Methods in Molecular Biology

Series Editor
John M. Walker
School of Life and Medical Sciences
University of Hertfordshire
Hatfield, Hertfordshire, AL10 9AB, UK

For further volumes:
http://www.springer.com/series/7651

Plant Chromatin Dynamics

Methods and Protocols

Edited by

Marian Bemer

Department of Molecular Biology, Wageningen University & Research, Wageningen, The Netherlands

Célia Baroux

Department of Plant and Microbial Biology, Zürich-Basel Plant Science Center, University of Zürich, Zürich, Switzerland

Humana Press

Editors
Marian Bemer
Department of Molecular Biology
Wageningen University & Research
Wageningen, The Netherlands

Célia Baroux
Department of Plant and Microbial Biology
Zürich-Basel Plant Science Center, University of Zürich
Zürich, Switzerland

ISSN 1064-3745 ISSN 1940-6029 (electronic)
Methods in Molecular Biology
ISBN 978-1-4939-7317-0 ISBN 978-1-4939-7318-7 (eBook)
DOI 10.1007/978-1-4939-7318-7

Library of Congress Control Number: 2017951723

Printed on acid-free paper

This Humana Press imprint is published by Springer Nature
The registered company is Springer Science+Business Media LLC
The registered company address is: 233 Spring Street, New York, NY 10013, U.S.A.

Preface

The nucleus provides eukaryotic cells with a unique functional compartment to protect and process the genetic information, offering both versatility and robustness to fine-tune gene expression and genome function. Inside the nucleus, the DNA is wrapped around histone proteins in a highly organized manner to form chromatin, which provides packaging and regulatory functions to control gene expression, genome replication, repair, and recombination. First discovered as a stainable substance in the cell during the nineteenth century, chromatin is now recognized as a highly dynamic macromolecular assembly that integrates developmental and environmental cues to adjust the transcriptional program, to respond to damage, and to facilitate DNA replication. Chromatin dynamics are driven by a complex interplay of events at the microscopic, nanoscopic, molecular, and biochemical levels. There is a growing interest in understanding how dynamic changes in chromatin organization and the epigenetic landscape direct and respond to physiological and developmental processes, and with this, the variety and power of methods to analyze plant chromatin have increased accordingly. For example, the rise of high-throughput techniques to analyze chromatin fractions has made it possible to elucidate the genome-wide composition and distribution of nucleosomes, histone variants, histone modifications and DNA modifications in relation to gene accessibility, local chromatin structure, gene expression, and DNA replication and repair activities. In addition, novel experimental designs have facilitated the identification of spatial interactions between genomic regions within the nuclear space, generating models for interchromosomal associations and functional, subnuclear chromatin domains.

Considerable progress has been made in obtaining genome-wide snapshots of chromatin states, but the effect of these different states on processes such as transcription and replication and on the 3D organization of chromatin is largely unknown. Extensive effort is therefore being spent on understanding these processes by using a combination of classical biochemistry and genetics, together with state-of-the-art protein complex identification methods. Recent progress in methods to assess 3D chromatin architecture has fuelled the emerging view that the spatial organization of chromatin has a major role in the regulation of gene expression. Studies on inter- and intrachromosomal chromatin contacts, gene positions relative to distinct chromatin compartments, and the functional compartmentalization of chromatin domains in different cell types are valuable approaches to elucidate the functional relevance of chromatin organization.

Understanding chromatin dynamics and functionality in plants requires the development of optimized protocols that solve or circumvent the technical challenges posed by the complexity of plant genomes, the plant cell wall, and the relative inaccessibility of specific cell types inside complex tissues and organs. In this edition, we provide a comprehensive collection of protocols that can be exploited to study plant chromatin structure and composition (Section I), to investigate the regulation of chromatin features in relation to the epigenetic regulation of gene expression and function (Section II), and to explore the interaction between chromatin modifications, gene regulation, and the 3D spatial organization of the chromatin inside the nucleus (Section III). In addition to presenting a large number of state-of-the-art protocols, each section is prefaced by a Review summarizing the current state of knowledge on the topic and the various approaches that can be used to study the specific aspects of plant chromatin dynamics. Moreover, three Technical Reviews offer an overview of specific methods together with a balanced discussion of their benefits and limitations, with the aim of providing practical advice to the reader.

This volume contains a comprehensive collection of detailed protocols and (technical) reviews that will assist the plant chromatin community in understanding the relationships between 3D chromatin architecture, local chromatin structure, epigenetic modifications, and functional genomics. We are very grateful to all authors for sharing their knowledge, practical experience, and theoretical considerations by providing detailed protocols presented in a pragmatic and accessible manner. We believe that this collective effort will greatly contribute to exciting discoveries in the field of plant chromatin research.

Wageningen, The Netherlands ***Marian Bemer***
Zürich, Switzerland ***Célia Baroux***

Contents

Contributors

PHILIPPE ANDREY • *Institut Jean-Pierre Bourgin, INRA, AgroParisTech, CNRS, Université Paris-Saclay, Versailles, France*
GERCO C. ANGENENT • *Laboratory of Molecular Biology, Wageningen University & Research, Wageningen, The Netherlands; Business Unit Bioscience, Wageningen Plant Research, Wageningen University & Research, Wageningen, The Netherlands*
JAVIER ARPÒN • *Institut Jean-Pierre Bourgin, INRA, AgroParisTech, CNRS, Université Paris-Saclay, Versailles, France*
MARIAMAWIT S. ASHENAFI • *Department of Plant and Microbial Biology, Zürich-Basel Plant Science Center, University of Zürich, Zürich, Switzerland*
DAPHNÉ AUTRAN • *Epigenetic Regulation and Seed Development Group, UMR 232 DIADE, IRD Institut de Recherche Pour le Développement—Université de Montpellier, Montpellier, France*
VÁCLAV BAČOVSKÝ • *Department of Plant Developmental Genetics, Institute of Biophysics, Czech Academy of Sciences, Brno, Czech Republic*
MARKO BAJIC • *Graduate Program in Genetics and Molecular Biology, Emory University, Atlanta, GA, USA; Department of Biology, Emory University, Atlanta, GA, USA*
FREDY BARNECHE • *Département de Biologie, IBENS, Ecole Normale Supérieure, CNRS, INSERM, PSL Research University, Paris, France*
CÉLIA BAROUX • *Department of Plant and Microbial Biology, Zürich-Basel Plant Science Center, University of Zürich, Zürich, Switzerland*
MARIAN BEMER • *Department of Molecular Biology, Wageningen University & Research, Wageningen, The Netherlands*
MOUSSA BENHAMED • *Institute of Plant Sciences Paris-Saclay (IPS2), UMR 9213/ UMR1403, CNRS, INRA, Université Paris-Sud, Université d'Evry, Université Paris-Diderot, Sorbonne Paris-Cité, Orsay, France*
TILL DAVID BEY • *Plant Development and (Epi)Genetics, Swammerdam Institute for Life Sciences, University of Amsterdam, Amsterdam, The Netherlands*
CLARA BOURBOUSSE • *Département de Biologie, IBENS, Ecole Normale Supérieure, CNRS, INSERM, PSL Research University, Paris, France*
ERWANN CAILLIEUX • *Centre National de la Recherche Scientifique (CNRS) UMR8197, Institut de Biologie de l'Ecole Normale Supérieure, Institut National de la Santé et de la Recherche Médicale (INSERM) U1024, Ecole Normale Supérieure, Paris Cedex 05, France*
CRISTEL C. CARLES • *LPCV, CEA, CNRS, INRA, Université Grenoble-Alpes, BIG, Grenoble, France; Laboratoire Physiologie Cellulaire Végétale, BIG, CEA, Grenoble Cedex 9, France*
MARIE-CHRISTINE CARPENTIER • *Laboratoire Génome et Développement des Plantes, CNRS, UMR5096, Perpignan, France; Laboratoire Génome et Développement des Plantes, Univ. Perpignan Via Domitia, UMR5096, Perpignan, France*
GIACOMO CAVALLI • *Institute of Human Genetics, Centre National de la Recherche UMR9002, Montpellier, France*
YUN-RU CHEN • *School of Life Sciences, The Chinese University of Hong Kong, Shatin, Hong Kong, China*

LUSIK CHERKEZYAN • *Department of Biomedical Engineering, Northwestern University, Evanston, IL, USA*
VINCENT COLOT • *Centre National de la Recherche Scientifique (CNRS) UMR8197, Institut de Biologie de l'Ecole Normale Supérieure, Institut National de la Santé et de la Recherche Médicale (INSERM) U1024, Ecole Normale Supérieure, Paris Cedex 05, France*
ROGER B. DEAL • *Department of Biology, Emory University, Atlanta, GA, USA*
SOPHIE DESSET • *Université Clermont Auvergne, CNRS, INSERM, GReD, Clermont-Ferrand, France*
BÉNÉDICTE DESVOYES • *Centro de Biología Molecular Severo Ochoa, CSIC-UAM, Madrid, Spain*
SUSANNE EDELMANN • *Biocenter Klein Flottbek, University of Hamburg, Hamburg, Germany*
JULIA ENGELHORN • *LPCV, CEA, CNRS, INRA, Université Grenoble-Alpes, BIG, Grenoble, France; School of Life Science, University of Warwick, Coventry, UK*
TOHNYUI NDINYANKA FABRICE • *Department of Plant and Microbial Biology, Basel-Zürich Plant Science Center, University of Zürich, Zürich, Switzerland*
SILVIA FARINATI • *Department of Agronomy, Food, Natural resources, Animals and Environment (DAFNAE) Agripolis, University of Padova, Legnaro (PD), Italy*
CRISTIAN FORESTAN • *Department of Agronomy, Food, Natural resources, Animals and Environment (DAFNAE) Agripolis, University of Padova, Legnaro (PD), Italy*
PAUL FRANSZ • *Plant Development and (Epi)Genetics, Swammerdam Institute for Life Sciences, University of Amsterdam, Amsterdam, The Netherlands*
MARCELINA GARCÍA-AGUILAR • *Laboratory of Genetics and Epigenetics of Seed Development, Laboratorio Nacional de Genómica para la Biodiversidad, Irapuato, Guanajuato, Mexico*
VALÉRIE GAUDIN • *Institut Jean-Pierre Bourgin, INRA, AgroParisTech, CNRS, Université Paris-Saclay, Versailles, France*
STEFAN GROB • *Institute of Human Genetics, Centre National de la Recherche UMR9002, Montpellier, France*
UELI GROSSNIKLAUS • *Department of Plant and Microbial Biology, Zürich-Basel Plant Science Center, University of Zürich, Zürich, Switzerland*
CRISANTO GUTIERREZ • *Centro de Biología Molecular Severo Ochoa, CSIC-UAM, Madrid, Spain*
LARS HENNIG • *Department of Plant Biology, Linnean Center for Plant Biology, Swedish University of Agricultural Sciences, Uppsala, Sweden*
ROMAN HOBZA • *Department of Plant Developmental Genetics, Institute of Biophysics, Czech Academy of Sciences, Brno, Czech Republic*
YANNICK JACOB • *Department of Molecular, Cellular and Developmental Biology, Yale University, New Haven, CT, USA*
SURAJ JAMGE • *Laboratory of Molecular Biology, Wageningen University & Research, Wageningen, The Netherlands*
ANDRZEJ JERZMANOWSKI • *Laboratory of Systems Biology, Faculty of Biology, University of Warsaw, Warsaw, Poland; Institute of Biochemistry and Biophysics, Polish Academy of Sciences, Warsaw, Poland*
KERSTIN KAUFMANN • *Institute for Biology, Plant Cell and Molecular Biology, Humboldt-Universität zu Berlin, Berlin, Germany*
JOOST J.B. KEURENTJES • *Laboratory of Genetics, Wageningen University and Research, Wageningen, The Netherlands*

KAMAL KISHORE • *Center for Genomic Science of IIT@SEMM, Fondazione Istituto Italiano di Tecnologia (IIT), Milan, Italy; Cancer Research UK Cambridge Institute, University of Cambridge, Cambridge, UK*
MARIA KOINI • *Plant Development and (Epi)Genetics, Swammerdam Institute for Life Sciences, University of Amsterdam, Amsterdam, The Netherlands*
MACIEJ KOTLIŃSKI • *Laboratory of Systems Biology, Faculty of Biology, University of Warsaw, Warsaw, Poland*
KATHRIN LAUSS • *Swammerdam Institute for Life Sciences, University of Amsterdam, Amsterdam, The Netherlands*
FUQUAN LIU • *School of Biological Sciences, Queen's University Belfast, Belfast, Northern Ireland, UK*
ALICE LUNARDON • *Department of Agronomy, Food, Natural resources, Animals and Environment (DAFNAE) Agripolis, University of Padova, Legnaro (PD), Italy*
SOFIA MADEIRA • *Centro de Biología Molecular Severo Ochoa, CSIC-UAM, Madrid, Spain*
KELSEY A. MAHER • *Department of Biology, Emory University, Atlanta, GA, USA; Graduate Program in Biochemistry, Cell, and Developmental Biology, Emory University, Atlanta, GA, USA*
WALID MAHREZ • *Department of Plant Biology, Linnean Center for Plant Biology, Swedish University of Agricultural Sciences, Uppsala, Sweden*
BENOIT MERMAZ • *Department of Life Sciences, Imperial College London, London, UK*
ANA KARINA MORAO • *Centre National de la Recherche Scientifique (CNRS) UMR8197, Institut de Biologie de l'Ecole Normale Supérieure, Institut National de la Santé et de la Recherche Médicale (INSERM) U1024, Ecole Normale Supérieure, Paris Cedex 05, France*
JOSE M. MUIÑO • *Department of Computational Molecular Biology, Max-Planck Institute for Molecular Genetics, Berlin, Germany*
ALICE PAJORO • *Laboratory of Molecular Biology, Wageningen University & Research, Wageningen, The Netherlands*
MATTIA PELIZZOLA • *Center for Genomic Science of IIT@SEMM, Fondazione Istituto Italiano di Tecnologia (IIT), Milan, Italy*
ARIADNA PICART-PICOLO • *Laboratoire Génome et Développement des Plantes, CNRS, UMR5096, Perpignan, France; Laboratoire Génome et Développement des Plantes, Univ. Perpignan Via Domitia, UMR5096, Perpignan, France*
NUNO D. PIRES • *Department of Plant and Microbial Biology, Zürich-Basel Plant Science Center, University of Zürich, Zürich, Switzerland*
FRÉDÉRIC PONTVIANNE • *Laboratoire Génome et Développement des Plantes, CNRS, UMR5096, Perpignan, France; Laboratoire Génome et Développement des Plantes, Univ. Perpignan Via Domitia, UMR5096, Perpignan, France*
AXEL POULET • *Université Clermont Auvergne, CNRS, INSERM, GReD, Clermont-Ferrand, France; Department of Biological and Medical Sciences, Oxford Brookes University, Oxford, UK*
ALINE V. PROBST • *GReD, Université Clermont Auvergne, CNRS, INSERM, Clermont-Ferrand, France*
CHRISTOPH RINGLI • *Department of Plant and Microbial Biology, Basel-Zürich Plant Science Center, University of Zürich, Zürich, Switzerland*
STEFANIE ROSA • *Institute of Biochemistry and Biology, Plant Physiology, University of Potsdam, Potsdam-Golm, Germany*

François Roudier • *Centre National de la Recherche Scientifique (CNRS) UMR8197, Institut de Biologie de l'Ecole Normale Supérieure, Institut National de la Santé et de la Recherche Médicale (INSERM) U1024, Ecole Normale Supérieure, Paris Cedex 05, France; Laboratoire Reproduction et Développement des Plantes, Univ Lyon, ENS de Lyon, UCB Lyon 1, CNRS, INRA, Lyon, France*

Stefan Scholten • *Biocenter Klein Flottbek, University of Hamburg, Hamburg, Germany; Institute for Plant Breeding, Seed Science and Population Genetics, University of Hohenheim, Stuttgart, Germany*

Veit Schubert • *Leibniz Institute of Plant Genetics and Crop Plant Research (IPK) Gatersleben, Seeland, Germany*

Joana Sequeira-Mendes • *Centro de Biología Molecular Severo Ochoa, CSIC-UAM, Madrid, Spain*

Wenjing She • *Department of Plant and Microbial Biology, Zürich-Basel Plant Science Center, University of Zürich, Zürich, Switzerland*

Lauriane Simon • *GReD, Université Clermont Auvergne, CNRS, INSERM, Clermont-Ferrand, France*

Jie Song • *Department of Life Sciences, Imperial College London, London, UK*

Maike Stam • *Laboratory of Plant Development and Epigenetics, Swammerdam Institute for Life Sciences, University of Amsterdam, Amsterdam, The Netherlands*

Christophe Tatout • *Université Clermont Auvergne, CNRS, INSERM, GReD, Clermont-Ferrand, France*

Serena Varotto • *Department of Agronomy, Food, Natural resources, Animals and Environment (DAFNAE) Agripolis, University of Padova, Legnaro (PD), Italy*

Zaida Vergara • *Centro de Biología Molecular Severo Ochoa, CSIC-UAM, Madrid, Spain*

Philipp Voigt • *Wellcome Trust Centre for Cell Biology, The University of Edinburgh, Edinburgh, UK*

Boris Vyskot • *Department of Plant Developmental Genetics, Institute of Biophysics, Czech Academy of Sciences, Brno, Czech Republic*

Blaise Weber • *Laboratory of Plant Development and Epigenetics, Swammerdam Institute for Life Sciences, University of Amsterdam, Amsterdam, The Netherlands*

Frank Wellmer • *Smurfit Institute of Genetics, Trinity College Dublin, Dublin, Ireland*

Sheng Yu • *School of Life Sciences, The Chinese University of Hong Kong, Shatin, Hong Kong, China*

Silin Zhong • *School of Life Sciences, The Chinese University of Hong Kong, Shatin, Hong Kong, China*

Part I

Profiling of Chromatin States

Chapter 1

Profiling Developmentally and Environmentally Controlled Chromatin Reprogramming

Clara Bourbousse, Moussa Benhamed, and Fredy Barneche

Abstract

Dynamic reshuffling of the chromatin landscape is a recurrent theme orchestrated in many, if not all, plant developmental transitions and adaptive responses. Spatiotemporal variations of the chromatin properties on regulatory genes and on structural genomic elements trigger the establishment of distinct transcriptional contexts, which in some instances can epigenetically be inherited. Studies on plant cell plasticity during the differentiation of stem cells, including gametogenesis, or the specialization of vegetative cells in various organs, as well as the investigation of allele-specific gene regulation have long been impaired by technical challenges in generating specific chromatin profiles in complex or hardly accessible cell populations. Recent advances in increasing the sensitivity of genome-enabled technologies and in the isolation of specific cell types have allowed for overcoming such limitations. These developments hint at multilevel regulatory events ranging from nucleosome accessibility and composition to higher order chromatin organization and genome topology. Uncovering the large extent to which chromatin dynamics and epigenetic processes influence gene expression is therefore not surprisingly revolutionizing current views on plant molecular genetics and (epi)genomics as well as their perspectives in eco-evolutionary biology. Here, we introduce current methodologies to probe genome-wide chromatin variations for which protocols are detailed in this book chapter, with an emphasis on the plant model species Arabidopsis.

Key words Chromatin, Histone, DNA methylation, Epigenome, Methodology

1 Introduction

In multicellular organisms, intrinsic and environmental signals are transduced by multiple molecular pathways that control cell proliferation and specialization. As sessile organisms, plants are particularly prone to phenotypic plasticity of cell identity, possibly hinting at their great capacity for physiological and morphological adaptive responses. These steps involve the fine-tuning of gene expression that depends on multiple regulatory layers converging onto "chromatin-based" controls. Indeed, in eukaryotic cells, genomic DNA is not naked in the nucleoplasm but is structurally organized with myriads of proteins altogether forming the complex structure of chromatin. This information superimposes with DNA sequence

Marian Bemer and Célia Baroux (eds.), *Plant Chromatin Dynamics: Methods and Protocols*, Methods in Molecular Biology, vol. 1675, DOI 10.1007/978-1-4939-7318-7_1, © Springer Science+Business Media LLC 2018

information to determine genome readout and notably gene expression programs (reviewed in [1–3]). Chromatin-level information has the remarkable property to be metastable, being reversible, yet in some cases epigenetically transmitted through cell division or meiosis [4]. While all somatic cells from a multicellular organism share an identical genome, modulation of gene expression patterns by signaling components and transcription factors in different cells comes with distinct flavors of the chromatin landscape. Consequently, chromatin reprogramming events are hypothesized to determine or facilitate the establishment of specific transcriptional programs. Uncovering the large extent to which chromatin dynamics and epigenetic processes influence gene expression has recently revolutionized current views and future perspective in plant cell and evolutionary biology.

The elemental repeat unit of chromatin is the nucleosome, which consists of 146/147 base pair (bp) of DNA wrapped around a histone core particle containing two copies each of positively charged histone H2A, H2B, H3, and H4 [5, 6]. The stretch of DNA lying in between two neighboring nucleosomes is defined as linker DNA, whose length can vary from a few up to 80 bp with an average length of 30–40 bp in *O. sativa* and *A. thaliana* plant species [7]. Nucleosome units are lined up to form a beads-on-a-string structure, which can further be compacted into higher-order structures and a 30 nm fiber through the assistance of linker histone H1 (reviewed in [8]). Nucleosomal density is commonly representative of chromatin condensation and activity, and its regulation is critical for most DNA-related processes [9]. Local and large-scale variations in chromatin condensation have first been studied at the single cell level using cytogenetic methods, notably assessing the functional partitioning into gene-rich "euchromatin" and highly condensed and silent "heterochromatin" (Fig. 1). Heterochromatin contains the majority of centromeric and pericentromeric repeats and other dispersed silent repeats, including transposable elements (TEs), which can aggregate to form conspicuous nuclear foci referred to as "chromocenters" [10, 11].

Nucleosome positioning is first influenced in *cis* by DNA sequence and can be modulated by ATP-dependent nucleosome remodelers, helicases and transcription factors (TFs) among others [12, 13]. Notably, the presence of nucleosome-free (or depleted) regions (NFRs/NDRs) coinciding with transcriptional regulatory regions and start sites hints at the influence of DNA accessibility on transcription initiation [13]. Different classes of transcription factors differ in their sensitivity to the presence of nucleosome for binding to their target sites. It has been proposed that "pioneer" transcription factors can bind DNA in a nucleosomal context and subsequently deplete the region from nucleosomes via the recruitment of remodeler complexes, thereby facilitating the subsequent binding of other TFs [14]. An ever-increasing number of "digital

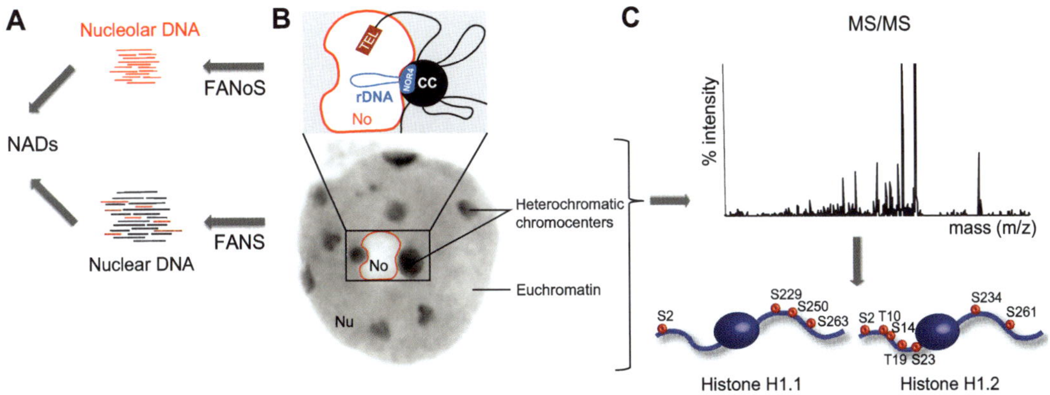

Fig. 1 Chromatin in a nuclear context. (**a**) Diagram representing a NADs detection procedure. Fluorescence-activated nuclear sorting (FANS) of cell homogenates allows for isolating intact nuclei. In parallel, FANoS is used to isolate membrane-free nucleoli from disrupted nuclei. Comparative analysis of the nuclear and nucleolar fractions allows for identifying nucleolar-enriched genomic domains, proteins, and transcripts. (**b**) Representative nucleus of Arabidopsis leaf mesophyll cells stained with DAPI (6-diamidino-2-phenylindole) displaying a characteristic nucleolus (No, circled in *red*) and ten conspicuous chromocenters in which centromeres, pericentromeres, and most heterochromatin are compacted. The *inset* schematically represents how ribosomal DNA euchromatic units loop out from NOR-containing chromocenters into the nucleolus where RNA Pol I transcription and rRNA processing events occur. (**c**) Identification of chromatin-associated histone types and PTMs by MS/MS. The diagram exemplifies the detection of phosphorylated peptides from histone H1 by nanoLC-MS/MS upon trypsin digestion of purified histones. The residues found to be phosphorylated in histone H1.1 and H1.2 by Kotlinski et al. [85] are depicted

epigenomic techniques" have been developed to precisely determine the positioning of nucleosomes and TFs (reviewed in [15]). These techniques interrogate DNA accessible domains by nuclease footprinting using deoxyribonuclease I (DNase I) or micrococcal nuclease (MNase), or by exploiting the preferential integration of transposons into nucleosome-free regions by ATAC (assay for transposase-accessible chromatin; Fig. 2). Specialized algorithms are applied to the sequencing data allowing for translating accessibility signals into nucleosome positions and TF footprints.

Plant genomes contain multiple genes for each histone type (e.g., 13 histone H2A genes in *A. thaliana*), which commonly encode subtypes of different length and amino-acid sequence with either similar or distinct functions. Several of the Arabidopsis H1, H2A, H2B, and H3 subtypes correspond to histone variants with distinct spatiotemporal expression patterns and biochemical properties, assigning them to specialized functions in the nucleus such as gene specific H2A.Z and heterochromatin specific H2A.W (recently reviewed in [16]; Fig. 2a). An additional layer of complexity in the chromatin landscape is determined by myriads of post-translational modifications (PTMs) deposited at specific residues of chromatin proteins either before or after their incorporation into nucleosomes [17]. PTMs are enriched over the tail regions of

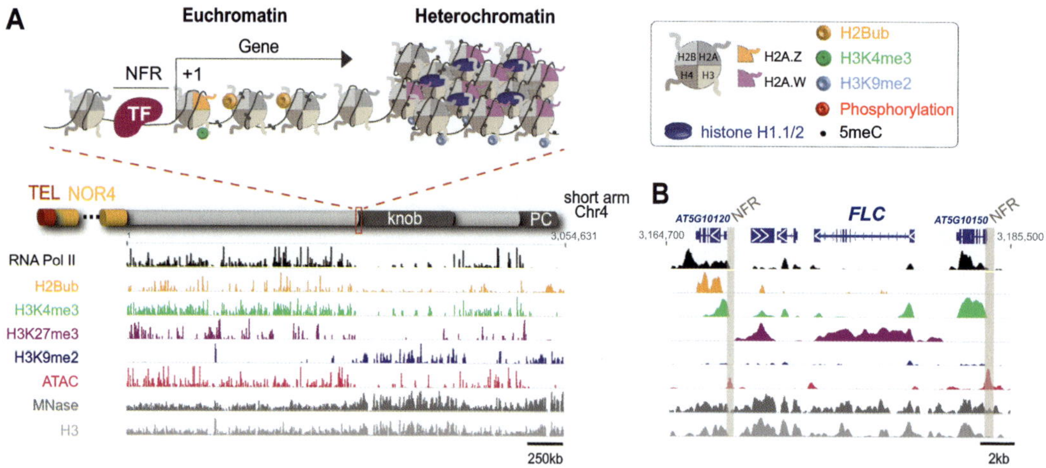

Fig. 2 Profiling the chromatin landscape along the Arabidopsis genome. (**a**) Schematic representation of the chromatin landscape on the Arabidopsis chromosome 4 short arm showing euchromatic–heterochromatic boundaries around the knob and the pericentromeric (PC) domains. Heterochromatin is typically characterized by high nucleosomal occupancy (histone H3 level), weak MNase cleavage and Tn5 "tagmentation" by ATAC-seq, and harbors the H3K9me2 hallmark [97]. Euchromatin displays typical chromatin features associated with gene expression control as detailed in (**b**). (**b**) Enlarged view around the *FLC* gene on chromosome 5 showing H3K27me3 enrichment, a characteristic *PRC2*-mediated repressive chromatin status. In contrast, the neighboring gene *AT5G10120* displays typical hallmarks of transcriptionally active chromatin with elevated RNA Pol II levels, a H3K4me3 peak at the +1 nucleosome, histone H2B monoubiquitination along the gene body and a putative nucleosome-free region (NFR) located immediately upstream the Transcription Start Site (TSS) that is reflected by an ATAC-seq peak. An ATAC peak at the 3′ end of the *FLC* gene presumably corresponding to *COOLAIR* antisense transcription can also be noted. In (**a**) and (**b**), RNA Pol II, nonmodified H3 and H3K9me2 ChIP-seq data as well as ATAC-seq data are from M. Benhamed et al., while H2Bub, H3K4me3 and H3K27me3 ChIP-seq data are from F. Barneche et al. (unpublished data). Datasets were generated using entire *A. thaliana* Col-*0* seedlings grown under standard laboratory conditions. *Nu* nucleus, *No* nucleolus, *TEL* telomeres, *NFR* nucleosome-free region, *NOR* nucleolus organizing region. Genomic coordinates correspond to TAIR10

histone proteins that protrude from the nucleosomal structure and are presumably more accessible to histone readers [18], the most extensively studied PTMs being arginine/lysine mono/di/tri-methylation and acetylation as well as serine/threonine phosphorylation. Several other histone PTMs such as monoubiquitination of histone H2A and H2B occur on residues in the nucleosomal inner region [19]. Some of the multiple PTMs display an intrinsic capacity to modify the chromatin structure. For example, both acetylation and monoubiquitination can impair nucleosome-array folding and facilitate DNA accessibility by loosening histone–DNA charge contacts or by modifying the nucleosome structure, respectively [20–22]. Other histone PTMs might primarily serve as anchoring points for effector proteins harboring specific histone-binding domains or by recruiting histone chaperones that bring specific activities [1, 23].

The DNA itself can be modified by cytosine methylation in different sequences contexts (CG, CHG where H is A, T, or C, and CHH), which can be mapped at base-pair resolution using bisulfite-sequencing ([24]; detailed in Chapter 2 by Chen et al. [25]). In Arabidopsis as in other plant species, DNA methylation is essential for the epigenetic transmission of several types of silent chromatin states through cell division [4], notably by the maintenance of symmetric CG and CHG methylation and the reestablishment of CHH methylation by RNA-dependent DNA methylation (RdDM) after semiconservative DNA replication (reviewed in [26]).

Combined together, histone PTMs, DNA methylation, and incorporation of specific histone variants contribute to organize the genome into functionally distinct domains that influence transcriptional outcomes to coordinate cell responses to internal and external signals (Fig. 2). Controlled regulation of the chromatin landscape contributes to orchestrate developmental phases during the plant's life cycle, notably by their prominent role in the differentiation of new cell types from meristematic stem cells and in cell specialization. Extensive transitions between different chromatin states occur during reproduction to establish germ line- and zygote-specific programs, but also in somatic cells subjected to intrinsic signals or to environmental variations, such as prolonged heat or suboptimal light conditions (reviewed in [27–30]). In both reproductive and vegetative phases, gene expression reprogramming frequently coincides with multilevel modification of the chromatin landscape, from local changes around individual genes to wide rearrangements of the nuclear architecture, with major waves of variations being shared in a given cell type or organ. Deciphering the extent, the driving forces and the functional impact of chromatin dynamics in terms of variations of the epigenome landscape and in terms of changes in the chromatin spatial organization has recently gained an enormous interest (reviewed in [31–37]).

Profiling of histone variants, PTMs, and associated proteins has allowed for shedding light on multiple regulatory pathways that control epigenome variations. Among them, *Polycomb*-mediated activity has profound impacts on the establishment and the maintenance of cell specialization and adaptations. Chromatin dynamics are also at play in plant developmental switches that involve morphologic and metabolic adaptations of somatic cells in differentiated organs without stem cell differentiation (such as germination, photomorphogenesis, or senescence). Among such transitions, epigenome profiling during Arabidopsis seedling photomorphogenesis over a short period of time (1–6 h) has revealed extensive chromatin state dynamics, which occur independently of cell division in multiple cells of the seedling (recently reviewed in [38, 39]). These studies unveiled a combination of events acting downstream the light signaling pathways, which rely on nucleosome occupancy/positioning dynamics as well as on loss or deposition of histone PTMs over hundreds of genes, such as trimethylation of H3K27 by PRC2; acetylation and deacetylation of H3K9 by the opposite action of the GCN5 acetyltransferase and histone deacetylase 1; and H2B ubiquitination by the HUB1/HUB2 ubiquitin ligase [40–43].

Several studies also shed light on the impact of controlled histone exchange in Arabidopsis transcriptional programs. For example, control of H2A.Z occupancy is a key player during Arabidopsis adaptive responses to heat (recently reviewed in [44]). H2A.Z replacement by canonical H2A in the +1 nucleosome of temperature-responsive genes increases DNA accessibility to TFs and RNA Polymerase II [45], consequently upregulating their transcriptional activity [46, 47]. H2A.Z is frequently enriched over either Transcriptional Start Sites (TSS) or gene bodies, but its presence correlates with low expression levels and high gene responsiveness only in the latter context [48]. Comparative analyses of histone and DNA methylation profiles uncovered a strict anticorrelation between H2A.Z and DNA methylation coverage over gene bodies, suggesting that genic DNA methylation may exclude H2A.Z from genes leveled for constitutive expression [48].

In contrast to mammals [49], the genome-wide extent of internally controlled variations of the chromatin landscape has just begun to be uncovered in plants. Among them, circadian clock controlled chromatin dynamics are expected to impact a large part of the epigenome, as suggested by the recent observation that enrichment levels of H3K9ac, H3K27ac and H3S28p fluctuate over hundreds of genes between dusk and dawn in *A. thaliana* seedlings exposed to short day conditions [50]. Accordingly, the periodic repression of central clock oscillator genes relies on the removal of transcriptionally permissive histone marks by histone deacetylation and demethylation, and also from histone exchange by chaperones independently from the polycomb repressive complexes 2 (PRC2) pathway (reviewed in [38, 51]).

2 Current Methodologies to Probe Variations in Chromatin Composition and Epigenome Profiles

Over the past 15 years, knowledge on chromatin composition, regulation and activity has been considerably extended with the rapid methodological progresses allowing for the detection of a large repertoire of histone PTMs by mass spectrometry. Epigenomic approaches now also provide DNA sequence information on chromatin properties such as condensation and accessibility as well as differential enrichments of histone types and PTMs, DNA modification, TFs, RNA polymerases, and physically associated factors. In a first intention, chromatin immunoprecipitation (ChIP) has allowed for performing combinatorial analyses of multiple chromatin marks along the genome, shedding light on the general principles of plant chromatin organization (detailed below in Subheading 2.2). This notably unveiled that chromosomal distribution of histone and DNA modifications largely falls into a few prevalent sets of chromatin signatures in metazoans [52] and plants [53–55]. This approach has generated reference epigenomic maps for *A. thaliana* seedlings in standard laboratory growth conditions [53, 54]. Genome-enabled approaches also aim at characterizing three-dimensional (3D) chromatin organization and genome topology by providing quantitative information on short and long distance interactions analyzed by locus-specific chromosome conformation capture (3C) and its genome-wide derivatives 4C (that is one-against-all) and Hi-C (all-against-all; reviewed in [56]). Recent studies on Arabidopsis have unraveled a great overlap between the information gathered using chromatin profiling and microscopic analyses of nuclear architecture, such as the prevalence of intrachromosomal interactions and the 3D association between heterochromatic and *Polycomb* transcriptionally silenced domains [37, 57–61]. Novel developments of 3C-derived approaches such as the chromatin interaction analysis by paired-end tag-sequencing (ChIA-PET) aim at capturing genomic regions linked to a protein of interest, which are enriched following a chromatin immunoprecipitation step. For example, ChIA-PET analysis of RNA Pol II has allowed for unveiling the complexity of short distance "transcriptional interactomes" that link promoter and distal enhancer regions in zebrafish [62]. To our knowledge this methodology has not been reported for plant species so far. Intense efforts are nonetheless devoted to determining the genomic domains that preferentially associate with specific subnuclear compartments such as the nucleolus-associated domains (NADs; [63–65]), and possibly also lamin-associated domains (LADs; [66]).

Due to the inherent difficulty of performing functional analyses of chromatin activity in single plant cells, a strategy consists in using biological systems in which endogenous heterogeneity is minimized. Examples include studies considering (1) a synchronous physiological transition in response to developmental or environmental signals [41, 42, 46], (2) an endogenously synchronized process such as clock-controlled chromatin dynamics (that follows a hierarchical and uniform control in most cells) [67, 68], or (3) artificially synchronizing a developmental process such as the elegant floral synchronization system presented in Chapter 16 by Engelhorn et al. [69]. Dedicated methodologies have emerged more recently to profile the chromatin landscape of specific cell types or subpopulations, most notably the INTACT (isolation of nuclei tagged in specific cell types) strategy [70]. INTACT has first allowed for determining specificities in histone methylation profiles and their relationship to gene expression in hair and non-hair root epidermis cells [71]. The original INTACT protocol relies on biotin labeling of a nuclear envelope protein in individual cell types followed by streptavidin-mediated selective capture of biotinylated nuclei. In vivo biotinylation is achieved by transgenic expression of the bacterial BirA enzyme acting on the biotin ligase recognition peptide (BLRP) of the nuclear targeting fusion protein (NTF), which also contains a nuclear envelop-targeting domain and a fluorescent moiety (GFP or mCherry) for visualization purposes. As detailed in Chapter 8 by Morao et al. [72], a modified version of the initial protocol, aiming at improving the procedure in terms of yield, purity, and time, involves affinity-purifying nuclei using magnetic beads coated with an antibody directed against the GFP or mCherry of the original NTF protein.

The functional significance of epigenome variations can be assessed by probing gene expression patterns using quantitative RT-PCR or transcriptomics. Yet, in the absence of time-resolved experimental design, the quantification of steady-state levels of both short and long-lived RNA species can impair efficient comparisons with chromatin dynamics. A better proxy of transcriptional activity is attained by ChIP profiling of RNA polymerase occupancy along the genome as described in Chapter 8 [72]. Reliable profiling of the transcriptional activity is also obtained using either GRO-seq (Global nuclear Run-On sequencing) or 5′ GRO-seq that captures 7meG-capped transcripts. The latter methods rely on the BrUTP labeling of the neosynthesized transcripts performed ex vivo in isolated nuclei [73, 74]. In addition, a complementary strategy to probe global changes in transcriptional activity consists in quantifying the absolute and relative contents of the elongating (S2-P) and nonphosphorylated forms of RNA Pol II signals in isolated nuclei extracted from different cell populations signals using microscopy or cytometry methodologies [75]. Here, we introduce molecular and biochemical procedures to profile chromatin status as well as its variations along the genome for which protocols are detailed in this book section.

2.1 Determination of Chromatin Composition by Mass Spectrometry

Mass spectrometry-based techniques interrogate the whole repertoire of known histone proteins and PTMs, and can also detect previously uncharacterized chromatin marks in abundant plant samples [76–78]. Standard biochemical techniques have not easily been adapted to Arabidopsis vegetative tissues, which have relatively low amounts of histones per cell. In addition, they contain relatively high amounts of charged polycarbohydrates, which can interfere with isolation of basic proteins such as histones by standard methods applied to animal tissues. An efficient strategy to circumvent these issues consists in proliferative tissues. For instance, inflorescences from other Brassicaceae species such as *Brassica oleracea* (cauliflower) provide a prolific, Arabidopsis-related and chloroplast-free material enabling the extraction of abundant histone amount. As presented in Chapters 9 and 10 [79, 80], their highly basic nature allows for efficient extraction of both nucleosomal and linker H1 histones using sulfuric or perchloric acid followed by reverse phase or ion-exchange chromatography, respectively, before in-depth proteomic analyses.

Early MS analyses of plant chromatin composition have shed light on the partial evolutionary conservation of histone PTMs between mammals, yeast, and plants. The first detailed reports of Arabidopsis histone isoforms/variants and their modifications [81, 82] revealed that numerous core histone PTMs are conserved between plants and mammals, while others have so far only been detected in plants (such as H4K20Ac). Reciprocally, H3K79me3 was not found in Arabidopsis [82], in line with absence of homologs of the cognate histone methyltransferase Dot1/DOT1L present in mammals, drosophila, and *S. cerevisiae* [83, 84].

Recent optimization of a novel efficient multistep procedure has allowed for isolation and quantitative assessment of posttranslational signatures of H1.1 and H1.2, the two major linker histone subtypes in Arabidopsis ([85]; detailed in Chapter 10 by Kotlinski and Jerzmanowski [80]). These include phosphorylation in both N and C-terminal H1.1 and H1.2 domains, suggesting that function of histone H1 is modulated by PTMs like core histones (Fig. 1c). Interestingly, their distribution over the conserved H1 globular domains is similar in Arabidopsis and in mammals while it differs in the C-terminal domain. Such PTMs might be involved in the control of chromatin fiber condensation, possibly reflecting a diversification of histone H1 functions during evolution [85]. Linker histones play an important role in packing and repressing repetitive transposable elements (TEs) in Arabidopsis as well as in other species [86–88], and the influence of H1 PTMs dynamics on gene expression patterns remains poorly understood.

Unlike ChIP methodologies, MS does not rely on antibody availability for a specific protein or peptide. It also provides precise information on each type of PTM bore by specific histone subtypes or variants. For example, in Arabidopsis, the cell cycle regulated canonical H3.1 is selectively monomethylated on Lys-27 while the constitutively expressed histone H3.3 variant is not, due to subtle amino-acid sequence divergence in their N-terminal tails that impair recognition by the histone methyltransferases ATXR5 and ATXR6 [88]. Moreover, MS approaches can allow for detecting co-occurring PTMs on the same histone molecule, as observed for example for the H3K27me2 and H3K36me1 modifications on the same histone H3 molecule [89]. This can represent a great asset of MS-based approaches, since different combinations of chromatin marks can lead to different functional outputs. Sequential ChIP (Re-ChIP) allows for identifying two histone marks present on the same region of a chromatin fiber, or even on the same nucleosome if performed using mononucleosomes ([54, 90], as detailed in Chapter 6 by Desvoyes et al. [91]). Yet Re-ChIP methods cannot ascertain whether a "bivalent" nucleosomal status corresponds to co-occurrence on a single histone protein.

Most studies of the plant's histone proteome use trypsin digestion before MS analysis, a "bottom-up" approach that generates small peptides from lysine/arginine-rich histone amino-terminal tails [78]. This drawback can also generate confounding signals when distinct PTMs on the same peptide have a similar mass (e.g., trimethylation and acetylation). An alternative consists in using proteases with different cleavage specificities, such as Arg-C or thermolysin, as described in Chapter 10 by Kotlinski and Jerzmanowski [80]. The inherent technical properties of the approach also limit the capacity for internal normalization in label-free experiments. This needs to be considered before attempting quantitative comparisons of histone PTMs between different plant samples. More robust quantification can be achieved upon in vivo labeling using N^{15}/C^{13} lysine/arginine by SILAC (Stable Isotope Labeling with Amino acids in Cell culture), which, for example, has been successfully used to detect a global switch between trimethylated H3K27 and acetylated H3K27 in mammalian embryonic stem cells lacking *Polycomb* Repressive Complex 2 (PRC2) activity [92].

2.2 Profiling Genome-Wide Changes of Chromatin Proteins and PTMs

Locus-specific and genome-wide profiling of chromatin proteins is commonly achieved using ChIP-derived methods that consist in co-immunoprecipitating associated DNA fragments followed by either quantitative PCR, microarray hybridization (ChIP-on-chip; [93]) or high-throughput sequencing (ChIP-seq; *see* Chapter 5 by Desvoyes et al. [94]). ChIP requires the availability of an antibody for one or more epitopes that allow for specifically detecting a whole protein, a peptide or a modified peptide (e.g., mono/di/trimethylated, phosphorylated, or branched with a ubiquitin moiety). As recommended by the ENCODE program and several studies, the specificity of each antibody should be ascertained carefully, for example using histone peptide arrays or by competition with corresponding peptides [95, 96]. An alternative is to use an epitope tagged histone or chromatin-associated protein, after evaluating whether its expression pattern and function recapitulates the endogenous protein properties.

In most instances a sufficient ChIP resolution is attained by sonication of cross-linked chromatin in which proteins are covalently bound to DNA into 150–500 bp fragments (*see* Chapter 5 by Desvoyes et al. [94]), but complete degradation of linker DNA by MNase digestion of native chromatin further allows for reaching a mononucleosomal resolution [97]. In both cases, deep sequencing analysis of the immunoprecipitate allows for refining the epigenomic profiles to a nearly single nucleosome resolution. Important guidelines and caveats about ChIP deep-sequencing and subsequent bioinformatics analysis are available at the ENCODE website (http://encodeproject.org/ENCODE/experiment_guidelines.html) including concerns about the determination of sequencing depth and quality criteria and standards for reporting data [95]. Another caveat in ChIP analyses is the lack of an internal normalization standard to quantitatively compare site occupancy in different chromatin samples. Indeed, the inherent principles of (1) normalizing ChIPed DNA to input levels, and (2) adjusting genome-wide enrichment levels around a similar value in all samples are well suited to probe local variations at a subset of sites but drastically buffer global variations [98–101]. Consequently, massive chromatin changes can be missed or are difficult to analyze when they are distributed over a large portion of the genome. A procedure has recently been developed that consists, before immunoprecipitation, in spiking in a fixed amount of foreign chromatin with little DNA sequence similarity to the studied organism or a commercially available modified peptide coupled to an artificial DNA sequence [98–101]. These exogenous epitopes serve as a robust internal standard for qPCR or sequencing after immunoprecipitation. A spike adjustment factor is determined either by counting the number of reads mapped to the foreign reference genome or by normalizing peak values of plant chromatin to the foreign chromatin.

ChIP-seq has for long required a large amount of starting material for robust peak detection, therefore hampering its application to limited cell populations. Indeed, amplification of immunoprecipitated DNA during library preparation is limited to avoid the generation of clonal reads upon sequencing. Library amplification biases can also result in significant over- or underrepresented regions, which can be corrected by normalizing the data to the input chromatin. Recently, genome-wide profiling of H3K4me3 and H3K27me3 histone PTMs in specific root epidermis cell types has been reported in Arabidopsis using INTACT [70, 71]. A method combining efficient ChIP-seq of H3K4me3 and RNA Pol II after INTACT of root apical meristem cells is detailed in Chapter 8 by Morao et al. [72]. Successful INTACT notably depends on the availability of a promoter specific to the desired cell type that drives sufficient expression of the nuclear marker.

2.3 Profiling Genome-Wide Changes of Nucleosome Positioning and Chromatin Condensation

Nucleosomal occupancy can be assessed by ChIP profiling of canonical or variant histones irrespective of PTMs, as performed for example to compare the relative distribution of Arabidopsis histones H3.1 and H3.3 [102, 103]. These studies notably showed that H3.1 is frequently found on heterochromatic loci, whereas H3.3 is prevalently found over the 3′ end of actively transcribed genes. However, the ChIP procedure presents a limited resolution that can be overcome by dedicated approaches to reach a single base-pair resolution of nucleosome positioning. These methods take advantage of nonspecific endonucleases such as micrococcal nuclease (MNase) and deoxyribonuclease I (DNase I) that cleave accessible DNA, triggering the elimination of all linker DNA upon complete digestion (Reviewed in [15]). Interestingly, limited digestion with MNase or DNase I allows for distinguishing different degrees of chromatin accessibility, delineating hypersensitive sites (e.g., DHSs) that commonly represent *cis*-regulatory DNA and TF binding sites (Fig. 2). The quality of the assay is therefore dependent on nuclease digestion efficiency, which should be experimentally set using time series for each sample type as detailed in Chapter 11 by Pajoro et al. [104]. Careful adjustments and controls are also essential when making comparisons between different samples or experiments. Although MNase is not sequence-specific, a higher efficiency over A/T-rich loci has been observed in the maize genome, possibly generating sequencing biases that may be taken into account for genome-wide comparisons [105]. Similar to histone H3 enrichment, MNase-seq has shown that long heterochromatic TEs have higher nucleosome occupancy than short ones and protein-coding genes, and further depend on the DDM1 SWI/SNF chromatin remodeler to facilitate the access to DNA methyltransferases [86]. Using these methods, genome-wide nucleosome positions have been investigated in various plants, such as Arabidopsis [106, 107], rice [108] and maize [109, 110] and revealed strong relationships between chromatin condensation and transcript levels [13].

As highlighted above, a remarkable study monitoring DNase I hypersensitive site variations during Arabidopsis responses to heat or light has identified ≈700,000 sites of TF occupancy at the nucleotide resolution in vivo [40]. The photomorphogenic response was shown to involve more than 700 dynamic DHSs, many of which reside in the proximity of light-responsive genes. Furthermore, clustering of DHSs generated in the different conditions identified different classes of *cis*-DNA regulatory networks matching occupancy sites of known photomorphogenic TFs such as HY5 and PIFs, altogether unveiling that chromatin of etiolated plants is largely poised for the first exposure to light. This study more generally showed that network motif topology in *A. thaliana* follows similar rules as the previously described *C. elegans* neuronal and human TF systems [40]. DNase I/MNase-based profiling has also been used for functional studies on nucleosome dynamics, unveiling for example that heterochromatin decondensation and loss of nucleosomes act as a molecular determinant of TE reactivation under prolonged heat stress, which relies on the CAF-1 (chromatin assembly factor-1) histone H3 chaperone complex for efficient restoration of silencing [111]. As a last example, case studies using loss-of-function lines for the Brahma-related chromatin remodeler BAF60 (SWP73B) resulted in nucleosome alterations over 5′ domains of hundreds of genes, including the *FLOWERING LOCUS C* gene (*FLC*), which frequently correspond to *cis*-regulatory regions involved in developmental regulations [112, 113].

Several other procedures based on chromatin condensation variations have been set to probe nucleosome occupancy and *cis*-regulatory DNA regions, such as the FAIRE-seq (formaldehyde-assisted isolation of regulatory elements [114]) and ATAC-seq methods. ATAC has been established for *S. cerevisiae* cells [115], before being adapted to plant species as detailed in Chapter 12 by Bajic et al. [116]. This methodology is originally a derivate from a standard procedure used in sequencing library preparation by DNA fragmentation and ligation of adapters (the so-called tagmentation), exploiting the preferential integration of a hyperactive form of the prokaryotic Tn5 transposon into nucleosome-free regions [115]. Transposition is performed in vivo on isolated nuclei and the extracted DNA is directly used to generate sequencing libraries. ATAC-seq is therefore a fast two-step method to profile accessible chromatin regions, from which nucleosome positioning, TF occupancy and chromatin compaction can also be inferred at a high resolution [117]. However, in contrast to ChIP experiments that can be performed either on fresh plants or on formaldehyde cross-linked samples, MNase/DNase I-seq and ATAC-seq are performed on intact nuclei, potentially limiting them to specific scopes. Both DNase I/MNase-seq and ATAC-seq can be challenging, notably to avoid saturating DNA cleavage.

While DNase/MNase-seq commonly require high amounts of material (typically 1×10^6 cells), ATAC-seq has been applied to single human cells captured through a programmable microfluidics platform, notably showing combinations of *trans*-acting factors associated with either induction or suppression of cell-to-cell variability [118]. ATAC-seq has recently been adapted to plants with the successful mapping of open chromatin and TF-binding sites using 500–50,000 Arabidopsis seedling nuclei [119]. Such small amounts offer the possibility to couple ATAC-seq with nuclei sorting to study chromatin accessibility in specific cell types. Nuclei sorting also helps for improving subsequent mapping steps by reducing the number of reads originating from the plastid genome. While peaks of open chromatin are detected using the entire sequence tags, TF footprints are predicted by mapping Tn5 cut positions and read orientation to define footprint boundaries at high resolution using the pyDNase software [119, 120]. Similarly to MNase-seq, ATAC-seq profiling is sensitive to some sequence biases, which can be detected by profiling Tn5 integration in naked genomic DNA [119].

2.4 Profiling DNA Methylation Variations

Initial approaches to profile DNA methylation patterns were based on the immunoprecipitation of 5-methylcytosine (MeDIP) or on methylation sensitive restriction enzymes and have extensively been used to study plant epigenomes. DNA methylome profiling has subsequently been revolutionized by selective conversion of non methylated cytosines using sodium bisulfite [121] coupled to whole-genome sequencing (WGBS, also referred to as MethylC-seq or BS-seq). Quantitative estimation of DNA methylation profiles at a single base-pair resolution by WGBS was first used for the study of the Arabidopsis methylome [122, 123]. Comparison of the DNA methylation patterns from various Arabidopsis mutant lines revealed correlations between the establishment, maintenance, and removal of DNA methylation, with complex interplays and possible feedback controls between these distinct pathways (reviewed in [26, 124]). This notably helped the investigation of the mechanisms involved in gene imprinting and transgenerational epigenetic inheritance in plants (reviewed in [125, 126]). DNA methylation reprogramming plays a central role during gametogenesis to ensure stable silencing of TEs in the future embryo and across generations. An emerging model depicts active DNA demethylation and loss of silencing in the male and female companion cells through the activity of DNA glycosylases, which may allow for specifically targeting DNA methylation of TEs and repeats in the male and female gametes through a mobile signal consisting of siRNAs [127–129]. This model was validated for male gametogenesis by a recent study using both fluorescent silencing of reporter genes and of a 21–24 nt siRNAs sequester gene under the control of promoters specific to either the pollen companion cell or the sperm cell [130]. WGBS has also been used to compare DNA methylation variation between diverse *Arabidopsis* species and natural accessions showing that patterns of DNA methylation along genes and TEs is highly variable in natural populations [131–134]. The potential contribution of this epigenetic trigger on adaptive capacity and plant fitness is currently a flourishing field of research.

Detailed protocols for WGBS are presented in Chapters 2 and 3 [25, 135]. Advantages and pitfalls of these different methods have been contrasted in [136] and are summarized in Table 1. Estimating the nonconversion rate is the first important checkpoint in analyzing WGBS data. A first indication can be obtained by measuring the fraction of nonconverted cytosines in the unmethylated plastid genome, a rate that should not exceed a few percent. Yet, for cell samples with no or few plastids, another option consists in including a nonconversion filter to discard reads with more than three consecutive methylated CHHs [122]. These are known to be more likely due to an inefficient conversion than to the rare natural cases. Yet, this presents the risk of losing information on highly methylated CHH regions. More generally, defining differentially methylated regions (DMRs) can represent a major difficulty. DMRs can be called on single cytosines or larger regions using diverse statistical methods. Also, whereas cytosines in a CG context often

Table 1
Main advantages and disadvantages of commonly used methods for the profiling of plant chromatin composition and activity

Method	Purpose	Advantages	Pitfalls
BS-seq	5-meC profiles at base-pair resolution	Identifies meC sequence context; quantitative	Requires complete bisulfite conversion; bioinformatics difficulties to map reads corresponding to repeated loci and for calling relevant DMRs due to statistical and intrinsic variability
ChIP-seq	Genomic profiles of chromatin marks, TFs, RNA Polymerases and other chromatin associated proteins	Whole-genome information; quantitative; easily allows for comparing the profiles in multiple samples; RNA Pol II ChIP allows for estimating transcriptional activity	Requires specific antibody; can present important biases when comparing samples with major differences in the global level of the analyzed mark
INTACT	Isolation of specific cell populations	Potentially cell type specific; isolation of homogeneous cell subsets; suited to multiple downstream applications on isolated nuclei or on nuclear components	Promoter specificity; may require large amounts of material for certain cell types
MNase-seq	Genomic profile of DNA accessibility	Delineate *cis*-regulatory domains	Requires large amounts of material and tight controls of MNase digestion efficiency to maximize the sensitivity
ATAC-seq	Genomic profile of accessible chromatin regions	Requires more limited amounts of material than MNase-seq and DNase I-seq	May require a tight control of tagmentation frequency to maximize the sensitivity
NAD-seq	Determination of nucleolus-associated genomic regions	Provides information on nuclear spatial and functional organization	May enrich repeated NADs over less represented domains; requires independent validation of subnuclear localization by alternative approaches
MS/MS	Global profiling of chromatin protein composition	Can determine protein PTMs and their combinatorial forms on single histone molecules; can identify novel PTMs	Large amounts of material required; multiple drawbacks for quantitative comparisons of lysine/arginine-enriched PTMs
GRO-seq/ 5′ GRO-seq	Genome-wide profiling of nascent transcripts in isolated nuclei	Quantitative measure of transcriptional profiles	Large amounts of material required; ex vivo labeling of neosynthesized transcripts; may be inappropriate for monitoring transcriptional changes in intact tissues or in response to cellular/external signals

More details are found in the corresponding chapters

present an "all-or-nothing" methylated rate, this is not the case for CHG and CHH contexts that require a higher coverage to determine a refined methylation level. Finally, DMRs can arise even between wild-type isogenic lines [137] and some regions of the genome are prone to high levels of variation in their methylation profiles [138]. Hence, natural variation has to be taken into account not to overestimate differential DNA methylation, especially when working with plant lines bearing subtle (weak or highly locus specific) DMRs. This bias can be minimized by only considering reproducible DMRs from independent biological replicates. Finally, like for all other genomic approaches, an inherent difficulty resides in mapping short sequence reads to a reference sequence, especially for repeated elements that give multiple hits over the genome. This difficulty is somehow reinforced in WGBS with the combined effects of increased mapping complexity after bisulfite conversion of unmethylated cytosines and of specific interests on TEs and other repeated genetic elements that are strongly regulated by DNA methylation. Chapter 4 by Kishore and Pelizolla [139] is dedicated to WGBS data mining. Finally, because obtaining sufficient amounts of genomic DNA can be limiting (e.g., after INTACT) or cost-effective, Chapter 3 by Edelmann and Scholten [135] details guidelines for performing reduced representation bisulfite sequencing (RRBS) with small DNA amounts.

2.5 Mapping Chromatin Spatial Organization

The linear information provided by chromatin profiling has just begun to converge with 3-dimensional information obtained by microscopy and by genome-enabled methodologies such as chromosome conformation capture (3C; [140]), which is detailed in the Section II of this book. Cytological and genomic approaches have refined at the genetic and molecular levels the euchromatic and heterochromatic states identified microscopically in the early twentieth century [141, 142], most notably revealing that they correspond to distinct chromosomal domains displaying drastically different epigenomic landscapes [53, 54, 97, 143]. In *A. thaliana* interphase nuclei, the chromosome arms occupy distinct territories [144]. Most of the heterochromatic pericentromeric and TE repeats are aggregated close to the centromeres within a few chromocenters located close to the nuclear periphery or to the nucleolus for those containing large silent domains of the Nucleolus Organizing Regions [143, 145]. Strikingly, TEs of the five Arabidopsis chromosomes also tend to interact within another nuclear substructure referred to as the KNOT, which may contribute to safeguard the genome from their invasive capacity [146]. Nuclear compartmentalization is exemplified in mouse by the extreme cases of transcription factories and by the aggregation of transcriptionally silent heterochromatic domains within highly condensed chromocenters [147, 148]. In *A. thaliana*, chromocenters can easily be visualized in most cell types, but their organization is subject to natural variation in their regulatory factors and can also decondensate following developmental or abiotic signals (e.g., temperature and light) or pathogenic aggressions [28, 149–152].

3C-based methodologies allow for a quantitative assessment of nuclear architecture by probing intrachromosomal and interchromosomal interactions between distant genomic loci at a single gene resolution [57–61, 146] (detailed in Chapter 14 by Grob and Cavalli [153]). Arabidopsis Hi-C maps match cytogenetic observations with heterochromatin and euchromatin mostly forming two distinct 3C interactomes. Refined examination of the euchromatin interactome further indicated that actively transcribed genes tend to weakly associate with each other [60], possibly acting as insulator domains. This is in agreement with the scattered distribution of RNA polymerase II foci in the Arabidopsis nucleoplasm, suggesting that transcription sites do not cluster in large domains such as the macromolecular transcription factories in animal cells [154]. This may also relate to the observation that chromatin signatures are markedly indexed to single genetic elements such as genes or transposable elements (TEs) with a given transcriptional status (reviewed in [155]). Accordingly, a set of transcriptionally repressed genes enriched in the *Polycomb*-associated H3K27me3 histone mark shows a tendency for long-range interactions [60]. This observation is substantiated cytogenetically by the cold-induced association of *FLC* alleles in the nuclear space, a dynamic property that temporally correlates with their transcriptional repression and H3K27me3 marking during vernalization [156]. The dynamic properties and biological significance of *Polycomb*-repressed gene clustering in plants remain to be assessed.

At a smaller scale, hundreds of chromatin loops formed by interactions between the 5′ and 3′ ends of highly expressed genes have been detected over the Arabidopsis genome [60]. Investigation of gene looping on the *FLC* locus by 3C-based analysis showed that its disruption is an early step required for switching off this gene by *Polycomb* complexes [157]. Interestingly, the process involves the SWI/SNF complex subunit BAF60, revealing that ATP-dependent chromatin remodeling is at play [113]. Taken together, these findings hint at functional links between spatial organization of the chromatin and transcriptional competency. The precise influence of *FLC* loop formation on RNA Polymerase II recruitment and activity constitute interesting mechanistic and temporal aspects to assess in future studies.

Possible interplays between genome topology and transcriptional status is also sustained by the observation that several light-induced genes are subject to long-distance repositioning from the nuclear interior to the nuclear periphery during cotyledon de-etiolation [158]. Similarly, artificially tethering a transgenic locus to the nuclear periphery of leaf and inflorescence cells can enhance its expression [159]. An emerging picture therefore discriminates two antagonistic effects of peripheral localization in Arabidopsis. On one hand, proximity with the nuclear membrane could favor gene expression, for example by facilitating mRNA export to the cytosol. On the other hand, vicinity to some lamina-like components may contribute to heterochromatin organization and result in many chromocenters being localized at the nuclear periphery. This second observation relates to the finding that mammal Lamin Associated Domains (LADs) mainly correspond to silent genomic regions (reviewed in [160]). Despite lacking clear homologs of lamin proteins, the plant nuclear matrix appears to be functionally similar to the animal lamina (reviewed in [161]). Dynamic variations of plant LAD counterparts in response to developmental or environmental cues remain to be determined.

Arabidopsis nucleolus-associated domains (NADs) have recently been identified [65]. Intact plant nucleoli can be isolated by gradient fractionation as reported in human cells [63, 64, 162], or, as detailed in Chapter 7 by Carpentier et al. [163], by fluorescence-activated cell sorting (FACS) using a YFP-tagged nucleolar protein, FIBRILLARIN 2 (Fig. 1a). Deep sequencing showed that Arabidopsis NADs are expectedly enriched in actively transcribed ribosomal DNA genes and, interestingly, they also appear to encompass several heterochromatic regions [65]. These include multiple TEs as well as inactive protein-coding genes and subtelomeric regions located close to the nucleolar organizing regions (NORs), which may be sequestered away from RNA Pol II. These findings highlight the importance of the nucleolus in anchoring or confining silent chromatin domains. Future studies should contribute to a better understanding of how subnuclear partitioning contributes to organize the plant cell's transcriptional pattern.

3 Perspectives

An ever-increasing number of studies are shedding light on the role of histone modification and replacement, DNA methylation, and nucleosome repositioning, which together contribute to epigenome reorganization into functionally distinct domains to control transcriptional outcomes. Dynamic variations of the epigenome in response to internal or external signals is thought to facilitate the establishment of a novel transcriptional program or, in some instances, to define a chromatin status that is poised for anticipated changes [40, 164, 165]. Recent findings also hint at a combinatorial effect between local chromatin signatures, genome topology through intrachromosomal/interchromosomal long-distance chromatin interactions, and partitioning of chromatin domains within nuclear substructures such as the nuclear envelope or the nucleolus. Spatial positioning and/or physical proximity of different genomic regions in the nuclear volume may trigger (or favor) the establishment of local chromatin contexts that are either permissive or repressive to transcription (reviewed in [166]). However, the importance of nuclear positioning for local transcriptional activity is yet unclear. This will hopefully be further elucidated by the methodologies discussed here. Studies on cell identity and on allele-specific gene regulation (e.g., maternally or paternally imprinted) have long been impaired by technical challenges in generating single-cell epigenome profiles. Recent advances in increasing the sensitivity of genome-enabled technologies and in the isolation of specific cell types such as the development of INTACT methods should now allow for overcoming these limitations. In future studies, integrating multilevel information on epigenomic and nuclear architecture dynamics should also allow for establishing comprehensive models of chromatin function and of their impact plant developmental and environmental adaptations from the single cell to the whole plant.

Acknowledgments

The authors thank Chris Bowler for constant support, Vincent Colot (IBENS, Paris France) and François Roudier (ENS, Lyon France) for helpful discussions and sharing unpublished data. They are also grateful to Damarys Loew (Curie Institute, Paris France) and Julie Law (Salk Institute for Biological Studies, San Diego USA) for helpful discussions. Work by the authors is supported by the CNRS, ANR-11-JSV2-003-01, Investissements d'Avenir Labex MEMOLIFE ANR-10-LABX-54 to FB, by PSL Research University to FB and CB, and by Université Paris-Saclay to MB.

References

1. Berger SL (2007) The complex language of chromatin regulation during transcription. Nature 447:407–412. doi:10.1038/nature05915
2. Kouzarides T (2007) Chromatin modifications and their function. Cell 128:693–705. doi:10.1016/j.cell.2007.02.005
3. Li G, Zhu P (2015) Structure and organization of chromatin fiber in the nucleus. FEBS Lett 589:2893–2904. doi:10.1016/j.febslet.2015.04.023
4. Richards EJ (2006) Inherited epigenetic variation—revisiting soft inheritance. Nat Rev Genet 7:395–401
5. Richmond TJ, Finch JT, Rushton B, Rhodes D, Klug A (1984) Structure of the nucleosome core particle at 7 [angst] resolution. Nature 311:532–537. doi:10.1038/311532a0
6. Luger K, Mader AW, Richmond RK, Sargent DF, Richmond TJ (1997) Crystal structure of the nucleosome core particle at 2.8 A resolution. Nature 389:251–260. doi:10.1038/38444
7. Jiang J, Zhang T, Zhang W (2015) Genome-wide nucleosome occupancy and positioning and their impact on gene expression and evolution in plants. Plant Physiol 168(4):1406. doi:10.1104/pp.15.00125
8. Cutter AR, Hayes JJ (2015) A brief review of nucleosome structure. FEBS Lett 589:2914–2922. doi:10.1016/j.febslet.2015.05.016

9. Jiang C, Pugh BF (2009) Nucleosome positioning and gene regulation: advances through genomics. Nat Rev Genet 10:161–172. doi:10.1038/nrg2522
10. Fransz P, Soppe W, Schubert I (2003) Heterochromatin in interphase nuclei of *Arabidopsis thaliana*. Chromosome Res 11:227–240
11. Fransz P, de Jong H (2011) From nucleosome to chromosome: a dynamic organization of genetic information. Plant J Cell Mol Biol 66:4–17. doi:10.1111/j.1365-313X.2011.04526.x
12. Struhl K, Segal E (2013) Determinants of nucleosome positioning. Nat Struct Mol Biol 20:267–273. doi:10.1038/nsmb.2506
13. Liu M-J, Seddon AE, Tsai ZT-Y, Major IT, Floer M, Howe GA, Shiu S-H (2015) Determinants of nucleosome positioning and their influence on plant gene expression. Genome Res 25(8):1182. doi:10.1101/gr.188680.114
14. Zaret KS, Carroll JS (2011) Pioneer transcription factors: establishing competence for gene expression. Genes Dev 25:2227–2241. doi:10.1101/gad.176826.111
15. Zentner GE, Henikoff S (2012) Surveying the epigenomic landscape, one base at a time. Genome Biol 13:250. doi:10.1186/gb-2012-13-10-250
16. Jiang D, Berger F (2016) Histone variants in plant transcriptional regulation. Biochim Biophys Acta. doi:10.1016/j.bbagrm.2016.07.002
17. Huang H, Sabari BR, Garcia BA, Allis CD, Zhao Y (2014) SnapShot: histone modifications. Cell 159:458–458.e1. doi:10.1016/j.cell.2014.09.037
18. Kornberg RD, Lorch Y (1999) Twenty-five years of the nucleosome, fundamental particle of the eukaryote chromosome. Cell 98:285–294. doi:10.1016/S0092-8674(00)81958-3
19. Feng J, Shen WH (2014) Dynamic regulation and function of histone monoubiquitination in plants. Front Plant Sci 5:83. doi:10.3389/fpls.2014.00083
20. Shogren-Knaak M, Ishii H, Sun J-M, Pazin MJ, Davie JR, Peterson CL (2006) Histone H4-K16 acetylation controls chromatin structure and protein interactions. Science 311:844–847. doi:10.1126/science.1124000
21. Robinson PJ, An W, Routh A, Martino F, Chapman L, Roeder RG, Rhodes D (2008) 30 nm chromatin fibre decompaction requires both H4-K16 acetylation and linker histone eviction. J Mol Biol 381:816–825
22. Fierz B, Chatterjee C, McGinty RK, Bar-Dagan M, Raleigh DP, Muir TW (2011) Histone H2B ubiquitylation disrupts local and higher-order chromatin compaction. Nat Chem Biol 7:113–119
23. Smith E, Shilatifard A (2010) The chromatin signaling pathway: diverse mechanisms of recruitment of histone-modifying enzymes and varied biological outcomes. Mol Cell 40:689–701
24. Krueger F, Kreck B, Franke A, Andrews SR (2012) DNA methylome analysis using short bisulfite sequencing data. Nat Methods 9:145–151. doi:10.1038/nmeth.1828
25. Chen Y-R, Sheng Y, Zhong S (2017) Profiling DNA methylation using bisulfite sequencing (BS-Seq). In: Bemer M, Baroux C (eds) Plant chromatin dynamics: methods and protocols. Springer, New York, NY. doi:10.1007/978-1-4939-7318-7_2
26. Law JA, Jacobsen SE (2010) Establishing, maintaining and modifying DNA methylation patterns in plants and animals. Nat Rev Genet 11:204–220
27. Feng S, Jacobsen SE (2011) Epigenetic modifications in plants: an evolutionary perspective. Curr Opin Plant Biol 14:179–186
28. Baroux C, Raissig MT, Grossniklaus U (2011) Epigenetic regulation and reprogramming during gamete formation in plants. Curr Opin Genet Dev 21:124–133. doi:10.1016/j.gde.2011.01.017
29. Jullien PE, Susaki D, Yelagandula R, Higashiyama T, Berger F (2012) DNA methylation dynamics during sexual reproduction in Arabidopsis thaliana. Curr Biol 22:1825–1830. doi:10.1016/j.cub.2012.07.061
30. Bourc'his D, Voinnet O (2010) A small-RNA perspective on gametogenesis, fertilization, and early zygotic development. Science 330:617–622. doi:10.1126/science.1194776
31. Berger F, Gaudin V (2003) Chromatin dynamics and Arabidopsis development. Chromosome Res 11:277–304
32. Crevillen P, Dean C (2010) Regulation of the floral repressor gene FLC: the complexity of transcription in a chromatin context. Curr Opin Plant Biol 14:38–44
33. Berr A, Shafiq S, Shen W-H (2011) Histone modifications in transcriptional activation during plant development. Biochim Biophys Acta 1809:567–576

34. He G, Elling AA, Deng XW (2011) The epigenome and plant development. Annu Rev Plant Biol 62:411–435
35. Grimanelli D, Roudier F (2013) Epigenetics and development in plants: green light to convergent innovations. Curr Top Dev Biol 104:189–222. doi:10.1016/B978-0-12-416027-9.00006-1
36. Patel DJ, Wang Z (2013) Readout of epigenetic modifications. Annu Rev Biochem 82:81–118. doi:10.1146/annurev-biochem-072711-165700
37. Liu C, Weigel D (2015) Chromatin in 3D: progress and prospects for plants. Genome Biol 16:170. doi:10.1186/s13059-015-0738-6
38. Barneche F, Malapeira J, Mas P (2014) The impact of chromatin dynamics on plant light responses and circadian clock function. J Exp Bot 65:2895–2913. doi:10.1093/jxb/eru011
39. Perrella G, Kaiserli E (2016) Light behind the curtain: photoregulation of nuclear architecture and chromatin dynamics in plants. New Phytol 212:908–919. doi:10.1111/nph.14269
40. Sullivan AM, Arsovski AA, Lempe J, Bubb KL, Weirauch MT, Sabo PJ, Sandstrom R, Thurman RE, Neph S, Reynolds AP, Stergachis AB, Vernot B, Johnson AK, Haugen E, Sullivan ST, Thompson A, Neri FV 3rd, Weaver M, Diegel M, Mnaimneh S, Yang A, Hughes TR, Nemhauser JL, Queitsch C, Stamatoyannopoulos JA (2014) Mapping and dynamics of regulatory DNA and transcription factor networks in A. thaliana. Cell Rep 8:2015–2030. doi:10.1016/j.celrep.2014.08.019
41. Charron J-BF, He H, Elling AA, Deng XW (2009) Dynamic landscapes of four histone modifications during deetiolation in Arabidopsis. Plant Cell 21:3732–3748
42. Bourbousse C, Ahmed I, Roudier F, Zabulon G, Blondet E, Balzergue S, Colot V, Bowler C, Barneche F (2012) Histone H2B monoubiquitination facilitates the rapid modulation of gene expression during Arabidopsis photomorphogenesis. PLoS Genet 8:e1002825. doi:10.1371/journal.pgen.1002825
43. Benhamed M, Bertrand C, Servet C, Zhou DX (2006) Arabidopsis GCN5, HD1, and TAF1/HAF2 interact to regulate histone acetylation required for light-responsive gene expression. Plant Cell 18:2893–2903
44. Quint M, Delker C, Franklin KA, Wigge PA, Halliday KJ, van Zanten M (2016) Molecular and genetic control of plant thermomorphogenesis. Nat Plants 2:15190. doi:10.1038/nplants.2015.190
45. March-Díaz R, Reyes JC (2009) The beauty of being a variant: H2A.Z and the SWR1 complex in plants. Mol Plant 2(7):565–577. doi:10.1093/mp/ssp019
46. Kumar SV, Wigge PA (2010) H2A.Z-containing nucleosomes mediate the thermosensory response in Arabidopsis. Cell 140:136–147
47. Kumar SV, Lucyshyn D, Jaeger KE, Alos E, Alvey E, Harberd NP, Wigge PA (2012) Transcription factor PIF4 controls the thermosensory activation of flowering. Nature 484:242–245. doi:10.1038/nature10928
48. Coleman-Derr D, Zilberman D (2012) Deposition of histone variant H2A.Z within gene bodies regulates responsive genes. PLoS Genet 8:e1002988. doi:10.1371/journal.pgen.1002988
49. Koike N, Yoo SH, Huang HC, Kumar V, Lee C, Kim TK, Takahashi JS (2012) Transcriptional architecture and chromatin landscape of the core circadian clock in mammals. Science 338:349–354. doi:10.1126/science.1226339
50. Baerenfaller K, Shu H, Hirsch-Hoffmann M, Fütterer J, Opitz L, Rehrauer H, Hennig L, Gruissem W (2016) Diurnal changes in the histone H3 signature H3K9ac|H3K27ac|H3S28p are associated with diurnal gene expression in Arabidopsis. Plant Cell Environ 39:2557–2569. doi:10.1111/pce.12811
51. Seo PJ, Mas P (2015) STRESSing the role of the plant circadian clock. Trends Plant Sci 20:230–237. doi:10.1016/j.tplants.2015.01.001
52. Filion GJ, van Bemmel JG, Braunschweig U, Talhout W, Kind J, Ward LD, Brugman W, de Castro IJ, Kerkhoven RM, Bussemaker HJ (2010) Systematic protein location mapping reveals five principal chromatin types in Drosophila cells. Cell 143:212–224
53. Roudier F, Ahmed I, Bérard C, Sarazin A, Mary-Huard T, Cortijo S, Bouyer D, Caillieux E, Duvernois-Berthet E, Al-Shikhley L (2011) Integrative epigenomic mapping defines four main chromatin states in Arabidopsis. EMBO J 30:1928–1938
54. Sequeira-Mendes J, Araguez I, Peiro R, Mendez-Giraldez R, Zhang X, Jacobsen SE, Bastolla U, Gutierrez C (2014) The functional topography of the arabidopsis genome is organized in a reduced number of linear motifs of chromatin states. Plant Cell 26:2351–2366. doi:10.1105/tpc.114.124578

55. Baker K, Dhillon T, Colas I, Cook N, Milne I, Milne L, Bayer M, Flavell AJ (2015) Chromatin state analysis of the barley epigenome reveals a higher-order structure defined by H3K27me1 and H3K27me3 abundance. Plant J 84:111–124. doi:10.1111/tpj.12963
56. Bonev B, Cavalli G (2016) Organization and function of the 3D genome. Nat Rev Genet 17:661–678. doi:10.1038/nrg.2016.112
57. Grob S, Schmid MW, Grossniklaus U (2014) Hi-C analysis in Arabidopsis identifies the KNOT, a structure with similarities to the flamenco locus of Drosophila. Mol Cell 55:678–693. doi:10.1016/j.molcel.2014.07.009
58. Feng S, Cokus SJ, Schubert V, Zhai J, Pellegrini M, Jacobsen SE (2014) Genome-wide Hi-C analyses in wild-type and mutants reveal high-resolution chromatin interactions in Arabidopsis. Mol Cell 55:694–707. doi:10.1016/j.molcel.2014.07.008
59. Veluchamy A, Jégu T, Ariel F, Latrasse D, Mariappan KG, Kim S-K, Crespi M, Hirt H, Bergounioux C, Raynaud C, Benhamed M (2016) LHP1 regulates H3K27me3 spreading and shapes the three-dimensional conformation of the arabidopsis genome. PLoS One 11:e0158936. doi:10.1371/journal.pone.0158936
60. Liu C, Wang C, Wang G, Becker C, Zaidem M, Weigel D (2016) Genome-wide analysis of chromatin packing in Arabidopsis thaliana at single-gene resolution. Genome Res 26 (8):1057. doi:10.1101/gr.204032.116
61. Wang C, Liu C, Roqueiro D, Grimm D, Schwab R, Becker C, Lanz C, Weigel D (2014) Genome-wide analysis of local chromatin packing in Arabidopsis thaliana. Genome Res 25(2):246. doi:10.1101/gr.170332.113
62. Zhang Y, Wong C-H, Birnbaum RY, Li G, Favaro R, Ngan CY, Lim J, Tai E, Poh HM, Wong E, Mulawadi FH, Sung W-K, Nicolis S, Ahituv N, Ruan Y, Wei C-L (2013) Chromatin connectivity maps reveal dynamic promoter-enhancer long-range associations. Nature 504:306–310. doi:10.1038/nature12716
63. van Koningsbruggen S, Gierliński M, Schofield P, Martin D, Barton GJ, Ariyurek Y, den Dunnen JT, Lamond AI (2010) High-resolution whole-genome sequencing reveals that specific chromatin domains from most human chromosomes associate with nucleoli. Mol Biol Cell 21:3735–3748. doi:10.1091/mbc.E10-06-0508
64. Németh A, Conesa A, Santoyo-Lopez J, Medina I, Montaner D, Péterfia B, Solovei I, Cremer T, Dopazo J, Längst G (2010) Initial genomics of the human nucleolus. PLoS Genet 6:e1000889. doi:10.1371/journal.pgen.1000889
65. Pontvianne F, Carpentier M-C, Durut N, Pavlištová V, Jaške K, Schořová Š, Parrinello H, Rohmer M, Pikaard CS, Fojtová M, Fajkus J, Sáez-Vásquez J (2016) Identification of nucleolus-associated chromatin domains reveals a role for the nucleolus in 3D organization of the A. thaliana genome. Cell Rep 16:1574–1587. doi:10.1016/j.celrep.2016.07.016
66. Pickersgill H, Kalverda B, de Wit E, Talhout W, Fornerod M, van Steensel B (2006) Characterization of the Drosophila melanogaster genome at the nuclear lamina. Nat Genet 38:1005–1014. doi:10.1038/ng1852
67. Malapeira J, Khaitova LC, Mas P (2012) Ordered changes in histone modifications at the core of the Arabidopsis circadian clock. Proc Natl Acad Sci U S A 109:21540–21545. doi:10.1073/pnas.1217022110
68. Takahashi N, Hirata Y, Aihara K, Mas P (2015) A hierarchical multi-oscillator network orchestrates the arabidopsis circadian system. Cell 163(1):148–159. doi:10.1016/j.cell.2015.08.062
69. Engelhorn J, Wellmer F, Carles CC (2017) Profiling histone modifications in synchronised floral tissues for quantitative resolution of chromatin and transcriptome dynamics. In: Bemer M, Baroux C (eds) Plant chromatin dynamics: methods and protocols. Springer, New York, NY. doi:10.1007/978-1-4939-7318-7_16
70. Deal RB, Henikoff S (2011) The INTACT method for cell type-specific gene expression and chromatin profiling in Arabidopsis thaliana. Nat Protoc 6:56–68. doi:10.1038/nprot.2010.175
71. Deal RB, Henikoff S (2010) A simple method for gene expression and chromatin profiling of individual cell types within a tissue. Dev Cell 18(6):1030. doi:10.1016/j.devcel.2010.05.013
72. Morao K, Caillieux E, Colot V, Roudier F (2017) Cell type-specific profiling of chromatin modifications and associated proteins. In: Bemer M, Baroux C (eds) Plant chromatin dynamics: methods and protocols. Springer, New York, NY. doi:10.1007/978-1-4939-7318-7_8
73. Erhard KF, Talbot J-ERB, Deans NC, McClish AE, Hollick JB (2015) Nascent transcription affected by RNA polymerase IV in

Zea mays. Genetics 199:1107–1125. doi:10.1534/genetics.115.174714

74. Hetzel J, Duttke SH, Benner C, Chory J (2016) Nascent RNA sequencing reveals distinct features in plant transcription. Proc Natl Acad Sci 113:12316–12321. doi:10.1073/pnas.1603217113
75. Bourbousse C, Mestiri I, Zabulon G, Bourge M, Formiggini F, Koinig M, Spencer CB, Fransz P, Bowler C, Barneche F (2015) Heterochromatin reorganization during photomorphogenic reprogramming of plant development. Proc Natl Acad Sci U S A 112: E2836–E2844. doi:10.1073/pnas.1503512112
76. Huang H, Lin S, Garcia BA, Zhao Y (2015) Quantitative proteomic analysis of histone modifications. Chem Rev 115:2376–2418. doi:10.1021/cr500491u
77. Sidoli S, Cheng L, Jensen ON (2012) Proteomics in chromatin biology and epigenetics: elucidation of post-translational modifications of histone proteins by mass spectrometry. J Proteomics 75(6):3419–3433. doi:10.1016/j.jprot.2011.12.029
78. Zheng Y, Huang X, Kelleher NL (2016) Epiproteomics: quantitative analysis of histone marks and codes by mass spectrometry. Curr Opin Chem Biol 33:142–150. doi:10.1016/j.cbpa.2016.06.007
79. Mahrez W, Hennig L (2017) Mapping of histone modifications in plants by tandem mass spectrometry. In: Bemer M, Baroux C (eds) Plant chromatin dynamics: methods and protocols. Springer, New York, NY. doi:10.1007/978-1-4939-7318-7_9
80. Kotliński M, Jerzmanowski A (2017) Histone H1 purification and PTM profiling by high-sensitive MS approaches (Orbitrap). In: Bemer M, Baroux C (eds) Plant chromatin dynamics: methods and protocols. Springer, New York, NY. doi:10.1007/978-1-4939-7318-7_10
81. Bergmuller E, Gehrig PM, Gruissem W (2007) Characterization of post-translational modifications of histone H2B-variants isolated from Arabidopsis thaliana. J Proteome Res 6:3655–3668
82. Zhang K, Sridhar VV, Zhu J, Kapoor A, Zhu J-K (2007) Distinctive core histone post-translational modification patterns in arabidopsis thaliana. PLoS One 2:e1210
83. Shanower GA, Muller M, Blanton JL, Honti V, Gyurkovics H, Schedl P (2005) Characterization of the grappa gene, the Drosophila histone H3 lysine 79 methyltransferase. Genetics 169:173–184. doi:10.1534/genetics.104.033191
84. Feng Q, Wang H, Ng HH, Erdjument-Bromage H, Tempst P, Struhl K, Zhang Y (2002) Methylation of H3-lysine 79 is mediated by a new family of HMTases without a SET domain. Curr Biol 12:1052–1058. doi:10.1016/S0960-9822(02)00901-6
85. Kotliński M, Rutowicz K, Kniżewski Ł, Palusiński A, Olędzki J, Fogtman A, Rubel T, Koblowska M, Dadlez M, Ginalski K, Jerzmanowski A (2016) Histone H1 variants in arabidopsis are subject to numerous post-translational modifications, both conserved and previously unknown in histones, suggesting complex functions of H1 in plants. PLoS One 11:e0147908. doi:10.1371/journal.pone.0147908
86. Zemach A, Kim MY, Hsieh PH, Coleman-Derr D, Eshed-Williams L, Thao K, Harmer SL, Zilberman D (2013) The Arabidopsis nucleosome remodeler DDM1 allows DNA methyltransferases to access H1-containing heterochromatin. Cell 153:193–205. doi:10.1016/j.cell.2013.02.033
87. Rutowicz K, Puzio M, Halibart-Puzio J, Lirski M, Kroten MA, Kotlinski M, Knizewski L, Lange B, Muszewska A, Sniegowska-Swierk K, Koscielniak J, Iwanicka-Nowicka R, Zmuda K, Buza K, Janowiak F, Joesaar I, Laskowska-Kaszub K, Fogtman A, Zielenkiewicz P, Tiuryn J, Kollist H, Siedlecki P, Ginalski K, Swiezewski S, Koblowska M, Archacki R, Wilczynski B, Rapacz M, Jerzmanowski A (2015) A specialized histone H1 variant is required for adaptive responses to complex abiotic stress and related DNA methylation in Arabidopsis. Plant Physiol 169 (3):2080. doi:10.1104/pp.15.00493
88. Jacob Y, Bergamin E, Donoghue MT, Mongeon V, LeBlanc C, Voigt P, Underwood CJ, Brunzelle JS, Michaels SD, Reinberg D, Couture JF, Martienssen RA (2014) Selective methylation of histone H3 variant H3.1 regulates heterochromatin replication. Science 343:1249–1253. doi:10.1126/science.1248357
89. Johnson L, Mollah S, Garcia BA, Muratore TL, Shabanowitz J, Hunt DF, Jacobsen SE (2004) Mass spectrometry analysis of Arabidopsis histone H3 reveals distinct combinations of post-translational modifications. Nucleic Acids Res 32:6511–6518. doi:10.1093/nar/gkh992
90. Luo C, Sidote DJ, Zhang Y, Kerstetter RA, Michael TP, Lam E (2013) Integrative analysis of chromatin states in Arabidopsis identified potential regulatory mechanisms for

natural antisense transcript production. Plant J 73:77–90. doi:10.1111/tpj.12017

91. Desvoyes B, Sequeira-Mendes J, Vergara Z, Madeira S, Gutierrez C (2017) Sequential ChIP protocol for profiling bivalent epigenetic modifications (ReChIP). In: Bemer M, Baroux C (eds) Plant chromatin dynamics: methods and protocols. Springer, New York, NY. doi:10.1007/978-1-4939-7318-7_6
92. Pasini D, Malatesta M, Jung HR, Walfridsson J, Willer A, Olsson L, Skotte J, Wutz A, Porse B, Jensen ON, Helin K (2010) Characterization of an antagonistic switch between histone H3 lysine 27 methylation and acetylation in the transcriptional regulation of Polycomb group target genes. Nucleic Acids Res 38:4958–4969. doi:10.1093/nar/gkq244
93. Gendrel A-V, Lippman Z, Martienssen R, Colot V (2005) Profiling histone modification patterns in plants using genomic tiling microarrays. Nat Methods 2:213–218. doi:10.1038/nmeth0305-213
94. Desvoyes B, Vergara Z, Sequeira-Mendes JO, Madeira S, Gutierrez C (2017) A rapid and efficient ChIP protocol to profile chromatin binding proteins and epigenetic modifications in bulk Arabidopsis tissue. In: Bemer M, Baroux C (eds) Plant chromatin dynamics: methods and protocols. Springer, New York, NY. doi:10.1007/978-1-4939-7318-7_5
95. Landt SG, Marinov GK, Kundaje A, Kheradpour P, Pauli F, Batzoglou S, Bernstein BE, Bickel P, Brown JB, Cayting P, Chen Y, DeSalvo G, Epstein C, Fisher-Aylor KI, Euskirchen G, Gerstein M, Gertz J, Hartemink AJ, Hoffman MM, Iyer VR, Jung YL, Karmakar S, Kellis M, Kharchenko PV, Li Q, Liu T, Liu XS, Ma L, Milosavljevic A, Myers RM, Park PJ, Pazin MJ, Perry MD, Raha D, Reddy TE, Rozowsky J, Shoresh N, Sidow A, Slattery M, Stamatoyannopoulos JA, Tolstorukov MY, White KP, Xi S, Farnham PJ, Lieb JD, Wold BJ, Snyder M (2012) ChIP-seq guidelines and practices of the ENCODE and modENCODE consortia. Genome Res 22:1813–1831. doi:10.1101/gr.136184.111
96. Egelhofer TA, Minoda A, Klugman S, Lee K, Kolasinska-Zwierz P, Alekseyenko AA, Cheung M-S, Day DS, Gadel S, Gorchakov AA, Gu T, Kharchenko PV, Kuan S, Latorre I, Linder-Basso D, Luu Y, Ngo Q, Perry M, Rechtsteiner A, Riddle NC, Schwartz YB, Shanower GA, Vielle A, Ahringer J, Elgin SCR, Kuroda MI, Pirrotta V, Ren B, Strome S, Park PJ, Karpen GH, Hawkins RD, Lieb JD (2011) An assessment of histone-modification antibody quality. Nat Struct Mol Biol 18:91–93. doi:10.1038/nsmb.1972
97. Bernatavichute YV, Zhang X, Cokus S, Pellegrini M, Jacobsen SE (2008) Genome-wide association of histone H3 lysine nine methylation with CHG DNA methylation in Arabidopsis thaliana. PLoS One 3:e3156. doi:10.1371/journal.pone.0003156
98. Bonhoure N, Bounova G, Bernasconi D, Praz V, Lammers F, Canella D, Willis IM, Herr W, Hernandez N, Delorenzi M, The CycliX Consortium (2014) Quantifying ChIP-seq data: a spiking method providing an internal reference for sample-to-sample normalization. Genome Res 24:1157–1168. doi:10.1101/gr.168260.113
99. Orlando DA, Chen MW, Brown VE, Solanki S, Choi YJ, Olson ER, Fritz CC, Bradner JE, Guenther MG (2014) Quantitative ChIP-Seq normalization reveals global modulation of the epigenome. Cell Rep 9:1163–1170. doi:10.1016/j.celrep.2014.10.018
100. Egan B, Yuan C-C, Craske ML, Labhart P, Guler GD, Arnott D, Maile TM, Busby J, Henry C, Kelly TK, Tindell CA, Jhunjhunwala S, Zhao F, Hatton C, Bryant BM, Classon M, Trojer P (2016) An alternative approach to ChIP-Seq normalization enables detection of genome-wide changes in histone H3 lysine 27 trimethylation upon EZH2 inhibition. PLoS One 11:e0166438. doi:10.1371/journal.pone.0166438
101. Grzybowski AT, Chen Z, Ruthenburg AJ (2015) Calibrating ChIP-Seq with nucleosomal internal standards to measure histone modification density genome wide. Mol Cell 58:886–899. doi:10.1016/j.molcel.2015.04.022
102. Stroud H, Otero S, Desvoyes B, Ramírez-Parra E, Jacobsen SE, Gutierrez C (2012) Genome-wide analysis of histone H3.1 and H3.3 variants in Arabidopsis thaliana. Proc Natl Acad Sci 109:5370–5375. doi:10.1073/pnas.1203145109
103. Wollmann H, Holec S, Alden K, Clarke ND, Jacques P-É, Berger F (2012) Dynamic deposition of histone variant H3.3 accompanies developmental remodeling of the arabidopsis transcriptome. PLoS Genet 8:e1002658. doi:10.1371/journal.pgen.1002658
104. Pajoro A, Muiño J, Angenent G, Kaufmann K (2017) Profiling nucleosome occupancy by MNase-seq: experimental protocol and computational analysis. In: Bemer M, Baroux C (eds) Plant chromatin dynamics: methods and protocols. Springer, New York, NY. doi:10.1007/978-1-4939-7318-7_11

105. Gent JI, Madzima TF, Bader R, Kent MR, Zhang X, Stam M, McGinnis KM, Dawe RK (2014) Accessible DNA and relative depletion of H3K9me2 at maize loci undergoing RNA-directed DNA methylation. Plant Cell 26 (12):4903. doi:10.1105/tpc.114.130427
106. Chodavarapu R, Feng S, Bernatavichute Y, Chen P, Stroud H, Yu Y, Hetzel J, Kuo F, Kim J, Cokus S, Casero D, Bernal M, Huijser P, Clark A, Kramer U, Merchant S, Zhang X, Jacobsen S, Pellegrini M (2010) Relationship between nucleosome positioning and DNA methylation. Nature 466:388–392
107. Li G, Liu S, Wang J, He J, Huang H, Zhang Y, Xu L (2014) ISWI proteins participate in the genome-wide nucleosome distribution in Arabidopsis. Plant J 78:706–714. doi:10.1111/tpj.12499
108. Wu Y, Zhang W, Jiang J (2014) Genome-wide nucleosome positioning is orchestrated by genomic regions associated with DNase I hypersensitivity in rice. PLoS Genet 10: e1004378. doi:10.1371/journal.pgen.1004378
109. Fincher JA, Vera DL, Hughes DD, McGinnis KM, Dennis JH, Bass HW (2013) Genome-wide prediction of nucleosome occupancy in maize reveals plant chromatin structural features at genes and other elements at multiple scales. Plant Physiol 162:1127–1141. doi:10.1104/pp.113.216432
110. Vera DL, Madzima TF, Labonne JD, Alam MP, Hoffman GG, Girimurugan SB, Zhang J, McGinnis KM, Dennis JH, Bass HW (2014) Differential nuclease sensitivity profiling of chromatin reveals biochemical footprints coupled to gene expression and functional DNA elements in maize. Plant Cell 26:3883–3893. doi:10.1105/tpc.114.130609
111. Pecinka A, Dinh HQ, Baubec T, Rosa M, Lettner N, Mittelsten Scheid O (2010) Epigenetic regulation of repetitive elements is attenuated by prolonged heat stress in Arabidopsis. Plant Cell 22:3118–3129. doi:10.1105/tpc.110.078493
112. Sacharowski SP, Gratkowska DM, Sarnowska EA, Kondrak P, Jancewicz I, Porri A, Bucior E, Rolicka AT, Franzen R, Kowalczyk J, Pawlikowska K, Huettel B, Torti S, Schmelzer E, Coupland G, Jerzmanowski A, Koncz C, Sarnowski TJ (2015) SWP73 subunits of arabidopsis SWI/SNF chromatin remodeling complexes play distinct roles in leaf and flower development. Plant Cell 27:1889–1906. doi:10.1105/tpc.15.00233
113. Jegu T, Latrasse D, Delarue M, Hirt H, Domenichini S, Ariel F (2014) The BAF60 subunit of the SWI/SNF chromatin-remodeling complex directly controls the formation of a gene loop at FLOWERING LOCUS C in Arabidopsis. Plant Cell 26 (2):538. doi:10.1105/tpc.113.114454
114. Simon JM, Giresi PG, Davis IJ, Lieb JD (2012) Using formaldehyde-assisted isolation of regulatory elements (FAIRE) to isolate active regulatory DNA. Nat Protoc 7:256–267. doi:10.1038/nprot.2011.444
115. Gangadharan S, Mularoni L, Fain-Thornton J, Wheelan SJ, Craig NL (2010) DNA transposon Hermes inserts into DNA in nucleosome-free regions in vivo. Proc Natl Acad Sci 107:21966–21972. doi:10.1073/pnas.1016382107
116. Bajic M, Maher KA, Deal RB (2017) Identification of open chromatin regions in plant genomes using ATAC-Seq. In: Bemer M, Baroux C (eds) Plant chromatin dynamics: methods and protocols. Springer, New York, NY. doi:10.1007/978-1-4939-7318-7_12
117. Buenrostro JD, Giresi PG, Zaba LC, Chang HY, Greenleaf WJ (2013) Transposition of native chromatin for fast and sensitive epigenomic profiling of open chromatin, DNA-binding proteins and nucleosome position. Nat Methods 10:1213–1218. doi:10.1038/nmeth.2688
118. Buenrostro JD, Wu B, Litzenburger UM, Ruff D, Gonzales ML, Snyder MP, Chang HY, Greenleaf WJ (2015) Single-cell chromatin accessibility reveals principles of regulatory variation. Nature 523:486–490. doi:10.1038/nature14590
119. Lu Z, Hofmeister BT, Vollmers C, RM DB, Schmitz RJ (2017) Combining ATAC-seq with nuclei sorting for discovery of cis-regulatory regions in plant genomes. Nucleic Acids Res 45(6):e41. doi:10.1093/nar/gkw1179. gkw1179
120. Piper J, Elze MC, Cauchy P, Cockerill PN, Bonifer C, Ott S (2013) Wellington: a novel method for the accurate identification of digital genomic footprints from DNase-seq data. Nucleic Acids Res 41:e201–e201. doi:10.1093/nar/gkt850
121. Frommer M, McDonald LE, Millar DS, Collis CM, Watt F, Grigg GW, Molloy PL, Paul CL (1992) A genomic sequencing protocol that yields a positive display of 5-methylcytosine residues in individual DNA strands. Proc Natl Acad Sci 89:1827–1831. doi:10.1073/pnas.89.5.1827
122. Cokus SJ, Feng S, Zhang X, Chen Z, Merriman B, Haudenschild CD, Pradhan S, Nelson SF, Pellegrini M, Jacobsen SE (2008) Shotgun bisulphite sequencing of the Arabidopsis

genome reveals DNA methylation patterning. Nature 452:215–219. doi:10.1038/nature06745

123. Lister R, O'Malley RC, Tonti-Filippini J, Gregory BD, Berry CC, Millar AH, Ecker JR (2008) Highly integrated single-base resolution maps of the epigenome in arabidopsis. Cell 133:523–536

124. Stroud H, Greenberg MV, Feng S, Bernatavichute YV, Jacobsen SE (2013) Comprehensive analysis of silencing mutants reveals complex regulation of the *Arabidopsis methylome*. Cell 152:352–364

125. Heard E, Martienssen RA (2014) Transgenerational epigenetic inheritance: myths and mechanisms. Cell 157(3):95–109. doi:10.1016/j.cell.2014.02.045

126. Becker C, Weigel D (2012) Epigenetic variation: origin and transgenerational inheritance. Curr Opin Plant Biol 15(5):562. doi:10.1016/j.pbi.2012.08.004

127. Slotkin RK, Vaughn M, Borges F, Tanurdzic M, Becker JD, Feijó JA, Martienssen RA (2009) Epigenetic reprogramming and small RNA silencing of transposable elements in pollen. Cell 136:461–472. doi:10.1016/j.cell.2008.12.038

128. Calarco JP, Borges F, Donoghue MT, Van Ex F, Jullien PE, Lopes T, Gardner R, Berger F, Feijo JA, Becker JD, Martienssen RA (2012) Reprogramming of DNA methylation in pollen guides epigenetic inheritance via small RNA. Cell 151:194–205. doi:10.1016/j.cell.2012.09.001

129. Ibarra CA, Feng X, Schoft VK, Hsieh TF, Uzawa R, Rodrigues JA, Zemach A, Chumak N, Machlicova A, Nishimura T, Rojas D, Fischer RL, Tamaru H, Zilberman D (2012) Active DNA demethylation in plant companion cells reinforces transposon methylation in gametes. Science 337:1360–1364. doi:10.1126/science.1224839

130. Martínez G, Panda K, Köhler C, Slotkin RK (2016) Silencing in sperm cells is directed by RNA movement from the surrounding nurse cell. Nat Plants 2:16030. doi:10.1038/nplants.2016.30

131. Schmitz RJ, Schultz MD, Urich MA, Nery JR, Pelizzola M, Libiger O, Alix A, McCosh RB, Chen H, Schork NJ, Ecker JR (2013) Patterns of population epigenomic diversity. Nature 495:193–198. doi:10.1038/nature11968

132. Dubin MJ, Zhang P, Meng D, Remigereau M-S, Osborne EJ, Paolo Casale F, Drewe P, Kahles A, Jean G, Vilhjálmsson B, Jagoda J, Irez S, Voronin V, Song Q, Long Q, Rätsch G, Stegle O, Clark RM, Nordborg M (2015) DNA methylation in Arabidopsis has a genetic basis and shows evidence of local adaptation. Elife 4:e05255. doi:10.7554/eLife.05255

133. Willing E-M, Rawat V, Mandáková T, Maumus F, James GV, Nordström KJV, Becker C, Warthmann N, Chica C, Szarzynska B, Zytnicki M, Albani MC, Kiefer C, Bergonzi S, Castaings L, Mateos JL, Berns MC, Bujdoso N, Piofczyk T, de Lorenzo L, Barrero-Sicilia C, Mateos I, Piednoël M, Hagmann J, Chen-Min-Tao R, Iglesias-Fernández R, Schuster SC, Alonso-Blanco C, Roudier F, Carbonero P, Paz-Ares J, Davis SJ, Pecinka A, Quesneville H, Colot V, Lysak MA, Weigel D, Coupland G, Schneeberger K (2015) Genome expansion of Arabis alpina linked with retrotransposition and reduced symmetric DNA methylation. Nat Plants 1:14023. doi:10.1038/nplants.2014.23

134. Quadrana L, Bortolini Silveira A, Mayhew GF, LeBlanc C, Martienssen RA, Jeddeloh JA, Colot V (2016) The Arabidopsis thaliana mobilome and its impact at the species level. Elife 5:e15716. doi:10.7554/eLife.15716

135. Edelmann S, Scholten S (2017) Bisulphite sequencing using small DNA amounts. In: Bemer M, Baroux C (eds) Plant chromatin dynamics: methods and protocols. Springer, New York, NY. doi:10.1007/978-1-4939-7318-7_3

136. Suzuki MM, Bird A (2008) DNA methylation landscapes: provocative insights from epigenomics. Nat Rev Genet 9:465–476. doi:10.1038/nrg2341

137. Becker C, Hagmann J, Muller J, Koenig D, Stegle O, Borgwardt K, Weigel D (2011) Spontaneous epigenetic variation in the Arabidopsis thaliana methylome. Nature 480:245–249. doi:10.1038/nature10555

138. Schmitz RJ, Schultz MD, Lewsey MG, O'Malley RC, Urich MA, Libiger O, Schork NJ, Ecker JR (2011) Transgenerational epigenetic instability is a source of novel methylation variants. Science 334(6054):369. doi:10.1126/science.1212959

139. Kishore K, Pelizzola M (2017) Identification of differentially methylated regions in the Arabidopsis thaliana genome. In: Bemer M, Baroux C (eds) Plant chromatin dynamics: methods and protocols. Springer, New York, NY. doi:10.1007/978-1-4939-7318-7_4

140. Lieberman-Aiden E, Berkum NL, Williams L, Imakaev M, Ragoczy T, Telling A (2009)

Comprehensive mapping of long-range interactions reveals folding principles of the human genome. Science 326(5950):289. doi:10.1126/science.1181369

141. Baccarini J (1908) Sulle cinesi vegitative de Cynomorium coccineum L. Nuovo Giorn Botan Ital 15:189–203

142. Heitz E (1928) Das Heterochromatin der Moose. Jahrb Wiss Bot 69:762–818

143. Fransz P, De Jong JH, Lysak M, Castiglione MR, Schubert I (2002) Interphase chromosomes in Arabidopsis are organized as well defined chromocenters from which euchromatin loops emanate. Proc Natl Acad Sci U S A 99:14584–14589

144. Pecinka A, Schubert V, Meister A, Kreth G, Klatte M, Lysak M, Fuchs J, Schubert I (2004) Chromosome territory arrangement and homologous pairing in nuclei of *Arabidopsis thaliana* are predominantly random except for NOR-bearing chromosomes. Chromosoma 113:258–269

145. Chandrasekhara C, Mohannath G, Blevins T, Pontvianne F, Pikaard CS (2016) Chromosome-specific NOR inactivation explains selective rRNA gene silencing and dosage control in Arabidopsis. Genes Dev 30 (2):177. doi:10.1101/gad.273755.115

146. Grob S, Schmid MW, Luedtke NW, Wicker T, Grossniklaus U (2013) Characterization of chromosomal architecture in Arabidopsis by chromosome conformation capture. Genome Biol 14:R129

147. Guenatri M, Bailly D, Maison C, Almouzni G (2004) Mouse centric and pericentric satellite repeats form distinct functional heterochromatin. J Cell Biol 166:493–505

148. Mitchell JA, Fraser P (2008) Transcription factories are nuclear subcompartments that remain in the absence of transcription. Genes Dev 22:20–25

149. Tessadori F, van Zanten M, Pavlova P, Clifton R, Pontvianne F, Snoek LB, Millenaar FF, Schulkes RK, van Driel R, Voesenek LA (2009) Phytochrome B and histone deacetylase 6 control light-induced chromatin compaction in Arabidopsis thaliana. PLoS Genet 5:e1000638

150. Del Prete S, Arpon J, Sakai K, Andrey P, Gaudin V (2014) Nuclear architecture and chromatin dynamics in interphase nuclei of Arabidopsis thaliana. Cytogenet Genome Res 143:28–50. doi:10.1159/000363724

151. Probst AV, Mittelsten Scheid O (2015) Stress-induced structural changes in plant chromatin. Curr Opin Plant Biol 27:8–16. doi:10.1016/j.pbi.2015.05.011

152. van Zanten M, Tessadori F, McLoughlin F, Smith R, Millenaar FF, van Driel R, Voesenek LACJ, Peeters A, Fransz PF (2010) Photoreceptors CRYTOCHROME 2 and Phytochrome B control chromatin compaction in Arabidopsis thaliana. Plant Physiol 154 (4):1686

153. Grob S, Cavalli G (2017) Technical review: a Hitchhiker's guide to chromosome conformation capture. In: Bemer M, Baroux C (eds) Plant chromatin dynamics: methods and protocols. Springer, New York, NY. doi:10.1007/978-1-4939-7318-7_14

154. Schubert V (2014) RNA polymerase II forms transcription networks in rye and Arabidopsis nuclei and its amount increases with endopolyploidy. Cytogenet Genome Res 143:69–77. doi:10.1159/000365233

155. Roudier F, Teixeira FK, Colot V (2009) Chromatin indexing in Arabidopsis: an epigenomic tale of tails and more. Trends Genet 25:511–517. doi:10.1016/j.tig.2009.09.013

156. Rosa S, De Lucia F, Mylne JS, Zhu D, Ohmido N, Pendle A, Kato N, Shaw P, Dean C (2013) Physical clustering of FLC alleles during Polycomb-mediated epigenetic silencing in vernalization. Genes Dev 27:1845–1850. doi:10.1101/gad.221713.113

157. Crevillen P, Sonmez C, Wu Z, Dean C (2013) A gene loop containing the floral repressor FLC is disrupted in the early phase of vernalization. EMBO J 32(1):140. doi:10.1038/emboj.2012.324

158. Feng CM, Qiu Y, Van Buskirk EK, Yang EJ, Chen M (2014) Light-regulated gene repositioning in Arabidopsis. Nat Commun 5:3027. doi:10.1038/ncomms4027

159. Smith S, Galinha C, Desset S, Tolmie F, Evans D, Tatout C, Graumann K (2015) Marker gene tethering by nucleoporins affects gene expression in plants. Nucleus 6:471–478. doi:10.1080/19491034.2015.1126028

160. Amendola M, van Steensel B (2014) Mechanisms and dynamics of nuclear lamina–genome interactions. Curr Opin Cell Biol 28 (6):61–68. doi:10.1016/j.ceb.2014.03.003

161. Poulet A, Probst AV, Graumann K, Tatout C, Evans D (2017) Exploring the evolution of the proteins of the plant nuclear envelope. Nucleus 28:1–46. doi:10.1080/19491034.2016.1236166

162. Andersen JS, Lam YW, Leung AKL, Ong S-E, Lyon CE, Lamond AI, Mann M (2005) Nucleolar proteome dynamics. Nature 433:77–83. doi:10.1038/nature03207

163. Carpentier M-C, Picart-Picolo A, Pontvianne F (2017) A method to identify nucleolus-associated chromatin domains (NADs). In: Bemer M, Baroux C (eds) Plant chromatin dynamics: methods and protocols. Springer, New York, NY. doi:10.1007/978-1-4939-7318-7_7

164. Ding Y, Fromm M, Avramova Z (2012) Multiple exposures to drought "train" transcriptional responses in Arabidopsis. Nat Commun 3:740. doi:10.1038/ncomms1732

165. Sani E, Herzyk P, Perrella G, Colot V, Amtmann A (2013) Hyperosmotic priming of Arabidopsis seedlings establishes a long-term somatic memory accompanied by specific changes of the epigenome. Genome Biol 14: R59. doi:10.1186/gb-2013-14-6-r59

166. Sequeira-Mendes J, Gutierrez C (2016) Genome architecture: from linear organisation of chromatin to the 3D assembly in the nucleus. Chromosoma 125:455–469. doi:10.1007/s00412-015-0538-5

Chapter 2

Profiling DNA Methylation Using Bisulfite Sequencing (BS-Seq)

Yun-Ru Chen, Sheng Yu, and Silin Zhong

Abstract

DNA cytosine methylation is one of the most abundant epigenetic marks found in the plant nuclear genome. Bisulfite sequencing (BS-Seq) is the method of choice for profiling DNA cytosine methylation genome-wide at a single nucleotide resolution. The basis of this technique is that the unmethylated cytosine can be deaminated to uracil by sodium bisulfite, while the methylated cytosine is resistant to the treatment. By deep sequencing of the bisulfite converted genomic DNA, the methylation level of each mappable cytosine position in the genome could be measured. In this chapter, we present a detailed 2-day protocol for performing a BS-Seq experiment and a simple bioinformatic workflow for wet lab biologists to visualize the methylation data.

Key words DNA methylation, Bisulfite sequencing

1 Introduction

Bisulfite sequencing (BS-Seq) is the most efficient method for the analysis of methylation status at single nucleotide resolution [1, 2]. As Sanger sequencing and most of the second generation sequencing methods cannot distinguish methylated from unmethylated cytosines, a bisulfite treatment is required to convert the unmethylated cytosine to uracil prior to sequencing. However, third generation sequencing such as PacBio is capable of detecting DNA modifications based on the pausing time of the polymerase when encountering a modified base, and could therefore eventually render the BS-Seq method obsolete.

The success of a BS-Seq experiment depends on the near complete conversion of the unmethylated cytosine to uracil. A small amount (1%) of exogenous DNA, such as lambda phage DNA, is often added before the bisulfite treatment, and the conversion rate can then be calculated from the spike-in unmethylated phage DNA. Plant DNA extract is often contaminated with organelle genomic DNA such as the chloroplast DNA, which lacks 5mC. Hence the

Marian Bemer and Célia Baroux (eds.), *Plant Chromatin Dynamics: Methods and Protocols*, Methods in Molecular Biology, vol. 1675, DOI 10.1007/978-1-4939-7318-7_2, © Springer Science+Business Media LLC 2018

conversion rate could also be inferred from the chloroplast cytosine positions. For plant draft genomes without a chloroplast reference sequence, it might be necessary to include a lambda phage DNA spike-in. In addition to the conversion rate, sufficient sequence coverage is also essential for a successful BS-Seq experiment. Only cytosine positions with sufficient sequencing coverage will provide useful information and allow the determination of the methylation status, where a T indicates that the cytosine is unmethylated and a C that it is methylated. A typical BS-Seq experiment requires a sequencing depth of 10× to 20× genome coverage per strand, which is generally considered sufficient for calling differentially methylated regions (DMRs). However, nonunique alignments, such as the reads mapped to transposons and repeats, are discarded. Hence, in large repetitive plant genomes, the whole-genome coverage is often an underestimate of the actual cytosine coverage.

In this chapter, we provide a 2-day protocol for BS-Seq library preparation that is based on previously published methods [3] with some modifications to optimize for the current sequencing standards. In short, genomic DNA is first sonicated and the 300–500 bp fragments are recovered from agarose gel. The DNA fragments are end-repaired, dA-tailed and ligated to a fully methylated Illumina Y-shape adapter. Finally, the adapter ligated DNA fragments are bisulfite converted and PCR amplified prior to sequencing. In addition, we also provide a simple data analysis workflow that wet-lab biologist can perform to visualize the data. For advanced bioinformatic analysis, please refer to Chapter 4 [4] for details.

2 Materials

2.1 DNA Fragmentation and Size Selection

1. Agarose.
2. SYBR-Safe dye with compatible blue light transilluminator.
3. Optional: lambda phage DNA.
4. DNA loading buffer.
5. 1 kb DNA marker.
6. Gel running device.
7. Sonicator (e.g., Covaris M220 and Diagenode Bioruptor).
8. Purification kit for PCR product.
9. Purification kit for DNA extraction from agarose gel.

2.2 End Repair

1. DNA End Repair kit (e.g., from Enzymatics or NEB).
2. 10× T4 ligase buffer with ATP (e.g., from Enzymatics or NEB).

3. 10 mM dNTP w/o dCTP (10 mM each dATP, dGTP, and dTTP).
4. Thermocycler.
5. Magnetic stand for 0.2 ml PCR tubes.
6. AMPure XP purification beads.
7. 80% EtOH.

2.3 dA-Tailing

1. 10× T4 ligase buffer with ATP.
2. 10 mM dATP.
3. Klenow 3′–5′ exo- (e.g., from Enzymatics or NEB).
4. Thermocycler.
5. Magnetic stand for 0.2 ml PCR tubes.
6. AMPure XP purification beads.
7. 80% EtOH.

2.4 Adapter Ligation and Double AMPure XP Cleanup

1. 25 μM TruSeq 5mC double-stranded adapter:

   Adapter Oligos:

   Methylated TruSeq adapter oligo A (100 μM in TE buffer):

   A[5Me~dC]A[5Me~dC]T[5Me~dC]TTT[5Me~dC][5Me~dC][5Me~dC]TA[5Me~dC]A[5Me~dC]GA[5Me~dC]G[5Me~dC]T[5Me~dC]TT[5Me~dC][5Me~dC]GAT[5Me~dC]*T

   * indicates the phosphorothioate modification.

   Methylated TruSeq adapter oligo B (100 μM in TE buffer):

   [Phos]GAT[5Me~dC]GGAAGAG[5Me~dC]A[5Me~dC]A[5Me~dC]GT[5Me~dC]TGAA[5Me~dC]T[5Me~dC][5Me~dC]AGT[5Me~dC]A[5Me~dC]

   Anneal the adapter oligos first, to obtain a double-stranded adapter: Mix 25 μl of both adapter oligos (100 μM) in a 0.2 ml PCR tube. Heat to 90 °C in a thermocycler, and slowly chill to room temperate at a ramp rate of −1 °C/min. The annealed adapter (50 μM) can then be diluted to 25 μM with TE and stored in small aliquots at −20 °C.

2. 2× ligase buffer with ATP and PEG.
3. T4 Ligase HC (high concentration; e.g., from Enzymatics or NEB).
4. Thermocycler.
5. Magnetic stand for 0.2 ml PCR tubes.
6. AMPure XP purification beads.
7. 80% EtOH.
8. TE buffer: 10 mM Tris–HCl, 1 mM EDTA.

2.5 Bisulfite Conversion

1. Fluorometer (e.g., Invitrogen Qubit)
2. Fluorescent dye (e.g., Invitrogen Qubit dsDNA HS Assay Kit Q32851 or Promega QuantiFluor dsDNA Dye E258A)
3. Bisulfite conversion kit (e.g., Invitrogen MethylCode Bisulfite Conversion Kit or ZYMO EZ DNA Methylation-Gold Kit)
4. Thermocycler.
5. 100% EtOH.
6. Nuclease-free water.

2.6 PCR Amplification

1. Thermocycler.
2. DNA polymerase (e.g., KAPA HiFi Uracil+ or NEB Epimark Hotstart Taq).
3. TruSeq Adapter and index PCR primers:

   Primers:

   TruSeq universal PCR primer 1.0

   AATGATACGGCGACCACCGAGATCTACACTCTTTCCC TACACGACGCTCTTCCGATC*T
   TruSeq index PCR primer 2.x
   CAAGCAGAAGACGGCATACGAGAT[NNNNNN] GTGACTGGAGTTCAGACGTGTGCTCTTCCGATC*T
   * indicates the phosphorothioate modification.
   [NNNNNN] indicates the 6 nucleotide index sequence in the TruSeq PCR primer 2.
4. Thermocycler.
5. Magnetic stand for 0.2 ml PCR tubes.
6. AMPure XP purification beads.
7. 80% EtOH.
8. Nuclease-free water.

2.7 Analysis of BS-Seq Data

To perform BS-Seq data analysis, a computer with Linux/UNIX operating system is required, as well as installation of the here listed software [5, 6]. The instruction on how to download and install these packages can be found via the provided URLs.

1. BSMAP (https://github.com/zyndagj/BSMAP).
2. IGV Browser (http://software.broadinstitute.org/software/igv/).
3. Tabix (https://github.com/samtools/tabix).
4. R version 3.0.2 or higher (https://www.r-project.org/).
5. Bioconductor R package DSS (http://bioconductor.org/packages/release/bioc/html/DSS.html).
6. Samtools (http://www.htslib.org/).

3 Methods

3.1 DNA Fragmentation and Size Selection (TIMING: ~2 h)

Fragmentation of DNA creates small inserts for library preparation, and is the first step in next generation sequencing workflows. The current sequencing read length for the Illumina HiSeq X ten machine is paired end 150 bp. Hence, the ideal BS-Seq library insert should be slightly over 300 bp.

1. Prepare a 1% agarose gel with SYBR-Safe dye in advance.
2. Determine the concentration of the DNA sample; dilute 2 μg of DNA into 130 μl of water.
3. Optional: add 2 ng of lambda phage DNA as spike-in.
4. Fragment the DNA in a sonicator. We use the Covaris M220 sonicator, and select the 300 bp fragment peak size protocol (*see* **Note 1**).
5. Transfer the sample into a snap cap microTUBE and sonicate the DNA.
6. Optional: confirm the sonication efficiency by running 50 ng of the sonicated DNA in an agarose gel.
7. Concentrate the DNA using a PCR purification kit and elute in 25 μl elution buffer.
8. Perform gel electrophoresis and recover the DNA fragments in the range between 300 and 500 bp using a gel purification kit. Elute the DNA with 25 μl water.
9. Determine the DNA concentration, around 300 ng is required.

3.2 End Repair (TIMING: ~50 min)

The fragmented DNA contains different types of overhangs, and these need to be blunt-ended and phosphorylated for adapter ligation. This is achieved by the combined action of the T4 DNA polymerase and the T4 polynucleotide kinase in the End Repair enzyme mix.

1. Assemble the end repair reaction in a PCR tube as follows:

Components	Volume (μl)
300–500 ng of purified DNA in water	21
10× T4 Ligase buffer with ATP	2.5
10 mM dNTP w/o dCTP (*see* **Note 2**)	0.5
End Repair Enzyme Mix	1

2. Mix by pipetting, incubate in a thermocycler at 20 °C for 30 min.

3. Add 25 μl AMPure XP beads into the end-repair reaction, mix by pipetting (at least ten times) and incubate at room temperature for 10 min.
4. Place the End-Repair reaction with AMPure XP on a magnetic stand for ~1 min or until the beads are fully attached to the magnetic side of the tube. Carefully pipette out the solution without touching the beads. Do not remove the tube from the magnet.
5. Add 150 μl 80% EtOH without disturbing the beads and wait for 30 s, carefully remove the ethanol using the pipette.
6. Repeat the ethanol wash one more time. Do not remove the tube from the magnet throughout the process.
7. Use a new pipette tip to remove all residual ethanol from the bottom and the side of the PCR tube. Do not overdry the AMPure XP beads.
8. Add 20.5 μl of water to elute the end-repaired DNA, mix well by pipetting.
9. Place the PCR tube on the magnetic stand for ~1 min or until the beads are fully attached to the magnetic side of the tube. Carefully pipette out the end-repaired DNA to a new PCR tube without touching the beads.

3.3 dA-Tailing (TIMING: ~50 min)

To prevent ligation between the blunt and phosphorylated DNA fragments and also self-circularization, a 3′ dA overhang is added by the Klenow Fragment (3′ exo-) of the *E. coli* DNA polymerase I.

1. Prepare the dA-tailing reaction as follows:

Components	Volume (μl)
End-repaired DNA in water	20.5
10× T4 Ligase buffer with ATP	2.5
10 mM dATP	1
Klenow 3′–5′ exo-	1

2. Place the tube in a thermocycler and incubate at 37 °C for 30 min.
3. Add 25 μl AMPure XP beads into the tube, mix by pipetting (at least ten times) and incubate at room temperature for 10 min.
4. Place the tube on a magnetic stand for ~1 min or until the beads are fully attached to the magnetic side of the tube. Carefully pipette out the solution without touching the beads. Do not remove the tube from the magnet.

5. Add 150 μl 80% EtOH without disturbing the beads and wait for 30 s, carefully remove the ethanol using the pipette.
6. Repeat the ethanol wash one more time. Do not remove the tube from the magnet throughout the process.
7. Use a new pipette tip to remove all residual ethanol from the bottom and the side of the PCR tube. Do not overdry the AMPure XP beads.
8. Add 22 μl of water to elute the dA-tailed DNA, mix well by pipetting.
9. Place the PCR tube on the magnetic stand for ~1 min or until the beads are fully attached to the magnetic side of the tube. Carefully pipette out all the DNA to a new PCR tube without touching the beads.

3.4 Adapter Ligation and Double AMPure XP Clean Up (TIMING: ~2 h or Overnight)

After dA-tailing, the DNA fragment with 3′ dA overhang is ready to be ligated to the Illumina adapter, which contains a complementary 3′ dT overhang. Compared to the regular Illumina adapter, the cytosines in the BS-Seq adapter are methylated, and will not be bisulfite converted to uridine.

1. Prepare the ligation reaction as follows:

Components	Volume (μl)
dA-tailed DNA in water	22
2× Ligase buffer with ATP and PEG	25
25 μM TruSeq 5mC double stranded adapter	2
T4 Ligase	1

2. The solution containing 2× ligase buffer is very viscous, pipette up and down at least ten times to mix after adding the T4 DNA Ligase.
3. Incubate the tube in a thermocycler at 25 °C for 15 min and hold at 4 °C for at least 1 h. Alternatively, perform the ligation overnight at 16 °C (*see* **Note 3**).
4. Add 50 μl AMPure XP beads into the tube, mix by pipetting (at least ten times) and incubate at room temperature for 10 min.
5. Place the tube on a magnetic stand for ~1 min or until beads are fully attached to the magnetic side of the tube. Carefully pipette out the solution without touching the beads. Do not remove the tube from the magnet.
6. Add 150 μl 80% EtOH without disturbing the beads and wait for 30 s, carefully remove the ethanol using the pipette.

7. Repeat the ethanol wash one more time. Do not remove the tube from the magnet during the process.
8. Use a new pipette tip to remove all residual ethanol from the bottom and the side of the PCR tube. Do not overdry the AMPure XP beads.
9. Add 50 μl of TE to elute the adapter ligated DNA, mix well by pipetting.
10. Repeat the AMPure XP cleanup to achieve very pure DNA. After drying the beads, add 21 μl of TE to elute the adapter ligated DNA, mix well by pipetting.
11. Place the PCR tube on the magnetic stand for ~1 min or until the beads are fully attached to the magnetic side of the tube. Carefully pipette out all the DNA to a new PCR tube without touching the beads.

3.5 Bisulfite Conversion (TIMING: 3 h)

Sodium bisulfite can convert unmethylated cytosine into uracil, which can be subsequently converted to thymine during PCR. The 5-methylcytosines in the adapter and the DNA insert are protected from this conversion reaction. The following protocol describes a bisulfite conversion using the Invitrogen MethylCode Bisulfite Conversion Kit. We have also used conversion kits from QIAGEN and ZYMO with success following the condition recommended by the manufacturer.

1. Use 1 μl of the library for DNA concentration measurement on a Qubit fluorometer. The expected DNA amount is around 300–500 ng (*see* **Note 4**).
2. Set up the bisulfite conversion reaction as follows:

Components	Volume (μl)
Library DNA in TE	20
Freshly prepared CT conversion reagent	130

3. Mix the reaction carefully and incubate it in the PCR machine using the following cycle conditions:

Step	Time	Temp (°C)
Denature	10 min	98
Incubation	2.5 h	64
Hold	Indefinite	4

4. Purify the bisulfite converted library according to the kit's manual, except that the elution is performed using 2 × 20 μL water instead of the EB buffer, which interferes with PCR.

5. **Critical step**: the purified library is unstable, so it should be PCR-amplified immediately. Do not store the purified libraries at −20 °C. If storage is required, it can be kept in a 4 °C fridge for 1 day without significant loss.

3.6 PCR Amplification (TIMING: ~120 min)

Cytosine deamination occurs naturally inside the cell, and is a major promutagenic event. Archaeal DNA polymerase such as *Pfu* can recognize the uracil in the template strand and stall polymerization. Hence, uracil tolerant polymerase must be used to amplify the bisulfite-converted library.

1. Assemble the PCR reaction as follows (*see* **Note 5**):

Components	Volume (μl)
BS-treated library	20
2× KAPA HiFi Uracil+ ReadyMix	25
TruSeq universal primer (10 μM)	2.5
TruSeq index primer (10 μM)	2.5

2. Perform the PCR in a thermocycler using the following program.

Cycle	Denature	Anneal	Extend	Hold
1	94 °C for 1 min	–	–	–
6–8	98 °C for 10 s	65 °C for 20 s	72 °C for 30 s	–
1	–	–	72 °C for 3 min	–
1	–	–	–	4 °C

3. Purify the PCR product using 1 volume of AMPure XP as previously described. The library DNA is now ready for sequencing (*see* **Note 6**).

3.7 Analysis of BS-Seq Data: Read Mapping and Methylation Calling

In this example, we assume that there is a paired-end *Arabidopsis* BS-Seq Illumina dataset from sample 'At_C1'. The fastq files are named At_C1_R1.fq.gz and At_C1_R2.fq.gz. We also assume that the reference genome sequence file TAIR10.fa is stored in the same directory.

The following script will invoke BSMAP to align the BS-Seq reads against the Arabidopsis genome. The number following option '-p' in the script specifies the number of CPU cores to be used during the BSMAP mapping. The alignment BAM file is then used by the python program methratio.py supplied by BSMAP to call the methylation level for each cytosine position. The result (a bigwig file named AtC1_methratio.bw) can be directly viewed in the IGV browser.

Create a shell script (in the Unix-Shell command line interpreter):

```
$ cat >run_bsmap.sh
#!/bin/bash
sample=At_C1
ref=./TAIR10.fa
bsmap -a ${sample}_R1.fq.gz -b ${sample}_R2.fq.gz -d $ref -o $sample.bam -r 0 -p
8 -v 5 -q 10 2>${sample}_bsmap.err
methratio.py -o ${sample}_methratio.txt -w ${sample}_methratio.wig -b 1 -d $ref
-z -r $sample.bam 2>${sample}_methratio.err
samtools view -H $sample.bam | grep '^@SQ' | cut -f 2- | sed -e 's/SN://' -e 's/
LN://' >genome.size
wigToBigWig ${sample}_methratio.wig genome.size ${sample}_methratio.bw
bgzip ${sample}_methratio.txt
tabix -b 2 -e 2 -S 1 ${sample}_methratio.txt.gz
Press Ctrl-C to save and exit, then execute the script:
$ nohup bash run_bsmap.sh &
```

3.8 Analysis of BS-Seq Data: Differential Methylation Calling

Once the mapping and methylation calling are finished, one can call differentially methylated regions (DMRs) between two samples using DSS. In the following scenario, we assume that BS-Seq data from the control and treatment groups (with one biological replicate) have been processed as indicated above. The methylation file for two control samples, (At_C1_methratio.txt.gz and At_C2_methratio.txt.gz) and the treatment samples, (At_T1_methratio.txt.gz and At_T2_methratio.txt.gz) are in the same path.

Create an R script:

```
$ cat >run_dss.R
library(DSS)
require(bsseq)
readData <- function(file, context="CG", skip=1){
 if(! file.exists(file)){stop(file, " doesn't exist !!!")}
 classes = c("character", "integer", "character", "character", "NULL", "NULL",
"integer", "integer", rep("NULL", 4))
 names = c("chr", "pos", "strand", "context", "NULL", "NULL", "X", "N", rep
("NULL", 4))
 data <- read.table(file, colClasses=classes, col.names=names, skip=skip)
 data <- data[data$context==context, c("chr", "pos", "N", "X")]
}
sample.c1 <- "At_C1"
sample.c2 <- "At_C2"
sample.t1 <- "At_T1"
```

```
sample.t2 <- "At_T2"

sample.c <- "At_C"
sample.t <- "At_T"

file.c1 <- paste(sample.c1, "methratio.txt.gz", sep="_")
file.c2 <- paste(sample.c2, "methratio.txt.gz", sep="_")
file.t1 <- paste(sample.t1, "methratio.txt.gz", sep="_")
file.t2 <- paste(sample.t2, "methratio.txt.gz", sep="_")

context.all <- c("CG", "CHG", "CHH")

for(context in context.all){
 data.c1 <- readData(file.c1, context)
 data.c2 <- readData(file.c2, context)
 data.t1 <- readData(file.t1, context)
 data.t2 <- readData(file.t2, context)
 BSobj <- makeBSseqData(list(data.c1, data.c2, data.t1, data.t2), c(sample.c1,
sample.c2, sample.t1, sample.t2))
 rm(data1, data2)

 dml <- DMLtest(BSobj, group1=c(sample.c1, sample.c2), group2=c(sample.t1,
sample.t2), smoothing=F)

 dml.r <- callDML(dml, delta=0.1, p.threshold=0.05)
 dmr <- callDMR(dml, delta=0.1, p.threshold=0.05, minCG=4, dis.merge=100, pct.
sig=0.75)

 prefix <- paste(paste(sample.c, sample.t, sep="-"), context, sep=".")
 save(dml, file=paste(prefix, "Rdata", sep="."))
 write.table(dml.r, file=paste(prefix, "DMC.txt", sep="."), sep="\t", quote=F,
row.names=F)
 write.table(dmr, file=paste(prefix, "DMR.txt", sep="."), sep="\t", quote=F,
row.names=F)
}
```

Type Ctrl-C to save and exit.
Then execute the script:

```
$ nohup Rscript run_dss.R &
```

The output will be two tables containing the DMR and DMC information that can be viewed directly in MS Excel.

4 Notes

1. We often use 1–2 μg genomic DNA in 130 μl as starting material. Some polysaccharide-rich DNA, such as those prepared from fruits, need to be diluted and sonicated in multiple batches. In this protocol, we choose to purify the 300 bp DNA fragments because we use HiSeq X ten for sequencing and the current read length is paired end 150 bp.

2. Due to the lack of dCTP, 5′ overhang containing guanosine cannot be filled in, and, hence, those DNA fragments cannot be converted to library. If the starting DNA amount is limited (e.g., DNA from laser capture microdissection), we could add 1 μl of mung bean nuclease and incubate for 10 min to remove the remaining 5′ overhang after the end-repair reaction. One could also include dCTP in the end-repair reaction. The unmethylated dCTP added to the DNA ends could be excluded from the analysis, and the BSMAP package described in this chapter have an option "–trim-fillin N" supporting this type of end-repaired library.
3. Although a 15 min room temperature ligation reaction is often sufficient for RNA-Seq or ChIP-Seq library preparation, we found that for BS-Seq library preparation, an extended ligation at 4 °C or even overnight at 16 °C could significantly improve the library yield.
4. We do not recommend using more than 500 ng library DNA in each bisulfite reaction, as too much DNA can reduce conversion rate. If more DNA has to be used, the samples can be split into multiple bisulfite conversion reactions. On the other hand, too few DNA could lead to over bisulfite conversion, degradation of DNA fragments with low 5mC level and loss in the post-bisulfite column purification step. If DNA sample is limited, one could add some raw lambda DNA, which could not be PCR-amplified, to the bisulfite conversion reaction to bring the final DNA amount in the bisulfite reaction to the optimum range (300–500 ng).
5. The PCR cycle number depends on the input DNA amount, the adapter ligation efficiency and the loss during bisulfite conversion. In our hands, 100–200 ng of pre-bisulfite DNA requires no more than 8 cycles of PCR. The choice of a different polymerase could also affect the PCR cycle condition. For example, 2–3 more cycles are often required if Pfu Turbo Cx (Stratagene) is used. We found that the low cost NEB EpiMark Taq polymerase is as efficient as the KAPA Uracil+ enzyme. However, it will also amplify the adapter dimer if the AMPure XP cleanup step after adapter ligation failed to remove all adapters, and the PCR amplified libraries can be gel purified.
6. Each Illumina sequencing service provider has its preferred library format. The sequencing core facility we use requires that each library is to be submitted in TE solution with a concentration no <2 ng/μL and in a volume of no <10 μL. Some service providers prefer the user to ethanol precipitate 50–100 ng of library DNA with glycogen as carrier and post the library in the form of a dry pellet.

References

1. Frommer M, LE MD, Millar DS, Collis CM, Watt F, Grigg GW, Molloy PL, Paul CL (1992) A genomic sequencing protocol that yields a positive display of 5-methylcytosine residues in individual DNA strands. Proc Natl Acad Sci 89:1872–1831
2. Clark SJ, Harrison J, Paul CL, Frommer M (1994) High sensitivity mapping of methylated cytosines. Nucleic Acids Res 22:2990–2997
3. Lister R, O'Malley RC, Tonti-Filippini J, Gregory BD, Berry CC, Millar AH, Ecker JR (2008) Highly integrated single-base resolution maps of the epigenome in Arabidopsis. Cell 133:523–536
4. Kishore K, Pelizzola M (2017) Identification of differentially methylated regions in the Arabidopsis thaliana genome. In: Bemer M, Baroux C (eds) Plant chromatin dynamics: methods and protocols. Springer, New York, NY. doi:10.1007/978-1-4939-7318-7_4
5. Xi Y, Li W (2009) BSMAP: whole genome bisulfite sequence mapping program. BMC Bioinformatics 10:232
6. Wu H, Xu T, Feng H, Chen L, Li B, Yao B, Qin Z, Jin P, Conneely KN (2015) Detection of differentially methylated regions from whole-genome bisulfite sequencing data without replicates. Nucleic Acids Res 43:e141

Chapter 3

Bisulfite Sequencing Using Small DNA Amounts

Susanne Edelmann and Stefan Scholten

Abstract

Bisulfite sequencing (BS-seq) enables the detection of DNA methylation at cytosine residues (5mC) at single-nucleotide resolution. For many applications, a limiting factor of conventional BS-seq protocols is the high amount of DNA required, since the treatment with bisulfite causes severe DNA fragmentation. Here, we describe a post-bisulfite tagging method that accounts for this problem. Illumina-compatible BS-seq libraries can be obtained from as little as five single haploid maize cells, enabling whole genome BS-seq (WGBS) for the generation of genome-wide, cell-type specific DNA methylation profiles. The method can also be used to analyze defined fractions of genomes from limited samples by Reduced Representation Bisulfite Sequencing (RRBS). This involves restriction digestion, gel separation and fragment elution prior to BS-seq library preparation to enrich certain areas of the genome. This reduction of represented genomic regions lowers the sequencing cost considerably while providing an accurate assessment of total genome-wide DNA methylation levels and assessment of DNA methylation in categorical genomic regions.

Key words DNA methylation, Cell-type specific, Whole-genome bisulfite sequencing, Reduced representation bisulfite sequencing

1 Introduction

Cytosine methylation is a DNA modification with important regulatory functions in plants. The DNA methylation patterns in higher plants are particularly dynamic during early zygotic embryogenesis, and the epigenetic pattern established in this phase is maintained in the adult plant [1]. In addition, DNA demethylation is required for proper gene expression in the endosperm after fertilization [2]. To study DNA methylation in sexual plant reproduction, a method for comprehensive, genome-wide 5mC profiling of gametes, early embryogenesis and endosperm development is required, but naturally, the sample material is very limited, complicating 5mC profiling. In female gametes this limitation restricted 5mC profiling to individual genes in the past [3, 4]. The method we describe here is suitable for low sample amounts and recently proved to be effective for genome-wide 5mC profiles of female *Arabidopsis* and

Marian Bemer and Célia Baroux (eds.), *Plant Chromatin Dynamics: Methods and Protocols*, Methods in Molecular Biology, vol. 1675, DOI 10.1007/978-1-4939-7318-7_3, © Springer Science+Business Media LLC 2018

rice gametes [5]. At the same time it is very cost-effective, which is in particular interesting for breeding, since epigenetic variation is increasingly recognized as an important component of phenotypic variation [6] and crop yield [7], raising the demand for comprehensive genome-wide 5mC profiling of large breeding populations.

The "gold standard" for DNA methylation analysis is bisulfite sequencing, because it allows for the quantification of 5mC at individual cytosine residues. The basis for bisulfite sequencing is the treatment of genomic DNA with bisulfite, which converts cytosine to uracil while leaving 5mC unaffected. In common BS-seq protocols, the DNA is fragmented and sequencing adapters are ligated before bisulfite conversion. This strategy results in a loss of ligated DNA fragments due to DNA degradation by the bisulfite treatment and thus the need for relatively high DNA amounts to start with. To minimize DNA loss, we modified post-bisulfite adapter tagging methods, which were developed for subnanogram quantities of DNA [8] or single mammalian cells [9].

In our approach, the bisulfite treatment is performed first, leading to simultaneous DNA fragmentation and conversion of unmethylated cytosines to uracil. Complementary strand synthesis is primed by 5′ biotin modified oligonucleotides containing Illumina-compatible adapter sequences and a stretch of nine random nucleotides at the 3′ end. After capturing the tagged strands with streptavidin-coated magnetic beads, a second adapter is integrated in the same way, and PCR amplification is performed with indexed primers. This method can be combined with reduced representation bisulfite sequencing (RRBS) [10], in which only a fraction of the genome is used for sequencing, reducing the costs. The protocol we present here for bisulfite sequencing of small DNA amounts can either be used for whole-genome bisulfite sequencing (WGBS) or for the cost-effective RRBS [10]. The RRBS variant is in particular useful in case the genome under investigation is large or if many samples need to be analyzed. Through combination of our post-bisulfite adapter tagging protocol with modified techniques for DNA fragmentation and size selection [11–13], we developed a cost-effective RRBS method that is applicable to very low amounts of plant genomic DNA.

If RRBS is the method of choice, the protocol starts with DNA restriction with the methylation-insensitive Taq$^{\alpha}$I (restriction site 5′-T|CGA-3′) [14], followed by fragment size selection by common agarose gel electrophoresis and a custom elution protocol. If WGBS will be performed, the first steps of the protocol (Subheadings 3.1–3.3) can be skipped. We generated and successfully sequenced WGBS libraries from as little as 5 single haploid maize cells (approximately 12.5 pg genomic DNA) and RRBS libraries with approximately 375 ng genomic DNA from 7-day-old maize

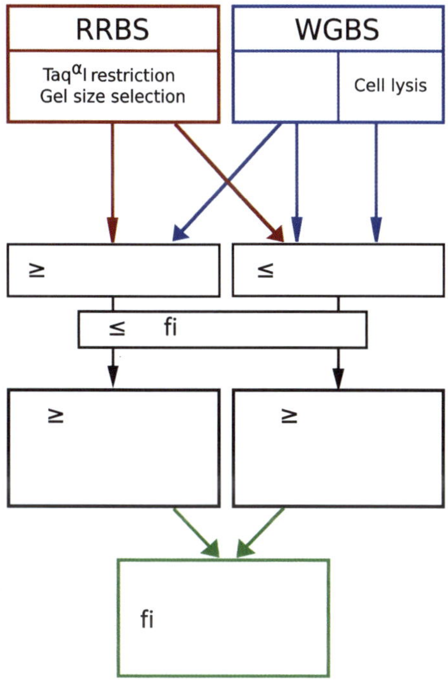

Fig. 1 Schematic overview of the alternative methods for BS-seq library preparation depending on the aim

embryos as input. Together, our protocol provides robust and cost-effective methods that are well adapted to various needs on BS-seq, which range from single cell-type to population-wide DNA methylation analyses (Fig. 1).

2 Materials

Prepare all solutions using ultrapure water and molecular biology grade reagents if not specified otherwise. Sterilize/decontaminate all buffers after preparation through filters with a pore size of 0.45 μm. Primers should be at least HPLC purified. In different steps, ethanol absolute to prepare washing solutions, ultrapure H_2O and 1 M Tris–HCl pH 8.5 is needed. For 1 M Tris, dissolve 121.14 g Tris in 800 mL H_2O. Adjust the pH to 8.5 by adding concentrated HCl. Adjust the volume to 1 L with H_2O. Sterilize the working space with a decontamination solution for the removal of nucleic acids contaminations (e.g., DNA-Exitus, Applichem; for alternatives please refer to [15]).

2.1 Taq$^{\alpha}$I Restriction

1. Taq$^{\alpha}$I restriction endonuclease (10 units/μL) and corresponding 10× Buffer.
2. Magnetic beads for purification or PCR purification kit (optional).

2.2 Gel Size Selection

1. 50× TAE buffer: Dissolve 242 g Tris in 500 mL H_2O for a 50× stock solution. Add 100 mL 0.5 M Na_2EDTA (pH 8.0) and 57.1 mL glacial acetic acid. Adjust volume to 1 L with H_2O. Dilute the 50× TAE to 1× before use.
2. Agarose (molecular grade).
3. 3 M NaOAc pH 5.2: Dissolve 408.24 g sodium acetate ($NaOAc*3H_2O$) in 800 mL H_2O. Adjust the pH to pH 5.2 with glacial acetic acid and add H_2O until 1 L.

2.3 Cell Lysis and Bisulfite Conversion

1. 2× Lysis/Binding buffer: 200 mM Tris–HCl (pH 7.5), 1 M LiCl, 20 mM EDTA (pH 8.0), and 2% lithium dodecyl sulfate (LiDS).
2. 0.5 mg/mL BSA: Dissolve 5 mg bovine serum albumin (BSA) in 10 mL H_2O. Store aliquots at −20 °C.
3. 40 mg/mL Pronase (e.g., Roche). Store aliquots at −20 °C.
4. Molecular Kit for bisulfite DNA conversion and purification (e.g., Imprint DNA Modification Kit, Sigma-Aldrich).

2.4 Klenow Reactions

1. 10 mM dNTPs.
2. Klenow exo- (High concentration, 3′-5′ exo-, 50 U/μL) and corresponding 10× Buffer.
3. 100 μM Biotin-labeled Oligo1:

   5′-(Biotin)CTACACGACGCTCTTCCGATCTNNNNNNN NN-3′.
4. 100 μM Oligo2:

   5′-TGCTGAACCGCTCTTCCGATCTNNNNNNNNN-3′.

2.5 Klenow Reaction Cleanup

1. Exonuclease I (40 U/μL).
2. Magnetic beads for purification (e.g., HighPrep PCR, MagBio, or Agencourt AMPureXP, Beckman Coulter).
3. Magnetic beads for isolating biotinylated nucleic acids, ca. 1–3 μm diameter (e.g., M-280 Streptavidin Dynabeads, Invitrogen or Streptavidin Magnetic Beads, NEB).
4. 5× Streptavidin Beads Binding Buffer: 25 mM Tris–HCl pH 7.5, 2.5 mM EDTA, and 5 M NaCl.
5. 0.1 N NaOH: Weight one pellet of NaOH directly in a tube. Add for each mg 12.5 μL H_2O for a stock solution of 2 N NaOH. From this stock solution prepare 0.1 N freshly before use.

2.6 Library Amplification

1. Amplification Kit for bisulfite-modified template DNA (e.g., EPIK Amplification Kit, Bioline).
2. 10 μM PE1.0 forward primer: 5′-AATGATACGGCGAC CACCGAGATCTACACTCTTTCCCTACACGACGCTCTT CCGATC*T-3′.

Table 1
Used Index sequences [16]

Oligo Name	Single correcting, double and shift detecting octamers	Sequence obtained
iPCRtagT1	AACGTGAT	ATCACGTTAT
iPCRtagT2	AAACATCG	CGATGTTTAT
iPCRtagT3	ATGCCTAA	TTAGGCATAT
iPCRtagT4	AGTGGTCA	TGACCACTAT
iPCRtagT5	ACCACTGT	ACAGTGGTAT
iPCRtagT6	ACATTGGC	GCCAATGTAT
iPCRtagT7	CAGATCTG	CAGATCTGAT
iPCRtagT8	CATCAAGT	ACTTGATGAT
iPCRtagT9	CGCTGATC	GATCAGCGAT
iPCRtagT10	ACAAGCTA	TAGCTTGTAT

3. 10 μM Index reverse primers, e.g., iPCRtagT1: 5′-CAAGCAGAAGACGGCATACGAGATAACGTGATGAGATCGGTCTCGGCATTCCTGCTGAACCGCTCTTCCGATC*T -3′. The * indicates phosphorothioate bonds; the underlined sequence indicates the eight nucleotide index sequence. Please refer to Table 1 or to [16] for differently indexed primers or for the primer to read the index sequence.

3 Methods

Set up all enzymatic reactions on ice except for the cell lysis (Subheading 3.4). The beads and column purification protocols should be carried out at room temperature. Use low binding/low retention tubes and tips. When working with limited sample amounts, it is extremely important to avoid contamination. The work should ideally take place in a sterile or semisterile room. At least, clean every surface with a decontamination solution for the removal of nucleic acids contaminations (e.g., DNA-Exitus, Applichem, or alternatives [15]) and 70% ethanol before starting the experiment. Use a bench just for these single cell experiments and perform PCR product analyses in a separate room. To check for contamination, always perform one reaction without DNA or cells but with all solutions and reagents used in parallel.

Depending on the type of starting material and the library type to be generated, different steps of the protocol are used as starting points:

Cells or tissue in lysis buffer for WGBS:

Start with the cell lysis (Subheading 3.4).

Isolated genomic DNA either for WGBS or RRBS:

For RRBS start with Taq$^{\alpha}$I restriction (Subheading 3.1).

For WGBS start with the cell lysis (Subheading 3.4) and proceed either with the one-step (Subheading 3.5.1) or the two-step (Subheading 3.5.2) imprint DNA modification procedure with more or less than 10 ng DNA, respectively. These limits also account for DNA fragments for RRBS. The one-step modification offers a more convenient procedure to be used with higher DNA input (>10 ng), whereas the two-step modification procedure is recommended for low DNA input (<10 ng) to increase conversion efficiency.

3.1 Taq$^{\alpha}$I Restriction

The gDNA can be extracted from any tissue (*see* **Note 1**), and should be purified by phenol/chloroform extraction and ethanol precipitation. The volume of water and gDNA depends on the sample concentration; we usually use 1–5 μg gDNA.

1. Mix 5 μL 10× CutSmart Buffer and 1 μL Taq$^{\alpha}$I (20 U/μL) with the sample and add ultrapure H_2O up to a volume of 50 μL.
2. Incubate at 65 °C for 90 min (*see* **Note 2**).

3.2 Restriction Cleanup (Optional)

The restriction purification is needed to visualize the digested DNA on the fragment analyzer (or similar) or if the samples are stored at >4 °C without enzyme inactivation (*see* **Notes 3–8**).

1. Add 50 μL magnetic beads to the sample (1:1) and mix well by pipetting. Incubate for 15 min at room temperature.
2. Wash 2× with freshly prepared 80% EtOH while leaving them on the magnetic stand with the pellet intact.
3. Dry the beads (ca. 3–5 min, *see* **Note 8**).
4. Add 43 μL 10 mM Tris–HCl pH 8.5 and mix well. Elute the restricted DNA for 5 min at room temperature.
5. Place the samples on the magnetic stand and wait for 5 min until the supernatant appears clear.
6. Transfer 42 μL of the supernatant to a new tube. We usually keep 2 μL for the quality check on the fragment analyzer (Fig. 2e) together with size selected DNA (Subheading 3.3).

3.3 Size Selection of Restricted DNA

1. Prepare a 1% agarose–EtBr gel by boiling 100 mL of 1× TAE buffer with 1 g agarose. Cool down to 60 °C, add 10 μL of 10 mg/mL EtBr and pour in an appropriate gel chamber tray.
2. Add 8 μL of 6× Loading Dye to the 40 μL of sample and load the entire volume onto the gel. We usually use the 1 kb DNA ladder as the marker. Then run the gel in 1× TAE buffer at 90 V for 45 min.

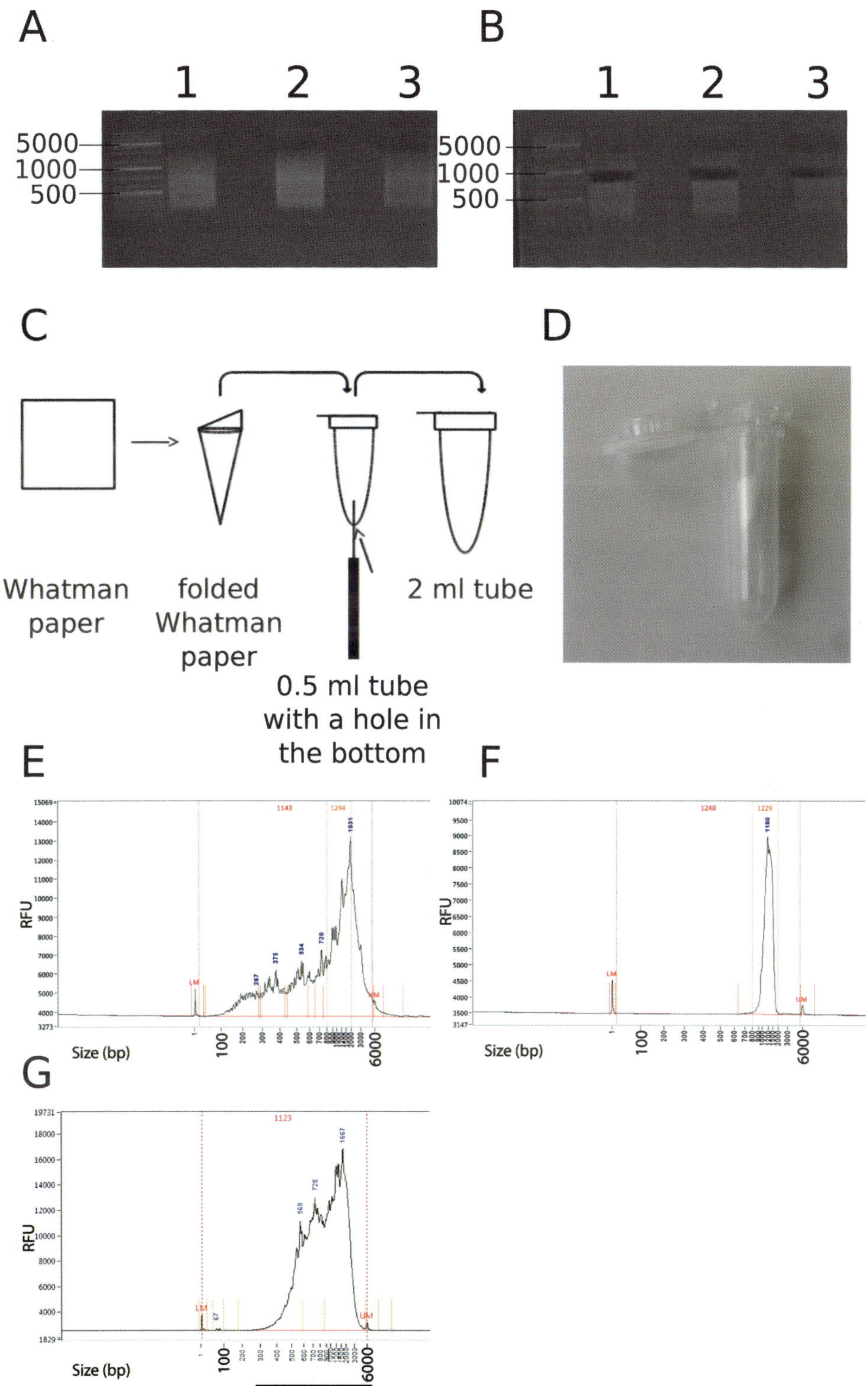

Fig. 2 Representative gel size selection of restricted canola DNA on a 1% agarose–ethidium bromide gel. Fragments were separated with 90 V for 45 min and size was determined with the GeneRuler 1 kb Plus DNA

3. Visualize the gel on a gel documentation instrument (Fig. 2a).
4. Use a clean razor or scalpel to excise a slice of gel from each sample lane corresponding to 1000–1200 bp (*see* **Note 9**, Fig. 2b).
5. Add the sample containing gel slice to the custom filter device (Fig. 2c, d) and centrifuge for 1 min at 16,000 rcf. There should be a clear DNA solution in the 2 mL tube.
6. On ice, add 0.1 Vol NaOAc and 2.5 Vol 100% EtOH to precipitate the DNA from the solution for at least 60 min at −20 °C (*see* **Note 10**).
7. Centrifuge for 20 min at >16,000 rcf at 4 °C.
8. Remove the supernatant and dry the pellet.
9. Resuspend the pellet in 14 μL 10 mM Tris–HCl pH 8.5. We analyze 2 μL for quality check and determination of the DNA concentration on the fragment analyzer together with the restricted DNA from Subheading 3.2 (Fig. 2f).
10. Proceed with Subheading 3.5.1 if the concentration is more than 1 ng/μL and with Subheading 3.5.2 if the concentration is less than 1 ng/μL.

3.4 Cell Lysis

1. Adjust the cells sampled in 1× Lysis/Binding buffer (*see* **Note 11**) with 0.5 mg/mL BSA to a volume of 11 μL (*see* **Note 12**).
2. Add 1 μL Pronase (40 mg/mL), mix thoroughly and incubate for 45 min at 37 °C.
3. Proceed with Subheading 3.5.2.

3.5 Bisulfite Conversion with Imprint DNA Modification Kit

Add 1.1 mL of DNA Modification Solution to 1 vial of DNA Modification Powder and mix well until the solution is clear. Add 40 μL Balance Solution and vortex briefly (Bisulfite Mix, *see* **Note 13**).

3.5.1 One-Step Modification Procedure

1. To 10 μL of gDNA (from Subheading 3.4) or size-selected DNA (from Subheading 3.3), add 110 μL Bisulfite Mix and vortex.

Fig. 2 (continued) ladder (Thermo Scientific). (**a**, **e**) Fragment length distribution after Taq$^{\alpha}$I restriction shows the main portion of 500–1500 bp fragments. (**b**, **f**) DNA fragments between 1000 and 1200 bp were cut from the gel. DNA was eluted from the gel via a "selfmade" column to elute DNA from an agarose gel. (**c**) The schematic overview shows the folded Whatman filter paper, which is located in a 0.5 mL tube with a subtle hole in the bottom. (**d**) The 0.5 mL tube is placed in a 2 mL tube to collect the DNA solution. (**e–g**) Fragment Analyzer electropherograms of 1 μg canola DNA that was digested with the restriction enzyme Taq$^{\alpha}$I (**e**) and size selected between 1000 and 1200 bp on a gel, which resulted in an actual fragment length distribution between 800 and 2000 bp (**f**). Fragment size selection by magnetic DNA purification beads revealed a broader distribution of fragment sizes (**g**)

2. Denature the DNA at 99 °C for 6 min and incubate at 65 °C for 90 min. Store the modified DNA at 7 °C overnight (*see* **Note 14**).

3.5.2 Two-Step Modification Procedure

1. Add 0.5 μL Balance solution to 12 μL gDNA/BSA solution (from Subheading 3.4, *see* **Note 15**) or to 12 μL size-selected DNA (from Subheading 3.3), mix and incubate for 10 min at 37 °C to denaturate the DNA.
2. Add 62.5 μL of Bisulfite Mix (0.5 reaction). Incubate in the thermocycler with a heated lid at 65 °C for 90 min, followed by a second denaturation at 95 °C for 3 min and extended modification at 65 °C for 20 min. Store the modified DNA at 7 °C overnight (*see* **Note 14**).

3.6 Desulfonation and Column Cleanup (Imprint DNA Modification Kit)

1. Prepare the following solutions according to the manufacturer's protocol: For the Ethanol-Diluted Cleaning Solution, add 8.2 mL of absolute EtOH to the bottle and mix. For 90% Ethanol Solution, add 500 μL ultrapure water to 4.5 mL absolute ethanol. For each sample, 400 μL 90% Ethanol is needed. For the Balance/Ethanol Wash Solution, mix 5 μL Balance Solution and 550 μL 90% EtOH. For each sample, 50 μL of this solution is needed.
2. Place a column in a 2 mL collection tube and add 300 μL of the Capture Solution.
3. Add the sample to the Capture Solution on the column and mix by pipetting. Centrifuge for 30 s at 16,000 rcf and discard the flow-through.
4. Add 200 μL of the Ethanol-Diluted Cleaning Solution to the column and centrifuge for 30 s at 16,000 rcf.
5. Add 50 μL of the freshly prepared Balance/Ethanol Wash Solution and incubate for 8 min at room temperature. After desulfonation, centrifuge for 30 s at 16,000 rcf and discard the flow-through.
6. Add 200 μL of the freshly prepared 90% EtOH to the column and centrifuge for 30 s at 16,000 rcf. Discard the flow-through.
7. Add 200 μL of the freshly prepared 90% EtOH and centrifuge for 1 min at 16,000 rcf to dry the column membrane. Then transfer the column to a fresh 1.5 mL tube.
8. Add 10 μL of prewarmed (60 °C) Elution Solution and incubate for 1 min.
9. Centrifuge for 30 s at 16,000 rcf to elute the bisulfite-converted DNA.
10. Repeat the elution with 10 μL Elution Solution (**steps 8** and **9**).

3.7 Klenow Reaction #1

1. Prepare the Klenow-Mastermix by combining 1 μL dNTPs (10 mM), 1 μL Oligo 1 (Biotin, 10 μM for a DNA input <10 ng, 100 μM for a DNA input >10 ng), and 2.5 μL Blue Buffer (10×) for each sample.
2. Aliquot 4.5 μL of the Klenow-Mastermix to a new PCR tube, add the entire sample and mix thoroughly. Incubate for 3 min at 65 °C, cool to 4 °C.
3. On ice, add 1 μL Klenow (3′- 5′ exo-, 50 U/μL) and mix thoroughly. Then incubate in a thermocycler at 4 °C for 5 min with a heated lid (set at 37 °C). After that, increase the temperature by 1 °C every 15 s up to 37 °C. Incubate at 37 °C for 30 min.
4. During the incubation time, prepare additional Klenow-Mastermix 1 for samples with a DNA quantity lower than 10 ng: Mix 2.5 μL ultrapure H_2O, 5 μL of Oligo 1 (Biotin, 10 μM), 0.5 μL dNTPs (10 mM), and 2.5 μL Klenow (3′- 5′ exo-; 50 U/μL) for each sample.
5. Incubate the samples at 95 °C for 1 min and put the samples immediately on ice.
6. Add 2.5 μL of the Klenow-Mastermix 1, mix and centrifuge to remove any liquid residues from the lid.
7. Incubate in a thermocycler with heated lid (set at 37 °C) at 4 °C for 5 min. After that, increase the temperature by 1 °C every 15 s up to 37 °C. Incubate at 37 °C for 30 min.
8. Repeat **steps 5–7** four times for a DNA input <10 ng. For a higher input amount, no repetition of the Klenow reaction #1 is needed. In this case, proceed after **step 3** to DNA cleanup and capture (Subheading 3.8) without denaturation.

3.8 DNA Cleanup and Capture

1. Add 3 μL Exonuclease I (20 U/μL) and incubate at 37 °C for 1 h.
2. Add 39 μL or 20.4 μL magnetic beads depending on whether the Klenow #1 reaction was five times or once, respectively. Mix thoroughly by pipetting. The ratio of sample–beads should be 1:1 (*see* **Notes 5** and **6**) .
3. Incubate for 15 min at room temperature.
4. Place the samples on a magnetic stand, wait until the supernatant is clear and discard it.
5. Wash the beads 2× with freshly prepared 80% EtOH while leaving them on the magnetic stand and keeping the pellet intact.
6. Dry the beads (3–5 min, *see* **Note 8**).
7. Elute the DNA from the beads in 41 μL 10 mM Tris–HCl pH 8.5.

8. Incubate for 5 min at room temperature. After incubation, separate the dissolved DNA from the beads via the magnetic stand.
9. During incubation time, prepare the Dynabeads M-280 Streptavidin as follows:
   (a) Add 12 μL per sample to an empty well/tube.
   (b) Place the tubes on a magnetic stand and remove the supernatant.
   (c) Wash the Streptavidin beads with 500 μL 10 mM Tris–HCl pH 8.5.
10. Transfer 40 μL of the DNA-containing supernatant to the Streptavidin beads and mix by pipetting.
11. Add 10 μL of 5× Streptavidin Beads Binding Buffer to the mixture and mix.
12. Incubate the Streptavidin beads for 20 min at room temperature under constant movement to avoid settlement of the beads (e.g., on a roller).
13. Wash 2× with 100 μL freshly prepared 0.1 N NaOH.
14. Wash 2× with 100 μL 10 mM Tris–HCl pH 8.5.

3.9 Klenow Reaction #2

1. Prepare the Klenow-Mastermix 2 by combining 41 μL ultrapure H_2O, 5 μL Blue Buffer (10×), 2 μL of Oligo 2 (10 μM for a DNA input <10 ng, 100 μM for a DNA input >10 ng), and 2 μL dNTPs (10 mM) for each sample.
2. Resuspend the Streptavidin beads with captured DNA from Subheading 3.8 in 48 μL of the Klenow-Mastermix 2.
3. Incubate samples at 95 °C for 45 s. Put samples immediately on ice.
4. Add 2 μL Klenow (3′- 5′ exo-, 50 U/μL), mix thoroughly and centrifuge to remove any liquid residues from the lid.
5. Incubate in a thermocycler with a heated lid (set at 37 °C) at 4 ° C for 5 min. After that, increase the temperature by 1 °C every 15 s up to 37 °C. When the cycler reached this temperature, incubate at 37 °C for 90 min.
6. Wash the beads once with 100 μL 10 mM Tris–HCl pH 8.5.

3.10 PCR Amplification Step

1. During incubation time of the Klenow reaction 2, prepare a PCR master mix of 21 μL ultrapure H_2O, 25 μL EPIK Amplification mix (2×) and 2 μL Primer PE1.0 for (10 μM) for each sample.
2. Add 2 μL of a 10 μM Index Primer rev to the sample.
3. Add 48 μL of the PCR master mix to the sample and mix by pipetting.

4. Amplify the library in a thermocycler with a heated lid: 95 °C for 2 min, denaturation at 95 °C for 30 s, annealing at 65 °C for 30 s and extension at 72 °C for 30 s. Cycle in total 10–13 times (*see* **Note 16**). The final extension is at 72 °C for 3 min.

3.11 Final Cleanup

1. Provide 40 μL magnetic beads per sample (0.8:1) in a new tube (*see* **Notes 5** and **6**).
2. Place the samples on a magnetic stand to accumulate the Streptavidin beads and transfer the supernatant to the magnetic purification beads in a new tube (*see* **Note 17**).
3. Mix thoroughly by pipetting and incubate for 15 min at room temperature.
4. Wash the beads 2× with freshly prepared 80% EtOH while leaving them on the magnetic stand and leaving the pellet intact.
5. Dry the beads (3–5 min, *see* **Note 8**).
6. Resuspend the beads in 21 μL 10 mM Tris–HCl pH 8.5.
7. Incubate for 5 min at room temperature.
8. Place the samples on a magnetic stand and wait 5 min until the supernatant appears clear.
9. Transfer 20 μL of the supernatant, containing the final library, to a new tube.

3.12 Quality Control and Quantity Measurement

We use a Fragment Analyzer (Advanced Analytical) to analyze the quality and quantity of the methylation libraries. Two microliter of each sample is needed for the High Sensitivity NGS Fragment Analysis Kit. A good library should show a fragment length distribution from 300 to 500 bp (Fig. 3). For successful sequencing on a HiSeq4000 machine (*see* **Notes 18** and **19**), the concentration should be at least 10 nM and contamination with primer dimers should not exceed 5–10% of the entire library concentration.

4 Notes

1. For restriction, we use purified gDNA. The most appropriate extraction method is depending on the organism and tissue.
2. The restriction pattern of gDNA with a particular restriction enzyme is species specific. The restriction enzyme Taq$^{\alpha}$I gave an appropriate RRBS fragment length distribution for rapeseed and maize. Enzyme inactivation (20 min at 80 °C) is not necessary if the reaction is purified immediately after restriction.
3. Size selection is also possible via magnetic beads (for a detailed protocol, *see* [17, 18]), but this is not as accurate and specific as

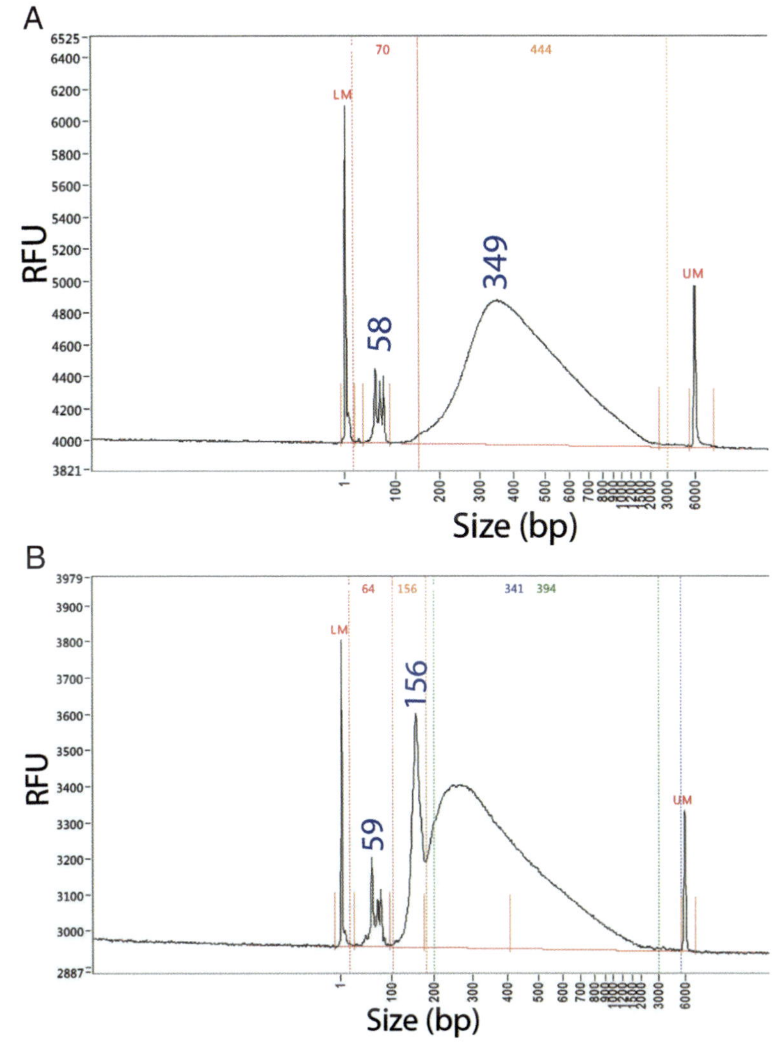

Fig 3 A representative methylation library (**a**) prepared from 30 pg gDNA from lysed maize cells has a fragment length distribution of about 200–800 bp with a main length of 350 bp as shown on the Fragment Analyzer electropherogram. A high adapter contamination (**b**), which is shown in a representative methylation library prepared from 50 pg gDNA from lysed *Arabidopsis* cells, leads to loss of informative reads during sequencing

gel size selection (Fig. 2f, g). For a better fragment separation, we run the gel rather slow.

4. We preferably use Agencourt AMPure XP beads (Beckman Coulter), which resulted in a faster workflow, because they bind faster to the magnetic stand, whereas HighPrep PCR beads (MagBio Genomics) take longer to bind because of their smaller size. The final result, however, is identical.

5. The ratio between beads and sample volumes influences the fragment size of the outcome. The general guideline is: the less the beads volume, the larger the fragment size.
6. It is very important to mix well after adding beads to the sample; the solution should appear homogenous. This ensures proper binding of a high amount of fragments to the beads.
7. We usually leave small volumes (1–2 μL) in the tube while washing the beads or transferring supernatants, even if we work with a small DNA input amount, to increase yield at a high quality.
8. While drying the beads at room temperature to remove any ethanol residues, one should take care not to overdry the beads. A good state is a non-shiny pellet without flaws.
9. While size selection for RRBS libraries is usually between 150 and 300 bp, we cut slices with fragment sizes between 1000 and 1500 bp. In our protocol, the size selection takes place before the bisulfite conversion, which is known to additionally degrade the DNA through the bisulfite treatment.
10. DNA precipitation can be faster at −70 °C (≈20 min) or even through dipping the tube in liquid nitrogen until the bubbles disappear. Especially for low DNA amounts, we recommend to precipitate overnight.
11. We sample single or few cells manually by using fine glass needles to remove the surrounding tissue under an inverted microscope and transfer one or more cells subsequently to a tube containing lysis buffer. The cells are then immediately frozen in liquid nitrogen and stored at −70 °C until processing [19]. Alternatively, cells or nuclei of individual cell types can be isolated by FACS (fluorescence activated cell sorting) [20], INTACT (isolation of nuclei tagged in specific cell types) [21], or an INTACT-derivative method (*see* Chapter 8 of this issue by Morao et al.).
12. The BSA-water solution will improve DNA recovery with low input concentrations. Ideally 5–6 μL BSA is used to keep the DNA protected by SDS while adding useful amounts of BSA. Thus, try to sample the cells in finally 5–6 μL 1× Lysis/Binding buffer.
13. The prepared DNA modification solution can be stored at −20 °C up to 2 weeks. Preparation and thawing is described in the manufacturer's protocol.
14. To ensure a proper DNA conversion, we usually extend the bisulfite reaction at 7 °C overnight.
15. The gDNA/BSA solution should approximately contain 5–6 μL 0.5 mg/mL BSA solution when using less than 10 ng gDNA. The total volume needs to be exactly 12 μL.

16. For amplification of a methylation library made of a DNA input amount >10 ng/μL, 10 cycles in total are sufficient. 12–13 cycles are necessary to amplify a library prepared from a smaller DNA amount.
17. Before opening tubes containing PCR products, change the room to avoid contamination of future prePCR samples with previous PCR products.
18. It is important to know for bioinformatic analysis that libraries prepared with our method are nondirectional and that the first nine nucleotides of the reads have to be removed. Please refer to [8] for details.
19. It is always worth to do test sequencing to check for the conversion rate and mapping efficiencies of the individual libraries. The use of different index primers allows for sequencing many libraries on one lane for this purpose.

Acknowledgments

This work was supported by a Heisenberg Scholarship of the German Research Foundation (DFG) to S. Scholten.

References

1. Jullien PE, Susaki D, Yelagandula R et al (2012) DNA methylation dynamics during sexual reproduction in Arabidopsis thaliana. Curr Biol 22:1825–1830. doi:10.1016/j.cub.2012.07.061
2. Rodrigues JA, Zilberman D (2015) Evolution and function of genomic imprinting in plants. Genes Dev 29:2517–2531. doi:10.1101/gad.269902.115
3. Gutiérrez-Marcos JF, Costa LM, Dal Prà M et al (2006) Epigenetic asymmetry of imprinted genes in plant gametes. Nat Genet 38:876–878. doi:10.1038/ng1828
4. Jahnke S, Scholten S (2009) Epigenetic resetting of a gene imprinted in plant embryos. Curr Biol 19:1677–1681. doi:10.1016/j.cub.2009.08.053
5. Park K, Kim MY, Vickers M et al (2016) DNA demethylation is initiated in the central cells of Arabidopsis and rice. Proc Natl Acad Sci USA 113(52):15138–15143. doi:10.1073/pnas.1619047114
6. Kawakatsu T, Huang S-SC, Jupe F et al (2016) Epigenomic diversity in a global collection of Arabidopsis thaliana accessions. Cell 166:492–505. doi:10.1016/j.cell.2016.06.044
7. Rodríguez López CM, Wilkinson MJ (2015) Epi-fingerprinting and epi-interventions for improved crop production and food quality. Front Plant Sci 6:397. doi:10.3389/fpls.2015.00397
8. Miura F, Enomoto Y, Dairiki R, Ito T (2012) Amplification-free whole-genome bisulfite sequencing by post-bisulfite adaptor tagging. Nucleic Acids Res 40:e136. doi:10.1093/nar/gks454
9. Smallwood SA, Lee HJ, Angermueller C et al (2014) Single-cell genome-wide bisulfite sequencing for assessing epigenetic heterogeneity. Nat Methods 11:817–820. doi:10.1038/nmeth.3035
10. Gu H, Smith ZD, Bock C et al (2011) Preparation of reduced representation bisulfite sequencing libraries for genome-scale DNA methylation profiling. Nat Protoc 6:468–481. doi:10.1038/nprot.2010.190
11. Meissner A (2005) Reduced representation bisulfite sequencing for comparative high-resolution DNA methylation analysis. Nucleic Acids Res 33:5868–5877. doi:10.1093/nar/gki901
12. Harris RA, Wang T, Coarfa C et al (2010) Comparison of sequencing-based methods to

profile DNA methylation and identification of monoallelic epigenetic modifications. Nat Biotechnol 28:1097–1105. doi:10.1038/nbt.1682

13. Hardcastle TJ (2013) High-throughput sequencing of cytosine methylation in plant DNA. Plant Methods 9:16. doi:10.1186/1746-4811-9-16
14. Lee Y, Jin S, Duan S et al (2014) Improved reduced representation bisulfite sequencing for epigenomic profiling of clinical samples. Biol Proced Online 16:1. doi:10.1186/1480-9222-16-1
15. Fischer M, Renevey N, Thür B, Hoffmann D, Beer M, Hoffmann B (2016) Efficacy assessment of nucleic acid decontamination reagents used in molecular diagnostic laboratories. PLoS One 11(7):e0159274. doi:10.1371/journal.pone.0159274
16. Quail MA, Otto TD, Gu Y et al (2012) Optimal enzymes for amplifying sequencing libraries. Nat Methods 9:10–11. doi:10.1038/nmeth.1814
17. Edwards D (2012) PCR Purification: AMPure and Simple. http://www.keatslab.org/blog/pcrpurificationampureandsimple. Accessed 1 Nov 2016
18. Stephenson K (2013) AMPure and not-so-simple. http://www.keatslab.org/blog/ampureandnot-so-simple. Accessed 1 Nov 2016
19. Kranz E, Scholten S (2007) In vitro fertilization: analysis of early post-fertilization development using cytological and molecular techniques. Sex Plant Reprod 21:67–77. doi:10.1007/s00497-007-0060-x
20. Birnbaum K, Shasha DE, Wang JY et al (2003) A gene expression map of the Arabidopsis root. Science 302:1956–1960. doi:10.1126/science.1090022
21. Deal RB, Henikoff S (2010) The INTACT method for cell type-specific gene expression and chromatin profiling in Arabidopsis thaliana. Nat Protoc 6:56–68

Chapter 4

Identification of Differentially Methylated Regions in the Genome of *Arabidopsis thaliana*

Kamal Kishore and Mattia Pelizzola

Abstract

DNA methylation profiling in the epigenome of *Arabidopsis thaliana* (Arabidopsis) has provided great insights in the role of this epigenetic mark for the regulation of transcription in plants, and is often based on high-throughput sequencing. The analysis of these data involves a series of steps including quality checks, filtering, alignment, identification of methyl-cytosines, and the identification of differentially methylated regions. This chapter outlines the computational methodology required to profile genome-wide differential methylation patterns based on publicly available Arabidopsis base-resolution bisulfite sequencing data. The *methylPipe* Bioconductor package is adopted for the identification of the differentially methylated regions, and all the steps from the raw data to the required input are described in detail.

Key words DNA methylation, Bisulfite sequencing, Computational biology, DMRs, methylPipe, Arabidopsis

1 Introduction

This chapter describes a protocol that can be used for the identification of differentially methylated regions in Arabidopsis using the methylPipe R/Bioconductor package [1] [http://bioconductor.org/packages/methylPipe/], including methyl-cytosines in the non-CpG context. The whole procedure, from processing of raw data up to the identification of differentially methylated regions, is illustrated step-by-step reporting the specific code that has been used. The data used in this protocol are methylC-seq data of Arabidopsis (ecotype Col-0) immature floral tissue and mutant plants deficient in the maintenance of CpG DNA methylation (*met1-3* mutant, referred to as *met1*) [2]. These data allow quantifying the base-resolution methylation status of most of the cytosines in the 135 Mbp Arabidopsis genome. The methylC-seq data result from the sequencing of bisulfite converted DNA, in which bisulfite was used to convert cytosines (but not methyl-cytosines) into uracil, thus allowing the base-resolution identification of the

Marian Bemer and Célia Baroux (eds.), *Plant Chromatin Dynamics: Methods and Protocols*, Methods in Molecular Biology, vol. 1675, DOI 10.1007/978-1-4939-7318-7_4, © Springer Science+Business Media LLC 2018

methylation events by high-throughput sequencing [3]. The methylC-seq data were retrieved from NCBI Gene Expression Omnibus (GEO, http://www.ncbi.nlm.nih.gov/geo) through the GEO Series accession number GSE10877. The data is available in the form of sra files that were converted to fastq files using the fastq-dump tool [https://trace.ncbi.nlm.nih.gov/Traces/sra/].

2 Materials

The analysis can be performed on a standard computer, on which the below listed software packages have to be installed. There are no specific hardware requirements (especially for Arabidopsis DNA methylomes, which are significantly smaller than mammalian methylomes), but the installation of some of these tools (mostly described in Subheadings 2.1.2–2.1.4) is restricted to Linux or Mac operating systems.

2.1 Software to be Installed (in No Particular Order)

1. FastQC: http://www.bioinformatics.bbsrc.ac.uk/projects/fastqc.
2. FASTX-Toolkit: http://hannonlab.cshl.edu/fastx_toolkit.
3. Bismark aligner: http://www.bioinformatics.babraham.ac.uk/projects/bismark/.
4. Samtools and TABIX: http://www.htslib.org/download/.
5. methylPipe: http://bioconductor.org/packages/methylPipe/.
6. rtracklayer: http://bioconductor.org/packages/rtracklayer/.

2.2 Additional Reference URLs

1. Phred scores: https://en.wikipedia.org/wiki/Phred_quality_score.
2. Download genome sequence: TAIR10 genome http://plants.ensembl.org/Arabidopsis_thaliana/Info/Index.
   Download genome sequence: Araport 11
   https://www.araport.org/data/araport11.
3. FASTQ format: https://en.wikipedia.org/wiki/FASTQ_format.

3 Methods

3.1 Data Processing

Cleaning and filtering of the raw sequencing data is an important processing step. The FastQC tool can be used to evaluate the quality of the raw sequencing data. The resulting report will illustrate the quality of the raw data from different perspectives, and will help deciding if the data have the required quality for downstream analyses (*see* **Note 1**). While the FastQC report can help deciding about including or discarding individual samples, other tools can be used to remove low quality portions of the data, thus improving the overall quality of the remaining data. In this protocol we will filter, trim and mask bad quality reads using the tools available in the FASTX-Toolkit.

Given the intrinsic lower quality of the base call at the ends of the sequencing reads, the reads were first trimmed from the ends removing bases with quality score lower than 20 (Phred scores, *see* **step 1** of Subheading 2.2. for more info), indicating base calls of particularly low quality. Individual reads will be completely discarded if only less than 10 bases remained after the trimming process. This filter stops trimming the reads as soon as one base of sufficient quality is encountered, and, thus, it does not have an effect on the internal low quality bases. The remaining internal low-quality bases (quality score less than 20) were masked with N so that they will not influence the subsequent reads alignment (*see* **Note 2** for details on the –Q33 setting).

```
fastq_quality_trimmer -Q33 -t 20 -l 10 -v -i col0.fastq -o
col0_trim.fastq
fastq_quality_trimmer -Q33 -t 20 -l 10 -v -i met1.fastq -o
met1_trim.fastq
fastq_masker -Q33 -q 20 -r N -v -i col0_trim.fastq -o col0_
final.fastq
fastq_masker -Q33 -q 20 -r N -v -i met1_trim.fastq -o met1_
final.fastq
```

3.2 Data Alignment

The processed fastq files can then be aligned to the Arabidopsis reference genome (TAIR10 version). The Bismark aligner, which was developed specifically for the alignment of bisulfite converted DNA, should be used for this purpose [4]. The first step is to download a reference genome and place it in a specific folder. We obtained the Arabidopsis.thaliana.TAIR10 genome from the Ensembl website. Recently, a new annotated version of the Col-0 reference genome became available, which is named Araport 11 (*See* "Download DNA sequence" at **step 2** of Subheading 2.2.).

The Bismark genome preparation script is first ran to prepare the reference genome for the alignment of bisulfite converted sequences. The genome has to be in fasta format with either .fa or .fasta extension, with single or multiple sequence (chromosomes) entries per file (*see* **Note 3**).

```
/bismark_v0.16.3/bismark_genome_preparation -path_to_bowtie/
path_to_bowtie/
 -verbose /path_to_refgenome/
```

The subsequent alignment step requires the path to the folder containing the indexed genome (as generated above) and the file(s) containing the read sequences to be aligned (in either fastq or fasta format). Bismark saves the alignment results in the BAM/SAM format, which are the most common binary and text format used to store aligned reads, respectively. The bisulfite sequencing methylation library can be constructed in a directional or nondirectional manner. In directional libraries only reads from the forward or reverse strands are present, while in the nondirectional libraries reads from complementary strands are also present. Hence, specifying "–non_directional" instructs Bismark to use all four alignment outputs (*see* **Note 4**).

```
/bismark_v0.16.3/bismark -q --phred33-quals --non_direc-
tional--path_to_bowtie
 /path_to_bowtie/ /path_to_refgenome/ col0_final.fastq
/bismark_v0.16.3/bismark -q --phred33-quals --non_directional
--path_to_bowtie
 /path_to_bowtie/ /path_to_refgenome/ met1_final.fastq
```

3.3 Data Processing by methylPipe

The next step is to extract methylation information corresponding to each cytosine in the genome (covered by sequencing), so that the location of the methyl-cytosines can be defined, together with the quantification of the events (typically representative of their frequency within the cell population) and their significance. The SAM files generated by Bismark and containing the aligned reads were sorted (using the samtools [5] sort command) and processed by methylPipe [1] by the *meth.call* function. The read.context parameter is specified as "All" resulting in the extraction of the methylation information for cytosines in all contexts (CpG and non-CpG, *see* **Note 5**). To minimize the amount of stored data, methylPipe only saves cytosine positions supported by at least one methylated read (reads where a given cytosine remained unconverted after bisulfite treatment), and genomic regions not covered by sequencing are saved as GRanges object (the R/Bioconductor representation of genomic ranges, that in the minimal version are analogous to BED files). Hence, the *meth.call* function of methylPipe produces a text file with the methylation calls and a file listing the uncovered genomic regions for each sample in the output folder (*see* **Note 6**). The combination of this information allows to determine for each genomic cytosine if it is (1) not covered by sequencing, (2) not methylated, or (3) methylated (and at which level and significance).

```
library(methylPipe)
require(BSgenome.Athaliana.TAIR.TAIR10)
meth.call(files_location=/path_to_the_data/, output_folder=/
path_to_the_data/, no_overlap=F, read.context="All", Nproc=2)
```

methylPipe adopts TABIX compressed indexing to compress the methylation call text file to reduce the disk space and to guarantee fast access to these data [6]. The methylPipe *BSprepare* function takes care of this process (thus requiring a local installation of the TABIX software available within the HTSlib package), and finalizes the methylation call by applying binomial tests on the putative methyl-cytosines. The binomial test is performed taking into account both the bisulfite conversion rate (provided through the "bc" argument) and the sequencing error rate. The bisulphite conversion rate is typically calculated by sequencing of an unmethylated spike-in (such as the lambda phage [3]) or by determining the conversion rate of reads from the chloroplast genome, which is not methylated. These binomial *p*-values are added to the tabix-compressed file.

```
BSprepare(files_location=file_loc, output_folder=file_loc,
tabixPath=/path_to_tabix/, bc=0.5/100)
```

The processed base-resolution data are stored in the form of objects of class BSdata for each sample. In case of multiple samples, which are required for the identification of differentially methylated regions, multiple BSdata objects can be conveniently stored in an object of class BSdataSet, by specifying group name for each sample either as" C″ (control) or" E" (Experiment) (*see* **Note 7**).

```
seqnames(Athaliana) <- c("chr1" ,"chr2" ,"chr3", "chr4"
,"chr5")
col0 <- "/path_to_the_data/col0_chr4_tabix_out.txt.gz"
met1 <- "/path_to_the_data/met1_chr4_tabix_out.txt.gz"
load("/path_to_the_data/col0_uncov.Rdata")
load("/path_to_the_data/met1_uncov.Rdata")
### Creation of BSdata
col0.db <- BSdata(file=col0, uncov=col0_uncov, org=Athaliana)
met1.db <- BSdata(file=met1, uncov=met1_uncov, org=Athaliana)
### Creation of BSdataSet
col_met.set <- BSdataSet(org=Athaliana, group=c("C","E"),
col0=col0.db, met1=met1.db)
```

3.4 Identification of Differentially Methylated Regions

The methylPipe *findDMR* function is used here to identify the differentially methylated regions (DMRs) for methyl-cytosines in the CpG and non-CpG sequence context for chromosome 4. The non-CpG methylation occurs at CHG and CHH sites (where H is A, C, or T). The *findDMR* function adopts different statistical tests according to the number of groups to be compared. Specifically, the Wilcoxon-rank test or Kruskal–Wallis nonparametric statistical tests are used in case of pairwise or multisample comparisons, respectively. The evaluated regions are reported in the output along with their corresponding statistical significance. The change in DNA methylation is reported in terms of percentage difference between the mean methylation of experiment and control (MethDiffPerc) and log2 of mean methylation of experiment over control (log2Enrichment, *see* **Note 8**). The settings of the findDMR function require choosing a few parameters, i.e., the number of methyl-cytosines within each tested region, the size of the regions, and optionally the binsize that is used to smooth (or average) the information for more scattered events (such as non-CpG methyl-cytosines in human). Typically for non-CpG methylation in human, a dmrSize of 50, a dmrBp of 50,000, and a binsize of 1000 are used. For CpG methylation in human and both CpG and non-CpG methylation in plants the default settings are usually fine. Additional arguments not used here can be conveniently tuned to filter the regions before applying the statistical tests, to ease the multiple testing issue (*see* **Note 9**).

Thereafter, the ***consolidateDMRs*** function can be applied to correct the DMRs for multiple-testing and consolidate them according to their relative distance, by joining DMRs closer than a given threshold. In addition, you can also specify the type of DMRs (hypomethylated or hypermethylated in the baseline) and thresholds of methylation differences to further filter the significant DMRs. This function results in a GRanges object with the final set of DMRs, including their adjusted *p*-value (possibly combined between adjacent DMRs using the Fisher's Method), methylation difference and log enrichment.

Non-CpG DMRs (in chromosome 4) were identified by comparing ***met1*** to Col-0 mCHG and mCHH smoothed methylation profiles. The average methylation level of methyl-cytosines in the mCHG and mCHH sequence context was determined in 1 kb windows (binsize). The genome was scanned considering groups of at least 50 adjacent windows (dmrSize) over a distance less than 50 kb. The set of 50 smoothed values for mCHG in the ***met1*** sample was compared to a similar set in the Col-0 sample using the Wilcoxon-rank test. The resulting *p*-values were corrected with the Benjamini–Hochberg method. Regions with adjusted *p*-value lower than 0.01 and adjacent windows were joined. The same procedure was repeated for the methyl-cytosines in the CHH sequence context. This results in a final set of 60 CHG-DMRs and 43 CHH-DMRs.

```
col_met.dmrs.CHG <- findDMR(object= col_met.set, ROI=NULL,
Nproc=1, MCClass="mCHG", dmrSize=50, dmrBp=50000,
binsize=1000)
col_met.dmrs.CHH <- findDMR(object= col_met.set, ROI=NULL,
Nproc=1, MCClass="mCHH", dmrSize=50, dmrBp=50000,
binsize=1000)
### consolidating the DMRs
```

```
CHGdmrsWG <- consolidateDMRs(col_met.dmrs.CHG, pvThr=0.01,
GAP=100, correct=TRUE)
CHHdmrsWG <- consolidateDMRs(col_met.dmrs.CHH, pvThr=0.01,
GAP=100, correct=TRUE)
```

For the DMRs in the CpG context, the genome was scanned looking for groups of 20 CpGs within 1 kbp max, with a methylation difference greater than 30% between conditions. The Wilcoxon-rank paired test was used to score each region by comparing methylation levels of those CpGs in *met1* and Col-0-conditions. Regions with adjusted *p*-value lower than 0.01 and that were closer than 100 bp were combined, resulting in a final set of 3564 CG-DMRs.

```
col_met.dmrs.CG <- findDMR(object= col_met.set, ROI=NULL,
Nproc=1, MCClass="mCG", dmrSize=20, dmrBp=1000,eprop=0.3)
### consolidating the DMRs
CGdmrsWG <- consolidateDMRs(col_met.dmrs.CG, pvThr=0.01,
GAP=100, correct=TRUE)
```

The final DMRs can easily be exported to view in a genome browser as bed files using the export.bed function of the rtracklayer Bioconductor package.

```
library(rtracklayer)
export.bed(CHGdmrsWG,con="CHGdmrsWG.bed")
```

4 Notes

1. The FastQC report contains useful general statistics, such as the number of reads, and various plots illustrating the base-level and overall quality of the reads, their duplication level, and several measures estimating the presence of contaminants and sequencing artifacts. In particular, relevant sections of the report include: "basic statistics" (will give general information about number and length of reads), "overrepresented sequences" (will inform on the residual presence of sequencing adapters) and "per base sequence quality" reporting the distributions of base-level quality scores of the reads. Most plots and tables include the quantification of the evaluated feature, such as the level of duplication. Please refer to the FastQC documentation for hints on the interpretation of those values.
2. The sequence quality score encoding is specified by the '-Q33' parameter, used to define how the base quality scores were converted into ASCII characters contained in the fastq files

(for details *see* the Encoding information under link 3 of Subheading 2.2.).

3. Bismark is internally based on the Bowtie aligner, which is not specific for bisulfite converted DNA sequences. The "verbose" argument specifies if the software will run in a quiet mode, or if it will prompt logs and various information about the progress and each specific step. The notation "/path_to_XX" has to be replaced with the path at which a given tool (XX) can be found. Please note that use of common and back-slashes in this protocol is tailored for UNIX-based systems, such as Linux or Mac OS. They will likely have to be reverted (and possibly complemented by indication on the drive unit) in Windows-based systems.
4. The Bismark genome preparation script creates two individual folders for a C-T and a G-A converted genome. During the alignment, bisulfite-converted forward (C->T) and reverse reads (G->A) are then aligned to both converted versions of the genome. These genomes, once created, can be reused unless the alignment against a different genome is desired.
5. Methyl-cytosines are often found in the CpG sequence context. In human and mammals, cytosine methylation in other sequence contexts (non-CpG) is present at low frequency in various tissues and can be found at high frequency in pluripotent cells and in brain [7–9]. In contrast, methyl-cytosines in the non-CpG context are common in plants, and their functional role is relatively well characterized [2].
6. Within the ***meth.call*** function you could set the "no_overlap" argument to TRUE so that overlapping paired-end read pair information is counted only once for methylation calling. For parallel processing of multiple samples, "Nproc" argument allows specifying the number of processors to use.
7. In BSdataSet consisting of multiple samples only two groups can be specified at present, either as "C" (control) or "E" (Experiment). In case of multiple groups, the group labeling is not relevant, as the statistical test (a nonparametric version of ANOVA) will look for DMRs in the whole set of samples and groups, independently from the group membership of each sample. The chromosome notation should be consistent to the one present in the output of the ***meth.call*** function, and should agree with the official notation present in the corresponding BSgenome object.
8. Log2Enrichment and MethDiffPerc are two complimentary measures of differential methylation. While the former ensures that there is a substantial variation between the conditions, the latter informs on absolute difference between them. For example, consider two regions: (i) a region with methylation levels of

0.01 and 0.1 in two different conditions, and (ii) a region with methylation levels 0.1 and 1. For both regions the log2Enrichment will report that they are ten times more methylated, but only the second region is associated with a relevant difference between conditions. In practice, one could decide to filter only based on the MethDiffPerc, since the data are constrained in the [0,1] range.

9. The lower the number of tests performed by the *findDMR* function, the less the resulting *p*-values will be penalized by the multiple testing correction performed by the *consolidateDMRs* function. For this reason, the eprop argument of *findDMR* can be used to skip cytosines with a difference between the considered conditions lower than a given amount (default set at 0.3, indicating that a minimum 30% difference is required for the statistical test to be applied).

References

1. Kishore K, de Pretis S, Lister R, Morelli MJ, Bianchi V, Amati B et al (2015) methylPipe and compEpiTools: a suite of R packages for the integrative analysis of epigenomics data. BMC Bioinformatics 16:1–11
2. Lister R, O'Malley RC, Tonti-Filippini J, Gregory BD, Berry CC, Millar AH et al (2008) Highly integrated Single-Base resolution maps of the Epigenome in Arabidopsis. Cell 133:523–536
3. Urich MA, Nery JR, Lister R, Schmitz RJ, Ecker JR (2015) MethylC-seq library preparation for base-resolution whole-genome bisulfite sequencing. Nat Protoc 10:475–483. http://dx.doi.org/10.1038/nprot.2014.114
4. Krueger F, Andrews SR (2011) Bismark: a flexible aligner and methylation caller for bisulfite-Seq applications. Bioinformatics 27:1571–1572
5. Li H, Handsaker B, Wysoker A, Fennell T, Ruan J, Homer N et al (2009) The sequence alignment/map format and SAMtools. Bioinformatics 25:2078–2079
6. Li H (2011) Tabix: fast retrieval of sequence features from generic TAB-delimited files. Bioinformatics 27:718–719
7. Lister R, Pelizzola M, Dowen RH, Hawkins RD, Hon G, Tonti-Filippini J et al (2009) Human DNA methylomes at base resolution show widespread epigenomic differences. Nature 462:315–322
8. Lister R, Mukamel EA, Nery JR, Urich M, Puddifoot CA, Nicholas D et al (2013) Global Epigenomic Reconfi guration during mammalian brain development. Science 341:629–643
9. Schultz MD, He Y, Whitaker JW, Hariharan M, Mukamel EA, Leung D et al (2015) Human body epigenome maps reveal noncanonical DNA methylation variation. Nature 523:212–216

Chapter 5

A Rapid and Efficient ChIP Protocol to Profile Chromatin Binding Proteins and Epigenetic Modifications in Arabidopsis

Bénédicte Desvoyes, Zaida Vergara, Joana Sequeira-Mendes, Sofia Madeira, and Crisanto Gutierrez

Abstract

Chromatin immunoprecipitation (ChIP) is a widely used and very powerful procedure to identify the proteins that are associated with the DNA to regulate developmental processes. These proteins can be transcription factors, or specific histone variants and modified histones, which are all crucial for gene regulation. In order to obtain reliable results, ChIP must be carried out under highly reproducible conditions. Here, we describe a simple and fast ChIP protocol adapted for Arabidopsis seedlings, which can serve as a basis for other species, organs or more sophisticated procedures, such as the sequential ChIP. We also provide user-oriented troubleshooting to increase the chances of successful applications.

Key words Chromatin immunoprecipitation (ChIP), Histone marks, Epigenetics, Chromatin-binding protein, Chromatin modification, Antibody, Cross-linking

1 Introduction

Plant development relies on the establishment of gene regulatory networks, which are spatially and temporally controlled. This is primarily achieved by highly specific interactions of transcription factors and associated proteins with DNA at target genomic sites. In addition, the presence of a plethora of chromatin modifications, such as acetylation, methylation, ubiquitylation, and phosphorylation, among others [1], is associated with gene activity [2–4]. Finally, gene expression during plant development is also intimately coordinated with genome replication, which is in turn highly dependent on chromatin modifications [5]. Therefore, the identification of protein–DNA interactions and chromatin modifications is of primary importance to understand developmental processes.

Bénédicte Desvoyes and Zaida Vergara contributed equally to this work.

Marian Bemer and Célia Baroux (eds.), *Plant Chromatin Dynamics: Methods and Protocols*, Methods in Molecular Biology, vol. 1675, DOI 10.1007/978-1-4939-7318-7_5, © Springer Science+Business Media LLC 2018

Chromatin immunoprecipitation (ChIP) is probably one of the most powerful and reliable approaches of choice [6–9]. There are various reports describing ChIP protocols, which can be subsequently combined with qPCR and sequencing procedures. While at first sight ChIP strategies look straightforward, it is a challenging procedure due to some crucial steps that contribute significantly to sensitivity, reproducibility, and reliability of the final results. Here, we describe in detail a relatively fast procedure, modified from several currently available protocols [10–15], that delivers robust results using a simplified procedure. We also pay special attention to highlight the critical steps to solve common problems.

2 Materials

General equipment: a rotary wheel, a mini centrifuge, a centrifuge for 50 ml and 15 ml tubes with a fixed angle rotor, and a ThermoMixer. Common reagents are: autoclaved milliQ water, PMSF (100 mM stock: 1.74 g of phenylmethylsulfonyl fluoride dissolved in 100 ml of 100% ethanol, stored at 4 °C), and protease inhibitor cocktail for plant tissues (100× stock solution; P9599, Sigma). The indicated quantities are based on handling four samples simultaneously.

2.1 Cross-Linking

1. Seedlings (4–12 days old) or parts of them (roots, aerial part, or other tissues).
2. 10× PBS: 80 mM Na_2HPO_4, 20 mM KH_2PO_4 pH 7.4, 1.37 M NaCl, and 27 mM KCl.
3. 36.5% formaldehyde.
4. 2 M glycine (10 ml freshly prepared).
5. Miracloth (22–25 μm pore size).
6. Vacuum pump and desiccator.
7. Liquid nitrogen.

2.2 Chromatin Preparation

1. Liquid nitrogen.
2. Mortar and pestle.
3. Extraction buffer (200 ml): 10 mM Tris–HCl pH 8.0, 0.25 M sucrose, 10 mM $MgCl_2$, and 1% Triton X-100. Add 1 mM PMSF and 1× protease inhibitor cocktail for plant tissues before use.
4. Dounce homogenizer device with a loose pestle (0.114 ± 0.025 mm clearance) and a tight pestle (0.05 ± 0.025 mm clearance).
5. Miracloth (22–25 μm pore size).

6. Glass funnels.
7. Lysis buffer (50 ml): 50 mM Tris–HCl pH 8.0, and 10 mM EDTA. Add 1 mM PMSF and 1× protease inhibitor cocktail for plant tissues before use.
8. 10% SDS.
9. Sonicator (we use Bioruptor Plus from Diagenode, but there are alternative devices).
10. 1.5 ml TPX microtubes from Diagenode (or equivalent for your sonication device).

2.3 Analysis of Chromatin, DNA Purification and Quantification

1. 5 M NaCl.
2. RNase A at 2 mg/ml and RNase T1 at 5000 U/ml, DNase and protease free.
3. 1 M Tris–HCl, pH 6.8.
4. 0.5 M EDTA, pH 8.0.
5. 10 mg/ml proteinase K.
6. Phenol–chloroform–isoamyl alcohol (25:24:1).
7. Phase lock gel tubes. Both heavy and light type can be used with similar performance.
8. Glycogen (20 mg/ml), free of nucleases and proteases.
9. 3 M CH_3COONa, pH 5.2.
10. 100% ethanol and 70% ethanol.
11. TE buffer: 10 mM Tris–HCl pH 8.0, 1 mM EDTA pH 8.0.
12. A fluorimetric double-stranded DNA (dsDNA) quantification system and dsDNA assay kit (we use a system from Thermo Scientific but alternative devices may be available).

2.4 Immunoprecipitation

1. ChIP Dilution buffer (100 ml): 16.7 mM Tris–HCl pH 8.0, 167 mM NaCl, 1.2 mM EDTA, and 1.1% Triton X-100. Add before use 1 mM PMSF and 1× protease inhibitor cocktail for plant tissues.
2. Protein A or G agarose beads (*see* **Note 1**).
3. ChIP grade antibodies (also for positive and negative control, *see* **Note 2**). We have been successfully using anti-H3K4me3 (2 μg/sample; Ab8580, Abcam) and anti-H3K27me3 (5 μg/sample; Ab6002, Abcam).
4. Liquid nitrogen.

2.5 Recovery of Immune Complexes and Washing Steps

1. Protein A or G agarose beads.
2. ChIP Dilution buffer (*see* **step 1** of Subheading 2.4).
3. Low Salt Wash Buffer (100 ml): 20 mM Tris–HCl pH 8.0, 2 mM EDTA, 150 mM NaCl, 0.1% SDS, and 1% Triton X-100.

Add before use 1 mM PMSF and 1× protease inhibitor cocktail for plant tissues.

4. High Salt Wash Buffer (100 ml): 20 mM Tris–HCl pH 8.0, 2 mM EDTA, 500 mM NaCl, 0.1% SDS, and 1% Triton X-100. Add before use 1 mM PMSF and 1× protease inhibitor cocktail for plant tissues.
5. LiCl Wash Buffer (100 ml): 10 mM Tris–HCl pH 8.0, 1 mM EDTA, 0.25 M LiCl, 1% Igepal CA-630, and 1% sodium deoxycholate. Add before use 1 mM PMSF and 1× protease inhibitor cocktail for plant tissues.
6. TE Buffer (100 ml): 10 mM Tris–HCl pH 8.0, and 1 mM EDTA. Add before use 1 mM PMSF and 1× protease inhibitor cocktail for plant tissues.
7. Elution Buffer (freshly prepared): 1% SDS, 0.1 M $NaHCO_3$.

3 Methods

3.1 Cross-Linking (~1 h)

To fix the protein–DNA interactions, the tissue of choice has to be cross-linked:

1. Harvest 1 g of the desired plant material in 50 ml tubes with 35.5 ml ice cold 1× PBS. Place the tubes on ice to keep all the material and solutions cold during the entire cross-linking step.
2. Add formaldehyde to a final concentration of 1% and use a piece of miracloth mesh soaked in cold 1× PBS as a lid to keep all the material submerged in the cross-linking solution. Vacuum infiltrate (85,000 Pa) three rounds, 6 min each, incubating for 4 min without vacuum between the rounds. To eliminate air bubbles that appear during vacuum infiltration, move the desiccator up and down against the bench. Release the vacuum slowly (*see* **Notes 3** and **4**).
3. Add glycine to a final concentration of 125 mM to stop the cross-linking reaction and vacuum infiltrate 5 min.
4. Wash three times with ice-cold milliQ water.
5. Pat-dry the material on tissue paper and freeze in liquid nitrogen. The plant material can be stored in the −80 °C freezer for up to 10 months.

3.2 Chromatin Preparation (~5 h)

To isolate the chromatin, the plant material has to be grinded and homogenized, followed by isolation of the nuclei. To disrupt the nuclei and isolate the chromatin, sonication is performed in lysis buffer. For an efficient immunoprecipitation (IP) and possible downstream sequencing procedures, the chromatin has to be fractionated by sonication into 100–500 bp fragments.

1. Grind at least 1 g of material in liquid nitrogen using a pre-cooled mortar and pestle (*see* **Note 5**).
2. Resuspend each gram of cross-linked material in 10 ml of extraction buffer.
3. Incubate for 30 min on a rotary shaker at 4 °C.
4. Homogenize in an ice-cold mortar (*see* **Note 6**).
5. Use a Dounce homogenizer device to increase the yield of isolated nuclei. Use twice (up and down) the loose pestle and twice the tight one (*see* **Note** 7).
6. Filter through a double miracloth mesh (moistened with extraction buffer) employing a glass funnel to a new 50 ml tube.
7. Spin for 20 min at 3,000 × *g* and 4 °C.
8. Discard the supernatant and keep the nuclear pellet.
9. Resuspend the pelleted nuclei in 1 ml of lysis buffer per gram of cross-linked material, pipetting up and down being careful not to form any foam.
10. Once the nuclear pellet is well resuspended, measure the volume and add SDS to a final concentration of 1% from a 10% stock solution.
11. Incubate for 15–30 min on a rotary shaker at 4 °C.
12. Sonicate the chromatin. We use a Bioruptor Plus device and 1.5 ml TPX tubes (*see* **Note 8**). Sonication is carried out at high power mode for 30 cycles (sonication cycle: 30 s ON, 30 s OFF). If you are using other types of sonicators, it will be necessary to determine experimentally the numbers of sonication cycles needed to get fragments between 100 and 500 bp (*see* Fig. 1).
13. Centrifuge for 5 min at 19,000 × *g* and 4 °C in a microcentrifuge to separate the soluble and insoluble chromatin.

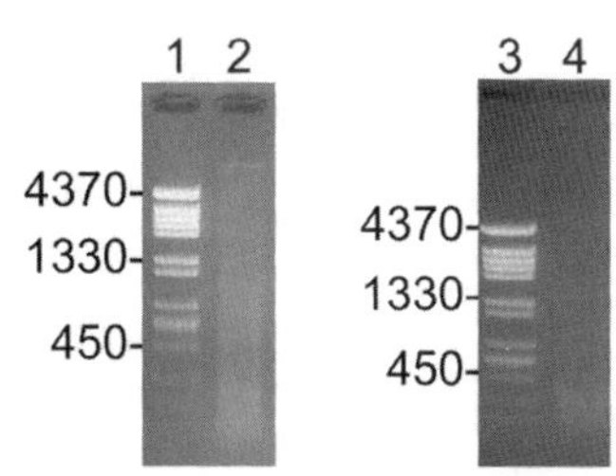

Fig. 1 Size analysis of the sonicated chromatin. The sonicated DNA is expected to have a size of 100–500 bp and no RNA contamination Samples correspond approximately to 50 mg of seedlings. *Lanes 1* and *3*: DNA marker (Φ29 DNA digested with Hind III); *Lane 2*: chromatin prior to sonication; *Lane 4*: chromatin after sonication

Transfer an aliquot of 30–60 μl of the supernatant (soluble chromatin) to another tube to assess the sonication efficiency and quantify the chromatin (Subheading 3.3). Keep the remaining sample on ice in the cold chamber until the immunoprecipitation step (Subheading 3.4).

3.3 Chromatin Fractionation Analysis and Quantification

It is important to check the quality and fragment sizes of the chromatin before proceeding to the IP reaction. The quantity of DNA necessary for the IP can then be calculated, and additional sonication could be performed on the sample if the fragments do not yet have the desired sizes. However, it is possible to skip this step and only assess the quality and quantity after the IP, with the risk that quantity and/or quality are not sufficient.

3.3.1 Reversion of Cross-Links (Overnight)

1. Add milliQ water to the aliquot of **step 13** from Subheading 3.2 to achieve a volume of 200 μl.
2. Add 8 μl of 5 M NaCl.
3. Incubate at 65 °C and 1000 rpm overnight in a ThermoMixer.

3.3.2 Purification of the Sonicated DNA (~6 h)

1. Digest the RNA by adding 4 μl of the RNase A/T1 mix and 8 μl of 1 M Tris–HCl, pH 6.8. Incubate for 30 min–1 h at 37 °C and 500 rpm in a ThermoMixer (*see* **Note 9**).
2. To digest the proteins, add 4 μl of 0.5 M EDTA and 2 μl of 10 mg/ml proteinase K. Incubate for 2 h at 37 °C and 500 rpm in a ThermoMixer.
3. Extract the DNA with phenol–chloroform–IAA using phase lock gel tubes. First spin the empty tubes 30 s, then add the sample and finally one volume of phenol–chloroform–IAA. Mix gently and centrifuge for 5 min at 19,000 × *g* at room temperature.
4. Collect the supernatant into a new tube and precipitate the DNA with 20 μg of glycogen, 0.1 volume of 3 M CH_3COONa pH 5.2, and 2.5 volumes of cold absolute ethanol. Incubate for 30–60 min at −80 °C.
5. Spin down for 30 min at 19,000 × *g* and 4 °C.
6. Wash once with 70% ethanol.
7. Air-dry the pellet.
8. Resuspend the pellet in 30–60 μl of TE. Incubate for at least 30 min on ice before quantification.

3.3.3 Quantification (~45 min)

1. Quantify the DNA with the Qubit system. Use 1–5 μl of DNA and the Qubit dsDNA HS assay kit. In order to calculate the concentration of the original sample of chromatin take into account the dilution factor (*see* **Note 10**).

2. To check the size of the sonicated chromatin, fractionate 0.5–1 μg of DNA in a 1.5% agarose gel. The sonicated DNA is expected to have a size of 100–500 bp (for sequencing) and no RNA contamination (*see* **Note 11** and Fig. 1).

3.4 Immuno-precipitation (~1.5/overnight)

In the IP reaction, the DNA associated with a protein of interest (e.g., modified histones) is specifically bound to an antibody against that protein. The workflow of the IP procedure including the subsequent washing, elution, and cross-linking steps is visualized in Fig. 2.

1. Calculate the quantity of chromatin for each ChIP reaction (normally corresponding to 1–8 μg of DNA).
2. Centrifuge the chromatin stored on ice (from **step 13** of Subheading 3.2) for 10 min at 19,000 × *g* at 4 °C. Transfer the supernatant to a new tube.
3. Dilute the soluble chromatin at least ten times in ChIP dilution buffer to decrease the SDS concentration to 0.1% maximum. Use up to 1 ml of diluted chromatin per ChIP (*see* **Notes 12** and **13**).
4. Wash 30 μl of protein A or G coated agarose beads per each ml of diluted chromatin with 1 ml of ChIP Dilution Buffer (*see* **Note 1**). Spin down at 1000 × *g* for 2 min at room temperature. Discard the supernatant and repeat.
5. Preclear the diluted chromatin: Add 30 μl of washed beads to 1 ml of diluted chromatin.
6. Incubate for 1 h at 4 °C on a rotary shaker.
7. Spin down the beads for 5 min at 1000 × *g* and 4 °C, collect the supernatant.
8. Set aside 10% of the diluted and precleared chromatin as input sample and store it at −20 °C.
9. Add the antibodies (negative, positive, and the specific modified histone) to each tube of diluted chromatin (*see* **Note 2**).
10. Incubate overnight at 4 °C on a rotary shaker.
11. Make aliquots with the remaining soluble chromatin and snap-freeze them on liquid nitrogen. Store at −80 °C up to 6 months.

3.5 Recovery of Immune Complexes and Washing Steps (~5 h)

In this step, the antibody-associated chromatin is bound to agarose beads, and weakly bound and unbound proteins are washed away during several washing steps. The specifically bound chromatin is then eluted.

1. Pull down the ChIP with 50 μl of protein A or G agarose (previously washed twice with ChIP Dilution Buffer). Incubate for 2 h at 4 °C on a rotary shaker.

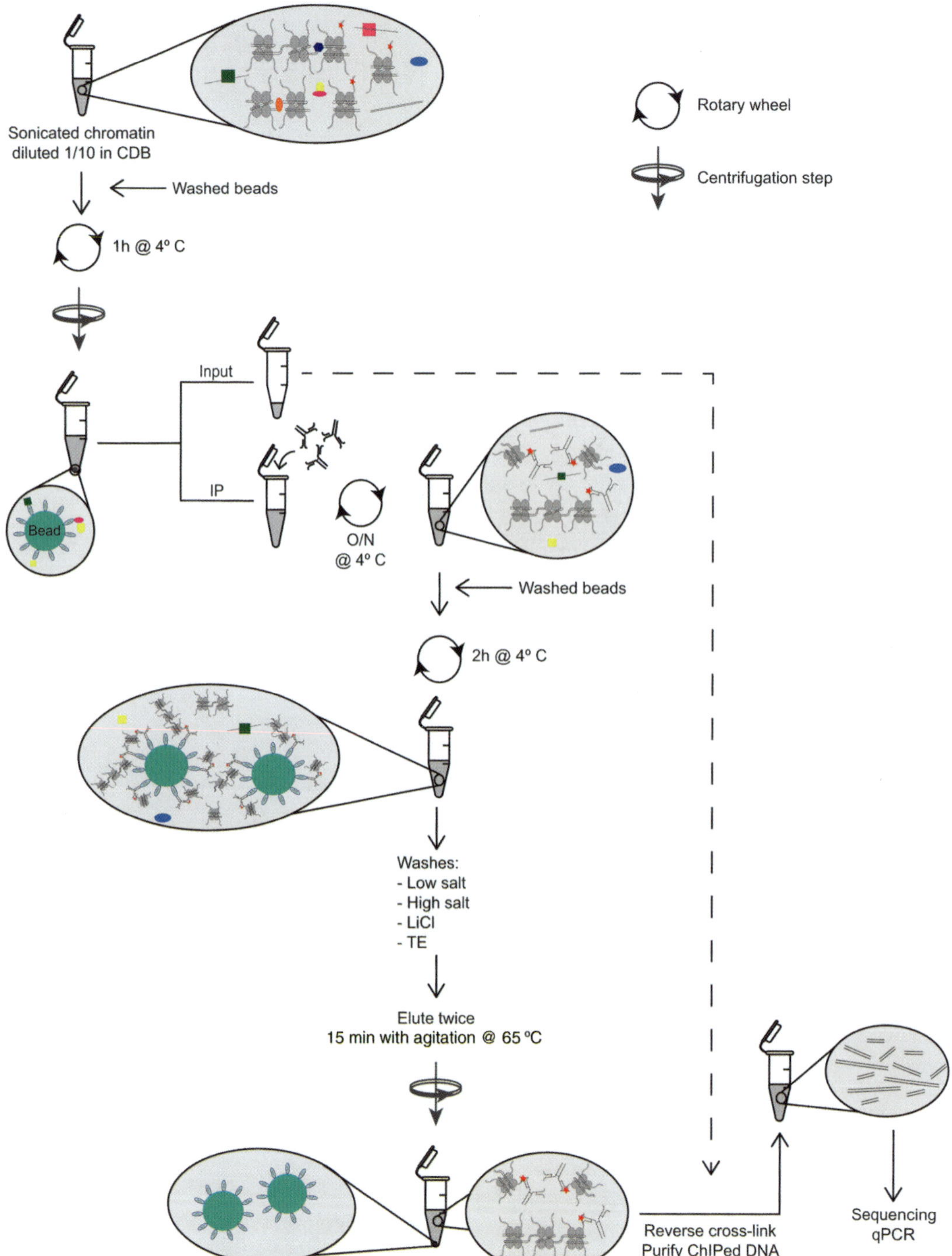

Fig. 2 Schematic diagram of the ChIP protocol. Diluted chromatin is precleared with protein A or G agarose coated beads to remove nonspecific binding. Then, the chromatin is immunoprecipitated with specific antibodies. The immune complexes are pulled down with protein A or G agarose beads and the unbound chromatin is washed off. After elution, both ChIPed DNA and input DNA are purified for further analysis. CDB: ChIP dilution buffer; IP: immunoprecipitation, Ab: antibody

2. Prepare the elution buffer and keep it at 65 °C.
3. Wash the beads twice with 1 ml of the following washing buffers:
   (a) Low salt wash buffer.
   (b) High salt wash buffer.
   (c) LiCl wash buffer.
   (d) TE buffer.

   For each buffer, first perform a quick wash and then incubate for 5 min at 4 °C with rotation. To spin down the beads in each wash, centrifuge the tubes for 2 min at 1000 × *g* at 4 °C, and then remove the supernatant.
4. To elute the immune complexes from the beads, add 200 μl of the elution buffer (preheated at 65 °C). Vortex briefly and incubate at 65 °C with agitation for 15 min in a ThermoMixer.
5. Spin down the beads for 1 min at 1000 × *g* and transfer the supernatant to a new tube.
6. Repeat **steps 4** and **5**.
7. Adjust the volume of the input sample to 400 μl with elution buffer.

3.6 Purification of the ChIPed DNA, Negative Controls, and Input Samples

3.6.1 Reversion of Cross-Links (Overnight)

1. Add 16 μl of 5 M NaCl to each eluted sample, (positive and negative ChIP reactions) and input sample.
2. Incubate at 65 °C with agitation overnight in a ThermoMixer.

3.6.2 Purification of the ChIPed DNA, Negative Control, and Input Sample (~5 h)

1. Digest the RNA by adding 8 μl of the RNase A/T1 mix and 16 μl of 1 M Tris–HCl, pH 6.8. Incubate for 30–60 min at 37 °C, at 500 rpm in a ThermoMixer (RNA degradation step is optional for validations using qPCR).
2. To digest the proteins, add 8 μl of 0.5 M EDTA and 2 μl of 10 mg/ml proteinase K. Incubate for 2 h at 37 °C and 500 rpm in a ThermoMixer.
3. Extract the DNA with phenol–chloroform–IAA using phase lock gel tubes. First spin the empty tubes 30 s, then add the sample and finally one volume of phenol–chloroform–IAA. Mix gently and centrifuge for 5 min at 19,000 × *g* and room temperature.
4. Collect the supernatant to a fresh tube and precipitate the DNA with 20 μg of glycogen, 0.1 volumes of 3 M CH_3COONa, pH 5.2, and 2.5 volumes of cold absolute ethanol. Incubate for 1 h at −80 °C.

5. Centrifuge for 30 min at 19,000 × *g* and 4 °C.
6. Wash once with 70% ethanol.
7. Air-dry the pellet.
8. Resuspend in 100 μl of TE. Incubate for at least 30 min on ice.

3.6.3 Analyze the ChIP

For sequencing: Quantify 10% of the ChIPed DNA and the input sample using the fluorimetric dsDNA quantification assay. Proceed to the library preparation.

For validations using qPCR: Dilute the input ten times to 1% input chromatin. Depending on the target and the ChIPed DNA you may dilute also the ChIPed DNA sample.

4 Notes

1. Proteins A or G have varying affinities for the antibodies depending on the species and the immunoglobulin subclass (heavy chain properties). Consider in every ChIP reaction which of the coated beads (protein A or G) is the most suitable one. When using magnetic beads instead of agarose beads, a preclearing step is not necessary.
2. For the negative control it is recommended to use an IgG antibody of the same species as the antibody used for the ChIP. For the positive control use an antibody against the corresponding unmodified histone (e.g., use anti-total H3 as a positive control for the study of H3K4me3). The positive control sample should later also be used to correct for differences in total H3 deposition.
3. Formaldehyde should be handled with caution in a fume hood. All residues should be discarded appropriately according to the toxic waste regulations.
4. The cross-linking process is one of the most crucial steps in the ChIP procedure. It can be affected by different factors such as temperature, percentage of formaldehyde, salt concentration, pH, or cross-linking time. The reaction can be performed in buffers such as PBS or HEPES, but buffers containing primary amines (e.g., Tris) should be avoided because they will quench the reaction.
5. Good grinding of the material is essential to obtain enough nuclei. Collect the powder into a new 50 ml tube using a liquid nitrogen-cooled spatula.
6. This step is optional but recommended to better resuspend the powder.
7. This step helps to release more nuclei. If you do not have a Dounce homogenizer you can perform **step 3** of Subheading 3.2 for a longer time until an apparently homogenous sample is obtained.

8. TPX microtubes are thinner and recommended for efficient sonication with the Bioruptor Plus system. The sample volume should be comprised between 100 and 300 μl. Check the specific minimum and maximum sample volume to be used with your sonication device.
9. There are several types of RNases that hydrolyze the RNA at specific sites. The RNase A specifically cuts RNA at C and U residues, while the RNase T1 degrades RNA at G nucleotides. The combination of these two RNA degradation activities leads to small fragments of RNA that are undetectable on an agarose gel.
10. To check the amount of sheared chromatin, it is necessary to use a highly sensitive method such as the fluorimetric-based dsDNA quantification system. Regular UV spectrophotometric methods are not sensitive enough and tend to overestimate the quantity of chromatin.
11. If the chromatin is not well sheared, you should mix the chromatin sample stored at 4 °C (soluble and insoluble chromatin) and resonicate the material until it reaches the appropriate fragment size. Check again the quantity and size of the fragments.
12. ChIP will not work in a buffer with more than 0.1% SDS.
13. When performing several IP reactions, dilute the chromatin ten times in a 15 ml tube and preclear all together (use 30 μl of beads per ml of diluted chromatin and scale-up). After the preclearing, separate the chromatin in tubes according to the number of ChIP reactions.

Acknowledgments

Z.V. and S.M. were recipients of Fellowships BES-2010-037158 (MINECO; Spain) and SFRH/BD/105550/2014 (FCT, Portugal), respectively. This work has been supported by grants BFU2012-34821 (MINECO), BIO2013-50098-EXP (MINECO) and BFU2015-68396-R (MINECO-FEDER), and by an institutional grant from Fundación Ramón Areces to the Centro de Biología Molecular Severo Ochoa.

References

1. Kouzarides T (2007) Chromatin modifications and their function. Cell 128:693–705
2. Strahl BD, Allis CD (2000) The language of covalent histone modifications. Nature 403:41–45
3. Jenuwein T, Allis CD (2001) Translating the histone code. Science 293:1074–1080
4. Bannister AJ, Kouzarides T (2011) Regulation of chromatin by histone modifications. Cell Res 21:381–395
5. Sequeira-Mendes J, Gutierrez C (2015) Links between genome replication and chromatin landscapes. Plant J 83:38–51
6. Kuo MH, Allis CD (1999) In vivo cross-linking and immunoprecipitation for studying dynamic protein:DNA associations in a chromatin environment. Methods 19:425–433
7. Orlando V (2000) Mapping chromosomal proteins in vivo by formaldehyde-crosslinked-chromatin immunoprecipitation. Trends Biochem Sci 25:99–104
8. Hoffman EA, Frey BL, Smith LM, Auble DT (2015) Formaldehyde crosslinking: a tool for the study of chromatin complexes. J Biol Chem 290:26404–26411
9. Hyun BR, McElwee JL, Soloway PD (2015) Single molecule and single cell epigenomics. Methods 72:41–50
10. Bowler C, Benvenuto G, Laflamme P, Molino D, Probst AV, Tariq M, Paszkowski J (2004) Chromatin techniques for plant cells. Plant J 39:776–789
11. Villar CB, Kohler C (2010) Plant chromatin immunoprecipitation. Methods Mol Biol 655:401–411
12. Malapeira J, Khaitova LC, Mas P (2012) Ordered changes in histone modifications at the core of the Arabidopsis circadian clock. Proc Natl Acad Sci U S A 109:21540–21545
13. Henaff E, Vives C, Desvoyes B, Chaurasia A, Payet J, Gutierrez C, Casacuberta JM (2014) Extensive amplification of the E2F transcription factor binding sites by transposons during evolution of Brassica species. Plant J 77:852–862
14. Malapeira J, Mas P (2014) ChIP-seq analysis of histone modifications at the core of the Arabidopsis circadian clock. Methods Mol Biol 1158:57–69
15. Komar DN, Mouriz A, Jarillo JA, Piñeiro M (2016) Chromatin immunoprecipitation assay for the identification of Arabidopsis protein-DNA interactions. J Vis Exp 107:e53422

Chapter 6

Sequential ChIP Protocol for Profiling Bivalent Epigenetic Modifications (ReChIP)

Bénédicte Desvoyes, Joana Sequeira-Mendes, Zaida Vergara, Sofia Madeira, and Crisanto Gutierrez

Abstract

Identification of chromatin modifications, e.g., histone acetylation and methylation, among others, is widely carried out by using a chromatin immunoprecipitation (ChIP) strategy. The information obtained with these procedures is useful to gain an overall picture of modifications present in all cells of the population under study. It also serves as a basis to figure out the mechanisms of chromatin organization and gene regulation at the population level. However, the ultimate goal is to understand gene regulation at the level of single chromatin fibers. This requires the identification of chromatin modifications that occur at a given genomic location and within the same chromatin fiber. This is achieved by following a sequential ChIP strategy using two antibodies to distinguish different chromatin modifications. Here, we describe a sequential ChIP protocol (Re-ChIP), paying special attention to the controls needed and the required steps to obtain meaningful and reproducible results. The protocol is developed for young Arabidopsis seedlings but could be adapted to other plant materials.

Key words Sequential chromatin immunoprecipitation (Re-ChIP), ChIP, Histone marks, Chromatin modification, Epigenetics, Chromatin-binding protein, Antibody, Cross-linking

1 Introduction

Chromatin organization and gene regulation depend on the occurrence of specific modifications in DNA and histones. In the case of histones, multiple modifications, primarily acetylation, methylation, phosphorylation, and ubiquitylation, have been identified [1]. The combination of different chromatin modifications leads to an extreme combinatorial complexity. Efforts to simplify have allowed the identification of different chromatin states that largely reflect the structural and, in many cases, functional organization of the genome in plants, flies, and humans [2, 3].

Bénédicte Desvoyes and Joana Sequeira-Mendes contributed equally to this work.

Marian Bemer and Célia Baroux (eds.), *Plant Chromatin Dynamics: Methods and Protocols*, Methods in Molecular Biology, vol. 1675, DOI 10.1007/978-1-4939-7318-7_6, © Springer Science+Business Media LLC 2018

The presence of chromatin modifications is now widely determined by chromatin immunoprecipitation (ChIP) procedures [4–8]. A standardized, simple, and fast ChIP protocol for Arabidopsis seedlings, which allows for the purification of DNA material useful for sequencing or qPCR analysis, is published in this book (Chapter 5) [9]. ChIP renders results of chromatin modifications occurring in all cells of the organ or tissue under analysis. However, gene regulation occurs at the single cell level and, consequently, it is of primary importance to identify chromatin modifications that occur in the same chromatin fiber either at the locus of interest or genome wide. Moreover, it is often a specific combination of chromatin modifications that defines gene activity, which cannot be identified using regular ChIP protocols. This also accounts for the identification of the so-called bivalent chromatin observed in poised promoters, which refers to chromatin fibers that possess histone modifications typical of both active and repressed states in the same chromatin fiber, e.g., H3K4me3 and H3K27me3 [10, 11].

The identification of multiple chromatin modifications at the same locus requires the application of sequential ChIP strategies. Thus, the material immunoprecipitated with one antibody that recognizes a given histone modification is subjected to a second ChIP step with another antibody that recognizes a different histone modification. In plants, this sequential ChIP (Re-ChIP) strategy has been used to define the bivalent state of flowering genes [12] and to identify a genome-wide set of genomic loci of potential bivalent nature [2]. Here, we describe a Re-ChIP protocol using standard reagents used in ChIP protocols with emphasis on the critical steps and controls needed to render reproducible and meaningful results. The first part of the protocol presented here has been adopted from Chapter 5 [9].

2 Materials

Several reagents and materials are used in many steps. These are: autoclaved milliQ water, PMSF (100 mM stock: 1.74 g of phenylmethylsulfonyl fluoride dissolved in 100 ml of 100% ethanol, stored at 4 °C), protease inhibitor cocktail for plant tissues (100× stock solution; P9599, Sigma), low binding 1.5 ml tubes, a mini centrifuge, a fixed-angle rotor centrifuge for 15 ml and 50 ml tubes, a rotary wheel, and a ThermoMixer. The volumes indicated below are based on handlingfour samples simultaneously.

2.1 Cross-Linking

1. Seedlings (4–12 days old) or parts of them (roots and aerial part). Other plant tissues can also be used, but the cross-linking conditions may have to be adjusted.

2. 10× PBS: 80 mM Na_2HPO_4, 20 mM KH_2PO_4 pH 7.4, 1.37 M NaCl, and 27 mM KCl.
3. 36.5% formaldehyde.
4. 2 M glycine (10 ml freshly prepared).
5. Miracloth (22–25 μm pore).
6. Vacuum pump and desiccator.
7. Liquid nitrogen.

2.2 Chromatin Preparation

1. Liquid nitrogen.
2. Mortar and pestle.
3. Extraction buffer (200 ml, stored at 4 °C): 10 mM Tris–HCl pH 8.0, 0.25 M sucrose, 10 mM $MgCl_2$, and 1% Triton X-100. Add 1 mM PMSF and 1× protease inhibitor cocktail for plant tissues before use.
4. Dounce homogenizer device with a loose pestle (0.114 ± 0.025 mm clearance) and a tight pestle (0.05 ± 0.025 mm clearance).
5. Miracloth (22–25 μm pore).
6. Glass funnels.
7. Lysis buffer (50 ml): 50 mM Tris–HCl pH 8.0 and 10 mM EDTA. Add 1 mM PMSF and 1× protease inhibitor cocktail for plant tissues before use.
8. 10% SDS.
9. Sonicator (we use Bioruptor Plus from Diagenode, but there are alternative devices).
10. 1.5 ml TPX microtubes from Diagenode (or equivalent for your sonication device).

2.3 Chromatin Analysis and Reversion of Cross-Links

1. 5 M NaCl.
2. RNase A at 2 mg/ml and RNase T1 at 5000 U/ml, DNase and protease free.
3. 1 M Tris–HCl, pH 6.8.
4. 0.5 M EDTA, pH 8.0.
5. 10 mg/ml proteinase K.
6. Phenol–chloroform–isoamyl alcohol (25:24:1).
7. Phase lock gel tubes. Both heavy and light type can be used with similar performance.
8. Glycogen (20 mg/ml), free of nucleases and proteases.
9. 3 M CH_3COONa pH 5.2.
10. 100% ethanol and 70% ethanol.
11. TE buffer: 10 mM Tris–HCl pH 8.0 and 1 mM EDTA pH 8.0.

12. A fluorimetric double-stranded DNA (dsDNA) quantification system and dsDNA assay kit (we use the Qubit system and Qubit dsDNA HS assay kit from Thermo Scientific but alternative devices may be available).

2.4 First and Second Immunoprecipitation Steps

1. ChIP Dilution buffer (100 ml stored at 4 °C): 16.7 mM Tris–HCl pH 8.0, 167 mM NaCl, 1.2 mM EDTA, and 1.1% Triton X-100. Add before use 1 mM PMSF and 1× protease inhibitor cocktail for plant tissues.
2. Protein A or G agarose beads (*see* **Note 1**).
3. ChIP grade antibodies (also for the controls (*see* **Note 2**). We have been successfully using anti-H3K4me3 (2 μg/sample; Ab8580, Abcam) and anti-H3K27me3 (5 μg/sample; Ab6002, Abcam).
4. Low Salt Wash Buffer (100 ml, store at 4 °C): 20 mM Tris–HCl, pH 8.0, 2 mM EDTA, 150 mM NaCl, 0.1% SDS, and 1% Triton X-100. Add before use 1 mM PMSF and 1× protease inhibitor cocktail for plant tissues.
5. High Salt Wash Buffer (100 ml, store at 4 °C): 20 mM Tris–HCl pH 8.0, 2 mM EDTA, 500 mM NaCl, 0.1% SDS, and 1% Triton X-100. Add before use 1 mM PMSF and 1× protease inhibitor cocktail for plant tissues.
6. LiCl Wash Buffer (100 ml, store at 4 °C): 10 mM Tris–HCl pH 8.0, 1 mM EDTA, 0.25 M LiCl, 1% Igepal CA-630, and 1% sodium deoxycholate. Add before use 1 mM PMSF and 1× protease inhibitor cocktail for plant tissues.
7. TE buffer (100 ml, store at 4 °C): 10 mM Tris–HCl pH 8.0 and 1 mM EDTA pH 8.0.
8. 1 M DTT.
9. 25% Triton X-100.
10. 0.2 ml microtubes.

2.5 Recovery of Immunoprecipitated DNA

1. First ChIP elution Buffer (freshly prepared): 50 mM Tris–HCl pH 8.0, 5 mM EDTA, 20 mM DTT, and 1% SDS.
2. HiPPR™ Detergent Removal Resin (Thermo Scientific).
3. Prepacked columns SEPADEXTRANS™ 25 (ABT).
4. Column buffer: 15 mM Tris–HCl pH 8.0, 150 mM NaCl, and 1 mM EDTA.
5. ReChIP elution buffer (freshly prepared): 0.1 M NaHCO3 and 1% SDS.

3 Methods

3.1 Cross-Linking (~1 h)

To fix the protein–DNA interactions, the chromatin in the tissue of choice has to be cross-linked:

1. Harvest 1 g of the desired plant material in 50 ml tubes with 35.5 ml ice cold 1× PBS. Place the tubes on ice to keep all the material and solutions cold during the entire cross-linking step.
2. Add formaldehyde to a final concentration of 1% and use a piece of miracloth mesh soaked in cold 1× PBS as a lid to keep all the material submerged in the cross-linking solution. Vacuum-infiltrate (85,000 Pa) three rounds, 6 min each, incubating 4 min without vacuum between the rounds. To eliminate air bubbles that appear during vacuum infiltration, move the desiccator up and down against the bench. Release the vacuum slowly (*see* **Notes 3** and **4**).
3. Add glycine to a final concentration of 125 mM to stop the cross-linking reaction and vacuum-infiltrate for 5 min.
4. Wash three times with ice-cold milliQ water.
5. Pat-dry the material on tissue paper and freeze in liquid nitrogen. The plant material can be stored in the −80 °C freezer for up to 10 months.

3.2 Chromatin Preparation (~5 h)

Isolation of the chromatin requires thorough grinding of the tissue and homogenization in extraction buffer, followed by isolation of the nuclei. Disruption of the nuclei to isolate the chromatin is then performed by sonication in lysis buffer. For an efficient immunoprecipitation (IP) and possible downstream sequencing procedures, the chromatin has to be fractionated into 100–500 bp fragments.

1. Grind at least 1 g of material in liquid nitrogen using a precooled mortar and pestle (*see* **Note 5**).
2. Resuspend each gram of cross-linked material in 10 ml of extraction buffer supplemented with protease inhibitors.
3. Incubate for 30 min on a rotary shaker at 4 °C.
4. Homogenize in an ice-cold mortar (*see* **Note 6**).
5. Use a Dounce homogenizer device to increase the yield of isolated nuclei. Use twice (up and down) the loose pestle and twice the tight one (*see* **Note 7**).
6. Filter through a double miracloth mesh (moistened with extraction buffer) employing a glass funnel to a new 50 ml tube.
7. Spin 20 min at 3000 × *g* and 4 °C.
8. Discard the supernatant and keep the nuclear pellet.

9. Resuspend the pelleted nuclei in 1 ml of lysis buffer per gram of cross-linked material, pipetting up and down being careful to not form any foam.
10. Once the nuclear pellet is well resuspended, measure the volume and add SDS to a final concentration of 1% from a 10% stock solution.
11. Incubate for 15–30 min on a rotary shaker at 4 °C.
12. Sonicate the chromatin. We use a Bioruptor Plus device and 1.5 ml TPX tubes (*see* **Note 8**). Sonication is carried out at high power mode for 30 cycles (sonication cycle: 30 s ON, 30 s OFF). If you are using other types of sonicators, it will be necessary to determine experimentally the numbers of sonication cycles needed to get fragments between 100 and 500 bp.
13. Centrifuge for 5 min at 19,000 × *g* and 4 °C in a mini centrifuge to separate the soluble and insoluble chromatin. Transfer an aliquot of 30–60 μl of the supernatant (soluble chromatin) to another tube to assess the sonication efficiency and quantify the chromatin (Subheading 3.3). Keep the remaining sample on ice in the cold chamber until the first immunoprecipitation step (Subheading 3.4).

3.3 Analysis of Chromatin and Quantification

Checking the quantity and fragment sizes of the chromatin at this point allows the precise quantification of the chromatin for the first IP and the adjustment of fragment sizes by further sonication if necessary (*see* **Note 9**).

3.3.1 Reversion of Cross-Links (Overnight)

1. Add milliQ water up to 200 μl and 8 μl of 5 M NaCl.
2. Incubate at 65 °C and 1000 rpm, overnight in a ThermoMixer.

3.3.2 Purification of the Sonicated DNA (~6 h)

1. Digest RNA by adding 4 μl of the RNase A/T1 mix and 8 μl of 1 M Tris–HCl, pH 6.8. Incubate from 30 min to 1 h at 37 °C and 500 rpm in a ThermoMixer (*see* **Note 10**).
2. To digest the proteins, add 4 μl of 0.5 M EDTA and 2 μl of 10 mg/ml proteinase K. Incubate for 2 h at 37 °C and 500 rpm in a ThermoMixer.
3. Extract the DNA with Phenol–chloroform–IAA using phase lock gel tubes. First spin the empty tubes 30 s, then add the sample and finally one volume of Phenol–chloroform–IAA. Mix gently and centrifuge for 5 min at 19,000 × *g* and room temperature.
4. Collect the supernatant into a new tube and precipitate it with 20 μg of glycogen, 0.1 volumes of 3 M CH_3COONa, pH 5.2 and 2.5 volumes of cold absolute ethanol. Incubate for 30–60 min at −80 °C.

5. Spin down for 30 min at 19,000 × *g* and 4 °C.
6. Wash once with 70% ethanol.
7. Air-dry the pellet.
8. Resuspend the pellet in 30 μl of TE. Incubate for at least 30 min on ice before quantification.

3.3.3 Quantification (~45 min)

1. Quantify the DNA with the Qubit system. Use 1–5 μl of DNA and the Qubit dsDNA HS assay kit. In order to calculate the concentration of the original sample of chromatin take into account the dilution factor (*see* **Note 11**).
2. To check the size of the sonicated chromatin, fractionate 0.5–1 μg of DNA in a 1.5% agarose gel. The sonicated DNA is expected to have a size of 100–500 bp (for sequencing) and no RNA contamination (*see* **Note 9**).

3.4 First Immunoprecipitation (IP1) (~1.5 h/Overnight)

In the IP1 reaction, the DNA associated with the first protein of interest (modified histone) is specifically bound to an antibody against that protein. The workflow of the first IP procedure including the subsequent washing, elution and cross-linking steps is visualized in Fig. 1.

1. Centrifuge the chromatin solution (from **step 13** of Subheading 3.2) to remove insoluble debris for 10 min at 19,000 × *g*, at 4 °C.
2. Use a chromatin amount corresponding to 3–5 μg of DNA for the negative IgG control and four times more (12–20 μg) for the ChIP with Ab1 (*see* **Note 12**).
3. Dilute the chromatin 1/10 with ChIP dilution buffer in order to lower the SDS concentration from 1 to 0.1% (*see* **Note 13**).
4. Preclear the chromatin with protein A or G agarose beads (*see* **Note 1**). Calculate the amount of beads required (30 μl of protein A or G coated agarose beads per each ml of diluted chromatin). Wash the beads as follows: (1) add 1 ml of ChIP Dilution Buffer to the protein beads, (2) spin down at 1000 × *g* for 2 min at room temperature, and (3) discard the supernatant. Repeat the washing step twice. Add 1 ml of diluted chromatin to the washed beads. Incubate for 1 h on a rotary wheel at 4 °C.
5. Centrifuge for 1 min at 1000 × *g* to pellet the beads. Transfer the supernatant to a new 1.5 ml tube.
6. Save 10% of the diluted chromatin as input sample and keep it at −20 °C until **step 8** of Subheading 3.6.
7. Add the antibodies to each tube of diluted chromatin (IP1). Check the supplier's information to determine the right amount of antibody. Use one volume of diluted chromatin

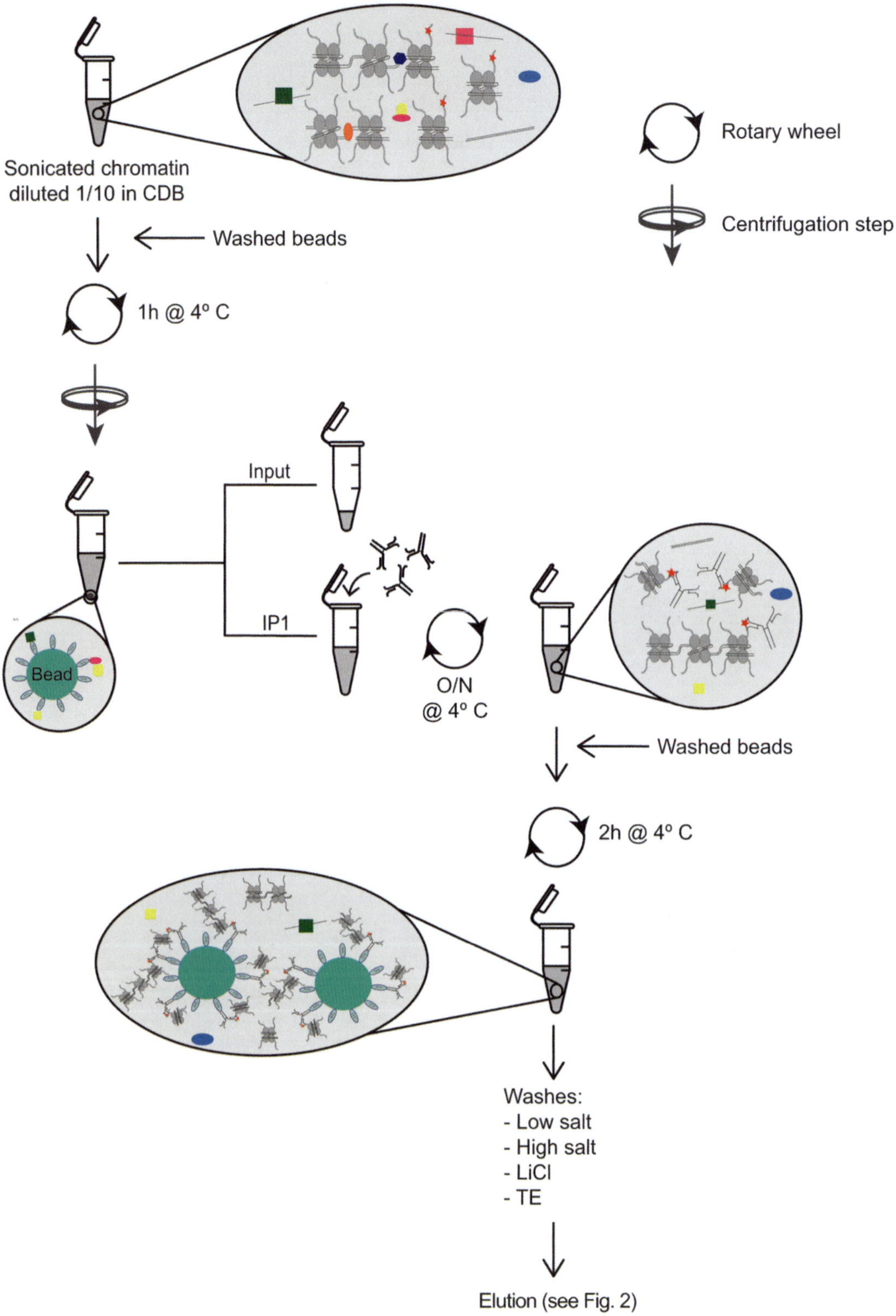

Fig. 1 Schematic diagram of the first immunoprecipitation (IP1) of the ReChIP protocol. Diluted chromatin is precleared with protein A or G agarose coated beads to remove nonspecific binding. Then, the chromatin is immunoprecipitated with Ab1. The immune complexes are pulled down with protein A or G agarose beads and the unbound chromatin is washed off. CDB: ChIP dilution buffer; IP: immunoprecipitation; Ab: antibody

for the negative control IP (*see* **Note 2**) and four volumes for the IP with antibody 1 (Ab1) (*see* **Notes 12** and **14**).

8. Incubate overnight on a rotary wheel at 4 °C.

3.5 Recovery and Elution of Immunoprecipitated Chromatin with Ab1 (~6 h)

To specifically recover the antibody-associated chromatin, the sample is mixed with agarose beads, which bind the antibodies, and washed several times to remove the unbound proteins. After elution, the sample needs to be prepared for the second immunoprecipitation by removal of the SDS and DTT (*see* Fig. 2).

1. Wash 20 and 80 μl of protein A or G agarose beads according to **step 4** of Subheading 3.4, for the IgG control and ChIP sample, respectively. Incubate for 2 h at 4 °C on a rotary wheel (*see* **Note 1**).
2. Centrifuge for 1 min at 1000 × *g* and discard the supernatant.
3. Wash the beads twice with 1 ml of the following washing buffers:
   - Low salt wash buffer.
   - High salt wash buffer.
   - LiCl wash buffer.
   - TE buffer.

   For each buffer, first perform a quick wash and then incubate for 5 min at 4 °C with rotation. To spin down the beads after each wash, centrifuge the tubes for 2 min at 1000 × *g* at 4 °C, and then remove the supernatant.
4. Elute twice the bound chromatin and IgG control with 100 μl of first ChIP elution buffer (Fig. 2; *see* **Note 15**). Incubate at room temperature 15 min under agitation. Spin down the beads for 2 min at 1000 × *g* at RT, transfer the supernatant to a fresh tube and repeat. Keep the eluted chromatin at RT to avoid SDS precipitation. Reserve the IgG control on ice until **step 9** of Subheading 3.6.
5. Remove the SDS from the eluted chromatin as follows: pipet 300 μl of HiPPRTM detergent removal resin onto the disposable column, centrifuge the column 1 min at 1500 × *g* to remove the storage buffer. Wash the resin three times with 300 μl column buffer. Discard the flow-through.
6. Insert the bottom plug on the column and place it on a new collection tube.
7. Load the eluted chromatin on top of the resin and vortex gently (*see* **Note 16**). Incubate at room temperature for 10 min. Repeat the gentle vortexing every 2 min.

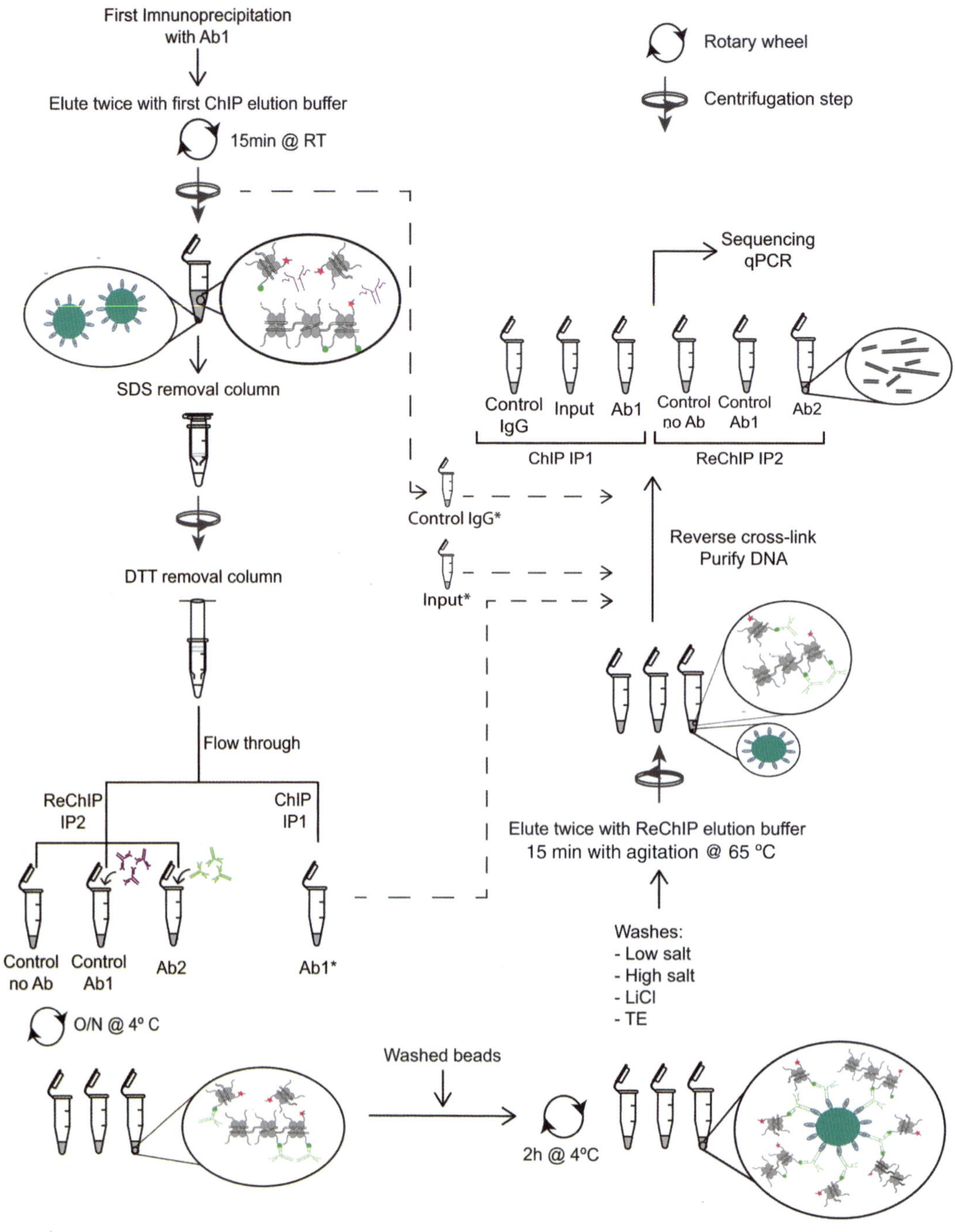

Fig. 2 Schematic diagram of the second immunoprecipitation (IP2) for the ReChIP protocol. DTT and SDS removal are performed after elution of the first immunoprecipitation (IP1). Chromatin is then split into four tubes: one tube (ChIP IP1_Ab1) as a control of the first IP, and three to carry out IP2 (Re-ChIP). Re-ChIP is performed by adding antibody 2 to one of the tubes (ReChIP IP2_Ab2) and by making two controls: one negative control without antibody (ReChIP_control no Ab), and one positive control by adding again antibody 1 (ReChIP_control Ab1). Immune complexes are pulled down after the addition of beads and the unbound chromatin is washed off. Following elution, cross-links are reversed in all IP samples, as well as in the Ab1, Ab2, and IgG control samples and the input sample. DNA is purified for further analysis. *Asterisks* indicate the samples (ChIP IP1 Ab1, Input, and IgG control) set aside until the cross-linking reversal step for DNA purification

8. Remove the bottom plug and recover the sample by centrifugation 1 min at 1500 × *g*. Keep the sample on ice (*see* **Note 17**).
9. Remove the DTT using a prepacked desalting column (Sepadextrans™ 25 medium SC, or equivalent). Equilibrate the column with at least 5 ml of column buffer supplemented with 1 mM DTT (*see* **Note 18**).
10. Load the 200 μl of eluted chromatin onto the column. When the sample has penetrated the gel, add 150 μl of column buffer. Place the column onto a new low binding tube and elute with 350 μl of column buffer supplemented with 1 mM DTT (*see* **Note 19**).
11. Add 16 μl of 25% Triton X-100. Adjust the volume to 400 μl with column buffer supplemented with 1 mM DTT and add 1× protease inhibitors cocktail for plant tissues and 1 mM PMSF.

3.6 Second Immunoprecipitation (IP2) (Overnight/~3 h)

The second immunoprecipitation procedure is similar to the first one, except for the composition of the elution buffer. Importantly, the sample from IP1 is split into four samples here (*see* **steps 1** and **2**), of which three have to be processed. The entire workflow of the IP2, starting from the elution of IP1, is visualized in Fig. 2.

1. Separate the eluted chromatin in four samples of 100 μl each in 0.2 ml microtubes. Keep one tube on ice as input for the second ChIP (ChIP IP1_Ab1) until **step 7**.
2. The other three samples are the following:
   (a) ReChIP negative control: do not add any antibodies.
   (b) ReChIP positive control: add antibody 1.
   (c) ReChIP experiment: add antibody 2.
3. Incubate overnight at 4 °C on a rotary wheel.
4. Add 20 μl of protein A or G agarose beads previously washed three times with ChIP dilution buffer (according to **step 4** of Subheading 3.4) and incubate for 2 h at 4 °C on a rotary wheel.
5. Wash the beads as described in **step 3** of Subheading 3.5.
6. Prepare the ReChIP elution buffer and keep it at 65 °C.
7. Elute twice the immune complexes by incubating the beads each time with 100 μl of ReChIP elution buffer for 15 min at 65 °C under agitation. Spin down the beads for 2 min at 1000 × *g* at room temperature and transfer the supernatants to a fresh tube.
8. Take the input samples of the first and second ChIP and adjust the volume to 200 μl with ReChIP elution buffer.
9. Take the IgG control sample.

3.7 Reverse Cross-Linking and DNA Purification (Overnight/~5 h)

1. To revert the cross-links, add 8 μl of 5 M NaCl to all samples and incubate overnight at 65 °C under gentle agitation.
2. RNA digestion (this step is optional for validations using qPCR): add 8 μl of 1 M Tris–HCl, pH 6.8 and 4 μl of the RNase A/T1 mix. Incubate for 1 h at 37 °C under gentle agitation.
3. Protein digestion: add 4 μl of 0.5 M EDTA pH 8.0 and 2 μl of 10 mg/ml proteinase K. Incubate for 2 h at 37 °C under gentle agitation.
4. Purify the DNA with 1 volume of Phenol–chloroform–IAA using phase lock gel tubes. Mix gently and centrifuge for 5 min at 19,000 × *g* at room temperature.
5. Transfer the supernatant to a new tube and precipitate the DNA with 20 μg of glycogen, 1/10 volume of 3 M CH_3COONa, pH 5.2 and 2.5 volumes of cold absolute ethanol. Incubate for 1 h at −80 °C.
6. Centrifuge for 30 min at 19,000 × *g* at 4 °C.
7. Wash with 1 ml 70% ethanol.
8. Air-dry the pellet.
9. Resuspend the pellet in 50 μl of TE buffer for at least 30 min on ice.

3.8 ChIP Analysis

For sequencing: Quantify the ChIPed DNA and the input samples using the fluorimetric dsDNA quantification assay (e.g., Qubit). Proceed to the library preparation.

For validations using qPCR: Dilute the input ten times to 1% input chromatin. Compare the results of the ChIPed sample to the input sample using primers of the loci of interest and positive and negative control primers (Fig. 3).

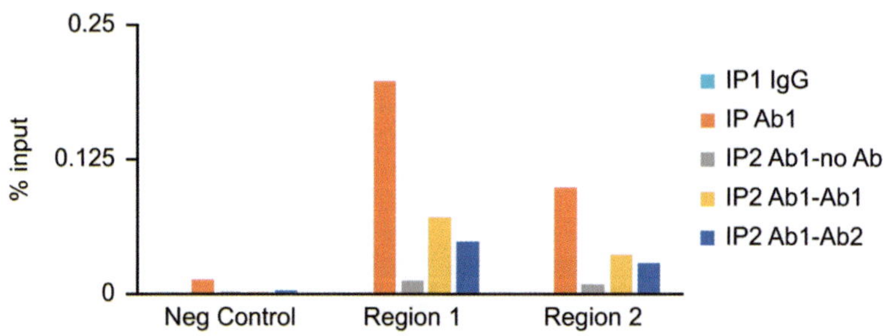

Fig. 3 qPCR ReChIP analysis. The experiment was performed with chromatin extracted from 10-day-old seedlings (3 μg/ChIP). We used anti-H3K4me3 (2 μg; Ab8580, Abcam) as Ab1 and anti-H3K27me3 (5 μg; Ab6002, Abcam) as Ab2. Primers used for the qPCR were designed based on [2]. Region 1 and 2 are bivalent loci and the negative control (Neg Control) corresponds to a region depleted of H3K4me3

4 Notes

1. The use of protein A or G depends on the isotype of the antibody and the animal species in which it was developed. Check the binding properties of your antibodies. When using magnetic beads instead of agarose beads, a preclearing step is not necessary.
2. The first and second ChIP experiments both require specific antibodies and positive and negative antibody controls. For the negative control, it is recommended to use an IgG antibody of the same species as the antibody used for the ChIP. For the positive control you can use for IP1 an antibody against the corresponding unmodified histone (e.g., use anti-total H3 as a positive control for the study of H3K4me3). This positive control sample can later also be used to correct for differences in total H3 deposition between the samples in the case of ChIP-seq. For IP2, the positive control consists of the first antibody, but if Ab2 is against a different histone and the samples will be sequenced, an unmodified histone positive control (anti-total H2A, H2B, or H4) is also advisable. In that case, adjust the number of the IP2 samples accordingly.
3. Formaldehyde should be handled with caution in a fume hood. All residues should be discarded appropriately according to the toxic waste regulations.
4. The cross-linking process is one of the most crucial steps in the ChIP and Re-ChIP procedures. It can be affected by different factors such as temperature, percentage of formaldehyde, salt concentration, pH, or cross-linking time. The reaction can be performed in buffers such as PBS or HEPES, but buffers containing primary amines (e.g., Tris) should be avoided because they will quench the reaction.
5. Good grinding of the material is essential to obtain enough nuclei. Collect the powder into a new 50 ml tube using a liquid nitrogen-cooled spatula.
6. This step is optional but recommended to better resuspend the powder.
7. This step helps to release more nuclei. If you do not have a Dounce homogenizer you can perform **step 3** of Subheading 3.2 for a longer time until an apparently homogenous sample is obtained.
8. TPX microtubes are thinner and recommended for efficient sonication with the Bioruptor Plus system. The sample volume should be between 100 and 300 μl. Check the specific minimum and maximum sample volume to be used with your sonication device.

9. If the chromatin is not well sheared, you should mix the chromatin sample stored at 4 °C (soluble and insoluble chromatin) and resonicate the material until it reaches the appropriate fragment size. Check again the quantity and size of the fragments.
10. There are several types of RNases that hydrolyze the RNA at specific sites. The RNase A specifically cuts RNA at C and U residues, while the RNase T1 degrades RNA at G nucleotides. The combination of these two RNA degradation activities leads to small fragments of RNA that are undetectable on an agarose gel.
11. To check the amount of sheared chromatin, it is necessary to use a highly sensitive fluorimetric-based dsDNA quantification system (such as the Qubit system). Regular UV spectrophotometric methods are not sensitive enough and tend to overestimate the quantity of chromatin.
12. Note that immunoprecipitated chromatin with antibody 1 (Ab1) will be divided in four samples to proceed with the second ChIP with antibody 2 (Ab2). Use four times the amount of chromatin for IP1 with Ab1 than for the negative IgG control. The four samples will be ChIP IP1; ReChIP IP2 No Ab; ReChIP IP2 with Ab1; and ReChIP IP2 with Ab2.
13. ChIP will not work in a buffer with more than 0.1% SDS.
14. The two antibodies used for the first and second IP may have different affinities for their specific targets that will influence the elution step. It is recommended to repeat the ReChIP by exchanging the antibodies used in each step.
15. The 1 M DTT stock solution used to prepare the elution buffer should be fresh.
16. It is very important to avoid forming any foam when loading and mixing the sample with the resin.
17. Check the efficiency of SDS removal by placing your sample on ice. It should not precipitate. In case that SDS is still present, repeat the detergent removing column step.
18. 1 mM DTT is used in order to maintain reducing conditions preventing the reformation of disulfide bonds of Ab1, but is compatible with the second IP.
19. The volumes of column buffer used can vary depending on the characteristics of the desalting column used and the sample volume. Follow the recommendations of the manufacturer.

References

1. Kouzarides T (2007) Chromatin modifications and their function. Cell 128:693–705
2. Sequeira-Mendes J, Araguez I, Peiro R, Mendez-Giraldez R, Zhang X, Jacobsen SE, Bastolla U, Gutierrez C (2014) The functional topography of the arabidopsis genome is organized in a reduced number of linear motifs of chromatin states. Plant Cell 26:2351–2366
3. Sequeira-Mendes J, Gutierrez C (2016) Genome architecture: from linear organisation of chromatin to the 3D assembly in the nucleus. Chromosoma 125:455–469
4. Bowler C, Benvenuto G, Laflamme P, Molino D, Probst AV, Tariq M, Paszkowski J (2004) Chromatin techniques for plant cells. Plant J 39:776–789
5. Villar CB, Kohler C (2010) Plant chromatin immunoprecipitation. Methods Mol Biol 655:401–411
6. Malapeira J, Khaitova LC, Mas P (2012) Ordered changes in histone modifications at the core of the Arabidopsis circadian clock. Proc Natl Acad Sci U S A 109:21540–21545
7. Orlando V (2000) Mapping chromosomal proteins in vivo by formaldehyde-crosslinked-chromatin immunoprecipitation. Trends Biochem Sci 25:99–104
8. Yamaguchi N, Winter CM, Wu MF, Kwon CS, William DA, Wagner D (2014) PROTOCOLS: chromatin immunoprecipitation from arabidopsis tissues. Arabidopsis book 12:e0170
9. Desvoyes B, Vergara Z, Sequeira-Mendes J, Madeira S, Gutierrez C (2017) A rapid and efficient ChIP protocol to profile chromatin binding proteins and epigenetic modifications in Arabidopsis. Methods Mol Biol 1675. doi:10.1007/978-1-4939-7318-7_5
10. Bernstein BE, Mikkelsen TS, Xie X, Kamal M, Huebert DJ, Cuff J, Fry B, Meissner A, Wernig M, Plath K, Jaenisch R, Wagschal A, Feil R, Schreiber SL, Lander ES (2006) A bivalent chromatin structure marks key developmental genes in embryonic stem cells. Cell 125:315–326
11. Roh TY, Cuddapah S, Cui K, Zhao K (2006) The genomic landscape of histone modifications in human T cells. Proc Natl Acad Sci U S A 103:15782–15787
12. Jiang D, Wang Y, Wang Y, He Y (2008) Repression of FLOWERING LOCUS C and FLOWERING LOCUS T by the arabidopsis polycomb repressive complex 2 components. PLoS One 3:e3404

Chapter 7

A Method to Identify Nucleolus-Associated Chromatin Domains (NADs)

Marie-Christine Carpentier, Ariadna Picart-Picolo, and Frédéric Pontvianne

Abstract

The nuclear context needs to be taken into consideration to better understand the mechanisms shaping the epigenome and its organization, and therefore its impact on gene expression. For example, in Arabidopsis, heterochromatin is preferentially localized at the nuclear and the nucleolar periphery. Although chromatin domains associating with the nuclear periphery remain to be identified in plant cells, Nucleolus Associated chromatin Domains (NADs) can be identified thanks to a protocol allowing the isolation of pure nucleoli. We describe here the protocol enabling the identification of NADs in Arabidopsis. Providing the transfer of a nucleolus marker as described here in other crop species, this protocol is broadly applicable.

Key words Nucleolus-associated chromatin domains, NADs, Nucleolus, Nucleus, Chromatin, FACS

1 Introduction

Gene accessibility to transcription factors and RNA polymerases is a key step in gene transcriptional regulation. This accessibility depends on local chromatin structure, but also on specific localization within the nucleus (i.e., the nuclear context) [1]. Global genome-wide approaches are routinely used to identify the precise chromatin context for each gene, revealing the existence of several chromatin states that correlate with the expression level of genes that they contain. In *Arabidopsis thaliana* (Arabidopsis), up to nine different chromatin states, characterized by specific combination of chromatin modifications, have been defined [2, 3]. However, the position of a gene with respect to a nuclear pore, nuclear lamina, and the nucleolus, as well as interchromosome or intrachromosome interactions, may also affect its transcriptional regulation, so they have to be taken into consideration too [1].

In the nucleoplasm, the largest nuclear body is the nucleolus, which can represent up to half of the nuclear volume in a certain cell

Marian Bemer and Célia Baroux (eds.), *Plant Chromatin Dynamics: Methods and Protocols*, Methods in Molecular Biology, vol. 1675, DOI 10.1007/978-1-4939-7318-7_7, © Springer Science+Business Media LLC 2018

type. The nucleolus is a direct consequence of ribosome biogenesis, but it is also a plurifunctional body implicated in other mechanisms including stress sensing, cell cycle progression, viral replication, and RNP biogenesis [4, 5]. Similar to the nuclear periphery, a thick heterochromatin layer surrounds the nucleolus, constraining chromatin domain mobility within the nucleoplasm [6]. In mammalian cells, nucleolus-associated chromatin domain (NAD) identification showed that all 23 human chromosomes possess at least one region associated with the nucleolus [7, 8]. In Arabidopsis leaf cells, all five chromosomes also have regions associating with the nucleolus [9, 10]. In both cases, NADs are composed of genomic regions enriched in silent chromatin modifications associated with transcriptional repression, such as DNA and/or Lysine 9 Histone 3 methylation. The identification of NADs is a complementary approach that allows to directly take into consideration the local environment of a locus. For example, RNA polymerase II tends to be much less abundant within the nucleolus. Thus, whether the association of a gene with the nucleolus influences its transcriptional regulation remains an open question, which requires additional studies, and particularly, the analysis of chromatin states in NADs.

Here we provide a detailed protocol allowing the identification of NADs in plant cells, and more specifically in Arabidopsis (Fig. 1). Four major steps are required and presented in this chapter: (1) nuclei and nucleoli isolation by fluorescence-activated cell sorting, (2) nuclear and nucleolar DNA recovery, (3) DNA-seq library preparation and sequencing, and (4) bioinformatic analyses of the sequences to identify NADs, as well as their respective chromatin states. These steps are indicated in the flowchart in Fig. 2.

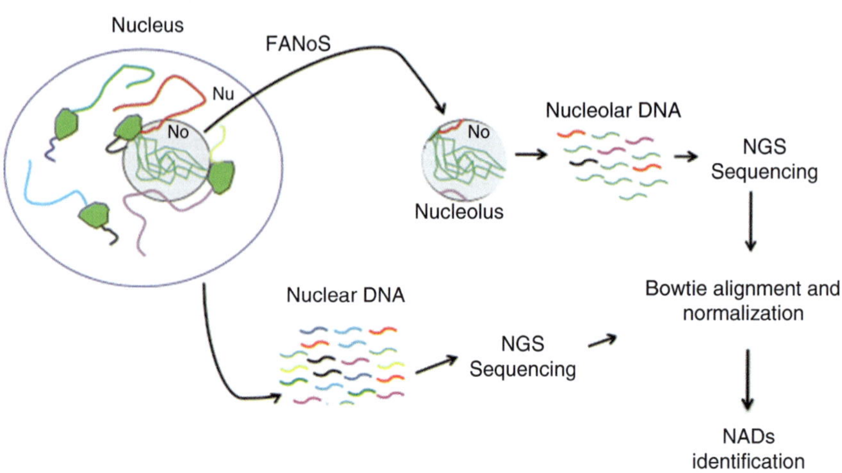

Fig. 1 NADs ID general strategy. Strategy used to purify and sequence total nuclear (Nu) and nucleolar (No) DNA in order to identify NADs by high throughput sequencing

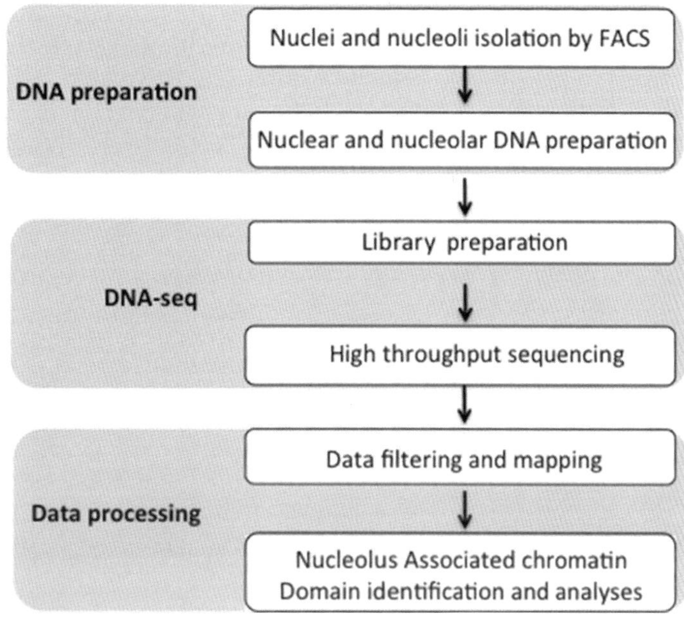

Fig. 2 Flowchart showing a schematic overview of the entire procedure

2 Materials and Reagents

2.1 Materials

1. Plants expressing the chimeric protein YFP fused to FIBRILLARIN 2 to mark the nucleoli (*see* **Note 1**).
2. Vertically sloped filters with a 30 μm mesh size (*see* **Note 2**).
3. Gentle sonicator device for DNA shearing (*see* **Note 3**).
4. Flow cytometer/Cell sorter: Fluorescence-Activated Cell Sorting (FACS) is carried out by flow cytometry with a cell sorter. YFP or GFP is excited with a 488 nm blue laser of 100 mW and signal emission is captured with a 505 long pass filter and a 530/30 (515–545 nm) filter. The sorting procedure is carried out in sterile conditions and the sheath fluid is 1× PBS.
5. A fluorimetric DNA quantification device distinguishing single stranded and double stranded DNA (e.g., Qubit).
6. Bioanalyzer system to control the size, the quantity and quality of DNA.
7. High-throughput DNA sequencer device.

2.2 Reagents

1. TEN buffer: 10 mM Tris–HCl pH 7.5, 10 mM Na-EDTA pH 8, and 100 mM NaCl. Has to be freshly prepared from stock solution (1 M Tris–HCl pH 8, 500 mM Na-EDTA pH 8, and 5 M NaCl, all kept at room temperature).
2. 36% formaldehyde solution.

3. FACS Solution (or Galbraith's buffer): 45 mM $MgCl_2$, 20 mM 3-(N-morpholino)propanesulfonic acid (MOPS), 30 mM sodium citrate, 0.1% Triton X-100 adjusted to pH 7 with NaOH. Can be prepared in advance, in 5 mL aliquot and stored at −20 °C for several months.
4. 10× Phosphate buffer saline (PBS): 1370 mM sodium chloride (NaCl), 27 mM potassium chloride (KCl), 100 mM disodium hydrogen phosphate (Na_2HPO_4), and 18 mM potassium dihydrogen phosphate (KH_2PO_4). Can be stored are room temperature for several months.
5. DNAse-free RNAse A, used at a concentration of 10 μg/mL.
6. DNAse-free Proteinase K, used at a concentration of 10 μg/mL.
7. A kit allowing the purification and the concentration and of low amount of DNA (*see* **Note 4**).
8. A kit enabling the preparation of a DNA library compatible for NGS sequencing from a low amount of DNA (*see* **Note 5**).

2.3 Software and Databases

1. Bowtie2.
2. Bedcoverage.
3. R.
4. Unix system and awk command line.
5. Computing cluster.
6. TAIR10 reference genome sequence database (*see* **Note 6**): https://www.arabidopsis.org/download_files/Genes/TAIR10_genome_release/TAIR10_chromosome_files/TAIR10_chr_all.fas.

3 Methods

3.1 Nuclei and Nucleoli Isolation by Fluorescence-Activated Cell Sorting

First, nuclear and nucleolar DNA have to be purified from isolated nuclei or nucleoli respectively. Two main protocols are available to isolate nuclei and nucleoli: (1) a protocol based on successive centrifugation steps with a sucrose gradient and variable salt concentrations or (2) a protocol based on the utilization of a fluorescence-activated cell sorter (FACS). However, in (1), protoplasts have to be the starting material [11], while the entire organism and any plant tissue can be used in (2).

Thus, we recommend to purify nuclei or nucleoli by FACS using the approaches named FANS for fluorescence-activated nucleus sorting and FANoS for fluorescence-activated nucleolus sorting [12]. This purification is based on the labeling of nucleoli via a fluorescent marker that delimits the nucleolar position in the cell of interest *in planta*. Hence, plants expressing the nucleolus-specific protein Fibrillarin 2 (FIB2) coding sequence from Arabidopsis (At4G25630), with the Yellow Fluorescent Protein (YFP) at its C-terminal part and under the control of the ubiquitous promoter 35S (YFP:FIB2 plants) are used here (*see* **Note 1**) [13]. The tissues is prepared for FACS as follows:

1. Fix 1 g tissue in 10 mL of TEN buffer with 4% formaldehyde on a rotor in a cold room for 20 min.
2. Wash twice for 10 min in 10 mL of ice-cold TEN buffer.
3. Mince the tissue with a razor blade in 1 mL FACS buffer (*see* **Note 7**).
4. Filter the homogenate through a 30 μm mesh diameter nylon membrane into a 1.5 mL tube (*see* **Note 8**).
5. To sort nuclei by FANS (*see* **Note 9**), the filtered homogenate can be directly subjected to FACS.
6. To sort nucleoli using FANoS, the filtered homogenate should be sonicated four times for 5 min at medium power using a Bioruptor to liberate the nucleoli, and can then be subjected to FACS.
7. Sorting is carried out with a fluorescent-activated cell sorter (FACS). To facilitate the setting of the sorting window, it is better to first analyse the sample containing the fluorescent marker, then the negative control. Because the nucleolar marker used may not be ubiquitously expressed, we recommend to always perform nuclear sorting using the fluorescent signal of YFP:FIB2.
8. Nucleolar sorting is performed using the YFP:FIB2 signal (*see* **Note 10**). As for nuclear sorting, the negative control sample (without a nucleolar marker) is used to determine the area where no YFP signal is detected from the particle cloud.

   For NADs identification by high throughput sequencing, at least 500,000 nuclei and 1 million nucleoli are necessary as starting material.

3.2 Nuclear and Nucleolar DNA Recovery

1. To homogenize the samples, sonicate the sorted nuclei five times for 5 min (with a 30 s ON and 30 s OFF interval) at medium power, using a Bioruptor (*see* **Note 3**). In parallel, sonicate the sorted nucleoli one time for 5 min using the same settings (*see* **Note 11**).
2. To disrupt the nuclei or nucleoli, place the samples for 10 min at 95 °C.
3. To deteriorate the RNA present in the samples, incubate them for 30 min at 37 °C with RNAse A at a concentration of 10 μg/mL.
4. To degrade the proteins, incubate for 30 min at 45 °C with Proteinase K at a concentration of 10 μg/mL.

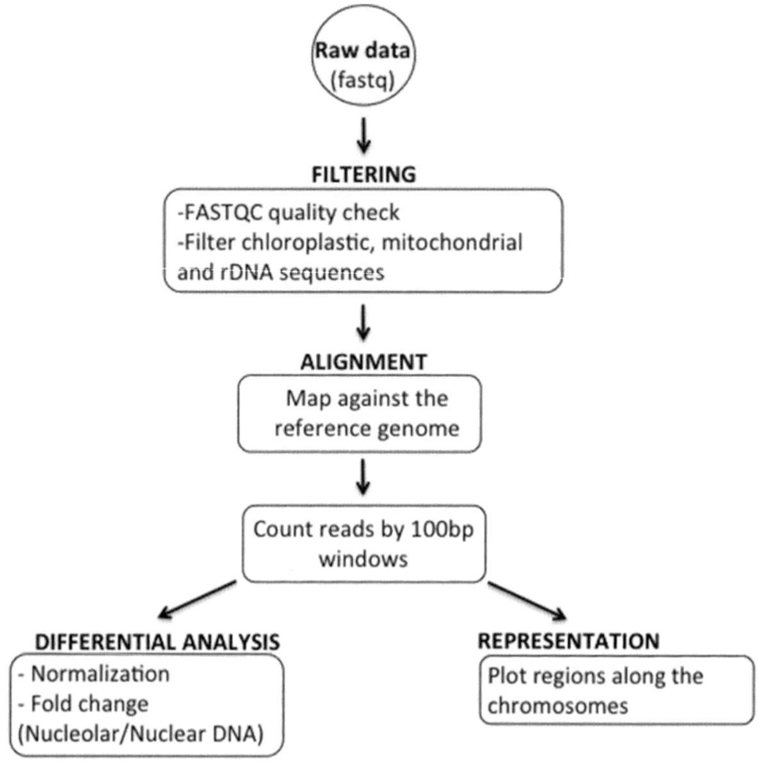

Fig. 3 NADs ID pipeline. Bioinformatic pipeline used to identify NADs from fastQ reads

5. Boil the samples for 10 min at 99 °C to inactivate Proteinase K.
6. For high-throughput DNA sequencing, nuclear and nucleolar DNA is purified and concentrated using the ChIP DNA Clean & Concentrator kit according to the manufacturer's instructions.
7. The Quant-iT PicoGreen dsDNA kit can be used to determine the concentration of the samples, according to the manufacturer's instructions.

3.3 DNA-Seq Library Preparation and Sequencing

The libraries are generated via the Nextera XT DNA library preparation kit or equivalent and dosed with for example the Agilent High Sensitivity DNA kit, according to the manufacturer's instructions. DNA samples are then subjected to high-throughput paired-end sequencing (*see* **Note 12**).

3.4 DNA-Seq Data Analysis

3.4.1 Processing of the Data

The pipeline described below is also presented in the Fig. 3.

1. Download the TAIR10 genome reference from the TAIR website (*see* **Note 13**).
2. Separate the organellar (chloroplasts and mitochondria) and ribosomal DNA from the nuclear chromosomes by generating

two different files (For example *chrC_M_rDNA.fa and TAIR10_5chr.fasta respectively*).

3. Create an index with chloroplastic and mitochondrial TAIR10 chromosomes and rDNA (*chrC_M_rDNA.fa*) sequences:

```
bowtie2-build index chrC_M_rDNA.fa chrC_M_rDNA
```

4. Filter the nonnuclear sequences from the raw data by mapping the fastq reads against the sequences above, i.e., chrC_M_rDNA, on the cluster, in parallel, with 12 threads per sample:

```
bowtie2 –sensitive -p 12 –un-conc $out-filtered.fastq
–x chrC_M_rDNA -1 $fq1 -2 $fq2 -S $out.sam
```

5. Keep the pairs that did not align with the index (in fastq format).

3.4.2 Dataset Alignment on the Reference Genome

1. Create an index with the five chromosomes of the TAIR10 database.

```
bowtie2-build TAIR10_5chr.fasta TAIR10_5chr
```

2. Map the filtered fastq libraries against the Arabidopsis chromosomes:

```
bowtie2 –sensitive-local -p 12 -x TAIR10_5chr -1
$fq1_A2 -2 $fq2_A2 -S $out.sam
```

3.4.3 Quantification

1. Segment the TAIR10 chromosomes by 100 bp windows:

```
bedtools make windows -g size_chr_TAIR10.txt -w 100 >
TAIR10_100bpwindows.bed
```

2. Convert the sam alignment files in sorted bam files*:*

```
samtools view -bS $in.sam -o $out.bam and samtools
sort $out.bam $out.sort
```

3. Count for each library, the number of reads that map against each 100 bp window:

```
coveragebed -abam $bam -b TAIR10_100bpwindows.bed >
$out.100pb.bed.coverage
```

4. Reformat the quantification table obtained with bedcoverage for data treatment in R:

```
awk -F « \t » '{print $1 « _ »$2 « _ »$3 « \t »$4'}
$out.100pb.bed.coverage
```

3.4.4 Differential Analysis in R

In the following commands, the initial file names used are *Col0_N* (nuclear DNA) and *Col0_No* (nucleolar DNA).

1. Normalize the counted data by the sequence coverage of each library.

```
tmpRESC$Col0_N_norm ←tmpRESC$Col0_N/sum(tmpRESC$Col0_N)
```

2. Filter the uncovered windows, i.e., keep the 100 bp windows that have a minimum of 10 aligned reads from each DNA sample

```
FRESC<-tmpRESC[which(tmpRESC$mCol0_N>10),]
```

3. For each 100 bp window, calculate the mean number of mapped reads between the nuclear (N) or nucleolar (No) DNA replicates of the same condition

```
tmpRESC$mCol0_N<-rowMeans(tmpRESC[,2:3])
```

4. Determine the fold change (No/N) for each condition.

```
FRESC$FC < −FRESC$Col0_No_n/FRESC$Col0_N_n
```

5. A region is defined as a NAD when the window is differentially covered, i.e., fold change ≥2 or fold change ≤0.5.

```
FRESC$DC←ifelse((FRESC$FC>=2 | FRESC$FC<=0.5 ),TRUE,FALSE)
```

3.4.5 Plot Regions Along the Chromosomes in R

For each chromosome, plot the log2 fold change (log2 No/N) of 100 kb windows according to their chromosome position. 100 kb windows are used here to allow a better visualization of the NADs along the chromosomes. To obtain them, restart the analyses from Subheading 3.4.3, using 100 kb instead of 100 bp windows. Every 100 kb window is represented by a black dot; nucleolus enriched genomic regions above the threshold (NADs) are shown in grey.

```
plot(log2(No/N),ylim=c(-1.5,3.5),col=ifelse(log2(No/N)>0.5,
« gray», « black »,pch=16)
```

An example of a chromosome plot is presented in Fig. 4.

3.4.6 Define the chromatin states of the NADs

1. Create a bed file for each chromatin state in TAIR10 reference genome starting from the Supplemental Data Set 2 published in [3]:

```
awk -F "\t" '{print "chr"$1"\t"$2"\t"$3"\tstate1"}'
State1_TAIR.csv > reformat-State1_TAIR.txt
cat reformat-State* > reformat-State_all_TAIR.bed
bedtools sort -i reformat-State_all_TAIR.bed > reformat-
State_all_TAIR.sort.bed
```

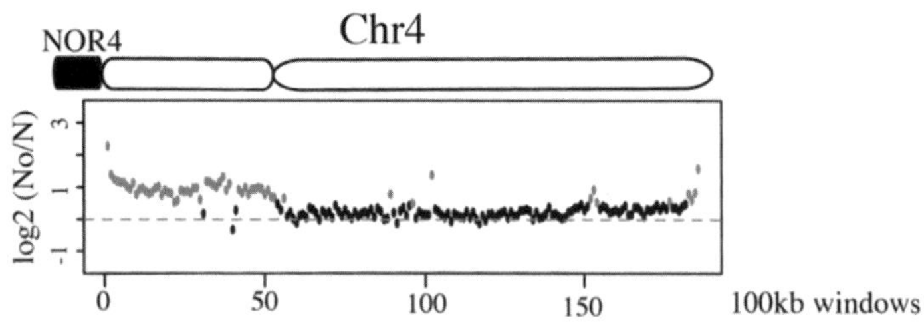

Fig. 4 Visualisation of NADs position along the chromosome. The relative enrichment of a given genomic segment in the nucleolus can be visualized using a chromosome plot. In this example, nucleolar and nuclear DNA from wild-type Col-0 plants were used. The *y axis* displays the fold change ratio No/N. Each dot represents a 100-kb window along chromosome 4. Nucleolus enriched genomic regions above the threshold are shown in *grey*

2. Intersect the NADs with the chromatin state bed files:

```
bedtools intersect -a $TAIR10_states.bed -b NADS_re-
gions.bed -wa -wb
```

3. Sum for each chromatin state the number of NADs from the reference genome:

```
with home-made perl program
perl parse_state.pl > sum_state_TAIR10.tab
```

4. Create a histogram showing the proportion of sequences in each chromatin state:

```
barplot((ref$V1/sum(ref$V1)*100),col="lightblue",
border="blue",names.arg=ref$V2,las=2,ylim=c(0,30))
```

4 Notes

1. The chimeric protein YFP::FIB2 is expressed under control of the 35S promoter, as described in [9]. However, other promoters including cell-specific promoters can be used to isolate nucleoli from specific cell-types.
2. Depending on the nozzle size in the FACS machine, 50 μm mesh sizes can also be used.
3. Because the sonication procedure depends on the device used, we describe here the setting used for the Bioruptor standard sonicator device (Diagenode). If an alternative sonication device is used, make sure to adapt the settings properly.
4. We used the Nextera XT DNA library preparation kit (Illumina, USA) at this stage, but alternative kits could also be used here.

5. For instance, the ChIP DNA Clean & Concentrator™ kit (Zymo Research, USA) could be used at this stage.
6. The reference genome from Araport, 11 can also be used here: https://www.araport.org/data/araport11
7. Nuclei isolation by FACS can be performed using the fluorescent signal of the YFP:FIB2 marker. Alternatively, FANS can be performed after DNA labeling with a DNA marker like DAPI or propidium iodide.
8. To mince the samples, you can use a plastic weight boat or the interior of a petri dish.
9. It is important to use low retention eppendorf tubes because nucleoli tend to stick on the surface of regular tubes.
10. The size of the nucleoli detected also varies because they tend to aggregate with each other once released from the nucleus.
11. When sorted by FACS, each nucleus or nucleolus is isolated in an approximately 3 nL drop. Therefore, a centrifugation step can be added here: 100 g 15 min. Keep only 100 μL of the supernatant and resuspend the nuclei or nucleoli in this volume.
12. For a better alignment of the reads along the reference genome, paired-end sequencing of at least 100 bp is recommended.
13. Of course, any reference genome could be used here.

Acknowledgments

This work was supported by the CNRS and ANR JCJC NucleoReg (ANR-15-CE12–0013-01) to FP.

References

1. Bickmore WA, van Steensel B (2013) Genome architecture: domain organization of interphase chromosomes. Cell 152:1270–1284. doi:10.1016/j.cell.2013.02.001
2. Roudier F, Ahmed I, Berard C et al (2011) Integrative epigenomic mapping defines four main chromatin states in Arabidopsis. EMBO J 30:1928–1938. doi:10.1038/emboj.2011.103
3. Sequeira-Mendes J, Araguez I, Peiro R et al (2014) The functional topography of the Arabidopsis genome is organized in a reduced number of linear motifs of chromatin states. Plant Cell 26:2351–2366. doi:10.1105/tpc.114.124578
4. Boulon S, Westman BJ, Hutten S et al (2010) The nucleolus under stress. Mol Cell 40:216–227. doi:10.1016/j.molcel.2010.09.024
5. Pederson T (2011) The nucleolus. Cold Spring Harb Perspect Biol. doi:10.1101/cshperspect.a000638
6. Ferreira J, Paolella G, Ramos C, Lamond AI (1997) Spatial organization of large-scale chromatin domains in the nucleus: a magnified view of single chromosome territories. J Cell Biol 139:1597–1610
7. van Koningsbruggen S, Gierlinski M, Schofield P et al (2010) High-resolution whole-genome sequencing reveals that specific chromatin

domains from most human chromosomes associate with nucleoli. Mol Biol Cell 21:3735–3748. doi:10.1091/mbc.E10-06-0508

8. Nemeth A, Conesa A, Santoyo-Lopez J et al (2010) Initial genomics of the human nucleolus. PLoS Genet 6:e1000889. doi:10.1371/journal.pgen.1000889
9. Pontvianne F, Carpentier M-C, Durut N et al (2016) Identification of nucleolus-associated chromatin domains reveals a role for the nucleolus in 3D organization of the A. thaliana genome. Cell Rep 16:1574–1587. doi:10.1016/j.celrep.2016.07.016
10. Picart C, Pontvianne F (2016) Plant nucleolar DNA: green light shed on the role of Nucleolin in genome organization. Nucleus Austin Tex 8:11–16. doi:10.1080/19491034.2016.1236167
11. Hacot S, Coute Y, Belin S et al (2010) Isolation of nucleoli. Curr Protoc Cell Biol Chapter 3: Unit3.36. doi:10.1002/0471143030.cb0336s47
12. Pontvianne F, Boyer-Clavel M, Saez-Vasquez J (2016) Fluorescence-activated nucleolus sorting in Arabidopsis. Methods Mol Biol 1455:203–211. doi:10.1007/978-1-4939-3792-9_15
13. Pontvianne F, Blevins T, Chandrasekhara C et al (2013) Subnuclear partitioning of rRNA genes between the nucleolus and nucleoplasm reflects alternative epiallelic states. Genes Dev 27:1545–1550. doi:10.1101/gad.221648.113

Chapter 8

Cell Type-Specific Profiling of Chromatin Modifications and Associated Proteins

Ana Karina Morao, Erwann Caillieux, Vincent Colot, and François Roudier

Abstract

Progression of a cell along a differentiation path is characterized by changes in gene expression profiles. Alterations of these transcriptional programs result from cell type-specific transcription factors that act in a dynamic chromatin environment. Understanding the precise contribution of these molecular factors during the differentiation process requires accessing specific cell types within a developing organ. This chapter describes a streamlined and alternative version of INTACT, a method enabling the isolation of specific cell populations by affinity-purification of tagged nuclei and the subsequent analysis of gene expression, transcription factor binding profiles, as well as chromatin state at a genome-wide scale. In particular, modifications of the nuclei isolation, capture, and purification procedures are proposed that improve time scale, yield, and purity. In addition, the combination of different tags enables the analysis of distinct cell populations from a single transgenic line and the subtractive purification of subpopulations of cells, including those for which no specific promoter is available. Finally, we describe a chromatin immunoprecipitation protocol that has been successfully used to profile histone modifications and other chromatin-associated proteins such as RNA Polymerase II in different cell populations of the Arabidopsis root, including the quiescent center of the stem cell niche.

Key words Chromatin, Histone modification, INTACT, RNA polymerase II, Cell type, Nucleus, Epigenome, Transcriptome

1 Introduction

Understanding how cells acquire and maintain the distinct identities and properties necessary for tissue function is a central question in developmental biology. While cell type-specific transcription factors (TFs) have a key role in instructing the gene expression profiles that govern cell differentiation, the chromatin organization in which these TFs act provides an essential context to orchestrate and stabilize transcriptional programs. Different methods have been used to isolate specific cell populations from a multicellular organism in order to investigate gene expression and TF binding profiles as well as chromatin state at a genome-wide scale. These techniques rely on fluorescence-activated cell sorting (FACS), laser

Marian Bemer and Célia Baroux (eds.), *Plant Chromatin Dynamics: Methods and Protocols*, Methods in Molecular Biology, vol. 1675, DOI 10.1007/978-1-4939-7318-7_8, © Springer Science+Business Media LLC 2018

capture microdissection (LCM) or affinity-based methods such as isolation of nuclei tagged in specific cell types (INTACT). The INTACT strategy was originally described in *Arabidopsis thaliana* [1, 2] and extended to *Solanum* spp. [3], *Caenorhabditis elegans* and *Drosophila melanogaster* [4, 5], as well as *Xenopus laevis* [6]. The principle is to mark in vivo the nuclei of a cell population of interest with a genetically encoded tag expressed under a cell type-specific promoter, and to purify labeled nuclei from a crude preparation, thanks to the tag-specific binding affinity.

This chapter describes a streamlined version of INTACT with improved nuclei isolation, capture and purification procedures, which was used to analyze gene expression [7] and chromatin features in both abundant and rare cell populations in the Arabidopsis root. Unlike the original INTACT approach [1, 2], this method does not rely on a two-component system to achieve streptavidin-mediated capture of biotinylated nuclei. Instead, nuclei are affinity-purified using magnetic beads coated with an antibody directed against the fluorescent moiety (GFP or mCherry) of the nuclear tagging fusion (NTF), which also contains a nuclear envelop-targeting domain [1, 2]. GFP- and mCherry-NTF versions were cloned in a gateway system compatible with the compendium of cell type-specific promoters recently published [7]. This alternative approach, which has been used in other systems [5, 7–9], also offers the possibility to combine multiple NTFs in a single line and to purify subpopulations of related cells, including those for which no specific promoter is available, as described below. In addition, nuclei capture is performed in batch rather than with a column, in a manner similar to what was recently proposed [10]. This significantly speeds up the procedure and results in better yield with higher purity. Following this purification method, we describe a chromatin immunoprecipitation protocol that has been successfully used to measure the enrichment of histone modifications, such as trimethylated lysine 4 of histone H3 (H3K4me3) that is associated with transcriptional activity [11], and other chromatin-bound proteins such as RNA Polymerase II in different cell populations of the Arabidopsis root, including the quiescent center of the stem cell niche.

2 Materials

2.1 Preparation of Biological Material

1. Gateway-compatible donor and destination vectors as well as transgenic lines expressing the NTF under different promoters are available at the European Arabidopsis Stock Centre (NASC, http://arabidopsis.info/CollectionInfo?id=156, sets N2106369, N2106365 and N2106369, respectively) [7], or available upon request (Gateway-compatible mCherry-NTF cassette and additional promoters). The original INTACT

GFP-NTF and the mCherry-NTF cassettes were a gift from R. Deal (Department of Biology, O. Wayne Rollins Research Center, Emory University, 1510 Clifton Road NE, Atlanta, GA, 30,322, USA) and D. Grimanelli (Epigenetic Regulations and Seed Development, UMR232, Institut de Recherche pour le Développement (IRD), Université de Montpellier, 34,394 Montpellier, France), respectively.

2. Square 120 × 120 mm petri dishes.
3. Murashige and Skoog basal medium.
4. Nylon sifting fabric (NITEX, mesh opening 100 μm).

2.2 Extraction and Purification of Tagged Nuclei

All buffers are prepared freshly using molecular biology-grade H_2O and commercial, sterilized or freshly prepared stocks solutions. The volume required for a single purification experiment is indicated for each buffer.

In **items 1–7**, the stock solutions that can be prepared for multiple experiments are listed:

1. 1 M MOPS: Dissolve 10.46 g of MOPS in 50 mL H_2O, adjust the pH to 7.0 with NaOH and store 10 mL aliquots at −20 °C.
2. 5 M NaCl: Dissolve 146.1 g of NaCl in H_2O. Make up the volume to 500 mL with H_2O, autoclave the solution and keep it at room temperature for up to 1 year.
3. 2 M KCl: Dissolve 14.9 g of KCl in H_2O. Make up the volume to 100 mL with H_2O, autoclave the solution and keep it at room temperature for up to 1 year.
4. 0.5 M EDTA pH 8.0: Dissolve 14.61 g of EDTA in H_2O. Under agitation, gradually add ~2 g of NaOH to adjust the pH to 8.0. Make up the volume to 100 mL with H_2O, autoclave the solution and keep it at room temperature for up to 1 year.
5. 0.5 M EGTA: Dissolve 19 g of EGTA in H_2O. Adjust the pH to 8.0 using NaOH. Make up the volume to 100 mL with H_2O, autoclave the solution and keep it at room temperature for up to 1 year.
6. 2 M Spermidine: dissolve 2.904 g of spermidine powder in 10 mL of H_2O and store 1 mL aliquots at −20 °C.
7. 200 mM Spermine: dissolve 0.405 g of spermine powder in 10 mL of H_2O and store 1 mL aliquots at −20 °C.
8. Nuclei Purification Buffer (NPB): 20 mM MOPS pH 7, 40 mM NaCl, 90 mM KCl, 2 mM EDTA, 0.5 mM EGTA, 0.5 mM spermidine, 0.2 mM spermine, 1× protease inhibitors (e.g., Complete protease inhibitors, Roche). Prepare 150 mL, keep on ice and use immediately.
9. Nuclei Purification Buffer with 1% formaldehyde (NPBf): 20 mM MOPS pH 7, 40 mM NaCl, 90 mM KCl, 2 mM

EDTA, 0.5 mM EGTA, 1% (v/v) formaldehyde (from a 37% commercial solution stabilized with methanol). Prepare 120 mL, keep at room temperature and use immediately. Work with formaldehyde should be performed under a fume hood.

10. 2 M Glycine: dissolve 1.5 g of glycine powder in 10 mL H_2O, keep at room temperature and use immediately.
11. DAPI stock solution (1 mg/mL): dissolve 1 mg of 4′,6-diamidino-2-phenylindole (DAPI) powder in 1 mL of H_2O and store 100 μL aliquots at −20 °C in the dark.
12. Nuclei Purification Buffer with DAPI (NPBd): Supplement NPB with 4 μg/mL DAPI. Prepare 1 mL, keep on ice, protect from light and use immediately.
13. Nuclei Purification Buffer with 1% Bovine Serum Albumin (BSA) (NPBb): Supplement NPB with 1% (v/v) BSA (from a 10% stock solution). Prepare 5 mL, keep on ice and use immediately.
14. BSA stock solution (10%): dissolve 10 g of BSA powder (heat shock fraction) in 100 mL of H_2O and store 10 mL aliquots at −20 °C.
15. Nuclei Purification Buffer with 0.5% Bovine Serum Albumin (BSA) and 0.1% Triton X-100 (NPBbt): Supplement NPB with 0.5% (v/v) BSA (from a 10% stock solution) and 0.1% (v/v) Triton X-100. Prepare 80 mL, keep on ice and use immediately.
16. Vacuum pump and chamber for tissue cross-linking.
17. Porcelain 50–100 mL mortars and pestles.
18. Liquid nitrogen.
19. 50 and 15 mL plastic tubes.
20. 1.5 mL plastic tubes.
21. Dounce tissue grinder 40 mL (e.g., Wheaton).
22. 70 and 40 μm cell strainers (e.g., Fisher Scientific) or equivalent.
23. Paramagnetic beads with recombinant Protein A and G (*see* **Note 1**).
24. Antibodies against GFP and mCherry (*see* **Note 2**).
25. Magnetic rack for bead separation with inlets for 15 mL tubes and a rack for 1.5 mL tubes (*see* **Note 1**).
26. Rotating wheel for 15 and 1.5 mL plastic tubes in a 4 °C cold room.
27. Refrigerated centrifuge with rotor for 50 mL tubes and bench-top centrifuge for 1.5 mL tubes.

28. Cell counting chamber (hemocytometer, *see* **Note 3**).
29. 4 °C cold room.
30. Epifluorescence microscope.

2.3 Chromatin Immunoprecipitation and Sequencing

All buffers are prepared freshly using molecular biology-grade H_2O and commercial, sterilized or freshly prepared stocks solutions. Buffer volumes are indicated for up to three ChIP experiments.

1. Nuclei lysis buffer: 50 mM Tris–HCl pH 8, 10 mM EDTA pH 8, 0.1% (w/v) SDS, and 1× protease inhibitor. Prepare fresh 5 mL, keep on ice and use immediately.
2. 1 M Tris–HCl pH 8.0: Dissolve 121.14 g of Tris in 800 mL H_2O. Adjust the pH of the solution to 8.0 using HCl. Make up the volume to 1 L with H_2O, autoclave the solution and keep it at room temperature for up to 1 year.
3. Sonicator suited for handling small volumes (< 1 mL) and appropriate microtubes (*see* **Note 4**).
4. ChIP dilution buffer (CDB): 1.1% (v/v) Triton X-100, 1.2 mM EDTA pH 8, 16.7 mM Tris–HCl pH 8, 167 mM NaCl, 0.1% (w/v) SDS. Prepare 10 mL and keep on ice.
5. Appropriate ChIP-grade antibodies. We successfully used antibodies for H3K4me3 (Millipore 07-473) and for RNA polymerase II (Abcam, Ab817). Alternative sources of ChIP-grade antibodies might work equally well.
6. Paramagnetic beads with recombinant Protein A and G (*see* **Note 1**).
7. Low-salt wash buffer: 20 mM Tris–HCl pH 8, 150 mM NaCl, 0.1% (w/v) SDS, 1% (v/v) Triton X-100, and 2 mM EDTA pH 8. Prepare 50 mL and keep on ice. Solution can be stored at 4 °C up to 3 months.
8. High-salt wash buffer: 20 mM Tris–HCl pH 8, 500 mM NaCl, 0.1% (w/v) SDS, 1% (v/v) Triton X-100, and 2 mM EDTA pH 8. Prepare 50 mL and keep on ice. Solution can be stored at 4 °C up to 3 months.
9. 4 M LiCl stock solution: Dissolve 17 g of LiCl in H_2O. Make up the volume to 100 mL with H_2O, autoclave the solution and keep it at room temperature for up to 1 year.
10. LiCl wash buffer: 10 mM Tris–HCl pH 8, 250 mM LiCl, 1% (w/v) sodium deoxycholate, 1% (v/v) NP-40, 1 mM EDTA pH 8. Prepare 50 mL and keep on ice. Solution can be stored at 4 °C up to 3 months.
11. TE: 10 mM Tris (pH 8), 1 mM EDTA (pH 8). Prepare 50 mL and keep on ice. Solution can be stored at 4 °C up to 3 months.

12. Elution buffer: 100 mM $NaHCO_3$, 1% (w/v) SDS. Prepare 10 mL, keep at room temperature and use immediately.
13. High-grade PCR purification kit (*see* **Note 5**).
14. DNA quantification assay kit and fluorimetric measurement device (*see* **Note 6**).
15. ChIP-seq library preparation kit working with low amounts (1 ng or less) of DNA (*see* **Note 7**).
16. Low-binding 1.5 mL tubes.
17. Real-time quantitative PCR cycler.

3 Methods

3.1 Preparation of Biological Material (11 Days +60 Min for Tissue Collection and Cross-Linking)

This section describes the procedure to generate cross-linked root tip material in sufficient amount to purify nuclei from abundant as well as rare cell populations (*see* **Note 8** and Fig. 2).

1. Surface-sterilize the appropriate amount of seeds. Typically, 0.5 g of seeds (~30,000 seeds of *Arabidopsis thaliana* accession Col-0) will generate about 1.5 g of 10-mm-long root tips (*see* **Note 8**). Stratify the seeds in sterile H_2O for 3 days in the dark at 4 °C in order to get a synchronized germination.
2. Sow the seeds over 0.5× MS plates overlaid with a nitex nylon mesh in four lines (2–3 seeds wide, ~1000 seeds per plate; Fig. 1a). Seal the plates with Parafilm, place them vertically in a growth chamber and grow the seedlings at 22 °C for 7 days in long-day conditions (Fig. 1a).
3. To collect the root tips, use a scalpel to cut the appropriate root segments (*see* **Note 8**) along each row of seedlings (Fig. 1a). For each plate, gather together the root segments into a small stack and put the cover back to prevent dehydration. Once all plates have been processed, transfer all root segments swiftly to a cell strainer floating in a petri dish filled with cold H_2O and kept on ice (*see* **Note 9**). Proceed immediately with crosslinking (*see* **Note 10**).
4. Transfer the cell strainer with the roots segments to a new petri dish containing a sufficient volume of NPBf to cover all the biological material (120 mL is appropriate for a square 120 × 120 mm petri dish). **Steps 4–6** should be performed under a fume hood.
5. Proceed with cross-linking by incubating the samples for 15 min under vacuum at room temperature.
6. Stop the cross-linking reaction by adding to the formaldehyde solution a volume of 2 M glycine required to reach 0.125 M final concentration. Incubate for 5 min with gentle mixing.

Fig. 1 Collection of root tips and nuclei purification. (**a**) Image of 7-day-old Arabidopsis seedlings before (*left panel*) and after (*right panel*) dissection of the root tips. (**b**) Epifluorescence microscope image of the nuclei suspension during (*top panel*) and at the end (*bottom panel*) of the purification procedure. Nuclei stained with DAPI are shown in *blue*; autofluorescent beads and debris appear in *green*. *White arrows* indicate isolated nuclei coated with magnetic beads. *Scale bar*: 20 μm

7. Transfer the cell strainer to a new petri dish filled with H_2O to rinse the tissues for 1 min with gentle mixing.
8. Remove all the liquid by placing the strainer on a paper towel, collect the root segments in a plastic tube and freeze in liquid N_2. The cross-linked material can be stored at −80 °C for a few weeks.

3.2 Extraction of Tagged Nuclei (50 Min)

This section describes the procedure to achieve a thorough extraction of nuclei from the cross-linked tissues, which is critical for the final yield and representative sampling of the targeted population. We do not recommend increasing the amount of starting material beyond 1.5 g without adjusting all buffer volumes proportionally (*see* **Note 8**). All steps are carried out at 4 °C using precooled materials unless stated otherwise. All the pipetting of the nuclei suspension is done by gently pipetting up and down with cut tips. Parallel processing of multiple samples is not recommended and should be restricted to proficient users.

1. Before starting the isolation of nuclei, wash the appropriate amount of Protein A or G paramagnetic beads (depending on the tag-antibody isotype) with 1 mL of NPB (*see* **Note 11**). Collect the beads using a magnetic rack and resuspend in 500 μL NPB. Add the appropriate amount of anti-tag antibody (Ab) (*see* **Note 11**) and proceed with binding for 30 min at 4 °C. This beads–Ab coupling step should be carried out

during the nuclei extraction to minimize the time required to proceed to the purification step (Subheading 3.3).

2. In a mortar filled with liquid N_2, grind to a fine powder 1.5 g of root tips using a pestle. Resuspend the tissue powder in 20 mL of ice-cold NPB buffer.
3. Homogenize the suspension in a 40 mL Dounce tissue grinder with 7 strokes using the loose pestle and 7 strokes with the tight pestle. This step is important to ensure that nuclei are properly released and individualized from the tissues.
4. Recover the nuclei suspension in a 50 mL tube by filtering through a 70 μm nylon cell strainer. Rinse the filter with 5 mL NPB. Filter again using a 40 μm nylon strainer and rinse the filter with 5 mL NPB.
5. Spin down the nuclei at 1000 × *g* for 7 min at 4 °C. Carefully remove the supernatant.
6. Resuspend the pellet of nuclei and debris in 1 mL NPBd, transfer the suspension to a 1.5 mL tube and incubate on ice for 3 min in the dark.
7. Spin down the nuclei at 1000 × *g* for 7 min at 4 °C. Discard the supernatant, resuspend the pellet in 1 mL NPBb and keep on ice. Proceed immediately to the purification step (Subheading 3.3).

3.3 Purification of Tagged Nuclei (60 Min)

This section describes the procedure to purify the population of interest to a high yield and purity. We do not recommend increasing the amount of starting nuclei if working with rare cell populations (*see* **Note 8**). Parallel processing of multiple nuclei suspension is not recommended to guarantee the swiftness of this critical procedure. All steps are carried out at 4 °C using precooled materials unless stated otherwise. All the pipetting of the nuclei suspension is done by gently pipetting up and down with cut tips.

1. Transfer the nuclei suspension to a 15 mL tube and resuspend gently by adding 3 mL NPBb and 4 mL NPB.
2. Collect the antibody-bound beads from **step 1** of Subheading 3.2 using a magnetic rack and discard the supernatant. Resuspend the beads in 0.5 mL of NPB and add to the nuclei suspension. Gently mix on a rotating wheel at 4 °C for 30 min.
3. Transfer the bead–nuclei mixture to a new 15 mL tube and adjust the volume to 12 mL with NPBbt. Place the tube vertically in a precooled magnetic rack for 15 mL tubes for at least 2 min.
4. Remove carefully the supernatant by pipetting slowly in the center of the tube using a 1 mL pipetman. Avoid touching the sides of the tube, disturbing the beads or the

bead–nuclei–debris pellet that gradually accumulates at the bottom of the tube (*see* **Note 12**).

5. Resuspend the bead–nuclei mixture in 12 mL of NPBbt, mix by gently inverting the tube a few times. Place the tube in the magnetic rack for 2 min.
6. Repeat **steps 4** and **5** at least twice or until all the beads remain along the tube walls after removal of the supernatant (*see* **Note 12**).
7. Resuspend the nuclei–beads in 1 mL NPBbt. Take a 10 μL sample for microscopic observation to determine whether the bead-bound nuclei suspension is sufficiently purified (Fig. 1b, bottom panel) or whether extra round of washes are needed (Fig. 1b, top panel).
8. Once satisfactorily washed, transfer the bead–nuclei suspension into a 1.5 mL tube. Take a 20 μL sample for counting and purity estimation (*see* **step 9**). Place the remaining bead–nuclei suspension on a magnetic rack and discard supernatant. Proceed immediately with the chromatin immunoprecipitation procedure or store the beads–nuclei at −80 °C for a few days.
9. Using a hemocytometer under an epifluorescence microscope, estimate the yield and purity of the target nuclei using the sample from **step 8** by counting the number of purified (bead-bound) and contaminating (free) nuclei. Expected purity ranges from 90 to 100%. In order to visualize the beads, the nuclei and the hemocytometer grid at the same time, use simultaneously the DAPI fluorescence channel and dim white field illumination.

3.4 Chromatin Immunoprecipitation from Small Amounts of Nuclei (4 Days)

This section describes the procedure to achieve efficient chromatin immunoprecipitation from relatively low amounts of nuclei. It includes the steps of chromatin shearing and immunoprecipitation as well as of DNA purification for sequencing library preparation (ChIP-seq). Keep samples and buffers on ice unless stated otherwise. The use of low binding tubes is recommended at all stages.

3.4.1 Chromatin Shearing and Validation of Sonication Efficiency (1.5 Day)

1. Resuspend the nuclei–beads pellet in 110 μL of Nuclei Lysis Buffer.
2. If working with at least 2×10^5 nuclei, keep a 5 μL aliquot aside as the unsonicated control, which is used later on to determine the sonication efficiency (if working with a lower amount of nuclei, *see* **Note 13**).
3. Sonicate the nuclei lysate (*see* **Note 4**). Appropriate chromatin shearing to 200–700 bp fragments can be achieved using a Covaris S220 ultra-sonicator with the following parameters:

9 min with 5% duty cycle, 105 W peak power and 200 cycles per burst. These parameters need to be determined if another type of sonicator is used (*see* below for checking sonication efficiency).

4. If starting with at least 2×10^5 nuclei, keep 10 μL of the sonicated chromatin to check sonication efficiency together with the unsonicated sample (if working with less nuclei, *see* **Note 13**). The rest of the sonicated chromatin can be stored at −80 °C for a few days.
5. Add 0.5 μL of 5 M NaCl to the unsonicated and sonicated aliquots from **steps 2** and **4**.
6. Seal the tubes with Parafilm and incubate at 65 °C for at least 6 h up to overnight to reverse the formaldehyde cross-linking.
7. To each tube, add 82 μL of H_2O, 5 μL of 1 M Tris–HCl pH 6.5, 2.5 μL of 0.5 M EDTA and 1.0 μL of RNAse A (20 mg/mL). Incubate 30 min at 42 °C with gentle agitation.
8. Add 0.5 μL of Proteinase K (20 mg/mL) and incubate 90 min at 42 °C.
9. Recover the DNA by classical phenol–chloroform extraction followed by sodium acetate/ethanol precipitation.
10. Resuspend the pellets in 15 μL of H_2O. Determine the sonication efficiency by agarose gel electrophoresis. The size of the DNA fragments should range between 200 and 700 bp. Use an aliquot to measure the amount of DNA using a spectrophotometer.

3.4.2 Chromatin Immunoprecipitation (1.5 Day)

Keep samples and buffers on ice unless stated otherwise. The use of low binding tubes is recommended at all stages.

1. Centrifuge the sonicated chromatin solution (from **step 3** of Subheading 3.4.1) at 12,000 × *g* for 5 min at 4 °C to pellet the debris. Transfer the supernatant to a new 1.5 mL tube. Take 10% of the chromatin solution for a total DNA control (INPUT) and store at −20 °C.
2. Dilute the remaining sonicated chromatin with ChIP Dilution Buffer (CDB) to a final volume of 1.2 mL.
3. Wash the required amount of magnetic beads coupled to protein A or G (depending on the antibody isotype) with CDB in a 1.5 mL tube. Typically, 10 μL of beads is used for chromatin preclearing (*see* **step 6** and **Note 14**) and 25 μL for each ChIP experiment including a no-antibody control.
4. Capture the beads using a magnetic rack, discard the supernatant and resuspend the beads in their initial volume with CDB.
5. Set aside 25 μL of washed beads for the no-antibody control.

6. Add 10 μL of washed beads to the diluted chromatin for preclearing. Incubate for 1 h at 4 °C on a rotating wheel (Optional, *see* **Note 14**).
7. In parallel to chromatin preclearing, resuspend the appropriate amount of washed beads in 400 μL of CDB and add the appropriate amount of antibody (1–2 μg per ChIP experiment). Proceed with bead–antibody coupling by incubating for 1 h at 4 °C on a rotating wheel.
8. After the preclearing of the chromatin, remove the beads using a magnetic rack and transfer the cleared chromatin to a new 1.5 mL tube. Split the chromatin into three aliquots (400 μL each) in 1.5 mL tubes.
9. Following the bead–antibody coupling step, collect the bead-coupled antibodies using a magnetic rack and remove the supernatant.
10. Resuspend the bead-coupled antibodies in 25 μL of CDB.
11. Add the 25 μL of bead-coupled antibodies to the 400 μL of precleared chromatin.
12. Include a no-antibody control by adding the 25 μL of washed beads set aside in step 5 to 400 μL of precleared chromatin.
13. Add 600 μL of CDB to each tube to reach a final volume of 1 mL.
14. Seal each tube with Parafilm to avoid leakage.
15. Incubate overnight at 4 °C on a rotating wheel.
16. Spin the tubes for a few seconds using a minicentrifuge to collect the solution caught in the cap and capture the beads using a magnetic rack. Discard the supernatant.
17. Remove the tubes from the magnet and resuspend the beads in 1 mL of the low-salt wash buffer by inverting the tubes.
18. Spin the tubes for a few seconds using a minicentrifuge to collect the solution caught in the cap and capture the beads using a magnetic rack. Discard the supernatant.
19. Remove the tubes from the magnet, resuspend the beads in 1 mL of the low-salt wash buffer by inverting the tube and incubate for 5 min at 4 °C on a rotating wheel.
20. Spin the tubes for a few seconds using a minicentrifuge to collect the solution caught in the cap and capture the beads using a magnetic rack. Discard the supernatant.
21. Repeat **steps 17–20** using the high-salt, LiCl, and TE wash buffers (in this order).
22. Following the second wash with TE, transfer the bead–nuclei suspension to new 1.5 mL tubes.

23. Collect the beads using a magnetic rack and discard the supernatant.
24. Remove the tubes from the magnet and add 125 μL of elution buffer preheated at 65 °C to the pelleted beads.
25. Vortex for a few seconds and incubate at 65 °C for 15 min.
26. Mix the tube contents repeatedly by inversion during the incubation time.
27. Spin the tubes for a few seconds using a minicentrifuge to collect the solution caught in the cap, capture the beads using a magnetic rack and transfer the eluate to a new 1.5 mL tube.
28. Repeat the elution by removing the tubes from the magnet and adding 125 μL of elution buffer preheated at 65 °C to the pelleted beads.
29. Vortex for a few seconds and incubate at 65 °C for 15 min.
30. Spin the tubes for a few seconds using a minicentrifuge to collect the solution caught in the cap, capture the beads using a magnetic rack and combine the eluate with the first one (**step 27**).
31. Resuspend the INPUT chromatin (total DNA control) from **step 1** in 250 μL of elution buffer.

3.4.3 DNA Purification (1.5 Day)

1. Add 20 μL of 5 M NaCl to the eluted and INPUT chromatin and reverse the formaldehyde cross-linking by incubating at 65 °C for at least 6 h and up to overnight.
2. Add 10 μL 1 M Tris–HCl pH 6.5, 5 μL of 0.5 M EDTA, and 2 μL of 10 mg/mL proteinase K to the samples and incubate for 1 h at 45 °C.
3. Recover and purify the DNA using a PCR purification kit following the manufacturer's instructions.
4. Measure the DNA concentration using a fluorimetric quantitation assay (*see* **Note 6**).
5. Validate the ChIP efficiency and purity by quantitative real-time PCR using a set of known positive and negative control regions, if available (*see* for example Figs. 2b, d and 3d). Include the INPUT and the no-antibody control DNA to estimate the enrichment levels and background noise, respectively.
6. Use 0.5–1 ng of immunoprecipitated and INPUT DNA to prepare sequencing libraries (*see* **Note 7**).

3.5 Sequential Purification of Distinct Nuclei Populations from a Single Transgenic Line

The use of different tags (GFP, mCherry) offers the possibility to combine NTF transgenes within a single line and purify more than one nuclei population from a single experiment, in a sequential manner. In addition, this "double INTACT" procedure can provide access to a subpopulation for which no specific promoter is

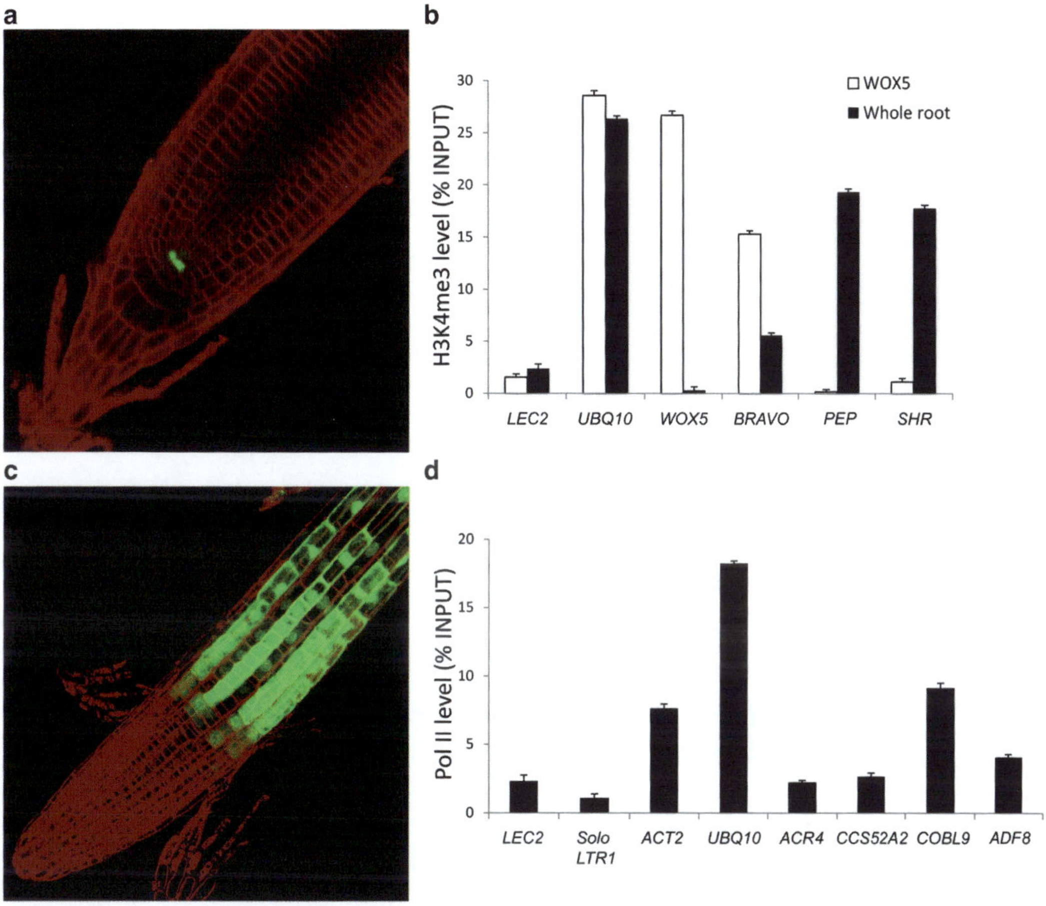

Fig. 2 Analyses of H3K4me3 and RNA Pol II enrichment in nuclei of the QC and the root maturation zone. Confocal microscopy images of root tips counterstained with FM4–64 (*red*) and expressing either the GFP-NTF tag in the quiescent center (*pWOX5-NTF*) (**a**) or in cells of the postmitotic, maturation zone (*pCC552A1-NTF*) (**b, c**) ChIP-qPCR analysis of H3K4me3 enrichment levels in QC-purified (WOX5) versus whole root nuclei at loci selected based on their expression pattern [15]. *LEC2* and *UBQ10* are used as negative and positive controls, respectively. *WOX5* and *BRAVO*, which are preferentially expressed in the QC, show higher H3K4me3 enrichment in the QC population compared to the whole root. *PEP* and *SHR*, which are preferentially expressed in the cortex and stele respectively, display background levels of H3K4me3 in QC nuclei. (**d**) ChIP-qPCR analysis of RNA polymerase II enrichment at selected loci in the maturation zone nuclei (*pCCS52A1-NTF*). *LEC2* and *SoloLTR1* are used as negative controls; *ACT2* and *UBQ10* are used as positive controls. *ACR4* and *CCS52A2*, which are preferentially expressed outside the maturation zone, show background levels of RNA Pol II, whereas a significant enrichment is detected at the *COBL9* and *ADF8* loci, both of which are expressed in the purified population analyzed. Error bars correspond to standard deviations between replicates. *See* Table 1 for primer sequences

available. For instance, in the case of cell populations partially overlapping, an initial capture based on a first tag can be used to remove unwanted, double-tagged cells from the subpopulation of interest that is then recovered thanks to the second tag. As a proof of concept for this subtraction-based approach, we purified epidermal

Table 1
List of primer pairs

ID	Name	Sequence
AT1G28300	LEC2_F	CCTGTTGATCCTTGCCATCT
	LEC2_R	TGAATCCTCAGCCGGTTTAC
AT4G05320	UBQ10_F	AACAATTGGAGGATGGTCGT
	UBQ10_R	GTGTCGGAGCTTTCCACTTC
AT3G11260	WOX5_F	TTCTCCGTGAAAGGTCGAAG
	WOX5_R	CTGAATCTGATCAGTTGTTGGAG
AT5G17800	BRAVO_F	TCCCTAAACCCAGAAACATGA
	BRAVO_R	CCCATGTTCATCTTCCGACT
AT4G26000	PEP_F	TTGAATCAGCCCGATAGCTT
	PEP_R	TACGAGCACGAGTCTCCTCA
AT4G37650	SHR_F	ACCTCAAACTCCTCCGTCCT
	SHR_R	GCGTCCATAGGATTTGTTGC
AT5TE35950	SoloLTRl_F	TGCATTACAAAAACCTTCTGATTG
	SoloLTRl_R	GAAAAAGAGAAGAGAAAGAGAAAGCA
AT3G18780	ACT2_F	GCCATCCAAGCTGTTCTCTC
	ACT2_R	CCCTCGTAGATTGGCACAGT
AT1G69040	ACR4_F	GCCTGAAGCTTGTTTCTCCA
	ACR4_R	GCTGCTCTGGTGTTGTGTGT
AT4G11920	CCS52A2_F	AGAAGATGGGGCTGGTTCTT
	CCS52A2_R	CTTCCTCGGCGACTTAACAG
AT5G49270	COBL9_F	AGCTCAATCTCAGCCGTCAT
	COBL9_R	GACAGAGCTTCCTGCTTTGG
AT4G00680	ADF8_F	TGTGTGTGTTGTGCAGGCTA
	ADF8_R	CATCGGGGATAGAGCTGGTA
AT 1G12560	EXP7_F	CGAGTTCGTCGCTGGATACT
	EXP7_R	TGTCTGGAAACGTTGGATCA
AT1G25220	ASB1_F	GACAATGGCGGCTTCTACAT
	ASB1_R	TACGACTCTTCCCCAAAACG

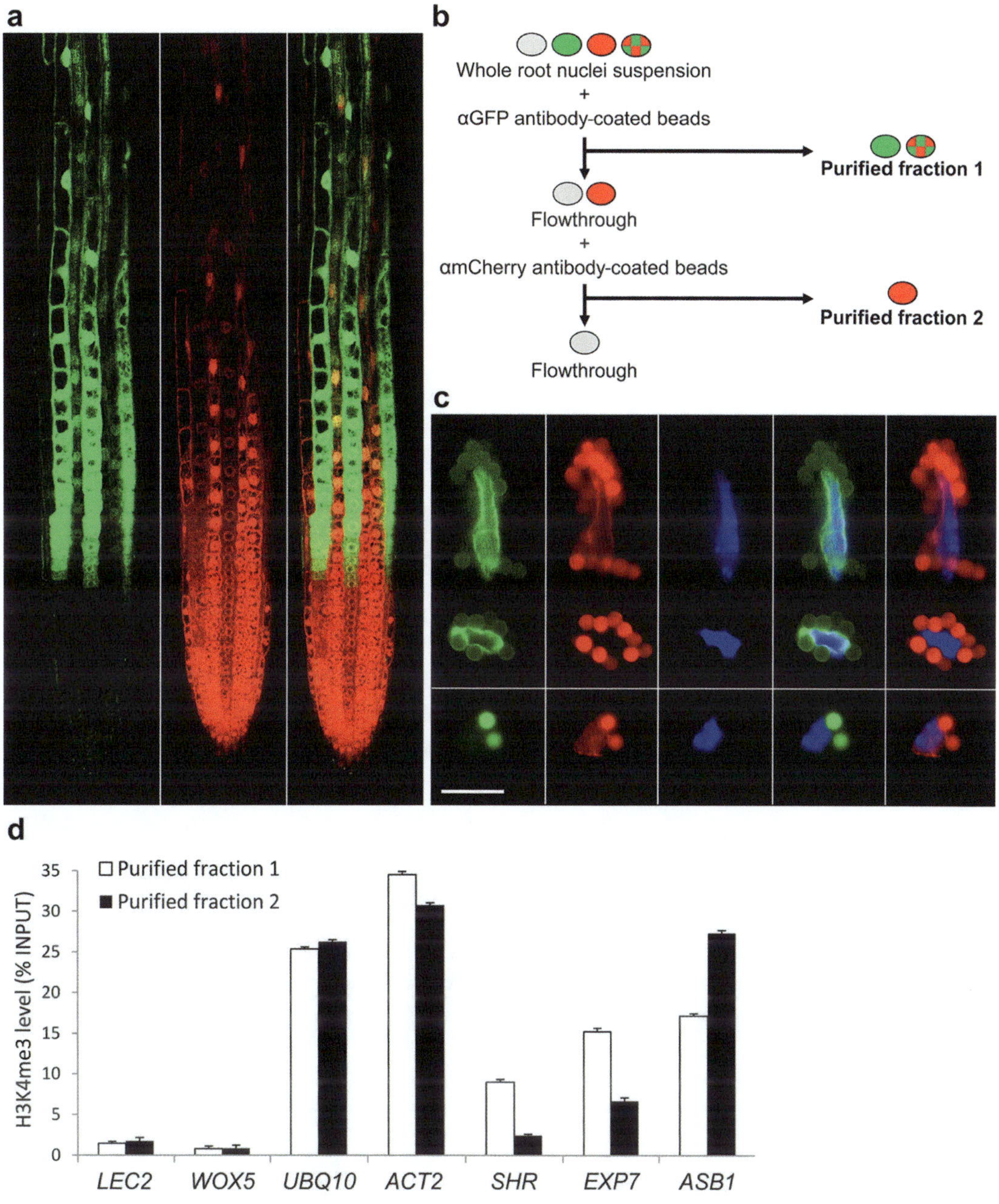

Fig. 3 Principle and validation of the double-INTACT procedure. (**a**) Confocal microscopy images of root tips expressing GFP-NTF in cells of the postmitotic zone (*pCCS52A1-GFP-NTF*, *green*) and mCherry-NTF in epidermal cells (*pWER-mCherry-NTF*, *red*). (**b**) Schematic diagram of the double INTACT procedure. (**c**) Epifluorescence micrographs of nuclei purified using the double INTACT procedure. *Top panel*: nuclei from purified fraction one tagged with both GFP and mCherry (epidermal cell from the maturation zone) as well as GFP only (non-epidermal cell from the maturation zone). Bottom panel: nucleus from purified fraction two tagged with mCherry only (meristematic epidermal cell). Staining with DAPI is shown in *blue*. Merge images of the tag and DAPI fluorescence are provided. *Scale bar*: 10 μm. (**d**) ChIP-qPCR analysis of H3K4me3 enrichment levels at selected loci in the two nuclei subpopulations purified following the double INTACT procedure. *LEC2* and *WOX5* are used as negative controls and *UBQ10* and *ACT2* as positive controls. *SHR* (expressed in the vasculature) and *EXP7* (expressed in root hair cells from the maturation zone) show higher enrichment levels in purified fraction 1. In contrast, ASB1, which is preferentially expressed in the meristematic zone, show higher H3K4me3 levels in purified fraction 2. Error bars correspond to standard deviations between replicates. *See* Table 1 for primer sequences

cells of the root meristematic zone by combining, on the one hand, a GFP-NTF expressed in cells of all root tissue layers in the post-mitotic, maturation zone under the *CELL CYCLE SWITCH PROTEIN 52 A1* (*CCS52A1*) promoter [12] and, on the other hand, an mCherry-NTF expressed only in the root epidermis under the *WEREWOLF* (*WER*) promoter [13, 14] (Fig. 3a). Thus, in a first step, beads coupled with an anti-GFP antibody enabled the capture of *pCCS52A1-NTF-GFP* positive nuclei (all cell layers in the maturation zone including the epidermis). Then in a second step, beads coupled with an anti-mCherry antibody are used to purify the meristematic epidermal cells from the nuclei suspension depleted in GFP-positive nuclei (Fig. 3b).

1. Proceed with the steps described in Subheadings 3.1, 3.2 and 3.3. In Subheading 3.2, **step 1** use anti-GFP antibodies. The amount of starting material needs to be adjusted according to the abundance of the second, final population of nuclei to be purified (*see* **Note 15**).
2. In **steps 4–6** of Subheading 3.3, collect the flow-through and washes of the first, GFP-based purification in 15 mL tubes.
3. Place these tubes on the magnetic rack to ascertain that all bead–nuclei complexes are captured and transfer the supernatant to a 50 mL tube (*see* **Note 16**). Proceed immediately with the second, mcherry-based nuclei purification.
4. Centrifuge the collected flow-through and wash fractions at 1000 × *g* for 7 min at 4 °C and discard the supernatant.
5. Resuspend the nuclei/debris pellet with 1 mL NPBbt. Take a 10 μL sample to verify the absence of GFP-positive nuclei using a fluorescence microscope.
6. Add 7 mL of NPBt to make up the volume of the nuclei suspension to 8 mL.
7. Prepare the appropriate amount of protein G-linked magnetic beads coupled with anti-mCherry antibodies as described in **step 1** of Subheading 3.2. This beads–Ab coupling step should be carried out during **step 2** of Subheading 3.3 to minimize the time required to proceed to the second purification step.
8. Continue as indicated in **steps 2–9** of Subheading 3.3.
9. Proceed to chromatin immunoprecipitation as described in Subheading 3.4.

4 Notes

1. Paramagnetic beads and related accessories can be purchased from several suppliers. We have good experience with the 2.8 μm Dynabeads Protein A and G and the magnetic racks DynaMag 15 and DynaMag 2 from Thermofisher Scientific.

2. Antibodies directed against GFP or mCherry. We have good experience with Abcam, Ref. ab290 and ThermoFisher Scientific Ref. A-11122 and PA5–34974.
3. Use Malassez, Nageotte, Neubauer-improved counting cell or an equivalent hemocytometer.
4. Use a sonicator capable of handling small volumes (less than 1 mL) and that shears chromatin into 200–700 bp fragments. We have good experience with the Covaris S220 Focused-ultrasonicator (Covaris) equipped with 130 μL microtubes with adaptive Focused Acoustics (AFA) fiber (Covaris). *See* Subheading 3.4.1 for checking sonication efficiency.
5. We have good experience with the Qiagen MinElute PCR purification kit (Qiagen).
6. Fluorimetric methods enabling the accurate measurement of double-stranded DNA in concentrations ranging between 10 and 100 pg/μL should be preferred. We have good experience with the Quant-it PicoGreen dsDNA quantitation assay or Qubit dsDNA high sensitivity (HS) Assay Kits (ThermoFisher Scientific) using a NanoDrop (ND3300, Thermo Scientific) or a Qubit (ThermoFisher Scientific) fluorometers, respectively.
7. We have good experience with the MicroPlex library preparation kit v2 (Diagenode). Alternative products working with low amounts of DNA (1 ng or less) might work equally well.
8. The amount of starting biological material should be adjusted according to two main factors: the abundance of the cell type of interest, which determines the yield of purified nuclei, and the enrichment level of the chromatin mark or protein studied, based on which the number of nuclei required for an efficient ChIP should be calculated. The ploidy level of the population of interest is an additional parameter to consider as the chromatin content is higher in differentiated cells in comparison with meristematic ones. In addition, the tagged/nontagged nuclei ratio can be improved by dissecting root segment from different length depending on the cell population of interest (*see* below).

   For a relatively abundant cell population, such as all cells from the maturation zone (tagged using a *pCCS52A1-NTF* construct, Figs. 2a and 3a), 10^6 tagged nuclei are obtained from a single purification round starting with 1 g of 10-mm-long root tips. For rare cell types, such as the quiescent center (4 cells/meristem) labeled using a *pWOX5-NTF* construct (Fig. 2a), 1.5 g of 5 mm-root tip yields about 5×10^4 captured nuclei. To reach the minimal amount of nuclei required to obtain reproducible ChIP-seq results (about 10^5 nuclei), we do not recommend increasing the amount of starting material beyond 1.5 g without adjusting all buffer volumes proportionally. In our hands, pooling the outcome of two or more independent purifications performed in the conditions described in Subheadings 3.1 and 3.2 gives the highest yield.

9. Collecting root segments from several dozens of plates can be time-consuming. It is important to minimize this collection time in order to preserve tissue and cell integrity. Therefore, it might be advisable to seek for the help of colleagues to speed up the collection step. Alternatively, collection and cross-linking can be done by batches of 10–12 plates.
10. If using INTACT to analyze nuclear RNAs (or if native ChIP is to be performed), omit this cross-linking step and start directly at Subheading 3.2, **step 1**. In addition, to prevent RNA degradation during the procedure, we recommend adding RNasin Plus RNAse Inhibitor (Promega) at 60 U/mL to the NPB.
11. The optimal amount of beads and antibody (Ab) needed depends upon the Ab used and the amount of nuclei to capture. Typically, 2 μg of anti-GFP antibody coupled with 20 μL of beads are sufficient to recover 10^6 tagged nuclei.
12. Washes of the bead-bound nuclei fraction is a critical step: incomplete washing will lead to low purity with significant chromatin, nucleic acid and protein carry over, whereas too much washing can decrease the final yield due to nuclei leakage and bursting. As indicated in **step 4** of Subheading 3.3, a large pellet of debris, beads, and nuclei will inevitably form at the bottom of the tube during the first wash. This pellet is usually rich in bead–nuclei complexes that were either not efficiently attracted by the magnet or displaced from the tube wall when pipetting the supernatant because of the high debris concentration. It is therefore important to keep this initial heterogeneous mixture and dilute it during the subsequent washing steps. If this pellet persist after three washes, transferring it to a new tube and washing it separately can be helpful.
13. If working with less than 2×10^5 nuclei, use a high-sensitivity Bioanalyzer DNA chip to assay sonication efficiency.
14. This preclearing step is optional but frequently improves the signal-to-noise ratio of ChIP experiments, especially when working with low amounts of chromatin.
15. The amount of starting material needs to be adjusted according to the abundance of the second, final population of nuclei to be purified. In addition to the factors discussed in **Note 1**, one needs to take into account the fact that the two sequential purifications lead to an overall increase of nuclei loss in comparison to a single purification procedure. In the experiment described here (Fig. 3), two purifications of *pCCS52A1-NTF-GFP* nuclei were performed in parallel, each starting from 1 g of cross-linked root tips. After pooling the purified fractions,

2×10^6 GFP-positive nuclei were obtained and 2×10^5 mCherry-tagged nuclei were purified from the pooled flow-throughs.

16. As it is essential that all nuclei labeled with the first tag are removed from the flow-through nuclei suspension that will be used for the second purification step, incubation on the magnet is repeated twice. The absence of beads and nuclei associated with tag 1 is verified by observation of an aliquot of the first flow-through under a fluorescence microscope. The amount and purity of tag 2-labeled nuclei is counted using a hemocytometer as indicated in Subheading 3.3, **step 9**.

Acknowledgments

We would like to thank Dr. R. Deal (Emory University, USA) and Dr. D. Grimanelli (IRD, France) for the GFP- and mCherry-NTF cassettes, respectively. This work was supported by the European Union Seventh Framework Programme Network of Excellence EpiGeneSys (HEALTH-F4-2010-257082), the CNRS and the Agence Nationale de la Recherche (Investissements d'Avenir ANR-10-LABX-54 MEMOLIFE and ANR-14-CE19-0008). A. K.M. is the recipient of a grant from the French Ministère de la Recherche et de l'Enseignement Supérieur as well as from the Fondation pour la Recherche Médicale (FRM: FDT20160435513).

References

1. Deal RB, Henikoff S (2010) A simple method for gene expression and chromatin profiling of individual cell types within a tissue. Dev Cell 18 (6):1030–1040. doi:10.1016/j.devcel.2010.05.013
2. Deal RB, Henikoff S (2011) The INTACT method for cell type-specific gene expression and chromatin profiling in Arabidopsis Thaliana. Nat Protoc 6(1):56–68. doi:10.1038/nprot.2010.175
3. Ron M, Kajala K, Pauluzzi G, Wang D, Reynoso MA, Zumstein K, Garcha J, Winte S, Masson H, Inagaki S, Federici F, Sinha N, Deal RB, Bailey-Serres J, Brady SM (2014) Hairy root transformation using agrobacterium rhizogenes as a tool for exploring cell type-specific gene expression and function using tomato as a model. Plant Physiol 166 (2):455–469. doi:10.1104/pp.114.239392
4. Steiner FA, Talbert PB, Kasinathan S, Deal RB, Henikoff S (2012) Cell-type-specific nuclei purification from whole animals for genome-wide expression and chromatin profiling. Genome Res 22(4):766–777. doi:10.1101/gr.131748.111
5. Henry GL, Davis FP, Picard S, Eddy SR (2012) Cell type-specific genomics of Drosophila neurons. Nucleic Acids Res 40(19):9691–9704. doi:10.1093/nar/gks671
6. Amin NM, Greco TM, Kuchenbrod LM, Rigney MM, Chung MI, Wallingford JB, Cristea IM, Conlon FL (2014) Proteomic profiling of cardiac tissue by isolation of nuclei tagged in specific cell types (INTACT). Development 141(4):962–973. doi:10.1242/dev.098327
7. Marques-Bueno MM, Morao AK, Cayrel A, Platre MP, Barberon M, Caillieux E, Colot V, Jaillais Y, Roudier F, Vert G (2016) A versatile multisite gateway-compatible promoter and transgenic line collection for cell type-specific functional genomics in Arabidopsis. Plant J 85 (2):320–333. doi:10.1111/tpj.13099
8. Ma J, Weake VM (2014) Affinity-based isolation of tagged nuclei from Drosophila tissues

for gene expression analysis. J Vis Exp 85. doi:10.3791/51418

9. Mo A, Mukamel EA, Davis FP, Luo C, Henry GL, Picard S, Urich MA, Nery JR, Sejnowski TJ, Lister R, Eddy SR, Ecker JR, Nathans J (2015) Epigenomic signatures of neuronal diversity in the mammalian brain. Neuron 86 (6):1369–1384. doi:10.1016/j.neuron.2015.05.018
10. Wang D, Deal RB (2015) Epigenome profiling of specific plant cell types using a streamlined INTACT protocol and ChIP-seq. Methods Mol Biol 1284:3–25. doi:10.1007/978-1-4939-2444-8_1
11. Roudier F, Ahmed I, Berard C, Sarazin A, Mary-Huard T, Cortijo S, Bouyer D, Caillieux E, Duvernois-Berthet E, Al-Shikhley L, Giraut L, Despres B, Drevensek S, Barneche F, Derozier S, Brunaud V, Aubourg S, Schnittger A, Bowler C, Martin-Magniette ML, Robin S, Caboche M, Colot V (2011) Integrative epigenomic mapping defines four main chromatin states in Arabidopsis. EMBO J 30(10):1928–1938. doi:10.1038/emboj.2011.103
12. Vanstraelen M, Baloban M, Da Ines O, Cultrone A, Lammens T, Boudolf V, Brown SC, De Veylder L, Mergaert P, Kondorosi E (2009) APC/C-CCS52A complexes control meristem maintenance in the Arabidopsis root. Proc Natl Acad Sci U S A 106(28):11806–11811. doi:10.1073/pnas.0901193106
13. Lee MM, Schiefelbein J (1999) WEREWOLF, a MYB-related protein in Arabidopsis, is a position-dependent regulator of epidermal cell patterning. Cell 99(5):473–483. S0092-8674(00)81536-6
14. Bruex A, Kainkaryam RM, Wieckowski Y, Kang YH, Bernhardt C, Xia Y, Zheng X, Wang JY, Lee MM, Benfey P, Woolf PJ, Schiefelbein J (2012) A gene regulatory network for root epidermis cell differentiation in Arabidopsis. PLoS Genet 8(1):e1002446. doi:10.1371/journal.pgen.1002446. PGENETICS-D-11-01595
15. Brady SM, Orlando DA, Lee JY, Wang JY, Koch J, Dinneny JR, Mace D, Ohler U, Benfey PN (2007) A high-resolution root spatiotemporal map reveals dominant expression patterns. Science 318(5851):801–806. 318/5851/801

Chapter 9

Mapping of Histone Modifications in Plants by Tandem Mass Spectrometry

Walid Mahrez and Lars Hennig

Abstract

To get an insight into the mechanisms of gene expression regulation in eukaryotic organisms, it is necessary to decipher the connection between the different chemical modifications occurring on the chromatin, at both the DNA and the associated histone proteins. Histones are basic proteins, which pack the DNA into nucleosomes, and are hot spots for several posttranslational modifications. Elucidating combinatorial histone modifications co-occurring on the same histone protein will greatly contribute to our understanding of the mechanisms involved in the development of eukaryotes. The advancements in mass spectrometry technologies, including sensitivity, accuracy, and ionization strategies, have significantly contributed to the identification of novel single and combinatorial modifications on histones isolated from model organisms. In this chapter, we describe detailed protocols applied for the extraction, purification, and processing of histones for subsequent analysis by tandem mass spectrometry, using *Brassica oleracea* (cauliflower), a close relative of *Arabidopsis thaliana*.

Key words Cauliflower, *Arabidopsis thaliana*, Histone posttranslational modification, Epigenetics, Chromatin, RP-HPLC, Tandem mass spectrometry, Trypsin digestion

1 Introduction

In eukaryotic cells, the DNA is tightly packed in the nucleus. This packing is mediated by proteins that interact with the DNA and together constitute a nucleoprotein complex called chromatin. The nucleosome, the basic unit of the chromatin, is constituted of 146 nucleotides of DNA wrapped around an octamer of two copies each of four different highly basic proteins called histones (H2A, H2B, H3 and H4) [1].

Histones are highly evolutionarily conserved, small proteins with molecular masses in the range of 10–22 kDa. Both the globular core region and the protruding N-terminal tails of histones are subject to a large number of potentially heritable and reversible covalent posttranslational modifications such as acetylation,

Marian Bemer and Célia Baroux (eds.), *Plant Chromatin Dynamics: Methods and Protocols*, Methods in Molecular Biology, vol. 1675, DOI 10.1007/978-1-4939-7318-7_9, © Springer Science+Business Media LLC 2018

methylation, phosphorylation, ubiquitination, sumoylation, deamination, and proline isomerization [2]. These modifications are deposited by specific proteins named "writers." Once deposited, they facilitate recruitment of several proteins that recognize, bind and interpret them ("readers") or remove them ("erasers") [3, 4]. This leads either to relaxation or compaction of chromatin structure required for induction or repression of transcription, respectively, or to DNA replication, recombination and repair [5–7].

Comparing the four core histones, the N-terminal tail of histone H3 contains the highest number of known target sites for chemical modifications, with more than 14 residues. The sequential deposition and combination of histone modifications on these sites and on others (H4, H2A, and H2B) constitute a code that defines actual or potential transcription states, regulates gene expression, and plays an important role in development [2, 8].

Owing to high-precision mass spectrometry, biochemistry, and molecular tools, the genomic distribution patterns and functions of histone modifications are well studied in yeast, as well as in animals [9–11]. Histone modification sites and patterns in plant tissues is not exhaustive [12, 13, 16], one way to identify novel ones is to apply a robust approach, combining a large amount of biological material (provided by cauliflower (***Brassica oleracea***) inflorescences) with high resolution mass spectrometers.

In this chapter we describe a simplified protocol, which was successfully applied for the extraction and fractionation of histone H3 and for the characterization of different H3 modifications from cauliflower heads [16]. Taking the advantage of their highly basic nature, crude histones were extracted from cauliflower inflorescences using sulfuric acid and further purified by reverse phase HPLC [14]. Fractions corresponding to H3 as documented by western blotting were collected, pooled, and digested either with trypsin for a limited time or with ArgC protease. The digested peptides were then analyzed by liquid chromatography–tandem mass spectrometry (LC/MS/MS) on an Orbitrap XL instrument associated with an ETD (Electro Transfer Detector) to analyze longer peptides [15]. The resulting raw data were submitted for MASCOT searching for histones posttranslational modification sites. The results are then validated by manual inspection of the peaks [16]. An overview of these different steps is summarized in Fig. 1.

2 Materials

2.1 Histone Extraction

1. Laminar flow hood.
2. Liquid nitrogen.
3. Mortar and pestle.

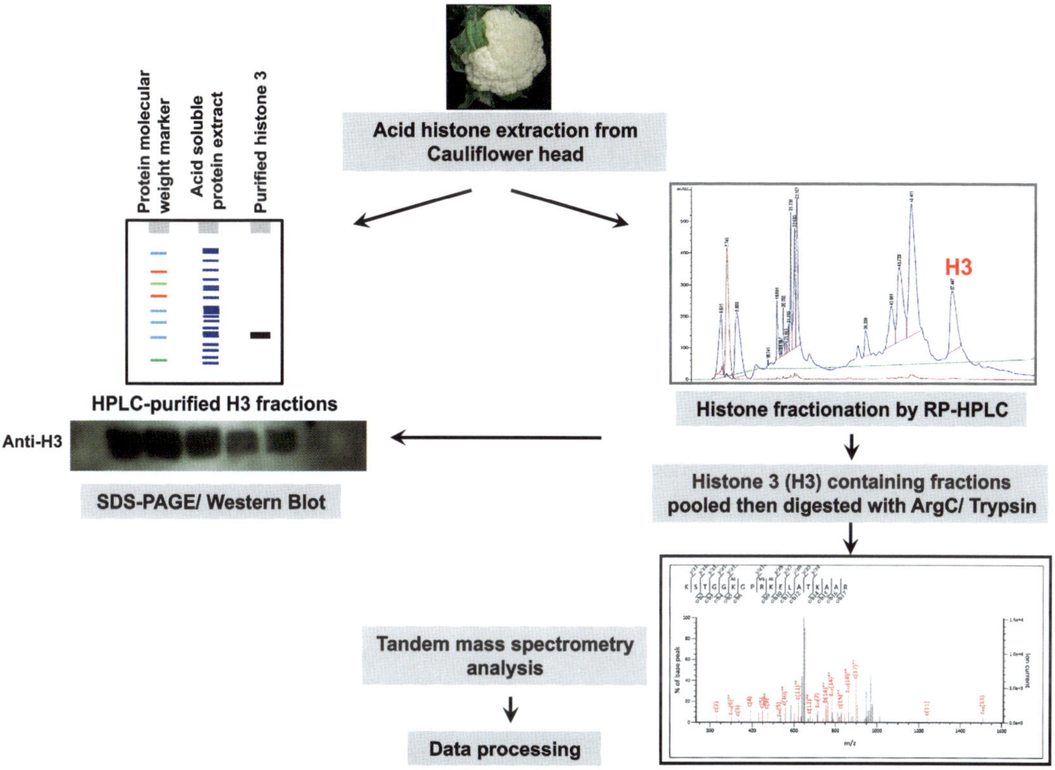

Fig. 1 A schematic overview of the different steps. Histones are extracted from cauliflower tissue, detected using western blot analysis and fractionated with HPLC. The histone containing fractions are then pooled, followed by histone digestion and purification. The purified peptides are subsequently analyzed with MS/MS

4. Nylon mesh (500 μm pore size) and fine nylon mesh (50 μm pore size) for filtration.
5. Spatula.
6. Scalpel.
7. 15 and 50 ml Falcon tubes.
8. 1.5 and 2 ml Microfuge tubes.
9. Single channel pipettes (e.g., Gilson).
10. Pipette tips (20, 200, and 1000 μl).
11. Refrigerated centrifuge with swinging bucket rotor suitable for 15 and 50 ml falcon tubes.
12. Refrigerated centrifuge with fixed-angle rotor suitable for 1.5 and 2 ml microfuge tubes.
13. Rotator at 4 °C.
14. pH-indicator paper.
15. Bath sonicator.

16. *Brassica oleracea* (cauliflower) heads are obtained from a local market (*see* **Note 1**).
17. Histone extraction buffer: 0.25 M sucrose, 1 mM $CaCl_2$, 15 mM NaCl, 60 mM KCl, 5 mM $MgCl_2$, 15 mM PIPES pH 7.0, 0.5% Triton X-100, 1× protease inhibitors cocktail, and 10 mM sodium butyrate.
18. Protease inhibitors cocktail (25×) (e.g., Roche): dissolve 1 tablet per 2 ml sterile distilled water, can be stored at −20 °C.
19. 100 mM sodium butyrate.
20. 100 mM PIPES dissolved in water, and the pH adjusted to 7.0 using 1 M KOH.
21. 0.2 N H_2SO_4 (*see* **Note 2**).
22. 100% (w/v) trichloroacetic acid (*see* **Note 2**).
23. 100% (v/v) acetone (*see* **Note 2**).
24. 0.33 M HCl (Dilute 37% concentrated HCl in water until a 1% (v/v) concentration is reached (*see* **Note 2**).
25. Acetone–HCl (add 0.1% HCl (0.033 M) to ice-cold acetone stored at −20 °C, prepare it fresh, before every extraction) (*see* **Note 2**).

2.2 SDS-PAGE

1. SDS-PAGE running buffer: 0.025 M Tris–HCl pH 8.3, 0.192 M glycine, 0.1% SDS.
2. SDS-PAGE running gel: 40% acrylamide–bisacrylamide (19:1), 1.5 M Tris–HCl pH 8.8, 10% SDS, glycerol 100%, TEMED, 10% APS (*see* **Note 3**).
3. SDS-PAGE stacking gel: 40% acrylamide–bisacrylamide (19:1), 1 M Tris–HCl pH 6.8, 10% SDS, TEMED, 10% APS.
4. Laemmli buffer (2×): 0.5 M Tris–HCl pH 6.8, 4.4% SDS, 20% Glycerol, 2% 2-mercaptoethanol, 0.1% Bromophenol Blue.
5. Glass plates for thin acrylamide gel (e.g., Mini PROTEAN® 3 System Glass plates, Bio-Rad).
6. plastic comb with ten wells.
7. SDS-PAGE gel chamber.

2.3 Western Blotting (WB)

1. Western blot transfer buffer: 0.025 M Tris–HCl pH 8.3, 0.192 M glycine, 20% methanol.
2. Tris-buffered saline (TBS 10×): 1.5 M NaCl, 0.1 M Tris–HCl pH 7.4.
3. Tris-buffered saline containing Tween (TBST): add 0.05% Tween20 to TBS.
4. Blocking solution: 5% fat free milk dissolved in TBS.
5. PVDF (polyvinylidenfluoride membrane).

6. Anti-Histone 3 primary antibody (CT, pan, rabbit polyclonal antibody, Millipore).
7. Horseradish peroxidase labeled, Anti-Rabbit IgG, secondary antibody (Bio-Rad).
8. Device for gel-blot transfer (e.g., Trans-Blot semi-dry transfer cell, Bio-Rad).

2.4 Histone Fractionation by Reverse Phase (RP): HPLC

1. Vortex mixer.
2. Heat block (100 °C).
3. Vacuum-based concentrator (e.g., SpeedVac system, ThermoFisher).
4. Buffer A: 5% acetonitrile (ACN) HPLC grade and 0.1% trifluoroacetic acid (TFA) spectroscopy grade (*see* **Notes 4** and **5**).
5. Buffer B: 90% ACN in 0.1% TFA (*see* **Note 5**).
6. Glass Hamilton syringe.
7. HPLC instrument configured for nanoflow delivery of solvents (e.g., Agilent HP1100 binary HPLC system, Agilent).
8. HPLC column appropriate to the HPLC instrument (e.g., ECLIPS XDB-C8 (4.6 × 150 mm), Agilent).

2.5 Histone Processing for Mass Spectrometry Analysis

1. Fluorimetric-based protein quantification assay (e.g., Qubit® 2.0 Fluorometer and Qubit®Protein Assay Kit *0.25–5 μg* Q33211, Invitrogen).
2. 50 mM Ammonium bicarbonate (NH_4HCO_3) pH 8.0 (prepare fresh).
3. Promega sequencing grade modified trypsin is prepared shortly before use by dissolving 20 μg of lyophilized powder in water to a final concentration of 1 μg/μl, dispensed into aliquots, and stored up to 6 months at −20 °C.
4. Endoproteinase ArgC stock solution is prepared by dissolving the lyophilized powder in water to a final concentration of 0.1 μg/μl, dispensed into aliquots, and stored up to 6 months at −20 °C.
5. 90% (w/v) formic acid (HPLC grade).
6. 50% acetonitrile (HPLC grade) dissolved in ddH_2O.
7. 1 M dithiothreitol (DTT): dissolve 154 mg of DTT in 1 ml of 100 mM NH_4HCO_3 (*see* **Note 5**).
8. 1 M iodoacetamide in 100 mM NH_4HCO_3: dissolve 37 mg of iodoacetamide in 200 μl of 100 mM NH_4HCO_3 (*see* **Note 5**).
9. C18 ZipTips (Millipore) are pipette tips packed with a bed of chromatography media at their end, and used for desalting, concentration, and purification of digested peptides before MS/MS analysis.

10. C18 ZipTips Wetting solution: 100% acetonitrile.
11. C18 ZipTips washing solution: 3% ACN + 0.1% TFA (*see* **Note 5**).
12. C18 ZipTips eluting solution: 60% ACN + 0.1% TFA (*see* **Note 5**).

2.6 Mass Spectrometry

1. Buffer A: 0.2% formic acid and 1% acetonitrile.
2. Buffer B: 0.2% formic acid and 80% acetonitrile.
3. Resolubilization buffer: 5% acetonitrile and 0.1% formic acid.
4. Tip column (75 μm × 70 mm) packed with reverse phase C18 material, prepared by a mass spectrometry qualified personal.
5. High-resolution, ion-trap mass analyzer, mass spectrometer instrument equipped with a CID/ETD (collision-induced fragmentation/Electron Transfer Dissociation) fragmentation, of which the high sensitivity and accuracy can distinguish between trimethylated and acetylated peptides ($\Delta m = 0.0364$ Da), (e.g., LTQ-Orbitrap XL, Thermo Fisher).

3 Methods

3.1 Histone Extraction from Plant Material

Histones isolation is based on acidic extraction of nuclear DNA binding proteins. The histone extraction steps are performed on ice using cooled tabletop centrifuges. The other steps are carried out at room temperature.

1. Use a new sharp scalpel, precleaned with ethanol, to shave the inflorescence tissues from the top of the cauliflower heads, 1–2 g of material should be used for histone extraction (*see* **Note 6**).
2. Precool mortar and pestle with liquid nitrogen, once the liquid nitrogen has evaporated, place the plant material in the mortar and submerge it with liquid nitrogen until the plant material is completely frozen. Grind the material quickly but carefully with the pestle to a fine powder.
3. Use a precooled spatula to transfer the ground tissue into a cold 50 ml falcon tube containing 15 ml of cold histone extraction buffer.
4. Place the falcon tube into a rotator, and leave the grounded material to dissolve slowly in the extraction buffer, at 4 °C for at least 20 min.
5. Filter the homogenate consecutively through a 500 μm nylon mesh, and then through a 50 μm filter into a fresh 50 ml tube (*see* **Note 7**).

6. Wash the mesh with 2 ml of cold histone extraction buffer, pass it through the 50 μm filter, and add it to the previous flow-through.
7. Centrifuge the flow-through at 4500 × *g* for 20 min at 4 °C.
8. Carefully pipet off the supernatant and discard.
9. Dissolve the pellet in 0.2 N H_2SO_4 (use 2 ml per 1 g starting material (*see* **Note 8**)).
10. Pipette up and down several times to dissolve the entire pellet, transfer the homogenate into three 1.5 ml microfuge tubes and vortex thoroughly to mix; this step can be performed either on the bench or under the fume hood.
11. Leave the homogenate to rotate at 4 °C for at least 30 min to overnight, to dissolve the basic histone proteins in the acid.
12. Centrifuge the homogenate for 10 min at 17000 × *g* in a cooled tabletop centrifuge at 4 °C to remove the insoluble material.
13. Carefully transfer the supernatant to new microfuge tubes and discard the pellets (*see* **Note 9**).
14. Precipitate the acid soluble histone proteins from the supernatant by adding 100% TCA to the cold sample to a final TCA concentration of 33% (v/v) (*see* **Notes 10** and **11**).
15. Keep the homogenate on ice for at least 1 h to allow the histone precipitation to occur.
16. Centrifuge the homogenate at full speed for 10 min at 4 °C.
17. Carefully pipette off the supernatant and wash the pelleted histone proteins with the freshly prepared, ice-cold acetone containing 0.1% HCl (0.033 M). Avoid disturbing the pellet.
18. Centrifuge the homogenate at full speed for 10 min and remove the supernatant with a pipette.
19. Wash the histone pellet one more time with acetone alone (*see* **Note 12**). Centrifuge the homogenate at full speed for 10 min and remove the supernatant with the pipette.
20. Air-dry the pellet until no liquid is visible in the tube anymore (*see* **Note 13**). Histones will be visible as a white smear on the tube's wall and bottom.
21. Depending on how dry and how concentrated it is, the pelleted total histone proteins can be either dissolved in 100 or 300 μl of ice-cold ddH_2O containing protease inhibitors (1×) and 100 mM sodium butyrate; Alternatively, the pelleted total histone proteins can be stored dry for up to several months at −20 °C.
22. To dissolve the histones, pipette up and down, and scratch the tube wall, without making bubbles. Sonicating the

homogenate for 1 min in a water bath sonicator facilitates the homogeneous dissolving of the histone pellets.

23. Spin the homogenate at 10,000 × *g* for 1 min at 4 °C.
24. Transfer the supernatant to a new 1.5 ml microfuge tube without disturbing the pellet.
25. Add 100 μl of cold ddH_2O containing protease inhibitors (1×) and 100 mM sodium butyrate to the undissolved histone protein pellet, mix well.
26. Centrifuge at 10,000 × *g* for 1 min.
27. Pool the supernatants containing solubilized histones and measure the histone concentration (*see* Subheading 3.2).
28. To investigate the presence of histones in the protein extract, 20 μl of water dissolved histones can be mixed with 1× Laemmli buffer and run onto SDS-PAGE followed by western blotting (*see* Fig. 1, **Note 14** and Subheading 3.4).

3.2 Histone Concentration Measurement

Due to the acid traces in the acid extracted histone proteins and because of the low number of aromatic residues, histone proteins are best quantified using a protein-specific fluorescence assay (e.g., using a Qubit Fluorometer, *see* Subheading 2.5). This method allows the accurate measurement of small protein quantities, using fluorescent dyes, which emit signals only when bound to specific target molecules. To determine the histone protein concentration before and after fractionation with the RP-HPLC, we used the Qubit®Protein Assay Kit following the manufacturer's recommendations, but alternative high precision fluorometric measurement methods can be used. An amount of more than 3 μg of dissolved histones is adequate for subsequent fractionation steps using RP-HPLC.

3.3 Histone Fractionation by RP-HPLC

To fractionate the different histone proteins from the bulk histones, homogeneously dissolved histones (*see* Subheading 3.1) were diluted to a concentration (≤1 μg/ml), then used for RP-HPLC purification.

1. Start the HPLC machine, connect the tubing, then perform a thorough cleaning of the system using buffer B then buffer A.
2. Adjust the pressure and the injection flow for the buffers.
3. Connect the chromatography column, and the sample loop.
4. Create and save the HPLC purification program for the histones, on the instrument's computer. The program is based on creating a gradient by combining two buffers A and B at different concentration, and different time periods, allowing the elution of the different histone fractions.
5. The program used for histone purification is as follow:

first step: 0–5 min, 0% Buffer B and 100% Buffer A; second step: 5–15 min, 0–35% Buffer B with simultaneous decrease of Buffer A (100–65%); third step 15–25 min, 35% Buffer B with 65% Buffer A, fourth step: 25–75 min, 35–65% Buffer B with simultaneous decrease of Buffer A (65–35%), all the steps should be run with a flow rate of 0.7 ml/min.

6. Equilibrate the column by passing buffer A through it for 10 min, and adjust the pressure and the flow again.
7. Standard proteins with known retention time should be run to confirm the stability of the signal.
8. Adjust the fractions collector, and load it with 1.5 ml microfuge tubes.
9. Load the diluted histone sample into the sample loop with a glass Hamilton syringe and inject the sample onto the chromatography column (ECLIPS XDB-C8 (4.6 × 150 mm)), attached to the HPLC system.
10. Run the HPLC purification program.
11. Collect the fractions from 12 to 80 min (90 fractions) (*see* Fig. 1), following the data on the chromatogram at 214 and 280 nm. Save 20 μl from each fraction into new microfuge tubes for histone detection by western blotting (*see* Subheading 3.4).
12. Lyophilize the remaining RP-HPLC fractions containing histone H3, using a vacuum concentrator. This step is important to remove any organic solvents; the samples are to be stored at −20 °C (*see* **Note 15**).

3.4 Histone Detection Using SDS-PAGE/Western Blotting

To assess the purity of the purified histone fractions and to locate the fractions that contain histone H3, aliquots of the fractionated histones have to be run on 15% SDS-PAGE, blotted and detected with a rabbit anti-H3 antibody.

1. Before running the SDS-PAGE, prepare the resolving gel by mixing [Tris–HCl pH 8.8, acrylamide, SDS, APS, and TEMED], and poor it in between a precasted glass plates.
2. Once the resolving gel solidified, the prepared stacking gel [Tris–HCl pH 6.8, 40% acrylamide, SDS, APS and TEMED] is layered on top of it, and the plastic comb is placed on it.
3. Once both gels are solidified, place the glass plates into the SDS-PAGE running chamber, and fill the whole chamber with running buffer.
4. Dissolve the extracted/purified histone aliquots (from Subheading 3.1 and 3.3) in 20 μl of 1× Laemmli buffer (*see* **Note 16**).

5. Boil the samples at 100 °C for 5 min to facilitate dissolving of the histone pellets.
6. Centrifuge at high speed for 1 min.
7. Load 10 μl of the supernatant on the SDS-PAGE gel, and let the gel run until a distinct separation of low molecular weight proteins is achieved as judged by the molecular weight markers.
8. Disassemble the electrophoresis unit, and transfer the SDS-PAGE gel onto a soaked and prewetted polyvinylidene difluoride (PVDF) membrane, in methanol and transfer buffer consecutively.
9. Put the SDS gel–PVDF membrane between two thick, prewetted filter papers, and place the sandwich in the Trans blot SD western blot device.
10. Blot at 15 V for 30 min.
11. Once blotting is finished, transfer the PVDF membrane into a clean plastic box.
12. Block the PVDF membrane, by soaking in the blocking solution under a gentle shaking at room temperature for 1 h.
13. Wash the PVDF membrane 3 times with TBST for 5 min each.
14. Incubate the membrane with the appropriate dilution of anti-H3 primary antibody, according to the antibody datasheet, with gentle shaking at 4 °C for 3 h to overnight.
15. Wash the membrane 3 times with TBST.
16. Incubate the PVDF membrane with horseradish peroxidase labeled secondary antibodies for an additional 1 h at 4 °C.
17. Detect the histone H3 signal on the gel using the western blotting developing kit and visualize it using a camera visualization system.
18. Observing single bands at a molecular weight size of nearly 15 kDa confirms the presence of histone proteins in the extracted homogenate/purified fractions (*see* Fig. 1).

3.5 Proteolytic Digestion of Histone Proteins Before Tandem Mass Spectrometry Analysis

The fractions containing H3, as deduced from western blotting (*see* Fig. 1), are dissolved in 50 mM ammonium bicarbonate pH 8.0, and then combined to be used for mass spectrometry analysis. Specific handling care is recommended (*see* **Note 17**).

1. Dissolve the lyophilized RP-HPLC fractions containing histone H3 (*see* Subheading 3.3) as deduced from the western blotting result, with 50 mM ammonium bicarbonate (AmBic) pH 8.0 (100 μl/Fraction) (*see* **Note 18**).
2. Add 10 mM DTT, and incubate for 30–60 min on a shaker at room temperature. This step allows a chemical reduction of the histones.

3. Add 40 mM iodoacetamide, and incubate for 60 min on a shaker at room temperature, and in the dark. This step allows the alkylation of the histone samples (*see* **Note 19**).
4. Add Trypsin (1:200 enzyme–substrate) or ArgC (1:50) and incubate for 2 h at 37 °C. This step allows the digestion of H3 (*see* **Note 20**).
5. Add 0.1% TFA, and incubate for 1 min, to stop the digestion reaction.
6. Dry the digested peptides using a vacuum concentrator, and keep them at −20 °C until use.

3.6 Sample Cleanup Before Tandem Mass Spectrometry Analysis

In order to obtain good quality spectra in a reproducible manner, the peptides should be concentrated and purified from contaminating salts. This is done by passing them over a pipette tip containing C_{18} reversed phase material "C18 ZipTips" as follows:

1. Resuspend the dried peptides in 20 μl of C18 ZipTips washing buffer.
2. Prewet the ZipTips by pipetting up and down, 10 μl of wetting solution for at least three times in order to activate the C18 matrix.
3. Wash the ZipTips with 10 μl of washing solution for three times.
4. Use the equilibrated ZipTips to pipette 10 μl of the digested histones for 3–10 times, this allows the peptides to bind to the filter matrix.
5. Wash the ZipTips four times with 10 μl of washing solution to discard unbound peptides and salt traces.
6. Elute the bound peptides with 5 μl of the elution solution for at least six times.
7. Dry the eluted samples with a vacuum concentrator to remove the acetonitrile. Use them immediately or store at −20 °C until analysis.

3.7 Sample Processing by Mass Spectrometry and Data Analysis

Samples are analyzed by liquid chromatography–tandem mass spectrometry (LC-MS/MS). We used an ion trapped mass analyzer LTQ-Orbitrap XL associated with an ETD instrument (Thermo Fisher Scientific) coupled to an Eksigent-Nano-HPLC system (Eksigent Technologies).

1. Resolubilize the dried peptides in 10 μl of resolubilization buffer and load them on a self-made tip column (75 μm × 70 mm) packed with reverse phase C18 material.
2. Elute the peptides with a flow rate of 200 nl per min by a gradient from 3 to 10% of Buffer B in 6 min, 35% B in 38 min, and 97% B in 45 min.

3. During sample processing by the tandem mass spectrometer, one scan cycle comprises of a full scan MS survey spectrum followed by up to 6 sequential CID and ETD MS/MS scans on the three most intense signals above a threshold of 500. Full-scan MS spectra (300–2000 *m/z*) are acquired in the FT-Orbitrap with a resolution of 60,000 at 400 *m/z* after accumulation to a target value of 200,000, while CID and ETD MS/MS spectra are recorded in the linear ion trap (target value of 1e4 for ion trap MSn scans). The ETD anion target value is set at 3e5 and activation time at 50 ms. Supplementary activation is employed to enhance the fragmentation efficiency for 2+ precursors and charge state dependent ETD time enabled. The ETD reaction time is 120 ms and isolation width 3 *m/z*. Charge state screening is enabled and singly charge states rejected. Precursor masses selected twice for MS/MS are excluded for further selection for 120 s. Samples are acquired using internal lock mass calibration set on *m/z* 429.0887 and 445.1200 (*see* Fig. 1).
4. The MS raw-files are converted into Mascot generic files (mgf) with Mascot Distiller software 2.4.2.0 (Matrix Science Ltd., London, UK).

The peak lists are searched using a Mascot Server 2.3 against the Arabidopsis TAIR10 protein database. The search parameters are: requirement for chymotryptic or ArgC ends, one missed cleavage allowed, peptide tolerance ±6 ppm, MS/MS tolerance ±0.5 Da. Carbamidomethylation of cysteine is set as fixed modification while oxidation (methionine), acetylation (N-term protein, lysine) monomethylation, dimethylation, and trimethylation (arginine, lysine) are set as variable. To assess the location of the post-translational modification sites, the site localization analysis provided by Mascot can be used. The presence of modification sites should be further validated by manual inspection of spectra.

4 Notes

1. Cauliflower inflorescence is commercially available year round as vegetable for consumption. Biologically speaking, this tissue exhibits a hypertrophic floral meristem formed upon indeterminate growth through mitotic division. The apical portion of the inflorescence consists of small cells and has a much higher DNA (chromatin) content than the underlying axial tissues. Because of being relatively free of chloroplasts and its close relationship to ***Arabidopsis thaliana***, the cauliflower inflorescence offers a good source of plant material for both soluble chromatin isolation and for different protein assays; it is

preferable to use small fresh and healthy heads for each new histone extraction experiment, to secure optimal yield.

2. Trichloroacetic acid, acetone, HCl, and H_2SO_4 are highly corrosive and harmful chemicals; it is recommended to wear lab coat and goggles. They should be handled and dispensed in a fume hood.
3. Acrylamide–bisacrylamide and TEMED are toxic and carcinogenic chemicals and should be handled and dispensed in a fume hood.
4. Acetonitrile (ACN) and trifluoroacetic acid (TFA) are toxic and highly corrosive chemicals that should be handled carefully and be dispensed in a fume hood. Besides, using tips with filters is also recommended.
5. Prepare all solutions shortly before use, using HPLC grade deionized water and analytical grade reagents. Prepare and store all reagents at 4 °C.
6. It is preferable to use fresh cauliflower heads for every histone extraction. However, plant material can be shock-frozen in liquid nitrogen and stored at −80 °C.
7. Avoid squeezing the nylon mesh too much after the first filtration, to prevent any additional cell debris from passing to the 50 μm filter.
8. The volume of the acid solution added should be adjusted depending on the amount of starting material; usually use 2 ml of 0.2 N H_2SO_4 for 1 g of starting material.
9. Sometimes, one acid extraction is not sufficient to release all the histones from the chromatin; thus, the pellet can be dissolved again in the acid solution for an additional 30 min. The supernatants can be pooled and subsequently used for histone precipitation.
10. TCA should be added dropwise to the supernatant while the mixture is vortexed in between to get a homogeneous mixture.
11. When adding the TCA, a milky solution is formed due to histone precipitation, the higher the intensity of the color, the higher the concentration of the precipitated proteins.
12. These washing steps are used to remove as much as possible traces of acid from the protein pellet. Excessive washing should be avoided, as this may lead to reduction of the histone yield.
13. Avoid to overdry the pellets as this prevents the full dissolving of the proteins into aqueous solutions.
14. Purity of histone extractions can be assessed by running an aliquot of the total histone extract on a 15% SDS-PAGE gel followed by Coomassie Blue staining.

15. The dried protein fractions can be stored for several months at −20 °C.
16. The dried samples from the total extracted histones or purified histone fractions may still contain some acid traces, resulting in the Laemmli buffer color turning yellow after dissolution. To avoid problems with the SDS-PAGE running behavior, 1 M Tris–HCl pH 8.0 is to be added until the blue color of the Bromophenol Blue reappears.
17. When handling samples for mass spectrometry analysis, gloves must be worn at all times, and rinsed occasionally to prevent contaminating the sample with human keratin and dust. All the steps should be performed in a laminar flow hood, and dedicated pipettes, tips, and tubes should be used.
18. It is very important to adjust the pH to 8.0, an appropriate pH for histone digesting enzymes. pH can be checked using pH paper, and adjusted to 8.0, by adding a few microliters of 100 mM ammonium bicarbonate, checking the pH increase in between.
19. Iodoacetamide is used for sulfhydryl group alkylation; it modifies the previously reduced sulfur groups by the DTT, preventing the formation or reformation of disulfide bonds.
20. As the histones contain a high percentage of lysine and arginine residues, digestion with trypsin results in a mixture of very small peptides, making the discovery of coexisting modifications on the same peptides difficult to achieve. In order to circumvent this problem, we recommend two alternatives: First, a different protease, such as ArgC, can be used. Alternatively, a partial tryptic digestion can be performed by diluting the Trypsin relative to the substrate and by reducing the digestion time to prevent complete cleavage.

References

1. Kornberg RD (1974) Chromatin structure: a repeating unit of histones and DNA. Science 184 (4139):868-871
2. Huang H, Sabari BR, Garcia BA, Allis CD, Zhao Y (2014) SnapShot: histone modifications. Cell 159(2):458–458.e451
3. Taverna SD, Li H, Ruthenburg AJ, Allis CD, Patel DJ (2007) How chromatin-binding modules interpret histone modifications: lessons from professional pocket pickers. Nat Struct Mol Biol 14(11):1025–1040
4. Jenuwein T, Allis CD (2001) Translating the histone code. Science 293(5532):1074–1080
5. Roudier F, Teixeira FK, Colot V (2009) Chromatin indexing in Arabidopsis: an epigenomic tale of tails and more. Trends Genet 25 (11):511–517
6. Bannister AJ, Kouzarides T (2011) Regulation of chromatin by histone modifications. Cell Res 21(3):381–395
7. Bowman GD, Poirier MG (2015) Post-translational modifications of histones that influence nucleosome dynamics. Chem Rev 115(6):2274–2295
8. Arnaudo AM, Garcia BA (2013) Proteomic characterization of novel histone post-translational modifications. Epigenetics Chromatin 6(1):24
9. Bowler C, Benvenuto G, Laflamme P, Molino D, Probst AV, Tariq M, Paszkowski J (2004)

Chromatin techniques for plant cells. Plant J 39(5):776–789

10. Garcia BA, Hake SB, Diaz RL, Kauer M, Morris SA, Recht J, Shabanowitz J, Mishra N, Strahl BD, Allis CD, Hunt DF (2007) Organismal differences in post-translational modifications in histones H3 and H4. J Biol Chem 282 (10):7641–7655
11. Garcia BA, Shabanowitz J, Hunt DF (2007) Characterization of histones and their post-translational modifications by mass spectrometry. Curr Opin Chem Biol 11(1):66–73
12. Zhang K, Sridhar VV, Zhu J, Kapoor A, Zhu JK (2007) Distinctive core histone post-translational modification patterns in Arabidopsis thaliana. PLoS One 2(11):e1210
13. Johnson L, Mollah S, Garcia BA, Muratore TL, Shabanowitz J, Hunt DF, Jacobsen SE (2004) Mass spectrometry analysis of Arabidopsis histone H3 reveals distinct combinations of post-translational modifications. Nucleic Acids Res 32(22):6511–6518
14. Tweedie-Cullen RY, Reck JM, Mansuy IM (2009) Comprehensive mapping of post-translational modifications on synaptic, nuclear, and histone proteins in the adult mouse brain. J Proteome Res 8 (11):4966–4982
15. Udeshi ND, Compton PD, Shabanowitz J, Hunt DF, Rose KL (2008) Methods for analyzing peptides and proteins on a chromatographic timescale by electron-transfer dissociation mass spectrometry. Nat Protoc 3 (11):1709–1717
16. Mahrez W, Arellano MS, Moreno-Romero J, Nakamura M, Shu H, Nanni P, Kohler C, Gruissem W, Hennig L (2016) H3K36ac is an evolutionary conserved plant histone modification that marks active genes. Plant Physiol 170 (3):1566–1577

Chapter 10

Histone H1 Purification and Post-Translational Modification Profiling by High–Resolution Mass Spectrometry

Maciej Kotliński and Andrzej Jerzmanowski

Abstract

It has proven particularly difficult to purify Linker (H1) histones from the model plant *Arabidopsis thaliana*. This is most likely due to its low nuclear DNA content and the abundance of substances that interfere with protein isolation. These problems have hindered the use of Arabidopsis for in-depth characterization of nuclear proteins by modern techniques based on mass spectrometry (MS). Here, we describe an improved methodology for preparing pure Arabidopsis H1s and separating them by HPLC into fractions corresponding to nonallelic variants. In addition, we outline basic approaches enabling the identification of posttranslational modifications of H1 by MS and their mapping by digestion with different proteases. We also discuss the analysis and interpretation of the acquired data.

Key words *Arabidopsis thaliana*, Linker (H1) histones, H1 purification, HPLC, Mass spectrometry, Posttranslational modifications of H1

1 Introduction

Linker (H1) histones are universal and abundant components of eukaryotic chromatin. They are mixed-domain proteins consisting of an evolutionarily conserved globular domain (GH1), flanked by highly basic unstructured N- and C-terminal domains. H1s are generally more variable than core histones. In both animals and plants, multiple nonallelic sequence variants of H1 are often present in the chromatin of a single cell. Linker histones are required for the stabilization of compact higher-order chromatin structures and are considered negative regulators of the accessibility of DNA to various regulatory factors [1]. However, H1s have also been shown to recruit at least 20 different proteins to chromatin, mostly via interactions with their basic C- and N-terminal domains [2]. Thus, in addition to restricting access to DNA, linker histones may also play a parallel and opposite role as mediators facilitating interactions between chromatin and certain nonhistone proteins. A crucial factor determining the in vivo dynamics and residence time of H1

Marian Bemer and Célia Baroux (eds.), *Plant Chromatin Dynamics: Methods and Protocols*, Methods in Molecular Biology, vol. 1675, DOI 10.1007/978-1-4939-7318-7_10,

in chromatin is the strong cooperativity between GH1 and the C-terminal domain [3]. There is growing evidence that H1s, similarly to core histones, participate in determining the pattern of chromatin epigenetic modifications. At the same time, in both animals [4] and plants [5], they are themselves the targets of numerous and highly diverse posttranslational modifications.

Due to the wide array of available experimental tools, such as extensive mutant collections and rich reference databases, Arabidopsis is the preferred model for studies on complex chromatin mechanisms underlying growth, development and the adaptive abilities of plants, including those dependent on linker histones. However, while the isolation of reasonable quantities of pure linker histones from plants such as tobacco and wheat is relatively straightforward, the small genome of Arabidopsis, resulting in a low level of histones per cell, seriously hinders the isolation of pure H1. This severely limits the usefulness of this model plant for studies on histones by MS-based proteomic methods, including characterization of their posttranslational modifications.

Here, we describe a modified procedure for the isolation of linker histones that has been optimized for Arabidopsis and yields protein suitable for various analytical studies. This protocol involves linker histone isolation, HPLC fractionation to purify the linker histones, and subsequent MALDI-TOF analysis to determine the nature of the linker histones in the different fractions. In addition, we also present a detailed methodology for the identification of posttranslational modifications of Arabidopsis H1s by high-resolution MS analysis. This includes different approaches to digest the linker histones into peptides of desirable length, identification of the peptides and their modifications by LC-MS/MS, and subsequent data analysis.

2 Materials

All solutions are prepared using deionized water (MilliQ or better) and analytical grade reagents, at room temperature unless stated otherwise. Standard molecular laboratory equipment (microcentrifuge, micropipettes, vortex, thermomixer or another type of shaker for 1.5 and 50 ml tubes, magnetic stirrer, beakers, graduated cylinders, spatulas, etc.) and plastic consumables (1.5 ml, 2 ml microcentrifuge tubes, 15 and 50 ml conical-bottomed centrifuge tubes, etc.) are required. Liquid nitrogen (LN_2) is used to freeze plant material and protein samples.

2.1 Plant Material

Arabidopsis thaliana Col-0 seeds are sown in soil and placed at 4 °C for 3 days without light. The pots (25 cm in diameter, 12 cm high) are then transferred to a greenhouse at 21 °C under natural light supplemented with sodium lamps, with a minimum day length of

16 h. After 7 days the seedlings are thinned to a density of about 100 plants per pot. Four weeks after sowing, the plants are harvested (without roots), frozen in LN_2, and stored at −80 °C until required (*see* **Note 1**).

2.2 Isolation of Linker Histones

1. 150 g of LN_2-frozen plants.
2. 5% perchloric acid (PCA)—1.5 l per 150 g of material. Store at room temperature. Stable for more than 1 year. Note that PCA is corrosive.
3. Rotor-stator- or a razor-blender-type homogenizer.
4. Miracloth.
5. Six centrifuge tubes of 30 ml (we use Corex-type round-bottomed glass tubes) with rubber adapters to fit the rotor for 50 ml tubes.
6. 100% (w/v) Trichloroacetic acid (TCA)—550 ml required for isolation from 150 g of material. Store at 4 °C for less than 1 month. Note that TCA is highly corrosive and can penetrate latex gloves.
7. Floor-standing high-speed refrigerated centrifuge (we use a Sorvall Superspeed centrifuge) with fixed angle rotor for 250 or 500 ml tubes (we use the Sorvall SLA-1000 4 × 250 ml rotor) and swinging-bucket rotor for 50 ml tubes (we use the Sorvall HB-6 6 × 50 rotor).
8. Twelve centrifuge tubes of 250 ml or eight centrifuge tubes of 500 ml that fit the rotor of the centrifuge.
9. Acetone (750 ml) cooled at −80 °C.
10. 4.5% G buffer: 4.5% (w/v) guanidine hydrochloride, 0.1 M potassium phosphate buffer, pH 6.8. The pH should be adjusted with 1 M H_3PO_4 after the addition of guanidine hydrochloride. Prepare 1.5 l, autoclave and store at 4 °C for up to 1 year.
11. 6% G buffer: 6% (w/v) guanidine hydrochloride, 0.1 M potassium phosphate buffer, pH 6.8. Prepare 50 ml, autoclave and store at 4 °C for up to 1 year.
12. 10% G buffer: 10% (w/v) guanidine hydrochloride, 0.1 M potassium phosphate buffer, pH 6.8. Prepare 50 ml, autoclave and store at 4 °C for up to 1 year.
13. 50% Bio-Rex 70 resin (Bio-Rad) in 4.5% G buffer. Prepare at least one day before isolation by equilibrating 10 g of resin in multiple changes of 4.5% G buffer (*see* **Note 2**). Suspend the resin immediately before use. Store at 4 °C, stable for 1 year.
14. 100 mM PMSF (phenylmethylsulfonyl fluoride) in ethanol. Store at room temperature or in a cold room on a magnetic stirrer for 1 year (100 mM PMSF stored in cold room without stirring will crystallize).

15. Glass chromatography column, 10 mm in diameter and ≥30 cm long. Column held in a holder/rack, fitted with an outlet valve and connected to 30–50 cm of flexible plastic tube (1–2 mm in diameter).
16. Dialysis tubing (cutoff ≤15 kDa; we use Spectrum-Labs Spectra/Por 1 membrane) and dialysis clips.
17. Acetic acid
18. Optional: formic acid or trifluoroacetic acid (TFA).

2.3 Fractionation by Reversed-Phase HPLC

1. HPLC buffer A: 0.1% (v/v) TFA in water. 1 l should be enough for an HPLC run plus equilibration and washing of the column.
2. HPLC buffer B: 90% acetonitrile, 0.1% (v/v) TFA. 1 l should be enough for an HPLC run plus equilibration and washing of the column.
3. 0.1% (v/v) TFA (10 ml).
4. HPLC column with C-18 resin: 25 mm × 4.6 mm column packed with 5 μm particle size resin or 15 mm × 4.6 mm column with 3 μm particle size resin. Pore size 300 Å.
5. HPLC system comprising pumps, solvent mixer and UV-Vis detector set to 220 nm or 220 and 280 nm.
6. Freeze-dryer.

2.4 MALDI-TOF Analysis of Whole Proteins

1. MALDI Matrix—Saturated solution of synapinic acid (SA) in 50% acetonitrile, 0.1% (v/v) TFA.
2. Target plate for MALDI mass spectrometer.
3. Matrix assisted laser desorption ionization-time of flight (MALDI-TOF) mass spectrometer working in linear mode (without ion reflector). We use Ultraflex Extreme (Bruker).

2.5 Identification of Posttranslational Modifications by Mass Spectrometry

1. Trypsin solution (100 ng/μl) in water (e.g., Sequencing grade modified porcine Trypsin, Promega). Minimize the time that the solution remains unfrozen and keep on ice. For storage, freeze in LN_2 and store at −80 °C for >2 years.
2. 10× trypsin digestion buffer: 1 M NH_4HCO_3 (Prepare 100 ml. Do not autoclave and store at room temperature for up to 2 years).
3. Arg-C solution (100 ng/μl) in water (Endoproteinase Arg-C sequencing grade, Roche). For storage, freeze in LN_2 and store at −80 °C for up to 2 years.
4. 5× Arg-C digestion buffer: 500 mM Tris–HCl pH 7.6, 100 mM $CaCl_2$, 50 mM DTT, 5% (v/v) acetonitrile. Prepare 1.5 ml fresh.

5. Thermolysin solution (0.5 μg/μl) in water (Thermolysin, Promega). For storage, freeze in LN_2 and store at −80 °C for up to 2 years.
6. 10× thermolysin digestion buffer: 500 mM Tris–HCl pH 8.0, 50 mM $CaCl_2$. Prepare 1.5 ml from stock solutions.
7. Immobilized Pepsin Agarose Resin (e.g., Thermo Fisher Scientific).
8. Pepsin digestion buffer: 20 mM sodium acetate, adjust the pH to 3.85 using HCl. Prepare 5 ml fresh.
9. 10% (v/v) Trifluoroacetic acid (TFA). Prepare 500 μl. Corrosive, store in the fume hood.
10. NanoHPLC buffer A: 0.1% (v/v) formic acid (FA) in water. Prepare 100 ml.
11. NanoHPLC buffer B: 0.1% (v/v) FA in acetonitrile. Prepare 50 ml.
12. Low retention glass sample vials that fit the autosampler of the nanoHPLC.
13. NanoHPLC precolumn: 180 μm × 20 mm column packed with 5 μm particles of C-18 resin or similar.
14. NanoHPLC column: 75 μm × 250 mm column packed with 1.7 μm particles of C-18 resin or similar.
15. NanoHPLC equipped with an automatic sample loader with sample loop and syringes that allow loading of 20 μl samples. HPLC system configured with sample loader rheodyne connected to a precolumn. The other side of the pre-column connected to a T-junction. Remaining ports of the T-junction connected to the waste valve and analytical column. The analytical column connected to the nanoelectrospray ion source of the mass spectrometer.
16. Mass spectrometer coupled with nanoHPLC. Preferably a high-resolution mass spectrometer with an in-beam collision-induced dissociation (CID) device, e.g., Orbitrap Velos, Elite, Fusion or Q-Exactive (utilizing HCD fragmentation). A modern quadrupole-time of flight (Q-TOF) mass spectrometer should also be suitable, but the mass accuracy will be lower.

2.6 Analysis of LC-MS Data

1. Software for acquisition of data and for controlling the HPLC system and mass spectrometer (e.g., Thermo Xcalibur, Waters MassLynx). Software should be configured during installation of the mass spectrometer.
2. Software allowing manual review of the raw data. Usually provided with mass spectrometer, e.g., Xcalibur for mass spectrometers produced by Thermo Finnigan, MasLynx for spectrometers from Waters, UK.

3. Software for processing raw data, e.g., Mascot Distiller, Matrix Science, UK or MaxQuant [6].
4. Protein identification software, e.g., Mascot server, Matrix Science, UK, MaxQuant [6], X!Tandem (http://www.thegpm.org/tandem/).
5. ProteinProspector service (http://prospector.ucsf.edu).
6. MS Expert GUI [7], now included in MaxQuant but; can be obtained as a stand-alone program.

3 Methods

3.1 Histone Isolation

Highly basic proteins like linker histones can be extracted in 5% perchloric acid (PCA), whereas core histones and other less basic proteins cannot [8]. Unlike in animals, numerous contaminants are coextracted from plant tissues using this method, and, therefore, the protein extract is further purified by ion-exchange chromatography. The histone H1 isolation procedure described here takes 3 days, with protein precipitation, dialysis and freeze-drying steps performed overnight (*see* **Note 3**).

1. Grind 150 g of frozen plant material in a mortar with LN_2. Alternatively, a LN_2-cooled orbital ball mill or LN_2-cooled coffee grinder can be used (only grinders with rotating blades and a metal bottom to the chamber can withstand LN_2). For high efficiency extraction, it is crucial to grind the material well.
2. Suspend the ground material in 1.5 l of 5% PCA in a beaker at room temperature. Use a spatula and/or vigorous shaking to rapidly suspend the material. Use 10 ml of 5% PCA per gram of plant material (*see* **Note 4**).
3. All subsequent isolation steps, including centrifugations, should be performed at 4 °C (preferably in a cold room).
4. Stir the suspended material on a magnetic stirrer in a cold room (4 °C). After 10 min, further homogenize the suspension using a rotor-stator homogenizer for 10 min, then stir the suspension for a further 90 min (*see* **Note 5**)
5. Transfer the suspension to 250 or 500 ml centrifuge tubes and centrifuge at 15,000–20,000 × *g* for 15 min.
6. Place pieces of folded Miracloth on top of new centrifuge tubes.
7. Gently decant the supernatants through the two layers of Miracloth into the new tubes.
8. Centrifuge for 15 min as above.
9. Place new pieces of folded Miracloth on top of a 2-l graduated cylinder.

10. Decant the supernatant into the cylinder through the Miracloth and note the volume.
11. To precipitate the proteins, add 1/3 of the noted volume of 100% TCA to the suspension (final TCA concentration—25%).
12. Transfer the suspension to a beaker and stir on a magnetic stirrer overnight.
13. Fill round-bottomed centrifuge tubes with the suspension and centrifuge for 45 min at 20,000–25,000 × *g* in a swing-out rotor. We use six 30 ml Corex-type glass tubes and a Sorvall HB-6 rotor. Attempts to harvest proteins in 250 ml flat-bottomed tubes have failed because the pellets do not stick to the bottom of the tubes.
14. Remove and discard the supernatants and add more suspension to fill the tubes.
15. Centrifuge again as above. Numerous centrifugations (usually 5) are required to collect the precipitate from the entire volume of suspension.
16. To remove residual TCA from the pelleted precipitate, add 30 ml of cold acetone to each tube and place at −20 °C for 10 min.
17. Centrifuge for 15 min as above.
18. Carefully remove the supernatants.
19. Repeat **steps 16–18** twice (i.e., three washes with acetone in total).
20. Dry the protein precipitates under vacuum (usually 10–30 min in a vacuum dryer), in a cold air flow, or leave the tubes open overnight at −20 °C. The dried precipitates can be stored at −20 °C for months.
21. Add 1 ml of 4.5% G buffer and 4 μl of 100 mM PMSF to each tube.
22. Place the tubes in a shaker for 60 min. We use a Thermomixer with a 50-ml tube block. The 30-ml centrifuge tubes fit perfectly into 50-ml Falcon-type tubes.
23. To unstick the pellets from the tube walls, sonicate the glass tubes in a sonicating bath for 15 min (we use a Diagenode BioRuptor working at maximal power for 50% of the time).
24. Transfer the suspensions of fragmented protein pellets from each the glass tubes to a new 2 ml tubes. Do not use thick-walled tubes which could block the ultrasound waves in subsequent steps.
25. Add 500 μl of 4.5% G buffer to each glass tube and shake for 15 min.

26. Transfer the solution with suspended pellet fragments to 2 ml tubes.
27. Add 250 μl of 4.5% G buffer to each glass tube and place the tubes on the shaker for 15 min.
28. Transfer the solution with suspended pellet fragments to 2 ml tubes.
29. Sonicate all 2 ml tubes for 15 min in a Diagenode BioRuptor set to maximal power, with the ultrasound source turned on for 50% of the time. The pellet clumps should be suspended after this step. When using an alternative sonication device, define the right settings to disrupt the pellet in advance.
30. Centrifuge the 2 ml tubes at 25,000 × *g* for 10 min.
31. Pool all supernatants in one new 50 ml tube and mix with 5 ml of 50% Bio-Rex 70 resin (*see* **Note 6**). Do not discard the 2 ml tubes.
32. Place the 50 ml tube on the rotator.
33. To increase the yield, the remaining pellets in the 2 ml tubes are again resuspended in 500 μl of 4.5% G buffer.
34. Sonicate the 2 ml tubes as above.
35. Centrifuge 2 ml tubes at 25,000 × *g* for 10 min.
36. Add the supernatants to the tube containing the Bio-Rex 70 resin.
37. Incubate the tube containing the resin for 4 h on the rotator.
38. Prepare the glass chromatography column.
39. Pour the resin suspension into the column and wait until the resin settles.
40. Put the tubing connected to the outlet at the base of the column into a new 50 ml tube to collect the eluted fractions containing unbound proteins. Open the valve and adjust the flow rate to ~0.5 ml/min by raising or lowering the end of the outlet tubing. The same eluate flow rate is maintained in all steps of chromatography.
41. If the volume of the resin suspension exceeds the column volume, close the valve in the outlet tubing once the liquid level has fallen and pour the remaining suspension into the column. Wait until the resin has settled and then open the valve again.
42. When almost all the liquid has entered the resin, close the valve to stop the flow. Be careful not to let the top of the resin dry out—leave a few millimeters of liquid over the resin. If the resin accidentally dries out, resuspend it and allow to settle again.
43. Replace the collection tube (fraction containing proteins from the first wash).

44. Gently add 10 ml of 4.5% G buffer to the column by layering, open the valve and wait until the solution has entered the resin.
45. Add another 10 ml of 4.5% G buffer to the column, open the valve and let it enter the resin.
46. Close the valve and replace the collection tube (fraction containing proteins eluted with 4.5% G buffer).
47. Add 10 ml of 6% G buffer to the column, open the valve and let it enter the resin.
48. Repeat **step 47**.
49. Close the valve and replace the collection tube (fraction containing H1).
50. Add 10 ml of 10% G buffer to the column, open the valve, and let it enter the resin.
51. Repeat this step twice (i.e., 3 × 10 ml of 10% G buffer in total) and remove the collection tube (fraction containing the linker histones).
52. Dialyze all eluate fractions against three changes of 5 l of 0.1% (v/v) acetic acid.
53. Freeze the dialyzed fractions in LN_2.
54. Vacuum-dry to dryness or until only 3–8 ml of solvent remains (which removes the need for resuspension).
55. If the sample was fully dried, resuspend in 2 ml of 0.1% (v/v) acetic, formic, or trifluoroacetic acid (*see* **Note 7**).
56. Freeze fractions in LN_2 and store at −80 °C (*see* **Note 8**).

3.2 HPLC Fractionation of Linker Histones

Reversed-phase HPLC is used to purify linker histones from other proteins that remain after column chromatography (mostly highly basic AT-hook motif-containing proteins) and to separate linker histone variants. A chromatographic column containing C-18 resin gives the best results, but C-8 or C-12 columns can also be used.

The procedure lasts approximately 3 h (centrifugation, HPLC separation) and is followed by vacuum-drying overnight plus 1 h the next day, followed by a further few hours of vacuum-drying (usually about 5 h).

1. Prepare the HPLC system with a column and buffers A and B. Set the flow rate to 1 ml/min. Equilibrate the column with buffer A. Prepare the HPLC gradient program: 0–2 min—buffer A; 5 min—35% buffer B; 105 min—43% buffer B; 125 min—63% buffer B; 130 min—100% buffer B. Set UV-Vis detector to 220 and 280 nm (280 nm is optional).
2. Thaw the sample and check if there is any visible precipitate. Centrifuge the sample (≥10,000 × g for 10 min) and take the

supernatant for HPLC. Avoid loading any precipitate onto the column.

3. Load the sample into the sample loop or directly onto the column (depending on the system configuration). If the sample loop is too small for the entire sample volume, repeat loading until all the sample has been applied to the column.
4. Prepare tubes to collect the fractions (50 ml tubes are best). Start the run and monitor the UV detector readings while collecting fractions eluted from the column. Linker histones do not absorb at 280 nm (aromatic amino acids), so the absorbance at 220 nm (peptide bond) is the most important indicator. Decide whether to collect all peaks or only the fractions containing histone H1, and whether or not to collect all non-allelic H1 variants as one fraction or attempt to separate them.
5. Strong peaks occurring between the 10th and 20th minute of the gradient run represent unbound and weakly bound proteins.
6. Linker histones should elute from the column between the 30th and 70th minute of the gradient run. Initially (30–40 min) numerous weak peaks are visible on the chromatogram. Stress-inducible variant H1.3, AT hook-containing proteins and various degradation products are present in this fraction. The most abundant linker histone variant H1.2 starts to elute between the 45th and 55th minute of the run. The second, less abundant, variant H1.1 elutes at around the 60th minute (Fig. 1) (*see* **Note 9**).

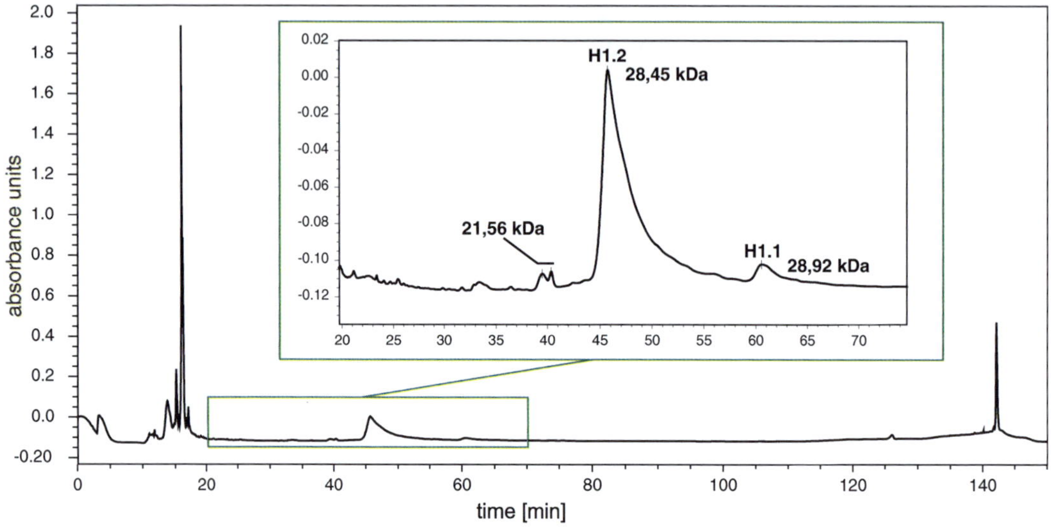

Fig. 1 Chromatogram for HPLC separation of linker histones from Arabidopsis, recorded at a wavelength of 220 nm

7. Freeze all collected fractions in LN_2 and vacuum dry in the freeze dryer.
8. Resuspend the fractions in 1 ml of 0.1% (v/v) TFA and transfer to new 1 ml tubes.
9. Freeze in LN_2 and vacuum dry in the freeze dryer.
10. Resuspend the samples in 100 μl of 0.1% (v/v) TFA (*see* **Note 10**).

3.3 MALDI-TOF Analysis of Whole Proteins

A MALDI-TOF mass spectrometer is used to measure the molecular masses of whole proteins. This method of establishing protein masses is more precise and requires a much smaller amount of protein than SDS-PAGE.

About 1 h is required to apply samples onto the MALDI target plate and then dry them. Manual measurement of one sample takes about 5 min (depending on the number of shots and the spectrometer).

1. For each sample, apply a 0.5 μl droplet of matrix onto a MALDI target plate, then immediately add 0.5 μl of the protein fraction. Mix the sample and matrix by gentle pipetting.
2. Allow the samples to dry on the bench.
3. Place the target plate in the MALDI-TOF mass spectrometer. Set the instrument to linear mode, positive ionization, with a detection range that includes at least masses from 10 to 50 kDa.
4. Run the MALDI-TOF MS and perform measurements. Try to acquire spectra using different laser power settings and from different places in the spot on the target plate. Peaks of H1.2 and H1.1 should be readily observable (Fig. 2) (*see* **Note 11**).

3.4 Digestion of Proteins for LC-MS/MS Analysis

Linker histones contain numerous lysines and arginines (>25% of all amino acid residues). This makes it impossible to produce large fragments for analysis using trypsin, since cleavage by this enzyme adjacent to K/R residues produces lots of very short nonspecific peptides. This problem is solved by using endoproteases with different sequence specificities. We use trypsin as well as Arg-C, cleaving near arginine, thermolysin, cleaving near hydrophobic residues, and pepsin, which is a less sequence-specific protease. It is worth noting that trypsin, despite digesting most histone molecules into very small peptides, still gives useful results because many posttranslational modifications of lysine and arginine reduce the efficiency of cleavage or even prevent digestion by this enzyme. Limited cleavage by trypsin at these sites means that this protease produces numerous modified peptides that can be detected after LC-MS analysis. Note also that histone H1 does not contain cysteine, so neither dithiothreitol (DTT) nor cysteine alkylating agents (e.g.,

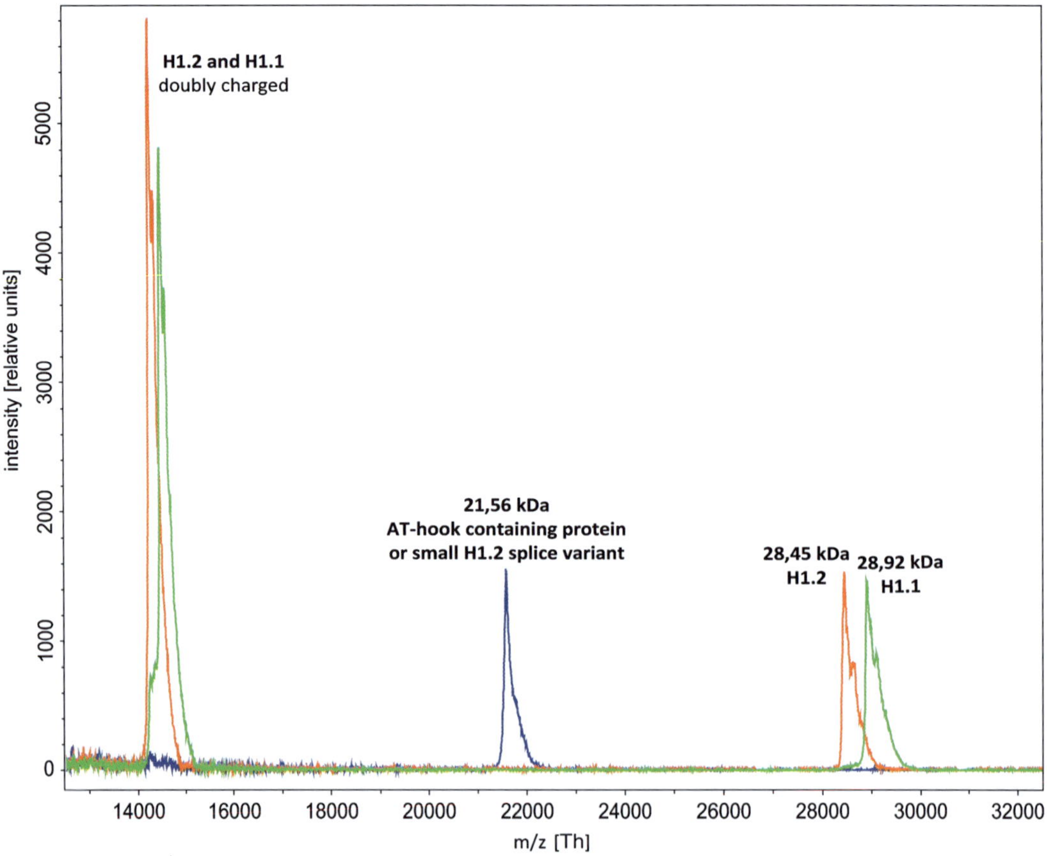

Fig. 2 MALDI mass spectrum of proteins from three fractions marked by *lines of different colors* in Fig. 1. Note that due to the basic nature of histone H1, the peaks originating from doubly charged protein ions have the highest intensity

iodoacetamide) need to be used (*see* **Note 12**). The digestion lasts between 1 h (thermolysin) and 8 h (trypsin). The whole procedure for pepsin takes less than 2 h.

3.4.1 Trypsin

1. Prepare 0.5 ml siliconized tubes for digestion of each protein fraction.
2. Mix 2.5 μl of 10 × trypsin digestion buffer, 15 μl of water, 2.5 μl of trypsin solution, and 5 μl of protein sample.
3. Place the tubes at 37 °C for 8 h. Incubation is conducted in an incubator or climatic chamber to ensure that the entire tube is maintained at the same temperature. Do not use a water bath, heat block, or thermomixer because water from the reaction mixture evaporates and condenses on the top of the tube, which hinders the digestion. A thermocycler with heated lid should also be suitable.
4. After digestion, freeze the samples in LN_2 and store at −80 °C.

3.4.2 Arg-C

1. Prepare 0.5 ml siliconized tubes for digestion of each protein fraction.
2. Mix 5 μl of 5× Arg-C digestion buffer, 14 μl of water, 1 μl of Arg-C solution, and 5 μl of protein sample.
3. Incubate at 37 °C for 4 h. Note that Arg-C also cleaves adjacent to lysines to some extent, and, therefore, prolonged digestion can produce tryptic-like peptides.
4. After digestion, freeze the samples in LN_2 and store at −80 °C.

3.4.3 Thermolysin

1. Prepare 0.5 ml siliconized tubes for digestion of each protein fraction.
2. Mix 2.5 μl of 10× Thermolysin digestion buffer, 17 μl of water, 0.5 μl Thermolysin solution and 5 μl of protein sample.
3. Incubate at 75 °C for 1 h. Use an incubator or thermocycler with heated lid to avoid water condensation on the top of the tube.
4. After digestion, freeze the samples in LN_2 and store at −80 °C.

3.4.4 Pepsin

1. Resuspend the resin with immobilized pepsin by vortexing shortly and transfer 30 μl to a new tube containing 500 μl of ice-cold pepsin digestion buffer.
2. Vortex and incubate for 10 min on ice.
3. Centrifuge for 5 min at 1000 × *g* at 4 °C.
4. Remove the supernatant and suspend the resin in 30 μl of ice-cold pepsin digestion buffer (for a maximum of ten samples).
5. Prepare 0.5 ml siliconized tubes for digestion of each protein fraction.
6. Mix 20 μl of digestion buffer, 3 μl of resin and 5 μl of protein sample. Work on ice.
7. Place the tubes in a Thermomixer set to 25 °C and incubate for 10 min with mixing (900 rpm).
8. Centrifuge for 5 min at 1000 × *g* at 4 °C.
9. Transfer the supernatants to new tubes.
10. To stop the digestion, add 3 μl of 1 M NH_4HCO_3, which increases the pH of the sample to 7.7.
11. Freeze the samples in LN_2 and store at −80 °C.

3.5 Analysis by HPLC-MS

The identification of proteins using a nanoHPLC system coupled with a mass spectrometer is the main tool employed in proteomic studies. For the investigation of linker histones, it is necessary to not only identify the peptides and proteins, but also to characterize the different types of posttranslational modifications they carry. These proteins can be modified by different chemical groups that often share similar masses. High resolution MS is employed to detect the masses of both pure amino acids and also residues modified in many different ways.

The preparation of samples and programming the LC-MS system takes less than one hour. Then automated analysis by LC-MS usually takes about 2 h per sample.

1. Acidify the samples digested with trypsin, Arg-C and thermolysin by adding 2.5 μl of 10% TFA. Do not acidify the samples digested with pepsin.
2. Transfer the digested protein samples to glass vials that fit the autosampler of the nanoHPLC system coupled to the mass spectrometer. Place the vials in the autosampler.
3. Program the HPLC system and mass spectrometer and start the analysis.

   HPLC: Samples loaded to precolumn with 100% buffer A at a flow rate of 10 μl/min for 3 min, with the waste valve open. After loading the precolumn, the waste valve is closed and the flow rate set to 250 nl/min. The gradient program is started, which ramps buffer B from 5% in the 1st minute to 35% in the 45th minute. The column is then washed by ramping the buffer B concentration to 90%.

   Mass spectrometer: Set to positive ionization data-dependent mode with dynamic exclusion and/or chromatographic peak apex prediction. Each MS scan is followed by 5–10 MS/MS scans (depending on the speed of the mass spectrometer), using in-beam fragmentation if possible: HCD device in Orbitraps, standard method in Q-TOF mass spectrometers (avoid CID fragmentation in ion trap which often gives poor results for phosphorylation and is affected by the 1/3 rule that disallows observation of daughter ions with m/z of $<1/3$ of the precursor m/z); m/z range for MS scans: 200–2000 Th, MS/MS scans: 100Th—mass of precursor. If an Orbitrap mass spectrometer is used, the resolution should be set to at least 30,000 for MS scans and 15,000 for MS/MS scans. We recommend setting the MS resolution to 60,000 and MS/MS resolution to 30,000 for Orbitrap Velos machines, and to 120,000 and 60,000, respectively, for faster and more precise Orbitrap Elite and Fusion machines. It is best to acquire profile (not centroided) mass spectra for both MS and MS/MS scans which yield much more information when analyzed manually. It is extremely important that the mass spectrometer is well calibrated to ensure that precise mass measurements are obtained. This is essential for distinguishing posttranslational modifications with similar masses and for characterizing unusual modifications.

3.6 Analysis of LC-MS Data

There are many different software tools for the analysis of mass spectrometric data, including those provided by the manufacturer of the mass spectrometer. Software is required for data acquisition, and in many cases for the initial processing. When analyzing post-translational modifications, this software is also useful for manual interpretation of raw data that is often required to check for the presence of particular modifications. Analysis of raw data usually permits the most accurate determination of masses, including mass shifts that are characteristic for particular modifications. Subsequent steps of data processing (especially regridding) usually cause deterioration of mass accuracy.

1. Check data generated by the mass spectrometer—review chromatograms and spectra. Check total ion current (TIC) chromatograms looking at signal strength and the stability of the signal (signal instability could indicate problems with ESI). Compare the strength of the signals in different fractions. The overall peak heights for a fraction should be comparable between digestions with different proteases (*see* **Note 13**).
2. Process raw data in order to obtain files for protein identification. Files with information concerning the m/z and charges of precursors and m/z of daughter ions are used for peptide and protein identification (Mascot Generic Files. MGF;. PKL files or similar). Some software packages, such as MaxQuant, perform both the initial data processing and protein/peptide identification. If there is the possibility of setting processing parameters (as with Mascot Distiller), set the regridding to at least 500 points/Da for MS data and 250 points/Da for MS/MS data. It is desirable to set even higher values, but this increases the time of processing. It is important to correctly set minimal peak widths—the software will not take into account peaks that are narrower than the limit (*see* **Note 14** for additional considerations regarding raw data processing).
3. Perform peptide/protein identification using software like Mascot Server or MaxQuant (which performs initial processing and identification in one run). Set sequence database (preferably a proteomic database for the selected species), the protease used (*see* **Note 15**), at least three missed cleavages (many modifications disable protein cleavage near the modified amino acid), but do not set fixed modification, i.e., the carbamidomethylation of cysteines (this is not important for H1 because it does not contain cysteines, but it affects other proteins that could be present). *See* **Notes 16** and **17** for additional information.
4. If a concrete modification is found, examine the mass error of the whole peptide (this should be <5 ppm on a well-calibrated Orbitrap) and the score (it should be significant), then review

the fragment masses found. In particular, look at the masses of fragments mapping the modification site (especially the smallest fragment of the series containing the modified amino acid and the largest one without modification). It is preferable to observe a whole series of fragments or part of a series mapping the modification and its neighborhood.

5. Look at the error plot of the identified peptide fragments. Errors should be distributed evenly across the whole *m/z* range. If mass errors form two separate distributions, where part of a distribution is shifted up or down by a few or more mmu, the peptide is probably modified by a group other than that identified. It is even possible that a peptide with a different sequence plus a different modification has been detected. Try to find modifications that do not show such nonuniform distribution.
6. Fragmentation data for a particular mother ion can be found in. MGF or. PKL files (text files) using the scan number and mass provided by Mascot server or other software (use a simple text editor like Notepad). The text data can be pasted into the MS-Product tool of ProteinProspector or MS Expert GUI. These tools can assist the visualization and interpretation of the data.
7. It is also worth inspecting the raw data, especially to check whether the monoisotopic *m/z* of a peptide was determined accurately by the software. Quite frequently the software does not locate the monoisotopic peak correctly, resulting in a 1 or 2 Da error in mass measurement.
8. Often, inspection of the raw data can identify more fragmentation peaks than are detected using the peptide identification software. Sometimes this permits the confirmation of uncertain identifications. It is also possible to find some typical neutral losses for different modifications, which confirms their occurrence. In the case of unexpected modifications, neutral losses and other information derived from MS/MS spectra can be useful in determining the nature of the modification. If the precise mass of the modification is required, we advise measuring modification mass shift of a few fragmentation ions in the MS/MS spectrum (*see* **Note 18**).
9. Prepare a map of posttranslational modifications found along the sequence of the analyzed proteins (here, linker histones). Include links to pages from Mascot with peptide identification results and the number of manually reviewed scans, because this permits the rapid reanalysis of the results. Check whether each modification is repeated at a particular site on peptides of longer or shorter sequences (including peptides obtained by digestion with different proteases). This additionally confirms the identification of modifications at a particular site.

4 Notes

1. The methods described here should also work for plants grown in-vitro and in hydroponic culture, although the yields of isolated histones may vary. The procedure can also be used for the isolation of H1 from other plant species. On testing these methods with tobacco, the efficiency of histone isolation was increased so that just 5–20 g of plant material gave comparable results.
2. Equilibrate the resin by suspending it in buffer and decanting the buffer again after the resin has settled down. Measure the pH of the buffer before and after suspension. When the pH of the decanted buffer is the same as that of the buffer before suspension, the resin is ready. To prepare 50% resin for H1 isolation, decant the buffer, check the volume of the resin and add the same volume of the buffer. Suspend the resin immediately before use.
3. The time of vacuum drying may differ considerably according to the conditions especially ambient light intensity.
4. The color of the suspension could be brown due to the presence of chlorophyll and a pink pigment in the mixture. After the first centrifugation, the color may change to pink or even cherry. The pigment responsible for this color is pH sensitive.
5. Before using the homogenizer, check that there are no lumps of ice in the sample. If ice is present, wait until it thaws.
6. The 50 ml tube is used to fit the solution added in subsequent protocol steps and to allow efficient mixing of the sample on the rotator.
7. If the suspension of proteins is not clear, increase the volume by adding more 0.1% acid.
8. It is advisable to also perform MALDI measurements before the HPLC separation of proteins. Measurements are made according to the instructions in Subheading 3.3.
9. Note that peaks of H1 are quite wide and often look like mixtures of different substances. This is probably due to the presence of numerous posttranslational modifications. Some of these modifications are likely to considerably alter the retention time, leading to the formation of separate peaks on the chromatogram. To fully analyze posttranslational modifications it is important to collect the eluate between the 30th and 65th minute of the gradient run. Note that the exact elution time depends on the HPLC setup employed, particularly the dead volume of the system and the column. The elution times stated here were recorded for a setup with a 150 mm × 4.6 mm

column packed with C-18 resin (particle size 3 μm, pore size 300 Å, ACE, UK).

10. It is highly advisable to check the isolation using a spectrophotometer, e.g., NanoDrop (H1 absorbs at a wavelength of 220 nm—peptide bonds).
11. It is possible to detect the Arabidopsis H1.2 and H1.1 variants using a MALDI mass spectrometer. However, we have been unable to identify the stress-inducible variant H1.3 (even when the plants are grown in conditions known to induce H1.3). Peaks of AT hook-containing proteins and alternative splice variants of H1 can also be observed. Histone H1 peaks are usually wide and contain subpeaks, which is probably due to the presence of numerous different posttranslational modifications.
12. Note that optimization of the digestion time may be necessary due to differences in protein concentration and possibly in protease activity between different enzyme lots. This optimization is most important for pepsin because this is the least specific of the recommended proteases.
13. If signals obtained with one protease are much weaker than those with others, and only ions of low mass are visible, the sample may be over-digested. This is especially common in the case of pepsin and less frequently so for thermolysin. It is advisable to repeat problematic digestions using a shorter incubation time and/or a decreased amount of enzyme. Digesting fractions with different amounts of pepsin and thermolysin (depending on the protein concentration and signals obtained in the first MS analysis) usually gives improved results.
14. Note that the resolution of Fourier Transform-based mass spectrometers (like the Orbitrap) is dependent on the *m/z* measured. The resolution is typically given for *m/z* values of around 450 Th. The resolution near 100 Th (the lowest mass acquired in MS/MS scans) is much higher (usually >4 times) than that at 450 Th. For the analysis of posttranslational modifications, a higher resolution is required. Therefore, the default settings normally used for initial processing are not suitable. Best practice is to check the peak half-width values in data acquired near the minimal *m/z* (usually 100 Th for MS/MS and 200 Th for MS) and set the minimal peak width 5–10 times lower. Also, check the peak width near the maximal *m/z* measured and set the maximal value a few times higher. Setting the minimal peak width too high during initial data processing can result in missing peaks of fragment ions below particular *m/z* values, and the software can lose more than half of the MS/MS data in this way.

15. Try both "pepsin" and "no enzyme" settings for pepsin digestion. It is advisable to test "no enzyme" settings for each protease to detect any nonspecific digestion products caused by contaminant proteases etc.
16. It is advisable to start with a 15 ppm mass error for MS scans and 50 mmu or 20 ppm error for MS/MS scans performed using an Orbitrap. The real error is usually much smaller but we have sometimes detected unexpected modifications with mass shifts similar to known modifications. Since each additional modification doubles the search time, it is not possible and practical to set all known histone modifications at once. It is necessary to search with different sets of possible modifications. Using the error-tolerant mode in Mascot enables one to test the single occurrence of one out of all modifications present in the database (typically a local copy of unimod.org containing >1000 different modifications) on each peptide. This is a good method for finding potential modifications that should be taken into account.
17. Surprisingly, it is quite common to find modifications that are not listed in Unimod and have been detected accidentally because a high mass error was set during the search. Such modifications are identified as different known modifications with a high mass error. Manual data analysis or additional experimental studies are necessary to determine the precise nature of a novel modification. Usually, multiple phosphorylation, acetylation, formylation, and crotonylation sites are readily identified in this step. Remember that the subsequent steps in the procedure are required to confirm their presence.
18. Try to use ions of lower m/z as the resolution of the MS analyzer is higher in this range. Then find the average or median of all mass shifts. The consistency of this data can be checked and the values compared with the mass shift calculated on the basis of the parent ion, to estimate the error. Remember that the analysis of raw data will give more accurate results. It is often possible to guess the molecular formula of the substance (here, the modification) but only on the basis of an accurate mass measurement. A tool for this type of analysis is available in the Qual Browser program from the Xcalibur package used with Orbitraps. Note that posttranslational modification can sometimes cause the removal of atoms from the amino acid. Usually, this is a single hydrogen, but other groups can also be lost.

Acknowledgments

We thank Prof. Michał Dadlez and Jacek Olędzki for fruitful discussions.

References

1. Zlatanova J, Caiafa P, Van Holde K (2000) Linker histone binding and displacement: versatile mechanism for transcriptional regulation. FASEB J 14:1697–1704
2. McBryant SJ, Lu X, Hansen JC (2010) Multifunctionality of the linker histones: an emerging role for protein-protein interactions. Cell Res 20:519–528. doi:10.1038/cr.2010.35
3. Stasevich TJ, Mueller F, Brown DT, McNally JG (2010) Dissecting the binding mechanism of the linker histone in live cells: an integrated FRAP analysis. EMBO J 29:1225–1234. doi:10.1038/emboj.2010.24
4. Wisniewski JR, Zougman A, Krüger S, Mann M (2007) Mass spectrometric mapping of linker histone H1 variants reveals multiple acetylations, methylations, and phosphorylation as well as differences between cell culture and tissue. Mol Cell Proteomics 6:72–87. doi:10.1074/mcp.M600255-MCP200
5. Kotliński M, Rutowicz K, Kniżewski Ł et al (2016) Histone H1 variants in Arabidopsis are subject to numerous post-translational modifications, both conserved and previously unknown in histones, suggesting complex functions of H1 in plants. PLoS One 11:e0147908. doi:10.1371/journal.pone.0147908
6. Tyanova S, Temu T, Cox J (2016) The MaxQuant computational platform for mass spectrometry-based shotgun proteomics. Nat Protoc 11:2301–2319. doi:10.1038/nprot.2016.136
7. Neuhauser N, Michalski A, Cox J, Mann M (2012) Expert system for computer-assisted annotation of MS/MS spectra. Mol Cell Proteomics 11:1500–1509. doi:10.1074/mcp.M112.020271
8. Johns EW (1964) Studies on histones. 7. Preparative methods for histone fractions from calf thymus. Biochem J 92:55–59

Chapter 11

Profiling Nucleosome Occupancy by MNase-seq: Experimental Protocol and Computational Analysis

Alice Pajoro, Jose M. Muiño, Gerco C. Angenent, and Kerstin Kaufmann

Abstract

Nucleosomes are the basic repeating units of eukaryotic chromatin. They play important roles in chromatin compaction and gene regulation. Therefore, it is important to profile the in vivo locations of nucleosomes in the genome. Here we illustrate how to profile nucleosome occupancy at genome-wide scale using micrococcal nuclease (MNase) digestion combined with high throughput Illumina sequencing (MNase-seq). Nucleosome-associated DNA is relatively insensitive to digestion by micrococcal nuclease (MNase). Upon mild MNase treatment, the undigested nucleosomal DNA can be purified and sequenced allowing a precise localization of in vivo nucleosomes at a genome-wide level.

Key words Nucleosome occupancy, MNase-seq, MNase digestion, Next-generation sequencing, Chromatin

1 Introduction

In the eukaryotic nucleus, DNA is packed into chromatin and therefore, all functions of the genome occur in the context of the chromatin, whose fundamental building block is the nucleosome. A nucleosome is constituted of 147 bp of DNA wrapped around a histone octamer, formed by two proteins each of the core histones H2A, H2B, H3 and H4, and one linker histone H1, which is critical for higher-order compaction of the chromatin [1]. Nucleosome stability and compaction are regulated by the combined effects of nucleosome-positioning sequences, histone chaperones, ATP-dependent nucleosome remodelers, histone variants and post-translational histone modifications. Dynamic post-transcriptional modifications of histones and histone tails, such as methylation, acetylation or ubiquitination, lead to a more or less compact chromatin structure that affects gene expression. ATP-dependent nucleosome remodelers can also modulate nucleosome position and local chromatin structure by sliding histone octamers, altering histone–DNA interactions or changing histone variants [2]. The

Marian Bemer and Célia Baroux (eds.), *Plant Chromatin Dynamics: Methods and Protocols*, Methods in Molecular Biology, vol. 1675, DOI 10.1007/978-1-4939-7318-7_11, © Springer Science+Business Media LLC 2018

use of endonuclease properties combined with next-generation sequencing allows detailed profiling of chromatin structure at a genome-wide scale [3]. Here we illustrate how to profile the distribution of nucleosomes using micrococcal nuclease (MNase) digestion combined with high throughput Illumina sequencing, MNase-seq.

Nucleosome-associated DNA is relatively insensitive to digestion by MNase, that makes this enzyme particularly useful for genome-wide profiling of nucleosome occupancy [4, 5]. MNase induces single-strand breaks and subsequently double stranded ones in DNA positions that are not covered by a histone octamer. It also has exonuclease activity, so after the double strand breaks are introduced, the DNA is digested until it reaches protected DNA, such as in a nucleosome [5]. Next, the protected DNA can be identified by high-throughput sequencing and the obtained sequences analyzed with appropriate bioinformatic tools to obtain the in vivo genome-wide location of nucleosomes [6–9].

We present a complete workflow for an experimental procedure and bioinformatics analysis to allow wet-lab biologists to perform and analyze their MNase-seq experiments (Fig. 1). This involves a nuclei isolation step to obtain native nuclear chromatin [10], followed by an MNase digestion step to obtain a partially digested sample. It is crucial to precisely determine the right amount of MNase enzyme, because a too extensive digestion of the chromatin will cause complete digestion of the DNA. Since the amount of DNA is difficult to quantify in native chromatin, it is impossible to determine a priori the right amount of enzyme, and we therefore perform an MNase titration step dividing the sample in multiple aliquots and adding increased amount of enzyme to each aliquot.

In our protocol, we select mononucleosomal DNA before sequencing. This step is important in case of a single-reads run to retrieve the position of a single nucleosome. If paired-end sequencing is used, the mononucleosomal DNA selection step can be omitted since the original fragment length can be extrapolated from the sequenced reads [11]. Another problem to take into account is the endonuclease DNA sequence preference. MNase, as many other endonucleases, has a sequence bias in DNA recognition and enzymatic activity [12]. To overcome this problem, we generate a "control sample" where we digest naked genomic DNA (DNA without any protein bound to it) with MNase. The "control sample" is then used in the data analysis step to correct for the sequence preference.

Finally, sequence reads obtained by an Illumina sequencer can be mapped to the reference genome using one of the many available mapping tools (e.g., Bowtie [13], SOAP [14], and BWA [15]). After mapping the reads to the reference genome using a mapping tool, we retrieve nucleosome position using DANPOS2 [7]. DANPOS2 also allows for quantitative comparison between samples;

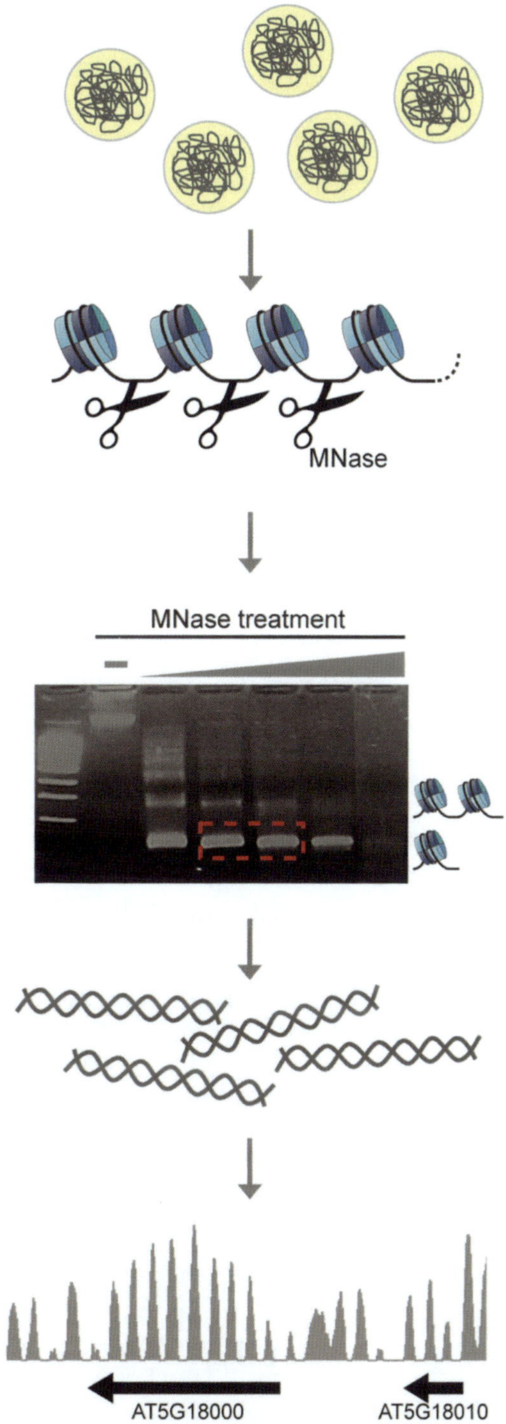

Fig. 1 Experimental workflow. After nuclei isolation the chromatin is subjected to MNase digestion. This enzyme preferentially cleaves linker DNA, leaving nucleosomal DNA intact. The chromatin is digested with different concentrations of MNase, the mononucleosomal DNA fraction is purified from gel and sequenced, leading to identification of nucleosome occupancy along the genome.

with this option, it is possible to study changes in nucleosome occupancy between difference genotypes, developmental stages or treatments.

2 Materials

2.1 Tissue Collection

1. Liquid nitrogen.
2. Nitrile gloves.
3. 50 ml centrifuge tubes.
4. Jeweler's forceps.

2.2 Nuclei Isolation

1. Mortar and pestle.
2. Liquid nitrogen.
3. Nylon mesh (55 μm pore size).
4. Nitrile gloves.
5. Glass funnel, recommended size 75 mm funnel top.
6. 50 ml centrifuge tubes.
7. 2.5 M sucrose.
8. Percoll.
9. HBM buffer: 25 mM Tris–HCl pH 7.5, 0.44 M sucrose, 10 mM $MgCl_2$, 10 mM β-mercaptoethanol, 0.1% Triton, and 2 mM spermine. Prepare 10 ml of buffer per sample freshly before use.
10. HBB buffer (fresh): 25 mM Tris–HCl pH 7.5, 0.44 M sucrose, 10 mM $MgCl_2$, 10 mM β-mercaptoethanol, and 0.1% Triton. Prepare 10 ml of buffer per sample freshly before use.
11. HBC buffer (fresh): 25 mM Tris–HCl pH 7.5, 0.44 M sucrose, 10 mM $MgCl_2$, 10 mM β-mercaptoethanol, 0.1% triton, 20% glycerol. Prepare 10 ml of buffer per sample freshly before use.

2.3 MNase Digestion

1. Nitrile gloves.
2. Wash buffer: 1 mM Tris–HCl pH 7.5, 5 mM $MgCl_2$, 60 mM KCl, 0.5 mM DTT, 15 mM NaCl, and 300 mM sucrose. Prepare 10 ml of buffer per sample freshly before use.
3. Reaction Buffer: 1 mM Tris–HCl pH 7.5, 1 mM CaCl, 60 mM KCl, 0.5 mM DTT, 15 mM NaCl, and 300 mM sucrose. Prepare 10 ml of buffer per sample freshly before use.
4. Lysis solution: 1% SDS, 50 mM Tris–HCl pH 8, 20 mM EDTA, and 10 mg/ml RNaseA. Prepare 10 ml of buffer per sample freshly before use.
5. Nuclease S7, Micrococcal nuclease, from *Staphylococcus aureus* (e.g., Roche).

6. 0.5 M EDTA pH 8.0 (adjusted to ~8.0 by the addition of NaOH).
7. Proteinase K (20 mg/ml) (e.g., Roche). This solution can be divided into 100 μl aliquots and stored at −20 °C.
8. Microcentrifuge tubes, 1.5 and 2 ml.

2.4 Mono-nucleosomal DNA Purification

1. Nitrile gloves.
2. Chloroform–isoamyl alcohol.
3. Isopropanol.
4. Ultrapure water.
5. Gel purification kit (e.g., MinElutegel extraction kit, QIAGEN).
6. PCR purification kit (e.g., QIAquick PCR purification kit, QIAGEN).
7. A multipurpose agarose.
8. 1× TE buffer: 1 mM EDTA pH 8 and 10 mM Tris–HCl pH 8.
9. Ethidium bromide 5 mg/ml.
10. Loading Dye.
11. Agarose gel running and visualization (UV) devices.
12. Smart Ladder small fragments (100 bp to 1 kb).
13. Sterile blades.
14. Low adhesion microcentrifuge tubes, 1.5 ml (e.g., Protein LoBind Tubes, Eppendorf).

2.5 DNA Library Preparation for Illumina Sequencing

1. Nitrile gloves.
2. Sequencing library preparation kit (e.g., TruSeq DNA sample preparation kit from Illumina).
3. Low adhesion microcentrifuge tubes, 1.5 ml (e.g., Protein LoBind Tubes, Eppendorf).

2.6 Illumina Sequencing

1. Nitrile gloves.
2. Fluorimetric-based DNA quantification assay (e.g., Qubit® dsDNA HS Assay Kit and fluorometer device, Invitrogen).
3. DNA quality check assay, device and reagents (ideally microcapillary-electrophoresis trace analyzer such as Agilent BioAnalyzer and Agilent BioAnalyzer DNA 1000 Kit, Agilent).
4. Elution buffer: 10 mM Tris–HCl pH 8.5 (e.g., from the MinElute Gel purification Kit, QIAGEN).
5. Next-Generation sequencing device and reagent kit (e.g., HiSeq from Illumina with Illumina sequencer reagent kit).

2.7 Equipment and Software for Bioinformatic Analysis

For the computational part of this protocol, a computer workstation running a Unix-based operating system or with access to a Unix-based server is needed. This protocol provides commands runnable in the Unix shell. Required software:

Bowtie 2: http://bowtie-bio.sourceforge.net/bowtie2/index.shtml.

DANPOS2: https://sites.google.com/site/danposdoc/.

3 Methods

3.1 Tissue Collection

Using a forceps, collect the plant tissue of interest into a 50 ml tube directly on liquid nitrogen. For one MNase-seq experiment generally 0.2–0.8 g of material is needed depending on the tissue used (for an example *see* **Note 1**).

3.2 Nuclei Isolation

The aim is to isolate high-quality, intact nuclei containing an intact chromatin. This requires careful handling at all stages particularly when nuclei are released from the cells: keep the samples on ice at all steps, avoid air bubbles, use cut-end tips, pipet slowly and carefully transfer the nuclei suspension into tubes (e.g., pipetting on the side walls by inclining the tubes), do not vortex the sample during the resuspension steps. Wear nitrile gloves for all steps. After step 1 (grinding), make sure to work in a fume hood since buffers contain β-mercaptoethanol. Before starting the protocol, cool-down all the buffers on ice and precool the centrifuge at 4 °C.

1. Grind the tissue in liquid nitrogen in a mortar until the tissue is completely homogeneous.
2. Transfer the ground tissue to a 50 ml tube (*see* **Note 2**).
3. Resuspend the homogeneous tissue in 2 ml of ice-cold HBM buffer (*see* **Note 3**).
4. Filter the resulting slurry through a 55 μm cloth mesh in a glass funnel, collect the flow-through in a 50 ml tube on ice. Wash the mesh with an additional 2 ml of ice-cold HBM buffer to collect all tissue (*see* **Note 4**).
5. Prepare 40% Percoll in HBM buffer (4 ml Percoll + 6 ml HBM buffer).
6. Prepare a 2.5 M sucrose–40% Percoll gradient in a 50 ml tube (*see* **Note 5**).
7. Pipet the filtrate gently on top of the sucrose–Percoll gradient, without disturbing the gradient (*see* Fig. 2).
8. Centrifuge at 1700 × *g* for 30 min at 4 °C.
9. The nuclei are now in the interface between the sucrose and the Percoll. Carefully remove and discard the Percoll phase leaving the pelleted nuclei undisturbed, then pipet the nuclei in a new 50 ml tube (*see* **Note 6**).

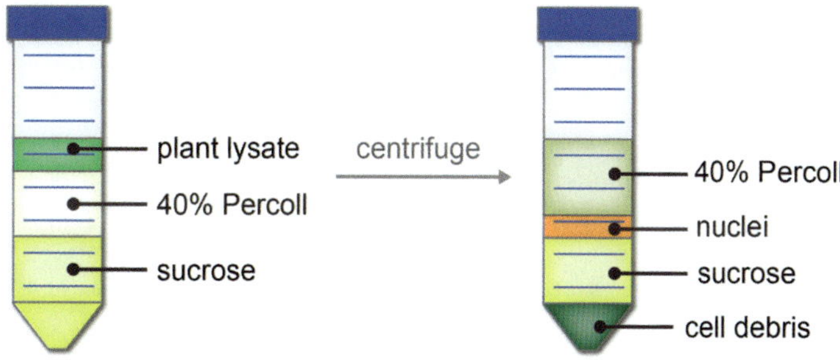

Fig. 2 Schematic drawing of a sucrose–Percoll gradient before and after the centrifugation step. After the centrifugation step the nuclei are collected in the interphase between the sucrose and the Percoll.

10. Resuspend the nuclear pellet in 10 ml of ice cold HBB buffer, and centrifuge at 1500 × *g* for 10 min at 4 °C.
11. Resuspend the nuclear pellet in 10 ml of ice cold HBC buffer, and centrifuge at 1500 × *g* for 10 min at 4 °C.

3.3 MNase Digestion

The aim is to obtain partially digested chromatin mainly consisting of mononucleosomal DNA. It is crucial to precisely determine the right amount of MNase enzyme, because a too extensive digestion of the chromatin will cause complete digestion of the DNA. Since the amount of DNA is difficult to quantify in native chromatin, it is impossible to determine a priori the right amount of enzyme, and we therefore perform an MNase titration step dividing the sample in multiple aliquots and adding increasing amounts of enzyme to each aliquot.

To avoid DNA degradation, wear nitrile gloves for all steps and be sure that your solutions are not contaminated with endonucleases. Resuspend the nuclei gently between the washing steps and do not vortex the sample. Before starting the protocol, cool-down all the buffers on ice. Precool the centrifuge at 4 °C before starting. Keep the samples on ice at all steps.

1. Resuspend the nuclei in 5 ml of ice-cold Wash buffer, and centrifuge at 1000 × *g* for 5 min at 4 °C.
2. Wash the nuclear pellet with 5 ml of ice-cold Reaction buffer, and centrifuge at 1000 × *g* for 5 min at 4 °C.
3. Resuspend the nuclei in 2 ml of ice-cold Reaction buffer.
4. Prepare a 1:20 dilution of S7 nuclease (5 μl S7 + 95 μl Reaction buffer).
5. Divide the sample into 10 aliquots of 200 μl each (*see* **Note 7**).
6. Add a different amount of diluted S7 nuclease to each aliquot: 0, 0.2, 0.4, 0.6, 0.8, 1, 1.5, 2, 3, or 5 μl. Keep the aliquots on ice while adding the enzyme.

7. Mix the samples gently.
8. Incubate at 37 °C for 10 min in a thermomixer gently shaking (500 rpm).
9. Stop the digestion by adding 23 μl of 0.5 M EDTA and incubate 5 min at room temperature.
10. Add 350 μl of Lysis solution and mix gently by inversion.
11. Add 10 μl of Proteinase K (20 mg/ml), mix gently and incubate at 37 °C overnight.

3.4 Mono-nucleosomal DNA Purification

The aim is to obtain pure mononucleosomal DNA for the following sequencing step. This step is crucial in case of a single end sequencing approach, while it can be skipped in case you adopt a paired end sequencing approach. In case of single end sequencing, it is essential to only have 150 bp length fragment (mononucleosomal DNA) to obtain a precise location of the nucleosomes in the sample. In case of a paired end sequencing approach, the original fragment length size can be retrieved from the sequencing output: for example, mononucleosomal DNA can be identified by selecting an insert size of 150 bp.

1. Centrifuge for 10 min at top speed in a microcentrifuge to spin down the cell debris.
2. Run 10 μl of each supernatant on a 1% agarose gel (*see* Fig. 1).
3. Select aliquots with mainly (80%) mononucleosomal DNA (*see* **Note 8**)
4. Transfer the supernatant (around 500 μl) to a new 1.5 ml tube.
5. Add 1 volume of chloroform–isoamyl alcohol and vortex briefly.
6. Centrifuge for 10 min at top speed.
7. Carefully transfer the upper phase to a 2 ml safe lock tube.
8. Precipitate the DNA by adding 0.9 volumes of isopropanol and incubate for 3 h at −20 °C.
9. Centrifuge for 30 min top speed at 4 °C.
10. Remove the supernatant and air dry the pellet for approximately 15 min, do not overdry the pellet, as this will reduce resuspension efficiency.
11. To resuspend the DNA, add 100 μl of ultrapure water.
12. Run 50 μl on a 2% agarose gel and cut the bands at 150 bp (mononucleosomal DNA size).
13. Purify the DNA from the gel using the purification kit. Pool all bands from the same sample in one tube for a single purification. Dissolve the gel at room temperature and after purification, elute the DNA with 18 μl of elution buffer in a 1.5 ml low-binding microfuge tube.

3.5 Preparation of the Control Sample

Most endonucleases exhibit a preference for particular DNA sequence contexts in their digestion activity. To correct for this bias in the enzyme digestion, we strongly advise to generate a control sample to be used in the data analysis. The control sample consists of naked DNA digested with MNase.

1. Perform a clean DNA extraction to obtain 2 μg of DNA (*see* **Note 9**).
2. Add 10 μl of Proteinase K (20 mg/ml) and incubate at 37 °C overnight (*see* **Note 10**).
3. Add 1 volume of Reaction buffer and divide into three aliquots.
4. Digest the DNA by adding 0 μl, 0.5 μl, and 1 μl of S7 nuclease 1:20.
5. Incubate at 37 °C for 5 min in a thermomixer.
6. Stop the digestion by adding 23 μl of 0.5 M EDTA and incubate 5 min at room temperature.
7. Purify the DNA using a nucleic acid purification kit. Elute in the smallest volume possible.
8. Load the purified DNA on a 2% agarose gel and cut a large band of 100–300 bp in size from the gel using a clean scalpel.
9. Purify the DNA from the gel using a gel purification kit. Dissolve the gel at room temperature and after purification elute the DNA with 18 μl of elution buffer in a 1.5 ml low-binding microfuge tube.

3.6 Library Preparation

We made use of the TruSeq DNA sample preparation kit from Illumina. However, any Illumina sequencing compatible kit can be used for the library preparation step. Work in a clean environment (e.g., treated with a nucleic acid decontamination solution or a laminar flow cabinet). Use nitrile gloves and filter tips at all steps to avoid contaminations. Prepare the sample for sequencing following the manufacturer's protocol (*see* **Note 11**).

3.7 Next-Generation Sequencing

Here we use an Illumina sequencing approach but alternatives are available. It is however critical to proceed with DNA concentrations and fragment size distributions determined with high accuracy. This is to date possible with fluorometer-based assays (e.g., QuBiT, Invitrogen) (spectrophotometer-based assays are not recommended) and a microcapillary electrophoresis trace analyzer (e.g., Bioanalyzer, Agilent), respectively. It is important to have libraries with equal concentration before combining them for the sequencing step; this will ensure a homogeneous distribution of sequencing reads between the different samples.

1. Measure the DNA concentration precisely (e.g., using the high-sensitive, HS Qubit, DNA fluorimetric assay) and the

average fragment size of the DNA (using the microfluidics chip, e.g., Agilent BioAnalyzer DNA chip). Calculate the molarity of the sample (*see* **Note 12**).

2. Dilute the sample to 10 nM using the elution buffer of the gel purification kit (*see* **Note 13**).
3. For a HiSeq2500 run, we advise to combine multiple samples into one lane. Pipet 5 μl of each 10 nM sample in a 1.5 ml low-binding microfuge tube.
4. Prepare the sequencing flow-cell following the provider's protocol (*see* **Note 14**).

3.8 Library Quality Check

Sequence reads obtained by an Illumina sequencer are normally provided in a FASTQ format. This format not only contains the read sequence but also gives information about the sequencing quality of the read. We advise to use the FastQC software [16] to check the sequence quality of the libraries. This software enables easy identification of common problems that can arise from either sequencing or library preparation. You can use FastQC for the sequence quality check in many Galaxy environments (e.g., http://galaxy.wur.nl/). For details on how to install and use the program, please refer to its manual: http://www.bioinformatics.babraham.ac.uk/projects/fastqc/Help/. Critical parameters that should be evaluated are the quality of the sequencing per nucleotide, and the percentage of Illumina adapters. In case of problems with the presence of adapter sequences, the reads need to be trimmed using software such as FASTAQ Trimmer in Galaxy. Low quality reads instead will not map to the reference genome.

3.9 Mapping the Sequencing Reads to a Reference Genome

Here we describe how to map the reads from an Illumina sequencing approach to a reference genome such as the Arabidopsis genome. If you conduct the experiment using a different species you can download the corresponding genome from Ensembl Plants.

1. Often the files obtained from the sequencer are compressed, generally as .gz files. In this case, unpack the file before further analysis: $ gunzip -c sample1A.fq.gz>sample1A.fq
2. Download the reference genome, if not already available. The Arabidopsis reference genome and other common plant genomes can be downloaded from Ensembl plants (http://plants.ensembl.org).
   - Go to the website of Ensembl Plant (http://plants.ensembl.org/).
   - Go to "Downloads" at the top of the page.
   - Select "Download data via FTP" under "Download databases & software".

- Search for the organism of interest (e.g., *Arabidopsis thaliana*).
- Click on "FASTA (DNA)" of the species of interest.
- Download the file ending with ".dna.genome.fa.gz"; this represents the unmasked genome sequence.
- Move the downloaded file (e.g., Arabidopsis_thaliana. TAIR10.dna.genome.fa.gz) to the working directory.
- Unzip the file.

3. There are many programs developed for fast and accurate alignment of reads to a reference genome, in our protocol, Bowtie2 [13] is used. Bowtie2 is commonly used software for Illumina sequence mapping and is often available in Galaxy environments. If you use the software in a Unix environment, please follow these steps:
   - Like other short read aligners, Bowtie2 uses index files of the reference genome for fast alignment of reads to the genome.

```
$ bowtie2-build Arabidopsis_thaliana.TAIR10.
dna.genome.fa Arabidopsis_thaliana.TAIR10.dna.genome.
fa.index
```

   - To perform the mapping to the reference genome use (*see* **Note 15**):

```
$ bowtie2 -x Arabidopsis_thaliana.TAIR10.dna.
genome.fa.index -U sample1A.fastq -S TAIR10_mapped_sam-
ple1A.bowtie
```

   - An alignment report will be available to check the percentage of mapped reads.

3.10 Determine Nucleosome Positions Using DANPOS2

After read mapping, we use DANPOS2 for detection of genome-wide nucleosome locations in each sample and comparison of nucleosome positions between samples. For details on how to install and use the program, please refer to: https://sites.google.com/site/danposdoc/. A description of all the optional arguments is available at the web page.

1. Group the mapping files from biological replicates of the same condition in a folder (e.g., sample1/; sample2/; control/).
2. If you would only like to define nucleosome positions for a single MNase-seq sample with a specific effect to be subtracted (e.g., the digested naked DNA sample, *see* Subheading 3.5), use the following command line code:

```
$ python danpos.py dtriple sample1 -b control -o
sample1_nucleosomes
```

3. If you would like to compare MNase-seq data between two samples (e.g., sample1 and sample2), where each sample has a specific effect to be subtracted (e.g., the digested naked DNA sample, *see* Subheading 3.5):

```
$ python danpos.py dtriple sample1:sample2 -b sample1:
control,sample2:control -o sample1_sample2_nucleosomes
```

4. In a new folder (specified by the optional argument *–o*), you will find the output of the analysis (*see* **Note 16**).

3.11 Nucleosome Profile Around the Genomic Features Using DANPOS2

The function "profile" of DANPOS2 can be used to obtain nucleosome profiles around genomic features, such as the transcription start site (TSS) or the transcription termination site (TTS), of a subset of genes of interest (*see* **Note 17**). For example, the following command can be used to obtain the average of nucleosome occupancy around the TSS of all genes annotated in the Arabidopsis genome (Fig. 3).

```
$ python danpos.py profile <wiggle_file_path>/sample1.wig -
genefile_paths TAIR10.txt -genomic_sites TSS
```

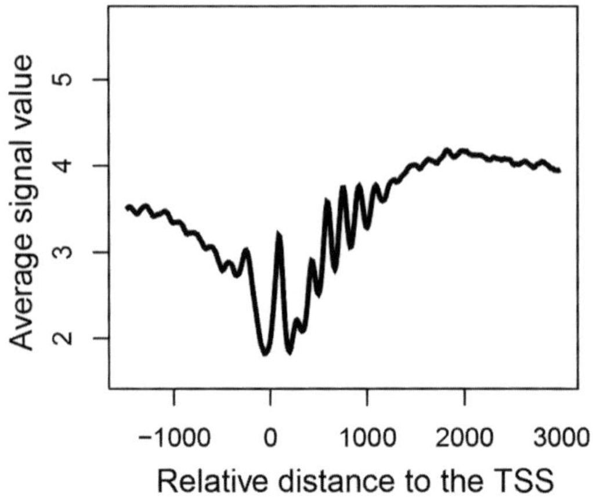

Fig. 3 Nucleosome density is plotted as a function of distance relative to the transcription start site (TSS), using the function "profile" of DANPOS2. When nucleosome density is plotted as a function of distance relative to the transcription start site (TSS), a − 1 nucleosome is positioned close to the TSS from −307 to −111 bp, a + 1 nucleosome is positioned from −5 to +144 bp and multiple nucleosomes are well positioned along the gene body. Moreover, two nucleosome free regions (NFR) are found before and after the +1 nucleosome.

4 Notes

1. For example, in our experiments we used 0.2 g of inflorescence meristem tissue of the ***ap1 cal*** mutant, while 0.8 g of tissue for wild type inflorescences. We advise to weight the tube before and after collection to estimate the amount of material. After this step, the tissue can be stored for several months at −80 °C.
2. Ground tissue can be stored at −80 °C for up to 2 days.
3. It may take some time (10 min) to resuspend the tissue, so keep the samples as cool as possible.
4. Carefully squeeze the membrane using a pipette tip to recover as much sample as possible.
5. The sucrose–Percoll gradient can be prepared in a 50 ml tube adding 10 ml of sucrose in the first layer and 10 ml of 40% Percoll in the top layer. We advise to pipet as slowly as possible to allow the formation of a clear gradient. In case of small amounts of starting material (0.2 g), the gradient can be prepared in a 10 ml tube with 3 ml of 2.5 M sucrose and 3 ml of 40% Percoll.
6. If the nuclei are not pelleted in the interphase between Percoll and sucrose, you may need to test different centrifugation speeds and/or times.
7. If only a small amount of starting material is available, we advise to resuspend the nuclei in 1 ml of reaction buffer and then to divide it into 5 aliquots.
8. The right aliquot is the aliquot where a band of dinucleosomal DNA is still present (*see* Fig. 1). More than one aliquot can be selected.
9. You can also use the 0 μl aliquot of **step 6** from Subheading 3.3 and perform a standard DNA purification with that. In that case, **step 2** can be omitted.
10. This step is important to ensure the digestion of all the proteins that may still be present on the DNA.
11. Make sure to select compatible indexed adapters if you want to pool together multiple samples in the sequencing step.
12. The molarity of the sample can be calculated using the following formula: {(concentration of the sample in pg/ul)/[(the average fragment length in bp) × 660]} × 10^3 = nM/l. Where 660 represents the Average weight of a DNA base pair.
13. The 10 nM library can be stored at −20 °C for several months.
14. For Arabidopsis, 50 bp single-end sequencing can be used. However, the optimal read-length and preferred sequencing

depth differ between species. Longer sequences or paired-end sequences will improve the mapping efficiency of the sample.

15. -U is used only to map single-end reads. For paired-ends reads, you must use the -1 and -2 options.
16. Nucleosome occupancy for the genomic locus of interest can be visualized loading the .wig file in a genome browser. We advise IGB (http://bioviz.org/igb/index.html).
17. For this a gene paths file is required (e.g., TAIR10.txt). This file contains a set of genomic elements with information about their location in the genome. The "TAIR10.txt" file was obtained from the Arabidopsis gene annotation file, using the following commands in R:

```
> gff< -read.table("Arabidopsis_thaliana.TAIR10.22.
gff3", header = FALSE, sep="\t", quote="")
> gff<-gff[gff$V3=="gene",]
> write.table(gff,file="TAIR10.txt")
```

The Arabidopsis gene annotation file (Arabidopsis_thaliana. TAIR10.22.gff3.gz) can be downloaded from Ensembl plants (http://plants.ensembl.org).

References

1. Noll M (1974) Subunit structure of chromatin. Nature 251(5472):249–251. doi:10.1038/251249a0
2. Clapier CR, Cairns BR (2009) The biology of chromatin remodeling complexes. Annu Rev Biochem 78:273–304. doi:10.1146/annurev.biochem.77.062706.153223
3. Zentner GE, Henikoff S (2014) High-resolution digital profiling of the epigenome. Nat Rev Genet 15(12):814–827. doi:10.1038/nrg3798
4. Axel R (1975) Cleavage of DNA in nuclei and chromatin with Staphylococcal nuclease. Biochemistry 14(13):2921–2925. doi:10.1021/Bi00684a020
5. Sulkowski E, Laskowski M (1962) Mechanism of action of micrococcal nuclease on deoxyribonucleic acid. J Biol Chem 237(8):2620
6. Becker J, Yau C, Hancock JM, Holmes CC (2013) NucleoFinder: a statistical approach for the detection of nucleosome positions. Bioinformatics 29(6):711–716. doi:10.1093/bioinformatics/bts719
7. Chen K, Xi Y, Pan X, Li Z, Kaestner K, Tyler J, Dent S, He X, Li W (2013) DANPOS: dynamic analysis of nucleosome position and occupancy by sequencing. Genome Res 23(2):341–351. doi:10.1101/gr.142067.112
8. Polishko A, Bunnik EM, Le Roch KG, Lonardi S (2014) PuFFIN – a parameter-free method to build nucleosome maps from paired-end reads. BMC Bioinformatics 15. doi:10.1186/1471-2105-15-S9-S11
9. Woo SS, Zhang XK, Sauteraud R, Robert F, Gottardo R (2013) PING 2.0: an R/bioconductor package for nucleosome positioning using next-generation sequencing data. Bioinformatics 29(16):2049–2050. doi:10.1093/bioinformatics/btt348
10. Zhang XY, Clarenz O, Cokus S, Bernatavichute YV, Pellegrini M, Goodrich J, Jacobsen SE (2007) Whole-genome analysis of histone H3 lysine 27 trimethylation in Arabidopsis. PLoS Biol 5(5):1026–1035. doi:10.1371/journal.pbio.0050129
11. Carone BR, Hung JH, Hainer SJ, Chou MT, Carone DM, Weng ZP, Fazzio TG, Rando OJ (2014) High-resolution mapping of chromatin packaging in mouse embryonic stem cells and sperm. Dev Cell 30(1):11–22. doi:10.1016/j.devcel.2014.05.024
12. Chung HR, Dunkel I, Heise F, Linke C, Krobitsch S, Ehrenhofer-Murray AE, Sperling SR, Vingron M (2010) The effect of micrococcal nuclease digestion on nucleosome positioning

data. PLoS One 5(12):e15754. doi:10.1371/journal.pone.0015754

13. Langmead B, Trapnell C, Pop M, Salzberg SL (2009) Ultrafast and memory-efficient alignment of short DNA sequences to the human genome. Genome Biol 10(3):R25. doi:10.1186/Gb-2009-10-3-R25
14. Li RQ, Yu C, Li YR, Lam TW, Yiu SM, Kristiansen K, Wang J (2009) SOAP2: an improved ultrafast tool for short read alignment. Bioinformatics 25(15):1966–1967. doi:10.1093/bioinformatics/btp336
15. Li H, Durbin R (2009) Fast and accurate short read alignment with Burrows-Wheeler transform. Bioinformatics 25(14):1754–1760. doi:10.1093/bioinformatics/btp324
16. Babraham Bioinformatics FASTQC: a quality control tool for high througput sequencing data. http://wwwbioinformaticsbabrahama-cuk/projects/fastqc/

Chapter 12

Identification of Open Chromatin Regions in Plant Genomes Using ATAC-Seq

Marko Bajic, Kelsey A. Maher, and Roger B. Deal

Abstract

Identifying and characterizing highly accessible chromatin regions assists in determining the location of genomic regulatory elements and understanding transcriptional regulation. In this chapter, we describe an approach to map accessible chromatin features in plants using the Assay for Transposase-Accessible Chromatin, combined with high-throughput sequencing (ATAC-seq), which was originally developed for cultured animal cells. This technique utilizes a hyperactive Tn5 transposase to cause DNA cleavage and simultaneous insertion of sequencing adapters into open chromatin regions of the input nuclei. The application of ATAC-seq to plant tissue has been challenging due to the difficulty of isolating nuclei sufficiently free of interfering organellar DNA. Here we present two different approaches to purify plant nuclei for ATAC-seq: the INTACT method (Isolation of Nuclei TAgged in specific Cell Types) to isolate nuclei from individual cell types of the plant, and tissue lysis followed by sucrose sedimentation to isolate sufficiently pure total nuclei. We provide detailed instructions for transposase treatment of nuclei isolated using either approach, as well as subsequent preparation of ATAC-seq libraries. Sequencing-ready ATAC-seq libraries can be prepared from plant tissue in as little as one day. The procedures described here are optimized for *Arabidopsis thaliana* but can also be applied to other plant species.

Key words ATAC-seq, INTACT system, Chromatin, Nucleus, Transposition, Nucleosome, Transcription factor, Enhancer

1 Introduction

Plants are sessile organisms that must precisely regulate their transcription in response to their environment, as well as for proper development, growth, and homeostasis. Transcription is associated with regions of relatively open chromatin, in which *cis*-regulatory elements such as enhancers and promoters can recruit transcription factors and RNA polymerase II to transcribe DNA [1]. Binding of transcription factors to DNA generally results in the depletion of nucleosomes, rendering these regions hypersensitive to nucleases. Characterizing such regulatory regions throughout the genome has therefore relied on methods that combine enzymatic digestion of nuclear DNA and high-throughput sequencing, such as

Marian Bemer and Célia Baroux (eds.), *Plant Chromatin Dynamics: Methods and Protocols*, Methods in Molecular Biology, vol. 1675, DOI 10.1007/978-1-4939-7318-7_12, © Springer Science+Business Media LLC 2018

microccocal nuclease sequencing (MNase-seq, *see* Chapter 10 [2]) and DNase I hypersensitivity sequencing (DNase-seq) [3, 4]. Alternatively, regulatory regions can be inferred by chromatin immunoprecipitation sequencing (ChIP-seq, *see* Chapter 5 [5]) where antibodies are used to pull down transcription factors or histone marks associated with active transcription [6].

An improved method for identifying accessible regions of chromatin and transcription factor binding is the Assay for Transposase-Accessible Chromatin with high-throughput sequencing (ATAC-seq) [7, 8]. This method uses a hyperactive Tn5 transposase to integrate preloaded sequencing adapters into regions of open chromatin (Fig. 1a). ATAC-seq is a fast protocol with simple library amplification steps and requires very small amounts of starting material, making it a vast improvement over alternative methods. However, a drawback of this protocol is that the hyperactive Tn5 transposase also targets sources of extranuclear genetic material, including the genomes of mitochondria and chloroplasts. This decreases the proportion of reads that map to the nuclear genome, reducing the amount of information that can be used to identify regulatory regions of open chromatin. Such extranuclear reads must be discarded at the start of the data analysis process, diminishing the efficiency of the assay both in terms of cost and in effective use of materials. To gain the maximum efficiency of this powerful procedure, input material free from extranuclear genetic material, such as purified nuclei, is the ideal input for ATAC-seq.

In this chapter, we describe the use of two different methods to isolate either total nuclei from tissues or nuclei from specific cell types of *Arabidopsis thaliana* (Fig. 1b). To isolate total nuclei from plant tissue we use extraction buffers with a non-ionic detergent to lyse organelles, followed by sucrose sedimentation to further purify the nuclei [9]. This method of nuclei isolation can be done in any lab on most plant tissues. However, these partially purified nuclei still contain some organellar DNA in addition to nuclear DNA, which reduces the efficiency of Tn5 transposition and results in fewer sequencing reads that map to nuclear DNA. In addition, we describe the Isolation of Nuclei TAgged in specific Cell Types (INTACT) method to isolate nuclei from tissue or from specific cell types [10]. This system uses two transgenes for nuclear targeting for affinity purification: (1) the Nuclear Tagging Fusion (NTF) construct, which encodes a fusion of WPP nuclear envelope-targeting domain, a green fluorescent protein (GFP), and the biotin ligase recognition peptide (BLRP); and (2) an *E. coli* biotin ligase (BirA), which biotinylates the BLRP tag. The BirA is expressed from a constitutive promoter while the NTF is expressed either from a constitutive or cell type-specific promoter. The specificity of the NTF promoter determines which cell types will have biotinylated nuclei, and can then be isolated by affinity purification with streptavidin-coated magnetic beads [11]. A key advantage of the

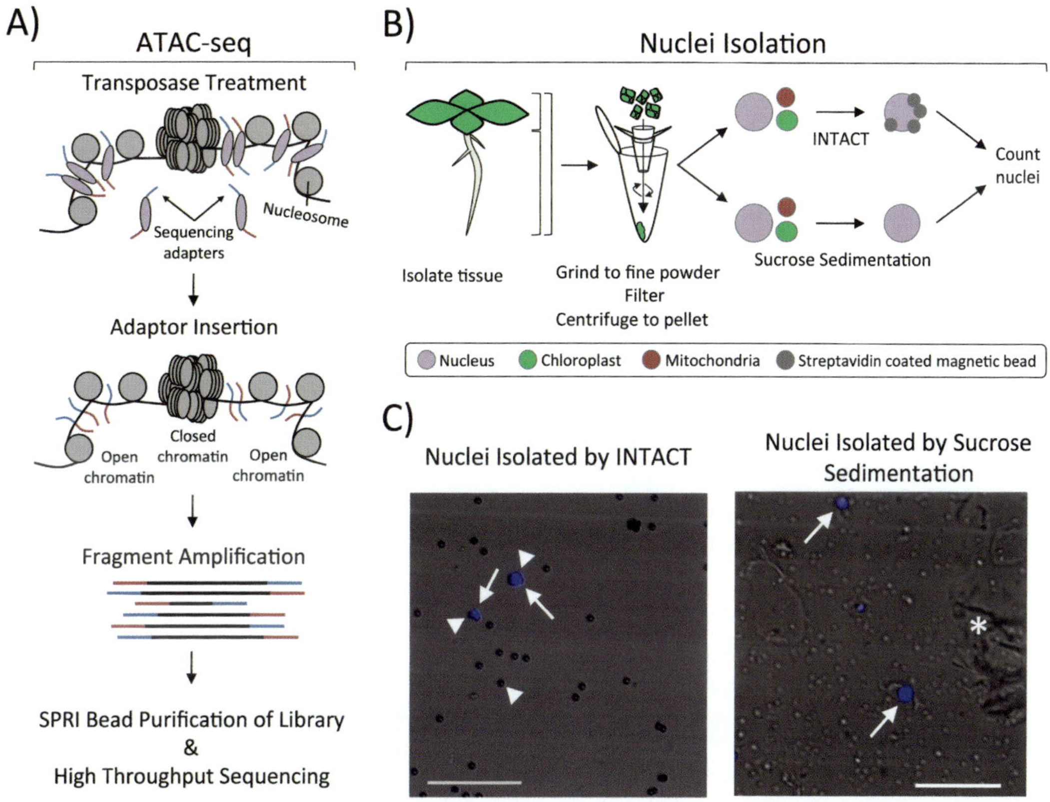

Fig. 1 ATAC-seq profiling using nuclei isolated by INTACT or sucrose sedimentation. (**a**) Overview of the ATAC-seq procedure. Nuclei are incubated with sequencing adapter-loaded Tn5 transposase, which diffuses into the nucleus to interact with chromatin. Sequencing adapters are inserted into open chromatin regions, and the fragmented DNA is amplified wherever the sequencing adapters were inserted. This generates a library of DNA fragments in which each end represents an insertion site. The amplified libraries are purified and sequenced with next-generation sequencing. (**b**) Two different methods for purifying nuclei from *Arabidopsis* can be used: (1) INTACT for isolating nuclei from specific cell types, and (2) sucrose sedimentation to isolate total nuclei from input tissue. The two methods have the same initial steps: tissue is collected from a specific part of the plant (root, leaf, or the entire plant), ground to a fine powder, resuspended, filtered, and centrifuged to pellet nuclei and cellular debris. Nuclei isolation using tissue that expresses INTACT transgenes uses streptavidin coated magnetic beads to affinity-purify biotinylated nuclei out of the resuspended pellet. This allows for the isolation of nuclei from specific cell types that express the Nuclear Tagging Fusion (NTF) protein and the biotin ligase BirA, resulting in very low contamination by organellar genomes. Alternatively, total nuclei can be isolated from tissue by resuspending the nuclei/debris pellet in a buffer with Triton X-100 to lyse organelles and centrifuging through a dense sucrose layer. Nuclei isolated from both procedures are stained with DAPI and quantified using a hemocytometer. (**c**) Fluorescent microscope images of nuclei (*white arrows*) stained with the DNA-binding dye DAPI (*blue*) isolated either through INTACT or sucrose sedimentation. INTACT isolated nuclei are identified by their DAPI fluorescence and binding to multiple beads (*white arrowhead*). Beads are easily visualized by increasing the transmission of white light while viewing the nuclei in the DAPI channel. Sucrose sedimentation isolated nuclei (*white arrows*) are DAPI-stained objects around 4–6 μm in diameter, although they can vary in size and shape depending on starting tissue. Much more cellular debris (*white asterisk*) is observed in sucrose sedimentation-isolated nuclei as compared to INTACT-purified nuclei, but this should not impact the procedure described here. Each picture contains a 50 μm scale bar shown at the bottom

INTACT approach is not only that the isolated nuclei have less organellar DNA contamination, but also that this method can be used to selectively isolate nuclei from specific cell types. While INTACT is a powerful technique, it does require that stable transgenic lines containing BirA and NTF cassettes for the cell type of interest are available, which are time consuming to generate and can be limiting for many species. Even so, the protocol described here, particularly ATAC-seq using sucrose sedimentation-purified nuclei, can readily be adapted for chromatin profiling in any plant species.

2 Materials

2.1 Equipment

1. Porcelain 50 mL mortar and pestle, or equivalent.
2. Liquid nitrogen.
3. Metal lab spoon.
4. Magnetic rack for 1.5 mL tubes (e.g., DynaMag 2, Life Technologies).
5. Magnetic rack for 15 mL tubes (e.g., DynaMag 15, Life Technologies).
6. 96-well magnetic separator plate (e.g., MagWell, EdgeBio).
7. Nylon cell strainers with 70 μm pores.
8. Long-stem analytical funnel.
9. Pipet-Aid.
10. Sterile 10 mL plastic serological pipettes.
11. Microfuge tubes, 1.5 mL (e.g., Eppendorf).
12. PCR tubes, 0.2 mL.
13. Falcon tubes, 15 mL and 50 mL.
14. Nutator platform rotator.
15. Hemocytometer (e.g., Hausser Bright Line hemocytometer, Fisher Scientific).
16. Microcentrifuge and refrigerated centrifuge with rotor for 15 mL tubes.
17. Cold room, 4 °C.
18. Molecular biology grade water.
19. Sterile disposable filter unit, 500 mL.
20. Sterile 0.2 μm syringe filter.
21. Sterile 10 mL plastic syringe.
22. Thermal cycler.
23. Real-time PCR machine.

24. A 64-bit computer with at least 1 TB hard disk and 16 Gb of memory for ATAC-seq data analysis.
25. Fluorescent microscope equipped for GFP detection and a 40× magnification objective or more.

2.2 Stock Solutions and Reagents

All solutions are done in sterile, ultrapure purified water unless indicated.

1. Complete, EDTA-free Protease Inhibitors (e.g., Roche).
2. Stock solution of 2 M spermidine. Prepare by dissolving 2.904 g spermidine powder in 10 mL water. Aliquot 1 mL per 1.5 mL Eppendorf tube and store at −20 °C.
3. Stock solution of 200 mM spermine. Prepare by dissolving 0.4047 g spermine powder in 10 mL of water. Aliquot 1 mL per 1.5 mL Eppendorf tube and store at −20 °C.
4. Stock solution (1 L) of incomplete Nuclei Purification Buffer (NPBi): 20 mM MOPS, 40 mM NaCl, 90 mM KCl, 2 mM EDTA, 0.5 mM EGTA, adjusted to pH 7 with 2 M KOH. Filter-sterilize the solution and degas under vacuum for 10 minutes. Store at 4 °C for up to 3 months.
5. Stock solution of 10% Triton X-100.
6. Stock solution of 10× DAPI. Prepare by dissolving 10 mg DAPI powder in 5 mL water, for a final concentration of 2 μg/μL. Filter-sterilize the solution and store at 4 °C in the dark for several months. To stain nuclei with DAPI, dilute the 10× DAPI solution to 1× using water (final concentration of 0.2 μg/μL), and use within 2–3 h.

2.3 Purification of Tagged Nuclei Using INTACT

1. Plant material: tissue from transgenic plants expressing both NTF and BirA in the cell type of interest. INTACT transgenic lines targeting the root epidermal hair and non-hair cell types, as well as INTACT plasmid vectors are available from the Arabidopsis Biological Resource Center at Ohio State University.
2. Magnetic beads coupled to Streptavidin, 2–3 μm diameter (e.g., M-280 Streptavidin Dynabeads, Life Technologies).
3. Protease inhibitor cocktail (e.g., Complete protease inhibitor, Roche).
4. Nuclei Purification Buffer (NPB): 20 mM, 40 mM NaCl, 90 mM KCl, 2 mM EDTA, 0.5 mM EGTA, 0.5 mM spermidine, 0.2 mM spermine, and 1× Roche Complete protease inhibitors, adjusted to pH 7 with 2 M KOH. Prepare by adding spermidine, spermine, and the protease inhibitors to NPBi just before starting the INTACT nuclei purification procedure. Keep solution on ice, and use within 1 h of preparation.

5. Nuclei Purification Buffer containing 0.1% Triton X-100 (NPBt): 20 mM MOPS pH 7, 40 mM NaCl, 90 mM KCl, 2 mM EDTA, 0.5 mM EGTA, 0.5 mM spermidine, 0.2 mM spermine, and 0.1% (v/v) Triton X-100. Prepare by adding spermidine, spermine, and Triton X-100 to NPBi just before starting the INTACT nuclei purification procedure. Keep solution on ice, and use within 1 day of preparation.

2.4 Purification of Total Nuclei Using Sucrose Sedimentation

1. Plant material: fresh or frozen plant tissue.
2. Stock solution of 1 M Tris–HCl pH 8.
3. Stock solution of 1 M $MgCl_2$.
4. Stock solution of 2 M sucrose.
5. Nuclei Purification Buffer (NPB): 20 mM MOPS pH 7, 40 mM NaCl, 90 mM KCl, 2 mM EDTA, 0.5 mM EGTA, 0.5 mM spermidine, 0.2 mM spermine, and 1× Roche Complete protease inhibitors. Prepare by adding spermidine, spermine, and Roche Complete protease inhibitors to NPBi just before starting the nuclei purification procedure. Keep solution on ice, and use within 1 h of preparation.
6. Nuclei Extraction Buffer 2 (NEB 2): 0.25 M Sucrose, 10 mM Tris–HCl pH 8, 10 mM $MgCl_2$, 1% Triton X-100, and 1× Roche Complete Protease Inhibitors. Prepare solution just before use, keep on ice, and use within 1 h of preparation.
7. Nuclei Extraction Buffer 3 (NEB 3): 1.7 M Sucrose, 10 mM Tris–HCl pH 8, 2 mM $MgCl_2$, and 0.15% Triton X-100, 1× Roche Complete Protease Inhibitors. Prepare solution just before use, keep on ice, and use within 1 h of preparation.

2.5 Tagmentation of Chromatin by Tn5 Transposase

1. Tagmentation-based library preparation kit (e.g., Nextera, Illumina, or equivalent).
2. PCR purification kit (e.g., MiniElute, Qiagen).
3. Nuclease-free water (e.g., Sigma, Ambion).
4. Nucleic acid decontamination detergent (e.g., DNA AWAY, or equivalent).

2.6 Sequencing Library Preparation

1. ATAC Primer 1 (AATGATACGGCGACCACCGAGATCTACACTCGTCGGCAGCGTCAGATGTG).
2. ATAC barcoded Primer 2 (CAAGCAGAAGACGGCATACGAGAT*NNNNNNNN*GTCTCGTGGGCTCGGAGATGT); N's indicate the 8-base index sequence. Each library to be pooled for sequencing should be amplified with a different barcoded primer 2. *See* Supplementary Table 1 of [7] for all primer sequences.
3. High-Fidelity PCR master mix for library amplification (e.g., NEBNext High-Fidelity 2× PCR Master Mix, NEB).

4. Nucleic acid fluorescent dyes for qPCR (e.g., Solution of 20× EvaGreen dye, Biotium).
5. Passive fluorescence dye for qPCR normalization (e.g., Solution of 50× ROX dye, Invitrogen).
6. PCR Purification kit (e.g., MinElute, Qiagen).
7. PCR purification magnetic beads (e.g., Agencourt AMPure XP, Beckman Coulter).
8. 100% ethanol.
9. Horizontal electrophoresis gel box and power source.
10. A 302 nm ultraviolet transilluminator.
11. Library Quantification kit (e.g., NEBNext, NEB).

3 Methods

Users should either begin at Subheading 3.1 for affinity purification of nuclei using INTACT, or at Subheading 3.2 for isolation of total nuclei. In either case, the purified nuclei are used for tagmentation by Tn5 transposase in Subheading 3.3. All procedures are carried out at room temperature (25 °C) unless otherwise specified.

3.1 Purification of Tagged Nuclei Using INTACT

Purification of nuclei by INTACT gives access to cell type-specific nuclei and provides the highest purity nuclei with the least organelle contamination. However, established transgenic lines are required.

1. Excise the tissue (3 g of roots or 0.5 g of leaves) into a weigh boat on ice until the desired amount has been collected and weighed.
2. Grind the tissue to a fine powder in liquid nitrogen using a mortar and pestle.
3. Using a nitrogen-cooled metal lab spoon, quickly transfer the frozen tissue powder to another mortar containing 10 mL of ice-cold Nuclei Purification Buffer (NPB).
4. Thoroughly resuspend the powder in NPB by grinding it with a new, clean pestle (*see* **Note 1**).
5. Place a 70 μm nylon cell strainer in the center of a long-stemmed funnel and above a 15 mL tube on ice.
6. Use a 10 mL serological pipette to draw up the tissue suspension and filter it through the strainer. Collect the flow-through in the 15 mL tube.
7. Spin down the nuclei at 1,200 × *g* for 10 min at 4 °C.
8. Use a 10 mL serological pipet and then a 1 mL pipette tip as needed to carefully remove as much of the supernatant as possible without disturbing the pellet.

9. Gently resuspend the pellet in 1 mL of ice-cold NPB. Transfer the crude nuclei suspension to a 1.5 mL tube. Keep on ice.
10. Streptavidin beads preparation: Wash the appropriate amount of magnetic beads (25 μL for nuclei from 3 g of roots or 10 μL for 0.5 g of leaves) with 1 mL of ice-cold NPB in a 1.5 mL tube. Collect the beads on the magnetic rack. Discard the supernatant and resuspend the beads with ice-cold NPB to their original volume (e.g., 25 μL). Keep on ice.
11. Bind the biotinylated nuclei to the streptavidin beads: Add the freshly prepared beads to the 1 mL of resuspended nuclei from **step 9**. Rotate on a nutator in a 4 °C cold room for 30 minutes. Work in the 4 °C cold room for **steps 12–23**.
12. Transfer the 1 mL bead–nuclei mixture to a 15 mL tube and slowly add 13 mL of ice-cold NPBt. Mix gently and place on a nutator for 30 s.
13. Place the 15 mL tube in the magnetic rack for 15 mL tubes for 2 min to capture the bead-bound nuclei along the walls of the tube.
14. Slowly remove the NPBt supernatant with a serological pipette, making sure not to disturb the beads on the side walls of the tube.
15. Gently resuspend the beads with 14 mL of ice-cold NPBt, mix gently, and place on a nutator for 30 s.
16. Place the 15 mL tube in the magnetic rack for 2 min to capture the nuclei and beads.
17. Repeat **steps 14** through **16** one more time, for a total of three washes.
18. Slowly remove the NPBt supernatant with a serological pipette.
19. Resuspend the beads in 1 mL of ice-cold NPBt.
20. Remove 25 μL of this nuclei–bead suspension and place in a 0.6 mL tube on ice for later quantification of captured nuclei with a hemocytometer (*see* **step 24**).
21. Transfer the remaining nuclei–bead suspension to an ice-cold 1.5 mL tube.
22. Place the 1.5 mL tube in the magnetic rack for microfuge tube to capture the beads along the walls of the tube.
23. Carefully remove the NPBt supernatant and resuspend the bead-bound nuclei in 20 μL of ice-cold NPB. Keep on ice until the nuclei are counted and ready for tagmentation (*see* **Note 2**).
24. To view and quantify nuclei under a light microscope, add 1 μL of diluted DAPI solution (0.2 μg/μL) to each 25 μL aliquot of nuclei from **step 20**. Mix well, and place on ice for 5 min in the dark.

25. Use a hemocytometer to count the DAPI-stained, bead-bound nuclei and determine the total yield. Purified nuclei should appear as shown in Fig. 1c (*see* **Note 3**).
26. Use the calculated total yield to determine the volume of resuspended nuclei from **step 23** needed to obtain 50,000 nuclei for the ATAC-seq reaction.
27. Transfer the volume containing 50,000 nuclei to a new 0.2 mL tube.
28. Place the 0.2 mL tube in the 96-well magnetic separator plate to capture the beads along the walls of the tube.
29. Remove the supernatant, and resuspend the bead-bound nuclei in 50 μL of ice-cold transposition reaction mix (*see* Subheading 3.3).
30. *Immediately* proceed to Subheading 3.3.

3.2 Purification of Total Nuclei Using Sucrose Sedimentation

This method for purification will provide nuclei suitable for ATAC-seq from *Arabidopsis* leaf or root tissue, and may also be effective for other plant species with modifications. Users should be aware that substantial organelle contamination will exist in nuclei samples prepared with sucrose sedimentation. As such, the sequencing depth of the ATAC-seq libraries will need to be deeper than those from INTACT-purified samples in order to account for this.

1. Excise the tissue (0.1–1 g of plant tissue) into a weigh boat on ice until the desired amount has been collected and weighed.
2. Grind the plant tissue to a fine powder in liquid nitrogen using a mortar and pestle (*see* **Note 4**).
3. Using a nitrogen-cooled metal lab spoon, quickly transfer the frozen tissue powder to another mortar containing 10 mL ice-cold Nuclei Purification Buffer (NPB).
4. Thoroughly resuspend the powder in NPB by grinding it with a new, clean pestle.
5. Place a 70 μm nylon cell strainer in the center of a long-stemmed funnel and above a 15 mL tube on ice.
6. Use a 10 mL serological pipette to draw up the tissue suspension and filter it through the strainer. Collect the flow-through in the 15 mL tube.
7. Spin down the nuclei at 1,200 × *g* for 10 min at 4 °C.
8. Use a 10 mL serological pipet and then a 1 mL pipette tip to carefully remove as much of the supernatant as possible without disturbing the pellet.
9. Gently resuspend the pellet in 1 mL of ice-cold Nuclei Extraction Buffer 2 (NEB2). Transfer this suspension to a new 1.5 mL tube.

10. Spin down the resuspended nuclei at 12,000 × *g* for 10 min at 4 °C.
11. Carefully remove the supernatant and resuspend the pellet thoroughly in 300 μL of Nuclei Extraction Buffer 3 (NEB3).
12. Add 300 μL of ice-cold NEB3 to a new 1.5 mL tube.
13. Carefully layer the resuspended pellet from **step 11** on top of the fresh NEB3 from **step 12**.
14. Spin down the two layers at 16,000 × *g* for 10 min at 4 °C (*see* **Note 5**).
15. Carefully remove the supernatant and resuspend the nuclei pellet in 1 mL of cold NPB. Keep on ice.
16. Remove 25 μL of this nuclei suspension and move to a fresh 0.6 mL tube on ice for quantification of isolated nuclei with a hemocytometer (*see* **step 18**).
17. Transfer the remaining isolated nuclei to an ice-cold 1.5 mL tube.
18. To view and quantify nuclei under a light microscope, add 1 μL diluted DAPI solution (0.2 μg/μL) to each 25 μL of nuclei from **step 16**. Mix well, and place on ice for 5 min in the dark.
19. Use a hemocytometer to quantify the DAPI-stained nuclei and determine the total yield. Purified nuclei should appear as shown in Fig. 1c (*see* **Note 6**).
20. Use the calculated total yield to determine the volume of resuspended nuclei from **step 17** needed to obtain 50,000 nuclei for the ATAC-seq reaction.
21. Transfer the volume containing 50,000 nuclei to a new 0.2 mL tube.
22. Spin down the nuclei at 1,500 × *g* for 7 min at 4 °C.
23. Remove the supernatant, and resuspend the nuclei in 50 μL of ice-cold transposition reaction mix (*see* Subheading 3.3).
24. *Immediately* proceed to Subheading 3.3.

3.3 Tagmentation with Tn5 Transposase

This step will fragment the chromatin while adding the adapter sequences needed for high-throughput sequencing.

1. Prepare the transposition reaction master mix in a 0.2 mL PCR tube on ice according to Table 1 and mix well. The volumes given in Table 1 are for a single reaction with 50,000 nuclei.
2. Place the nuclei resuspended in transposition reaction mix – either from Subheading 3.1, **step 29** if the nuclei were isolated using magnetic beads or from Subheading 3.2, **step 23** if the nuclei were isolated using sucrose sedimentation – in a thermal cycler block prewarmed to 37 °C.

Table 1
Transposition reaction mix

Component	Volume (μL)
2× TD buffer	25
Water	22.5
TDE1 Transposase	2.5
Total	50

3. Incubate for 30 min, gently mixing the reaction by hand every 5 min.
4. Purify the transposed DNA using the PCR purification kit according to the manufacturer's instructions.
5. Elute DNA in 11 μL of elution buffer (EB) provided in the kit. DNA can now be stored at −20 °C until future use, or used immediately for PCR amplification.

3.4 PCR Amplification of the DNA Library

After tagmentation, this step will increase the abundance of the library fragments while also allowing the libraries to be barcoded.

1. Prepare the PCR amplification mix in a 0.2 mL tube on ice according to Table 2.
2. Mix well, and perform PCR cycling as described in Table 3 (*see* **Note** 7).
3. Once the thermal cycler reaches 4 °C, remove the samples and place them on ice.
4. To determine the number of additional PCR cycles needed to adequately amplify the DNA library, prepare the qPCR Library Amplification Mix described in Table 4 in a 0.2 mL PCR tube. Keep the mixture on ice.
5. Perform thermal cycling in the qPCR machine according to Table 5.
6. To determine the optimal number of cycles needed to amplify the remaining 45 μL of each library from **step 2**, view the linear fluorescence versus cycle number plot on the qPCR machine once the reaction is finished. The cycle number at which the fluorescence for a given reaction is at 1/3 of its maximum is the number of additional cycles (*N*) that each library requires for adequate amplification (*see* **Note 8**).
7. Run the remaining 45 μL of each PCR reaction from **step 3** according to Table 6.

Table 2
Transposed DNA amplification mix

Component	Volume (μL)
Transposed DNA (from Subheading 3.3, **step 5**)	10
Water	10
25 μM ATAC Primer 1	2.5
25 μM ATAC barcoded Primer 2[a]	2.5
2× NEBNext High Fidelity PCR Mix	25
Total	50

[a]A different barcoded Primer 2 should be used for each library that is to be pooled into a single sequencing run.

Table 3
Thermal cycling conditions for transposed DNA amplification

Cycle number	Temperature (°C)	Time
1	72	5 min
	98	30 s
5 cycles	98	10 s
	63	30 s
	72	1 min
	4	Hold

Table 4
qPCR library amplification mix

Component	Volume (μL)
Amplified library (from Subheading 3.4, **step 3**)	5
Water	0.45
25 μM ATAC Primer 1	0.5
25 μM ATAC barcoded Primer 2	0.5
20× Evagreen dye	0.75
50× ROX dye[a]	0.30
2× NEBNext High Fidelity PCR Mix	7.5
Total	15

[a]ROX concentration may vary depending on the qPCR instrument. The amount described here is optimized for the ABI Step-One-Plus instrument.

Table 5
qPCR cycling conditions to determine additional library amplification cycles

Cycle number	Temperature (°C)	Time
1	98	30 s
20 cycles	98	10 s
	63	30 s
	72	1 min

Table 6
Final library amplification

Cycle number	Temperature (°C)	Time
1	98	30 s
N cycles	98	10 s
	63	30 s
	72	1 min
	4	Hold

8. Purify the libraries by mixing PCR purification magnetic beads with the reaction products at a 1.5:1 volume ratio of beads: PCR sample (*see* **Note 9**).
9. Incubate at room temperature for 5 min.
10. Place the 0.2 mL tube on the 96-well magnetic separator plate for 1 min to capture the PCR purification magnetic beads, and discard the supernatant.
11. With the tubes still in the magnetic plate, wash the beads twice for 30 s each with 200 μL of 80% ethanol, without disturbing the bead pellet.
12. After the last wash, allow the beads to dry for 5 min to remove all traces of ethanol (*see* **Note 10**).
13. Remove the tube from the magnet and resuspend the bead pellet in 20 μL 10 mM Tris pH 8.
14. Incubate at room temperature for 2 min.
15. Capture the beads on the magnet, and transfer the supernatant into a fresh 0.2 mL PCR tube on ice.
16. A small aliquot of the library, 1–2 μL, can be run on a 2% agarose gel to visualize the abundance and size distribution of amplified libraries (Fig. 2a) (*see* **Note 11**).

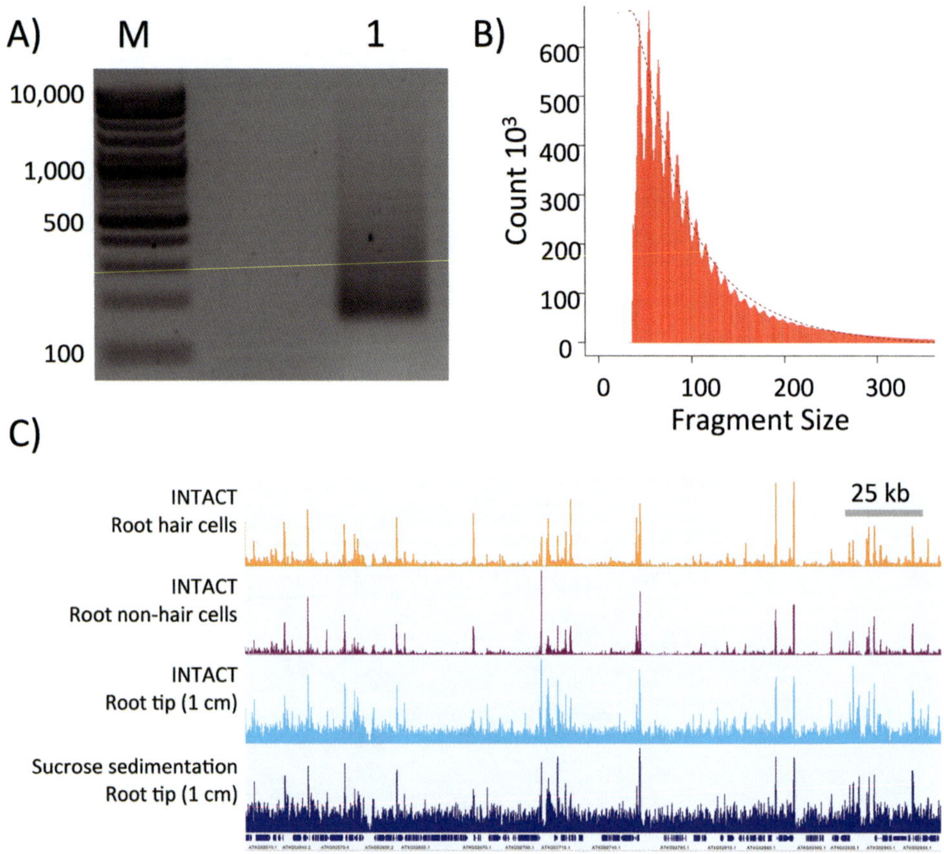

Fig. 2 ATAC-seq library preparation and high-throughput sequencing. (**a**) An amplified ATAC-seq library purified with AMPure XP beads (*lane* "1") was resolved in a 2% agarose gel stained with ethidium bromide. *Lane* "M" is the molecular weight marker lane. Amplified library fragments generally range in size from 180 bp to several kb in size. The size distribution of the resolved gel may vary somewhat, but the final product should be free of adapter dimers (distinct band around 125 bp) and primer dimers (distinct band around 80 bp) (*see* **Note 11**). (**b**) Insert sizes of ATAC-seq paired-end reads from 50,000 nuclei isolated by INTACT from non-hair cells calculated using the InsertSizeMetrics option from Picard Tools (*see* **Note 13**). The distribution shows periodicity of the helical pitch of DNA for fragments smaller than 200 bp. Fragments containing one or more nucleosomes, with insert sizes of 150 or 300 bp were not observed using the transposase: nuclei and bead: DNA ratios described in this protocol. (**c**) Integrated Genome Viewer snapshot of four different libraries sequenced on the Illumina platform. The tracks shown are of ATAC sequencing reads from INTACT isolated nuclei from root hair cells (*orange*), root non-hair cells (*purple*), root tip tissue (*cyan*), and sucrose sedimentation isolated nuclei from 1 cm root tip tissue (*navy*). Gene tracks are shown below the ATAC-seq tracks and a 25 kb scale bar is shown

17. The purified libraries can now be stored at −20 °C.
18. Quantify the molar concentrations of the libraries using the Library Quantification kit according to manufacturer's directions.
19. Once quantified, the libraries are ready for pooling and high-throughput sequencing on a next-generation sequencing platform (*see* **Note 12**).

20. Analysis of the quality of the sequencing reads, alignment to the genome, examination of the fragment size distribution (Fig. 2b), and further downstream analyses can be performed as described briefly in **Note 13**. A genome browser shot of the typical *Arabidopsis* ATAC-seq data from libraries made using the procedures described here can be seen in Fig. 2c.

4 Notes

1. This protocol is optimized for 3 g of root or 0.5 g of leaf tissue from *Arabidopsis*. Ground leaf tissue contains more debris relative to roots, and therefore necessitates a lower amount of starting material to obtain highly purified nuclei. INTACT may also be performed on fresh tissue by chopping the tissue in NPB as opposed to grinding to a fine powder using liquid nitrogen. However, this approach does require the use of fresh tissue. The number of samples that can be run through INTACT purification simultaneously is mainly limited by the capacity of the magnetic rack for 15 mL tubes used for nuclei capture. For example, up to four separate samples can be processed in parallel using one DynaMag 15 magnetic rack. Using an INTACT line with nuclei labeled in the root epidermal non-hair cell type, approximately 200,000 purified nuclei can be obtained from 3 g of roots. Larger amounts of tissue can be used for purifying nuclei from less abundant cell types, and this generally only requires adjustments to the amount of streptavidin beads used and the volume of solution used for bead capture. *See* [11] for more details on variations in the INTACT procedure.
2. After isolating the bead bound nuclei, keep the sample on ice while quantifying the nuclei from the aliquot in Subheading 3.1, **step 23**. *Do not* freeze the isolated nuclei before doing tagmentation and library preparation. Freezing and thawing of isolated nuclei can disrupt protein–DNA interactions.
3. After DAPI staining, nuclei purified by INTACT can be easily identified and counted using a hemocytometer. The ideal setup for visualizing nuclei is under a mix of dim white light and DAPI channel fluorescence. The dim white light allows for visualization of the hemocytometer grid and the beads, and the DAPI fluorescence allows for the visualization of nuclei. A sample image of isolated bead-bound nuclei is shown in Fig. 1c. A nucleus is identified as a punctate circle with strong DAPI fluorescence that has several beads clustered around it. Minimal cellular debris or contaminating unbound nuclei should be observed in the final product. These contaminants may be further reduced by using fewer beads and by increasing

the volumes of NPB and NPBt used during purification as described in **Note 1**. We have successfully used as few as 20,000 to as many as 200,000 INTACT-purified nuclei in this procedure without altering any other parameters of the protocol presented here.

4. This protocol is optimized for less than 1 g of root or 0.5 g of leaf tissue. Ground leaf tissue contains more debris relative to roots, and therefore necessitates a lower amount of starting material to obtain purified nuclei. As with the INTACT protocol, sucrose sedimentation of nuclei may also be performed on fresh tissue by chopping the tissue in NPB as opposed to grinding to a fine powder using liquid nitrogen. However, this approach does require the use of fresh tissue. We recommend starting with the minimum amount of tissue needed to obtain the required number of nuclei (e.g., 50,000 per ATAC-seq reaction).

5. Proper separation of nuclei from other cellular debris requires the nuclei to pass through the sucrose cushion during centrifugation. The NEB3 resuspended nuclei should therefore be placed gently on top of NEB3 layer present in the tube. After centrifugation, the contaminating organelles and debris may be visible at the top of the tube. If leaf tissue was used, the top layer will become greener after centrifugation and the pellet will become noticeably less green than it was prior to centrifugation.

6. After DAPI staining, nuclei purified by sucrose sedimentation can be identified and quantified using a hemocytometer. A mixture of DAPI-channel fluorescence and white light illumination allows the stained nuclei and the hemocytometer grid to be seen simultaneously. A sample image of isolated nuclei is shown in Fig. 1c. A nucleus is identified as a punctate circle with strong DAPI fluorescence. The nucleus is typically ~5 μm in size and can be easily identified at 200× and 400× magnifications. Cellular debris may be observed in the final preparation, but this generally does not affect the outcome of the ATAC-seq procedure. To reduce cellular debris contamination, starting tissue can be chopped with a razor blade (*see* **Note 4**) and/or additional NEB3 wash steps may also be done by repeating Subheading 3.2, **steps 11–14** for a second sucrose cushion centrifugation.

7. Ensure that all work surfaces, pipettes, and reagents needed for amplification and library preparation are free of DNA contamination by wiping them down with a nucleic acid decontamination detergent 10 min before starting work. For library amplification, unique barcoded primers are used for each sample if multiple libraries are to be sequenced in an individual flow cell lane. The sequences of all primers can be found in the supplementary material of [7].

8. The number of PCR cycles needed to amplify ATAC libraries is determined by the PCR reaction in Subheading 3.4, **step 6**. We recommend using the minimum number of cycles necessary to obtain a sufficient molar amount of library for Illumina sequencing. This must be determined empirically and will also depend on the number of libraries to be pooled for sequencing. Typically, when the libraries are amplified to 1/3 of their maximum fluorescence detected by qPCR the quantified molarities are from 50 to 300 nM. The variation in the final product is dependent on how reliably nuclei were counted in Subheading 3.1, **step 25** or Subheading 3.2, **step 19**.
9. The ratio of the PCR purification magnetic beads to the amplified library volume determines the size of the purified DNA fragments that are isolated. The 1.5 bead-to-amplified library reaction ratio results in the isolation of DNA fragments shown in Fig. 2a. Using ratios that have higher proportions of beads may result in purification of sequencing adapters and PCR primers, which can negatively affect sequencing.
10. A drying time of 5 min is generally sufficient to remove all traces of ethanol from the beads, but this time may vary based on humidity and room temperature. Ensure that all ethanol has evaporated before moving on to the next step. Do not allow beads to dry to the extent that the pellet begins to crack as this will decrease the purification yield.
11. Libraries can generally be visualized by agarose gel electrophoresis followed by ethidium bromide staining (Fig. 2b). Sensitivity can be greatly increased by staining the gel with a nucleic-acid specific green fluorescent dye (e.g., SYBR green stain, Qiagen, or equivalent) or using microfluidics-electrophoresis trace analysis (e.g., Agilent Bioanalyzer or equivalent instrument), if available. The libraries that we have prepared using this method generally present as a DNA smear starting at ~180 bp and ranging to greater than 1 kb, with peak intensity falling between ~180 and 500 bp (*see* Fig. 2a). The original publication on ATAC-seq [7] reported a nucleosome-like periodicity in the library size distribution, but we have not observed this phenomenon either by electrophoresis or by the estimation of fragment size distribution based on the distance between paired-end sequencing reads, as shown in Fig. 2b. This lack of observed nucleosome fractions may be due to the size selection of library fragments by PCR purification magnetic beads and the low transposase-to-nuclei ratio described in this protocol.
12. Paired-end sequencing is recommended in order to maximize the number of transposase integration events that can be observed in a given sample, and to allow measurement of the

length of the sequenced fragments (Fig. 2b). To identify open chromatin regions in *Arabidopsis*, users should aim to obtain at least 10–20 million reads per library that map to the nuclear genome. For transcription factor footprinting the number of nuclear genome-mapping reads should be increased to at least 60 million per library [12]. When using sucrose sedimentation for nuclei purification, users should expect ~50% of reads to map to the nuclear genome, while the use of INTACT purification will increase this number to >90%.

13. Sequencing reads are checked for overall quality using FastQC (http://www.bioinformatics.babraham.ac.uk/projects/fastqc/) or equivalent. The reads are aligned to the TAIR10 *Arabidopsis thaliana* genome (https://www.arabidopsis.org/download/index-auto.jsp?dir=%2Fdownload_files%2FGenes%2FTAIR10_genome_release) or the most recent genome annotation version using Bowtie2 (http://bowtie-bio.sourceforge.net/bowtie2/index.shtml). The resulting SAM file is converted to a binary BAM file, which is sorted and indexed using Samtools (http://samtools.sourceforge.net/). The quality of the resulting BAM file, including fragment size distribution, is analyzed using Picard Tools (https://broadinstitute.github.io/picard/). Alignment data is visualized using the Integrated Genome Viewer (http://software.broadinstitute.org/software/igv/). For ease of visualization, BAM files were converted to BigWig files using DeepTools BamPECoverage tool (http://deeptools.readthedocs.io/en/latest/index.html). Further downstream analyses of ATAC-seq data include calling peaks with HOMER (http://homer.salk.edu/homer/index.html), editing BED files with bedtools (http://bedtools.readthedocs.io/en/latest/), and identifying transcription factor footprints using pyDNase (http://pythonhosted.org/pyDNase/).

Acknowledgments

This work was supported by the National Science Foundation Grant No. IOS-1238243. We thank Paja Sijacic and Shannon Torres for helping to optimize the protocol for nuclei isolation and for suggestions on the manuscript.

References

1. Li B, Carey M, Workman JL (2007) The role of chromatin during transcription. Cell 128:707–719
2. Pajoro A, Muiño JM, Angenent GC, Kaufmann K (2017) Profiling nucleosome occupancy by MNase-seq: experimental protocol and computational analysis. In: Bemer M, Baroux C (eds) Plant chromatin dynamics: methods and protocols. Springer, New York, NY. doi:10.1007/978-1-4939-7318-7_11

3. Song L, Crawford GE (2010) DNase-seq: a high-resolution technique for mapping active gene regulatory elements across the genome from mammalian cells. Cold Spring Harb Protoc 2010:pdb.prot5384. doi:10.1101/pdb.prot5384
4. Ken Z (2005) Micrococcal nuclease analysis of chromatin structure. Curr Protoc Mol Biol 21 (1):1–17
5. Desvoyes B, Vergara Z, Sequeira-Mendes JO, Madeira S, Gutierrez C (2017) A rapid and efficient ChIP protocol to profile chromatin binding proteins and epigenetic modifications in bulk Arabidopsis tissue. In: Bemer M, Baroux C (eds) Plant chromatin dynamics: methods and protocols. Springer, New York, NY. doi:10.1007/978-1-4939-7318-7_5
6. Park PJ (2009) ChIP-seq: advantages and challenges of a maturing technology. Nat Rev Genet 10:669–680
7. Buenrostro JD, Giresi PG, Zaba LC, Chang HY, Greenleaf WJ (2013) Transposition of native chromatin for fast and sensitive epigenomic profiling of open chromatin, DNA-binding proteins and nucleosome position. Nat Methods 10:1213–1218
8. Buenrostro JD, Wu B, Chang HY, Greenleaf WJ (2015) ATAC-seq: a method for assaying chromatin accessibility genome-wide. Curr Protoc Mol Biol 109:21.29.1–21.29.9
9. Gendrel A, Lippman Z, Martienssen R, Colot V (2005) Profiling histone modification patterns in plants using genomic tiling microarrays. Nat Methods 2:213–218
10. Deal RB, Henikoff S (2010) A simple method for gene expression and chromatin profiling of individual cell types within a tissue. Dev Cell 18:1030–1040
11. Wang D, Deal RB (2015) Epigenome profiling of specific plant cell types using a StreamLined INTACT protocol and ChIP-seq. Mehods Mol Biol 1284:3–25
12. Lu Z, Hofmeister BT, Vollmers C, DuBois RM, Schmitz RJ (2016) Combining ATAC-seq with nuclei sorting for discovery of cis-regulatory regions in plant genomes. Nucleic Acids Res. doi:10.1093/nar/gkw1179

Part II

Chromatin Dynamics and Gene Regulation

Chapter 13

Unraveling the Complex Epigenetic Mechanisms that Regulate Gene Activity

Marian Bemer

Abstract

Our understanding of the epigenetic mechanisms that regulate gene expression has been largely increased in recent years by the development and refinement of different techniques. This has revealed that gene transcription is highly influenced by epigenetic mechanisms, i.e., those that do not involve changes in the genome sequence, but rather in nuclear architecture, chromosome conformation and histone and DNA modifications. Our understanding of how these different levels of epigenetic regulation interact with each other and with classical transcription-factor based gene regulation to influence gene transcription has just started to emerge. This review discusses the latest advances in unraveling the complex interactions between different types of epigenetic regulation and transcription factor activity, with special attention to the approaches that can be used to study these interactions.

Key words Epigenetics, Chromatin remodeling, Histone modification, Gene regulation, Transcription factors, Polycomb

1 Introduction

Gene expression is tightly regulated by a plethora of dynamic mechanisms. In addition to regulation by transcription factors (TFs) that bind to the promoter to repress or activate a gene, mechanisms such as nucleosome remodeling, histone modification and DNA methylation highly influence gene expression. These mechanisms do not involve changes in the DNA, but the effects they bring about can still be transferred to progeny cells and sometimes even be inherited by the offspring, and are therefore collectively referred to as "epigenetic". The development of high-throughput techniques in plants that allow a genome-wide analysis of nucleosome occupancy (*see* Chapters 11 and 12 [1, 2]), DNA methylation profiles (*see* Chapters 2–4 [3–5]), and histone modification profiles (*see* Chapters 5, 6, and 8 [6–8]) in combination with transcriptome profiling (*see* Chapter 16 [9]) has revealed the global dynamics of chromatin modifications and their importance for gene

Marian Bemer and Célia Baroux (eds.), *Plant Chromatin Dynamics: Methods and Protocols*, Methods in Molecular Biology, vol. 1675, DOI 10.1007/978-1-4939-7318-7_13,

expression. Unraveling the interactions between the different mechanisms however, as well as their link with TF activity and their effect on the transcriptional machinery, is highly challenging. In recent years, the development and refinement of various techniques for the plant field have facilitated studies on the interactions between different chromatin modifying mechanisms and TF activity, and their effect on the regulation of specific genes. These techniques are detailed in section II of this book, and include chromatin conformation capture (*see* Chapters 14 and 15 [10, 11]), immunoprecipitation followed by mass spectrometry (*see* Chapter 18 [12]), identification of sRNAs and their associated proteins (Chapters 17 and 19 [13, 14], and histone methyltransferase assays (*see* Chapter 20 [15]). In this review, recent advances in understanding the interactions between the different epigenetic mechanisms are discussed, with special emphasis on the methodologies that have been developed to study the local scale effects of chromatin remodeling on gene regulation. After a general introduction, common interactions between the different chromatin modifying mechanisms are discussed, followed by various examples from ***Arabidopsis thaliana*** (Arabidopsis) that illustrate the complexity of interactions at the gene level.

1.1 Global Chromatin Arrangement Influences Efficient Gene Silencing or Activation

DNA and proteins together constitute chromatin, a structure that both ensures the packaging of the DNA and guides gene regulation. The building blocks of chromatin are the nucleosomes, which consist of ca. 146 bp of DNA wrapped around a protein complex that contains two proteins each of histone H2A, H2B, H3, and H4. Nucleosomes facilitate chromosome compaction by fitting approximately 2 m of DNA inside the nucleus. Further compaction of the chromatin is regulated by modifications to the DNA itself, by posttranslational modifications of the histone tails in the nucleosomes, and by linker histones, which bind the DNA between the nucleosomes [16]. Actively transcribed genes are generally present in loosely packed euchromatin, which is accessible for regulatory protein complexes, while silenced genes and transposons are present in densely packed heterochromatin. Thus, in addition to allowing the packaging of the DNA into a small nucleus, chromatin structure is also of crucial importance for the regulation of gene expression. The 3D arrangement of the chromosomes and chromatin within the nucleus has been thoroughly investigated using different microscopic approaches (reviewed in Chapter 23 [17]). The currently available techniques allow simultaneous visualisation of specific DNA sequences and detection of proteins in the nucleus by combining fluorescence in situ hybridization (FISH) with immunolabeling (*see* Chapter 25 [18]). These approaches have recently been complemented with molecular approaches based on chromatin conformation capture (3C), where the chromatin is fixed, followed by DNA digestion and subsequent religation to detect the physically interacting fragments. While the 3C approach can detect the interaction between two specific fragments (*see* Chapter 15 [11] for a detailed protocol), variants of this technique, such as 4C and Hi-C, involve high-throughput sequencing and allow for the detection of interactions between a specific region and the rest of the chromatin (4C), or even between all chromosome fragments (Hi-C) (reviewed in Chapter 14 [10]). Combined application of microscopy and molecular approaches revealed that the 3D chromatin configuration brings similarly condensed regions close to each other, probably allowing a more efficient regulation of heterochromatin silencing or transcriptional regulation [19]. This 3D chromatin arrangement during interphase is not necessarily fixed, and repositioning of loci has been observed in Arabidopsis in response to light using a modified FISH protocol, and is associated with changes in gene expression [20]. This relocation may be linked to the global changes in chromatin condensation that are observed during photomorphogenesis [21]. How global chromatin architecture is regulated is still largely unknown, but recent studies have revealed distinct relationships between certain chromatin modifications and 3D chromosomal interactions, which are discussed below.

1.2 Local Chromatin Environment Is Regulated by a Complex Combination of Factors

The degree of DNA compaction is important for a gene's transcriptional status, and is regulated by a complex combination of different mechanisms involving nucleosome remodeling, histone modification and DNA methylation.

1.2.1 Nucleosome Remodeling

ATP-dependent nucleosome remodeling factors can reposition nucleosomes and thereby determine the distance between neighboring nucleosomes [22]. They are also involved in the incorporation of different histone variants, such as H3.3 or H2A.Z, which can affect the ability of nucleosomes to interact with each other or with the DNA and thereby have an effect on gene regulation [23]. For example, high H2A.Z levels in the gene body of stress-responsive genes correlate with the repression of these genes [24]. In addition to the core histones, each nucleosome can also be associated with a linker histone, histone H1, which guides further chromatin compaction. There are also three different H1 variants in Arabidopsis, of which the minor H1.3 variant is only incorporated in specific tissues and in response to stress [25]. H1.1 and H1.2 on the other hand, are ubiquitously expressed [25], but absent from the reproductive lineages [26, 27]. The removal of both linker histones is followed by a vast reprogramming of the epigenetic landscape in the meiocytes [26, 27]. There are four major subfamilies of ATP-dependent chromatin remodelers (INO80/SWR1, CHD, ISWI and SWI/SNF), of which the SWI/SNF-type chromatin remodeling factors PICKLE (PKL), PICKLE-RELATED 2 (PKR2), SPLAYED (SYD), and BRAHMA (BRM) have been shown to play major roles in Arabidopsis development [22, 28, 29].

1.2.2 Histone Modifications

Posttranslational modification (PTM) of the protein tails of the core histones, which affects the covalent interaction between the nucleosomes and the DNA, is another important mechanism involved in chromatin compaction. Linker histones are also subjected to PTMs, but the effect of these modifications on chromatin structure has not yet been ascertained [30]. The core histone tail modifications that play a major role in gene regulation in plants are trimethylation of histone H3 at lysine 4 (H3K4me3), H3K9me2, H3K9me3, H3K27me1, H3K27me3, and H3K36me3, as well as ubiquitination of histone H2A and histone H2B and acetylation of histone H3 at lysine 9 (H3K9ac), H3K14ac and H4K5ac. Some of these marks, which can either have a repressive or an activating effect on gene expression, co-occur in specific chromatin states, while others appear mutually exclusive [31, 32]. In euchromatic regions, active genes are often marked with H3K4me3, H3K36me3, H2Bub and H3K9ac, while repressed genes are marked with H3K27me3 and H2Aub. However, some loci display "bivalent" marks that are associated both with activation and repression. In particular H3K4me3 and H3K27me3 were found to co-occur at certain loci [32]. This observation might reflect a technical bias where two cell-populations in which the region was marked differently were sampled together, or suggest a mechanism for priming a gene for fast transcriptional activation and repression. These two options can be discriminated using a sequential chromatin immunoprecipitation (re-ChIP) approach, where the chromatin is first pulled down using an antibody against the first histone mark of interest, and subsequently subjected to an additional ChIP with the second antibody of interest (*see* Chapter 6 [7]). To determine whether the bivalent modifications actually occur on the same nucleosome, MNase treatment (Chapter 11 [1]) can be employed to fractionate the DNA, followed by extraction of mononucleosomal fragments prior to sequencing. In Table 1, an overview of the techniques that can be exploited to address different outstanding questions is presented.

1.2.3 Writers and Erasers of Histone Modifications

The deposition of the histone modifications occurs via histone methyltransferases, acetyltransferases, and ubiquitinases that function as epigenetic writers, whereas histone demethylases, deacetylases (HDACs), and deubiquitylating enzymes function as erasers that can remove these marks. In plants and animals, the highly conserved Polycomb Repressive Complex 2 (PRC2) is responsible for placing the repressive H3K27me3 mark at numerous loci. In Arabidopsis, PRC2 consists of four core subunits that are encoded by small gene families [29]. H3K27me3 repressed loci are often also marked with H2A monoubiquitination (H2AK119ub), which is deposited by Polycomb Repressive Complex 1 (PRC1), consisting of the RING-finger homologs AtRING1A, AtRING1B, AtBMIA, AtBMIB, and AtBMIC [37]. PRC1 was initially thought to recognize H3K27me3 and to strengthen and stabilize its silencing by depositing H2Aub, but more recent data showed that

Table 1
Overview of outstanding questions related to the epigenetic regulation of gene activity, and the approaches that can be employed to address these questions

Outstanding questions	Approach
What are the spatial and temporal dynamics of epigenetic writers, readers, and erasers when regulating gene activity?	– Use of synchronized tissues and/or induced expression of epigenetic writers, readers, or erasers, followed by temporal analysis of chromatin modifications and transcriptome dynamics (Chapter 16 [9, 33, 34]) – Studying specific cell lineages using the INTACT system (Chapter 8 [8])
Which chromatin remodelers function together in one complex?	– Immunoprecipitation of *in planta* complexes followed by mass spectrometry identification of the complex members (IP-MS, Chapter 18 [12])
How are chromatin remodeling complexes recruited to their different target loci?	– IP-MS to identify TFs associating with epigenetic writers, readers, or erasers (Chapter 18, [12]) – RNA immunoprecipitation (RIP) to identify RNAs associating with a specific chromatin remodeler/modifier (Chapter 19 [14])
Which combinations of histone modifications can co-occur on one nucleosome and how does this impact gene regulation?	– Sequential ChIP to identify histone modifications on the same chromatin fiber (Chapter 6 [7]), in combination with RNA-seq (Chapter 16 [9]) – Histone modification mapping using IP-MS/MS (Chapter 9 [45]) – Histone methyltransferase assays using different chromatin substrates (Chapter 20 [15]
How often does local-scale loop formation occur and what is the importance for gene activity?	– Chromatin Conformation Capture (3C) (Chapter 15 [11]), in combination with qPCR or DNA-seq
To what extent can epigenetic writers, readers, and erasers influence 3D chromatin conformation?	– Chromatin Conformation Capture (3C)-based strategies (Chapters 14 and 15 10, 11]), in different mutant backgrounds [35] or upon (induced) overexpression of epigenetic factors – Profiling nucleosome occupancy, and chromatin condensation and accessibility by MNase-seq (Chapter 11 [1], DNAse1-seq [36], or ATAC-seq (Chapter 12 [2]) in different backgrounds
How important is DNA methylation for the regulation of gene activity and gene function?	– Identify differentially methylated regions affecting the plant's phenotype by EpiQTL mapping (Chapter 22 [57]) – Identify parent-of-origin specific phenotypes using Bulk-seq (Chapter 21 [58]) – Identification of small RNAs involved in gene regulation using sRNA sequencing combined with mRNA sequencing (Chapter 17 [13])

H3K27me3 deposition can also depend on PRC1 activity, suggesting that PRC1 activity precedes PRC2 activity in these cases [37, 38]. The H3K4me2/3 and H3K36me3 marks can be deposited by different SET domain methyltransferases. The Arabidopsis genome contains at least 47 SET-domain group methyltransferases, with a putative role in histone methylation [39], of which up to ten have been identified as H3K4 methyltransferases [40]. Histone methyltransferases are counteracted by histone demethylases, which can remove the deposited marks and are therefore called "erasers." The Arabidopsis genome contains 21 JmJC-domain containing demethylases that are divided in subclasses based on the presence of additional domains, with specific demethylase activities [41, 42]. Histone lysine acetylation is regulated by histone acetyltransferase (HAT) writers, and histone deacetylase (HDAC) erasers. The Arabidopsis genome contains 12 HATs and 18 HDACs, which are further subdivided into different classes with specific activity [43]. In plants, acetylation marks can be deposited on lysine residues K9, K14, K18, K23, and K27 of histone H3, and K5, K8, K12, K16, and K20 of histone H4. Recently, a new histone modification, H3K36ac, was identified in plants using tandem mass spectrometry [44]. This technique (*see* Chapter 9 [45]), developed now for cauliflower inflorescences, is a valuable method to identify new histone modifications, as well as to study existing histone modifications and their relation to specific histone variants [44, 46]. In vitro histone methylation assays, such as described in Chapter 20 [15] are very useful to further elucidate the preference of histone methyltransferases for certain histone variants or histone modifications.

1.2.4 DNA Methylation

Finally, methylation of the DNA itself is also important for epigenetic regulation of transcription, being the major mechanism involved in the silencing of transposable elements (TEs). TE silencing is also accompanied by H3K27me1 and H3K9me2 histone modifications [32]. In plants, the DNA base cytosine can be methylated in three sequence contexts: CG, CHG and CHH (where H = A, T, or C), each regulated by different enzymes. DOMAINS REARRANGED METHYLTRANSFERASE 2 (DRM2) is a CHH methyltransferase that is guided by small RNAs in the RNA-directed DNA methylation (RdDM) pathway, and is required for de novo methylation in all sequence contexts. METHYLTRANSFERASE 1 (MET1) regulates the maintenance of CG methylation, while CHROMOMETHYLASE 3 (CMT3) is the main CHG methyltransferase [47]. The heterochromatic, condensed regions where most TEs are found contain ubiquitous DNA methylation in all sequence contexts associated with complete transcriptional silencing. However, a number of constitutively transcribed genes in open, euchromatic regions also contain CG methylation in their exons (gene body methylation), which is associated with gene expression and negatively correlated with the presence of the histone variant H2A.Z [48].

DNA methylation has been associated with specific histone modifications. The "reinforcing loop model" [47] has been used to describe the dependence of H3K9me2 deposition on the presence of non-CG methylation and vice versa [49–51]. In addition to binding H3K9me2, the SAWADEE HOMEODOMAIN HOMOLOG 1 (SHH1) protein, which is involved in the synthesis of siRNAs in the RdDM pathway, also specifically binds to unmethylated H3K4, suggesting that this active mark has to be removed prior to POL IV-dependent siRNA synthesis. Supporting this idea, H3K4me3 demethylases were identified as components of the RdDM pathway [52]. In addition, the histone deacetylase HDA6 has also been shown to function upstream of siRNA synthesis, providing more evidence that the active mark has to be removed prior to transcriptional silencing [47, 53]. Tang and coworkers [54] have recently shown that the DNA demethylase REPRESSOR OF SILENCING 1 (ROS1) antagonizes RdDM specifically at target TEs that are close to protein coding genes and that are marked with H3K18Ac and H3K27me3, but depleted of H3K27me1 and H3K9me2. This illustrates once more the strong link between DNA methylation and histone modifications.

While histone modifications such as H3K27me3 and H3K4me3 play a major role in the regulation of developmental programs during the plant's life cycle, DNA methylation is relatively stable throughout plant development. However, DNA methylation levels are dynamically regulated in response to several stresses, which is associated with differential expression of TEs and their neighboring genes [55]. In addition, DNA methylation is differentially established in the different cell types of the male and female germ lines, leading to differential DNA methylation profiles between maternal and paternal alleles, with an ensuing parent-of-origin effect on seed development. Although the effects of differentially methylated regions (DMRs) on gene expression and phenotype are often mild, they can explain a distinct part of the natural variation observed in plant traits [56]. These effects can best be studied using a forward epigenetic approach that takes into account the quantitative effects on the plant's phenotype. To identify Quantitative Trait Loci (QTLs) due to differential DNA methylation, a population of Recombinant Inbred Lines (RILs) generated using a DNA methylation mutant can be screened (*see* Chapter 22 [57]). Parent-of-origin-dependent QTLs that affect seed development can be identified using a modified DNA mapping approach termed Bulk-seq (*see* Chapter 21 [58]), which uses whole-genome sequencing of DNA pools to map causative loci.

1.3 Local 3D Chromatin Arrangement Depends on Histone and DNA Modifications

Comparison of Hi-C data with genome-wide histone methylation data revealed that certain histone modifications correlate with the local 3D arrangement of the chromatin [59, 60]. Heterochromatic interacting regions are enriched in H3K9me2, and loss of DNA methylation in the *met1* and *ddm1* mutants causes reduced interaction between densely compacted pericentromeric regions, strengthening the idea that the heterochromatic histone and DNA marks promote chromosome interactions, either directly or indirectly [19, 59, 60]. In addition, small interacting repressed regions were enriched in H3K27me3, a mark that was also found to be over-represented in long range promoter-promoter interactions [59, 61].

Interactions between decondensed euchromatic regions in plants appear to occur on a local rather than a global scale [61]. These local interactions typically result in chromatin looping and can greatly affect gene expression. Loops can either: (1) bring a distant enhancer in contact with the transcription start site (TSS), as for the maize *BOOSTER1* (*B1*) gene [62] and the floral regulator *FLOWERING LOCUS T* (*FT*, discussed below) [63]; (2) connect the TSS with the transcription termination site (TTS), to enhance gene transcription, for example at *FLOWERING LOCUS C* (*FLC*, discussed below) [64]; or (3) have a repressive effect on transcription by preventing access to the transcriptional machinery, such as at the *PINOID* (*PID*) locus in Arabidopsis [65]. The formation of the *PID* gene loop is regulated by the long noncoding RNA (lncRNA) APOLO, which is located 5 kb upstream of *PID*. The *APOLO* locus is highly methylated, and this methylation is required for loop formation, as the loop is absent in RdDM pathway mutants. *APOLO* and *PID* gene transcription is also correlated with the repressive histone marks H3K27me3 and H3K9me2, which are then reduced upon transcription [65]. More examples of gene loops that are important for gene regulation will likely be identified in the near future, now that robust plant 3C protocols are available (*see* Chapter 15 [11]). Combining Hi-C (*see* Chapter 14 [10]) with BS-seq (*see* Chapters 2–4 [3–5]) in methylation mutants might also shed more light on the general involvement of DNA methylation in the establishment of chromatin loops. Recently, the first plant Hi-C experiment was conducted at the single-gene resolution and revealed that local intrachromosomal interactions occur regularly between the 5′ and 3′ ends of actively transcribed genes [61]. This Hi-C study could not identify specific chromatin marks associated with the occurrence of loops in actively transcribed genes, but silent genes with local chromatin loops showed clear enrichment for the histone variant H3.3 at their 5′ and 3′ ends and were enriched for the repressive mark H3K27me3 [66], while being depleted of repressive heterochromatic marks in flanking regions [61]. Repressive histone and DNA marks appear to enhance long-range contacts between chromatin, in line with their proposed effect on chromatin compaction. Contacts within gene bodies might be regulated by other marks, such as histone H3.3, but this requires further investigation using high-resolution Hi-C (as performed by Liu and coworkers [61]), and more specific low-throughput gene-level studies using 3C or 4C.

2 Interactions Between the Different Chromatin Modifying Mechanisms

2.1 Epigenetic Writers and Epigenetic Readers

Chromatin remodeling proteins, histone modification proteins and DNA methylases together "write" the epigenetic code that determines the structural conformation of the chromatin. This code affects the compaction of the chromatin by altering histone–histone, histone–DNA, and nucleosome–nucleosome interactions. Modified histones can also serve as docking sites for proteins called "epigenetic readers" [19, 67], which can in turn affect gene regulation. Binding of these readers is thought to recruit or stabilize additional protein complexes that play a role in gene regulation or other processes such as DNA replication and repair [67]. The original concept of epigenetic writers and readers was clearly hierarchical and assumed that histone or DNA modifiers deposited an epigenetic mark that was subsequently interpreted by different reader proteins. However, recent advances in understanding the complex interactions between epigenetic marks and gene regulation have adjusted this hierarchical view. Several writer proteins have been shown to be capable of reading as well as writing, and either recognize their own deposited mark or interpret marks deposited by other chromatin modifying complexes [68]. In addition, writers can directly interact with other protein complexes to influence transcription. For example, H3K4/H3K36 methyltransferases interact with the mRNA cap-binding complex in *Arabidopsis*, thereby increasing the efficiency of mature mRNA production [69]. Moreover, transcription factors (TFs) not only function downstream of epigenetic writers but also play important roles in their recruitment [29, 70].

2.2 Recruitment of Writers, Readers, and Erasers

The core complexes of writers and erasers that deposit and remove histone methylation, ubiquitination and acetylation marks dynamically associate with other proteins that are involved in recruitment of the complexes, interpretation of the marks or cross talk between the different complexes. The identification of TFs that interact with chromatin remodelers and modifiers and recruit them to target sequences has greatly increased since the development of a sophisticated IP-MS approach (*see* Chapter 18 [12, 71]), in which the immunoprecipitation of nonfixed protein complexes by a specific antibody is followed by mass spectrometry and label-free quantification, allowing the identification of all proteins that are associated with the protein of interest *in planta*. This approach identified a number of stage- or tissue-specific interactors that will be discussed below.

Recruitment of PRC2 to specific loci has also been associated with lncRNAs in mammals and plants, but evidence for this function is weak [72]. However, Zhu et al. [73] found that the SWI/SNF subunit SWI3B physically interacts with the lncRNA-binding protein INVOLVED IN DE NOVO 2 (IDN2), and the SWI/SNF-induced opening of chromatin loops at certain gene loci has also been associated with lncRNAs (FLC and PID, [64, 65]). This suggests that lncRNAs might be involved in recruitment of nucleosome remodelers rather than in the recruitment of PRC2 as earlier proposed by the animal field [70]. However, much is still unclear about the function and abundance of lncRNAs, and elaborate protocols to identify lncRNAs in different species using RNA-Seq, such as described in Chapter 17 [13], and to identify the interaction between RNA and proteins using an RNA-immunoprecipitation (RIP) approach (Chapter 19 [14]), are very useful to identify more lncRNAs and unveil their association with regulatory proteins.

2.3 Interactions Between Epigenetic Writers, Readers, and the Transcriptional Machinery

In addition to stage- or tissue-specific interactors and lncRNAs that are involved in the recruitment of epigenetic complexes, there are also a few epigenetic regulators that are in general more important for the cross talk between different writer complexes and the interpretation of certain histone marks. The plant specific factors EMBRYONIC FLOWER 1 (EMF1) and LIKE HETEROCHROMATIN PROTEIN 1 (LHP1), can associate with chromatin and interact with both PRC1 and PRC2, and play important roles in regulating genes involved in flowering and flower development [23, 37, 74–78]. LHP1 can read the H3K27me3 mark and interact with the PRC2 subunit MSI1 [79], thereby probably regulating a reenforcing loop that is responsible for the spreading of the H3K27me3 mark over an entire locus. Initial deposition of the H3K27me3 mark may occur through recruitment by specific factors, while subsequent spreading is mediated by LHP1. LHP1 interacts with a DNA polymerase ε catalytic subunit [80] and requires the DNA polymerase α catalytic subunit INCURVATA 2 for the repression of target genes [81], pointing to a role for LHP1 in the reestablishment of H3K27me3 after DNA replication. Interestingly, the LHP1-mediated spreading of H3K27me3 also affects global chromatin interactions and loop formation, as was shown using Hi-C [35]. This points to an important role of LHP1 and H3K27me3 in the establishment of chromatin architecture. Recruitment of LHP1 to specific loci can occur via interaction with various transcription factors (as discussed below). In addition, LHP1 also interacts with the RNA-binding protein LIF2 [82, 83], suggesting a possible role of RNAs in LHP1 recruitment or function. Target gene analyses suggest that LHP1 and EMF1 can act together to regulate target genes [76, 84], although differences in their target genes and in their mutant phenotypes also point to independent functions for both factors [77, 84, 85].

The B3 transcription factors VAL1 (VP1/ABI3-like) and VAL2 play a major role during seed maturation. They have a chromodomain and a PHD domain, which are involved in chromatin binding, and can interact with both the PRC1 and PRC2 complexes, as well as with HDACs to repress seed maturation gene expression [37, 38, 86, 87]. In the *val1/2* double mutant, seed maturation genes are de-repressed, resulting in the development of embryo-like structures instead of true leaves [38, 88]. In addition, VAL1 was recently found to trigger histone deacetylation at FLOWERING LOCUS C (FLC), a key gene in vernalization-induced flowering, suggesting that the role of VAL is not restricted to the repression of seed maturation genes [89, 90]. Interestingly, VAL1 and VAL2 also interact with the CDK8 module of the Mediator complex [87]. This multiprotein complex, which consists of at least 33 subunits in Arabidopsis, functions as a link between transcription factors and RNA polymerase II [91]. The facultative CDK8 module of this complex is associated with transcriptional repression, which might be mediated by its interactions with VAL proteins and HDACs. In line with this, the CDK8 module is important for several developmental transitions in Arabidopsis, among which the regulation of flowering time via FLC [92].

Interestingly, activation of certain genes can be achieved via interaction of the Mediator subunit MED18 with the histone acetylase HOOKLESS1 (HSL1) [93], suggesting that the activity of the Mediator complex may more generally be associated with histone acetylation levels. H3K4 histone methyltransferases enhance transcription in a different way, since at least some of them, for example Arabidopsis Trithorax-like protein 1 (ATX1), can interact with a COMPASS-like complex [40]. This complex can be recruited by certain TFs and is required for accumulation of Pol II at promoters, and for H3K4me3-mediated efficiency of transcription elongation [94].

Recruitment of chromatin modifiers by general corepressors or coactivators appears to be a common phenomenon. The transcriptional corepressor LEUNIG can interact either through association with Mediator components or by directly interacting with HDA19 [95] and the TOPLESS-related corepressors (TPRs), interact with HDA19 to achieve transcriptional repression [96–98]. Similarly, the coactivator ANGUSTIFOLIA3 (AN3) associates with SWI/SNF-type chromatin remodelers to regulate transcription [99].

In conclusion, a number of chromatin mark readers and transcription-associated factors mediate the interaction between chromatin remodeling complexes on the one hand, and transcription factors on the other hand. The above described factors are generally involved in the mediation of these interactions, although tissue-specific effects are present. For example, PRC2 interacts with LHP1 to repress target genes involved in flowering, but with the PRC1 subunits AtBMI1 and AtRING1 to repress seed maturation genes [75]. The interactions of the different epigenetic modifiers are complex, as they interact with each other, with epigenetic readers, with transcription factors (TFs) and with the transcriptional machinery. In the next section, this complexity is illustrated on the basis of several gene specific examples, which is summarized in Fig. 1.

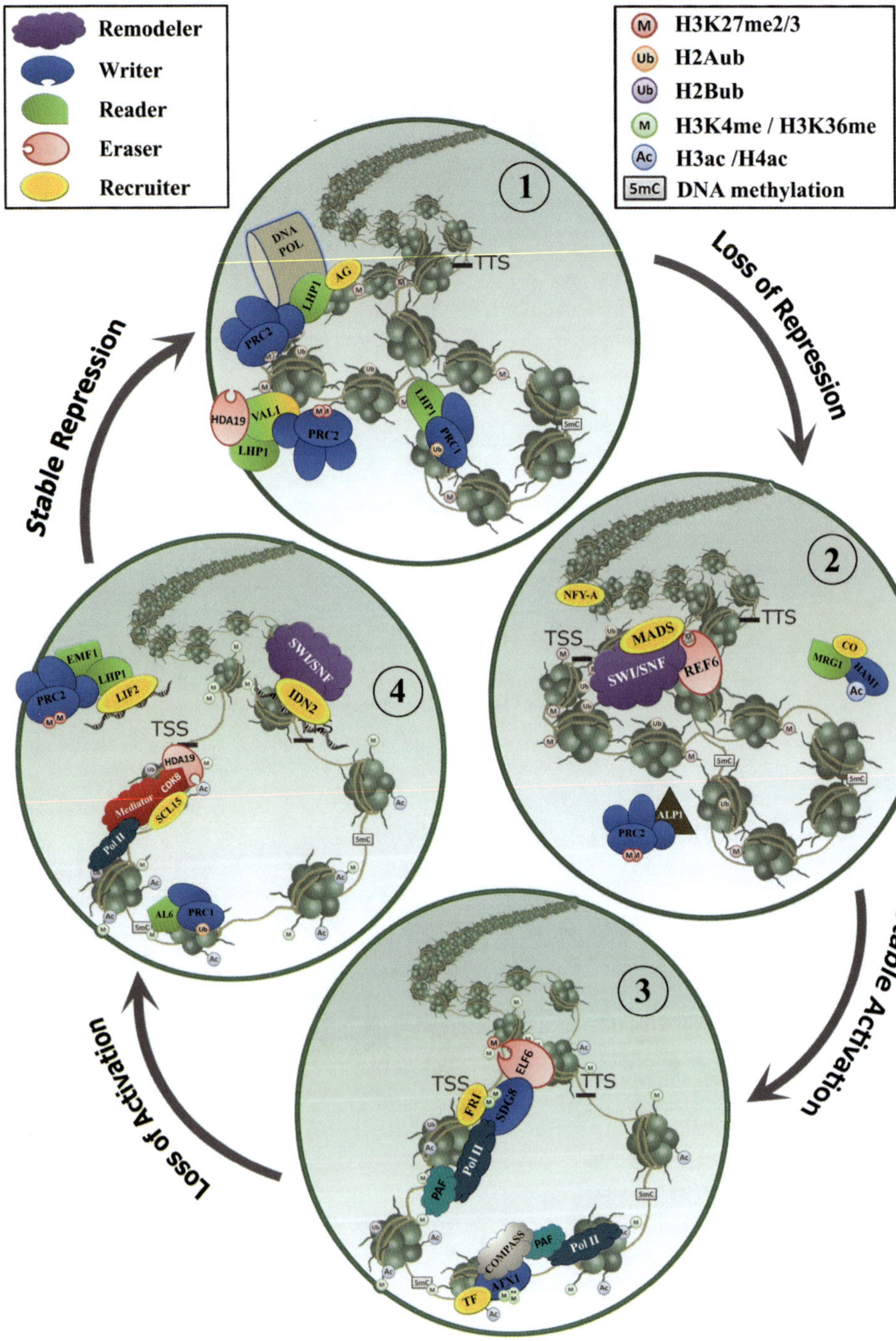

Fig. 1 Epigenetic complexes involved in the activation or repression of gene expression. The events that occur during the activation or repression of a hypothetical locus are shown in four different panels. (*1*) Stable Repression. H3K27me3 readers associate with the PRC2 complex to achieve spreading of the H3K27me3 signal. This complex is also associated with an HDAC, which removes acetylation marks from histones. Stable silencing of gene expression also involves deposition of H2Aub by PRC1. Inheritance of the transcriptionally repressed state is assured via interaction of LHP1 with subunits of the DNA polymerase complex. (*2*) Loss of

3 Complexity of Interactions at the Gene Level

Most epigenetic writers are broadly expressed and function in many cell types. However, the factors that recruit chromatin modifying complexes often show more specific expression, allowing only a subset of genes to change chromatin state. Epigenetic regulation of gene activity is particularly important during certain phase transitions, when cells change their identity. There are several transitions in the life cycle of plants that have been shown to be accompanied by major changes in the epigenome: (1) the somatic-to-reproductive transition followed by fertilization and embryogenesis, (2) the embryonic to vegetative transition during germination, (3) the transition from vegetative to reproductive development inducing flowering, and (4) flower development. Examples from Arabidopsis of these phase transitions are discussed here.

3.1 Fertilization and Seed Development

The plant germ cells are initiated from somatic cells in the anther or ovary, which marks the somatic-to-reproductive transition. This transition is accompanied by global chromatin modification events, such as chromatin decondensation, depletion of linker histones and changes in histone variants and histone modification patterns [26, 27]. The male and female germ cells each develop into small gametophytes that produce two sperm cells (male gametophyte), or an egg cell and a central cell (female gametophyte). Seed development starts with a double fertilization event, where the maternal central cell and egg cell each fuse with one sperm cell to give rise to the nourishing endosperm and the embryo, respectively.

In the central cells of Arabidopsis, rice, and maize, active DNA demethylation takes place [100], resulting in methylation differences between the maternal and paternal alleles in the endosperm. Similarly, differences in PRC2 activity between the male and female

Fig. 1 (continued) Repression. A SWI/SNF complex recruited by a TF, ejects a nucleosome close to the TSS and brings a distant enhancer region in contact with the promoter. This contact promotes the interaction between two TFs, of which one is bound to the distant enhancer element and the other to an H3K4me3 reader and a HAT. The activity of PRC2 complexes is blocked by interaction with ALP1. (*3*) Stable Activation. A TF recruits a histone H3K36me3 methyltransferase, associated with a H3K27me3 eraser to replace H3K27me3 with H3K36me3. This complex can also interact with the PAF complex, thereby facilitating transcriptional elongation. Similarly, H3K4me4 methyltransferases interact with a COMPASS-like complex that can also interact with PAF to induce transcriptional elongation. (*4*) Loss of Activation. A TF recruits the repressive unit of the Mediator complex, which interacts with an HDAC to repress transcription. lncRNAs recruit both a chromatin remodeling complex to disturb the active loop, and an LHP1-PRC2 complex for deposition of H3K27me3. Readers of the active mark H3K4me3 interact with PRC1 to facilitate deposition of H2Aub. *TSS* Transcriptional Start Site, *TTS* Transcriptional Termination Site, *HDAC* histone deacetylase, *HAT* histone acetylase, *TF* transcription factor

cells can also result in parent-of-origin dependent expression [101]. Epialleles that result in parent-of-origin effects are difficult to identify, because they generally have a mild effect on seed fitness. However, Pires et al. [102] recently described a method for the mapping of parent-of-origin effects using Bulk-seq (*see* Chapter 21 [58]), which revealed six new loci involved in seed fitness. The extent of endosperm proliferation is mainly regulated by the FIS-PRC2 complex, which ensures the repression of type I MADS box genes involved in endosperm cellularization [29, 103]. Cross talk between the maternal seed coat and the developing seed is necessary for proper seed development and involves the activity of FIS-PRC2, VRN/EMF-PRC2, the chromatin remodeler PKL, and the plant hormone auxin [104, 105]. How these factors interact with each other remains to be investigated. Several proteins interact with FIS-PRC2 subunits and their mutants display a fertilization-independent seed phenotype similar to the *fis-prc2* mutants. Applying the IP-MS approach, Derkacheva et al. [106] recently identified interactions between LHP1 and the H2A deubiquitinases UBP12 and UBP13 that are required for H3K27me3 deposition and repression of a subset of PcG target genes. The H2B deubiquitinase UBP26 has also been found to play a role in H3K27 trimethylation and repression of FIS-PRC2 targets [107], suggesting a role for both removal of H2A and H2B ubiquitination prior to PRC2-mediated H3K27 trimethylation. Interestingly, MSI1 can interact with the CUL4-DDB1 ubiquitin ligase complex involved in protein degradation, and this interaction is required for the repression of the FIS-PRC2 target *MEDEA* [108]. Similarly, association of MSI4 with CUL4-DDB1 revealed to be important for the H3K27me3-mediated repression of some flowering regulators [109], suggesting that in both cases, protein degradation is required for PRC2 activity.

3.2 Embryonic to Vegetative Transition

The transition from seed to seedling is accompanied by major chromatin modifications that promote the termination of the embryonic program. These modifications are initiated by EMF/VRN-PRC2, PRC1 as well as by HDACs, resulting in the repression of important seed maturation regulators such as *LEAFY COTYLEDON 1* (*LEC1*) and the B3 domain factors *LEC2*, *ABSCISIC ACID INSENSITIVE 3* (*ABI3*), and *FUSCA 3* (*FUS3*). As described above, the B3 domain transcription factors VAL1 and VAL2, which can also bind chromatin, are important for PRC1 and HDAC activity. Recruitment of the chromatin modifying complexes to their target loci can be regulated via VAL1/2, as they have a DNA-binding domain that recognizes a CATGCA motif in the regulatory regions of seed maturation genes [84, 88]. Recently, Merini et al. [84] showed that in the context of seed development, the PRC1 BMI subunits require VAL. The regulation of target genes by both factors is associated with the presence of the CATGCA motif and a GT-box element bound by the plant-specific trihelix DNA binding protein ARABIDOPSIS INTERACTING PROTEIN 1-LIKE 1 (ASIL1). In line with this, ASIL1 is involved in the repression of seed maturation genes after germination [110]. HDA19 can also be recruited to target genes via interaction with the GRAS domain factor SCARECROW-LIKE 15 (SCL15) [111], suggesting that both PRC1 and HDAC recruitment depends on VAL proteins as well as on other specific factors. The timely repression of active seed maturation genes, such as *ABI3* and the dormancy regulator *DELAY OF GERMINATION 1* (*DOG1*), also depends on the H3K4me3 readers ALFIN-LIKE 6 (AL6) and AL7. Both PHD-domain containing AL proteins can interact with the PRC1 components AtBMI1b and AtRING1a, and *al6 al7* double mutants display delayed seed germination and derepression of some seed maturation genes, suggesting that AL6 and AL7 are required for the recruitment of PRC1 to certain H3K4me3 marked targets, enabling the switch from H3K4me3 to H2Aub/H3K27me3 [112].

3.3 Flowering Time

The timely transition to flowering depends on the integration of both developmental and environmental cues via different pathways, including the photoperiod, autonomous, and vernalization pathways. Plants contain a few key regulators that integrate these different signals, known in Arabidopsis as FLOWERING LOCUS T (FT), CONSTANS (CO), SUPPRESSOR OF OVEREXPRESSION OF CONSTANS 1 (SOC1), and FLOWERING LOCUS C (FLC). The dynamic expression of these integrators is regulated by a plethora of different factors, among which many chromatin remodelers, allowing a robust regulation of flowering time. In particular the regulation of *FLC*, which controls the transition to flowering after a cold period (vernalization pathway), has been well-studied and has proven to be an excellent model to explore the complex epigenetic regulation that can influence gene expression [113].

FLC is active in the vegetative phase, and needs to be repressed by the autonomous pathway and/or by vernalization to allow upregulation of *FT* and *SOC1*, resulting in the transition to flowering. *FLC* is activated by the transcription factor FRIGIDA, which recruits the methyltransferase SDG8 to the *FLC* locus [114]. This activation is accompanied by SWI/SNF-type mediated incorporation of H2A.Z histone variants [115]. Although incorporation of H2A.Z in gene bodies has a negative effect on transcription, H2A.Z in +1 nucleosomes was recently found to be important for the maintenance of gene activity at some loci [24]. Active *FLC* is marked with H3K36me3 and H3K4me2/3, which is mediated by SDG8 and COMPASS-like complexes, respectively [40, 116]. In addition, SDG8 physically interacts with the H3K27me3 demethylase EARLY FLOWERING 6 (ELF6), suggesting that SDG8 actively opposes H3K27me3 deposition [113]. The increase in

FLC expression also requires the RNA polymerase II Associated Factor 1 Complex (PAF1C) [117]. PAF1C also interacts with SDG8, providing more evidence for the interaction between the transcriptional machinery and the deposition of active histone marks [113].

Repression of *FLC* via the autonomous pathway occurs through reduced transcription initiation and reduced Pol II elongation rate, mediated among others by the RNA-binding protein FCA and the H3K4me2 demethylase FLD [113, 118]. Repression of *FLC* via the vernalization pathway is more complex, and involves the switch from the active H3K36me3 state to the repressed H3K27me3 state, which is then stabilized by H3K27me3 spreading after prolonged cold. In the early phase of vernalization, a gene loop that brings the 5′ and 3′ flanking regions of *FLC* into contact and is controlled by the SWI/SNF chromatin remodeling complex [64], is disrupted. This loop disruption is accompanied by upregulation of the FLC COOLAIR antisense transcripts, and might enhance COOLAIR transcription [119]. The next phase of *FLC* silencing is the cold-induced upregulation of *VERNALIZATION INSENSITIVE 3* (*VIN3*), which encodes a PHD finger protein that associates with the PRC2 complex together with the other PHD finger proteins VRN5 and VEL1. Together with VAL1 and VAL2, the PHD finger proteins are required for the recruitment of PHD-PRC2 to the nucleation region at the border of the first exon/intron of *FLC*, resulting in the increase of H3K27me3 deposition on the *FLC* locus [89, 90, 120]. Prolonged cold ensures that all cells switch to an "*FLC*-off" state [121] that is maintained upon transfer to warm conditions by spreading of PHD-PRC2 and H3K27me3 over the *FLC* locus [120, 121]. Both VAL and PHD-PRC2 interact with HDA19 and LHP1, thereby probably mediating deacetylation and spreading of the H3K27me3 signal, respectively [89, 90]. A role for deacetylation in FLC-mediated repression was also suggested by the delayed flowering and upregulated expression of FLC in *hda5* mutants [122]. HDA5 can interact with HDA6, MSI4, and the histone deacetylase-associated protein FLOWERING LOCUS D (FLD) and also bind to the *FLC* locus. Thus, stable silencing of *FLC* involves disruption of the active chromatin loop, recruitment of PHD-PRC2 via VAL1/2 and removal of histone acetylation. The inheritance of the stable *FLC*-off state after mitosis is ensured via a cis-acting mechanism that involves the DNA polymerase alpha subunit ICU2 [81, 123].

Epigenetic regulation of the florigen *FT*, which functions as a master integrator of internal and external flowering signals downstream of FLC, is also complex. *FT* is repressed by the EMF2-PRC2 complex [124], and by an EMF1c complex, which consists of EMF1, LHP1, and the H3K4 demethylase JMJ14. The EMF1c complex is probably mediating the removal of H3K4 methylation together with spreading of H3K27me3 deposition, suggesting again that active H3K4me3 demethylation may be required for solid PcG repression [76]. Repression of floral initiation also involves the chromatin-binding PHD protein EARLY BOLTING IN SHORT DAYS (EBS), which recognizes H3K4me2/me3 and binds to *FT*. EBS interacts with the histone deacetylase HDA6, thereby preventing acetylation and the establishment of active chromatin [125]. Inheritance of the "*FT*-off" state is regulated by ICU2, as well as by the DNA polymerase ε subunit ESD7, which can recruit PRC2 during the replication process [126].

Activation of *FT* is associated with the formation of a chromatin loop that connects an enhancer located 5.3 kb upstream of *FT* with the TSS of *FT*. This loop is probably mediated by interaction between distantly bound NF-Y TFs and the *FT* activator CO, and disruption of the NF-Y binding site reduces *FT* promoter activity considerably [63, 127]. CO also needs to associate with the H3K4me3/H3K36me3 readers MORF RELATED GENE 1 (MRG1) and MRG2 to activate *FT* [128, 129]. In turn, MRG1/2 interacts with the histone 4 acetyltransferases HAM1/2, which leads to enhanced acetylation at the 5′ region of target loci [129]. Thus, the default silenced state of *FT* is regulated by both PcG complexes and histone deacetylases, which prevent premature deposition of active marks, and is enforced by flowering repressors such as FLC. Increase of *FT* transcription is initiated by the interaction between CO and NF-Y factors, resulting in loop opening, H3K4me3/H3K36me3 deposition and subsequent acetylation.

3.4 Flower Development

Repression of the floral homeotic genes prior to floral meristem determination is regulated by a CLF-containing PRC2 complex, which represses the MADS-domain floral organ specification genes *APETALA3* (*AP3*), *SEPALLATA 3* (*SEP3*) and *AGAMOUS* (*AG*) [75]. Comparison of H3K27me3 patterns in different PcG mutants by ChIP-seq revealed that the repression of floral homeotic genes by CLF-PRC2 is specifically associated with LHP1/EMF1 interaction and does not involve PRC1 [23, 75]. The interaction between PRC2 and LHP1/EMF1 can be blocked by ANTAGONIST OF LIKE HETEROCHROMATIN PROTEIN 1 (ALP1), which thereby acts as an inhibitor of PRC2 function [23]. Overrepresentation of MADS-box binding sites (CArG-boxes) and Homeobox motifs was observed in the promoters of CLF/LHP1 target genes, suggesting that the CLF-LHP1 PRC2 complex can be recruited to the promoters of target genes by MADS-domain or Homeobox TFs [75]. In agreement with this, AG is required for recruitment of PRC2 to the *WUS* locus, which is necessary for floral meristem termination [130]. Interestingly, this recruitment depends on the presence of DNA TOPOISOMERASE 1α (TOP1α), which can decrease nucleosome density at the 5′ end, allowing AG to bind and recruit PRC2 [131]. Despite the interaction between AG and PRC2, PRC2 members were not identified as interactors of floral MADS-domain proteins in IP-MS experiments [132]. Rather other epigenetic factors were identified, including the H3K27me3 demethylase RELATIVE OF EARLY FLOWERING 6 (REF6) and the chromatin remodelers SYD, BRM, CHR11 and CHR17 [28, 132], which are involved in gene activation. These data might indicate that, although MADS-domain proteins and CLF-LHP1 PRC2 bind to similar flower-associated targets they have opposite functions in the regulation of these targets. In line with this, IP-MS with CLF did not reveal any associated MADS-domain proteins [23]. To unravel the precise link between MADS-domain proteins and PcG complexes, more sophisticated experiments with a higher temporal resolution will be required. Methods that address the dynamics of histone modifications in combination with investigating actual gene regulation in synchronized floral tissues, such as described in Chapter 16 [9], will hopefully shed more light on the order of events that induce the de-repression of floral organ identity genes.

Activation of the floral organ identity genes is mediated by the chromatin-remodeling ATPases SYD and BRM, which can interact with the flower activators LEAFY (LFY), SEP3, AG, APETALA 1 (AP1), and AP3 [28, 132], and are recruited by these floral master regulators to the regulatory regions of target genes. Interestingly, BRM was also found to interact with the H3K27me3 demethylase REF6. REF6 can directly bind target DNA with a CTCTGYTY motif and this interaction enhances BRM occupancy at BRM-REF6 co-targets [133]. Since both BRM and REF6 are associated with MADS-domain complexes *in planta*, and most MADS proteins can bind to their own promoter [134], it is likely that a REF6-BRM-MADS complex boosts floral MADS protein expression via a positive feedback loop. Activation of the floral homeotic genes also depends on the activity of the BRM-like SWI/SNF factor SWI73B and the TrxG SAND-domain protein ULTRAPETALA1 (ULT1), as expression of the homeotic genes was reduced in *swi73b* mutants and in *35S:ULT1* leaves [135, 136]. In conclusion, the CLF-PRC2 mediated repression of the MADS-domain floral organ regulators, which has to be ensured throughout the vegetative phase, can be released during floral meristem development by chromatin remodelers. These are recruited to their target loci by specific transcription factors and by the DNA-binding REF6 protein, which also removes the H3K27me3 mark.

4 Conclusion and Perspectives

The case studies presented above illustrate the enormous complexity of epigenetic events that regulate gene activity. Although by far not exhaustive, these examples provide an overview of the multitude of mechanisms that are involved in the repression or induction of gene expression. The regulation of any transcription factor important for tissue identity in a certain cell lineage will probably involve complex epigenetic mechanisms. Stable gene activation can involve the recruitment of ATP-dependent nucleosome remodelers, the disruption of chromatin loops, the recruitment of H3K4me3/H3K36me3 methyltransferases and H3K27me3 demethylases, and the recruitment of histone acetylases (*see* Fig. 1). Whether there is a fixed order for these events is far from clear, but disruption of loops and repositioning of nucleosomes appear to be early events in all the cases described above. However, the various associations between different types of remodelers and modifiers, such as H3K27me3 demethylases and H3K36me3 methylases [113] and SWI/SNF type remodelers and H3K27me3 demethylases [132], suggest that a general order of events does not exist.

The recent development of sophisticated methods that enable the identification of *in planta* interactions between different epigenetic factors and transcription factors such as IP-MS (*see* Chapter 18 [12]), in combination with approaches that unravel the co-occurrence of histone modifications (Re-ChIP, *see* Chapter 6 [7]) as well as address the interdependency of histone and DNA modifications (Chapters 9 and 20 [15, 45]), will likely shed more light on the order of events necessary to stably activate or repress genes (*see* also Table 1). To better understand the role of the different complexes in the series of events that regulate the transcriptional activity of certain loci, it is essential to precisely monitor the chromatin modifications and gene expression changes that are induced by a specific complex. While experimentally challenging, the use of an inducible system (e.g., fusion of a TF to the glucocorticoid receptor (GR) domain), preferably in synchronized cells, can facilitate such an analysis (*see* Chapter 16 [9]). Alternatively, a better spatial and temporal resolution can be obtained by isolating a specific cell lineage using the INTACT system or FANS, followed by the analysis of chromatin state, TF binding profiles and transcriptome (*see* Chapters 7 and 8 [8, 137]).

In conclusion, unraveling the complex epigenetic networks that regulate gene activity is challenging and involves a broad range of molecular techniques. In this book, a comprehensive number of protocols is presented that allows to study all epigenetic aspects of gene regulation and to better understand the connections between the different epigenetic mechanisms and their connection with transcription factor-based gene regulation.

Acknowledgments

The author is very grateful to Kim Boutilier (WUR, Wageningen, the Netherlands) and Fredy Barneche (IBENS, Paris, France) for their critical reading of the manuscript and their helpful suggestions. This work was supported by a Dutch NWO-Veni grant.

References

1. Pajoro A, Muiño JM, Angenent GC, Kaufmann K (2017) Profiling nucleosome occupancy by MNase-seq: experimental protocol and computational analysis. In: Bemer M, Baroux C (eds) Plant chromatin dynamics: methods and protocols. Springer, New York, NY. doi:10.1007/978-1-4939-7318-7_11
2. Bajic M, Maher KA, Deal RB (2017) Identification of open chromatin regions in plant genomes using ATAC-Seq. In: Bemer M, Baroux C (eds) Plant chromatin dynamics: methods and protocols. Springer, New York, NY. doi:10.1007/978-1-4939-7318-7_12
3. Chen Y-R, Yu S, Zhong S (2017) Profiling DNA methylation using bisulfite sequencing (BS-Seq). In: Bemer M, Baroux C (eds) Plant chromatin dynamics: methods and protocols. Springer, New York, NY. doi:10.1007/978-1-4939-7318-7_2
4. Edelmann S, Scholten S (2017) Bisulphite sequencing using small DNA amounts. In: Bemer M, Baroux C (eds) Plant chromatin dynamics: methods and protocols. Springer, New York, NY. doi:10.1007/978-1-4939-7318-7_3
5. Kishore K, Pelizzola M (2017) Identification of differentially methylated regions in the genome of Arabidopsis thaliana. In: Bemer M, Baroux C (eds) Plant chromatin dynamics: methods and protocols. Springer, New York, NY. doi:10.1007/978-1-4939-7318-7_4
6. Desvoyes B, Vergara Z, Sequeira-Mendes J, Madeira S, Gutierrez C (2017) A rapid and efficient ChIP protocol to profile chromatin binding proteins and epigenetic modifications in Arabidopsis. In: Bemer M, Baroux C (eds) Plant chromatin dynamics: methods and protocols. Springer, New York, NY. doi:10.1007/978-1-4939-7318-7_5
7. Desvoyes B, Sequeira-Mendes J, Vergara Z, Madeira S, Gutierrez C (2017) Sequential ChIP protocol for profiling bivalent epigenetic modifications (ReChIP). In: Bemer M, Baroux C (eds) Plant chromatin dynamics: methods and protocols. Springer, New York, NY. doi:10.1007/978-1-4939-7318-7_6
8. Morao AK, Caillieux E, Colot V, Roudier F (2017) Cell type-specific profiling of chromatin modifications and associated proteins. In: Bemer M, Baroux C (eds) Plant chromatin dynamics: methods and protocols. Springer, New York, NY. doi:10.1007/978-1-4939-7318-7_8
9. Engelhorn J, Wellmer F, Carles CC (2017) Profiling histone modifications in synchronised floral tissues for quantitative resolution of chromatin and transcriptome dynamics. In: Bemer M, Baroux C (eds) Plant chromatin dynamics: methods and protocols. Springer, New York, NY. doi:10.1007/978-1-4939-7318-7_16
10. Grob S, Cavalli G (2017) A Hitchhiker's guide to chromosome conformation capture. In: Bemer M, Baroux C (eds) Plant chromatin dynamics: methods and protocols. Springer, New York, NY. doi:10.1007/978-1-4939-7318-7_14
11. Weber B, Jamge S, Stam ME (2017) 3C in maize and Arabidopsis. In: Bemer M, Baroux C (eds) Plant chromatin dynamics: methods and protocols. Springer, New York, NY. doi:10.1007/978-1-4939-7318-7_15
12. Jamge S, Angenent GC, Bemer M (2017) Identification of in planta protein-protein interactions using IP-MS. In: Bemer M, Baroux C (eds) Plant chromatin dynamics: methods and protocols. Springer, New York, NY. doi:10.1007/978-1-4939-7318-7_18
13. Lunardon A, Forestan C, Farinati S, Varotto S (2017) De novo identification of sRNA loci and non-coding RNAs by high-throughput sequencing. In: Bemer M, Baroux C (eds) Plant chromatin dynamics: methods and protocols. Springer, New York, NY. doi:10.1007/978-1-4939-7318-7_17
14. Mermaz B, Liu F, Song J (2017) RNA ImmunoPrecipitation protocol to profile RNA-binding proteins in Arabidopsis thaliana. In: Bemer M, Baroux C (eds) Plant chromatin dynamics: methods and protocols. Springer, New York, NY. doi:10.1007/978-1-4939-7318-7_19
15. Jacob Y, Voigt P (2017) In vitro assays to measure histone methyltransferase activity using different chromatin substrates. In: Bemer M, Baroux C (eds) Plant chromatin dynamics: methods and protocols. Springer, New York, NY. doi:10.1007/978-1-4939-7318-7_20
16. Over RS, Michaels SD (2014) Open and closed: the roles of linker histones in plants and animals. Mol Plant 7(3):481–491. doi:10.1093/mp/sst164
17. Probst AV (2017) A compendium of methods to analyze the spatial organization of plant chromatin. In: Bemer M, Baroux C (eds) Plant chromatin dynamics: methods and protocols. Springer, New York, NY. doi:10.1007/978-1-4939-7318-7_23
18. She W, Baroux C, Grossniklaus U (2017) Cell-type specific chromatin analysis in

whole-mount plant tissues by immunostaining. In: Bemer M, Baroux C (eds) Plant chromatin dynamics: methods and protocols. Springer, New York, NY. doi:10.1007/978-1-4939-7318-7_25

19. Rodriguez-Granados NY, Ramirez-Prado JS, Veluchamy A, Latrasse D, Raynaud C, Crespi M, Ariel F, Benhamed M (2016) Put your 3D glasses on: plant chromatin is on show. J Exp Bot. doi:10.1093/jxb/erw168
20. Feng C-M, Qiu Y, Van Buskirk EK, Yang EJ, Chen M (2014) Light-regulated gene repositioning in Arabidopsis. Nat Commun 5:3027–3027. doi:10.1038/ncomms4027
21. Bourbousse C, Mestiri I, Zabulon G, Bourge M, Formiggini F, Koini MA, Brown SC, Fransz P, Bowler C, Barneche F (2015) Light signaling controls nuclear architecture reorganization during seedling establishment. Proc Natl Acad Sci U S A 112(21): E2836–E2844. doi:10.1073/pnas.1503512112
22. Han S-K, Wu M-F, Cui S, Wagner D (2015) Roles and activities of chromatin remodeling ATPases in plants. Plant J 83(1):62–77. doi:10.1111/tpj.12877
23. Liang SC, Hartwig B, Perera P, Mora-García S, de Leau E, Thornton H, de Alves FL, Rapsilber J, Yang S, James GV, Schneeberger K, Finnegan EJ, Turck F, Goodrich J (2015) Kicking against the PRCs – a domesticated transposase antagonises silencing mediated by Polycomb group proteins and is an accessory component of Polycomb repressive complex 2. PLoS Genet 11(12):e1005660
24. Sura W, Kabza M, Karlowski WM, Bieluszewski T, Kuś-Slowinska M, Pawełoszek Ł, Sadowski J, Ziolkowski PA (2017) Dual role of the histone variant H2A.Z in transcriptional regulation of stress-response genes. Plant Cell. doi:10.1105/tpc.16.00573
25. Rutowicz K, Puzio M, Halibart-Puzio J, Lirski M, Kotliński M, Kroteń MĄ, Knizewski Ł, Lange B, Muszewska A, Śniegowska-Świerk K, Kościelniak J, Iwanicka-Nowicka R, Buza K, Janowiak F, Żmuda K, Jõesaar I, Laskowska-Kaszub K, Fogtman A, Kollist H, Zielenkiewicz P, Tiuryn J, Siedlecki P, Swiezewski S, Ginalski K, Koblowska M, Archacki R, Wilczynski B, Rapacz M, Jerzmanowski A (2015) A specialized histone H1 variant is required for adaptive responses to complex abiotic stress and related DNA methylation in Arabidopsis. Plant Physiol 169(3):2080–2101. doi:10.1104/pp.15.00493
26. She W, Grimanelli D, Rutowicz K, Whitehead MWJ, Puzio M, Kotliński M, Jerzmanowski A, Baroux C (2013) Chromatin reprogramming during the somatic-to-reproductive cell fate transition in plants. Development 140(19):4008
27. She W, Baroux C (2015) Chromatin dynamics in pollen mother cells underpin a common scenario at the somatic-to-reproductive fate transition of both the male and female lineages in Arabidopsis. Front Plant Sci 6:294. doi:10.3389/fpls.2015.00294
28. Wu M-F, Sang Y, Bezhani S, Yamaguchi N, Han S-K, Li Z, Su Y, Slewinski TL, Wagner D (2012) SWI2/SNF2 chromatin remodeling ATPases overcome Polycomb repression and control floral organ identity with the LEAFY and SEPALLATA3 transcription factors. Proc Natl Acad Sci U S A 109(9):3576–3581. doi:10.1073/pnas.1113409109
29. Bemer M, Grossniklaus U (2012) Dynamic regulation of Polycomb group activity during plant development. Curr Opin Plant Biol 15(5):523–529. doi:10.1016/j.pbi.2012.09.006
30. Kotliński M, Rutowicz K, Kniżewski Ł, Palusiński A, Olędzki J, Fogtman A, Rubel T, Koblowska M, Dadlez M, Ginalski K, Jerzmanowski A (2016) Histone H1 variants in Arabidopsis are subject to numerous post-translational modifications, both conserved and previously unknown in histones, suggesting complex functions of H1 in plants. PLoS One 11(1):e0147908. doi:10.1371/journal.pone.0147908
31. Roudier F, Ahmed I, Bérard C, Sarazin A, Mary-Huard T, Cortijo S, Bouyer D, Caillieux E, Duvernois-Berthet E, Al-Shikhley L, Giraut L, Després B, Drevensek S, Barneche F, Dèrozier S, Brunaud V, Aubourg S, Schnittger A, Bowler C, Martin-Magniette M-L, Robin S, Caboche M, Colot V (2011) Integrative epigenomic mapping defines four main chromatin states in Arabidopsis. EMBO J 30(10):1928–1938. doi:10.1038/emboj.2011.103
32. Sequeira-Mendes J, Aragüez I, Peiró R, Mendez-Giraldez R, Zhang X, Jacobsen SE, Bastolla U, Gutierrez C (2014) The functional topography of the Arabidopsis genome is organized in a reduced number of linear motifs of chromatin states. Plant Cell 26(6): 2351–2366. doi:10.1105/tpc.114.124578
33. Müller-Xing R, Clarenz O, Pokorny L, Goodrich J, Schubert D (2014) Polycomb-group proteins and FLOWERING LOCUS T maintain commitment to flowering in Arabidopsis thaliana. Plant Cell 26(6):2457–2471. doi:10.1105/tpc.114.123323
34. Pajoro A, Madrigal P, Muiño JM, Matus JT, Jin J, Mecchia MA, Debernardi JM, Palatnik

JF, Balazadeh S, Arif M, Ó'Maoiléidigh DS, Wellmer F, Krajewski P, Riechmann J-L, Angenent GC, Kaufmann K (2014) Dynamics of chromatin accessibility and gene regulation by MADS-domain transcription factors in flower development. Genome Biol 15(3): R41–R41. doi:10.1186/gb-2014-15-3-r41

35. Veluchamy A, Jégu T, Ariel F, Latrasse D, Mariappan KG, Kim S-K, Crespi M, Hirt H, Bergounioux C, Raynaud C, Benhamed M (2016) LHP1 regulates H3K27me3 spreading and shapes the three-dimensional conformation of the Arabidopsis genome. PLoS One 11(7):e0158936. doi:10.1371/journal.pone.0158936

36. Zhang W, Jiang J (2015) Genome-wide mapping of DNase I hypersensitive sites in plants. In: Alonso JM, Stepanova AN (eds) Plant functional genomics: methods and protocols. Springer, New York, pp 71–89. doi:10.1007/978-1-4939-2444-8_4

37. Merini W, Calonje M (2015) PRC1 is taking the lead in PcG repression. Plant J 83 (1):110–120. doi:10.1111/tpj.12818

38. Yang C, Bratzel F, Hohmann N, Koch M, Turck F, Calonje M (2013) VAL- and AtBMI1-mediated H2Aub initiate the switch from embryonic to postgerminative growth in Arabidopsis. Curr Biol 23(14):1324–1329. doi:10.1016/j.cub.2013.05.050

39. Ng DWK, Wang T, Chandrasekharan MB, Aramayo R, Kertbundit S, Hall TC (2007) Plant SET domain-containing proteins: structure, function and regulation. Biochim Biophys Acta 1769(5–6):316–329. doi:10.1016/j.bbaexp.2007.04.003

40. Jiang D, Kong NC, Gu X, Li Z, He Y (2011) Arabidopsis COMPASS-like complexes mediate histone H3 lysine-4 trimethylation to control floral transition and plant development. PLoS Genet 7(3):e1001330. doi:10.1371/journal.pgen.1001330

41. Chen X, Hu Y, Zhou D-X (2011) Epigenetic gene regulation by plant Jumonji group of histone demethylase. Biochim Biophys Acta 1809(8):421–426. doi:10.1016/j.bbagrm.2011.03.004

42. Xiao J, Lee U-S, Wagner D (2016) Tug of war: adding and removing histone lysine methylation in Arabidopsis. Curr Opin Plant Biol 34:41–53. doi:10.1016/j.pbi.2016.08.002

43. Pandey R, Müller A, Napoli CA, Selinger DA, Pikaard CS, Richards EJ, Bender J, Mount DW, Jorgensen RA (2002) Analysis of histone acetyltransferase and histone deacetylase families of Arabidopsis thaliana suggests functional diversification of chromatin modification among multicellular eukaryotes. Nucleic Acids Res 30(23):5036–5055

44. Mahrez W, Arellano MST, Moreno-Romero J, Nakamura M, Shu H, Nanni P, Köhler C, Gruissem W, Hennig L (2016) H3K36ac is an evolutionary conserved plant histone modification that marks active genes. Plant Physiol 170(3):1566–1577. doi:10.1104/pp.15.01744

45. Mahrez W, Hennig L (2017) Mapping of histone modifications in plants by tandem mass spectrometry. In: Bemer M, Baroux C (eds) Plant chromatin dynamics: methods and protocols. Springer, New York, NY. doi:10.1007/978-1-4939-7318-7_9

46. Berr A, Shafiq S, Shen W-H (2011) Histone modifications in transcriptional activation during plant development. Biochim Biophys Acta 1809(10):567–576. doi:10.1016/j.bbagrm.2011.07.001

47. Du J, Johnson LM, Jacobsen SE, Patel DJ (2015) DNA methylation pathways and their crosstalk with histone methylation. Nat Rev Mol Cell Biol 16(9):519–532. doi:10.1038/nrm4043

48. To TK, Saze H, Kakutani T (2015) DNA methylation within transcribed regions. Plant Physiol 168(4):1219–1225. doi:10.1104/pp.15.00543

49. Stroud H, Do T, Du J, Zhong X, Feng S, Johnson L, Patel DJ, Jacobsen SE (2014) Non-CG methylation patterns shape the epigenetic landscape in Arabidopsis. Nat Struct Mol Biol 21(1):64–72. doi:10.1038/nsmb.2735

50. Soppe WJJ, Jasencakova Z, Houben A, Kakutani T, Meister A, Huang MS, Jacobsen SE, Schubert I, Fransz PF (2002) DNA methylation controls histone H3 lysine 9 methylation and heterochromatin assembly in Arabidopsis. EMBO J 21(23):6549–6559. doi:10.1093/emboj/cdf657

51. Tran RK, Zilberman D, de Bustos C, Ditt RF, Henikoff JG, Lindroth AM, Delrow J, Boyle T, Kwong S, Bryson TD, Jacobsen SE, Henikoff S (2005) Chromatin and siRNA pathways cooperate to maintain DNA methylation of small transposable elements in Arabidopsis. Genome Biol 6(11):R90–R90. doi:10.1186/gb-2005-6-11-r90

52. Greenberg MVC, Deleris A, Hale CJ, Liu A, Feng S, Jacobsen SE (2013) Interplay between active chromatin marks and RNA-directed DNA methylation in Arabidopsis thaliana. PLoS Genet 9(11):e1003946. doi:10.1371/journal.pgen.1003946

53. Aufsatz W, Mette M, van der Winden J, Matzke M, Matzke AJM (2002) HDA6, a putative histone deacetylase needed to enhance DNA methylation induced by double-stranded RNA. EMBO J 21 (24):6832–6841. doi:10.1093/emboj/cdf663
54. Tang K, Lang Z, Zhang H, Zhu J-K (2016) The DNA demethylase ROS1 targets genomic regions with distinct chromatin modifications. Nat Plants 2:16169. doi:10.1038/nplants.2016.169
55. Dowen RH, Pelizzola M, Schmitz RJ, Lister R, Dowen JM, Nery JR, Dixon JE, Ecker JR (2012) Widespread dynamic DNA methylation in response to biotic stress. Proc Natl Acad Sci U S A 109(32):E2183–E2191. doi:10.1073/pnas.1209329109
56. Zhang Y-Y, Fischer M, Colot V, Bossdorf O (2013) Epigenetic variation creates potential for evolution of plant phenotypic plasticity. New Phytol 197(1):314–322. doi:10.1111/nph.12010
57. Lauss K, Keurentjes JJB (2017) epiQTL mapping in Arabidopsis thaliana. In: Bemer M, Baroux C (eds) Plant chromatin dynamics: methods and protocols. Springer, New York, NY. doi:10.1007/978-1-4939-7318-7_22
58. Pires ND, Grossniklaus U (2017) Identification of parent-of-origin-dependent QTLs using bulk-segregant sequencing (Bulk-Seq). In: Bemer M, Baroux C (eds) Plant chromatin dynamics: methods and protocols. Springer, New York, NY. doi:10.1007/978-1-4939-7318-7_21
59. Feng S, Cokus SJ, Schubert V, Zhai J, Pellegrini M, Jacobsen SE (2014) Genome-wide Hi-C analyses in wild type and mutants reveal high-resolution chromatin interactions in Arabidopsis. Mol Cell 55(5):694–707. doi:10.1016/j.molcel.2014.07.008
60. Grob S, Schmid Marc W, Grossniklaus U (2014) Hi-C analysis in Arabidopsis identifies the KNOT, a structure with similarities to the flamenco locus of Drosophila. Mol Cell 55 (5):678–693. doi:10.1016/j.molcel.2014.07.009
61. Liu C, Wang C, Wang G, Becker C, Zaidem M, Weigel D (2016) Genome-wide analysis of chromatin packing in Arabidopsis thaliana at single-gene resolution. Genome Res 26 (8):1057–1068. doi:10.1101/gr.204032.116
62. Louwers M, Bader R, Haring M, van Driel R, de Laat W, Stam M (2009) Tissue- and expression level-specific chromatin looping at maize b1 Epialleles. Plant Cell 21 (3):832–842. doi:10.1105/tpc.108.064329
63. Cao S, Kumimoto RW, Gnesutta N, Calogero AM, Mantovani R, Holt BF (2014) A distal CCAAT/NUCLEAR FACTOR Y complex promotes chromatin looping at the FLOWERING LOCUS T promoter and regulates the timing of flowering in Arabidopsis. Plant Cell 26(3):1009–1017. doi:10.1105/tpc.113.120352
64. Jégu T, Latrasse D, Delarue M, Hirt H, Domenichini S, Ariel F, Crespi M, Bergounioux C, Raynaud C, Benhamed M (2014) The BAF60 subunit of the SWI/SNF chromatin-remodeling complex directly controls the formation of a gene loop at FLOWERING LOCUS C in Arabidopsis. Plant Cell 26(2):538–551. doi:10.1105/tpc.113.114454
65. Ariel F, Jegu T, Latrasse D, Romero-Barrios N, Christ A, Benhamed M, Crespi M (2014) Noncoding transcription by alternative RNA polymerases dynamically regulates an auxin-driven chromatin loop. Mol Cell 55 (3):383–396. doi:10.1016/j.molcel.2014.06.011
66. Wang C, Liu C, Roqueiro D, Grimm D, Schwab R, Becker C, Lanz C, Weigel D (2015) Genome-wide analysis of local chromatin packing in Arabidopsis thaliana. Genome Res 25(2):246–256. doi:10.1101/gr.170332.113
67. Musselman CA, Lalonde M-E, Côté J, Kutateladze TG (2012) Perceiving the epigenetic landscape through histone readers. Nat Struct Mol Biol 19(12):1218–1227. doi:10.1038/nsmb.2436
68. Zhang T, Cooper S, Brockdorff N (2015) The interplay of histone modifications – writers that read. EMBO Rep 16(11):1467–1481. doi:10.15252/embr.201540945
69. Li Z, Jiang D, Fu X, Luo X, Liu R, He Y (2016) Coupling of histone methylation and RNA processing by the nuclear mRNA cap-binding complex. Nat Plants 2:16015. doi:10.1038/nplants.2016.15
70. Hecker A, Brand LH, Peter S, Simoncello N, Kilian J, Harter K, Gaudin V, Wanke D (2015) The Arabidopsis GAGA-binding factor BPC6 recruits PRC1 component LHP1 to GAGA DNA-motifs. Plant Physiol. doi:10.1104/pp.15.00409
71. Smaczniak C, Li N, Boeren S, America T, van Dongen W, Goerdayal SS, de Vries S, Angenent GC, Kaufmann K (2012) Proteomics-based identification of low-abundance signaling and regulatory protein complexes in native plant tissues. Nat Protoc 7 (12):2144–2158

72. Davidovich C, Cech TR (2015) The recruitment of chromatin modifiers by long noncoding RNAs: lessons from PRC2. RNA 21 (12):2007–2022. doi:10.1261/rna.053918.115
73. Zhu Y, Rowley MJ, Böhmdorfer G, Wierzbicki Andrzej T (2013) A SWI/SNF chromatin-remodeling complex acts in noncoding RNA-mediated transcriptional silencing. Mol Cell 49(2):298–309. doi:10.1016/j.molcel.2012.11.011
74. Bratzel F, López-Torrejón G, Koch M, Del Pozo JC, Calonje M (2010) Keeping cell identity in Arabidopsis requires PRC1 RING-finger homologs that catalyze H2A monoubiquitination. Curr Biol 20 (20):1853–1859. doi:10.1016/j.cub.2010.09.046
75. Wang H, Liu C, Cheng J, Liu J, Zhang L, He C, Shen W-H, Jin H, Xu L, Zhang Y (2016) Arabidopsis flower and embryo developmental genes are repressed in seedlings by different combinations of Polycomb group proteins in association with distinct sets of Cis-regulatory elements. PLoS Genet 12(1): e1005771. doi:10.1371/journal.pgen.1005771
76. Wang Y, Gu X, Yuan W, Schmitz Robert J, He Y (2014) Photoperiodic control of the floral transition through a distinct Polycomb repressive complex. Dev Cell 28 (6):727–736. doi:10.1016/j.devcel.2014.01.029
77. Calonje M, Sanchez R, Chen L, Sung ZR (2008) EMBRYONIC FLOWER1 participates in Polycomb group-mediated AG gene silencing in Arabidopsis. Plant Cell 20 (2):277–291. doi:10.1105/tpc.106.049957
78. Kim SY, Lee J, Eshed-Williams L, Zilberman D, Sung ZR (2012) EMF1 and PRC2 cooperate to repress key regulators of Arabidopsis development. PLoS Genet 8(3):e1002512. doi:10.1371/journal.pgen.1002512
79. Derkacheva M, Steinbach Y, Wildhaber T, Mozgová I, Mahrez W, Nanni P, Bischof S, Gruissem W, Hennig L (2013) Arabidopsis MSI1 connects LHP1 to PRC2 complexes. EMBO J 32(14):2073–2085. doi:10.1038/emboj.2013.145
80. Del Olmo I, López-González L, Martín-Trillo MM, Martínez-Zapater JM, Piñeiro M, Jarillo JA (2010) EARLY IN SHORT DAYS 7 (ESD7) encodes the catalytic subunit of DNA polymerase epsilon and is required for flowering repression through a mechanism involving epigenetic gene silencing. Plant J 61 (4):623–636. doi:10.1111/j.1365-313X.2009.04093.x
81. Hyun Y, Yun H, Park K, Ohr H, Lee O, Kim D-H, Sung S, Choi Y (2012) The catalytic subunit of Arabidopsis DNA polymerase α ensures stable maintenance of histone modification. Development 140(1):156
82. Latrasse D, Germann S, Houba-Hérin N, Dubois E, Bui-Prodhomme D, Hourcade D, Juul-Jensen T, Le Roux C, Majira A, Simoncello N, Granier F, Taconnat L, Renou J-P, Gaudin V (2011) Control of flowering and cell fate by LIF2, an RNA binding partner of the Polycomb complex component LHP1. PLoS One 6(1):e16592. doi:10.1371/journal.pone.0016592
83. Molitor AM, Latrasse D, Zytnicki M, Andrey P, Houba-Hérin N, Hachet M, Battail C, Del Prete S, Alberti A, Quesneville H, Gaudin V (2016) The Arabidopsis hnRNP-Q protein LIF2 and the PRC1 subunit LHP1 function in concert to regulate the transcription of stress-responsive genes. Plant Cell 28 (9):2197–2211. doi:10.1105/tpc.16.00244
84. Merini W, Romero-Campero FJ, Gomez-Zambrano A, Zhou Y, Turck F, Calonje M (2016) The Arabidopsis Polycomb Repressive Complex 1 (PRC1) components AtBMI1A, B and C impact gene networks throughout all stages of plant development. Plant Physiol. doi:10.1104/pp.16.01259
85. Zhou Y, Hartwig B, James GV, Schneeberger K, Turck F (2016) Complementary activities of TELOMERE REPEAT BINDING proteins and Polycomb group complexes in transcriptional regulation of target genes. Plant Cell 28(1):87–101. doi:10.1105/tpc.15.00787
86. Zhou Y, Tan B, Luo M, Li Y, Liu C, Chen C, Yu C-W, Yang S, Dong S, Ruan J, Yuan L, Zhang Z, Zhao L, Li C, Chen H, Cui Y, Wu K, Huang S (2013) HISTONE DEACETYLASE19 interacts with HSL1 and participates in the repression of seed maturation genes in Arabidopsis seedlings. Plant Cell 25 (1):134–148. doi:10.1105/tpc.112.096313
87. Chhun T, Chong SY, Park BS, Wong ECC, Yin J-L, Kim M, Chua N-H (2016) HSI2 repressor recruits MED13 and HDA6 to down-regulate seed maturation gene expression directly during Arabidopsis early seedling growth. Plant Cell Physiol 57(8):1689–1706. doi:10.1093/pcp/pcw095
88. Suzuki M, Wang HHY, McCarty DR (2007) Repression of the LEAFY COTYLEDON 1/B3 regulatory network in plant embryo development by VP1/ABSCISIC ACID INSENSITIVE 3-LIKE B3 genes. Plant Physiol 143 (2):902–911. doi:10.1104/pp.106.092320

89. Qüesta JI, Song J, Geraldo N, An H, Dean C (2016) Arabidopsis transcriptional repressor VAL1 triggers Polycomb silencing at FLC during vernalization. Science 353(6298):485
90. Yuan W, Luo X, Li Z, Yang W, Wang Y, Liu R, Du J, He Y (2016) A cis cold memory element and a trans epigenome reader mediate Polycomb silencing of FLC by vernalization in Arabidopsis. Nat Genet, advance online publication. doi:10.1038/ng.3712
91. Yang Y, Li L, Qu L-J (2016) Plant Mediator complex and its critical functions in transcription regulation. J Integr Plant Biol 58 (2):106–118. doi:10.1111/jipb.12377
92. Gillmor CS, Silva-Ortega CO, Willmann MR, Buendía-Monreal M, Poethig RS (2014) The Arabidopsis Mediator CDK8 module genes CCT (MED12) and GCT (MED13) are global regulators of developmental phase transitions. Development 141 (23):4580–4589. doi:10.1242/dev.111229
93. Liao C-J, Lai Z, Lee S, Yun D-J, Mengiste T (2016) Arabidopsis HOOKLESS1 regulates responses to pathogens and abscisic acid through interaction with MED18 and acetylation of WRKY33 and ABI5 chromatin. Plant Cell 28(7):1662–1681. doi:10.1105/tpc.16.00105
94. Ding Y, Ndamukong I, Xu Z, Lapko H, Fromm M, Avramova Z (2012) ATX1-generated H3K4me3 is required for efficient elongation of transcription, not initiation, at ATX1-regulated genes. PLoS Genet 8(12): e1003111. doi:10.1371/journal.pgen.1003111
95. Gonzalez D, Bowen AJ, Carroll TS, Conlan RS (2007) The transcription corepressor LEUNIG interacts with the histone deacetylase HDA19 and mediator components MED14 (SWP) and CDK8 (HEN3) to repress transcription. Mol Cell Biol 27 (15):5306–5315. doi:10.1128/mcb.01912-06
96. Causier B, Ashworth M, Guo W, Davies B (2012) The TOPLESS interactome: a framework for gene repression in Arabidopsis. Plant Physiol 158(1):423–438. doi:10.1104/pp.111.186999
97. Krogan NT, Hogan K, Long JA (2012) APETALA2 negatively regulates multiple floral organ identity genes in Arabidopsis by recruiting the co-repressor TOPLESS and the histone deacetylase HDA19. Development 139 (22):4180–4190. doi:10.1242/dev.085407
98. Ryu H, Cho H, Bae W, Hwang I (2014) Control of early seedling development by BES1/TPL/HDA19-mediated epigenetic regulation of ABI3. Nat Commun 5:4138. doi:10.1038/ncomms5138
99. Vercruyssen L, Verkest A, Gonzalez N, Heyndrickx KS, Eeckhout D, Han S-K, Jégu T, Archacki R, Van Leene J, Andriankaja M, De Bodt S, Abeel T, Coppens F, Dhondt S, De Milde L, Vermeersch M, Maleux K, Gevaert K, Jerzmanowski A, Benhamed M, Wagner D, Vandepoele K, De Jaeger G, Inzé D (2014) ANGUSTIFOLIA3 binds to SWI/SNF chromatin remodeling complexes to regulate transcription during Arabidopsis leaf development. Plant Cell 26(1):210–229. doi:10.1105/tpc.113.115907
100. Park K, Kim MY, Vickers M, Park J-S, Hyun Y, Okamoto T, Zilberman D, Fischer RL, Feng X, Choi Y, Scholten S (2016) DNA demethylation is initiated in the central cells of Arabidopsis and rice. Proc Natl Acad Sci U S A 113(52):15138–15143. doi:10.1073/pnas.1619047114
101. Raissig MT, Bemer M, Baroux C, Grossniklaus U (2013) Genomic imprinting in the Arabidopsis embryo is partly regulated by PRC2. PLoS Genet 9(12):e1003862. doi:10.1371/journal.pgen.1003862
102. Pires ND, Bemer M, Müller LM, Baroux C, Spillane C, Grossniklaus U (2016) Quantitative genetics identifies cryptic genetic variation involved in the paternal regulation of seed development. PLoS Genet 12(1): e1005806. doi:10.1371/journal.pgen.1005806
103. Hehenberger E, Kradolfer D, Köhler C (2012) Endosperm cellularization defines an important developmental transition for embryo development. Development 139 (11):2031
104. Roszak P, Köhler C (2011) Polycomb group proteins are required to couple seed coat initiation to fertilization. Proc Natl Acad Sci U S A 108(51):20826–20831. doi:10.1073/pnas.1117111108
105. Carter B, Henderson JT, Svedin E, Fiers M, McCarthy K, Smith A, Guo C, Bishop B, Zhang H, Riksen T, Shockley A, Dilkes BP, Boutilier K, Ogas J (2016) Cross-talk between sporophyte and gametophyte generations is promoted by CHD3 chromatin remodelers in Arabidopsis thaliana. Genetics 203(2):817
106. Derkacheva M, Liu S, Figueiredo DD, Gentry M, Mozgova I, Nanni P, Tang M, Mannervik M, Köhler C, Hennig L (2016) H2A deubiquitinases UBP12/13 are part of the Arabidopsis polycomb group protein system. Nat

Plants 2:16126. doi:10.1038/nplants.2016.126

107. Luo M, Luo M-Z, Buzas D, Finnegan J, Helliwell C, Dennis ES, Peacock WJ, Chaudhury A (2008) UBIQUITIN-SPECIFIC PROTEASE 26 is required for seed development and the repression of PHERES1 in Arabidopsis. Genetics 180(1):229–236. doi:10.1534/genetics.108.091736
108. Dumbliauskas E, Lechner E, Jaciubek M, Berr A, Pazhouhandeh M, Alioua M, Cognat V, Brukhin V, Koncz C, Grossniklaus U, Molinier J, Genschik P (2011) The Arabidopsis CUL4–DDB1 complex interacts with MSI1 and is required to maintain MEDEA parental imprinting. EMBO J 30(4):731–743. doi:10.1038/emboj.2010.359
109. Pazhouhandeh M, Molinier J, Berr A, Genschik P (2011) MSI4/FVE interacts with CUL4–DDB1 and a PRC2-like complex to control epigenetic regulation of flowering time in Arabidopsis. Proc Natl Acad Sci U S A 108(8):3430–3435. doi:10.1073/pnas.1018242108
110. Gao M-J, Lydiate DJ, Li X, Lui H, Gjetvaj B, Hegedus DD, Rozwadowski K (2009) Repression of seed maturation genes by a trihelix transcriptional repressor in Arabidopsis seedlings. Plant Cell 21(1):54–71. doi:10.1105/tpc.108.061309
111. Gao M-J, Li X, Huang J, Gropp GM, Gjetvaj B, Lindsay DL, Wei S, Coutu C, Chen Z, Wan X-C, Hannoufa A, Lydiate DJ, Gruber MY, Chen ZJ, Hegedus DD (2015) SCARECROW-LIKE15 interacts with HISTONE DEACETYLASE19 and is essential for repressing the seed maturation programme. Nat Commun 6:7243. doi:10.1038/ncomms8243
112. Molitor AM, Bu Z, Yu Y, Shen W-H (2014) Arabidopsis AL PHD-PRC1 complexes promote seed germination through H3K4me3-to-H3K27me3 chromatin state switch in repression of seed developmental genes. PLoS Genet 10(1):e1004091. doi:10.1371/journal.pgen.1004091
113. Yang H, Howard M, Dean C (2016) Physical coupling of activation and derepression activities to maintain an active transcriptional state at FLC. Proc Natl Acad Sci U S A 113 (33):9369–9374. doi:10.1073/pnas.1605733113
114. Choi K, Kim J, Hwang H-J, Kim S, Park C, Kim SY, Lee I (2011) The FRIGIDA complex activates transcription of FLC, a strong flowering repressor in Arabidopsis, by recruiting chromatin modification factors. Plant Cell 23 (1):289–303. doi:10.1105/tpc.110.075911
115. Zhang C, Cao L, Rong L, An Z, Zhou W, Ma J, Shen W-H, Zhu Y, Dong A (2015) The chromatin-remodeling factor AtINO80 plays crucial roles in genome stability maintenance and in plant development. Plant J 82 (4):655–668. doi:10.1111/tpj.12840
116. Kim SY, He Y, Jacob Y, Noh Y-S, Michaels S, Amasino R (2005) Establishment of the vernalization-responsive, winter-annual habit in Arabidopsis requires a putative histone H3 methyl transferase. Plant Cell 17 (12):3301–3310. doi:10.1105/tpc.105.034645
117. He Y, Doyle MR, Amasino RM (2004) PAF1-complex-mediated histone methylation of FLOWERING LOCUS C chromatin is required for the vernalization-responsive, winter-annual habit in Arabidopsis. Genes Dev 18(22):2774–2784. doi:10.1101/gad.1244504
118. Wu Z, Ietswaart R, Liu F, Yang H, Howard M, Dean C (2016) Quantitative regulation of FLC via coordinated transcriptional initiation and elongation. Proc Natl Acad Sci U S A 113 (1):218–223. doi:10.1073/pnas.1518369112
119. Crevillén P, Sonmez C, Wu Z, Dean C (2013) A gene loop containing the floral repressor FLC is disrupted in the early phase of vernalization. EMBO J 32(1):140–148. doi:10.1038/emboj.2012.324
120. De Lucia F, Crevillen P, Jones AME, Greb T, Dean C (2008) A PHD-Polycomb Repressive Complex 2 triggers the epigenetic silencing of FLC during vernalization. Proc Natl Acad Sci U S A 105(44):16831–16836. doi:10.1073/pnas.0808687105
121. Angel A, Song J, Yang H, Questa JI, Dean C, Howard M (2015) Vernalizing cold is registered digitally at FLC. Proc Natl Acad Sci U S A 112(13):4146–4151. doi:10.1073/pnas.1503100112
122. Luo M, Tai R, Yu C-W, Yang S, Chen C-Y, Lin W-D, Schmidt W, Wu K (2015) Regulation of flowering time by the histone deacetylase HDA5 in Arabidopsis. Plant J 82 (6):925–936. doi:10.1111/tpj.12868
123. Berry S, Hartley M, Olsson TSG, Dean C, Howard M (2015) Local chromatin environment of a Polycomb target gene instructs its own epigenetic inheritance. Elife 4:e07205. doi:10.7554/eLife.07205
124. Jiang D, Wang Y, Wang Y, He Y (2008) Repression of FLOWERING LOCUS C and FLOWERING LOCUS T by the Arabidopsis Polycomb Repressive Complex 2 components. PLoS One 3(10):e3404. doi:10.1371/journal.pone.0003404

125. López-González L, Mouriz A, Narro-Diego L, Bustos R, Martínez-Zapater JM, Jarillo JA, Piñeiro M (2014) Chromatin-dependent repression of the Arabidopsis floral integrator genes involves plant specific PHD-containing proteins. Plant Cell 26(10):3922–3938. doi:10.1105/tpc.114.130781

126. Olmo ID, López JA, Vázquez J, Raynaud C, Piñeiro M, Jarillo JA (2016) Arabidopsis DNA polymerase ε recruits components of Polycomb repressor complex to mediate epigenetic gene silencing. Nucleic Acids Res 44 (12):5597–5614. doi:10.1093/nar/gkw156

127. Liu C, Weigel D (2015) Chromatin in 3D: progress and prospects for plants. Genome Biol 16(1):170. doi:10.1186/s13059-015-0738-6

128. Bu Z, Yu Y, Li Z, Liu Y, Jiang W, Huang Y, Dong A-W (2014) Regulation of Arabidopsis flowering by the histone mark readers MRG1/2 via interaction with CONSTANS to modulate FT expression. PLoS Genet 10 (9):e1004617. doi:10.1371/journal.pgen.1004617

129. Xu Y, Gan E-S, Zhou J, Wee W-Y, Zhang X, Ito T (2014) Arabidopsis MRG domain proteins bridge two histone modifications to elevate expression of flowering genes. Nucleic Acids Res 42(17):10960–10974. doi:10.1093/nar/gku781

130. Liu X, Kim YJ, Müller R, Yumul RE, Liu C, Pan Y, Cao X, Goodrich J, Chen X (2011) AGAMOUS terminates floral stem cell maintenance in Arabidopsis by directly repressing WUSCHEL through recruitment of Polycomb group proteins. Plant Cell 23 (10):3654–3670. doi:10.1105/tpc.111.091538

131. Liu X, Gao L, Dinh TT, Shi T, Li D, Wang R, Guo L, Xiao L, Chen X (2014) DNA topoisomerase I affects Polycomb group protein-mediated epigenetic regulation and plant development by altering nucleosome distribution in Arabidopsis. Plant Cell 26(7): 2803–2817. doi:10.1105/tpc.114.124941

132. Smaczniak C, Immink RGH, Muiño JM, Blanvillain R, Busscher M, Busscher-Lange J, Dinh QD, Liu S, Westphal AH, Boeren S, Parcy F, Xu L, Carles CC, Angenent GC, Kaufmann K (2012) Characterization of MADS-domain transcription factor complexes in Arabidopsis flower development. Proc Natl Acad Sci U S A 109 (5):1560–1565. doi:10.1073/pnas.1112871109

133. Li C, Gu L, Gao L, Chen C, Wei C-Q, Qiu Q, Chien C-W, Wang S, Jiang L, Ai L-F, Chen C-Y, Yang S, Nguyen V, Qi Y, Snyder MP, Burlingame AL, Kohalmi SE, Huang S, Cao X, Wang Z-Y, Wu K, Chen X, Cui Y (2016) Concerted genomic targeting of H3K27 demethylase REF6 and chromatin-remodeling ATPase BRM in Arabidopsis. Nat Genet 48(6):687–693. doi:10.1038/ng.3555

134. Smaczniak C, Immink RGH, Angenent GC, Kaufmann K (2012) Developmental and evolutionary diversity of plant MADS-domain factors: insights from recent studies. Development 139(17):3081

135. Monfared MM, Carles CC, Rossignol P, Pires HR, Fletcher JC (2013) The *ULT1* and *ULT2* trxG genes play overlapping roles in *Arabidopsis* development and gene regulation. Mol Plant 6(5):1564–1579. doi:10.1093/mp/sst041

136. Sacharowski SP, Gratkowska DM, Sarnowska EA, Kondrak P, Jancewicz I, Porri A, Bucior E, Rolicka AT, Franzen R, Kowalczyk J, Pawlikowska K, Huettel B, Torti S, Schmelzer E, Coupland G, Jerzmanowski A, Koncz C, Sarnowski TJ (2015) SWP73 subunits of Arabidopsis SWI/SNF chromatin remodeling complexes play distinct roles in leaf and flower development. Plant Cell 27(7):1889–1906. doi:10.1105/tpc.15.00233

137. Carpentier M-C, Picart-Picolo A, Pontvianne F (2017) A method to identify nucleolus-associated chromatin domains (NADs). In: Bemer M, Baroux C (eds) Plant chromatin dynamics: methods and protocols. Springer, New York, NY. doi:10.1007/978-1-4939-7318-7_7

Chapter 14

Technical Review: A Hitchhiker's Guide to Chromosome Conformation Capture

Stefan Grob and Giacomo Cavalli

Abstract

The introduction of chromosome conformation capture (3C) technologies boosted the field of 3D-genome research and significantly enhanced the available toolset to study chromosomal architecture. 3C technologies not only offer increased resolution compared to the previously dominant cytological approaches but also allow the simultaneous study of genome-wide 3D chromatin contacts, thereby enabling a candidate-free perspective on 3D-genome architecture. Since its introduction in 2002, 3C technologies evolved rapidly and now constitute a collection of tools, each with their strengths and pitfalls with respect to specific research questions. This chapter aims at guiding 3C novices through the labyrinth of potential applications of the various family members, hopefully providing a valuable basis for choosing the appropriate strategy for different research questions.

Key words Chromosome Conformation Capture, Chromosome Conformation Carbon Copy, Hi-C, ChIA-PET, 3D genome organization, HiChIP, Capture Hi-C, Targeted Chromatin Capture

1 Introduction

In recent years, the study of 3D-genome architecture became one of the leading fields in genome research, providing novel insights into 3D genome organization and its effect on the cell's functions. As the field is rather young, our knowledge on the influence of the 3D-genome on genome functions, such as transcription, replication, and epigenetic inheritance of cellular states is only emerging, but an increasing number of studies are focusing on 3D-genome architecture and its functional consequence. Alongside with microscopy approaches, Chromosome Conformation Capture (3C) technologies are the most important drivers of this research field. To date, the 3C collection comprises six main members, varying in their suitability for specific research questions. 3C technologies are based on a common experimental protocol, which involves formaldehyde cross-linking of native chromatin to capture 3D chromosomal interactions, followed by either enzymatic or

Marian Bemer and Célia Baroux (eds.), *Plant Chromatin Dynamics: Methods and Protocols*, Methods in Molecular Biology, vol. 1675, DOI 10.1007/978-1-4939-7318-7_14, © Springer Science+Business Media LLC 2018

physical chromatin fragmentation and proximity ligation. This leads to the production of hybrid DNA molecules, termed as 3C templates, which are indicative for specific 3D-chromatin contacts. 3C templates typically include two DNA fragments, which were in contact in the 3D-space. The relative abundance of a given 3C template is then used to determine an averaged contact frequency across the cell population. The fragment size ultimately defines the resolution with which 3D-contacts can be analyzed. The downstream sample preparation and, most importantly, the readout of each assay distinguish the various members, which are discussed below. Hence, depending on the desired information, a certain 3C method is preferred for a given research question.

A number of 3C technologies require sophisticated bioinformatics approaches for data analysis. A comprehensive list and description of tools can be found at https://omictools.com/3c-4c-5c-hi-c-chia-pet-category.

1.1 Chromosome Conformation Capture (3C)

- *Type*: One-to-one.
- *Application*: Analyzing specific pairwise chromosomal 3D-contacts.
- *Protocols*: [1–5] and *see* Chapter 15 [6].

3C [7], the founding member of the 3C family, is employed to analyze the contact frequency of a pair of chromosomal regions (e.g., an enhancer and a promoter region) and, thus, represents a candidate approach to study specific chromatin interactions (*see* Fig. 1). 3C templates representing the regions of interest are PCR amplified using primers specific to each of the candidate regions (*see* Fig. 2). The primers are typically designed in parallel to each other (i.e., facing in the same direction) to ensure that only valid 3C templates, which underwent digestion and religation, can be amplified. The contact intensity is determined either semiquantitatively by the intensity of a band on an agarose gel or quantitatively by qPCR [5]. The obtained values are then typically compared to pairs of neighboring genomic regions, which are thought to exhibit lower contact frequencies (e.g., a region next to the enhancer element or adjacent to the promoter).

The correct application of 3C may be challenging. As the primers used to amplify the 3C template of interest and the primers used to amplify control 3C templates are not identical, utmost caution is essential to ensure appropriate PCR conditions and comparable primer efficiency [4]. Especially ensuring equal primer efficiency can be very tedious, as 3C PCR primers can neither be assessed on 3C DNA (as the amount of template is unknown), nor on genomic DNA (as 3C primers are designed in parallel and, thus, cannot amplify genomic sequences). Hence, known amounts of 3C-like templates have to be generated. This can be accomplished

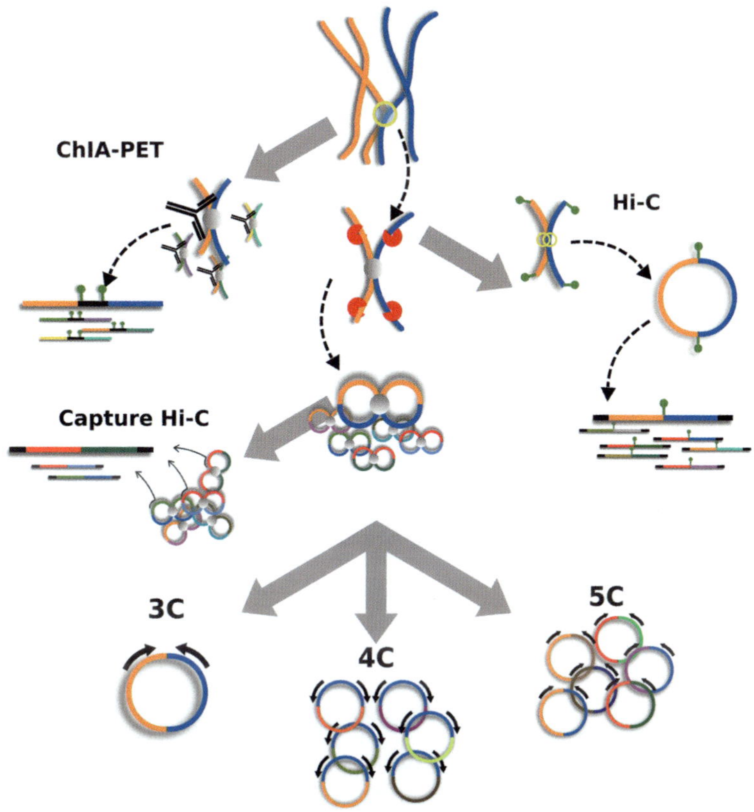

Fig. 1 Overview of the set of 3C technologies

by cloning of the expected 3C templates or by digestion and random religation of bacterial artificial chromosomes (BACs), covering the chromosomal region of interest. Furthermore, a number of control 3C templates originating from 3D-contacts adjacent to the candidate pair need to be included, leading to a substantial number of primer pairs, whose efficiency must be faithfully assessed using the above described strategies. However, once the experimental setup is established, various experimental conditions, such as comparisons between tissues or wild type and mutant genetic backgrounds can be investigated simultaneously without the need of sophisticated statistical analysis of the retrieved data.

In summary, performing standard 3C can be rather time consuming, thus, it is best applied to strong candidate regions rather than to screen for chromosomal contacts across large genomic domains. Since the downstream analysis of the retrieved data does not require profound knowledge on bio-statistics, 3C can be performed by any experienced molecular biologist without the need to engage bioinformatics specialists.

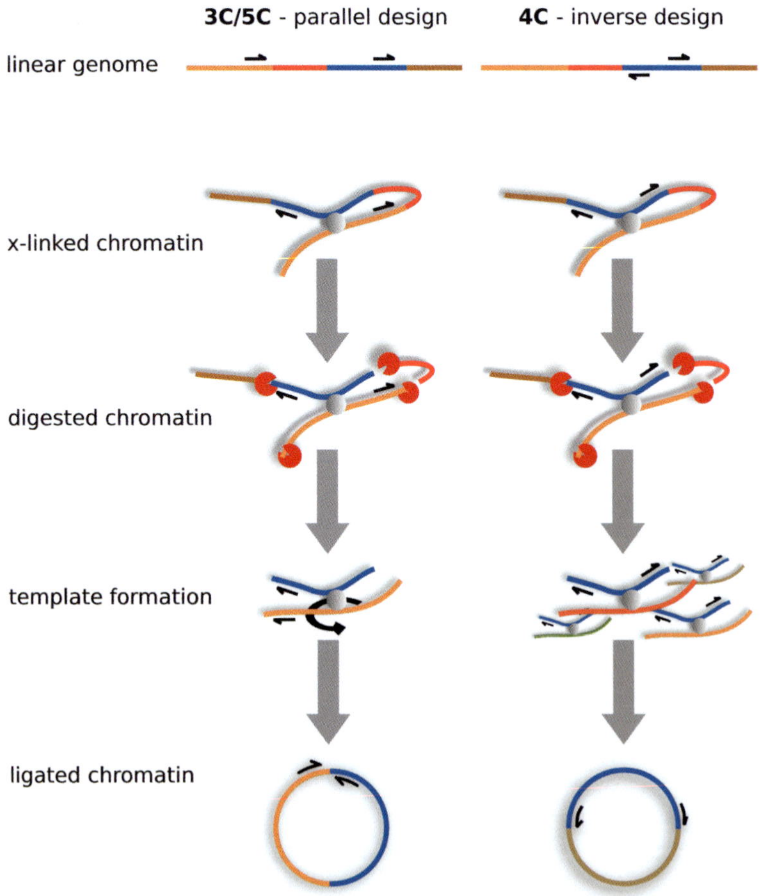

Fig. 2 Primer design. 3C, 4C, and 5C template amplification requires specific primer design, allowing specific amplification of informative 3C templates and preventing amplification of genomic DNA templates, which occurs in most 3C preparations due to incomplete chromatin digestion. Note in step "template formation": The cross-linking site acts like a hinge, around which the two fragments can turn and, thus, all fragment ends can ligate to each other

In plants, 3C has proven its power by important discoveries. In its first application in plants, 3C has been employed to describe the involvement of 3D contacts in the regulation of paramutagenic loci in maize [8]. Later, 3C allowed the identification of a vernalization-dependent gene loop linking 5′ and 3′ flanking regions of the flowering regulator *FLC* [9]. Additionally, 3C has allowed the characterization of chromatin loop involved in the regulation of the noncoding RNA *APOLO* expression [10].

1.2 Chromosome Conformation Carbon Copy (5C)

- *Type*: Many-to-many.
- *Application*: Comprehensive assessment of all 3D-contacts within a chromosomal region of interest.
- *Protocols*: [11–13].

5C [14] is the method of choice to comprehensively characterize all 3D-contacts of a chromosomal region of limited size up to a few megabases (*see* Fig. 1). Chromosomal regions, such as gene clusters, can engage in complex 3D-looping structures, connecting several regulatory and effector regions in 3D-space. Hence, the detailed analysis of all possible 3D-contacts in such a region can be of special interest. Methodologically, 5C is closely related to 3C. However, instead of using single primer pairs to amplify specific 3C templates characteristic for specific 3D-interactions, all 3C templates originating from the region of interest are amplified and later quantified using either microarray or next-generation sequencing technology. To amplify the entire interactome of the region of interest, primers specifically annealing to each restriction fragment are designed. To avoid amplification of circularized self-ligated fragments, primers are designed in parallel (i.e., pointing in the same direction) in order to only amplify head-to-head 3C ligation products (*see* Fig. 2). Furthermore, the specific primers carry tails common to all forward or reverse primers. This design subsequently allows for ligation-mediated amplification (LMA) of 3C templates within the region of interest. Similar to 3C, primer annealing efficiency has to be tightly controlled, as it is likely to vary among the pool of primers employed in the experiment. Control PCR amplification is achieved using digested and randomly religated BAC clones covering the region of interest.

Importantly, the complexity of the obtained 5C data increases exponentially with the number of fragments located in the region of interest. Analyzing a region of 100 fragments will generate 10,000 contact frequencies, whereas a region of 1000 fragments will yield 1,000,000 possible individual contacts. Hence, to obtain a satisfactory sequencing coverage, the number of fragments occurring in the region of interest has to be taken into consideration. Various bioinformatics tools were engineered to analyze 5C contact data (e.g., [13, 15]). In principle, most Hi-C (*see* below) data analysis tools can also be employed to process 5C data.

5C can generate 3D-contact maps of specific regions at unparalleled resolution. However, the design and generation of a large number of primers can be cost-intensive. Therefore, especially in case of relatively small genomes such as those of yeast, ***Drosophila melanogaster***, or ***Arabidopsis thaliana***, it may be advisable to perform Hi-C technology (*see* below) at high resolution, which may yield comparable resolution and cost for the whole genome. Furthermore, 5C technology is today in competition with a novel variant of Hi-C, termed Capture Hi-C (*see* below), which may produce comparable output at significantly lower cost.

Although widely employed in metazoen models, to our knowledge there has been no study using 5C published to date in the plant field. However, 5C may be beneficial to unravel complex folding principles of certain genomic loci. To date, little is known about enhancer–promoter interactions. 5C or Capture Hi-C (*see* below) may significantly advance how local chromatin folding contributes to transcriptional regulation in plants.

1.3 Circular Chromosome Conformation Capture/ Chromosome Conformation Capture on Chip (4C)

- *Type*: One-to-all.
- *Application*: Assessment of genome-wide chromosomal contacts of a candidate region.
- *Protocols*: [16–19].

4C [20, 21] is used to screen the genome for regions that specifically contact a candidate region in 3D-space (*see* Fig. 1). The search for interacting regulatory regions, such as enhancers, for a given chromosomal region of interest (e.g., a promoter) is one of the most prominent applications of 4C.

4C technology involves specific amplification of all 3C templates associated with a given region of interest (*see* Fig. 2). Thereby, an inverse PCR scheme is designed using primers specific for a candidate fragment, termed "bait" or "viewpoint." The primers anneal to the viewpoint's fragment ends pointing "outwards." This allows for amplification of all "prey" DNA fragments that interact with the viewpoint in 3D. As many 3C templates that were generated by a 6-cutter restriction enzyme may be comprised of prey fragments too large for efficient amplification, the 3C templates are subject to a second fragmentation and religation step, typically using a 4-cutter restriction enzyme. The composition of the final PCR products are then quantified using either microarrays or, more commonly, high-throughput sequencing. The obtained data are used to create a 4C-profile, which is typically represented by peaks corresponding to the contact frequency along the genome. The resolution of genome-wide 3D-contacts depends on sequencing coverage, which in turn is determined by the number of fragments generated during the primary 3C template generation (which is determined by the size of the genome and the frequency of restriction sites within). Hence, assuming a genome size of 1Gb, using state-of-the-art sequencing technology (e.g., one lane of Illumina sequencing) usually provides single fragment resolution, which is sufficient to analyze most biologically relevant chromosomal interactions.

A variant of 4C, termed enhanced 4C (e4C) [18], reduces library complexity by specific enrichment for a subset of chromatin associations and, thus, either significantly increases resolution at constant sequencing depth, or allows employing inexpensive microarray technology to retrieve the 4C experimental output.

The downstream assessment of 4C sequencing reads requires sequencing data processing and statistics and, thus, some skills in computational data analysis. Today, various computational tools are available (e.g., [22, 23]), which can also be used by scientists with relatively little experience in computational biology. To extract biologically significant 3D-contacts, the data sets have to be normalized for elevated contact frequencies that rely on linear proximity on the DNA. Neighboring genomic regions generally exhibit high 3D-contact frequencies, from which little biological significance can be inferred. However, they may mask biologically significant 3D-contacts, hence the need to normalize for linear distance-dependent 3D-contact frequencies. This normalization may be challenging especially in close proximity to the viewpoint, making it difficult to identify prey regions in close linear proximity to the viewpoint. Thus, 4C may not be suitable to analyze short-range interactions of less than approximately 50 kb.

Similar to 3C, 4C can easily be applied to study 3D-genome architecture in different experimental conditions. For many research questions, 4C data from various conditions can be compared using robust statistical software, designed for differential analysis of expression data (e.g., [24]). 4C technology is robust and can also be applied simultaneously to various viewpoints, which can be investigated in the same 4C sample using the primer sequences as bar codes. Although 4C can provide a 3D-contact profile in a rather short time and with limited financial requirements, it is not optimally suited to analyze 3D-contacts over short linear distances.

Despite the wide range of potential applications of 4C in plants, to our knowledge, only one report employing 4C has been published to date [17]. In a rather atypical application, general features of chromosomal architecture have been described based on 4C data. This study shows that 4C cannot only be beneficial to detect specific interactors of a given viewpoint, but, moreover, can also be deployed to reveal basic principles of chromosomal organization, such as the relationship of heterochromatin and euchromatin in 3D.

1.4 Genome-Wide 3C (Hi-C)

- *Type*: All-to-all.
- *Application*: Characterization of global chromatin contacts, assembly of genomes.
- *Protocols*: [25–29].

Hi-C [30] is designed to generate 3D-contact maps of entire genomes and catches all possible genome-wide chromatin interactions (*see* Fig. 1). Using Hi-C, basic organizational principles of genome organization were revealed [30, 31], significantly advancing the chromosomal architecture field. The identification of genome-wide contact frequencies can also assist in de novo genome assembly [32–35], using the fact that contigs in linear genomic neighborhood theoretically exhibit the highest contact frequencies among each other. Hence, by incorporation of the acquired contact frequencies, contigs of previously unknown genomic position can be put into linear relation.

In the Hi-C experimental protocol, successfully digested and religated fragments are marked using biotinylated nucleotides. Subsequently, 3C templates are fragmented and enriched for 3C-fragments carrying a biotin tag and, thus, representing hybrid molecules informative for a given 3D-chromatin contact. After ligation of adapter sequences to the fragmented 3C templates, the entire pool of enriched fragments is sequenced using high-throughput sequencing technology. A number of Hi-C variants have been established, varying the number of cells used as input material [36], the digestion procedure [29], and the enrichment [27] strategy.

Obtaining a genome-wide 3D contact map, typically represented by a contact frequency matrix, requires elaborate computational data processing and also significant computational resources. Various tools which preprocess and analyze complex Hi-C data are available today (see website cited above); however, to be able to fully understand and explore the obtained data, skills in computational biology are still required. Furthermore, faithful calling of 3D-contacts relies on the quality of the respective reference genome. Flaws in reference genomes can easily lead to artifacts, whose misinterpretation may lead to wrong conclusions. Hence, previous experience in analyzing genomics data represents an advisable prerequisite for successful application of Hi-C technology.

The obtained sequencing data is of high complexity, proportional to the square of the number of primary restriction fragments found in the entire genome. Small genomes, such as the one of ***Drosophila melanogaster*** or ***Arabidopsis thaliana*** contain approximately 30,000 fragments (given a six-cutter restriction enzyme), which yield a combinatorial space of nearly one billion contacts. Hence, achieving fragment size resolution requires considerable sequencing depth. Larger genomes, such as mammalian genomes and certain plant genomes, require exponentially more sequencing depth, and, hence, achieving fine-grain resolution using Hi-C may be challenging and cost-intensive.

Hi-C may appear as the gold-standard technology of the 3C family and, indeed, it is an extremely powerful technique not only to characterize global chromosomal architecture, but also to reveal previously unsuspected chromosomal contacts of biological significance. However, despite its power and relative ease to perform (especially compared to 3C and 5C), considerable limitations have to be taken into consideration. Identification of enhancer–promoter interactions may be challenging using Hi-C, as the required resolution can only be achieved with considerable investment in sequencing depth. Hence, although perfectly suited to acquire a global comprehension of 3D organization, investigations on specific chromosomal contacts maybe easier conducted using 3C or 4C.

Hi-C represents of the most widely used methods of the 3C collection in plants. In several studies, Hi-C has been used to characterize general chromosomal architecture [37–39] and specific chromosomal 3D conformations, such as the *KNOT*, positive strips [39] and chromatin loops [40]. Furthermore, the effects of various mutants on chromosomal architecture have been assessed by Hi-C [37, 38, 41, 42]. These studies provided a better understanding of the folding principles of the Arabidopsis genome and, importantly, revealed previously unknown genomic features, which can be related to gene regulation [40] and genome defense mechanisms [38].

1.5 Chromatin Interaction Analysis by Paired-End Tag Sequencing (ChIA-PET) and HiChIP: Protein-Centered Chromatin Interaction Approaches

- *Type*: All-to-all (bound by the same protein factor).
- *Application*: Detection of chromatin contacts, mediated by specific proteins.
- *Protocols*: [43–45].

ChIA-PET [46] resembles Hi-C technology, as it allows to detect genome-wide chromosomal contact frequencies (*see* Fig. 1). However, ChiA-PET does not represent a candidate-free approach, as chromatin contacts mediated by a protein factor of interest are enriched during the experimental protocol with the help of immunoprecipitation. Hence, a protein-specific interactome can be generated. Using this technique, it was shown that human estrogen receptor α (ER-α) mediates long-range contacts between regulatory ER-α binding sites and target gene promoters [46].

The ChIA-PET procedure significantly differs from all other experimental protocols of the 3C family. Cross-linked chromatin is physically fragmented by sonication and genomic fragments bound by a given protein factor are immunoprecipitated using specific antibodies. Linker sequences are ligated to the free DNA ends of the DNA–protein complexes. These linkers are subsequently integrated during proximity ligation that will create hybrid DNA molecules, consisting of protein factor bound genomic regions. With the help of the integrated linker sequences paired-end tag sequencing is performed to identify and quantify protein factor-mediated chromatin contacts. Due to the enrichment of a small set of genomic regions and enzyme-free fragmentation of the chromatin, ChIA-PET can yield a high resolution, allowing the pinpointing of biologically relevant chromatin interactions.

To date, ChIA-PET technology has not been as widely used as other 3C methods. However, in the presence of strong candidate protein factors, ChiA-PET represents a powerful method to identify genome-wide chromatin contacts of biological relevance.

Recently, another protein centered approach, termed HiChIP, has been established, which promises both higher yield of informative sequencing reads and lower requirements for the input material [45]. Whereas ChiA-PET requires up to 100 million cells, HiChIP has been reported to provide comparable output with as little as 1 million cells. HiChIP methodology shares more similarities to the other 3C members: In a first step, biotinylated Hi-C templates are generated. These templates are subsequently sheared by sonication followed by an immunoprecipitation with an antibody against the protein factor of interest. All downstream steps, including library generation, generally follow the standard Hi-C protocol. Computational pipelines to specifically analyze HiChIP are not implemented to date. However, standard Hi-C analysis programs are compatible with HiChIP data and may be sufficient to answer a wide range of research questions.

Hence, HiChIP promises to replace ChIA-PET in a foreseeable future, as it may generate more informative data output and, importantly, involves less complex methodology. Due to its novelty, independent experts of the field could not yet confirm the reliability of HiChIP. Thus, potential pitfalls of HiChIP may come to light with more researchers performing HiChIP.

HiChIP and ChIA-PET are promising methods to unravel fundamental questions on plant 3D chromatin organization. Thereby, a special focus could be laid on large regulatory protein complexes, such as Polycomb group (PcG) complexes. PcG complexes may play an important role in shaping 3D interactions and, hence, may represent an obvious target in order to establish HiChIP in plants.

1.6 Capture Hi-C (CHi-C)/Targeted Chromatin Capture (T2C)

- *Type*: Many-to-many/Many-to-all.
- *Application*: Determination of chromatin contacts of a specific set of chromosomal regions.
- *Protocols*: [47–50].

CHi-C [47, 49, 50] and T2C [48] technologies are recent members of the 3C collection. They allow identifying genome-wide chromosomal contacts of a specific set of genomic loci of interest, such as poised and active promoters with genome-wide interactors of those regions (*see* Fig. 1). Capture Hi-C methods are based on generation of a canonical 3C library; however, genomic regions of interest are enriched employing various capture sequencing strategies, such as the use of biotinylated RNA bait libraries. Using this strategy, capture Hi-C methods can either generate data similar to 5C, yielding a detailed description of chromosomal contacts within a genomic region of interest, or data similar to multiple 4C experiments, simultaneously characterizing the genome-wide interaction profiles of a large number of viewpoints (*see* Fig. 3). Hence, major

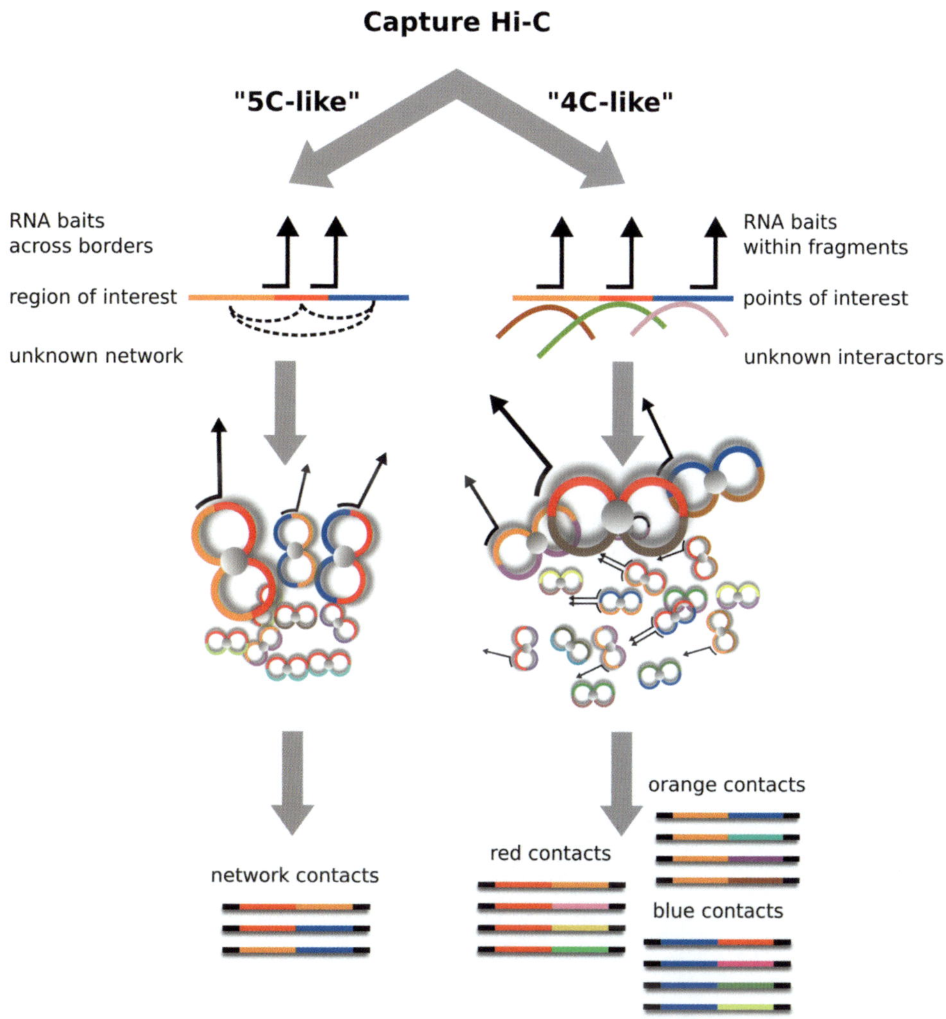

Fig. 3 Applications of Capture Hi-C. Capture Hi-C can generate two separate experimental outputs, depending on the design of biotinylated RNA baits. By probing single fragments, multiple "4C-like" datasets can be generated. By designing RNA baits across fragment borders, 5C like data can be generated, allowing the characterization of interaction networks of chromosomal regions of interest

limitations of Hi-C are overcome, namely, the high complexity of Hi-C libraries, which renders it difficult or very cost intensive to achieve high resolution. Furthermore, yielding a similar experimental output as 5C and multiple 4C, it eliminates the necessity to generate a large set of primers, which is not only cost-intensive but also a latent source of experimental biases, due to imbalanced PCR amplification efficiency.

For various research questions, a candidate-free approach is not required; hence, Capture Hi-C offers impressive resolution and limits amplification steps, which are a constant source of experimental noise that needs to be tightly controlled. As all candidate approaches however, it does not allow detection of previously unanticipated chromosomal contacts.

Related to 5C in their experimental output, CHi-C and T2C may be employed to answer a variety of biological questions, ranging from identification of potential promoter–enhancer interactions to detailed and unbiased characterization of 3D structuring of large genomic loci.

2 Conclusion

The collection of 3C technologies is highly proliferative, leading to the appearance of novel 3C-based approaches nearly every year over the last decade. Hence, even for experts of the field, it proves to be difficult to follow up on the technological advances and to assess the pros and cons of different 3C variants for specific research questions. However, with the current set of 3C variants at hand, we are finally offered a wide range of opportunities to study chromosomal architecture from various angles. These rapid technological advances are currently leading to a flourishing research field, which has been previously received insufficient attention, caused by technical inaccessibility of 3D chromatin structures. Hence, in recent years we rapidly gained a deepened understanding of plant 3D genome organization. Most of the published studies employing 3C technology in plants, to date, are of descriptive nature, setting a base line for future work (reviewed in [51, 52]). This work could now be aimed at understanding how the 3D genome influences fundamental cellular processes, such as transcription, differentiation, genome defense, and, last but least, genome evolution.

3C technologies represent the key to explore a fascinating level of genome organization. This may not only serve to more precisely define previously studied biological processes, but may lead to the discovery of novel biological phenomena as well.

References

1. Hagège H, Klous P, Braem C et al (2007) Quantitative analysis of chromosome conformation capture assays (3C-qPCR). Nat Protoc 2:1722–1733. doi:10.1038/nprot.2007.243
2. Louwers M, Splinter E, van Driel R et al (2009) Studying physical chromatin interactions in plants using Chromosome Conformation Capture (3C). Nat Protoc 4:1216–1229. doi:10.1038/nprot.2009.113
3. Miele A, Dekker J (2008) Mapping cis-and trans-chromatin interaction networks using chromosome conformation capture (3C). In: Hancock R. (ed) The nucleus: volume 2: chromatin, transcription, envelope, proteins, dynamics, and imaging. Methods in molecular biology. Humana, Totowa 464:105–121. doi:10.1007/978-1-60327-461-6_7
4. Dekker J (2006) The three "C" s of chromosome conformation capture: controls, controls, controls. Nat Methods 3:17–21. doi:10.1038/nmeth823
5. Comet I, Schuettengruber B, Sexton T, Cavalli G (2011) A chromatin insulator driving three-dimensional Polycomb response element (PRE) contacts and Polycomb association with the chromatin fiber. Proc Natl Acad Sci U S A 108:2294–2299. doi:10.1073/pnas.1002059108

6. Weber B, Jamge S, Stam M (2017) 3C analysis in Maize and Arabidopsis. Methods Mol Biol
7. Dekker J (2002) Capturing Chromosome Conformation. Science 295:1306–1311. doi:10.1126/science.1067799
8. Louwers M, Bader R, Haring M et al (2009) Tissue- and expression level-specific chromatin looping at Maize b1 Epialleles. Plant Cell 21:832–842. doi:10.1105/tpc.108.064329
9. Crevillen P, Sonmez C, Wu Z, Dean C (2012) A gene loop containing the floral repressor FLC is disrupted in the early phase of vernalization. EMBO J 32:140–148. doi:10.1038/emboj.2012.324
10. Ariel F, Jegu T, Latrasse D et al (2014) Non-coding transcription by alternative RNA polymerases dynamically regulates an auxin-driven chromatin loop. Mol Cell 55:383–396. doi:10.1016/j.molcel.2014.06.011
11. Dostie J, Zhan Y, Dekker J (2007) Chromosome conformation capture carbon copy technology. Curr Protoc Mol Biol Chapter 21:Unit 21.14. doi: 10.1002/0471142727.mb2114s80
12. Dostie J, Dekker J (2007) Mapping networks of physical interactions between genomic elements using 5C technology. Nat Protoc 2:988–1002. doi:10.1038/nprot.2007.116
13. Lajoie BR, van Berkum NL, Sanyal A, Dekker J (2009) My5C: web tools for chromosome conformation capture studies. Nat Methods 6:690–691. doi:10.1038/nmeth1009-690
14. Dostie J, Richmond TA, Arnaout RA et al (2006) Chromosome Conformation Capture Carbon Copy (5C): a massively parallel solution for mapping interactions between genomic elements. Genome Res 16:1299–1309. doi:10.1101/gr.5571506
15. Sauria ME, Phillips-Cremins JE, Corces VG, Taylor J (2015) HiFive: a tool suite for easy and efficient HiC and 5C data analysis. Genome Biol:1–10. doi:10.1186/s13059-015-0806-y
16. Simonis M, Klous P, Homminga I et al (2009) High-resolution identification of balanced and complex chromosomal rearrangements by 4C technology. Nat Methods 6:837–842. doi:10.1038/nmeth.1391
17. Grob S, Schmid MW, Luedtke NW et al (2013) Characterization of chromosomal architecture in Arabidopsis by chromosome conformation capture. Genome Biol 14:R129. doi:10.1186/gb-2013-14-11-r129
18. Sexton T, Kurukuti S, Mitchell JA et al (2012) Sensitive detection of chromatin coassociations using enhanced chromosome conformation capture on chip. Nat Protoc 7:1335–1350. doi:10.1038/nprot.2012.071
19. Grob S (2017) Circular chromosome conformation capture in plants. Methods Mol Biol 1610:73–92. doi:10.1007/978-1-4939-7003-2_6
20. Zhao Z, Tavoosidana G, Sjölinder M et al (2006) Circular chromosome conformation capture (4C) uncovers extensive networks of epigenetically regulated intra- and interchromosomal interactions. Nat Genet 38:1341–1347. doi:10.1038/ng1891
21. Simonis M, Klous P, Splinter E et al (2006) Nuclear organization of active and inactive chromatin domains uncovered by chromosome conformation capture-on-chip (4C). Nat Genet 38:1348–1354. doi:10.1038/ng1896
22. van de Werken HJG, Landan G, Holwerda SJB et al (2012) Robust 4C-seq data analysis to screen for regulatory DNA interactions. Nat Methods 9:969–972. doi:10.1038/nmeth.2173
23. Klein FA, Pakozdi T, Anders S et al (2015) FourCSeq: analysis of 4C sequencing data. Bioinformatics 31:3085–3091. doi:10.1093/bioinformatics/btv335
24. Robinson MD, McCarthy DJ, Smyth GK (2009) edgeR: a Bioconductor package for differential expression analysis of digital gene expression data. Bioinformatics 26:139–140. doi:10.1093/bioinformatics/btp616
25. Grob S, Grossniklaus U (2017) Chromatin Conformation Capture-based analysis of nuclear architecture. Methods Mol Biol 1456:15–32. doi: 10.1007/978-1-4899-7708-3_2
26. van Berkum NL, Lieberman-Aiden E, Williams L et al (2010) Hi-C: a method to study the three-dimensional architecture of genomes. J Vis Exp:1–7. doi:10.3791/1869
27. Kalhor R, Tjong H, Jayathilaka N et al (2011) Genome architectures revealed by tethered chromosome conformation capture and population-based modeling. Nat Biotechnol 30:90–98. doi:10.1038/nbt.2057
28. Belton J-M, McCord RP, Gibcus JH et al (2012) Hi-C: a comprehensive technique to capture the conformation of genomes. Methods 58:268–276. doi:10.1016/j.ymeth.2012.05.001
29. Ramani V, Cusanovich DA, Hause RJ et al (2016) Mapping 3D genome architecture through in situ DNase Hi-C. Nat Protoc 11:2104–2121. doi:10.1038/nprot.2016.126
30. Lieberman-Aiden E, van Berkum NL, Williams L et al (2009) Comprehensive mapping of long-range interactions reveals folding

principles of the human genome. Science 326:289–293. doi:10.1126/science.1181369

31. Sexton T, Yaffe E, Kenigsberg E et al (2012) Three-dimensional folding and functional organization principles of the Drosophila genome. Cell 148:458–472. doi:10.1016/j.cell.2012.01.010

32. Korbel JO, Lee C (2013) Genome assembly and haplotyping with Hi-C. Nat Biotechnol 31:1099–1101. doi:10.1038/nbt.2764

33. Kaplan N, Dekker J (2013) High-throughput genome scaffolding from in vivo DNA interaction frequency. Nat Biotechnol 31:1143–1147. doi:10.1038/nbt.2768

34. Adey A, Patwardhan RP, Qiu R et al (2013) Chromosome-scale scaffolding of de novo genome assemblies based on chromatin interactions. Nat Biotechnol 31:1119–1125. doi:10.1038/nbt.2727

35. Lesne A, Riposo J, Roger P et al (2014) 3D genome reconstruction from chromosomal contacts. Nat Methods 11:1141–1143. doi:10.1038/nmeth.3104

36. Nagano T, Lubling Y, Stevens TJ et al (2013) Single-cell Hi-C reveals cell-to-cell variability in chromosome structure. Nature 502:59–64. doi:10.1038/nature12593

37. Feng S, Cokus SJ, Schubert V et al (2014) Genome-wide Hi-C analyses in wild-type and mutants reveal high-resolution chromatin interactions in Arabidopsis. Mol Cell 55:694–707. doi:10.1016/j.molcel.2014.07.008

38. Grob S, Schmid MW, Grossniklaus U (2014) Hi-C analysis in Arabidopsis identifies the KNOT, a structure with similarities to the flamenco locus of Drosophila. Mol Cell 55:678–693. doi:10.1016/j.molcel.2014.07.009

39. Wang C, Liu C, Roqueiro D et al (2014) Genome-wide analysis of local chromatin packing in Arabidopsis thaliana. Genome Res 25:246–256. doi:10.1101/gr.170332.113

40. Liu C, Wang C, Wang G et al (2016) Genome-wide analysis of chromatin packing in Arabidopsis thaliana at single-gene resolution. Genome Res 26:1057–1068. doi:10.1101/gr.204032.116

41. Moissiard G, Cokus SJ, Cary J et al (2012) MORC family ATPases required for heterochromatin condensation and gene silencing. Science 336:1448–1451. doi:10.1126/science.1221472

42. Veluchamy A, Jegu T, Ariel F et al (2016) LHP1 regulates H3K27me3 spreading and shapes the three-dimensional conformation of the Arabidopsis Genome. PLoS One 11: e0158936–e0158925. doi:10.1371/journal.pone.0158936

43. Fullwood MJ, Han Y, Wei C-L et al (2010) Chromatin interaction analysis using paired-end tag sequencing. Curr Protoc Mol Biol Chapter 21:Unit 21.15.1–25. doi: 10.1002/0471142727.mb2115s89

44. Goh Y, Fullwood MJ, Poh HM et al (2012) Chromatin interaction analysis with paired-end tag sequencing (ChIA-PET) for mapping chromatin interactions and understanding transcription regulation. J Vis Exp. doi: 10.3791/3770

45. Mumbach MR, Rubin AJ, Flynn RA et al (2016) HiChIP: efficient and sensitive analysis of protein-directed genome architecture. Nat Methods 13:919–922. doi:10.1038/nmeth.3999

46. Fullwood MJ, Liu MH, Pan YF et al (2009) An oestrogen-receptor-α-bound human chromatin interactome. Nature 461:58–64. doi:10.1038/nature08497

47. Dryden NH, Broome LR, Dudbridge F et al (2014) Unbiased analysis of potential targets of breast cancer susceptibility loci by Capture Hi-C. Genome Res 24:1854–1868. doi:10.1101/gr.175034.114

48. Kolovos P, van de Werken HJ, Kepper N et al (2014) Targeted Chromatin Capture (T2C): a novel high resolution high throughput method to detect genomic interactions and regulatory elements. Epigenetics Chromatin 7:10–17. doi:10.1186/1756-8935-7-10

49. Mifsud B, Tavares-Cadete F, Young AN et al (2015) Mapping long-range promoter contacts in human cells with high-resolution capture Hi-C. Nat Genet 47:598–606. doi:10.1038/ng.3286

50. Jäger R, Migliorini G, Henrion M et al (2015) Capture Hi-C identifies the chromatin interactome of colorectal cancer risk loci. Nat Commun 6:6178. doi:10.1038/ncomms7178

51. Liu C, Weigel D (2015) Chromatin in 3D: progress and prospects for plants. Genome Biol 16:1–6. doi:10.1186/s13059-015-0738-6

52. Grob S, Grossniklaus U (2017) Chromosome Conformation Capture-based studies reveal novel features of plant nuclear architecture. Curr Opin Plant Biol 36:149–157

Chapter 15

3C in Maize and Arabidopsis

Blaise Weber, Suraj Jamge, and Maike Stam

Abstract

With Chromosome Conformation Capture (3C), the relative interaction frequency of one chromosomal fragment with another can be determined. The technique is especially suited for unraveling the 3D organization of specific loci when focusing on aspects such as enhancer–promoter interactions or other topological conformations of the genome. 3C has been extensively used in animal systems, among others providing insight into gene regulation by distant *cis*-regulatory elements. In recent years, the 3C technique has been applied in plant research. However, the complexity of plant tissues prevents direct application of existing protocols from animals. Here, we describe an adapted protocol suitable for plant tissues, especially *Arabidopsis thaliana* and *Zea mays*.

Key words Chromosome Conformation Capture, 3C protocol, Plant, *Arabidopsis thaliana*, *Zea mays*

1 Introduction

Since the first microscopic observation of nuclei, it has become clear that chromatin is not randomly organized [1]. A specific 3D architecture of the genome is established in each and every cell's nucleus to ensure proper regulation of gene expression [2, 3]. Such architecture includes large-scale chromatin domains as well as specific enhancer–promoter interactions that together shape the cell's transcriptome and its fate [4, 5]. Chromosome Conformation Capture (3C) allows to study chromatin organization and helps to understand how the spatiotemporal organization of chromatin influences gene expression [6]. 3C technology was first implemented in yeast [7] and rapidly adapted to other organisms including mice, human, and Drosophila [8–10]. The 3C technique provides the attractive opportunity to study chromosomal interactions at a resolution that was previously difficult to achieve with cytological methods [6]. The method relies on the proximity ligation concept (Fig. 1). In this method, chromatin from fixed nuclei is first subjected to

Blaise Weber and Suraj Jamge contributed equally to this work.

Marian Bemer and Célia Baroux (eds.), *Plant Chromatin Dynamics: Methods and Protocols*, Methods in Molecular Biology, vol. 1675, DOI 10.1007/978-1-4939-7318-7_15, © Springer Science+Business Media LLC 2018

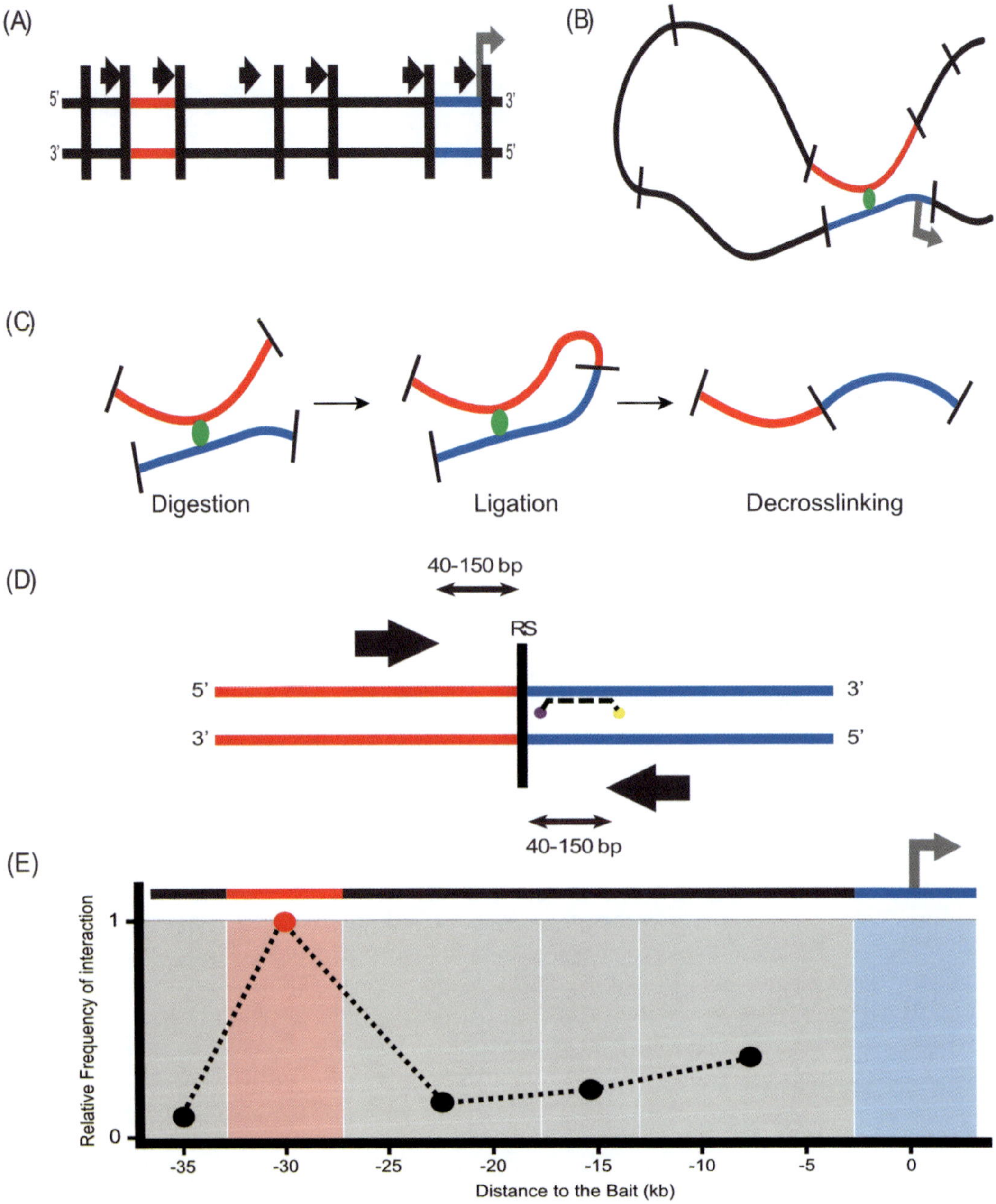

Fig. 1 Graphical overview of the 3C technique. (**a**) Schematic representation of a potential locus of interest. Primers (*black arrows*) are designed on one and the same strand for all fragments monitored. The bait is indicated in *blue*. (**b**) Schematic representation of a hypothetical chromosome conformation at the potential locus of interest. (**c**) Fixed chromatin is digested, ligated in a large volume and then decrosslinked. The products of these processes are shown. (**d**) Schematic representation of a 3C template. *Black vertical bar* indicates the ligated restriction site (RS). Primers (*black arrows*) anneal on each side of the RS and now form a primer pair. A TaqMan probe can be used for more specific quantification by qPCR (*see* **Note 14**) and should be designed on the bait fragment, on the strand complementary to the strand on which the bait primer is designed. (**e**) Graphical representation of a 3C plot. The relative frequency of interaction of each fragment with the bait is plotted on the *y*-axis (*black and red circles*). The distances to the middle of the bait fragment are reported on the *x*-axis. *Vertical boxes* are used to indicate the different fragments monitored

digestion with a methylation insensitive restriction enzyme, followed by ligation in conditions favoring intramolecular ligations. Hereby, regions that are in close contact in 3D have a higher chance of being ligated together. Subsequently, interaction frequencies of specific ligation events are quantified, providing insight into the 3D organization of chromatin at the genomic locus of interest. Frequencies of interaction are measured using quantitative PCR (qPCR) in combination with primers specific for each interaction one desires to examine. The use of primers recognizing specific fragments makes 3C an hypothesis-driven approach. Since the first publication in 2002, variants of the 3C method have been developed (e.g., 4C, 5C, and Hi-C, *see* Chapter 14) that allow a more systematic analysis of chromosomal interactions [11–14]. These methods identify many more interactions at the same time (one to all, many to many, and all to all) by including deep sequencing techniques. Hence, they are associated with higher costs, and given the complexity and amount of data generated, also time consuming data analysis pipelines. Therefore, 3C remains a method of choice when special focus is given to one specific locus. It allows faster results and often offers better resolution.

The specific nature of plant cells hampers the direct application of published 3C protocols from other species. Therefore, 3C on plant tissues requires plant specific steps. In literature, 3C protocols have been described for Arabidopsis and Maize [15, 16]. In this chapter, we provide a step-by-step bench protocol, starting from the design and setup of a 3C experiment, up to the analysis and interpretation of the 3C data. In addition, we provide critical notes on different aspects that need to be adapted when applying this method to other plants or tissues of interest.

2 Materials

Prepare all solutions using autoclaved milliQ water and analytical grade reagents. Sterilization by autoclaving is performed at 121 °C for 10 min unless indicated otherwise. Diligently follow all waste disposal regulations when disposing of waste materials.

2.1 Tissue Fixation and Nuclei Isolation

1. 1× phosphate buffered saline (PBS): add 800 mL of milliQ water into a graduated cylinder together with a magnetic stirrer. Weigh 8 g of NaCl, 0.2 g of KCl, 1.44 g of Na_2HPO_4, and 0.24 g of KH_2PO_4 and add it to the measuring column. Adjust the pH to 7.4 with HCl and add milliQ water up to 1 L. Autoclave the solution and store at room temperature (RT).
2. 4% paraformaldehyde (PFA) in PBS: Prepare this solution in the fume hood. Pour 100 mL of autoclaved 1× PBS in a 250 mL glass bottle. Weigh 4 g of powdered paraformaldehyde

and transfer it to the PBS-filled bottle. Adjust the pH to 9 with KOH, close the bottle and transfer it to a 65 °C water bath. Shake the bottle from time to time until the PFA is completely dissolved. Transfer it back to the fume hood and allow the bottle to cool down before adjusting the pH back to 7–7.5 with HCl. Prepare aliquots of 10 mL to be stored at −80 °C.

3. 2 M glycine: Add 80 mL of milliQ water to a glass graduated cylinder or beaker, together with a magnetic stirrer. Weigh 15 g of glycine and add it into the graduated cylinder or beaker. Dissolution can be enhanced by raising the temperature. After complete homogenization, adjust the volume to 100 mL with milliQ water and transfer the solution to a 100 mL glass bottle. Autoclave and store at 4 °C.
4. 20% Triton X-100: pipette twice 1 mL of Triton X-100 with a cut pipette tip and add to a 15 mL tube containing 8 mL of autoclaved milliQ water. Allow all Triton to get out of the tip by pipetting up and down. Avoid foaming as much as possible. Shield the tube from light with opaque tape or aluminum foil. Place the tube in a rotating wheel at RT overnight to allow complete homogenization. 20% Triton X-100 can be stored shielded from light at RT for a month.
5. 100 mM phenylmethylsulfonyl fluoride (PMSF): weigh 174 mg of PMSF and dissolve it in 10 mL of isopropanol. Prepare aliquots of 100 μL and store at −20 °C.
6. Complete protease inhibitor (Roche): dissolve 1 complete protease inhibitor tablet in 2 mL of autoclaved milliQ water by vortexing vigorously. Dissolved tablets can be stored for 1–2 weeks at 4 °C, or up to 6 weeks at −20 °C.
7. Nuclei extraction buffer (100 mL, prepare fresh): 2 mL 1 M Hepes pH 8, 25 mL 1 M sucrose, 0.1 mL 1 M $MgCl_2$, 0.5 mL 1 M KCl, 46 mL 87% glycerol, 1.25 mL 20% Triton X-100, 26 mL autoclaved milliQ, 100 μL 100 mM PMSF, 100 μL beta-mercaptoethanol.
8. 50 mL tubes.
9. Miracloth.
10. Sefar Nitek nylon filter 50 μm pore size.
11. Morter and pestle.
12. Liquid nitrogen.
13. Cooling centrifuge for 50 mL tubes.
14. Cooling microcentrifuge.

2.2 Digestion and Ligation of 3C and Control Samples

1. BAC-clone or other large plasmid that contains the region of interest (*see* Subheading 3.2.1).
2. Suitable restriction enzyme (*see* Table 1 and Subheading 3.1) with 10× restriction buffer.
3. Phenol–chloroform–isoamyl alcohol (25:24:1 v/v): under the fume hood, pipette 25 mL of Phenol equilibrated with 10 mM Tris–HCl, 1 mM EDTA, pH 8 and transfer it to a 50 mL glass bottle. Add 24 mL of chloroform and 1 mL of isoamyl alcohol. Close the lid and mix. Allow the phases to separate, protect the bottle from light, and store at 4 °C.
4. Chloroform–isoamyl alcohol (24:1 v/v): under the fume hood, pipette 24 mL of chloroform and transfer it to a 50 mL glass bottle. Add 1 mL of isoamyl alcohol. Close the lid and mix. Protect the bottle from light and store at 4 °C.
5. 2 M sodium acetate (NaOAc) pH 5.6: Add 80 mL of milliQ water to a graduated cylinder. Weigh 16.4 g of Anhydrous sodium acetate and dissolve it in the 80 mL of milliQ water. Adjust pH to 5.6 with HCl and bring volume up to 100 mL with milliQ water. Autoclave and store at RT.
6. Glycogen 20 mg/mL.
7. 96% and 70% ethanol.
8. 10 mM Tris–HCl pH 7.5.
9. 1 M Tris–HCl pH 7.8: add 800 mL of milliQ water into a graduated cylinder, together with a magnetic stirrer. Weight 157.6 g of Tris and gradually add it to the column while stirring. Set the pH to 7.8 with HCl and adjust the volume to 1 L with milliQ water. Autoclave solution and store at RT.

Table 1
List of REs regularly used in 3C experiments

Enzyme	Heat inactivation[a]	References
*Hind*III	Yes	[17, 18]
*Eco*RI	Yes	[7, 18]
*Bgl*II	No	[18–20]
*Bam*HI	No	[18, 20, 21]
*Dpn*II	Yes	[22, 23]
*Mfe*I	No	[24]
*Nla*III	Yes	[25]
*Xho*I	Yes	[18]

[a]This column indicates if an enzyme can be heat-inactivated without the addition of SDS

10. 1 M DL-dithiothreitol (DTT): dissolve 1.53 g of DTT powder into 10 mL of autoclaved milliQ water. Prepare aliquots of 100 μL and store at −20 °C.
11. 10× ligase buffer for 3C samples (1 mL, prepare freshly): 300 μL 1 M Tris–HCl pH 7.8, 100 μL 1 M $MgCl_2$, 100 μL 1 M DTT, 6 mg ATP (final concentration 10 mM), 500 μL autoclaved milliQ water.
12. Highly concentrated T4 DNA ligase.
13. 20% sodium dodecyl sulfate (SDS): open SDS container under the fume hood and weigh 10 g of SDS. Still under the fume hood, transfer SDS to a 50 mL glass bottle filled with 35 mL of autoclaved milliQ water. Ensure complete homogenization and bring volume to 50 mL with autoclaved milliQ water. Store bottle at RT.
14. Agarose.
15. Gel running device and UV transilluminator.
16. Water bath.
17. Heat block.
18. Cooling microcentrifuge.
19. Cooling centrifuge for 50 mL tubes.
20. 50 mL tubes.

2.3 Decrosslinking, DNA Purification, and qPCR Analysis

1. 10 mg/mL Proteinase K: weight 100 mg of proteinase K and dissolve in 10 mL of autoclaved milliQ water. Prepare aliquots of 100 μL and store at −20 °C.
2. 10 mg/mL RNase A: for a final volume of 1 mL, add 900 μL of 10 mM NaOAc pH 5.6 to a 1.5 mL eppendorf. Add 10 mg of RNase A. Mix until complete dissolved and place the tube into boiling water for 15 min. Allow solution to cool down at RT and add 100 μL of Tris–HCl pH 7.4. Store at −20 °C.
3. Phenol–chloroform–isoamyl alcohol (25:24:1 v/v), *see* **item 4** of Subheading 2.2.
4. Chloroform–isoamyl alcohol (24:1 v/v), *see* **item 5** of Subheading 2.2.
5. Agarose.
6. Gel running device and UV transilluminator.
7. 10 μM primers for the ligation products to be tested (*see* **step 3** of Subheading 3.1).
8. 10 mM dNTPs.
9. Quantitative PCR machine.
10. Reagents for real–time PCR analysis (e.g., SYBR Green Mastermix).

3 Methods

3.1 Study Design

1. *Define your region of interest*: 3C can be performed at any specific locus or gene of interest for which the DNA sequence is known. In general, one should select a restriction enzyme (RE) that will generate a restriction pattern compatible with intramolecular ligation (fragments smaller than 300 bp are more difficult to ligate) and allows the verification of hypothesized interactions (regions of interest should be located in fragments >300 bp). Define a "bait" or "viewpoint" (*see* Fig. 1) by selecting the fragment for which you will quantify the frequency of interactions with other fragments.
2. *Selection of the restriction enzyme*: REs used in the 3C procedure need to efficiently digest cross-linked chromatin, which is challenging. Different REs or combinations of REs that allow digestion of fixed chromatin have been reported in published 3C protocols and studies. Using such REs (*see* Table 1) is a safe choice. However, if the restriction patterns generated by these enzymes are not compatible with one's hypothesis at the locus of interest, newly selected enzymes should fulfill a number of requirements. First, the selected RE should be methylation insensitive, since methylation sensitive enzymes may result in partial digestion and thus introduce a bias. Second, the selected RE should ideally display optimal efficiency for digesting fixed chromatin at 37 °C, and preferably maintain its activity over a long period of time (e.g., overnight). In case of a short survival time, aliquots of restriction enzymes can be added sequentially during the digestion time. Thirdly, high ligation efficiency is crucial for a successful 3C experiment. Thus, favor enzymes generating sticky ends (the larger the overhang, the better), and ensure that the religation efficiency is high without the need of adding any macromolecular crowding agent such as polyethylene glycol (PEG).

   Once the digestion is complete, to stop digestion the RE needs to be inactivated by elevated temperatures. In addition, for some enzymes detergents need to be added for inactivation (*see* Table 1). Note that the addition of SDS is associated with negative effects on ligation efficiency.
3. *Type of tissue and number of nuclei*: chromosome conformations are mostly cell type-dependent [8, 19, 26]. Therefore, ideally, for 3C one uses fresh (*see* **Note 1**), homogeneous and synchronized cell populations. Plant tissues have an heterogeneous cell type composition, and their cells are not synchronized. The heterogeneity of plant tissues does not preclude the use of 3C technology, however, one should keep in mind that the results obtained will reflect an average of the chromosomal

interactions occurring in different and unsynchronized cell types. Combining the 3C procedure with methods that allow the isolation of specific cell types (e.g., FACS or INTACT) could, although technically very challenging, allow studying cell type-specific interactions [27–29].

The type of plant tissues used for 3C analysis is dependent on the research question. Theoretically, most tissues are compatible with 3C analysis, however, tissues that are highly lignified or have a high starch content pose difficulties at grinding and/or downstream process. In addition, the amount of starting material needs to be sufficient to allow isolating one to ten million nuclei (to assess the number of nuclei *see* **Note 2**). Tissues that are difficult to harvest in bulk, such as meristems, are therefore challenging.

4. *Primer design*: Specific primers need to be designed for all DNA fragments one wishes to study. 3C primers need to be orientated unidirectionally, meaning that all primers are designed complementary to the same strand. This way, only ligation events between two different fragments will result in amplicon formation (*see* Fig. 1). Primers are usually designed 40–150 bp away from the restriction sites. Furthermore, they are preferably 18–27 bp long with a GC content of about 50%, a Tm between 57 °C and 63 °C and no more than 2 °C difference in Tm between them. Primers need to be very specific, which should be determined using BLAST (High Throughput Genome Sequence database). Select primers only if one perfect match is found; homology to sequences elsewhere in the genome should be less than 75% of the primer length and exclude the 3′ end of the primer. Primers should be tested experimentally on control template (Random Ligation Library; *see* Subheading 3.2.1) and genomic DNA to ensure that only the correct amplicon size is amplified when using the Random Ligation Library.

3.2 Controls in the 3C Procedure

The correct interpretation of 3C results demands the use of a number of controls. For instance, controls are used to ensure an optimal digestion efficiency (a nondigested 3C sample and digested gDNA), correct for primer efficiency (Random Ligation Library), and also account for technical and biological variation between 3C libraries (endogenous locus).

3.2.1 Random Ligation Library (RL-Library)

To determine the relative frequency of interaction of one fragment with another, one needs to ensure that no quantification bias arises due to primer pairs with different amplification efficiencies. Therefore, to normalize for the primer efficiency, a random ligation library (RL-Library) is prepared that consists of all possible ligation products that need to be analyzed by qPCR (*see* Subheading 3.7).

The RL-Library can be used as a template to first test primer efficiency and specificity. An RL-Library can be generated in different ways. One option is by digesting a bacterial artificial chromosome (BAC) or other large plasmid that contains the locus of interest, followed by religation. The digestion of the BAC is performed with the same RE selected to generate the 3C library but in a smaller volume compared to the ligation step for the 3C library. This allows all fragments to randomly ligate with one another. Alternative to using a BAC-derived RL-Library in the qPCR experiments, one can PCR amplify all potential ligation products from a 3C library, and mix those in equimolar amounts. Finally, note that a different RL-Library has to be prepared for every locus one examines (locus of interest and the endogenous locus).

1. Prepare a 1.5 mL eppendorf tube with 10 μg of template (e.g., BAC) in a final volume of 100 μL with a twofold excess of restriction enzyme (e.g., 1 μL of 20,000 units/mL *Hin*dIII) and its recommended restriction buffer at a final concentration of 1×.
2. Incubate for 5 h at 37 °C.
3. Check the digestion efficiency by running 10 μL of the digestion mixture on a 0.8% agarose gel. Satellite bands should be visible.
4. Add 210 μL of sterile milliQ to bring the volume up to 300 μL.
5. Add 1 Volume of Phenol–chloroform–isoamyl alcohol (25:24:1) to the digestion mixture. Mix thoroughly.
6. Centrifuge at full speed for 10 min at RT.
7. Transfer the top aqueous phase to a new 1.5 mL eppendorf tube and add 1 volume of chloroform–isoamyl alcohol (24:1). Mix thoroughly.
8. Repeat the centrifugation step at full speed for 10 min at RT.
9. Transfer the top (aqueous) phase to a new 1.5 mL eppendorf tube and precipitate the DNA with 1/10 volume of 2 M NaOAc (pH 5.6), 2 volumes of 96% Ethanol and 10 μL of Glycogen (20 mg/mL).
10. Store the tube at −80 °C for 2 h to overnight.
11. Centrifuge at full speed for 30 min at 4 °C.
12. Gently pour off the supernatant and wash the pellet with 1 mL of 70% ethanol.
13. Centrifuge at full speed for 10 min at 4 °C.
14. Gently pour off the supernatant and use a pipette tip to remove the remaining droplets.
15. Let the pellet air-dry for 2–5 min and resuspend the DNA in 22 μL of 10 mM Tris–HCl pH 7.5.

16. Transfer 20 μL of the digested DNA to a new eppendorf tube. Add 5 μL of fresh 10× ligation buffer, sterile milliQ water and 20 Units of T4 DNA ligase up to a final volume of 50 μL (*see* **Note 3**). Store the remaining 2 μL of digested BAC DNA at −20 °C.
17. Incubate the ligation reaction overnight at 16 °C.
18. Bring the volume to 300 μL by addition of 250 μL of sterile milliQ water.
19. Add 1 volume of phenol–chloroform–isoamyl alcohol (25:24:1) to the ligation mixture. Mix thoroughly.
20. Centrifuge at full speed for 10 min at RT.
21. Transfer the top phase to a new 1.5 mL eppendorf tube and add 1 volume of chloroform–isoamyl alcohol (24:1). Mix thoroughly.
22. Centrifuge at full speed for 10 min at RT.
23. Transfer the top phase to a new 1.5 mL eppendorf tube and precipitate the DNA with 1/10 volume of 2 M NaOAc (pH 5.6), 2 volumes of 96% Ethanol and 10 μL of Glycogen (20 mg/mL).
24. Store the tube at −80 °C for 2 h to overnight.
25. Centrifuge at full speed for 30 min at 4 °C.
26. Gently pour off the supernatant and wash the pellet with 1 mL of 70% ethanol.
27. Centrifuge at full speed for 10 min at 4 °C.
28. Gently pour off the supernatant and use a pipette tip to remove the remaining droplets.
29. Let the pellet air dry for 2–5 min and resuspend the DNA in 22 μL of 10 mM Tris–HCl pH 7.5.
30. Prepare a 0.8% agarose gel and mix both the 2 μL of the RL-Library, and the 2 μL of digested 3C sample (Subheading 3.6, **step 16**) with 7 μL of sterile milliQ and 1 μL of 10× loading buffer.
31. Store the RL-Library at −20 °C.
32. Run the digested and ligated samples on the prepared 0.8% agarose gel to check the ligation efficiency.
33. Prepare 1/10, 1/100, 1/1000 and 1/10,000 dilutions of the RL-Library in sterile milliQ, and add nondigested gDNA to each dilution (final concentration of 50 ng gDNA/μL). Addition of nondigested gDNA mimics the PCR conditions with the 3C library as a template. Measure the Ct values for the different primer pairs of interest using qPCR. For each 20 μL PCR reaction prepare a qPCR mix according to your own setup. Use 1 μL of template (concentration depends on the

dilution) and 1 μL of each primer (10 μM). When analyzing the 3C library for the first time with qPCR, use the different dilution series of your RL-Library complemented with gDNA. Subsequently, for normalization use the dilution showing Ct values in the range of the Ct values obtained for the 3C library. In later qPCR analyses, one can use only the relevant dilution(s) of the RL-Library.

3.2.2 Endogenous Control

Another crucial 3C control, an endogenous control locus, accounts for technical and biological variation between samples. Technical variation hereby refers to differences in quantity and quality of the sample, biological variation refers to differences in interaction frequencies at the locus of interest between different tissues. To this end, an endogenous locus is chosen that shows similar RNA expression levels across the different tissues examined by 3C. A similar RNA expression level indicates a similar chromatin conformation. Typically, genic loci referred to as housekeeping genes are known to be similarly expressed and can be assumed to show similar chromatin conformation in different tissues. The frequency of interactions at such loci can therefore be used to normalize the data between biological samples.

1. Identify a proper endogenous control locus (e.g., *SAM* (GRMZM2G154397) in *Z. mays* or *TIP41* (AT4G34270.1) in *A. thaliana*). Do this by checking if RNA transcript levels are similar in the tissues of interest.
2. Design primers complementary to multiple restriction fragments at the endogenous control locus and its flanking sequences.
3. Using qPCR and your 3C library as a template, measure the frequencies of interaction between the selected bait fragment and the other fragments of the endogenous locus. Usually the bait fragment consists of the fragment that contains the TSS of the selected endogenous locus. Select a primer pair displaying a frequency of interaction comparable to the mean of your frequency of interactions at the studied locus and take this primer pair along at subsequent 3C experiments for normalization.

3.2.3 Positive Digestion Control

The positive digestion control is used as a reference to determine the efficiency of digestion. The positive digestion control consists of fully digested gDNA. In this sample, the pattern of digestion (size range of the smear and satellite bands) should be clearly visible.

1. Prepare a 1.5 mL eppendorf tube with 10 μg of gDNA template in a final volume of 100 μL with a twofold excess of restriction enzyme and its recommended buffer at a final concentration of 1×.

2. Incubate overnight at 37 °C.
3. Check the digestion efficiency by running 10 μL of the digestion mixture on a 0.8% agarose gel. Satellite bands should be visible. If the pattern of digestion is not clearly visible, extend the incubation time.
4. Add 200 μL of sterile milliQ to bring the volume up to 300 μL.
5. Add 1 volume of phenol–chloroform–isoamyl alcohol (25:24:1) to the digestion mixture. Mix thoroughly.
6. Centrifuge at full speed for 10 min at RT.
7. Transfer the top aqueous phase to a new 1.5 mL eppendorf tube and add 1 volume of Chloroform–isoamyl alcohol (24:1). Mix thoroughly.
8. Repeat the centrifugation step at full speed for 10 min at RT.
9. Transfer the top (aqueous) phase to a new 1.5 mL eppendorf tube and precipitate the DNA with 1/10 volume of 2 M NaOAc (pH 5.6), 2 volumes of 96% ethanol and 10 μL of Glycogen (20 mg/mL).
10. Store the tube at −80 °C for 2 h to overnight.
11. Centrifuge at full speed for 30 min at 4 °C.
12. Gently pour off the supernatant and wash the pellet with 1 mL of 70% ethanol.
13. Centrifuge at full speed for 10 min at 4 °C.
14. Gently pour off the supernatant and use a pipette tip to remove the remaining droplets.
15. Let the pellet air-dry for 2–5 min and resuspend the DNA in 50 μL of 10 mM Tris–HCl pH 7.5.
16. Store the tube at −20 °C.
17. Load 5 μL of the positive digestion control on a 0.8% agarose gel when checking the digestion efficiency of the 3C library (*see* Subheading 3.6, **step 16**).

3.3 Plant Tissue Fixation and Nuclei Isolation

The following procedure describes the handling of one biological sample. Multiple samples can be handled at the same time. In our hands, working with more than four samples at once is cumbersome and might result in suboptimal 3C library quality.

1. Prepare nuclei extraction buffer (NEB) and place on ice. Also, precool the centrifuge for 50 mL tubes (swing-out) and eppendorf centrifuge to 4 °C.
2. Fill a 50 mL centrifuge tube (preferably PPCO tubes—Poly Propylene COpolymer) and fill it with 10 mL of 1× PBS. Place the tube on ice.

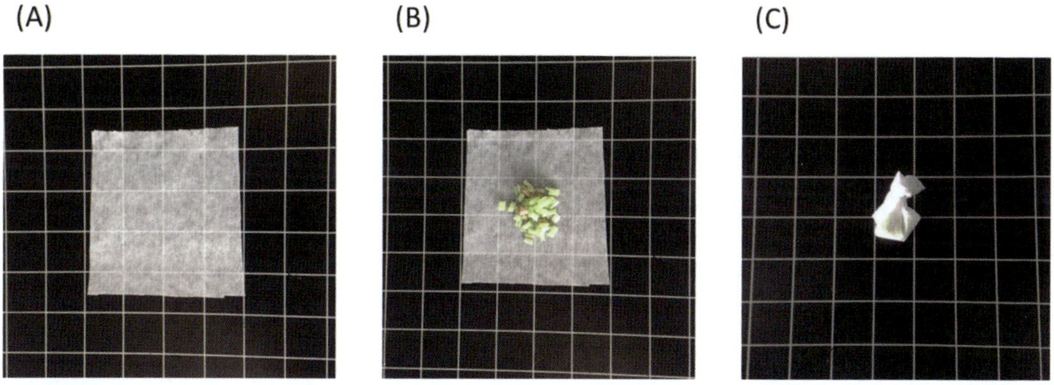

Fig. 2 Preparation of a "tea bag" from Miracloth. *One black square* is 4 by 4 cm. (**a**) Cut a 12 by 12 cm piece of Miracloth. (**b**) Place the plant tissue at the center of the piece of Miracloth. (**c**) Fold by joining all corners and staple them together

3. Harvest tissue of interest (1–3 g; *see* **Note 2**) and place it in a large petri dish on ice. If necessary, cut tissue into pieces with a sharp scalpel to improve penetration of the fixative (*see* **Note 4**). Place the tissue on top of a 12 × 12 cm piece of Miracloth (*see* **Note 5**). Enclose the tissue into the Miracloth by folding and stapling the corners, generating a "tea bag" (*see* Fig. 2). Completely submerge the bag into the PBS solution.
4. Place the tube under the fume hood. Add 10 mL of 4% PFA.
5. Vacuum infiltrate the tissue at RT (to determine the time of fixation, *see* **Note 4**). During fixation turn vacuum OFF and ON again 3 times to ensure good penetration of the fixative.
6. Stop fixation by adding 1.25 mL of ice-cold 2 M glycine (final concentration is 0.125 M) and vacuum infiltrate for 5 min.
7. Place the tube back on ice under the fume hood and discard PFA solution. Add autoclaved milliQ water to the tube, close the lid and shake vigorously to wash the tissue. Repeat this step twice. Discard PFA solution and milliQ water used for the washes according to your waste disposal regulations for fixative.
8. Place the tea bag in between two stacks of paper towels and press to dry the tissue. Repeat this process with new paper towels until the tissue is sufficiently dry (*see* **Note 6**).
9. Open the tea bag and place the tissue into a prechilled mortar containing liquid nitrogen. Grind the tissue into a fine powder. Avoid thawing the ground material.
10. Add sufficient NEB to submerge all the ground material (usually 10–20 mL). The NEB may freeze upon addition into the frozen mortar. Wait for it to thaw and mix from time to time. Avoid the suspension to warm up higher than 4 °C.

11. Place a new 50 mL tube on ice with a funnel on top.
12. Prepare a 12 by 12 cm piece of Sefar Nitek nylon filter (50 μm pore size) and of Miracloth. First place the Sefar Nitek filter in the funnel and then cover it with the piece of Miracloth, resulting in a two-layer filter.
13. Pipette the tissue suspension (from **step 10**) onto the two-layer filter and allow it to flow through by gravity. Rinse the mortar with an additional 5–10 mL of NEB and pipette it on top of the filter. Do not compress the filter! Upon squeezing you also obtain undesirable debris. Let gravity do its work. The filtrate contains your nuclei.
14. Centrifuge filtered nuclei at 1900 × *g* for 15 min at 4 °C.
15. Promptly place the tube back on ice and gently pour off the supernatant. Resuspend the pellet in 1 mL of NEB and transfer the nuclei suspension into a 1.5 mL eppendorf tube. At this step the number and quality of extracted nuclei can be determined (*see* **Note 2**).
16. Centrifuge at 1900 × *g* for 5 min at 4 °C.
17. Promptly place the eppendorf tube back on ice and gently pipette off the supernatant. Resuspend the pellet in 1 mL of NEB and repeat the centrifugation step (1900 × *g* for 5 min at 4 °C).
18. Promptly place the tube back on ice.

3.4 Digestion

1. Gently pipette off the supernatant and resuspend the nuclei into 400 μL of 1.2× restriction buffer (refer to the manufacturer's instructions for the optimal restriction buffer).
2. Centrifuge at 1900 × *g* for 5 min at 4 °C.
3. Promptly place the tube back on ice and gently pipette off the supernatant. Resuspend the nuclei into 500 μL of 1.2× restriction buffer.
4. Add 7.5 μL of 20% SDS (final concentration 0.3%) to permeabilize the nuclei and inactivate endogenous nucleases. Incubate at 65 °C for 40 min in a shaker at 900 rpm (*see* **Note 7**).
5. Place the tube at 37 °C for 20 min, still shaking at 900 rpm.
6. Add 50 μL 20% Triton X-100 (final concentration is 2%). Incubate at 37 °C for 1 h while shaking at 900 rpm. The Triton X-100 will sequester the SDS, preventing a negative impact on the digestion efficiency. The susceptibility to SDS varies from one restriction enzyme to another. In case of poor digestion efficiency, the final concentration of SDS might have to be adjusted (*see* **Note 8**).
7. For a nondigested control sample: transfer 28 μL of the nuclei suspension to a new 1.5 mL eppendorf tube containing 272 μL

of milliQ water. Store the tube at −20 °C until all samples will be decrosslinked (*see* **step** 7 of Subheading 3.5).

8. Add 400 Units of Restriction Enzyme and incubate overnight at 37 °C while shaking at 900 rpm (*see* **Note 9**).

3.5 Intramolecular Ligation and Decrosslinking

The ligation of fragments cross-linked together needs to be favored. Therefore, the volume of ligation needs to be sufficiently large to favor intramolecular ligations. At the same time, a too large volume will result in low DNA recovery. Thus, genome size-specific adaptations are required. For small-genome organisms such as *A. thaliana* the volume of ligation needs to be decreased compared to the volume used for large-genome organisms such as *Z. mays*. The following part of the protocol describes volumes based on the *Z. mays* genome size. To determine in which volume intramolecular ligation should be performed for other organisms *see* **Note 10** and Table 2.

1. Prepare 10× ligation buffer and store at RT (*see* **Note 11**).
2. Inactivate the restriction enzyme either by heat inactivation (refer to the manufacturer's instructions, shake at 900 rpm) or by addition of 40 μL of 20% SDS (final concentration is 1.6%) followed by incubation for 25 min at 65 °C, 900 rpm.
3. Transfer the digested sample to a 50 mL tube and add 7 mL of 1× ligation Buffer (700 mL of 10× ligation buffer plus 6.3 mL of sterile milliQ water).
4. Sequester the SDS by addition of 375 μL of 20% Triton X-100 (final concentration is 1%) and incubation for 1 h at 37 °C.
5. For a digested control sample, pipette 300 μL of the digested sample into a 1.5 mL eppendorf tube and store the tube at −20 °C until the decrosslinking step (*see* **step** 7).
6. Add 100 units of highly concentrated T4 DNA ligase to the ligation mix and incubate for 5 h at 16 °C, followed by 45 min at RT (*see* **Note 3**).

Table 2
Reported volumes of ligation reactions in different 3C protocols and organisms with their respective haploid genome size

Organism	*S. cerevisiae*	*A. thaliana*	*Z. mays*	*M. musculus*	*H. sapiens*
Genome size (Mbp)	~12,5	~135	~2100	~2800	~3300
Volume ligation reaction (mL)	0.8	2	7	7.5	7.5
References	[7]	[16]	[15, 19]	[25]	[18]

7. Add 30 μL of 10 mg/mL proteinase K to the ligation mix, and 5 μL to the nondigested and digested samples previously stored at −20 °C (*see* **step 7** of Subheading 3.4 and **step 5** of Subheading 3.5).
8. Incubate all tubes overnight in a 65 °C water bath.

3.6 DNA Purification

1. Place the Phenol–chloroform–isoamyl alcohol and chloroform–isoamyl alcohol solutions under the fume hood at RT at least 2 h before starting (*see* **Note 12**).
2. Add 30 μL of 10 mg/mL RNase A to the ligation sample, and 5 μL to the nondigested and digested samples.
3. Incubate all tubes at 37 °C for 30–45 min.
4. Place all tubes under the fume hood and add 10 mL of phenol–chloroform–isoamyl alcohol to the ligated sample and 300 μL to the nondigested and digested samples. Close all lids tightly and shake vigorously.
5. Spin all tubes at 4500 × *g* for 10 min at RT.
6. Transfer the aqueous phase to a new 50 or 1.5 mL tube.
7. Precipitate the DNA by first doubling the volume with milliQ, followed by addition of 1/10 volume of 2 M NaOac (pH 5.6) and 2 volumes of 96% ethanol. To promote high precipitation efficiency the addition of glycogen is strongly recommended. Typically, 20–40 μL (20 mg/mL) is added to the ligated sample and 5 μL to the nondigested and digested samples.
8. Incubate all tubes at −80 °C for at least 2 h.
9. Cool the centrifuge for 50 mL tubes (swing-out) and eppendorf centrifuge to 4 °C.
10. Centrifuge all tubes at 4500 × *g* for 60 min at 4 °C.
11. Pour off the supernatant and wash the pellet of the ligated sample with 10 mL and the nondigested and digested samples with 1 mL of 70% Ethanol.
12. Centrifuge all tubes at 4500 × *g* for 20 min at 4 °C.
13. Promptly place the tubes back on ice and gently pour off the supernatant. With a pipette tip, remove the residual droplet of ethanol and let the pellet air-dry for 2–5 min. Alternatively, when dealing with multiple samples one can use a pump to remove the supernatant and then dry the pellet. Using a pump is certainly faster, but be cautious as loose pellets might get lost.
14. Resuspend the ligated sample in 150 μL, and the nondigested and digested samples in 15 μL Tris–HCl pH 7.5.
15. Incubate all tubes at 4 °C overnight to optimally resuspend the DNA pellet.

16. To assay the digestion and ligation efficiency, mix 2 μL of the generated 3C library with 1.5 μL of 10× loading buffer and 11.5 μL of milliQ water. Load the sample, together with the entire 15 μL of the nondigested and digested samples, and a positive control for digestion consisting of digested gDNA (*see* **step 3** of Subheading 3.2), all with 10× loading buffer, on a 0.8% agarose gel. For the expected results, *see* Fig. 3 and **Note 13**.
17. The 3C library concentration can be estimated using a Qubit fluorometer. Alternatively, the concentration can be estimated on gel by comparison to a dilution series of a gDNA sample of known concentration. For more accuracy, load several dilutions of the 3C library on gel (1:1, 1:2, 1:4, and 1:8). Note that NanoDrop measurement is not reliable for quantification of complex DNA samples such as 3C libraries.

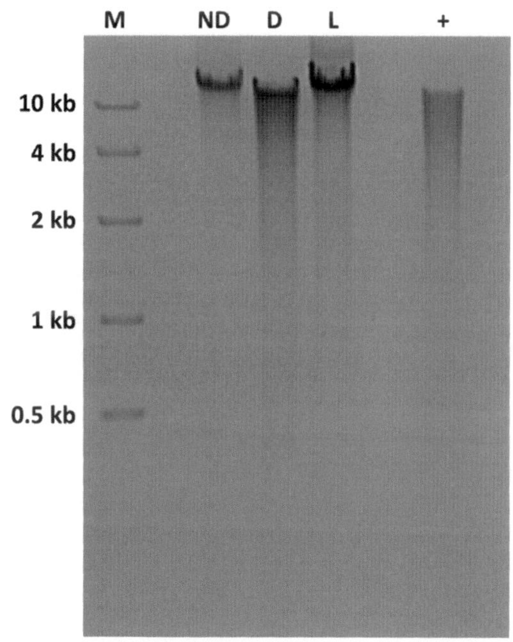

Fig. 3 Nondigested (ND), digested (D), and ligated 3C library (L) samples on a 0.8% agarose gel. The ND and L samples show one high molecular weight band, indicating intact DNA (ND sample) and efficiently ligated DNA (L sample). The presence of low molecular weight products in the ND and L samples would indicate DNA degradation and inefficient ligation, respectively. The digestion (D) sample has to show a similar banding pattern, including signs of satellite bands, as the positive digestion control (+), which consists of digested genomic DNA. *M* indicates the size marker lane

3.7 Quantification of Interaction Frequencies and Data Analysis

In 3C experiments the Relative Frequency of Interaction (RFI) is measured between a fixed fragment, the bait or viewpoint, and another fragment at the locus of interest. To estimate the RFI of a given fragment (e.g., the red fragment in Fig. 1) with the bait fragment (blue fragment in Fig. 1), one needs to compare it to the RFI of other fragments (black fragments in Fig. 1) with the bait. Below, we provide step-by-step qPCR and RFI quantification methods. The qPCR protocol is based on using SYBR green. The use of TaqMan probes is an alternative to SYBR green technology. For this, please *see* **Note 14**.

1. Prepare the qPCR mix, excluding primers, according to your own setup. The final volume for each reaction should be 20 μL.
2. Use 50 ng of 3C library as a template per reaction. Prepare similar reactions for the locus of interest and the endogenous control locus.
3. For the RL-Library controls, use the previously prepared dilution(s) complemented with gDNA (*see* **step 33** of Subheading 3.2.1). Prepare similar reactions for the RL-Library dilution of the locus of interest and the endogenous control locus.
4. Add 2 μL of primer (1 μL bait primer + 1 μL fragment primer from 10 nM stock) to each well.
5. For each primer pair, perform a triplicate qPCR reaction on the 3C library and RL-Libraries.
6. For each primer pair, the RFI is calculated as follows:

$$\text{RFI} = \left(2^{-(Ct_{3Ci}-Ct_{RLi})}\right)/\left(2^{-(Ct_{3Ce}-Ct_{RLe})}\right).$$

Ct_{3Ci} = mean value from the 3C triplicate for a given primer pair *i*.

Ct_{RLi} = mean value from the RL-Library triplicate for primer pair *i*.

Ct_{3Ce} = mean value from the 3C triplicate for the primer pair of the endogenous locus.

Ct_{RLe} = mean value from the RL-library triplicate for the endogenous locus.

7. Normalize all RFIs to the highest RFI value and plot the normalized RFIs as a function of their distance to the bait (Fig. 1e).

4 Notes

1. Use always fresh plant material if possible. Fresh material results in the most optimal digestion of fixed chromatin. In case material collection and the 3C protocol cannot be performed

simultaneously, we advise to store fixed, dried material rather than fixed, ground material. This appeared more effective in our hands.

2. The amount of tissue to be processed per sample ranges between 1 and 3 g of fresh plant material. This range should not be exceeded as too much tissue will affect the efficiency of fixation. At the same time, the amount of tissue used in each experiment should yield a sufficient number of nuclei (1 to 10×10^6). To estimate the number of nuclei isolated, take a 2 μL aliquot after the first resuspension of nuclei in 1 mL of NEB (**step 15** of Subheading 3.3). The nuclei have the tendency to sink to the bottom of the tube. Therefore, invert the tube with the resuspended nuclei gently three times before pipetting to avoid underestimating the actual yield. Add 2 μL of DAPI stain to the sample (final DAPI concentration of 2 μg/mL; Dilution Factor of 2) and count the nuclei on a hemocytometer using fluorescence microscopy (10–20× magnification, use DAPI filter). The total number of nuclei is estimated with the following formula: Total number of nuclei = (total number of counted nuclei × Dilution Factor × 10^4)/(number of chambers counted). In case of a low number of nuclei per gram of tissue, one could consider generating multiple tissue samples and subsequently pool the nuclei together when resuspending the nuclear pellets (**step 15** of Subheading 3.3).
3. Crowding reagents such as PEG cannot be used to increase the efficiency of the 3C ligation as it compromises the intramolecular nature of the ligation. Note that the addition of PEG 4000 (10% final concentration) can be used to increase the ligation efficiency of the RL-library.
4. Some tissue types can be used directly for fixation, other types need to be cut in smaller pieces. For instance, relatively permeable tissue like Arabidopsis rosettes can be used directly, while maize inner stem and husk tissue needs to be cut in ~1 cm^2 pieces before fixation. For each tissue type the optimal fixation time needs to be determined. Under-fixation will negatively affect the ability to detect chromosomal interactions. Overfixation will negatively affect digestion and decrosslinking efficiency, but also increase the background level of interactions. A good indication that tissue gets fixed is when it gets a translucent appearance. To define the optimal fixation time, one should perform a time series, fixing tissue samples for different time periods, and proceed with nuclei isolation and digestion (from **step 6** of Subheading 3.3 to **step 8** of Subheading 3.4), followed by decrosslinking and DNA precipitation (from **step 7** of Subheading 3.5 to **step 14** of Subheading 3.6). Run the DNA samples on a 0.8% agarose gel in parallel with an unfixed, decrosslinked sample (no fixative added, no

vacuum infiltration, positive control for DNA isolation) and a fixed, non-decrosslinked sample (negative control). At the optimal time of fixation, samples display efficient digestion and a relatively high DNA recovery after decrosslinking and DNA isolation. Alternatively, one could test the effect of different formaldehyde concentrations while using a fixed incubation time.

5. For efficient fixation, wrap the tissue or pieces of tissue in a 12 by 12 cm piece of Miracloth, and close the Miracloth with a staple, generating a "tea bag" (*see* Fig. 2). The tea bag ensures complete submersion of the tissue during fixation, prevents spilling of tissue, and allows easier handling of the sample during subsequent washes.
6. To efficiently dry the tissue, place the Miracloth "tea bag" containing the tissue between a stack of paper towels and use a bottle or something similar as a roller to remove water. Repeat this procedure till the paper towels do not get wet anymore. Then the tissue is considered dry. Remaining water negatively impacts the grinding efficiency and hence the yield of nuclei.
7. The incubation of the chromatin for 40 min at 65 °C prior to digestion is crucial for inactivation of endogenous nucleases that would otherwise become active during the digestion step at 37 °C.
8. Digestion efficiency is sensitive to the chemicals present in a solution, including the SDS used to permeabilize the nuclei (**step 4** of Subheading 3.4). In case of poor digestion efficiency, we recommend testing the effect of different SDS concentrations (0.05–0.3% final concentration) on the digestion efficiency of the restriction enzyme.
9. Efficient digestion is key to a successful 3C experiment. For enzymes with a low performance over a long incubation period, we recommend to add fractions of the total number of enzyme units (400) at different time points during the digestion procedure. This helps to maintain high digestion efficiency.
10. Good intramolecular ligation conditions are important to ensure that only fragments cross-linked with one another are being ligated. To favor intramolecular ligation events, the ligation reaction is carried out in a large volume. This volume needs to be adapted to the genome size, as a low DNA concentration hampers an efficient precipitation of the ligation products, while a too high DNA concentration also allows intermolecular ligation events. In Table 2 we report volumes of intramolecular ligation used in different published 3C protocols for different organisms. Note that the appropriate

ligation volume (**step 3** of Subheading 3.5) is influenced by the concentration of SDS present before the ligation step. If one wishes to lower the ligation volume, the volume of digestion should be adapted such that the SDS concentration will not hamper ligation efficiency. For instance, when working with *Arabidopsis thaliana*, in our hands a ligation volume of 2 mL (*see* Table 2) requires a digestion volume of 250 μL (**step 3** of Subheading 3.4) followed by addition of SDS to a final concentration of 0.2% (**step 4** of Subheading 3.4).

11. We highly recommend preparing a fresh 10× ligation buffer each time to ensure efficient ligation. Addition of extra ATP (0.6 mg/mL) after 2–3 h of incubation of the ligation reaction can further improve the ligation efficiency.
12. In case of low DNA recovery after the ligation step: make sure that the phenol–chloroform–isoamyl alcohol and chloroform–isoamyl alcohol solutions are well equilibrated at RT before adding them to the samples. Cold phenol–chloroform–isoamyl alcohol and chloroform–isoamyl alcohol are more prone to phase inversion. Increasing the amount of glycogen added and incubation of the precipitation reaction overnight at −80 °C can help to increase precipitation efficiency.
13. Digestion efficiency can be evaluated on an agarose gel as indicated at **step 16** of Subheading 3.6 (*see* Fig. 3). A more accurate evaluation of digestion efficiency can be achieved by designing a few primer pairs spanning restriction sites, followed by qPCR. The digestion efficiency of the 3C sample should be evaluated by comparison to a nondigested and digested gDNA template. Ideally the digestion efficiency is above 80%. To check for variation in the amount of each template, use a primer pair amplifying a region not cut by the restriction enzymes used.
14. Plant genomes can harbor a very high density of repetitive elements (*see* e.g., *Zea mays*) [30]. Specific amplification of the desired amplicons can therefore be challenging. When analyzing the qPCR results, always perform melting curve analyses for all primer pairs to check amplicon specificity. If obtaining specific primers appears to be very difficult, the design and use of a TaqMan probe (Fig. 1) can help to increase signal specificity. With a TaqMan probe one does not rely on the use of a non-sequence-specific fluorescent dye such as SYBR. A TaqMan probe should be designed for the bait fragment, on the opposite strand of the bait primer (*see* Fig. 1). In this way, the quencher from the probe can only be removed when a new strand is synthesized using the primer annealed at the ligated fragment. TaqMan probes are usually designed as an approximately 30 bp oligo with a Tm 7–10 °C above the Tm of the primers [25].

Acknowledgments

We gratefully acknowledge the support of the European Commission Seventh Framework-People-2012-ITN Project EpiTRAITS, GA-316965 (Epigenetic regulation of economically important plant traits) for Blaise Weber and Suraj Jamge. We thank Iris Hövel for her advice on 3C experiments on plant material.

References

1. Heitz E (1928) Das heterochromatin der moose. Jahrb Wiss Bot 69:762–818
2. Manuelidis L (1990) A view of interphase chromosomes. Science 250:1533–1540. doi:10.1126/science.2274784
3. Gorkin DU, Leung D, Ren B (2014) The 3D genome in transcriptional regulation and pluripotency. Cell Stem Cell 14:762–775. doi:10.1016/j.stem.2014.05.017
4. Sakabe NJ, Savic D, Nobrega MA (2012) Transcriptional enhancers in development and disease. Genome Biol 13:238. doi:10.1186/gb-2012-13-1-238
5. Kolovos P, Knoch TA, Grosveld FG, Cook PR, Papantonis A (2012) Enhancers and silencers: an integrated and simple model for their function. Epigenetics Chromatin 5:1. doi:10.1186/1756-8935-5-1
6. de Wit E, de Laat W (2012) A decade of 3C technologies: insights into nuclear organization. Genes Dev 26:11–24. doi:10.1101/gad.179804.111
7. Dekker J, Rippe K, Dekker M, Kleckner N (2002) Capturing chromosome conformation. Science 295:1306–1311. doi:10.1126/science.1067799
8. Tolhuis B, Palstra RJ, Splinter E, Grosveld F, de Laat W (2002) Looping and interaction between hypersensitive sites in the active beta-globin locus. Mol Cell 10:1453–1465
9. Vernimmen D, Gobbi MD, Sloane-Stanley JA, Wood WG, Higgs DR (2007) Long-range chromosomal interactions regulate the timing of the transition between poised and active gene expression. EMBO J 26:2041–2051. doi:10.1038/sj.emboj.7601654
10. Comet I, Schuettengruber B, Sexton T, Cavalli G (2011) A chromatin insulator driving three-dimensional Polycomb response element (PRE) contacts and Polycomb association with the chromatin fiber. Proc Natl Acad Sci 108:2294–2299. doi:10.1073/pnas.1002059108
11. Zhao Z, Tavoosidana G, Sjölinder M, Göndör A, Mariano P, Wang S, Kanduri C, Lezcano M, Singh Sandhu K, Singh U, Pant V, Tiwari V, Kurukuti S, Ohlsson R (2006) Circular chromosome conformation capture (4C) uncovers extensive networks of epigenetically regulated intra- and interchromosomal interactions. Nat Genet 38:1341–1347. doi:10.1038/ng1891
12. Grob S, Schmid MW, Luedtke NW, Wicker T, Grossniklaus U (2013) Characterization of chromosomal architecture in Arabidopsis by chromosome conformation capture. Genome Biol 14:R129. doi:10.1186/gb-2013-14-11-r129
13. Nora EP, Lajoie BR, Schulz EG, Giorgetti L, Okamoto I, Servant N, Piolot T, van Berkum NL, Meisig J, Sedat J, Gribnau J, Barillot E, Blüthgen N, Dekker J, Heard E (2012) Spatial partitioning of the regulatory landscape of the X-inactivation centre. Nature 485:381–385. doi:10.1038/nature11049
14. Lieberman-Aiden E, van Berkum NL, Williams L, Imakaev M, Ragoczy T, Telling A, Amit I, Lajoie BR, Sabo PJ, Dorschner MO, Sandstrom R, Bernstein B, Bender MA, Groudine M, Gnirke A, Stamatoyannopoulos J, Mirny LA, Lander ES, Dekker J (2009) Comprehensive mapping of long-range interactions reveals folding principles of the human genome. Science 326:289–293. doi:10.1126/science.1181369
15. Louwers M, Splinter E, van Driel R, de Laat W, Stam M (2009) Studying physical chromatin interactions in plants using Chromosome Conformation Capture (3C). Nat Protoc 4:1216–1229. doi:10.1038/nprot.2009.113
16. Hövel I, Louwers M, Stam M (2012) 3C technologies in plants. Methods 58:204–211. doi:10.1016/j.ymeth.2012.06.010
17. Palstra R-J, Tolhuis B, Splinter E, Nijmeijer R, Grosveld F, de Laat W (2003) The β-globin nuclear compartment in development and

erythroid differentiation. Nat Genet 35:190–194. doi:10.1038/ng1244

18. Stadhouders R, Kolovos P, Brouwer R, Zuin J, van den Heuvel A, Kockx C, Palstra R-J, Wendt KS, Grosveld F, van Ijcken W, Soler E (2013) Multiplexed chromosome conformation capture sequencing for rapid genome-scale high-resolution detection of long-range chromatin interactions. Nat Protoc 8:509–524. doi:10.1038/nprot.2013.018
19. Louwers M, Bader R, Haring M, van Driel R, de Laat W, Stam M (2009) Tissue- and expression level-specific chromatin looping at Maize b1 Epialleles. Plant Cell 21:832–842. doi:10.1105/tpc.108.064329
20. Crevillén P, Sonmez C, Wu Z, Dean C (2013) A gene loop containing the floral repressor FLC is disrupted in the early phase of vernalization. EMBO J 32:140–148. doi:10.1038/emboj.2012.324
21. Kang H, Wiedmer A, Yuan Y, Robertson E, Lieberman PM (2011) Coordination of KSHV latent and lytic gene control by CTCF-cohesin mediated chromosome conformation. PLoS Pathog 7:e1002140. doi:10.1371/journal.ppat.1002140
22. Kyrchanova O, Toshchakov S, Parshikov A, Georgiev P (2007) Study of the functional interaction between Mcp insulators from the Drosophila bithorax complex: effects of insulator pairing on enhancer-promoter communication. Mol Cell Biol 27:3035–3043. doi:10.1128/MCB.02203-06
23. Cao S, Kumimoto RW, Gnesutta N, Calogero AM, Mantovani R, Holt BF (2014) A distal CCAAT/NUCLEAR FACTOR Y complex promotes chromatin looping at the FLOWERING LOCUS T promoter and regulates the timing of flowering in Arabidopsis. Plant Cell 26(3):1009–1017. doi:10.1105/tpc.113.120352
24. Jégu T, Domenichini S, Blein T, Ariel F, Christ A, Kim S-K, Crespi M, Boutet-Mercey S, Mouille G, Bourge M, Hirt H, Bergounioux C, Raynaud C, Benhamed M (2015) A SWI/SNF chromatin remodelling protein controls cytokinin production through the regulation of chromatin architecture. PLoS One 10:e0138276. doi:10.1371/journal.pone.0138276
25. Hagège H, Klous P, Braem C, Splinter E, Dekker J, Cathala G, de Laat W, Forné T (2007) Quantitative analysis of chromosome conformation capture assays (3C-qPCR). Nat Protoc 2:1722–1733. doi:10.1038/nprot.2007.243
26. Mifsud B, Tavares-Cadete F, Young AN, Sugar R, Schoenfelder S, Ferreira L, Wingett SW, Andrews S, Grey W, Ewels PA, Herman B, Happe S, Higgs A, LeProust E, Follows GA, Fraser P, Luscombe NM, Osborne CS (2015) Mapping long-range promoter contacts in human cells with high-resolution capture Hi-C. Nat Genet 47:598–606. doi:10.1038/ng.3286
27. Deal RB, Henikoff S (2011) The INTACT method for cell type-specific gene expression and chromatin profiling in Arabidopsis thaliana. Nat Protoc 6:56–68. doi:10.1038/nprot.2010.175
28. Wang D, Deal RB (2015) Epigenome profiling of specific plant cell types using a streamlined INTACT protocol and ChIP-seq. Methods Mol Biol 1284:3–25. doi: 10.1007/978-1-4939-2444-8_1
29. Bonn S, Zinzen RP, Girardot C, Gustafson EH, Perez-Gonzalez A, Delhomme N, Ghavi-Helm Y, Wilczyński B, Riddell A, Furlong EEM (2012) Tissue-specific analysis of chromatin state identifies temporal signatures of enhancer activity during embryonic development. Nat Genet 44:148–156. doi:10.1038/ng.1064
30. Schnable PS, Ware D, Fulton RS, Stein JC, Wei F, Pasternak S, Liang C, Zhang J, Fulton L, Graves TA, Minx P, Reily AD, Courtney L, Kruchowski SS, Tomlinson C, Strong C, Delehaunty K, Fronick C, Courtney B, Rock SM, Belter E, Du F, Kim K, Abbott RM, Cotton M, Levy A, Marchetto P, Ochoa K, Jackson SM, Gillam B, Chen W, Yan L, Higginbotham J, Cardenas M, Waligorski J, Applebaum E, Phelps L, Falcone J, Kanchi K, Thane T, Scimone A, Thane N, Henke J, Wang T, Ruppert J, Shah N, Rotter K, Hodges J, Ingenthron E, Cordes M, Kohlberg S, Sgro J, Delgado B, Mead K, Chinwalla A, Leonard S, Crouse K, Collura K, Kudrna D, Currie J, He R, Angelova A, Rajasekar S, Mueller T, Lomeli R, Scara G, Ko A, Delaney K, Wissotski M, Lopez G, Campos D, Braidotti M, Ashley E, Golser W, Kim H, Lee S, Lin J, Dujmic Z, Kim W, Talag J, Zuccolo A, Fan C, Sebastian A, Kramer M, Spiegel L, Nascimento L, Zutavern T, Miller B, Ambroise C, Muller S, Spooner W, Narechania A, Ren L, Wei S, Kumari S, Faga B, Levy MJ, McMahan L, Van Buren P, Vaughn MW, Ying K, Yeh C-T, Emrich SJ, Jia Y, Kalyanaraman A, Hsia A-P, Barbazuk WB, Baucom RS, Brutnell TP, Carpita NC, Chaparro C, Chia J-M, Deragon J-M, Estill JC, Fu Y, Jeddeloh JA, Han Y, Lee H, Li P, Lisch DR, Liu S, Liu Z,

Nagel DH, McCann MC, SanMiguel P, Myers AM, Nettleton D, Nguyen J, Penning BW, Ponnala L, Schneider KL, Schwartz DC, Sharma A, Soderlund C, Springer NM, Sun Q, Wang H, Waterman M, Westerman R, Wolfgruber TK, Yang L, Yu Y, Zhang L, Zhou S, Zhu Q, Bennetzen JL, Dawe RK, Jiang J, Jiang N, Presting GG, Wessler SR, Aluru S, Martienssen RA, Clifton SW, McCombie WR, Wing RA, Wilson RK (2009) The B73 maize genome: complexity, diversity, and dynamics. Science 326:1112–1115. doi:10.1126/science.1178534

Chapter 16

Profiling Histone Modifications in Synchronized Floral Tissues for Quantitative Resolution of Chromatin and Transcriptome Dynamics

Julia Engelhorn, Frank Wellmer, and Cristel C. Carles

Abstract

Covalent histone modifications and their effects on chromatin state and accessibility play a key role in the regulation of gene expression in eukaryotes. To gain insights into their functions during plant growth and development, the distribution of histone modifications can be analyzed at a genome-wide scale through chromatin immunoprecipitation assays followed by sequencing of the isolated genomic DNA. Here, we present a protocol for systematic analysis of the distribution and dynamic changes of selected histone modifications, during flower development in the model plant *Arabidopsis thaliana*. This protocol utilizes a previously established floral induction system to synchronize flower development, which allows the collection of sufficient plant material for analysis by genomic technologies. In this chapter, we describe how to use this system to study, from the same set of samples, chromatin and transcriptome dynamics during early stages of flower formation.

Key words Histone, Epigenetics, ChIP-seq, RNA-seq, Floral induction system, Flower development, Inflorescence meristem, Stem cell, Differentiation

1 Introduction

Floral primordia are initiated on the flanks of the inflorescence meristem after a plant has switched from the vegetative to the reproductive phase of development. Their formation is accompanied by dramatic changes in gene expression, with activation of gene sets required for flower development and repression of gene sets that are no longer needed. These expression changes are known to involve alterations in the plant's epigenome; however, the underlying mechanisms and the nature of prevalent chromatin states remain largely unknown. Genomic approaches such as chromatin immunoprecipitation coupled to next-generation DNA sequencing (ChIP-seq) and RNA-sequencing (RNA-seq) allow the systematic characterization of epigenomes and transcriptomes. However, application of these experimental strategies has been hampered for

Marian Bemer and Célia Baroux (eds.), *Plant Chromatin Dynamics: Methods and Protocols*, Methods in Molecular Biology, vol. 1675, DOI 10.1007/978-1-4939-7318-7_16,

flower development because primordia are initiated sequentially by the inflorescence meristem. As a consequence, all floral buds in an inflorescence are at a different developmental stage, making the collection of a sufficient amount of tissue for stage-specific epigenomic analyses extremely challenging. This problem is especially pronounced for early developmental stages, when floral primordia are minute and not easily accessible because they are surrounded by older, much larger flowers [1]. To circumvent this problem, a floral induction system (FIS) has been developed, which allows the synchronization of flower development [2]. This system is based on a double mutant in which the paralogous genes *APETALA1* (*AP1*) and *CAULIFLOWER* (*CAL*) are disrupted. *AP1* and *CAL* encode MADS domain transcription factors that control the onset of flower development in a partially redundant manner [3]. Concomitant inactivation of these genes in an ***ap1 cal*** double mutant leads to a significant delay in flower formation and a massive over-proliferation of inflorescence-like meristems [3]. In the floral induction system, this over-proliferation phenotype is being exploited to convert individual meristems into flowers. To this end, a transgene that allows a specific activation of AP1 is used. Several different strategies to mediate this activation have been established over the years [4]. The strategy that works best in our hands is based on expression of a fusion protein between AP1 and the hormone-binding domain of the rat glucocorticoid receptor (GR) in the ***ap1 cal*** background. This fusion protein can be activated by treating plants with the steroid hormone dexamethasone (Dex), resulting in its nuclear import. Expression of the AP1-GR fusion protein has been either driven by the constitutive Cauliflower mosaic virus 35S promoter or by the *AP1* promoter [2, 4].

In this chapter, we provide information on all experimental and analysis steps that make use of this inducible system for correlative and quantitative resolution of histone modifications and transcriptome dynamics (*see* **Note 1**). This information includes: (1) Instructions for proper induction and harvest of synchronized tissues that require specific cultivation conditions as well as precise tissue dissection. (2) A protocol for quantitative analysis of histone modification dynamics described from the wet-lab steps to bioinformatics throughout, including material cross-linking, preparation of nuclear extracts, fragmentation of chromatin, immunoprecipitation using antibodies against modified histones, reverse cross-linking and DNA cleanup, library preparation, sequencing, and pipeline for quantitative analysis of the sequencing output. (3) Analysis recommendations of transcriptome dynamics involving regular RNA preparation using a commercial kit and standard library preparation, sequencing, and data analysis. (4) Suggestions for correlative analysis between chromatin and transcriptome profiles, including comparison of populations of differentially

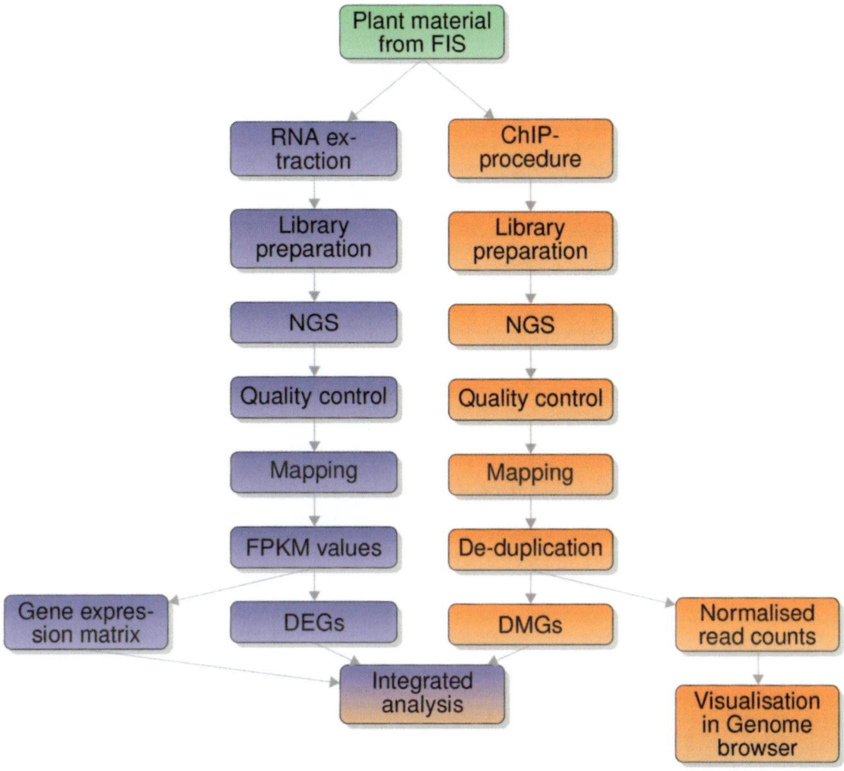

Fig. 1 Overview of the workflow for integrative analysis of RNA-seq and ChIP-seq data

expressed and differentially histone-modified genes, for which we discuss limitations and propose original illustration designs. This complete experimental setup is visualized in Fig. 1.

2 Materials

The wet-lab steps for both ChIP-seq and RNA-seq procedures require standard molecular biology laboratory equipment including a refrigerated microcentrifuge, a bench-top centrifuge for 50 mL tubes, a thermocycler, a set of micropipettes (*see* **Note 2**), Milli-Q water, agarose gel electrophoresis chambers, liquid nitrogen, a rotating wheel for microcentrifuge tubes and a heat block for microcentrifuge tubes. To avoid loss of nucleic acid material and prevent contamination, low binding filtered pipette tips and microcentrifuge tubes should be used throughout the protocol. For the ChIP procedure, all surfaces and nonautoclavable materials should be cleaned with detergents and 70% ethanol. Starting from Subheading 2.2.3, all buffers are prepared in Falcon tubes with sterile-filtered stock solutions. Final working concentrations of components in buffers are noted in brackets.

The computational analysis can be performed on standard up to date laptop computers. A minimum of 4 GB RAM is required, however, we would recommend to have at least 8 GB RAM. Most analyses are conducted using a UNIX command line.

2.1 Induction of Synchronized Tissues

1. *Arabidopsis thaliana* (Arabidopsis) plants of the following genotypes: wild-type (L*er*-1, CS1642), *ap1-1 cal-1* (Accession number CS6161) [3], *ap1-1 cal-1 35S::AP1-GR* (accession number CS67157) [2], *ap1-1 cal-1 pAP1::AP1-GR* [4], all in the Landsberg erecta ecotype.
2. Growth chamber with controlled temperature, light and humidity conditions (16 h day/8 h night, 100 μE cool white light, 70% humidity and a moderate temperature of 18 °C/16 °C to prevent leakage of the *35S::AP1-GR* construct, which may occur at higher temperatures).
3. Petri dishes containing sterile germination Murashige–Skoog (MS) medium supplemented with 0.3% sucrose.
4. Planting trays, pots (e.g., 8 cm × 8 cm pots to host four plants each).
5. Soil–vermiculite mixture (2:1).
6. Fertilizer: Nitrogen–phosphorus–potassium (1:1:1).
7. Fume hood.
8. Dexamethasone induction solution: Prepare a 10 mM dexamethasone stock solution in 100% EtOH (can be stored at −20 ° C for several years). Just before induction, mix 50 μL of 10 mM Dex (10 μM), 7.5 μL Silwet L-77 (0.015% (v/v) and adjust volume to 50 mL with Milli-Q water.

2.2 Quantitative Analysis of Histone Modification Dynamics

2.2.1 Tissue Harvest and Chromatin Fixation

1. Scintillation vials.
2. Tweezers.
3. Fine forceps (number 5).
4. Cross-linking buffer: For five samples, just before use, mix 13.69 g Sucrose (13.69% (w/v)), 1 mL 100 mM PMSF (1 mM), 1 mL 1 M Tris–HCL pH 8 (10 mM), 200 μL 0.5 M EDTA (1 mM), 2.7 mL 37% formaldehyde (1% (v/v)), and adjust volume to 100 mL with sterile Milli-Q water. Pour 20 mL of cross-linking buffer per sample, in scintillation vials.
5. Vacuum pump and desiccator.
6. 2 M glycine (a 40 mL stock can be prepared and filter-sterilized, for several uses).
7. Sieve (e.g., a tea sieve, fine enough to hold back small pieces of tissue, with approx. 0.5 mm pore size).
8. Whatman 3MM paper.
9. Aluminum foil.

2.2.2 Preparation of Nuclear Extracts

1. Mortars and pestles.
2. 50 mL Falcon tubes.
3. Honda buffer: Dissolve 0.48 g Hepes (20 mM) in 70 mL sterile Milli-Q water and adjust pH to 7.4 with KOH. Add 15.06 g sucrose (0.44 M), 1.25 g Ficoll (1.25% (w/v)), 2.5 g Dextran T40 (2.5% (w/v)), 1 mL of 1 M $MgCl_2$ solution (10 mM), and 500 μL of Triton X-100 (0.5% (v/v)) and stir until dissolved (the Honda buffer can be prepared the day before and kept at 4 °C). Just before use, add 500 μL of 1 M DTT solution (5 mM), 1 mL of 100 mM PMSF solution (1 mM), and 200 μL of Plant Protease Inhibitor Cocktail (0.2% (v/v)) (Sigma). Adjust the volume to 100 mL with sterile Milli-Q water.
4. 50 μm and 75 μm Nylon mesh.
5. Funnels, adaptable to the 50 mL Falcon tubes.

2.2.3 Fragmentation of Chromatin and Size Verification

1. Water-bath sonicator (e.g., Bioruptor, Diagenode).
2. Optional: DNA fragment distribution analyzer, such as a microcapillary electrophoresis trace analyzer (e.g., Bioanalyzer with corresponding High Sensitivity DNA Analysis Kit, Agilent).
3. Nuclei Lysis Buffer (NLB): Mix 0.25 mL of 1 M Tris–HCl pH 8 (50 mM), 100 μL of 0.5 M EDTA (10 mM), and 0.25 mL of 20% SDS (1%), and adjust the volume to 5 mL with sterile Milli-Q water (NLB can be prepared during the wash steps and kept at room temperature, do not store on ice to prevent SDS precipitation). Just before use, add 50 μL of 100 mM PMSF (1 mM) and 10 μL of Plant Protease Inhibitors (0.2%).

2.2.4 Immuno-precipitation of Modified Histones

1. Specific, ChIP-grade antibodies against the chromatin feature of interest, e.g., anti-H3K4me3 and anti-H3K27me3 antibodies (e.g., Millipore) (*see* **Note 3**).
2. Corresponding modified peptides, i.e., containing for example monomethylated, dimethylated, and trimethylated versions of H3K27 and H3K4 (e.g., Abcam, ab1780-1782, ab1340, ab1768, and ab1742 but alternatives are available).
3. Protein A agarose beads.
4. ChIP Dilution Buffer (prepare the same day): Mix 110 μL of 100%Triton X-100 (1.1%), 24 μL of 0.5 M EDTA (1.2 mM), 167 μL of 1 M Tris–HCl pH 8 (16.7 mM), and 334 μL of 5 M NaCl (167 mM), and adjust volume to 10 mL with sterile Milli-Q water. Just before use, add 20 μL of Plant Protease Inhibitors (0.2%).

5. Binding/Washing Buffer: Mix 1.5 mL of 5 M NaCl (150 mM), 1 mL of 1 M Tris–HCl pH 8 (20 mM), 200 μL of 0.5 M EDTA (2 mM), 500 μL of 100% Triton X-100 (1%), and 250 μL of 20% SDS (0.1%), and adjust volume to 50 mL with Milli-Q water (this can be done the day before, keep buffer at 4 °C). Just before use, add 500 μL of 100 mM PMSF (1 mM) and 100 μL of Plant Protease Inhibitors (0.2%).

2.2.5 Reverse Cross-Linking and DNA Cleanup

1. 1 M Tris–HCl pH 9.
2. Nucleic acid decontamination solution (e.g., RNase away, Roth).
3. Glycine Elution Buffer (GEB): Mix 1 mL 1 M glycine (0.1 M), 1 mL 5 M NaCl (0.5 M), and 5 μL Tween 20 (0.05% (v/v)), add sterile Milli-Q water to 10 mL, and adjust pH to 2.8 (using 37% HCl). Prepare a fresh 1 M glycine solution before every use. The pH Meter should be cleaned with nucleic acid decontamination solution and sterile water. Solution should be filter-sterilized.
4. RNase A (10 mg/mL).
5. Proteinase K (10 mg/mL).
6. PCR fragment purification kit (e.g., MiniElute Reaction Cleanup Kit, Qiagen).
7. 3 M NaAc (pH 5.2).

2.2.6 Quality Control of the DNA and Library Preparation for Sequencing

1. Microcapillary electrophoresis trace analyzer (e.g., Bioanalyzer, Agilent Technologies) and/or DNA quantification system: ideally, fluorometric-based (e.g., QubiT fluorometer, Thermo Fischer Scientifics and corresponding assay kit, Qubit HS DNA assay or Quantifluor dsDNA system, Promega) or alternative spectrophotometer (e.g., NanoDrop Instruments).
2. Library preparation kit for next generation DNA sequencing (e.g., TruSeq ChIP sample preparation kit, Illumina).

2.2.7 Quantitative Analysis of the Sequencing Output

Required software packages:

1. BWA [5].
2. SAMtools [6].
3. Picard-tools (picard.sourceforge.net).
4. bedtools [7].
5. SICER [8].
6. awk.
7. Optional: Generic Genome Browser (GBrowse 2.54) [9].
8. Optional: bedGraphToBigWig [10].

2.3 Analysis of Transcriptome Dynamics

2.3.1 Requirements for Wet Lab Procedure

1. Mortars and pestles.
2. RNA extraction kit (e.g., Qiagen RNeasy).
3. DNase (e.g., DNA-free™Kit (Ambion)).
4. Reverse transcriptase (e.g., SuperScript II).

2.3.2 Required Software

1. TopHat and Cufflinks [11].
2. Bowtie2 [12].

2.4 Correlation Between Chromatin and Transcriptome Profiles

Software*:* Heat map visualization software (e.g., Genesis, [13]).

3 Methods

3.1 Induction and Harvest of Synchronized Tissues

This section describes the cultivation of the plants for the floral induction system and the harvesting of tissue for the ChIP experiments and RNA extraction. The procedure takes about 10 weeks from seed to harvested material, but once harvested and fixed (*see* Subheading 3.2), the material is stable at −80 °C for several months.

3.1.1 Experimental Planning

Depending on the chromatin feature and the developmental program to be analyzed, different time points during floral development can be of interest. Despite every effort to achieve reproducible growth conditions, there will always be variation in plant developmental timing between different laboratories. It is therefore recommended to perform a first pilot analysis for both chromatin features and expression by qPCR. Based on the results of this experiment, a harvesting scheme for the main experiment can be scheduled (Fig. 2). The publication of Wellmer et al. (2006) [2] can be used as a guide to design this first experiment and gives a good overview on which gene expression states to expect at different time points during induction. Three main choices have to be considered (and eventually tested in the pilot experiment) for an effective experimental design:

1. The promoter for the floral induction system: either the native AP1-promoter or the constitutive 35S promoter. It might be worth growing both genotypes in the cabinet for one generation and visually inspect the formation of the "cauliflower" structures (Fig. 3). They should remain smooth and without any visible floral organs formed for several weeks if the construct is not leaking. We found that depending on the cabinet used, either one or the other promoter construct forms nicer structures.

Experiment 1	replicate 1	**induction with 10uM Dex** (0.1% v/v EtOH; 0.015% Silwet)

Calendar	Day 1 11am	Day 1 5pm	Day 3 11am	Day 6 11am
induction				
	ap1cal AP1-GR			
harvest for Chromatin & RNA (to ChIP-seq and RNA-seq)				
	ap1cal *ap1cal* AP1-GR **t=0** WT leaf WT inflorescence	*ap1cal* AP1-GR **t=6h**	*ap1cal* AP1-GR **t=2d**	*ap1cal AP1-GR* **t=5d**
harvest for Tissue & RNA (to assess correct induction by SEM, ish and qRT-PCR)				
	ap1cal *ap1cal* AP1-GR **t=0** WT leaf WT inflorescence			*ap1cal* AP1-GR **t=5d**
Number of Plants (All sown at the same time)	ca. 30 pl	ca. 15 pl	ca. 15 pl	ca. 30 pl

Fig. 2 Example of experimental setup using the floral induction system for analysis of chromatin and transcription dynamics. The 'tissue and RNA' harvest allows morphological verification of the correct development of induced floral buds (by scanning electron microscopy, SEM) and molecular verification of the correct spatial and temporal induction of floral genes (by RNA in situ hybridization (ish) and qRT-PCR)

Fig. 3 Developmental output of the floral induction system. (**a**) Macroscopic view. Bright-field pictures were taken 18 days after induction. (**b**) Microscopic view by scanning electron microscopy at t = 0 and t = 5 days after induction

2. The time points to be analyzed should span the time when the chromatin feature of interest changes in abundance and genes of interest change in expression. Transcriptional changes usually start to become visible at 6 h after induction (hai) and for histone marks we saw significant changes at 2 days after induction (dai) and flower development in the FIS stays synchronized until about 5 dai. Thus, a reasonable selection of time points for a pilot experiment would be t = 0, t = 6 h, t = 2 dai and t = 5 dai.
3. Additional controls outside of the FIS time points. For several biological questions it might be of interest to also know the state of certain chromatin features and expression in nonfloral tissues, e.g., as a baseline when floral organ identity genes are not expressed or when meristematic genes are not activated. For this reason, we usually include leaf tissue from wild type plants of the same age in our series. Another extreme time point can be fully developed wild type inflorescences. To check for leaking of the glucocorticoid receptor system, we also include *ap1cal* plants without the FIS construct to serve as control in pilot experiments for expression analyses (*see* Subheading 3.3.2).

3.1.2 Plant Cultivation

1. Sterilize approximately 120 seeds of each genotype (L*er* WT and *ap1cal 35S::AP1-GR* or *ap1cal pAP1::AP1-GR*, and *ap1cal* for control experiments) using your standard seed sterilization protocol and place on MS plates.
2. Stratify seeds at 4 °C for 2–4 days.
3. Transfer to a growth chamber providing long day conditions (16 h day/8 h night), 100 μE cool white light, 70% humidity and a moderate temperature (18 °C/16 °C), in order to prevent any leaking of the glucocorticoid receptor system.
4. After 1 week of growth, transfer seedlings to soil and place in the same growth cabinet. Count approx. 15 plants per sample per material analysis (i.e., for chromatin preparation or RNA preparation), except for WT, which requires approx. 25 plants. Also grow 15 more plants for extreme time-points (t = 0 and 5 dai) to analyze by scanning electron microscopy and RNA in situ hybridization/qRT-PCR, for verification of correct development after induction. It is recommended to perform this verification in preliminary experiments, before performing the ChIP (*see* **Note 4**). Thus, a typical experiment with four time points would require 90 plants containing the floral induction system (*ap1cal 35S::AP1-GR* or *ap1cal pAP1::AP1-GR*) and 30 wild-type plants (Fig. 2). Supplement once a week with fertilizer.

3.1.3 Cauliflower Induction and Tissue Harvest

1. After bolting, when stems have reached a height of ~5–10 cm, reinduction of floral meristem development is performed on the "cauliflower" inflorescences by application of dexamethasone induction solution. For this, under a fume hood, apply drops of induction solution directly onto each inflorescence using a 200 μL pipette. Completely cover the inflorescence in the drop and make sure it goes into the tissue. Let dry and repeat the procedure once. Induce 15 plants per desired time point plus 15 plants for later analysis and verification. Make sure that induced plants are fully dry before placing them back in the growth chamber and avoid contact with noninduced control plants.
2. After the desired time (e.g., 6 hai or 2 dai), harvest all material at approximately the same time. To harvest the samples of the floral induction system, cut the stem ca. 1 cm underneath the main inflorescence. Use this piece of stem to hold the structure while removing as much as possible of small leaf like structure that may form on the cauliflower structure. Afterward, remove as much stem and pedicel tissue as possible. From wild type plants, collect inflorescences (remove all flowers of stage 12 and later and as much stem tissue as possible) and rosette leaves. For RNA and chromatin, split material from the same sample in a way to get approximately 0.15–0.3 g of cauliflower tissue/ inflorescence tissue (1–1.3 g for leaves) for chromatin and 0.06 g for RNA. Material for RNA extraction should be directly frozen in liquid nitrogen, while chromatin samples should be harvested directly into cross-linking buffer (*see* below).
3. We recommend to harvest material for three independent biological replicates for both RNA-seq and ChIP-seq. In the case of ChIP-seq, the third sample may not be sequenced, but employed to confirm the data by semiquantitative or qRT-PCR.

3.2 Quantitative Analysis of Histone Modification Dynamics

This section describes how to fix plant material in formaldehyde to crosslink associated proteins to the DNA, prepare a nuclear extract, immunoprecipitate the chromatin associated with a histone mark of interest, isolate the DNA and perform next-generation sequencing analysis. Fixation has to be performed immediately after harvest, which can take a full day depending on the amount of plants. Nuclear extracts can be prepared in 1 day and are stable at −80 °C for several months. The ChIP procedure requires three consecutive days and the resulting DNA is stable at −20 °C. Performing a pilot DNA extraction (Subheading 3.2.3) to test the sonication efficiency takes one more day; we highly recommend this for the first experiment. We also recommend validation of the antibodies by dot blot, which takes a day (*see* **Note 3**).

Sequencing of the resulting DNA will require several days for library preparation and sequencing, but times vary depending on the machine and services employed and are mainly dependent on the queue time of the service facility, which can be up to several months and should be discussed in advance. Analysis of the data as described here takes a few days.

3.2.1 Tissue Harvest and Chromatin Fixation

1. Make 20 mL of cross-linking buffer per sample, fill in scintillation vials and place them on ice (*see* **Notes 5** and **6**).
2. Harvest plant material directly in the buffer (under the fume hood).
3. Infiltrate with vacuum on ice at least twice (6 min and 24 min).
4. Add 1 mL of 2 M Glycine per 20 mL fix and apply vacuum for 5 min.
5. Place the sample in a sieve and rinse with around 200 mL Milli-Q water.
6. Dry on Whatman 3MM paper and freeze in liquid nitrogen. Store at −80 °C.

3.2.2 Preparation of Nuclear Extracts

1. Grind cross-linked plant material in liquid nitrogen and transfer to a 50 mL tube.
2. Add 15 mL Honda Buffer and keep on ice.
3. Perform **steps 1** and **2** for all samples.
4. Place a 50 μm and a 75 μm Nylon mesh in a funnel in a 50 mL conical disposable centrifuge tube and wet it by adding 1 mL Honda Buffer, filter the solution through the meshes.
5. Wash the tube and meshes with 10 mL Honda Buffer, so that the total volume is around 25 mL.
6. Centrifuge at 3000 × *g*, 15 min, 4 °C.
7. Decant supernatant to a new tube (in case the pellet is too loose) and avoid losing bigger pieces of the pellet. Carefully resuspend the pellet in the remaining supernatant by pipetting and transfer to a microcentrifuge tube.
8. Spin the tube in a prechilled (4 °C) microcentrifuge at 1500 × *g* for 15 min.
9. Discard the supernatant and resuspend the pellet in 1–1.5 mL Honda buffer.
10. Spin the tube in a prechilled (4 °C) microcentrifuge at 1500 × *g* for 7 min. Discard the supernatant and resuspend the pellet in 1–1.5 mL Honda buffer.
11. Repeat **step 10** until the pellet is no longer green.
12. After the last centrifugation step, resuspend the pellet in 500 μL Nuclei Lysis Buffer and split to two tubes.

3.2.3 Fragmentation of Chromatin and Size Verification

1. Sonicate the chromatin: for instance in a Bioruptor filled with ice-cold water, 13× 30 s on, 1 min off, position H. After 10 cycles, put new ice to cool the water bath again. Wait until the ice has melted to continue (*see* **Note 7**).
2. Spin 10 min max. speed in a microcentrifuge at 4 °C, pool the supernatants of each sample to one tube and measure the DNA content. Concentrations of DNA are approximately 200–400 ng/μL for the leaf and inflorescence tissues and 800–1000 ng/μL for the cauliflower tissues.
3. Freeze the nuclear extract in liquid nitrogen and store at −80 °C.

   The following **steps** (**4**–7) allow assessing the size distribution of fragments on a small fraction of the nuclear extract. For ChIP-seq on chromatin marks, it is recommended to have the majority of fragments in a size range between 200 bp and 600 bp.
4. Take 12 μL of nuclear extract.
5. Add 150 μL Tris–HCl pH 9.0 and 300 μL Glycine Elution Buffer (*see* Subheading 2.2.4.) or 450 μL TE (pH 7.5–8).
6. De-crosslink and purify samples (*see* Subheading 3.2.5, **steps 2**–**12**).
7. Check the size distribution of the DNA fragments: either by gel electrophoresis using 10 μL or better, with the Bioanalyzer using 1 μL.

3.2.4 Immuno-precipitation of Modified Histones

1. Per immunoprecipitation (IP), take an aliquot of the nuclear extract corresponding to 20 μg DNA amount and dilute to 50 μL with Nuclear Lysis Buffer (*see* **Note 8**). From this, take 10% as an Input and freeze at −80 °C. Always run one IP without antibody as control.
2. Aliquot 45 μL to a separate tube and add 450 μL IP-dilution buffer.
3. Spin 10 min at max. speed, at 4 °C.
4. In the meantime, prepare new tubes and add the antibody (AB) to these tubes. Depending on the dot-blot results (*see* **Note 3**), use between for example 3 μL and 5 μL of antibody for H3K4me3 and 4 μL to 9 μL for H3K27me3).
5. Add the supernatant from **step 4** to the tubes with AB (normally no precipitate is observed after **step 3**.).
6. Incubate on a rotating wheel at 4 °C for 5 h at rotation speed 12 (secure tubes with snap locks).
7. After 4.5 h, start the preparation of the protein A beads: if n is the number of samples, take (n + 1) * 35 μL protein A beads. Shake the bottle before taking out the beads (*see* **Note 9**).

8. Wash protein A beads three times with 1 mL Binding/Washing Buffer (1500 × *g*, 2 min).
9. After the 3rd wash, adjust the volume of the beads to (n + 1) *100 μL.
10. Add 100 μL Protein A beads to the samples from **step 6**, resuspend the beads by inverting after each second pipetting.
11. Incubate on the rotating wheel at 4 °C overnight.
12. Spin samples at max. speed for 2 min at 4 °C.
13. Wash the beads four times by adding 1 mL Binding/Washing buffer, incubate 10 min on the rotating wheel and spin 4 min max. speed, all on ice or at 4 °C. The supernatant can be decanted, we first decant in a tube, to be safe. During these steps, start to prepare Glycine Elution Buffer.
14. After the last spin, add 1 mL Binding/Washing buffer at RT, incubate 10 min at RT and spin at RT.
15. Decant the supernatant and remove the remaining supernatant with a pipette.
16. Add 100 μL ice-cold GEB, vortex for 30 s.
17. Spin the samples 1 min, max. speed at RT.
18. Prepare new tubes with 150 μL of 1 M Tris–HCl pH 9.
19. Transfer the supernatant to these tubes.
20. Repeat **steps 16** and **17** twice and transfer the supernatants to the tubes from **step 18** (total volume is 450 μL).
21. Spin the pooled elutions for 2 min at maximum speed in a microcentrifuge and transfer the supernatant to a new tube.

3.2.5 Reverse Cross-Linking and DNA Cleanup

1. Thaw the input samples on ice and add 150 μL of 1 M Tris–HCl pH 9.0 and 300 μL of GEB to each input sample. From here on, the input samples are treated along with the ChIP samples. Always treat input samples last in each step to avoid contamination.
2. Add 1 μL RNase A (10 mg/mL) to each sample and incubate for 15 min at 37 °C.
3. Add 3 μL Proteinase K (10 mg/mL) to each sample and incubate overnight at 37 °C.
4. Add another 3 μL of Proteinase K (10 mg/mL) and incubate for 6 h at 65 °C.
5. Cleanup procedure for one sample (e.g., inputs and no-antibody controls):

   For DNA cleanup with, for instance, the MinElute kit, split the samples to two tubes (225 μL each). *See* **Note 10** for an alternative protocol for **steps 5–12**, if performing three technical replicates per sample.

6. Add 675 μL ERC buffer (e.g., from the MinElute kit) to each tube; samples will turn red.
7. Add 40 μL of 3 M NaAc (pH 5.2); samples will turn yellow.
8. Load 500 μL on the MinElute column; spin the column in a microcentrifuge at maximum speed for 1 min.
9. **Repeat 3** more times to load the complete sample.
10. Add 750 μL PE wash buffer to the column, spin the column in a microcentrifuge at maximum speed for 1 min.
11. Discard the flow-through and spin the column in a microcentrifuge at maximum speed for 1 min to dry the membrane.
12. Elute with 30 μL EB.

3.2.6 Quality Control of the DNA

1. Check the distribution of IP fragments, for instance with the Bioanalyzer and quantify the DNA using 1 μL.
   (a) The fluorometric assay should detect DNA in your post-IP samples (can detect DNA in samples as low as 0.1–0.25 ng/μL but linear range starts at 0.25 ng/μL).
   (b) The Bioanalyzer, or equivalent, gives supplementary information on the sizes of the post-IP DNA. You expect a size distribution similar to that of the input sample, but often only input gives enough signal and IP samples are hardly detected (*see* **Note 11**).
2. In order to test for enrichment, perform qRT-PCR or semi-qRT-PCR on IPs and no-AB control samples. Choose a pair of primers amplifying a region that is known to be positive for the respective chromatin mark (e.g., a constitutively active gene like *ACTIN7* for an active chromatin mark like H3K4me3) and, ideally, a region that is negative for the respective mark (e.g., a gene repressed in all analyzed tissues in case of an active chromatin mark). Primers generating fragments of 120–250 bp worked best in our hands. Make dilutions of the ChIP, input and no-AB control samples. The dilution factor will vary depending on the efficiency of the antibody and the chromatin feature analysed (*see* **Note 12**).

3.2.7 Library Preparation and Sequencing

Library preparation can be performed at any platform offering Illumina HiSeq sequencing according to the Illumina standard protocols, using the appropriate kit. Since these are constantly updated, it is the best to refer to Illumina brochure/protocols for details and clarify the options with your preferred sequencing platform. In our experience, quantitative analysis for Arabidopsis is best possible with around 20–25 million uniquely mapped reads. To achieve such numbers, around 40–50 million raw reads are required. Based on these numbers, multiplexing options should be discussed depending on the machine used. We usually perform 50 bp single read sequencing. A critical step in the library preparation is the size selection; we highly recommend to perform size selection (e.g., by performing gel extraction after library preparation if necessary), as otherwise long fragments will result in background. Make sure that the majority of the fragments in your samples is within the selected size range (ideally between 200 bp and 600 bp).

3.2.8 Quantitative Analysis on Sequencing Output

If not otherwise specified, all computational analyses are performed using standard software packages that are well documented. Therefore, we only mention the commands we employ and assume that the reader refers to the manual of the specific software to install and run the software. We assume to have a fastq file named test.fastq, a reference genome sequence ref.fa and a file containing chromosome names and length (tair10.dat) as an example. Perform **steps 1–6** for all ChIP and Input files.

1. Perform quality control of the received data using fastqc

   ```
   fastqc test.fastq
   ```

   Usually, our libraries pass most of the criteria, some warnings are tolerable. Libraries might fail the sequence duplication criterion if sequencing was too deep. Two reasons are possible here: the amount of material sequenced was very high, thus identical sequenced reads were indeed precipitated. This should lead to saturated peaks in the subsequent analysis (*see* **step 9**) or diversity or complexity of the library was low. This should lead to PCR artifacts being sequenced and should be indicated by a strong peak at one duplication level, depending on the number of PCR cycles used to prepare the library. If this leads to very low numbers of unique reads (below 10 million) quantitative detection of subtle differences will become difficult.

   For failures in other criteria, refer to the fastqc documentation or your local sequencing platform to discuss quality issues.

2. Map reads to the genome (currently latest version tair10) with BWA

   ```
   bwa index ref.fa  (has to be done only once)
   bwa aln -t 6 ref.fa test.fastq > test.sai
   bwa samse ref.fa test.sai test.fastq > test.sam
   ```

3. Convert to bam format and sort using samtools

   ```
   samtools view -bS test.sam > test.bam
   samtools sort test.bam test-sorted
   samtools index test-sorted.bam
   ```

4. Remove duplicated reads using picard tools

```
picard-tools SortSam I=test-sorted.bam O=test-sorted2.bam SO=coordinate VALIDATION_STRINGENCY=LENIENT
```

```
picard-tools MarkDuplicates I=test-sorted2.bam O=test-sorted2-rmdups.bam M=log.txt REMOVE_DUPLICATES=true VALIDATION_STRINGENCY=LENIENT
```

5. Index with samtools again

```
samtools index test-sorted2-rmdups.bam
```

6. Convert to bed format using bedtools

```
bamToBed-i test-sorted2-rmdups.bam > test.bed
```

7. Count mapped reads

```
samtools idxstats test.bam | awk 'BEGIN{samples=0} {samples=samples+$3} END{print samples}'
```

(returns the number of total mapped reads)

```
samtools idxstats test-sorted2-rmdups.bam | awk 'BEGIN{samples=0} {samples=samples+$3} END{print samples}'
```

(returns the number of mapped unique reads)

8. Optional: Convert data to bigwig format for visualization (normalized as reads per million).

```
c=`samtools idxstats test-sorted2-rmdups.bam | awk 'BEGIN{samples=0} {samples=samples+$3} END{print samples}' -`;
```

```
x=$( printf "%s%d\n" 'scale = 10; 1000000/' $c | bc); genomeCoverageBed -scale $x -split -bg -ibam test-sorted2-rmdups.bam -g tair10.dat > test.bedgraph
```

```
LC_COLLATE=C sort -k1,1 -k2,2n test.bedgraph > test_sorted_for_bw.bedgraph
```

```
bedGraphToBigWig test_sorted_for_bw.bedgraph tair10.dat test.bw
```

9. Visual inspection of the data:
   (a) Load the different ChIP samples in a genome browser (we recommend loading read count normalized .bw files (**step** 7) as a bigwig set into gbrowse (Fig. 4a)).
   (b) Go to a locus you expect to be strongly targeted by the respective chromatin feature and ideally unchanged during the time course (e.g., a housekeeping gene like ACTIN7 for H3K4me3).

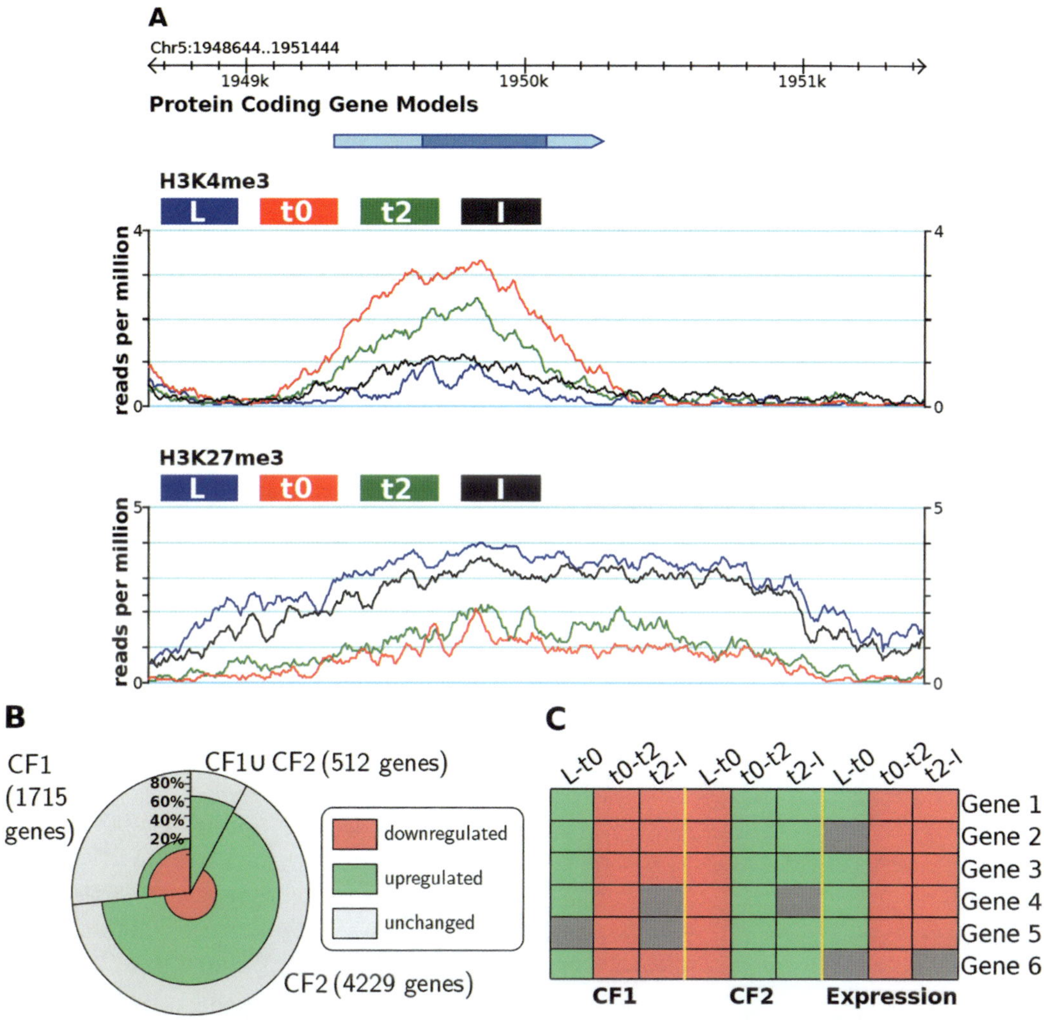

Fig. 4 Examples and suggestions for visualization of chromatin and transcriptome dynamics during flower development. (**a**) Example of a genome browser (gbrowse) view of H3K27me3 and H3K4me3 distribution in *ap1cal 35S::AP1-GR* at t0, t = 2 dai and in wild type leaves (L) and fully expanded inflorescences (I). The selected region contains one gene (shown as an arrow with *dark blue* showing the coding region and *light blue* UTR regions) and was found to show both significant elevation in H3K27me3 and significant reduction in H3K4me3 from t0 to t2 (t = 2 dai), with the analysis pipeline described in this chapter. The quantitative differences are small, however found to be significant in two biological replicates. Furthermore, both t0 and t2 are significantly different from either leaf or inflorescence tissue and also the difference in H3K27m3 between leaf and inflorescence samples is significant. The difference between leaf and inflorescence in H3K4me3 however, is not significant, as both peaks failed to pass the threshold for a positive region in one of the replicates. (**b**) Example pie-chart to represent the comparison of DEGs and DMGs in two chromatin features (CF) for one given comparison during the induction of flower development (e.g., t0 compared to t2 in the FIS). One pie would be drawn for each direction of change for the CFs, e.g., increase in CFs. The pie represents all genes that either change in CF1, in CF2 or both as one piece of the pie. The pieces themselves are then color-coded according to the number of differentially expressed genes in each group. This way, the proportions of DEGs on each list can be easily visualized. (**c**) Example heat map representation of significant changes in two

(c) At such a locus, all normalized files should show a similar pattern und very little variation in signal intensity. If there is a strong bias toward a higher signal for samples with lower read numbers, this will introduce bias in the quantitative analysis. Other normalization methods than normalizing by the total number of reads might help but usually cannot abolish the bias. Outside of the peak regions, the background signal should be low and equal among the different files. Differences in background intensity might also be a cause for unequal signal intensities among samples. In this case, we recommend to first identify the positive regions for the feature to be analyzed in both samples to be compared, and then to employ the filter_raw_tags_by_islands.py function from SICER as recommended in Zang et al. (2009) [8]. The resulting files contain only the reads on positive regions, which can then be employed for the quantitative analysis.

In case the peaks observed in a ChIP-seq experiment reach a plateau around a value of (2× read length/number of mapped unique reads in million), the peaks are saturated. In this case, removal of duplicated reads (**step 4**), should be omitted. To enhance the possibility that only true duplicates are kept, peaks can be first called with duplicates removed and quantitative differences determined only in positive regions with duplicates included.

10. Identification of differentially methylated regions (Sicer):

In the installation folder of Sicer, open the script SICER-df and set the parameter EFFECTIVEGENOME to 0.9.

Assuming two ChIP files to be compared (test1.bed and test2.bed) and two respective Input files (in1.bed and in2.bed):

```
$SICER/SICER-df.sh  test1.bed  in1.bed  test2.bed  in2.bed
200 400 1E-4 1E-4
```

where 200 is the window size, 400 is the gap size (windows with distance lower than this are merged) and 0.0001 is the FDR threshold for both significant positive regions and significant differences between the two samples (*see* **Note 13**).

The two files test1-W200-G400-increased-islands-summary-FDR1E-4 and test1-W200-G400-decreased-islands-summary-FDR1E-4 contain the regions with significantly

Fig. 4 (continued) CFs and expression over a time and tissue series containing leaf (L), t0 FIS, t2 FIS and inflorescence (I) samples. For each gene and comparison, a color code indicates whether the respective feature shows significant increase (*green*), decrease (*red*) or no change (*grey*). This representation allows a compact summary of the data for some selected genes of interest

elevated and reduced signal in test1 compared to test2 respectively. For further procedure, rename these files by adding the .bed file extension.

11. Determination of differentially marked genes:

    For this purpose, a genome annotation file in gff format is needed (download from arabidopsis.org). Below is an example command that returns all genes intersecting with the positive regions in the respective bed file, except for the pseudogenes:

```
intersectBed -wa -wb -a test1-W200-G400-increased-islands-summary-FDR1E-4.bed -b TAIR10_GFF3_genes.gff | grep 'gene' | grep -v 'pseudo' | cut -f 2 -d '=' | cut -f 1 -d ';' | grep -v '\.' > test1-W200-G400-increased-islands-summary-FDR1E-4_genes.txt
```

    Remove redundant entries of genes from the resulting list:

```
awk '!_[$1]++' test1-W200-G400-increased-islands-summary-FDR1E-4_genes.txt >test1-W200-G400-increased-islands-summary-FDR1E-4_genes_uni.txt
```

12. Determine genes with significant differential marking in all replicates.
13. The resulting lists of differentially marked genes (DMGs) can then be used for further analysis, e.g., functional categorization by Gene Ontology analysis.

3.3 Analysis of Transcriptome Dynamics

This section describes how to extract RNA from the floral induction system plants and analyze transcript abundance by RNA-seq. RNA extraction and DNase treatment can be performed within 1 day. Quality controls will take another day. As for ChIP-seq, library preparation and sequencing will take several days and queue times might apply depending on the facility. Data analysis takes a few days.

3.3.1 RNA Extraction

1. Grind the frozen plant material to a fine powder in liquid nitrogen.
2. Extract total RNA using a standard kit (e.g., RNeasy from Qiagen following manufacturer's instructions).
3. Treat 8 μg of RNA in 35 μL volume with a DNA-free Kit according to manufacturer's instructions, resulting in 36 μL final volume of DNase treated DNA
4. Take 20 μL for the RNA-seq experiment, store at −80 °C until further procedure, the remaining 16 μL will be used for quality control.

3.3.2 Quality Control and Control of the Floral Induction System

1. Load 4 μL of RNA on an agarose gel or Bioanalyzer machine to check the integrity of the RNA.
2. Perform reverse transcription on the remaining 12 μL using for example SuperScript® II Reverse Transcriptase.
3. Perform semiquantitative or qRT-PCR on the resulting cDNA to test the expression of the floral organ identity genes *AGAMOUS* and *APETALA3* using primers described in [14]. Transcripts of both genes should be absent in leaf and uninduced cauliflower structures. Slight expression should be observable 6 h after induction and strong expression at 2 dai and in wild type inflorescences.

3.3.3 Library Preparation, Sequencing and Data Analysis

1. Library preparation can be performed according to standard procedures for mRNA-seq. Multiplexing options should be discussed with the employed sequencing platform. We usually multiplex 4 samples per lane on a HiSeq 2000 machine, resulting in 40–50 million raw reads per sample. Quality control can be done by fastqc like for the ChIP-seq samples.
2. Differentially expressed genes can be determined using Tophat and Cufflinks with standard parameters as described in Trapnell et al. (2012) [11] using Bowtie2 to map to the tair10 genome annotation.
3. In a folder containing the reference genome file (e.g., ref.fa, the same genome file as for the ChIP-seq), the genome annotation file (e.g., TAIR10_GFF3_genes.gff) and the fastq files of the experiment, build the index for the genome mapper using:

```
bowtie2-build ref.fa ref_v2
```

4. Create a script to determine differentially expressed genes:

   Create a text document named "run_cuffdiff.sh" with the following content:

```
#!/bin/bash
FILELIST1=`ls tophat_${1}_*/accepted_hits.bam| awk
'{if(NR==1){printf("%s",$0);}else{printf(",%s",$0);}}'`
FILELIST2=`ls tophat_${2}_*/accepted_hits.bam| awk
'{if(NR==1){printf("%s",$0);}else{printf(",%s",$0);}}'`
cuffdiff TAIR10_GFF3_genes.gff -p 4 -o cuffdiff_${1}_${2}
$FILELIST1 $FILELIST2
```

5. Assuming two samples (*see* **Note 14**) with three replicates and resulting fastq files named for example sample1_rep1.fastq.gz, sample1_rep2.fastq.gz, sample1_rep3.fastq.gz and sample2_rep1.fastq.gz, sample2_rep2.fastq.gz, sample2_rep3.fastq.gz mapping can be performed as follows:

```
tophat -r 50 -G TAIR10_GFF3_genes.gff -p 4 -o
tophat_sample1_rep1 ref_v2 sample1_rep1.fastq.gz
```

   Repeat for all files.

6. Fragment per kilobase per million reads (FPKM) values and differentially expressed genes can then be determined by

```
./run_cuffdiff.sh sample1 sample2
```

7. From the resulting table file gene_exp.diff select the differentially expressed genes (they are marked with "yes" in the last column (significant)).
8. This procedure yields lists of differentially expressed genes for each pairwise comparison for the different tissues and time points. Furthermore, an expression matrix can be generated, which contains normalized mean expression values for the three replicates for each tissue per gene (from the file genes.fpkm_tracking containing FPKM values). Both files are necessary for further comparison with ChIP-seq results.

3.4 Correlation Between Chromatin and Transcriptome Profiles

This section describes how to combine the results of the ChIP-seq and RNA-seq analyses and discusses options for data visualization.

3.4.1 Comparison of DEG and DMGs

1. Perform pairwise comparisons of DEG and DMG for each pairwise comparison between tissues (Fig. 3). Pairwise comparisons between gene list can for example be performed using the Venn selector tool (http://bar.utoronto.ca/ntools/cgi-bin/ntools_venn_selector.cgi) of the BAR webservice [15]. This leads to four gene lists per comparison: chromatin feature elevated and expression elevated; chromatin feature reduced and expression reduced; chromatin feature elevated and expression reduced; and chromatin feature reduced and expression elevated.
2. Visualization suggestions:

   We found that for two chromatin features and expression data, usual bar chart representations or Venn diagrams are limited. We thus propose a pie chart representation, where the number of genes changed in one or both of the chromatin features in one tissue/time point comparison is represented by a pie of a certain size (Fig. 4b). This pie is then divided in slices according to the number of genes changing in one of the features or both. Each slice is subdivided according to the number of genes that are differentially expressed in each direction.

   In order to visualize DEGs and DMGs over all samples, it is useful to generate a matrix containing all this information for each gene in each sample. One can for example represent significantly upregulation of a gene and elevation in chromatin features by 1 and reduction by −1, while unchanged genes get a 0 value. From such a matrix, genes can then be extracted and their behavior in terms of both transcription and chromatin can be visualized in a heatmap using standard clustering programs such as Genesis (Fig. 4c).

3.4.2 Further Correlative Analysis

Determination of DEGs and DMGs both relies on *p*-value cutoff, thus false negatives cannot be excluded and genes not found in both lists can still be significantly changed in a chromatin feature, but just under the threshold in the expression analysis. To nevertheless see such tendencies for a limited number of genes of interest, e.g., floral organ identity genes, it can be useful to employ the gene expression matrix and visualize their expression in a heatmap.

Another possibility is to test for the correlation of fold-change value for expression and chromatin marking. However, such analyses are limited to chromatin features with narrow peaks like transcription factor binding sites or narrow chromatin marks like H3K4me3. For broad peaks, also the length of the changed region matters and fold change values would be very misleading. To circumvent this problem, again heatmaps can be employed. For example, DEGs in one condition can be sorted by their expression differences or fold-change values and then the distribution of a chromatin feature (e.g., the difference between two conditions) over each gene locus can be visualized in a heatmap. This representation allows observing correlations between differences in chromatin mark signal intensity, the length of the changed region and transcription.

4 Notes

1. Initially established for inflorescences of *ap1cal* inducible lines in Arabidopsis, the ChIP-seq protocol described in this chapter can also be used for the analysis of Arabidopsis leaves, inflorescences and young seedlings.
2. It is worth to reserve a set of micropipettes dedicated to ChIP-seq and RNA-seq procedures or autoclave before performing the protocol.
3. Information on the specificity of the commercially available antibodies directed against histone modifications can be found in The Histone Antibody Specificity Database at http://www.histoneantibodies.com [16]. However, upon reception of any new antibody lot and before performing the IP, it is highly recommended to verify the antibody efficiency and specificity by dot blot. For this, modified histone peptides are available on the market (*see* Subheading 3 for examples we used in the past). Note that for a same reference, we experienced important variations between production lots and even vials from the same lot may vary in performance.

4. Perform semiquantitative or qRT-PCR to test the expression of the floral organ identity genes *AGAMOUS* and *APETALA3* using primers described in [14]. Transcripts of both genes should be absent in leaf and uninduced cauliflower structures, slightly expressed 6 h after induction and strongly expressed at 2 dai and in wild type inflorescences. Expression of these genes in uninduced tissue indicates leakage of the inducible system. In this case, plants should be grown at lower temperatures and the experiment should be repeated.
5. Formaldehyde is very toxic. Never use it outside the fume hood, leave all papers and vials that were in contact with the buffer inside the fume hood for at least 24 h to allow remaining formaldehyde to evaporate. Discard formaldehyde waste according to your laboratory rules.
6. When performing the protocol on seedlings, use 40 mL cross-linking buffer and harvest 3 g of plant material. Perform infiltration 3× 10 min, 1× 5 min, otherwise the same protocol can be followed.
7. In case of a different sonicator, test various numbers of sonication cycles to find the appropriate sonication scheme required to achieve the desired fragment sizes (200–600 bp). Make sure to minimize foaming and avoid heating of the sample, as this might degrade the proteins.
8. In order to limit the impact of technical bias on the enrichment between samples and to gain sufficient DNA for sequencing to the depth needed for quantitative comparison, we recommend to perform three technical replicates per sample and antibody. These technical replicates will then be merged after DNA cleanup (*see* Subheading 3.2.5, **step 12** and **Note 10**), for preparation of the libraries. A Mastermix can thus be prepared (e.g., 80 μg of nuclear extract in a final volume of 200 μL NLB). For inflorescence samples, the concentration might be too low to gain 20 μg per replicate; smaller amounts, e.g., 10 μg per sample gave good results in our hands.
9. If n is higher than 13, split to several tubes. In this case, make one tube per AB, per replicate or whatever will be compared, to avoid bias due to slightly different amounts of beads between tubes.
10. Alternative protocol for MinElute column loading and elution volumes if 3 technical replicates are done for sequencing (**steps 5–12**):
    (a) Split 3 samples to four tubes: make one new tube, take out 112 μL out of each of the three tubes into the fourth (this is necessary since the added volume will otherwise exceed 1.5 mL in the next step).
    (b) Add 1 mL ERC Buffer to each tube, samples will turn red.

(c) Add 60 μL NaAc 3 M (pH 5.2), samples will turn yellow.

(d) Prepare two MinElute columns per three technical replicates and divide the first sample tube over both columns, spin 1 min max speed.

(e) Repeat 3 more times to load the complete sample of the remaining tubes.

(f) Wash the columns with 750 μL PE and spin 1 min max speed.

(g) Spin 1 min max speed to dry membrane.

(h) Elute the first column with 20 μL EB, place the 2nd column in the same tube and elute with 20 μL EB. This way, the total volume is around 37 μL, 30 μL can be used for sequencing and the rest for quality control.

11. These days, sequencing facilities accept samples for HiSeq with at least 20 ng (they use 10 ng per run and keep 10 ng aside in case). However, we never reached this amount and found that amounts around 1–5 ng were fine for sequencing as the protocol usually gives a very low background and low contamination.

12. Pools of three technical replicates usually give good results in 10× dilutions and 28–32 cycles. The dilution of the no-antibody control has to be adjusted accordingly, taking into account that it was eluted in 30 μL, not 40 μL, input samples usually give good results when diluted 10×. The PCR can be performed with standard Taq polymerase.

13. As window size, it is reasonable to choose a value close to the lower value of the size range sequenced. The gap size should be chosen based on the nature of the observed chromatin feature. For broad peaks like H3K27me3, 600 bp is a commonly used value, while for H3K4me3 200 bp might be sufficient. However, if both marks should be directly compared, we recommend to use a value suitable for both marks, hence our choice for 400 bp when comparing H3K27me3 and H3K4me3. Sicer produces various output files of which only two are important for quantitative comparison. However, it might be also of interest to know all target genes of a chromatin feature (not just the differentially marked genes/regions). For this purpose, the files test1-W200-G400-islands-summary-FDR1E-4.bed and test2-W200-G400-islands-summary-FDR1E-4.bed are generated. They contain all regions with significant enrichment of the chromatin feature of interest compared to the input sample. Genes in these regions can be determined as described for differentially marked regions.

14. To include more than two samples simply add more file lists to the run_cuffdiff.sh script, e.g.,

```
#!/bin/bash
FILELIST1='ls tophat_${1}_*/accepted_hits.bam| awk
'{if(NR==1){printf("%s",$0);}else{printf(",%s",$0);}}''
FILELIST2='ls tophat_${2}_*/accepted_hits.bam| awk
'{if(NR==1){printf("%s",$0);}else{printf(",%s",$0);}}''
FILELIST3='ls tophat_${3}_*/accepted_hits.bam| awk
'{if(NR==1){printf("%s",$0);}else{printf(",%s",$0);}}''
cuffdiff /PATH_TO/TAIR10_GFF3_genes.gff -p 4 -o
cuffdiff_${1}_${2}_${3} $FILELIST1 $FILELIST2 $FILELIST3
```

Acknowledgments

The authors would like to thank Sacha Preuss, Craig Pikkard, Kerstin Kaufmann, David Pose Padilla, and Markus Schmid whose ChIP protocols inspired the one described in this chapter. We further thank Laurent Journot and the team of the mgx sequencing platform in Montpellier for discussion on multiplexing procedures and for performing flexible library preparations and sequencing runs. We thank Laura Gregoire for input on visualization possibilities of multiple data sets. The work leading to this protocol was funded by the French National Agency Young Researcher grant for the ChromFlow project (ANR JCJC, project SVSE2-1206 01 to CCC), the Centre National de la Recherche Scientifique for a CNRS-Higher Education Chair (position 0428–64 to CCC), the Université Grenoble Alpes for an UGA-UJF Initiative Chair (to CCC), and a Marie Curie Intra-European Fellowship within the European Union's Seventh Framework Programme (FP7/2007-2013) under REA grant agreement n° 327377 (to JE).

References

1. Smyth DR, Bowman JL, Meyerowitz EM (1990) Early flower development in Arabidopsis. Plant Cell 2(8):755–767
2. Wellmer F, Alves-Ferreira M, Dubois A, Riechmann JL, Meyerowitz EM (2006) Genome-wide analysis of gene expression during early Arabidopsis flower development. PLoS Genet 2:e117
3. Bowman JL, Alvarez J, Weigel D, Meyerowitz EM, Smyth DR (1993) Control of flower development in Arabidopsis thaliana by APETALA 1 and interacting genes. Development 119:721–721
4. Ó'Maoiléidigh DS, Thomson B, Raganelli A, Wuest SE, Ryan PT, Kwaśniewska K, Carles CC, Graciet E, Wellmer F (2015) Gene network analysis of *Arabidopsis thaliana* flower development through dynamic gene perturbations. Plant J 83:344–358
5. Li H, Durbin R (2009) Fast and accurate short read alignment with Burrows-Wheeler transform. Bioinformatics 25:1754–1760
6. Li H, Handsaker B, Wysoker A, Fennell T, Ruan J, Homer N, Marth G, Abecasis G, Durbin R, Subgroup 1000 GPDP (2009) The sequence alignment/map format and SAMtools. Bioinformatics 25:2078–2079
7. Quinlan AR, Hall IM (2010) BEDTools: a flexible suite of utilities for comparing genomic features. Bioinformatics 26:841–842

8. Zang C, Schones DE, Zeng C, Cui K, Zhao K, Peng W (2009) A clustering approach for identification of enriched domains from histone modification ChIP-Seq data. Bioinformatics 25:1952–1958
9. Stein LD (2013) Using GBrowse 2.0 to visualize and share next-generation sequence data. Brief Bioinform 14:162–171
10. Kent WJ, Zweig AS, Barber G, Hinrichs AS, Karolchik D (2010) BigWig and BigBed: enabling browsing of large distributed datasets. Bioinformatics 26:2204–2207
11. Trapnell C, Roberts A, Goff L, Pertea G, Kim D, Kelley DR, Pimentel H, Salzberg SL, Rinn JL, Pachter L (2012) Differential gene and transcript expression analysis of RNA-seq experiments with TopHat and Cufflinks. Nat Protoc 7:562–578
12. Langmead B, Salzberg S (2012) Fast gapped-read alignment with Bowtie 2. Nat Methods 9:357–359
13. Sturn A, Quackenbush J, Trajanoski Z (2002) Genesis: cluster analysis of microarray data. Bioinformatics 18:207–208
14. Carles CC, Fletcher JC (2009) The SAND domain protein ULTRAPETALA1 acts as a trithorax group factor to regulate cell fate in plants. Genes Dev 23:2723–2728
15. Toufighi K, Brady SM, Austin R, Ly E, Provart NJ (2005) The botany array resource: e-Northerns, expression angling, and promoter analyses. Plant J 43:153–163
16. Rothbart SB, Dickson BM, Raab JR, Grzybowski AT, Krajewski K, Guo AH, Shanle EK, Josefowicz SZ, Fuchs SM, Allis CD, Magnuson TR, Ruthenburg AJ, Strahl BD (2015) An interactive database for the assessment of histone antibody specificity. Mol Cell 59 (3):502–511

Chapter 17

De Novo Identification of sRNA Loci and Non-coding RNAs by High-Throughput Sequencing

Alice Lunardon, Cristian Forestan, Silvia Farinati, and Serena Varotto

Abstract

Non-coding RNA transcripts, such as long non-coding RNAs, miRNAs, siRNAs, and transposon-originating transcripts, are involved in the regulation of RNA stability, protein translation, and/or the modulation of chromatin states. RNA-Seq can be used to catalog this diversity of novel transcripts and a joint analysis of these transcriptomic data can provide useful insights into epigenetic regulation of dynamic responses such as the stress response, which may not be deciphered from individual analysis of single transcript categories. Here, we present a protocol that allows the identification and analysis of small RNAs and long non-coding RNAs, together with the comparison of these species between different sample types.

Key words Epigenetic regulation, lncRNA, Non-coding RNA, RNA-Seq, sRNA, Transcriptome analysis, Transposable elements

1 Introduction

RNA-Seq approaches are widely used to study messenger RNA differential expression by comparing different tissues and/or treatments in a wide range of organisms. In addition, the more recent discovery of different groups of non-coding RNAs, such as long non-coding RNAs, transposon-originating transcripts, miRNAs and siRNAs has been greatly enhanced by RNA-Seq approaches. These non-coding transcripts revealed to be important for various gene regulatory processes, and can be strongly induced by stress conditions [1, 2]. Long non-coding RNAs (lncRNAs), which can be transcribed from intergenic regions or from the antisense strand of protein-coding regions, are emerging as important components of regulatory mechanisms involved in chromatin modification and epigenetic regulation [3]. The transcripts of class I transposable elements (TEs) can move in the genome via an RNA intermediate

Alice Lunardon and Cristian Forestan contributed equally to this work.

Marian Bemer and Célia Baroux (eds.), *Plant Chromatin Dynamics: Methods and Protocols*, Methods in Molecular Biology, vol. 1675, DOI 10.1007/978-1-4939-7318-7_17, © Springer Science+Business Media LLC 2018

called retrotransposon [4], and cause either mutation or influence the transcription of neighboring genes [5]. Finally, both miRNAs and siRNAs can associate with RISC complexes to recognize their targets via base-pairing and produce double-stranded RNA, which triggers silencing of their target regions [6]. In the case of miRNAs, these target regions are usually endogenous protein-coding mRNAs, while siRNAs can direct the methylation of previously unmodified cytosines in the DNA through the RNA-directed DNA methylation pathway (RdDM; reviewed in [7, 8]). The RdDM pathway ensures the silencing of TEs, which are located in compacted chromatin, accompanied by high levels of DNA methylation in every sequence context (mCG, mCHG, and mCHH, where H represents an A, T, or C), or are located in open chromatin in the flanking regions of genes. DNA methylation patterns can recruit chromatin enzyme complexes able to chemically modify the associated histones and vice versa, further affecting chromatin states and Pol II transcription. The non-coding transcriptome is thus very important for a proper regulation of gene expression and genome integrity, but classical RNA-Seq approaches fail to sufficiently identify this class of transcripts. Here, we present specific small RNA-Seq and RNA-Seq protocols for the identification of all species of non-coding RNAs, with emphasis on the analysis of siRNAs, lncRNAs, and TE transcripts from maize (*Zea mays*; [9, 10]). We describe the RNA extraction and library preparation protocols, as well as the extensive analysis of the transcripts that allow the identification and de novo annotation of different species of non-coding RNAs. Although we employed this protocol for maize, it can also be used for other species with small modifications in the bioinformatic procedures.

2 Materials

2.1 Equipment

1. RNase-free microcentrifuge tubes and pipette tips.
2. Mortar and pestle.
3. Liquid nitrogen.
4. Microcentrifuge (12,000 × g or higher).
5. Heat block or water bath.
6. Magnetic rack or stand for 1.5 mL tubes.
7. Nucleic acid quantification device (e.g., NanoDrop 2000 UV-Vis Spectrophotometer, Thermo Fisher).
8. Nuclei Acid profile analyzer (e.g., Agilent 2100 Bioanalyzer, Agilent).

2.2 Solutions, Chemicals and Buffers

1. 100% Ethanol.
2. RNase-free distilled water.
3. Nucleic acid surface decontamination solutions (e.g., RNase AWAY® Reagent, Thermo Fisher Scientific).
4. RNA extraction kit (e.g., Spectrum Plant Total RNA Kit from Sigma, *see* **Note 1**).
5. On-Column DNase I Digest Set (e.g., DNASE10 and DNASE70, Sigma).
6. Ribosomal RNA removal kit (e.g., Ribo-Zero™ Magnetic Kit (Plant), Illumina).
7. RNA purification kit (e.g., RNA Clean & Concentrator™-5 from Zymo Research).
8. Small RNA library preparation kit (e.g., TruSeQ® small RNA Sample Preparation Kit, Illumina).
9. Total RNA library preparation kit (e.g., TruSeq Stranded RNA Library Prep Kit, Illumina).
10. Next generation sequencing platform (e.g., Illumina Hiseq2000).

2.3 Data Analysis

1. Multicore computer (with at least 8 GB of RAM) or high-performance cluster with a Linux-based operating system.
2. Software for quality control of sequenced reads (e.g., FastQC tool, available at http://www.bioinformatics.babraham.ac.uk/projects/fastqc/).
3. Software for removal of adapter sequences, primers, poly-A tails, and other types of unwanted sequences from NGS sequencing reads (e.g., Cutadapt 1.2.1 [11], available at http://cutadapt.readthedocs.io/en/stable/index.html).
4. Software for read quality trimming and contamination filtering (e.g., Extended Randomized Numerical alignEr (ERNE) [12], available at http://erne.sourceforge.net/).
5. Software for annotation and quantification of small RNA loci (e.g., ShortStack 3.3 [13], available at https://github.com/MikeAxtell).
6. Software for unspliced read mapping (e.g., Bowtie2 [14], available at http://bowtie-bio.sourceforge.net/bowtie2/index.shtml).
7. Software for splice junction mapping of RNA-Seq reads (e.g., Tophat2 [15], available at https://ccb.jhu.edu/software/tophat/index.shtml).
8. Software for transcript assembly and differential expression analysis (e.g., Cufflinks package [16], available at http://cole-trapnell-lab.github.io/cufflinks/).

9. Software for transcript expression quantification (e.g., RNA-Seq by Expectation Maximization (RSEM) [17], available at https://deweylab.github.io/RSEM/).
10. Software for differential expression analysis (e.g., edgeR package [18], available at https://bioconductor.org/packages/release/bioc/html/edgeR.html).
11. Tools for manipulating read alignments in the SAM/BAM format (e.g., Samtools package [19], available at http://samtools.sourceforge.net/).
12. Tools for genomics analysis tasks (e.g., BEDTools package [20], available at http://bedtools.readthedocs.io/en/latest/).
13. We describe the RNA-Seq data analysis here based on the command-line interface, but many of the utilized tools are also implemented on the web-based workflow manager Galaxy (https://galaxyproject.org/).

3 Methods

Depending on the plant species and tissue type to be analyzed, growth conditions and RNA extraction procedures may vary. In general, it is important to sample all plant materials at the same time of the day to avoid diurnal variation, to randomize mutant and wild-type plants if possible, and to keep all conditions the same except for the treatment conditions. All samples should at least be processed in triplicate to allow for the identification of sRNAs that are significantly differentially expressed in different conditions. After harvesting, samples should be flash-frozen in liquid nitrogen as soon as possible to prevent RNA degradation, and can then be stored at −70 °C.

Since high RNA quality is essential for successful Illumina sequencing experiments, creating and maintaining an RNase-free work environment is critical for performing successful RNA isolation, rRNA removal reactions and library preparation. Using nucleic acid surface decontamination solutions (e.g., RNase AWAY® Reagent form Thermo Fisher Scientific) and good working practices are therefore crucial for a successful experiment.

3.1 RNA Extraction

Here we describe the extraction using the Spectrum Plant Total RNA Kit (Sigma) but alternatives are available. This kit usually allows the isolation of about 30–50 μg of total RNA from 100 mg of maize leaves. These yields are usually sufficient to perform required RNA quality controls and to prepare libraries for both small RNA-Seq and total RNA-Seq.

1. Grind the frozen tissue to a fine powder in liquid nitrogen using a mortar and pestle, avoiding thawing of the material.

2. Transfer 90–100 mg of the tissue powder in a 2-mL prechilled microcentrifuge tube and keep sample on liquid nitrogen or at −70 °C before lysis solution is added.
3. Add 500 μL of the Lysis Solution/2-ME Mixture to the tissue powder and vortex immediately and vigorously for at least 1 min.
4. Incubate the sample at 56 °C for 3–5 min.
5. Centrifuge the sample at maximum speed (12,000 × *g* or higher) at room temperature for 3 min to pellet cellular debris.
6. Pipette the lysate supernatant into a Filtration Column, close the cap and centrifuge at maximum speed for 1 min to remove residual debris.
7. Save the clarified flow-through lysate, add 750 μL of Binding Solution and mix immediately and thoroughly by pipetting at least five times or vortex briefly. The use of 750 μL of Binding Solution allows to recover the small-sized RNA molecules (such as microRNA and siRNA).
8. Pipette 700 μL of the mixture into a Binding Column, close the cap and centrifuge at maximum speed for 1 min to bind RNA. Discard the flow-through liquid, pipette the remaining mixture to the column and repeat the centrifugation and decanting steps.
9. Pipette 300 μL of Wash Solution I into the Binding Column, close the cap and centrifuge at maximum speed for 1 min. Discard the flow-through liquid.
10. For On-Column DNase Digestion, combine 10 μL of DNase I with 70 μL of DNase digestion buffer, mix gently by pipetting and pipette the mixture directly onto the center of the filter inside the Binding Column.
11. Incubate the sample at room temperature for 15 min.
12. Pipette 500 μL of Wash Solution 1 into the Binding Column and centrifuge at maximum speed for 1 min to remove the digested DNA. Decant the flow-through liquid.
13. Pipette 500 μL of the diluted Wash Solution 2 into the column and centrifuge at maximum speed for 30 s. Discard the flow-through liquid.
14. Pipette another 500 μL of the diluted Wash Solution 2 into the column, close the cap and centrifuge at maximum speed for 30 s. Discard the flow-through liquid and centrifuge again at maximum speed for 1 min to dry the column.
15. Transfer the column to a new, clean 2-mL Collection Tube. Pipette 50 μL of Elution Solution directly onto the center of the binding matrix inside the column. Close the cap and let the tube sit for 1 min. Centrifuge at maximum speed for 1 min to elute.

16. Purified RNA is now in the flow-through eluate and ready for quantification, while sample quality (integrity) is determined using a microcapillary-electrophoresis trace analysis, e.g., an Agilent Bioanalyzer (*see* **Note 2**).
17. Extracted RNA can be directly subjected to rRNA removal and small-RNA library preparation or stored at −70 °C.

3.2 rRNA Removal and RNA Purification

Ribosomal RNA (rRNA) constitutes over 90% of the total RNA in the cell. Typically, RNA-Seq libraries are prepared from total RNA using poly(A) enrichment of the mRNA (mRNA-Seq) to remove rRNA. However, this method fails to capture non-poly(A) transcripts, such as noncanonical transcripts derived from TEs expression or regulatory non-coding RNA transcribed by RNA Pol V. Being interested also in non-poly(A) transcripts, we used the Ribo-Zero™ Magnetic Kit (Plant Leaf) for high efficiency removal of cytoplasmic, mitochondrial and chloroplast ribosomal RNA prior to library preparation for total RNA-Seq. Nowadays, this depletion chemistry is integrated with the Illumina library preparation workflow (TruSeq Stranded Total RNA with Ribo-Zero Plant).

1. The first steps concern the washing and preparing of the Magnetic Beads. Slowly pipet 225 μL of Magnetic Beads into a 1.5 mL RNase-free microcentrifuge tube and place each tube on the magnetic stand for at least 1 min (until the solution appears clear).
2. Remove and discard the supernatant, and add 225 μL of RNase-free water to each tube to wash. Mix well by repeated pipetting or vortexing at medium speed. Magnetize, remove the solution and resuspend the beads in 65 μL Resuspension Solution. Mix well by repeated pipetting or vortexing at medium speed and add 1 μL of RiboGuard RNase Inhibitor
3. To perform rRNA removal, 5 μg total RNA is combined with 10 μL rRNA Removal Solution and 4 μL Reaction Buffer in 40 μL total volume.
4. Gently mix the reaction by pipetting and incubate at 68 °C for 10 min, and subsequently at room temperature for 5 min.
5. Add the treated RNA to the washed Magnetic Beads. Immediately and thoroughly mix the contents of the tube by pipetting at least ten times. Then, vortex the tube immediately at medium setting for 10 s and place at room temperature for 5 min.
6. Following incubation, mix the reactions by vortexing at medium speed for 5 s and then place at 50 °C for 5 min.
7. Place the tubes on a magnetic stand for at least 1 min (until the solution appears clear).

8. Carefully remove the supernatant (85–90 μL) containing the rRNA-depleted RNA and transfer to a 1.5-mL RNase-free microcentrifuge tube. Place the tube on ice and immediately proceed to the final purification steps.
9. The rRNA-depleted samples can be purified by different methods, such as ethanol precipitation, or column-based purification methods. For this purpose we used the RNA Clean & Concentrator™-5 columns (Zymo Research).
10. Add 180 μL of RNA Binding Buffer to the 90 μL of rRNA-depleted RNA and mix well.
11. Add 270 μL of 100% ethanol to the mixture and mix well.
12. Transfer the sample to the Zymo-Spin™ IC Column in a Collection Tube and centrifuge at ≥12,000 × *g* for 1 min and discard the flow-through.
13. Add 400 μL RNA Prep Buffer to the column and centrifuge at ≥12,000 × *g* for 1 min. Discard the flow-through.
14. Add 800 μL RNA Wash Buffer to the column and centrifuge at ≥12,000 × *g* for 30 s. Discard the flow-through. Repeat the wash step with 400 μL RNA Wash Buffer.
15. Centrifuge the Zymo-Spin™ IC Column in an emptied Collection Tube at ≥12,000 × *g* for 2 min to dry the column.
16. Remove the Zymo-Spin™ IC Column carefully from the Collection Tube and transfer it into an RNase-free tube.
17. Add 10 μL of DNase/RNase-free water directly to the column matrix and let stand for 1 min at room temperature. Centrifuge at 10,000 × *g* for 30 s.
18. The eluted RNA can be used immediately or stored at −70 °C.
19. Quantify the yield of rRNA-depleted RNA sample and estimate the efficiency of rRNA removal using a microcapillary-electrophoresis trace analysis, e.g., using an Agilent Bioanalyzer (*see* **Note 2**) prior to proceed to the RNA-Seq library preparation.

3.3 Illumina Library Preparation and Sequencing

The extracted and purified RNA should be converted to double-stranded complementary DNA (cDNA) and specific "adaptor" sequences have to be added at either end of each molecule prior to sequencing. Many specific Illumina kits and protocols are available for transcriptome analysis and others will probably be released in the coming years. The library preparation protocol details are therefore highly dependent on the strategy and specific sequencing platform that is chosen.

A key element of successful Illumina next-generation sequencing relies on high-quality library preparation (i.e., correct insert size, low or absent adapter concatenamers, and low duplication level).

Reproducibility of library preparation between replicates is another fundamental requisite to avoid biases in the interpretation of the sequencing data. To this end, library preparation can be carried-out by the NGS service provider, or in house using commercially available kit for experienced molecular biologists.

In our case sequencing libraries were directly prepared by the NGS service provider. The libraries for small RNA-Seq were prepared using the TruSeQ® small RNA Sample Preparation Kit (Illumina) and sequenced with a multiplexing level of 16, producing on average 9 million reads per sample. We also performed total RNA-Seq from rRNA-depleted RNA, to be able to link differential non-coding RNA expression to differential expression of coding genes or transposable elements. The total RNA was prepared using the TruSeq Stranded RNA Library Prep Kit (Illumina) and sequenced with a multiplex level of 4, producing on average 37 million of 50 bp single-end reads per library. The sequencing was performed using the Illumina HiSeq2000 platform.

3.4 Data Analysis

Sequenced reads obtained from the small RNA and the total RNA sequencing should be analyzed independently to characterize the two distinct populations of RNAs. In the following sections, we will describe the bioinformatic approach for the analysis of small and long RNA populations. Although this is based on the analysis we performed to identify and analyze the stress-responsive regulation in the B73 wild-type maize and the *rmr6* maize mutant [9], we also discuss the critical points that should be evaluated and adapted to the specific biological question of interest.

3.4.1 Small RNA-Seq Analysis

Because siRNA loci can show heterogeneous patterns in different samples, resulting in distinct sequences originating from the same functional locus, the sRNAs are preferably clustered in sRNA loci to facilitate annotation of the sRNA loci in the genome. If different treatments are applied, the sample number can increase considerably, as does the acquired data. For large sample numbers, we therefore employed the tool ShortStack, which performs well in terms of sensitivity and specificity of de novo sRNA loci identification, providing detailed description of the found loci [13]. This software includes a tool producing small RNA-seq alignments, where multimapped sRNAs tend to be placed near regions of confidently high density increasing the balance between precision and sensitivity. The alignment of multimapping reads is a critical step in the identification of genomic loci effectively producing sRNAs, especially in maize due to its highly repetitive genome.

1. Remove the 3′ and 5′ adapters from the sequences with Cutadapt 1.2.1 [11] ("-m 15 -discard-untrimmed" argument).
2. Remove low complexity sequences, containing only two different nucleobases, through a customized Perl script.

3. The FastQC tool can be used for checking the quality of the reads prior to and after the read cleaning steps, evaluating the sequence quality, GC content, the presence of adapters, overrepresented k-mers and duplicated reads. In each library, the Q score of the 90th percentile of reads should be $\geq$28 across all bases.
4. ShortStack version 3.3 [13], can be used to align the clean .fastq files to the maize genome (http://ensembl.gramene.org/Zea_mays/Info/Index) and to de novo annotate the sRNA loci ("-nostitch -mincov 20 -readfile [clean .fastq files]" arguments). When the input is the list of .fastq files of all samples, the software aligns each individual sample and then merges all the alignments into a single .bam file, and, therefore, the sRNA loci are found from the merging of all the alignments.
5. ShortStack produces a list of MIRNA loci, identified by following strict parameters aimed to reduce the false positives, and the rest of the identified sRNA loci are treated as siRNA loci. Among the siRNA loci, the phased loci can be identified with ShortStack and their *p*-values can be corrected for multiple testing using a Benjamini–Hochberg-adjusted significance level of 0.05.
6. With the same procedure, the sRNA loci can be identified in the TE exemplar sequences annotated in the maize TE database (http://maizetedb.org/~maize/), replacing the genome input with the .fasta file of the TE sequences.

3.4.2 Total RNA-Seq Data Analysis

The analysis of RNA-Seq data can be carried out following four consecutive steps: (1) filtering and cleanup of the sequenced reads prior to mapping to the reference genome, (2) mapping to the reference genome, (3) Reannotation of the maize transcriptome using the mapped reads (Subheading 3.4.3), and (4) Identification of reads representing lncRNAs or ascribable to transposable elements on the full set of maize transcripts (Subheading 3.4.5).

The first steps include sequence quality check, read processing and mapping:

1. A preliminary quality check of the sequenced reads can be performed with the FastQC tool, to evaluate the sequence quality, GC content, the presence of adaptors, overrepresented k-mers, and duplicated reads.
2. The sequenced reads can then be cleaned from adapters using Cutadapt 1.2.1 [11] and subsequently trimmed based on the quality scores.
3. Contaminant reads corresponding to rRNA can be filtered out with ERNE-FILTER 1.2 [12], using ribosomal RNA sequences retrieved from the "SILVA ribosomal RNA gene

database project" (http://www.arb-silva.de/) as contaminant reference sequences (*see* **Note 3**).

4. The high quality reads are then mapped to the reference genome (for the maize B73 reference genome, *see* http://ensembl.gramene.org/Zea_mays/Info/Index, most other maize genomes can be downloaded from http://maizegdb.org/) using Tophat2 [15], (*see* **Note 4**). The following modifications from default parameters should be used: maximum intron size, 60,000; minimum intron size, 5; up to three mismatches and gaps allowed. Strand-specific RNA-seq libraries have to be aligned in strand-specific mode (–library-type fr-firststrand argument).
5. After mapping, the percentage of uniquely mapped reads (assigned to only one position in the reference genome) and multimapped reads (reads mapped equally well to two or more genomic positions; *see* **Note 5**), were evaluated using Samtools [19], available at http://samtools.sourceforge.net/. Reads with MAPping Quality (MAPQ) smaller than 1 (corresponding to reads mapping to more than 10 different positions) should be filtered out using Samtools [19] together with PCR duplicates.

3.4.3 Total RNA-Seq Transcriptome Reannotation

The following steps can be performed for de novo transcriptome assembly and identification of novel transcripts (*see* Fig. 1 for an example of a Genome Browser overview of mapped RNA-seq data with newly identified sRNA loci):

1. Reassembly of transcriptome can be carried out with Cufflinks [16], available at http://cole-trapnell-lab.github.io/cufflinks/. Starting from the aligned reads, Cufflinks allows the identification of novel transcripts not included in the reference annotation: based on default parameters, annotation of novel transcripts requires the alignment of at least 10 reads in at least one library, and any new isoforms should represent at least 10% of the total gene abundance in at least one library.
2. Each BAM file should be singularly processed with Cufflinks: 2.2.1 RABT mode with –frag-bias-correct, –multi-read-correct, and –max-intron-length 60,000 options to produce single assemblies.
3. Merge the Cufflinks assemblies produced separately from each BAM file using Cuffmerge. The resulting GTF file can be edited to discard new annotations shorter than the shortest reference transcripts (57 bp in our case) and to resolve duplicated reference annotations.
4. Based on Cufflinks transfrag class codes, newly identified transcripts can be labeled as: (1) _j = potential novel isoform at known locus; (2) _O = generic exonic overlap with a reference

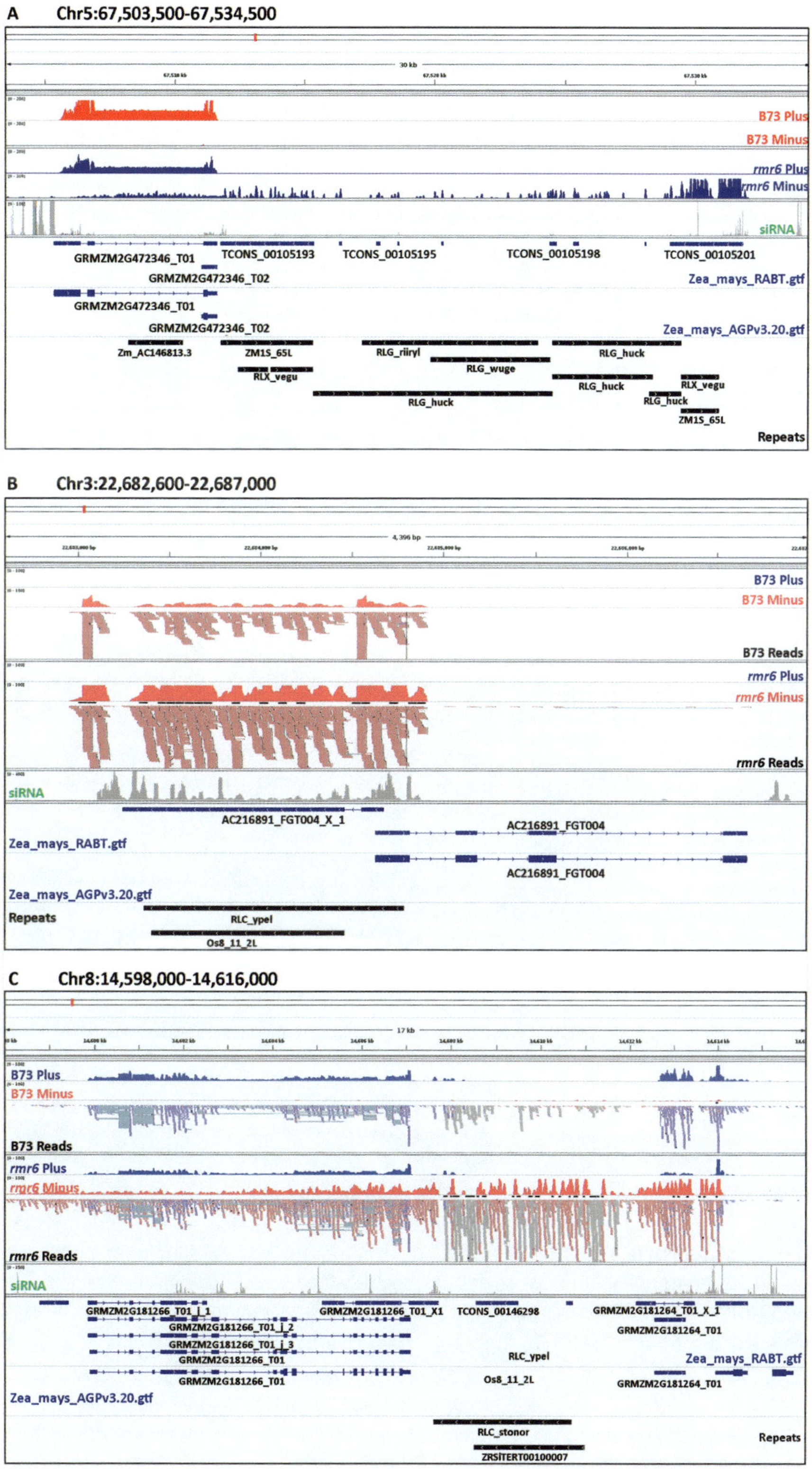

Fig. 1 Genome browser view of RNA-Seq and sRNA-Seq reads mapped at Class U and Class X newly identified loci. IGV—Integrative Genomics Viewer (http:/software.broadinstitute.org/software/igv/) snapshots reporting

transcript; (3) _X = transcript overlapping with reference on the opposite strand; (4) TCONS = unknown, intergenic transcript (also defined as Class U). Generic overlapping, antisense, and intergenic transcripts can be assigned to new loci named in the same way.

5. The accuracy of the new transcriptome annotation can be checked and validated using RSEM (RNA-Seq by Expectation Maximization; [17] software), remapping the cleaned reads against the reconstructed transcriptome and estimating isoform expression (*see* **Note 6**).

3.4.4 Identification of Long Non-coding RNAs (lncRNAs)

LncRNAs are defined as long functional ribonucleic acids not translated into proteins and are involved in the control of genome activity at chromatin level in eukaryotes. In particular, there are evidences that lncRNAs can affect chromatin structure and epigenetically regulate gene expression [1]. LncRNAs are usually grouped into different categories on the basis of positional features of their encoding sequence: long intergenic non-coding RNAs (lincRNAs), natural antisense transcripts (NATs) transcribed from the complementary DNA, and intronic RNAs (incRNAs; [21]). However, there is a rich variety of plant lncRNAs in terms of origin and function: for example in plants, most lncRNAs are transcribed by RNA Pol II (and can be polyadenylated) but also Pol IV and V can produce lncRNAs, which may function as scaffold in RdDM or other silencing pathways [22]. An increasing number of research works have shown that plant lncRNAs are expressed at low levels and differentially during development and in response to stressful environmental conditions [23]. A few lncRNAs have been well characterized in Arabidopsis for their role in the epigenetic regulation of flowering: COLDAIR, COOLAIR, and ASL participate in the epigenetic silencing of flowering repressor Flowering Locus C [24]. Different pipelines have been recently adopted for the genome wide characterization of maize lncRNA [25].

Fig. 1 (continued) mapped reads, annotated transcript (*blue boxes*) and repeat annotations (*black boxes*) for three maize genome regions in which new transcripts have been annotated. (**a**) Coverage of RNA directional sequencing reads (normalized to the total of mapped reads) for B73 (*blue*) and *rpd1/rmr6* mutant (*red*) showing a *rpd1/rmr6* 30 Kb transcriptional active region on chromosome 5 with several newly annotated Class U intergenic loci. siRNAs are particularly abundant at the boundaries of this region in which several RLG class I transposable elements are annotated. (**b**) Strand-specific coverage and mapped reads at a newly identified Class X locus, transcribed in both B73 and *rpd1/rmr6* on the opposite strand of the AC216891_FGT004 transcript. Antisense AC216891_FGT004_X_1 corresponds to an RLC class I transposable element, which is completely covered by siRNAs. (**c**) Strand-specific coverage and mapped reads on a chromosome 8 complex region. Blue mapped reads indicate transcription of the plus strand in both B73 and *rpd1/rmr6*, while red reads, mapped on the minus strand, are detectable only in the *rpd1/rmr6* mutant, resulting in the annotation of both antisense and intergenic new loci. Several peaks of small RNA reads are detectable in this region

1. To predict the potential long non-coding RNAs in the newly annotated transcriptome, a pipeline can be applied based on the following criteria to distinguish the long non-coding from coding transcripts: (1) length ≥200 bp; (2) the presence of an Open Reading Frame <120 amino acids; (3) when an ORF is present, the predicted protein must not match any protein in public databases [26]. This analysis will reveal potential long non-coding RNA (pot-lncRNAs), which can be further classified and characterized as follows:
2. Compare each long non-coding RNA with publicly available sequences of smallRNA precursors [27] and TE-elements (for maize: http://maizetedb.org/~maize/) and with your database of smallRNA identified through smallRNA-Seq, using blastn with the following parameters: e-value <0.01, 100% identity and word-size of 16. If transcripts have a match in the abovementioned databases, it suggests that they might be precursors of smallRNAs or expressed from transposable elements.
3. The remaining transcripts can be annotated as "truly" long non-coding RNAs and can, if possible, be compared with lncRNAs identified in previous studies (e.g., for maize, *see* ref. 25). This can be performed with the BEDTools function intersectBed.
4. The lncRNAs transcribed from the opposite strand of an annotated gene with at least a 10-nt overlapping sequence can be defined as long non-coding Natural Antisense (lncNAT) transcripts. Intersections of genomic coordinates between lncRNAs and coding transcripts were performed with the intersectBed tool.
5. To determine whether or not the remaining lncRNAs, classified as long-intergenic ncRNAs (lincRNAs), could regulate the expression of protein coding genes as long molecules, they can be tested for homology to CDS sequences using Blastn (more than 95% identity and minimum overlap of 100 bp) against the nucleotide sequence of the coding transcripts.

3.4.5 Identification of Transposon Transcripts

Transposable Elements (TEs) are DNA sequences capable of moving from one genome location to another one. TE insertions can promote genetic variability and affect gene expression by disrupting open reading frames or regulatory genomic sequences, such as promoter and enhancers. To maintain its stability, the genome reacts to TE movement by methylating their sequences in order to silence them and prevent further mobilization events. TE-induced methylation and chromatin mediated silencing can also be extended to TE flanking regions, potentially affecting the regulation and expression of neighboring genes. These epigenetically modified gene-coding sequences may be inherited to the offspring and represent in this case genetically transmitted epialleles. TEs are grouped into two principal classes, based on their transposition intermediate: (1) class I elements (also called retrotransposons), characterized by an RNA intermediate and a "copy and paste" mechanism; (2) class II elements, characterized by a DNA intermediate and by either "cut and paste" or "copy and paste" mechanisms. A further hierarchical classification that includes subclasses, orders, superfamily and family is commonly used for plant TEs [28]. During evolution, TEs have successfully colonized eukaryotic genomes and can represent more than 85% of the total genomic sequence in some plant species, such as maize [29, 30]. Two independent approaches can be followed to classify TE transcripts and identify differentially expressed sequences related to transposable elements (**steps 1–3** and **steps 4–6**):

1. Blast all transcript models against a TE database containing full-length sequences of curated, nonredundant TEs (for maize, *see* http://maizetedb.org/~maize/).
2. Identify TE-related transcripts using stringent criteria using the bit scores and coverage percentages of the alignments and classify the TEs in two subgroups: high-confident-TEs (HC-TEs: with Bit-score >500 and coverage >50%) and putative/relic-TEs (PR-TEs: with Bit-score >250 or coverage >30%).
3. Associate each TE-related transcript with its specific TE-family and superfamily to analyze the preferential transcription of specific TE-classes in different samples.
4. Alternatively, to identify TE members differentially expressed in certain samples compared to others, filtered reads can be mapped with Bowtie2 [14] against the full-length TE consensus sequences present in the database (for maize, *see* http://maizetedb.org/~maize/).
5. Use the BEDTools "multiBamCov" function [20] to extract a counts table of uniquely mapped reads for each TE accession starting from alignment files.
6. To identify differentially expressed TE families, the counts table can then be analyzed with the edgeR package [18] (log2 fold change ratio $\geq|2|$ and FDR- adjusted p value ≤ 0.01).

4 Notes

1. The appropriate RNA extraction protocol/kit should be chosen based on species and tissue/cell type. For difficult tissues, such as the *Zea mays* leaves, we used the Spectrum Plant Total RNA Kit from Sigma that facilitates the extraction of RNA from tissues with high levels of secondary metabolites and

allows also to recover small RNA molecules for subsequent sRNA-Seq analysis [9].

2. Prior to libraries preparation it is fundamental to know the quantity and the quality of the RNA starting material. The concentration of RNA should be determined by measuring the absorbance at 260 nm (A260) in a spectrophotometer, while RNA purity is estimated by measuring the A260/A280 ratio that should be approximately 1.8–2.0 (lower values indicate the presence of proteins in the extracted RNA). The integrity of the RNA is another important component of RNA quality: use of degraded RNA can result in overrepresentation of the 3′ ends of the RNA molecules or in the failure of library preparation. Moreover, degradation fragments could be later wrongly identified as small RNAs or differential degradation between samples could be mistaken for differential expression. RNA integrity can be rapidly evaluated by microfluidic analysis, e.g., using the Agilent Bioanalyzer, and the application of the RNA Integrity Number (RIN) algorithm to electrophoretic RNA trace. By comparing signal areas, intensities and ratios of different regions of the electropherogram, in particular at the level of rRNA peaks, the algorithm allows for the classification of eukaryotic total RNA, based on a numbering system from 1 to 10, with 1 being the most degraded profile and 10 being the most intact. The evaluation of RIN values for leafy plant tissues could be lower due to the presence of chloroplast rRNA peaks that are recognized as degradation products, giving lower RIN value than those of nongreen tissues, such as roots. Being valid the rule "the higher the better", the RIN should be higher than 6 for RNA extracted from leafy tissues and higher than 8 for other tissues. Comparison of RNA electropherogram profiles before and after the rRNA depletion also allows the evaluation of the reaction: peaks corresponding to rRNA should indeed not be present in rRNA-depleted RNA.

3. In our hands, the Ribo-Zero™ rRNA Removal Kit gave very good results: only 1–3% of the sequenced reads were ascribable to rRNA (depending on the library). These small percentages, however, resulted in several thousands of reads that could affect the downstream analysis and were thus filtered out. To do this, maize ribosomal RNA sequences (corresponding to cytoplasmic, mitochondrial, and chloroplast rRNA) were downloaded from the "SILVA ribosomal RNA gene database project" (http://www.arb-silva.de/) and used as reference for the ERNE-FILTER 1.2 [12] software to discard the sequenced reads corresponding to rRNAs.

4. Mapping to the reference genome, even when a transcriptome annotation is available, is required for the identification of

novel genes or transcripts and requires a specific splice junction mapper, as reads may span splice junctions. TopHat has been one of the most popular spliced mapper for years, but it is now largely superseded by HISAT2 which provides the same core functionality, in a more accurate and much more efficient way [31] . Similarly, Cufflinks and Cuffdiff tasks could be accomplished faster, with less memory and with more accurate overall results [32] by the recently released tools StringTie [33] and Ballgown [34].

5. These multireads are primarily due to paralogous genes and repetitive sequences, and account for about 60% of our reads (maize is an ancient allotetraploid species and has a highly repetitive genome, mainly consisting of TEs [35]. Since multireads contain important biological information, but also represent an important challenge for transcript identification and differential expression analysis, we decided to discard only those mapping on more than 10 different positions. Reads with MAPping Quality (MAPQ) smaller than 1 were therefore filtered out using Samtools [19] together with PCR duplicates.
6. The RSEM algorithm quantifies expression after read mapping to a reference transcriptome using an alternative strategy to effectively handle ambiguous/multiple-mapping reads, accurately estimating gene-level abundance starting from large numbers of short single-end reads. With this independent assay, 99% of Class X and Class U transcripts passed the expression arbitrary threshold of 1 RPKM (reads per kilobase per million mapped reads) in at least one condition. The percentage decreased slightly for Class O (79%) and Class J (66%) but it was twice as high as those of reference transcripts (35%).

Acknowledgments

The authors would like to thank Riccardo Aiese Cigliano and Walter Sanseverino (Sequentia Biotech) for their precious collaboration during the whole project. This work was supported by EC grant AENEAS and Italian MIUR-CNR EPIGEN Flagship Project to SV.

References

1. Chekanova JA (2015) Long non-coding RNAs and their functions in plants. Curr Opin Plant Biol 27:207–216
2. Zhao J, He Q, Chen G, Wang L, Jin B (2016) Regulation of non-coding RNAs in heat stress responses of plants. Front Plant Sci 7:1213
3. Wang H, Niu QW, Wu HW, Liu J, Ye J, Yu N, Chua NH (2015) Analysis of non-coding transcriptome in rice and maize uncovers roles of conserved lncRNAs associated with agriculture traits. Plant J 84:404–416
4. Grandbastien MA (2015) LTR retrotransposons, handy hitchhikers of plant regulation and stress response. Biochim Biophys Acta 1849:403–416

5. Makarevitch I, Waters AJ, West PT, Stitzer M, Hirsch CN, Ross-Ibarra J, Springer NM (2015) Transposable elements contribute to activation of maize genes in response to abiotic stress. PLoS Genet 11:e1004915
6. Carthew RW, Sontheimer EJ (2009) Origins and mechanisms of miRNAs and siRNAs. Cell 136:642–655
7. Matzke MA, Mosher RA (2014) RNA-directed DNA methylation: an epigenetic pathway of increasing complexity. Nat Rev Genet 15:394–408
8. Matzke MA, Kanno T, Matzke AJ (2015) RNA-directed DNA methylation: the evolution of a complex epigenetic pathway in flowering plants. Annu Rev Plant Biol 66:243–267
9. Lunardon A, Forestan C, Farinati S, Axtell M, Varotto S (2016) Genome-wide characterization of maize small RNA loci and their regulation in the required to maintain repression6-1 (rmr6-1) mutant and long-term abiotic stresses. Plant Physiol 170:1535–1548
10. Forestan C, Aiese Cigliano R, Farinati S, Lunardon A, Sanseverino W, Varotto S (2016) Stress-induced and epigenetic-mediated maize transcriptome regulation study by means of transcriptome reannotation and differential expression analysis. Sci Rep 6:30446
11. Martin M (2011) Cutadapt removes adapter sequences from high-throughput sequencing reads. EMBnetJ 17:10–12
12. Del Fabbro C, Scalabrin S, Morgante M, Giorgi FM (2013) An extensive evaluation of read trimming effects on Illumina NGS data analysis. PLoS One 8:e85024
13. Axtell MJ (2013) ShortStack: comprehensive annotation and quantification of small RNA genes. RNA 19:740–751
14. Langmead B, Salzberg SL (2012) Fast gapped-read alignment with bowtie 2. Nat Methods 9:357–359
15. Kim D, Pertea G, Trapnell C, Pimentel H, Kelley R, Salzberg SL (2013) TopHat2: accurate alignment of transcriptomes in the presence of insertions, deletions and gene fusions. Genome Biol 14:R36
16. Trapnell C, Roberts A, Goff L, Pertea G, Kim D, Kelley DR, Pimentel H, Salzberg SL, Rinn JL, Pachter L (2012) Differential gene and transcript expression analysis of RNA-seq experiments with TopHat and cufflinks. Nat Protoc 7:562–578
17. Li B, Dewey CN (2011) RSEM: accurate transcript quantification from RNA-seq data with or without a reference genome. BMC Bioinformatics 12:323
18. Robinson MD, McCarthy DJ, Smyth GK (2010) edgeR: a bioconductor package for differential expression analysis of digital gene expression data. Bioinformatics 26:139–140
19. Li H, Handsaker B, Wysoker A, Fennell T, Ruan J, Homer N, Marth G, Abecasis G, Durbin R, 1000 Genome Project Data Processing Subgroup (2009) The sequence Alignment/Map format and SAMtools. Bioinformatics 25:2078–2079
20. Quinlan AR, Hall IM (2010) BEDTools: a flexible suite of utilities for comparing genomic features. Bioinformatics 26:841–842
21. Ma L, Bajic VB, Zhang Z (2013) On the classification of long non-coding RNAs. RNA Biol 10:925–933
22. Wierzbicki AT, Haag JR, Pikaard CS (2008) Noncoding transcription by RNA polymerase pol IVb/Pol V mediates transcriptional silencing of overlapping and adjacent genes. Cell 135:635–648
23. Di C, Yuan J, Wu Y, Li J, Lin H, Hu L, Zhang T, Qi Y, Gerstein MB, Guo Y, ZJ L (2014) Characterization of stress-responsive lncRNAs in Arabidopsis thaliana by integrating expression, epigenetic and structural features. Plant J 80:848–861
24. Berry S, Dean C (2015) Environmental perception and epigenetic memory: mechanistic insight through FLC. Plant J 83:133–148
25. Li L, Eichten SR, Shimizu R, Petsch K, Yeh CT, Wu W, Chettoor AM, Givan SA, Cole RA, Fowler JE, Evans MM, Scanlon MJ, Yu J, Schnable PS, Timmermans MC, Springer NM, Muehlbauer GJ (2014) Genome-wide discovery and characterization of maize long non-coding RNAs. Genome Biol 15:R40
26. Paytuvi Gallart A, Hermoso Pulido A, Anzar Martinez de Lagran I, Sanseverino W, Aiese Cigliano R (2015) GREENC: a wiki-based database of plant lncRNAs. Nucleic Acids Res 44:D1161–D1166
27. Wang X, Elling AA, Li X, Li N, Peng Z, He G, Sun H, Qi Y, Liu XS, Deng XW (2009) Genome-wide and organ-specific landscapes of epigenetic modifications and their relationships to mRNA and small RNA transcriptomes in maize. Plant Cell 21:1053–1069
28. Wicker T, Sabot F, Hua-Van A, Bennetzen JL, Capy P, Chalhoub B, Flavell A, Leroy P, Morgante M, Panaud O, Paux E, SanMiguel P, Schulman AH (2007) A unified classification system for eukaryotic transposable elements. Nat Rev Genet 8:973–982
29. Baucom RS, Estill JC, Chaparro C, Upshaw N, Jogi A, Deragon JM, Westerman RP, Sanmiguel PJ, Bennetzen JL (2009) Exceptional

diversity, non-random distribution, and rapid evolution of retroelements in the B73 maize genome. PLoS Genet 5:e1000732

30. Eichten SR, Ellis NA, Makarevitch I, Yeh CT, Gent JI, Guo L, McGinnis KM, Zhang X, Schnable PS, Vaughn MW, Dawe RK, Springer NM (2012) Spreading of heterochromatin is limited to specific families of maize retrotransposons. PLoS Genet 8:e1003127
31. Kim D, Langmead B, Salzberg SL (2015) HISAT: a fast spliced aligner with low memory requirements. Nat Methods 12:357–360
32. Pertea M, Kim D, Pertea GM, Leek JT, Salzberg SL (2016) Transcript-level expression analysis of RNA-seq experiments with HISAT, StringTie and Ballgown. Nat Protoc 11:1650–1667
33. Pertea M, Pertea GM, Antonescu CM, Chang TC, Mendell JT, Salzberg SL (2015) StringTie enables improved reconstruction of a transcriptome from RNA-seq reads. Nat Biotechnol 33:290–295
34. Frazee AC, Pertea G, Jaffe AE, Langmead B, Salzberg SL, Leek JT (2015) Ballgown bridges the gap between transcriptome assembly and expression analysis. Nat Biotechnol 33:243–246
35. Schnable PS, Ware D, Fulton RS, Stein JC, Wei F, Pasternak S, Liang C, Zhang J, Fulton L, Graves TA, Minx P, Reily AD, Courtney L, Kruchowski SS, Tomlinson C, Strong C, Delehaunty K, Fronick C, Courtney B, Rock SM, Belter E, Du F, Kim K, Abbott RM, Cotton M, Levy A, Marchetto P, Ochoa K, Jackson SM, Gillam B, Chen W, Yan L, Higginbotham J, Cardenas M, Waligorski J, Applebaum E, Phelps L, Falcone J, Kanchi K, Thane T, Scimone A, Thane N, Henke J, Wang T, Ruppert J, Shah N, Rotter K, Hodges J, Ingenthron E, Cordes M, Kohlberg S, Sgro J, Delgado B, Mead K, Chinwalla A, Leonard S, Crouse K, Collura K, Kudrna D, Currie J, He R, Angelova A, Rajasekar S, Mueller T, Lomeli R, Scara G, Ko A, Delaney K, Wissotski M, Lopez G, Campos D, Braidotti M, Ashley E, Golser W, Kim H, Lee S, Lin J, Dujmic Z, Kim W, Talag J, Zuccolo A, Fan C, Sebastian A, Kramer M, Spiegel L, Nascimento L, Zutavern T, Miller B, Ambroise C, Muller S, Spooner W, Narechania A, Ren L, Wei S, Kumari S, Faga B, Levy MJ, McMahan L, Van Buren P, Vaughn MW, Ying K, Yeh CT, Emrich SJ, Jia Y, Kalyanaraman A, Hsia AP, Barbazuk WB, Baucom RS, Brutnell TP, Carpita NC, Chaparro C, Chia JM, Deragon JM, Estill JC, Fu Y, Jeddeloh JA, Han Y, Lee H, Li P, Lisch DR, Liu S, Liu Z, Nagel DH, McCann MC, SanMiguel P, Myers AM, Nettleton D, Nguyen J, Penning BW, Ponnala L, Schneider KL, Schwartz DC, Sharma A, Soderlund C, Springer NM, Sun Q, Wang H, Waterman M, Westerman R, Wolfgruber TK, Yang L, Yu Y, Zhang L, Zhou S, Zhu Q, Bennetzen JL, Dawe RK, Jiang J, Jiang N, Presting GG, Wessler SR, Aluru S, Martienssen RA, Clifton SW, McCombie WR, Wing RA, Wilson RK (2009) The B73 maize genome: complexity, diversity, and dynamics. Science 326:1112–1115

Chapter 18

Identification of In Planta Protein–Protein Interactions Using IP-MS

Suraj Jamge, Gerco C. Angenent, and Marian Bemer

Abstract

Gene regulation by transcription factors involves complex protein interaction networks, which include chromatin remodeling and modifying proteins as an integral part. Decoding these protein interactions is crucial for our understanding of chromatin-mediated gene regulation. Here, we describe a method for the immunoprecipitation of in planta nuclear protein complexes followed by mass spectrometry (IP-MS) to identify interactions between transcription factors and chromatin remodelers/modifiers in plants. In addition to a step-by-step bench protocol for immunoprecipitation and subsequent mass spectrometry, we provide guidelines and pointers on necessary controls and data analysis approaches.

Key words Arabidopsis, Protein–protein interactions, Immunoprecipitation, Mass spectrometry, Label free quantification, Chromatin remodelers

1 Introduction

Proteins interact with each other and form multimeric complexes that execute unique functions. Protein–protein interactions occur at different places in the cell and play crucial roles in a variety of processes, such as cell-to-cell signaling by effector proteins, chromatin organization by histones, and gene regulation by transcription factor proteins. Gene transcription is usually regulated by large protein complexes that can include transcription factors, transcriptional cofactors, and chromatin remodelers and modifiers [1]. While more and more functions of individual genes and their protein products have been elucidated, the interactions between the different factors in protein complexes and their role in gene regulatory networks are still far from understood. Although gene regulation was initially considered to be independently regulated by chromatin remodelers and transcription factors, recent data have revealed many physical interactions between these two types of proteins, suggesting that they act together in large regulatory complexes [1–3]. These regulatory protein modules are highly

Marian Bemer and Célia Baroux (eds.), *Plant Chromatin Dynamics: Methods and Protocols*, Methods in Molecular Biology, vol. 1675, DOI 10.1007/978-1-4939-7318-7_18, © Springer Science+Business Media LLC 2018

dynamic and different combinations of proteins are involved in the regulation of specific target genes in particular tissues or under specific conditions. To unravel the composition of the in planta complexes that regulate gene activity, the entire complex can be immunoprecipitated (IP) using a specific antibody against one of the proteins, or using an antibody against a tagged protein.

In the classical approach, this IP is followed by western blotting (co-IP). Although relatively easy to perform, a drawback of this approach lies in the need for specific antibodies against each of the potential complex partners, which requires prior knowledge of the putative interaction partners, and thereby prevents the identification of novel interactors. This problem has been overcome by technical advances in liquid chromatography coupled with mass-spectrometry (LC-MS), which have enabled the detection of low-abundance proteins, as well as high-throughput identification of hundreds of proteins from a single sample in a relatively short time. The additional availability of user-friendly data analysis tools and the choice of label free quantification (LFQ) make LC-MS an attractive option for protein interaction research. In this chapter, we describe a label-free method that enables the user to study the composition of in planta nuclear protein complexes and to identify the interactions between transcription factors and chromatin remodelers.

In this method (*see* Fig. 1), protein complexes are isolated from native plant tissue by immunoprecipitation. Subsequently, the

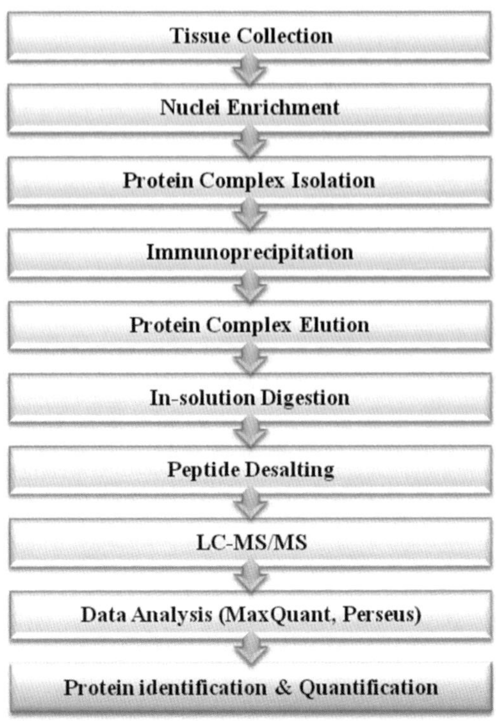

Fig. 1 Schematic workflow of the IP-MS/MS protocol

identification of all proteins requires a proteolytic step with trypsin, followed by purification of the sample. Digested peptides are then eluted and further injected into a mass spectrometer, where the molecules are ionized, accelerated, and separated based on their mass-to-charge ratio, enabling the deduction of peptide identity. In short, the peptide identification is done by liquid chromatography tandem-mass spectrometry (LC-MS/MS) followed by searches of the deduced peptides against a protein database, resulting in a list of proteins present in the IP sample. This is a straightforward procedure for Arabidopsis, because an exhaustive peptide database is available, but may be more challenging for other species. Here, we provide a detailed step-by-step protocol for performing IPs and sample preparation for LC-MS/MS analysis. In addition, guidelines on data analysis tools and label-free quantification as well as recommendations on necessary controls have been summarized.

2 Materials

The immunoprecipitation of in planta nuclear protein complexes can be achieved using a specific antibody against the protein of interest, or by using a specific antibody against a protein tag that has been added to the protein of interest. The advantage of the latter approach is that a specific antibody of high quality can be selected, but it requires the generation of stable transgenic lines that express the tagged protein of interest (*see* **Note 1**). Because the use of fluorophore-tagged proteins works best in our hands, this protocol describes the immunoprecipitation based on the use of magnetic anti-GREEN FLUORESCENT PROTEIN (GFP) microbeads. However, other combinations of antibodies and beads can also be used, but this will have to be optimized by the user (*see* **Note 2**). All the buffers and solutions described in this protocol are prepared with autoclaved milliQ water (before the IP) or HPLC water (after the IP). The recommended waste disposal regulations for each reagents should be followed.

2.1 Tissue Collection and Preparation

1. Mortar and pestle.
2. Liquid nitrogen
3. Nitrile gloves.
4. 50 mL centrifuge tubes.

2.2 Protein Complex Isolation

1. Liquid nitrogen.
2. Nitrile gloves.
3. Nylon mesh (pore size 55 μm).
4. Glass funnel.

5. 50 mL centrifuge tubes.
6. 100 mM sodium phosphate buffer pH 7.0: prepare 100 mM Na_2HPO_4 and 100 mM NaH_2PO_4. While measuring the pH, add 100 mM NaH_2PO_4 to the 100 mM Na_2HPO_4 solution until a pH of 7.0 is reached.
7. 20% Triton X-100: prepare fresh in advance, allowing the Triton to mix with the milliQ on a roller for several hours.
8. Complete Protease Inhibitor Cocktail tablets (e.g., from Roche).
9. M1 buffer (prepare freshly from stock solutions): 0.1 M NaCl, 10 mM sodium phosphate buffer pH 7.0, 1 M 2-methyl 2,4-pentanediol. Per 50 mL M1 buffer, add 35.4 μL 2-mercaptoethanol and 1 tablet of protease inhibitor cocktail (*see* **Note 3**). Bring up to a final volume of 50 mL using milliQ water.
10. M2 buffer (prepare freshly from stock solutions): 0.1 M NaCl, 10 mM sodium phosphate pH 7.0, 10 mM $MgCl_2$, 1 M 2-methyl 2,4-pentanediol, 0.5% Triton X-100 (from a fresh 20% stock solution). For 50 mL M2 buffer, add 35.4 μL and 1 tablet protease inhibitor cocktail (*see* **Note 3**). Bring up to a final volume of 50 mL using milliQ water.
11. M3 buffer (prepare freshly from stock solutions): 0.1 M NaCl, 10 mM sodium phosphate pH 7.0. For 10 mL buffer, add 7.1 μL 2-mercaptoethanol and ¼ tablet of protease inhibitor cocktail (*see* **Note 3**). Bring up to a final volume of 50 mL using milliQ water.
12. Lysis buffer (*see* **Note 4**): 1% Triton X-100, 50 mM Tris–HCl pH 8.0.
13. Benzonase (25 U/μL).
14. Sonicator.
15. 2 mL low protein-binding tubes.
16. 25× protease inhibitor mix: dissolve 1 tablet of Complete Protease Inhibitor Cocktail in 2 mL of HPLC water in a 2 mL tube. Store 100 μL aliquots in 1.5 mL tubes at −20 °C.

2.3 Protein Immunoprecipitation, Elution, and Tryptic Digestion

1. Nitrile gloves.
2. Because our proteins are tagged with GFP, we use anti-GFP magnetic microbeads for the IP. However, various other options are available.

   μMACS™ GFP Isolation Kit (Miltenyi Biotech) containing (*see* **Note 4**):

   (a) Anti-GFP microbeads.

   (b) Lysis buffer: 1% Triton X-100, 50 mM Tris–HCl pH 8.0.

   (c) Wash buffer 2: 20 mM Tris–HCl pH 7.5.

3. Magnetic stand for columns/tubes (e.g., μMACS Separator with MultiStand, Miltenyi Biotec).
4. Microbeads-binding matrix (e.g., μ-Columns, Miltenyi Biotec).
5. 8 M urea in HPLC water (fresh).
6. Rotating device.
7. 50 mM ammonium bicarbonate in HPLC water (fresh).
8. Low-binding microcentrifuge tubes (2 mL and 1.5 mL).
9. Low-binding tips.
10. 45 mM dithiothreitol in 50 mM ammonium bicarbonate (fresh).
11. 100 mM iodoacetamide in 50 mM ammonium bicarbonate (fresh).
12. Trypsin Gold, Mass Spectrometry Grade. Prepare 0.1 μg/μL trypsin in 50 mM acetic acid. Aliquots can be prepared in advance and stored at −80 °C.

2.4 Peptide Preparation

For the desalting of large numbers of samples simultaneously, we use the MultiScreen Vacuum Manifold system (Merck Millipore) in combination with a 96-wells Oasis HLB μElution plate. However, single samples can also be desalted using special desalting columns or tips, such as C18 ZipTips (Millipore).

1. Nitrile gloves.
2. 100% acetonitrile HPLC grade.
3. 100% formic acid HPLC grade.
4. 10% formic acid (in HPLC water).
5. 50% acetonitrile/5% formic acid (v/v).
6. HPLC water.
7. 96-well plate with sorbent for purification (e.g., Oasis HLB μElution plate, Waters).
8. Device for vacuum filtration of 96-well plates (e.g., Multi-screen HTS vacuum manifold apparatus, Millipore).
9. Vacuum pump.
10. Vacuum centrifuge concentrator (e.g., SpeedVac, Thermo Fisher Scientific).

2.5 LC-MS Equipment and Software for Data Analysis

1. Ion Trap-Orbitrap Mass Spectrometer such as LTQ-Orbitrap XL (Thermo Fisher Scientific).
2. MaxQuant, http://www.coxdocs.org/doku.php?id=:maxquant:start.
3. Perseus, http://www.coxdocs.org/doku.php?id=perseus:start.

3 Methods

At all times, work in a clean and sterile environment and use nitrile gloves during the handling of the samples. Based on the capacity of certain instruments and the handling and waiting time between steps, we recommend not to process more than six samples at once during the IP. For handling more than six samples, please make use of one of the pause points during which the samples can be stored for a few days or up to several weeks as stated. The trypsin digestion and desalting procedure can be performed on a large number of samples simultaneously if the 96-wells plate is used for purification.

3.1 Starting Material and Preparations

1. Harvest the tissue of interest, determine the weight and freeze immediately in liquid nitrogen. The tissue can be stored at −80 °C until use. The time of harvest, growth stage, amount of starting material, and number of replicates and controls are key factors to consider before performing the protocol (*see* **Notes 1**, **5–8**).
2. Using a chilled mortar and pestle, grind the harvested plant material using liquid nitrogen. Make sure the tissue is ground into a fine powder.
3. Transfer the powdered tissue with liquid nitrogen to a 50 mL centrifuge tube and allow the liquid nitrogen to vaporize. Once vaporized, the ground tissue can be used immediately or stored at −80 °C for up to 2–3 days. We recommend to grind the material a day before performing the protein extraction steps.

3.2 Protein Isolation and Sonication

In this step, the nuclei are purified to enrich for protein complexes that play a role in the nucleus, such as those containing transcription factors and chromatin modifiers (*see* **Note 6**). After isolation of the nuclei, the cells are disrupted by sonication in lysis buffer to release the proteins.

1. Freshly prepare all the buffers (M1, M2, and M3) on ice under a sterile fume hood.
2. Prepare the filter apparatus by placing an open 50 mL tube on ice, topped with a glass funnel, which contains a clean piece of cloth mesh (55 μm).
3. Take the grinded tissue sample from the −80 °C freezer (*see* **step 3** of Subheading 3.1) in a Dewar flask containing liquid nitrogen.
4. Take out the sample from the liquid nitrogen with a long forceps. Add approximately 20 mL M1 buffer per gram of starting material (weighed prior to grinding, *see* **Note 5**) and resuspend gently by shaking until it becomes homogeneous.

5. Once homogeneous, carefully pour the sample through the filter apparatus, in the glass funnel on ice. Allow the sample to flow through by gravity.
6. Pipet 5 mL of additional M1 buffer to wash the mesh of the filter apparatus and allow residual sample to flow through.
7. Centrifuge the filtrate at 1000 × *g* for 20 min at 4 °C.
8. Discard the supernatant in the appropriate waste container and place the sample back on ice.
9. To wash the cell pellet, add 5 mL of M2 buffer and resuspend the pellet gently by shaking. Do not vortex.
10. Centrifuge the resuspended pellet at 1000 × g for 10 min at 4 °C and discard the supernatant in the appropriate waste container and place the sample back on ice.
11. Repeat the washing step with 5 mL of M2 buffer four times.
12. After completing five washing steps with M2 buffer, resuspend the semipure nuclei gently shaking in 5 mL of M3 buffer.
13. Centrifuge the semipure nuclei at 1000 × *g* for 10 min at 4 °C.
14. Discard the supernatant in the appropriate waste container and place the tube back on ice.
15. Resuspend the crude nuclear pellet in 1 mL of lysis buffer with 45 μL of 25× protease inhibitor mix and 5 μL benzonase (to eliminate the DNA and RNA) and transfer the sample to a clean 2 mL tube.
16. Sonicate the nuclei for 10 s on ice (*see* **Note 9**). After sonication, invert the tube several times and place on ice for 45 s. Repeat the sonication step twice.
17. After 3 sonication cycles, place the sample on ice for 5 min.
18. Place the sample into a precooled microcentrifuge and centrifuge at maximum speed for 10 min at 4 °C.
19. Transfer the supernatant (soluble protein extract) to a new microcentrifuge tube and centrifuge again at maximum speed for 10 min at 4 °C. Repeat this step until no visible pellet is present anymore (*see* **Note 10**).
20. Upon final centrifugation, transfer the supernatant to a 2 mL low-protein binding tube and place on ice.
21. Optional: Take a small aliquot (50 μL) of the protein extract to check using Western blotting (*see* **Note 11**).

3.3 Protein Immunoprecipitation and Elution

In the IP step, the native protein complexes that contain the protein of interest will be pulled down. We describe here the pull down of a GFP-tagged protein using anti-GFP magnetic beads (*see* **Note 2**). For all the following steps in this protocol, work in a clean sterile environment (e.g., flow hood). Any contamination of the sample may affect the detection of immunoprecipitated proteins during the LC-MS/MS analysis. Therefore, while handling samples, the use of protective nitrile gloves is highly advised.

1. Resuspend the anti-GFP microbeads by vortexing and add 50 μL of anti-GFP microbeads to the soluble protein extract.
2. Incubate the sample on a rotating device for 1 h at 4 °C.
3. Place the μColumn onto the μMACS™ separator in the flow hood.
4. First calibrate the μColumn with 200 μL lysis buffer.
5. Load the immunoprecipitate (sample from **step 2**) onto the calibrated μColumn and allow it to flow through by gravity (*see* **Note 12**).
6. Wash the immobilized beads six times (*see* **Note 13**) with 200 μL μMACS lysis buffer.
7. To remove detergents, wash the immobilized beads twice with 200 μL of μMACS wash buffer 2.
8. Freshly prepare 1 mL 8 M Urea, preferably with HPLC water.
9. Add 20 μL (dead volume of the column) of 8 M Urea to the μColumn, which contains the washed immobilized beads.
10. Incubate the μColumn for 5 min at room temperature.
11. If a small droplet remains on the end of the μColumn, remove it using a pipette tip.
12. To collect the eluate, place a clean, labeled, 1.5 mL low-protein binding tube under the μColumn.
13. Add 50 μL of 8 M Urea on to the μColumn to elute the protein immunoprecipitate. Be patient and collect all the droplets during the elution.
14. The protein immunoprecipitate can be stored at −20 °C (for several weeks) until use.

3.4 Protein Digestion with Trypsin

For analysis of the proteins with LC-MS/MS, the proteins have to be digested into smaller peptides using the serine protease trypsin (*see* **Note 14**).

1. Freshly prepare all the solutions in 50 mM ammonium bicarbonate and continue working in a clean environment while wearing protective nitrile gloves.
2. Dilute the eluated proteins (i.e., the 50 μL protein immunoprecipitate from **step 13** of Subheading 3.3) four times by adding 150 μL of 50 mM ammonium bicarbonate to adjust the final concentration of urea to 2 M.
3. In order to reduce the number of protein disulfide bonds, add 10 μL 45 mM DTT to the sample.

4. Incubate at 37 °C for 30 min in a thermomixer at 550 rpm.
5. Allow the sample to cool down by leaving it at room temperature for 5 min.
6. Add 10 μL of 100 mM iodocetamide and place the sample in the dark at room temperature for 30 min. This will alkylate any free thiol groups.
7. Add 15 μL of 0.1 μg/μL trypsin (aliquots of trypsin can be prepared in advance and stored at −80 °C).
8. To digest the proteins into peptides, incubate the sample at 37 °C for 16–24 h.
9. The digested mixture can be stored at −80 °C for several days.

3.5 Peptide Cleanup

Prior to LC-MS/MS analysis, the peptides must be cleaned to remove salts and urea from the mixture. Freshly prepare all the solutions and work under the flow hood.

1. Add 26 μL 100% formic acid to the digested sample (final concentration of formic acid 10% v/v), vortex gently and briefly spin down.
2. Assemble the multiscreen HTS vacuum manifold apparatus with the vacuum pump and place the 96-well Oasis HLB μelution plate in it on top of a wash plate (96 well low-binding plate (*see* **Note 15**)). Mark the wells that are going to be used.
3. Turn on the vacuum pump and set the pressure to 2.5 inHg.
4. Equilibrate the HLB μelution plate by pipetting 150 μL of 100% acetonitrile in the marked wells and collect the flow-through into the wash plate. Increase the pressure to 10–12 inHg in order to allow all the solution to pass through. Reduce the pressure back to 2.5 inHg and stop the vacuum.
5. Repeat the wash with 150 μL of 100% acetonitrile under vacuum.
6. Discard the flow-through from the wash plate and place it back.
7. Start the vacuum pump and set the pressure to 2.5 inHg.
8. Wash the marked wells of the HLB μelution plate twice with 150 μL 10% formic acid. Vacuum at a high pressure (i.e., 10–12 inHg) in order to allow all the solution to pass through. Then reduce the pressure back to 2.5 inHg and then stop the vacuum. Discard the flow-through.
9. Switch on the vacuum pump and set the pressure to 2.5 inHg.
10. Load the sample into the marked well of the washed HLB μelution plate and allow the sample to flow through without increasing the pressure. Leaving the vacuum at low pressure provides maximum binding capacity to the column. Once all the sample mixture has run through, stop the vacuum and discard the waste.

11. Turn on the vacuum pump and set the pressure at 2.5 inHg.
12. Wash twice with 150 μL 10% formic acid. Increase the pressure to 10–12 inHg in order to allow all the solution to pass through. Then reduce the pressure back to 2.5 inHg and stop the vacuum to discard the waste.
13. Replace the wash plate from the apparatus with a collection plate (96-well low-binding plate).
14. Start the vacuum pump at a pressure of 2.5 inHg.
15. To elute the sample peptides, add 50 μL of 50% acetonitrile/5% formic acid (v/v). At the end of the elution the pressure can be increased to 15 inHg. Stop the vacuum after 1–2 min (when the eluate reached the collection plate). Repeat the elution step once. Transfer the eluate to a new 1.5 mL low-protein binding tube (*see* **Note 16**).
16. Place the peptide eluate in a SpeedVac to evaporate the elution solution and recover a dried peptide pellet.
17. The dried peptides are ready for LC-MS/MS analysis.

3.6 Peptide Sequencing

In reverse phase chromatography, the peptides are separated and then identified by tandem mass spectrometry (MS/MS) on an LTQ-Orbitrap XL [4]. Owing to the complexity of the peptide sample, a nano-LC setup with a 250 mm (or more) column of C18 particles should be used. For the separation, the dried peptides are dissolved in 50 μL 0.1% formic acid in water and 18 μL is injected onto a preconcentration column. The injection is performed at a constant pressure of 270 bar to obtain a constant flow of around 7 μL/min. The peptides are then eluted onto an analytical column with an acetonitrile gradient at a flow rate of 0.5 μL/min with a Proxeon EASY nanoLC. The MS/MS measurements are recorded with a gradient increase of 0.5% (v/v) acetonitrile per minute or less. FTMS spectrum between *m/z* 380 and 1400 is acquired by the orbitrap at a high resolution (60,000) with a target value of 1,000,000 or a maximum ion injection time of 500 ms. Data-dependent MS/MS spectra are obtained by the LTQ for the highest four multiple-charged peaks at a threshold target value of 5000 with an exclusion list of 500 *m/z* values and a 60-s exclusion duration (*see* **Note 17**).

3.7 Data Analysis

The LC-MS/MS raw data generated from the LTQ-orbitap XL can be analyzed with the MaxQuant software package, a tool for protein identification and quantification that is freely available [5]. A detailed description of the MaxQuant suite that summarizes the algorithms it uses, including those for peak detection, scoring peptides, and protein identification has been described by Cox and Mann [5], and a step-by-step protocol for the processing of LC-MS/MS datasets has been published by Smackzniak et al. [6] (*see* **Note 18**).

After the completion of the MaxQuant analysis, the output dataset can be further processed for statistical analysis with the Perseus package (*see* **Note 19**) [7]. Here we describe the key steps for filtering and statistical analysis.

1. Use the Generic matrix upload icon to load the proteinGroup.txt file of your MaxQuant output data in the Perseus software (version 1.5.3.2).
2. Selectively place the output data from the text file into the right categories as described below and click OK.
   (a) Expression (Main): Add "LFQ" and "iBAQ" intensities of all samples (both test and control samples).
   (b) Numerical Columns: Add "Unique peptides," "peptides," "number of proteins," and "iBAQ"
   (c) Text Columns: Add "Protein IDs" and "Fasta headers"
   (d) Categorical Columns: Add "Potential contaminants," "Reverse," and "Only identified by sites"
3. Using the 'Filter rows' tab, perform the following filtering:
   (a) Filter the rows based on categorical columns: filter for "Only identified by sites" and then for "Reverse." Optionally, one can also filter out the "potential contaminants" from the dataset. However, it is good to realize that your fluorophore, e.g., GFP and trypsin are also characterized as contaminants. These proteins should not be filtered out. Therefore, it is best to filter out contaminants using the "select rows manually" tab and remove contaminants excluding GFP and trypsin.
   (b) Filter the rows based on numerical/main columns: filter for "peptides" with at least two identified peptides per protein. Similarly filter for "unique peptides" with at least one unique peptide identified per protein.
4. Using the Basic tab, perform logarithmic transformation (log2) for the intensity (expression) data.
5. With the Imputation tab, replace missing values by a constant that is marginally lower than the lowest (Log) value measured.
6. Assemble the data set into Groups, e.g., test and control, and use an appropriate statistical test to calculate the relative protein abundance. Performing a *T*-test (e.g., $FDR = 0.01$ and $S_0 = 1$) allows to test the significant differences between the means.
7. In addition to data filtering, data imputation, and data normalization, Perseus offers several tools for quantification and visualization of the proteomics data, and the generation of volcano plots and scatter plots can be performed within the software package.
8. The analyzed data from Perseus can be exported into Excel format for further analysis.

4 Notes

1. We prefer to use fluorophore tagged lines and antibodies against the tag for the immunoprecipitation experiments. In our case, this approach gave better results than antibodies against the native protein. This requires the generation of transgenic lines that express the tagged protein of interest, preferably in the corresponding mutant background (e.g., *SEP3:GFP* in the *sep3* mutant background). If expressing the transgene in the mutant background, complementation should be observed. If this is not the case transferring the tag from the C-terminus to the N-terminus or vice versa may help.
2. When using an antibody against a native protein, test whether the antibody is specific for your protein of interest prior to the IP-MS. Subsequently, the antibody should be coupled to magnetic or agarose beads according to the manufacturer's description.
3. Dissolving a complete protease inhibitor tablet directly in the buffer will take some time. To facilitate the dissolution of the protease inhibitor tablet, the tablet can be disintegrated in a tube with 1 mL milliQ using tapping and up- and down pipetting and the complete 1 mL can be added to 50 mL buffer.
4. The buffers and wash solutions present in the μMACS™ GFP Isolation Kit may not be sufficient to process all samples. The user may have to prepare extra solutions as per the kit's instruction manual. The lysis buffer from the kit contains 1% Triton X-100. Depending on your experimental requirements, useful variations in lysis buffer can be performed. For example, lowering the detergent concentrations or switching to other detergents such as NP-40 or Digitonin may help to increase the signal to noise ratio for proteins that have higher or lower affinity for the antibody/beads than the average protein.
5. The amount and type of plant tissue used depends on the user's research question. This protocol has been successfully tested with 2 g of starting material and can be scaled down to as low as 0.5 g of tissue. This protocol is suitable for different Arabidopsis tissues and has been used for seedlings, rosette leaves, inflorescences, stems, cauline leaves, and siliques without modifications.
6. Moderate to high concentrations of the bait protein in the nucleus are essential. For proteins which are of low or variable abundance in the cell, the nuclear enrichment steps (**steps 1–14** of Subheading 3.2) can be skipped and IPs can be performed on crude protein extract.

7. We advise a minimum of 3–4 independent biological replicates for reliable reproducibility and estimation of protein abundance ratios between test and control samples. Replicates aid in normalization of the datasets and an improved comparison of protein abundance across multiple samples. The statistical power is increased with multiple biological replicates.
8. A reliable IP control is necessary for accurate comparison in quantitative proteomics. One possible control would be Arabidopsis wild-type plants for comparison against the transgenic GFP (or other fluorophore)- tagged line. This control can correct for nonspecific binding of GFP with other proteins. However, the best and most reliable control would be an Arabidopsis line expressing nuclear localized GFP under the same promoter as the bait protein.
9. Dependent on the type of sonicator used, or when using an ultrasonic bath, pulse times may need to be adjusted for proper cell disruption. However, the samples must be kept on ice during the entire procedure.
10. This step is critical as any debris/pellet may obstruct the column during IP steps.
11. To check if your protein of interest and/or protein tag can be detected in the sample, a western blot can be performed using the same antibody as used for the immunoprecipitation. At several points in the protocol, such as before the IP, and after the washing and elution steps, a small aliquot can be saved for western analysis. This analysis can be performed when the samples are stored after elution of the IP sample (**step 14** of Subheading 3.3).
12. If pipetting of the sample or the washing buffers on the column results in the formation of an air bubble, this can block the flow-through of the solutions (visible as a small circle in the column). This is probably the case if the flow through in one column is much slower than in the others. If an air bubble has formed, it can be removed by carefully pipetting up and down without touching the beads.
13. Depending on the type of nuclear protein, type of TF, strength of interactions etc. the number (lowering/increasing) of washing steps can be optimized. For example, washing can be reduced if you expect only weak interactions. For optimization, it can be useful to perform western blot analysis with the flow-through collected in different washing steps.
14. Depending on your experimental purpose, for example when the protein of interest is very lysine-rich, digestion with an alternative protease (e.g., Arg-C, thermolysin, and pepsin) can be an option. Choice of an alternative protease demands optimization by the user.

15. Correct assembly of the two plates, i.e., the 96-well Oasis HLB μelution plate and the 96-well wash plate on to the vacuum manifold apparatus and of the vacuum manifold apparatus to the vacuum pump is crucial. Gaps between the plates and the vacuum manifold will result in an incomplete vacuum, prohibiting efficient flow-through of the solutions. To fit the 96-well Oasis HLB μelution plate on top of the 96-well wash plate in to the vacuum manifold, we built an additional support. This in-house built support raised the platform of the 96-well wash plate, bringing it closer to the Oasis HLB μelution plate and closing any gaps. Before starting the cleanup of your sample using the HTS apparatus, 96-well wash plate and 96-well Oasis HLB μelution plate, it is recommended to first test whether the assembly is correctly set up using some test solutions. If only a few samples need to be processed, the cleanup can also be performed using desalting columns/tips such as C18 ZipTips (Millipore).
16. Droplets of 50% acetonitrile/5% formic acid (v/v) can stick to the sides of the wells and reduce the elution volume. Try to pipet the elution solution into the centre of the tube to avoid such droplets and if necessary, remove droplets from the walls with a pipette and then pipet them into the centre without making contact with the column. Typically, 50–80 μL of sample eluate is recovered.
17. Identification of post-translationally modified peptides, such as acetylated or phosphorylated peptides, is not standardly performed, but can be useful (although largely extending the time for the MS/MS run), as a considerable proportion of the proteome is post-translationally modified.
18. Additional insights and online training/tutorials on the use of these software packages have been made available by the MaxQuant developers at the following link:

    https://www.youtube.com/channel/UCKYzYTm1cnmc0CFAMhxDO8w
19. Statistical analysis can only be performed if three or more biological replicates have been performed. If certain peptides are not detected at all in the control sample (e.g., extracted from a line expressing a free GFP, ideally under control of the promoter of the gene of interest), they may not be recognized as statistically significant, although they can still represent important interactors. Thus, nonsignificant data may also be relevant. Therefore, researchers often present a table which summarizes the number of peptides identified and their abundance.

Acknowledgments

We thank Sjef Boeren and Twan America for helpful suggestions regarding the data analysis and Tom Brabbs and Richard Immink for critical reading of the manuscript. S.J. is supported by the European Commission Seventh Framework-People-2012-ITN Project EpiTRAITS, GA-316965 (Epigenetic regulation of economically important plant traits) and M.B. by an NWO-Veni grant.

References

1. Smaczniak C, Immink RGH, Muiño JM, Blanvillain R, Busscher M, Busscher-Lange J, Dinh QD, Liu S, Westphal AH, Boeren S, Parcy F, Xu L, Carles CC, Angenent GC, Kaufmann K (2012) Characterization of MADS-domain transcription factor complexes in Arabidopsis flower development. Proc Natl Acad Sci U S A 109 (5):1560–1565. doi:10.1073/pnas.1112871109
2. Liang SC, Hartwig B, Perera P, Mora-García S, de Leau E, Thornton H, de Alves FL, Rapsilber J, Yang S, James GV, Schneeberger K, Finnegan EJ, Turck F, Goodrich J (2015) Kicking against the PRCs – a domesticated Transposase Antagonises silencing mediated by Polycomb group proteins and is an accessory component of Polycomb repressive complex 2. PLoS Genet 11(12): e1005660
3. Olmo I, López JA, Vázquez J, Raynaud C, Piñeiro M, Jarillo JA (2016) Arabidopsis DNA polymerase ϵ recruits components of Polycomb repressor complex to mediate epigenetic gene silencing. Nucleic Acids Res 44 (12):5597–5614. doi:10.1093/nar/gkw156
4. Wendrich JR, Boeren S, Möller BK, Weijers D, De Rybel B (2017) In vivo identification of plant protein complexes using IP-MS/MS. In: Kleine-Vehn J, Sauer M (eds) Plant hormones: methods and protocols. Springer, New York, NY, pp 147–158. doi:10.1007/978-1-4939-6469-7_14
5. Cox J, Mann M (2008) MaxQuant enables high peptide identification rates, individualized p.p.b.-range mass accuracies and proteome-wide protein quantification. Nat Biotechnol 26 (12):1367–1372
6. Smaczniak C, Li N, Boeren S, America T, van Dongen W, Goerdayal SS, de Vries S, Angenent GC, Kaufmann K (2012) Proteomics-based identification of low-abundance signaling and regulatory protein complexes in native plant tissues. Nat Protoc 7(12):2144–2158
7. Tyanova S, Temu T, Sinitcyn P, Carlson A, Hein MY, Geiger T, Mann M, Cox J (2016) The Perseus computational platform for comprehensive analysis of (prote)omics data. Nat Methods 13(9):731–740

Chapter 19

RNA Immunoprecipitation Protocol to Identify Protein–RNA Interactions in *Arabidopsis thaliana*

Benoit Mermaz, Fuquan Liu, and Jie Song

Abstract

The role of RNA-binding proteins in the regulation of epigenetic processes has received increasing attention in the past decades. In particular noncoding RNAs have been shown to play a role in chromatin loop formation, recruitment of chromatin modifiers and RNA-dependent DNA methylation. In plants, the identification of specific RNA–protein interactions is now rising, facilitated by the development of specific approaches for plant tissues. Here, we present a simple one-day RNA immunoprecipitation (RIP) protocol adapted for *Arabidopsis*, suited for the identification of RNAs that are associated with a protein-of-interest *in planta*.

Key words RNA–protein interaction, RNA immunoprecipitation, RIP, lncRNA, Reverse transcription, Quantitative PCR

1 Introduction

It is well known that RNA Binding proteins (RBPs) play important roles in transcriptional regulation in eukaryotic cells, through regulation of transcriptional initiation, termination, and RNA splicing [1]. Increasing cases of RNA–protein interactions are also implicated in the gene-specific regulation of expression, for example via microRNA targeting, or by the action of long noncoding RNAs (lncRNAs) [2–4]. Recent studies have shown that lncRNAs can play a role in the recruitment of chromatin-modifying complexes [5] and the opening of chromatin loops [6]. In plants, only a few studies have investigated the role of lncRNAs in the regulation of specific gene expression so far, revealing a role for lncRNAs in the disruption of gene loops [7, 8].

A suitable approach to identify interactions between chromatin-modifying complexes and RNAs is RNA immunoprecipitation (RIP) [9]. RIP is developed to detect RNA–protein interactions *in vivo* and involves the immunoprecipitation (IP) of a target protein followed by purification of the associated RNA (Fig. 1). Key in this process is the antibodies, which should

Marian Bemer and Célia Baroux (eds.), *Plant Chromatin Dynamics: Methods and Protocols*, Methods in Molecular Biology, vol. 1675, DOI 10.1007/978-1-4939-7318-7_19, © Springer Science+Business Media LLC 2018

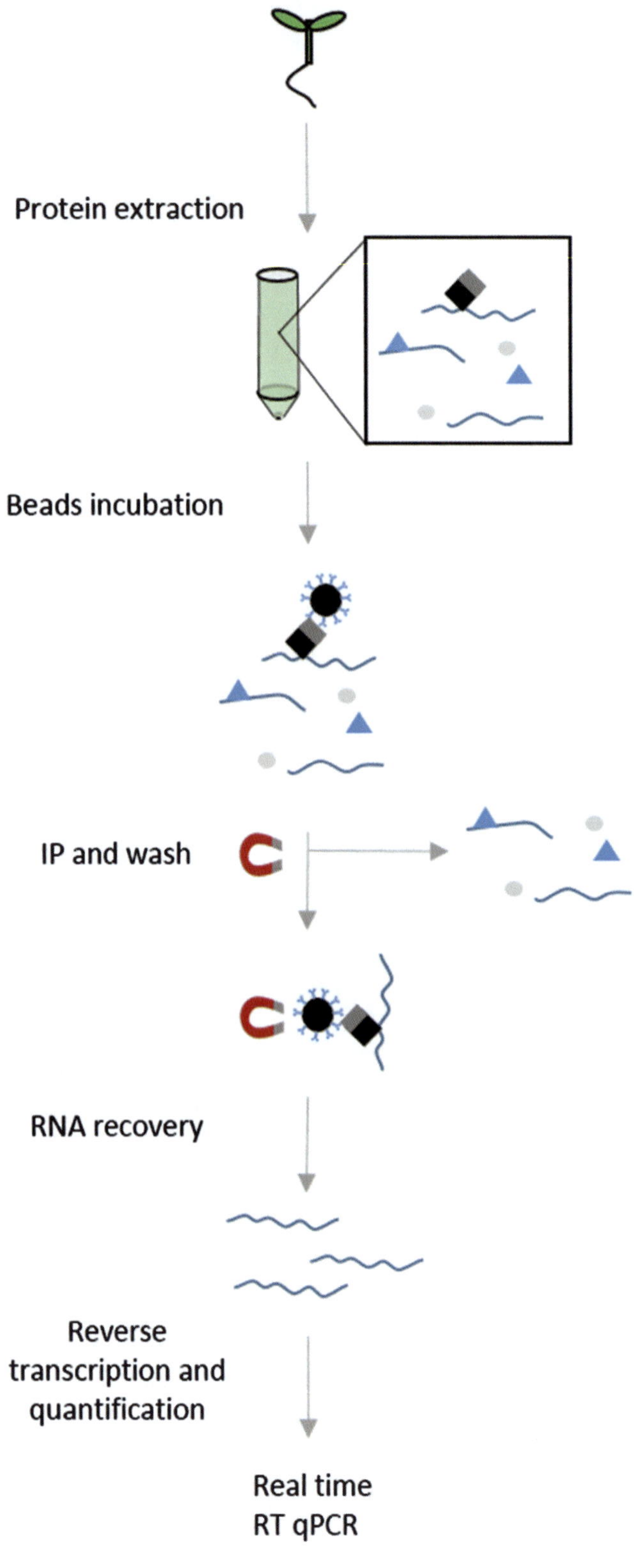

Fig. 1 Workflow of the RNA Immunoprecipitation (RIP). Proteins are extracted from seedlings by grinding the seedlings in liquid nitrogen and resuspending the powder in extraction buffer, followed by pelleting the debris.

specifically pull down the protein of interest. Because an antibody of sufficient quality is not always available, the IP is often performed using a translational fusion of the target protein with a peptide tag (e.g., GFP). The chimeric protein is then usually introduced into the corresponding mutant background to enable IP with commercially available (e.g., anti-GFP) antibodies [10]. If such an approach is adopted, the fusion protein needs to be assessed prior to RIP to confirm that it is fully functional and acts to replace the original protein. As peptide antibodies against native proteins are increasingly available from various companies, RIP may become more often possible without the need to generate transgenic lines. Here, we present the protocol with a GFP fusion protein as an example.

In contrast to chromatin immunoprecipitation (ChIP), where the protein–DNA interactions are usually cross-linked using a chemical fixative such as formaldehyde [10], we do not apply chemical fixation for the RIP samples. The RIP samples can be processed without any cross-linking (native conditions) or, to detect less stable interactions and to get a better indication of the binding site, the RNA–protein complexes can be preserved by UV cross-linking [11, 12]. However, because RIP from plant tissues is still a young technique, further optimisation is desirable to establish the best cross-linking conditions, and test different cross-linking reagents/procedures with different plant tissues [13].

Another important difference from a ChIP protocol is the stability of the targeted molecule. While DNA is not easily degraded, RNases are ubiquitous and are common contaminants in biological samples. Therefore, inhibition of RNase activity to protect the RNA from degradation is essential in this protocol. After immunoprecipitation, the RNAs have to be reverse-transcribed and can then be assessed by quantitative PCR (qPCR) [14], or high-throughput sequencing [15, 16]. A qPCR can be performed to test whether a particular RNA is enriched after the immunoprecipitation, and requires prior knowledge of the expected interactions [16]. The more expensive high throughput sequencing, on the other hand, provides a global picture of all transcripts that are immunoprecipitated with the target protein, and could thus lead to new discoveries [16].

Here, we present a simple protocol for RIP analysis that can be performed in 1 day. This protocol includes tissue sampling, protein–RNA extraction, immunoprecipitation and subsequent elution of the protein–RNA complexes. After recovery of the

Fig. 1 (continued) After extraction of the protein–RNA complexes, the proper combination of beads + antibody is incubated with the sample to bind the protein-of-interest. The immunoprecipitate is then isolated using a magnet and washed three times to get rid of the unbound proteins. Finally, the RNAs are eluted and purified before being analyzed by real-time RT-PCR

RNAs and reverse transcription, the enrichment of specific RNAs can be assayed by qPCR or high-throughput analysis (*see* Fig. 1 for a visualization of this workflow). We do not describe the quantification procedures in detail here, but give tips on how to perform solid quantification analysis.

2 Materials

Required equipment: microfuge tubes (1.5, 2 ml), low-binding tubes and tips, a rotator, a mobile minicentrifuge, a refrigerating centrifuge (allowing 20,000 × *g*) for 2 ml tubes or 50 ml tubes, and a thermoshaker. All solutions need to be freshly prepared with ultrapure DNase, RNase-, protease-free water and kept on ice, unless stated otherwise. We recommend RNase decontamination of lab ware and bench area (e.g., RNaseZap solution from Sigma).

2.1 Plant Material

1. ½ MS agar medium: half strength (2.2 g/l) Murashige and Skoog (MS) medium and 0.8% agar.
2. Transgenic plants with translational fusion of the protein-of-interest with a GFP tag and the background ecotype (e.g., Col-0) as a control. The ideal control is a transgenic plant expressing GFP alone under the control of the same promoter as for the protein-of-interest.
3. Optional: UV cross-linker (e.g., Stratalinker®).

2.2 Protein Extraction

1. Seedlings (10 day-old).
2. Liquid nitrogen (LN).
3. Mortar and pestle.
4. 1 M Tris–HCl pH 7.5.
5. 5 M NaCl.
6. 10% Triton X-100.
7. 2 M $MgCl_2$.
8. 80% Glycerol.
9. 1 M DTT.
10. DNase-, RNase-, protease-free, water.
11. Protease Inhibitor Cocktail Tablets, EDTA-free (e.g., SIGMA-FAST from Sigma) (*see* **Note 1**).
12. RNase inhibitor (e.g., RNaseOUT Recombinant Ribonuclease Inhibitor from Invitrogen).
13. Extraction buffer: 20 mM Tris–HCl pH 7.5, 150 mM NaCl, 2.5 mM $MgCl_2$, 0.5% Triton X-100, 10% glycerol, 0.5 mM DDT, 1× Protease Inhibitor Cocktail, and 20 U/ml RNase inhibitor (*see* **Note 2**).

14. Rotator.
15. Dounce homogenizer device with a tight pestle (0.05 ± 0.025 mm clearance).
16. 1.5 and 2 ml low adhesion microcentrifuge tubes (e.g., Protein LoBind Tubes from Eppendorf).
17. Dilution buffer: 20 mM Tris–HCl pH 7.5, 150 mM NaCl, 2.5 mM $MgCl_2$, 10% glycerol, 1× Protease Inhibitor Cocktail, and 20 U/ml RNase inhibitor.

2.3 Immunoprecipitation

1. Anti-GFP microbeads (e.g., GFP-Trap_MA beads from ChromoTek).
2. Dilution buffer.
3. Wash buffer: 20 mM Tris–HCl pH 7.5, 150 mM NaCl, 2.5 mM $MgCl_2$, 0.2% Triton X-100, 10% glycerol, 0.5 mM DDT, 1× Protease Inhibitor Cocktail, and 20 U/ml RNase inhibitor.
4. Magnetic separator for microcentrifuge tubes (e.g., Promega).
5. Magnetic separator for 15 ml and 50 ml Falcon tubes (e.g., Promega).

2.4 Elution, RNA Recovery, and cDNA Synthesis

1. 1 M Tris–HCl pH 8.0.
2. Protease buffer: 30 mM Tris–HCl pH 8.0.
3. Proteinase K (10 mg/ml).
4. Thermoshaker.
5. Magnetic separator for microcentrifuge tubes (e.g., Promega).
6. 0.5 M EDTA.
7. 10% Sodium Dodecyl Sulfate (SDS).
8. β-Mercaptoethanol.
9. Homogenisation buffer: 100 mM Tris–HCl pH 8.0, 5 mM EDTA pH 8.0, 100 mM NaCl, 0.5% SDS, and 0.01 volume β-mercaptoethanol (*see* **Note 2**).
10. Phenol–chloroform–isoamyl alcohol mixture (25:24:1) pH 4.0–5.0.
11. Isopropanol (2-propanol).
12. GlycoBlue (15 mg/ml), nuclease and protease-free.
13. 3 M NaAc, pH 5.2.
14. 75% Ethanol.
15. RNase-free DNase (e.g., RQ1 from Promega). RQ1 RNase-Free DNase is provided with the following reagents:
    (a) 10× RQ1 RNase-free DNase buffer.
    (b) RQ1 RNase-free DNase.
    (c) RQ1 DNase stop solution.

16. 10 mM dNTP.
17. 100 μM oligo d(T) and/or 50 μM random hexamers and/or specific primers against known RNA.
18. RNase inhibitor (e.g., RNaseOUT Recombinant Ribonuclease Inhibitor Invitrogen).
19. High-performance Reverse Transcriptase (e.g., SuperScript IV from Invitrogen). SuperScript IV is provided with the following reagents:
    (a) 5× SSIV buffer.
    (b) 0.1 M DTT.
    (c) SuperScript IV Reverse Transcriptase.

2.5 Enrichment Analysis by qPCR or Library Preparation

1. qPCR quantification Master mix (e.g., LightCycler 480 Probes Master SYBR Green Mix from Roche).
2. Real-time PCR system (e.g., LightCycler 480 System from Roche).
3. Appropriate primer set to detect the RNA/cDNA of interest as well as negative and positive controls if possible (*see* **Note 3**).
4. Optional for sequencing: fluorimetric dsDNA quantification system (e.g., Qubit, ThermoFisher).
5. Optional for sequencing: fragment analyzer (e.g., BioAnalyzer, Agilent).
6. Optional for sequencing: high-throughput sequencing device (e.g., Illumina) with corresponding library preparation kit.

3 Methods

The workflow for RIP is illustrated in Fig. 1. The whole procedure can be performed within a day but several pausing points are possible and mentioned below. Make sure to work in an RNase-free environment during the entire procedure and prevent any contamination with RNA or DNA.

3.1 Plant Material

For each sample, approximately 10 g of plant material from seedlings is required. We therefore usually grow 3000 seedlings per sample. Prior to the protein extraction (Subheading 3.2), the seedlings can either be directly harvested for native RIP or cross-linked with UV light to preserve the less stringent RNA–Protein interactions.

1. Start by sowing sterilized seeds on ½ MS agar medium (*see* **Note 4**). For UV cross-linking, put one layer of autoclaved Whatman paper on top of the MS agar medium and sow the seeds onto the Whatman paper.

2. Stratify the seeds at 4 °C for 2 days and then grow the Arabidopsis seedlings in a growth chamber for 10 days (*see* **Note 4**).

3.2 Protein Extraction

Precool the centrifuge at 4 °C, place the rotator at 4 °C (fridge/cold room), prepare all buffers and keep them on ice. Precool all tubes on ice prior to use.

1. Harvest 10 g of plant material from each treatment sample and the control in a 50 ml Falcon tube. Collect five portions of 2 g, which are directly frozen in liquid nitrogen.
   (a) For native RIP: collect the seedlings, remove excess of water from seedlings and freeze in liquid nitrogen (*see* **Note 5**). Proceed to **step 2** or store the sample at −80 °C.
   (b) For cross-linking RIP: cool down 2 l of sterile ddH_2O on ice. Remove the lid of the petri dish and put the petri dish into a UV cross-linker. Expose the seedlings three times to 4 Million Joules (MJ) of UV. Following each exposure, cool down the plants by adding ice-cold ddH_2O to the petri dish until the seedlings are just submerged and incubate for 1 min. After a full course of UV exposures, immediately harvest the seedlings, remove free water, and freeze in liquid nitrogen (*see* **Note 6**).
2. Grind the sample manually using a mortar and pestle with liquid nitrogen to a fine powder. Do not allow it to thaw (*see* **Note 7**).
3. Transfer the powder in a LN precooled 50 ml Falcon tube and add an equal volume of extraction buffer. Do NOT vortex or pipette but use a pipette tip to gently dislocate the powder pellet. Make sure all the powder is resuspended in the extraction buffer (*see* **Note 8**).
4. Incubate on a rotator at 4 °C for 10–15 min. The resulting solution should be clear of clumps.
5. Use a dounce homogenizer device with a tight pestle to further disrupt the cells and to release a maximum of proteins (up and down slow movements).
6. The next step requires a centrifuge able to reach 20,000 × *g*. If no such a centrifuge is available for 50 ml Falcon tube, transfer the protein solution into 2 ml low adhesion tubes instead (*see* **Note 9**).
7. Pellet the debris at max speed (>20,000 × *g*) for 10 min. Transfer the supernatant into fresh precooled 2 ml low adhesion tubes and repeat **step** 7 until the solution is debris-free (*see* **Note 10**).
8. After the last centrifugation, transfer the supernatant to a precooled 50 ml Falcon tube and dilute the proteins solution by adding 3 volumes of dilution buffer (*see* **Note 11**).

3.3 Immunoprecipitation

Here, we use anti-GFP beads to pull down the GFP-tagged protein-of-interest. However, many different types of beads are available for the immunoprecipitation of tagged proteins (either with a fluorophore such as GFP, or with for example a His- or FLAG-tag). If specific antibodies are used against the protein-of-interest, they will have to be bound to protein A or G agarose beads. In the case of other beads, please refer to the manufacturer's instructions (*see* **Notes 12** and **13**).

1. Use 50 μl of the appropriate resuspended beads per 10 g of samples. Here, GFP-Trap_MA beads are used for the immunoprecipitation of GFP-tagged proteins (*see* **Note 12**).
2. Equilibrate the beads three times using 500 μl of wash buffer each time. Collect the beads using a magnet and remove the supernatant.
3. After the last equilibration wash, resuspend the beads into 50 μl of wash buffer.
4. Add the beads into the diluted protein solution and incubate for 1 h at 4 °C on a rotator (*see* **Note 13**).
5. Collect the complexes (beads/antibody/protein-of-interest/RNA) using a magnet. Leave the tube 5 min on the magnet to collect all complexes on the side of the tube and remove the supernatant.
6. Resuspend the complexes into 1 ml of wash buffer and transfer to a fresh precooled 1.5 ml low adhesion tube.
7. Wash the complexes three times, 5 min on a rotator at 4 °C. The two first washes are done with 500 μl of wash buffer each time and the last one is performed with 500 μl of dilution buffer (no Triton X-100).

3.4 Elution, RNA Recovery, and cDNA Synthesis

1. Resuspend the complexes in 50 μl of protease buffer supplemented with 1 μl RNase inhibitor. Add 2 μl of proteinase K, mix by pipetting slowly and incubate at 55 °C for 30 min to elute protein and RNAs from the beads.
2. Remove the beads using a magnet.
3. Add homogenisation buffer to make up the volume to 400 μl and mix vigorously (*see* **Note 2**).
4. Extract the RNA by adding 400 μl of phenol–chloroform–isoamyl alcohol (pH 4–5), mix well and centrifuge at maximum speed for 15 min.
5. Precipitate the RNA by adding an equal volume of isopropanol, with 0.1 volume of 3 M NaAc, pH 5.2 and 2 μl GlycoBlue. Incubate for 1 h at −20 °C.
6. Centrifuge at 4 °C, max speed (>19,000 × *g*) for 20 min.

7. Rinse the pellet with 750 μl of 75% ethanol, mix by two inversions, centrifuge for 5 min and remove the supernatant.
8. Repeat **step 7**.
9. Air-dry the pellet.
10. Resuspend the pellet into 8 μl of DNase-, RNase-, and protease-free water. Carry on or store it at −80 °C.
11. Treat the RNAs with DNase following the manufacturer's protocol. We describe here the protocol for RQ1 DNase (Promega) (*see* **Note 14**).
    (a) Add 1 μl of RQ1 DNase 10× buffer, 1 μl of RQ1 DNase for 1 μg of RNA and make up the volume to 10 μl with water.
    (b) Incubate for 1 h at 37 °C.
    (c) Stop the reaction by adding 1 μl of RQ1 Stop solution and heat at 65 °C for 10 min to completely inactivate the DNase. Proceed immediately to the next step.
12. cDNA has to be synthesised with a reverse transcriptase (RT). We describe here the protocol for the use of SuperScript IV Reverse transcriptase following the manufacturer's protocol, but several alternatives are available. However, make sure that you use a reverse transcriptase with high performance. For each sample, run a "no reverse transcription" control (no SuperScript IV added).
    (a) Use approximately 500 ng of the RNA for the reverse transcription (RT) (*see* **Note 14**). Add 0.5 μl of 100 μM oligo d(T) or 1.0 μl of 50 μM of random hexamers or specific primers (*see* **Note 15**), 1 μl of 10 mM dNTP and complete the volume up to 13 μl with water.
    (b) Incubate the RNA mix at 65 °C for 5 min and transfer immediately on ice for at least 1 min.
    (c) In a separate tube prepare RT mix by adding 4 μl of 5× SSIV buffer, 1 μl of 0.1 M DTT, 1 μl of 40 U/μl RNase inhibitor, 0.5 μl of SuperScript IV Reverse Transcriptase, and 0.5 μl of water. Mix by slowly pipetting up and down.
    (d) Add the RT mix to the RNA mix and incubate at 55 °C for 15 min.
    (e) Inactivate the reaction by heating at 80 °C for 10 min.
13. Store the cDNA at −80 °C or use it immediately for qPCR analysis or library preparation for high-throughput sequencing.

3.5 Quantification by qPCR or Library Preparation

qPCR quantification: The RNA immunoprecipitate can be quantified with qRT-PCR to examine the enrichment of the RNAs of interest. Primers are designed to cover the sequences of interest and several controls (*see* **Note 3**). The raw data from the qPCR can be analyzed using the $2^{-[\Delta][\Delta] C_T}$ analysis method [17].

Sequencing: For high-throughput sequencing a library has to be constructed. A library can already be generated from the RNA pool (e.g., using the NEBNext Ultra Directional RNA Library Prep Kit from Illumina [16]), or from the cDNA after reverse transcription. From the cDNA, molecules between 300 bp and 500 bp should be selected, either before or after the library preparation. The samples have to be precisely quantified using a fluorimetric dsDNA quantification system (such as the Qubit), while fragment sizes can be precisely analyzed using a microcapillary-electrophoresis trace analyzer (such as the Agilent BioAnalyzer). After size selection and quantification, the library can be prepared with a kit (usually an Illumina sequencing kit). Because of the low RNA/cDNA concentrations, it is essential to work in a very clean environment.

4 Notes

1. Each Protease Inhibitor Cocktail Tablet should be dissolved in 10 ml of water (10× concentrated). Protease Inhibitor Cocktail is stable at pH 7.0 (in water) and can be stored at −20 °C for 4 weeks. After thawing frozen stock solutions of protease inhibitors, vortex briefly to mix and resuspend sediment particles. For working solutions with pH above 7.0, add protease inhibitor ideally just prior to use, or no longer than 30 min before use.
2. The following stock solutions can be prepared beforehand and stored at room temperature: 1 M Tris–HCl (pH 7.5), 1 M Tris–HCl (pH 8.0), 5 M NaCl, 2 M $MgCl_2$, 10% Triton X-100, and 80% glycerol. DTT (1 M) can be stored at −20 °C for up to a year. Homogenization buffer (w/o β-mercaptoethanol) can be prepared in advance and kept at room temperature. β-mercaptoethanol must be added prior to use in a fumehood.
3. To determine whether the RNA of interest is indeed enriched compared to the background, different primer sets have to be designed. It is important to design the primers in such a way that their annealing temperatures are in the same range and the amplified fragments will be of similar size (e.g., 80 bp) or no more than 150 bp. The primer efficiencies have to be carefully assayed before performing the actual enrichment qPCR. Differences in the efficiencies of the primer pairs can seriously affect the data. To check the primer efficiencies, a dilutions of the input sample can be used. Primers have to be designed for the RNAs/cDNAs of interest and for random reference RNAs that are not expected to bind to the protein of interest. The best is to randomly choose three other RNAs, for example

ACTIN2, *UBC21*, and *CBP20* [18] and check whether their abundance is more or less the same in the input sample and the IP sample. If the protein of interest is already known to bind a particular RNA, primers for this RNA can be included as a positive control.

4. Sow the seeds quite sparsely using 0.2% Agar and grow them in a 16 h light–8 h dark chamber for 10 days. Make sure that the seedlings are growing healthy (green, stress-free and without contamination).
5. Be aware of the risks when handling liquid nitrogen (LN) and ensure proper protection (lab coat, goggles, proper thermo gloves and use a spoon to manipulate LN). Be careful when flash-freezing the samples.
6. The UV exposure time needed to cross-link the sample will vary according to the tissues types or plant samples used and will need to be optimized. To do so, use a known RNA–protein interaction (e.g., ref. [19]). Good negative controls are non-cross-linked plants and plants that are deficient for the protein known to interact with the RNA. Analyze them in parallel with UV cross-linked samples with different exposure times during a RIP experiment. RNA enrichment should be observed with UV cross-linked sample compared to negative controls. The shortest exposure time that shows a good RNA immunoprecipitation result should be used to perform the UV cross-linking. Make sure to cool down the seedlings between the rounds of UV treatment. Not all RNA-binding proteins can be cross-linked by UV to their target RNAs. In these cases chemical cross-linker (e.g., formaldehyde) can be used [13].
7. RNases are naturally abundant in cells. To protect RNAs, it is important to keep RNases inactive by keeping the sample frozen during the grinding. Do not allow the powder to thaw. Use LN to refrigerate the mortar and the pestle before and during their usage.
8. If native RIP is performed (without cross-linking), vortexing or vigorous pipetting can break the bond between RNAs and the protein of interest and reduce the immunoprecipitation efficiency. Use a tip to gently dislocate the powder pellet and make sure that the pellet is soaked in extraction buffer.
9. Low adhesion tubes are specifically designed for use when the protein concentration tends to be very small. Significantly more protein can be recovered for downstream analyses.
10. The number of centrifugation steps required to pellet the debris will vary from samples to samples. After each centrifugation, we recommend to check the presence of debris left in suspension. If there is some, transfer the supernatant into a fresh tube and repeat the centrifugation step if necessary up to four times.

11. Triton X-100 is a detergent used to separate the beads from each other. If used in too high concentrations, it will denature proteins, antibody and reduce immunoprecipitation efficiency. We recommend to bring the Triton X-100 concentration to <0.2%.
12. GFP-Trap_MA beads are ready to use. They consist of magnetic beads coupled with small recombinant alpaca antibody fragments binding several types of GFP. GFP fusion proteins and their interacting factors can be isolated fast and efficiently by immunoprecipitation. Alternatively, anti-GFP antibody and protein A/G coated agarose beads can be used. Protein agarose beads are chosen according to their affinity for the antibody of interest, check https://www.neb.com/tools-and-resources/selection-charts/affinity-of-protein-ag-for-igg-types-from-different-species. We recommend the use of magnetic beads mainly for the ease of manually pipetting the solution from the beads without the need of centrifugation. No residual volume remains after the washes, which helps to reduce background.
13. ChromoTek recommends to incubate the GFP-Trap_MA beads for 1 h at 4 °C on rotator. Depending on the abundance of the GFP-tagged protein in the sample, beads can be incubated from 5 min to 2 h at 4 °C. For the use of protein A/G coated agarose beads, we advise to refer to manufacturer recommendations.
14. Prior to treating the RNA with DNase, RNA concentration can be determined. To perform the DNase treatment and the reverse transcription (RT), the manufacturer recommends to use 1 unit of DNase per microgram of RNA and to use no more than 500 ng per RT reaction. Higher amounts will reduce DNase and RT efficiencies. However, the quantity of the RNA after the IP is expected to be lower than 500 ng, and we therefore always use the entire RNA solution for the DNase treatment and for the RT, performing the DNase treatment with 1 ul DNase.
15. Oligo d(T) primer is suitable for use in first-strand cDNA synthesis with reverse transcriptase. The primer hybridizes to the poly(A) tail of mRNA. Alternatively if no information about the potential polyadenylation of the RNA interacting with the protein-of-interest is available, random hexamers can be used. They are short oligodeoxyribonucleotides of random sequence [$d(N)_6$] that anneal to random complementary sites on a RNA target to serve as primers for reverse transcriptase. If the RNAs interacting with the protein of interest are known, primers binding specifically these RNAs can be used for the reverse transcription.

References

1. Re A, Kulberkyte E, Joshi T, Workman CT (2015) RNA – protein Interactions : an overview RNA-protein interactions. Methods Mol Biol 1097:491–521
2. Au PCK, Helliwell C, Wang MB (2014) Characterizing RNA-protein interaction using cross-linking and metabolite supplemented nuclear RNA-immunoprecipitation. Mol Biol Rep 41:2971–2977
3. Wu L, Murat P, Matak-Vinkovic D, Murrell A, Balasubramanian S (2013) The binding interaction between long non-coding RNA HOTAIR and PRC2 proteins. Biochemistry 52:9519–9527
4. Meredith EK, Balas MM, Sindy K, Haislop K, Johnson AM (2016) An RNA matchmaker protein regulates the activity of the long noncoding RNA HOTAIR. RNA 22:995–1010
5. Kahlil AM, Guttman M, Huarte M, Garber M, Raj A, Morales DR, Thomas K, Presser A, Bernstein BE, van Oudenaarden A, Regev A, Lander ES, Rinn JL (2009) Many human large intergenic noncoding RNAs associate with chromatin-modifying complexes and affect gene expression. Proc Natl Acad Sci U S A 106:11667–11672
6. Názer E, Lei EP (2014) Modulation of chromatin modifying complexes by noncoding RNAs in trans. Curr Opin Genet Dev 25:68–73
7. Crevillén P, Sonmez C, Wu Z, Dean C (2013) A gene loop containing the floral repressor FLC is disrupted in the early phase of vernalization. EMBO J 32:140–148
8. Ariel F, Jegu T, Latrasse D, Romero-Barrios N, Christ A, Benhamed M, Crespi M (2014) Noncoding transcription by alternative RNA polymerases dynamically regulates an auxin-driven chromatin loop. Mol Cell 55:383–396
9. Rinn JL, Kertesz M, Wang JK, Squazzo SL, Xu X, Brugmann SA, Goodnough LH, Helms JA, Farnham PJ, Segal E, Chang HY (2007) Functional demarcation of active and silent chromatin domains in human HOX loci by noncoding RNAs. Cell 129:1311–1323
10. van Mourik H, Muino JM, Pajoro A, Angenent GC, Kaufmann K (2010) Characterization on in vivo DNA-binding events of plant transcription factors by ChIP-seq: experimental protocol and computational analysis. Plant Funct Genomics 380:93–121
11. Huppertz I, Attig J, D'Ambrogio A, Easton LE, Sibley CR, Sugimoto Y, Tajnik M, Konig J, Ule J (2014) iCLIP: protein-RNA interactions at nucleotide resolution. Methods 65:274–287
12. Jensen KB, Darnell RB (2008) CLIP: cross-linking and immunoprecipitation of in vivo RNA targets of RNA-binding proteins. Methods Mol Biol 488:85–98
13. Terzi LC, Simpson GG (2009) Arabidopsis RNA immunoprecipitation. Plant J 59:163–168
14. Berry S, Hartley M, Olsson TSG, Dean C, Howard M (2015) Local chromatin environment of a Polycomb target gene instructs its own epigenetic inheritance. elife 4:e07205
15. Moore MJ, Zhang C, Gantman EC, Mele A, Darnell JC, Darnell RB (2016) Mapping Argonaute and conventional RNA-binding protein interactions with RNA at single-nucleotide resolution using HITS-CLIP and CIMS analysis. Nat protoc 9:263–294
16. Xing D, Wang Y, Hamilton M, Ben-Hur A, Reddy ASN (2015) Transcriptome-wide identification of RNA targets of Arabidopsis SERINE/ARGININE-RICH45 uncovers the unexpected roles of this RNA binding protein in RNA processing. Plant Cell 27:3294–3308
17. Livak KJ, Schmittgen TD (2001) Analysis of relative gene expression data using real-time quantitative PCR and the 2(-Delta Delta C (T)) method. Methods 25:402–408
18. Czechowski T, Stitt M, Altmann T, Udvardi MK, Scheible W-R (2005) Genome-wide identification and testing of superior reference genes for transcript normalization in Arabidopsis. Plant Physiol 139:5–17
19. Zhang Y, Gu L, Hou Y, Wang L, Deng X, Hang R, Chen D, Zhang X, Zhang Y, Liu C, Cao X (2015) Integrative genome-wide analysis reveals HLP1, a novel RNA-binding protein, regulates plant flowering by targeting alternative polyadenylation. Cell Res 25:864–876

Chapter 20

In Vitro Assays to Measure Histone Methyltransferase Activity Using Different Chromatin Substrates

Yannick Jacob and Philipp Voigt

Abstract

In vitro histone modification (HM) assays are used to characterize the activity of chromatin-modifying enzymes. These assays provide information regarding the modification sites on histones, the product specificity, and the impact of other histone or nucleotide modifications on enzyme activity. In particular, histone methyltransferase (HMT) assays have been instrumental in elucidating the activity and site specificity of many plant HMT enzymes. In this chapter, we describe a general protocol that can be used to perform HMT assays using different chromatin substrates, detection methods, and enzymes directly purified from plant material or heterologous sources.

Key words Histone modifications, Histone variants, Histone lysine methylation, Histone peptides, Histone octamers, Nucleosomes

1 Introduction

The study of epigenetics is largely focused on understanding how different chemical modifications on DNA and histones affect chromatin-based biological processes. Identifying the specific activity of a chromatin-modifying enzyme is often the first step in elucidating the molecular function of a particular epigenetic mark. In vitro histone modification (HM) assays have long been the method of choice to precisely identify the substrate of histone methyltransferases, acetyltransferases, and kinases, as well as their antagonizing counterparts, the "erasers" of histone modifications [1]. These assays have significantly facilitated the study of plant chromatin since the discovery and characterization of the first plant histone methyltransferase (HMT) enzymes [2, 3]. Even though many chromatin-modifying enzymes have now been functionally characterized in terms of their substrate preferences, HM assays still

Yannick Jacob and Philipp Voigt contributed equally to this work.

Marian Bemer and Célia Baroux (eds.), *Plant Chromatin Dynamics: Methods and Protocols*, Methods in Molecular Biology, vol. 1675, DOI 10.1007/978-1-4939-7318-7_20, © Springer Science+Business Media LLC 2018

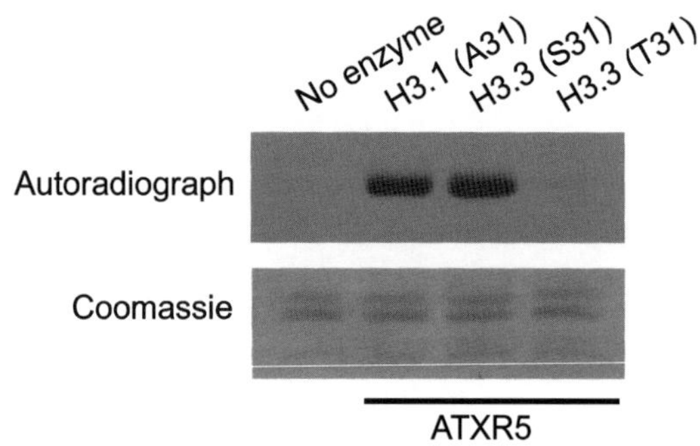

Fig. 1 HMT assay using the Arabidopsis histone H3 lysine 27 (H3K27) methyltransferase ATXR5. The enzyme with an N-terminal GST tag was expressed in E.coli and purified by affinity chromatography. The substrates used in the assay were plasmid-based nucleosomes containing different histone H3 variants: plant H3.1, mammalian H3.3, or plant H3.3. 3H-labeled SAM was used to detect the methylated histones. The result of the assay demonstrates the specific role of threonine 31 (T31) of plant H3.3 variants in inhibiting the activity of ATXR5

remain a very useful method in chromatin biology. For example, HM assays can be used to determine product specificity (e.g., monomethylation, dimethylation, or trimethylation) and the impact of histone variants and neighboring histone/DNA modifications on the enzymatic activity of histone-modifying enzymes (Fig. 1). Recent work in plant epigenetics on the histone H3 lysine 27 (H3K27) methyltransferases ATXR5 and ATXR6 underscores the usefulness of in vitro HM assays for understanding the functions of chromatin-modifying enzymes. HMT assays were used to show (1) that ATXR5 and ATXR6 do not methylate H3K4 (as predicted by sequence comparison) but rather H3K27 and (2) that the enzymatic activity of ATXR5 and ATX6 is specific for replication-dependent H3.1 variants [4, 5].

In vitro HM assays rely on setting up a chemical reaction between a histone-modifying enzyme and a chromatin substrate. The chromatin substrate used in these assays can be peptides, histone monomers, histone dimers/tetramers/octamers, or even complete nucleosomes. Many of these substrates are now commercially available. In addition, new techniques have made it possible to design and synthesize complex nucleosome substrates (e.g., including different histone/DNA modifications and/or histone variants) to answer precise questions about the interplay between different epigenetic marks [6, 7]. In this chapter, we describe a general protocol that can be used to perform HMT assays using different chromatin substrates. The addition of methyl groups can either be detected with radioactive assays, or by using antibodies against a specific histone mark. All these detection methods are described

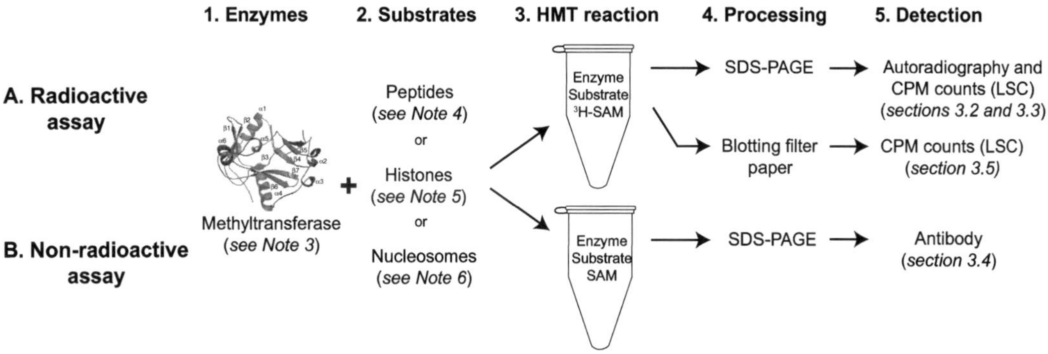

Fig. 2 Summary of the different methods described in this Chapter to perform HMT assays. *LSC* liquid scintillation counter

here in different sections of the protocol (Fig. 2). We seek to provide sufficient experimental details to help scientists adapt this assay to their specific scientific needs.

2 Materials

2.1 HMT Reactions

1. 5× methylation buffer: 250 mM Tris–HCl pH 8.5 (pH measured at 21 °C), 25 mM $MgCl_2$ (*see* **Note 1**).
2. ^3H-labeled S-adenosylmethionine (^3H-SAM, available from PerkinElmer, e.g., NET155H250UC, or in Europe also from Hartmann Analytic, e.g., ART0288), with the ^3H label present in the methyl group. Aliquot and store at −20 °C, avoid repeated freeze–thaw cycles (Radioactive assay only, *see* **Note 2**).
3. 0.2 M DTT (dithiothreitol) stock solution. Aliquot and freeze at −20 °C.
4. Nonradioactive SAM stock as a 32 mM stock solution. Alternatively, a 2.5 mM SAM stock solution can be created by dissolving 1 mg of SAM in 788 μl of 5 mM sulfuric acid, pH 2, 10% ethanol. Aliquot and freeze at −20 °C.
5. TE buffer: 10 mM Tris–HCl pH 8.0, 1 mM EDTA.
6. BC100 buffer: 20 mM Tris–HCl pH 8.0, 100 mM KCl, 0.2 mM EDTA, 20% glycerol, 1 mM DTT. Add DTT fresh before using.
7. 4× SDS-PAGE loading buffer: 200 mM Tris–HCl (pH 6.8), 400 mM DTT, 8% SDS, 0.4% bromophenol blue and 40% glycerol. Aliquot and store at −20 °C.
8. Histone methyltransferase preparation in suitable buffer: enzymes for these assays are typically obtained from recombinant expression in *E. coli* or Sf9 insect cells followed by

purification using affinity tags grafted onto the methyltransferase (*see* **Note 3**).

9. Methylation substrate: suitable substrates include histone peptides (*see* **Note 4**), recombinant or native histones (*see* **Note 5**), and mononucleosomes or oligonucleosomes (*see* **Note 6**). The choice of substrate depends on the specificity of the enzyme and experimental needs, however, the unavailability of more complex substrates may impose limitations in selecting the optimal substrate.
10. Heat block (with heated lid to prevent evaporation) or incubator set to 30 °C or reaction temperature of choice (*see* **Note 7**).
11. Laboratory space equipped and certified for the use of ^{3}H radioactive material (Radioactive assay only, *see* **Note 8**).

2.2 Detection of Methylated Histones by SDS-PAGE and Radioactivity

1. SDS-PAGE reagents and equipment for standard mini gels. Use of prestained protein marker is recommended to monitor transfer.
2. Equipment and reagents for wet or semidry transfer.
3. PVDF membrane.
4. Laboratory platform rocker.
5. Coomassie stain solution: 45% (v/v) methanol, 10% (v/v) acetic acid, 0.25% (w/v) Coomassie Brilliant Blue R. For 500 ml, dissolve 1.25 g of Coomassie Brilliant Blue R in 225 ml of methanol. Add distilled water up to a final volume of 450 ml. Add 50 ml glacial acetic acid. Filter to remove residual undissolved Coomassie dye.
6. Destain solution: 45% (v/v) methanol, 10% (v/v) acetic acid.
7. Camera or imaging system capable of taking white-light monochrome or color images. Alternatively, a scanner can be used.
8. Autoradiography enhancer (e.g., EN^{3}HANCE spray, Perkin Elmer).
9. Autoradiography film with high sensitivity for ^{3}H.
10. Autoradiography cassette and bag.
11. −80 °C freezer.
12. Darkroom with red safelight.
13. Automated X-ray developer or manual film development setup.
14. Hair dryer or laboratory heat gun capable of running in cool mode (recommended but not essential; *see* **Notes 12** and **13**).

If performing scintillation counting (Subheading 3.2, **steps 11–13**):

15. Scalpel or razor blade.

16. Liquid scintillation cocktail suitable for ^{3}H.
17. Scintillation vials.
18. Scintillation counter suitable for ^{3}H detection.

2.3 Detection of Methylated Peptides by SDS-PAGE and Radioactivity

1. Same materials as under Subheading 2.2 with the exception of blotting reagents and equipment, Coomassie stain solution, EN^{3}HANCE spray, and items for scintillation counting.
2. Autoradiography Enhancer (e.g., ENLIGHTNING Rapid, Perkin Elmer, or Amplify Fluorographic Reagent, Amersham).
3. Filter paper (e.g., Whatman 3MM Chr).
4. Thin clear wrap such as Saran Wrap.
5. Gel drying apparatus.

2.4 Detection of Methylated Histones or Peptides by SDS-PAGE and Antibodies

1. SDS-PAGE reagents and equipment for standard mini gels. Use of prestained protein ladder is recommended.
2. Equipment and reagents for semidry transfer.
3. Nitrocellulose or PVDF membrane.
4. Laboratory platform rocker.
5. Primary antibody against a specific histone mark. Follow the manufacturer's recommendation regarding the dilution to be used in a Western blot.
6. Secondary antibody coupled to horseradish peroxidase. Make sure that the secondary antibody used has specificity for the antibody species and isotype of the primary antibody. Refer to manufacturer's protocol for determining which dilution to use.
7. TBS: 20 mM Tris pH 7.5, 100 mM NaCl.
8. TBS-T: 20 mM Tris pH 7.5, 100 mM NaCl, 0.1% (v/v) Tween.
9. Blocking solution: 5% Nonfat Dry Milk in TBS-T solution.
10. Antibody solution: 2% Nonfat Dry Milk in TBS-T solution or follow the antibody manufacturer's recommendation.
11. ECL Western Blotting Detection Reagents.
12. X-ray film for Western blot.
13. Standard film cassette.
14. Automated X-ray developer or manual film development setup.

2.5 Detection of Methylated Peptides and Histones by Using Radioactivity and Filter Assays

1. Whatman P-81 filter papers (cellulose phosphate paper which is also a strong cation exchanger).
2. BD solution: 50 mM $NaHCO_3$ at pH 9.0. Prepare a fresh solution before use.
3. 1 l beaker.

4. Rocking shaker.
5. Liquid scintillation cocktail suitable for ^{3}H (e.g., Gold Star by Meridian) and scintillation vials.
6. Liquid scintillation counter suitable for ^{3}H detection.

3 Methods

3.1 HMT Reaction

This protocol can be adapted to a broad range of methyltransferases and substrates. Depending on the choice of substrate and detection method, subsequent steps differ and are outlined separately in the following sections (*see* Subheadings 3.2–3.5 and Fig. 2). If no activity is detectable, pointers for troubleshooting are provided at the end (*see* Subheading 3.6). If performing multiple assays with different substrates or enzyme preparations, it is advisable to prepare a master mix containing all common components.

1. On ice, combine distilled water (to make up a total reaction volume of 25 μl including substrate), 5 μl 5× methylation buffer, 0.5 μl 0.2 M DTT (4 mM final), 10 μM unlabeled SAM (e.g., 0.25 μl of a 1 mM predilution in water of the stock solution) or 25–75 kBq of ^{3}H-SAM (if relying on radioactivity-based detection methods), and methyltransferase preparation (in BC100, *see* **Note 3**). 0.5–1 pmol (25–50 ng for a 50 kDa protein) of purified enzyme is usually sufficient to detect activity, but higher amounts might be necessary for some methyltransferases. If higher reaction volumes are required, for instance due to diluted enzyme or substrate preparations, scale up 5× methylation buffer, DTT, and SAM accordingly to maintain their final concentrations.
2. Add substrate to start the reaction. If using peptides (*see* **Note 4**), add peptide to a final concentration of 0.1–1 mM (2.5–25 nmol). If using core histones (*see* **Note 5**), add 0.5–4 μg of histones (0.18–1.47 μM or 4.6–36.7 pmol for recombinant *Xenopus laevis* histones). If using nucleosomal substrates (*see* **Note 6**), use the same amounts as indicated for core histones.
3. Mix well by pipetting up and down. Centrifuge briefly if needed (*see* **Note 8**).
4. Incubate at 30 °C (*see* **Note** 7) for 1 h if doing end-point assays or for shorter time points if analyzing reaction kinetics to obtain enzymatic parameters (*see* **Note 9**).
5. Stop reactions by adding 8.3 μl of 4× SDS sample buffer (to reach 1× final) and boiling at 95 °C for 5 min. Centrifuge for 10 sec at full speed. Different detection procedures can be followed after this step. For detection of methylated histones

or methylated peptides using SDS-PAGE and ^{3}H fluorography, go to Subheading 3.2 or 3.3, respectively. For detection of methylated histones or peptides using antibodies, go to Subheading 3.4. If using filter assays to detect incorporation of radioactivity into histones or peptides, it is not necessary to stop the reaction here. Skip this step entirely and go directly to Subheading 3.5. Please refer to Fig. 2 for an overview of the different detection methods available and the sections of this protocol covering each method.

3.2 Detection of Methylated Histones by SDS-PAGE and ^{3}H Fluorography

This protocol outlines the steps required to detect methylation of histone proteins if radioactive ^{3}H-SAM has been used for the HMT reaction described in Subheading 3.1. It is applicable to all reactions performed on histone proteins, irrespective of the nature of the methyltransferase studied. ^{3}H-SAM incorporation is visualized by exposing an autoradiography film (**steps 1–10**). Quantitative data can be obtained subsequent to film exposure by liquid scintillation counting (optional **steps 11–13**).

1. Resolve samples on a 15% SDS polyacrylamide gel (*see* **Note 10**). Dispose of radioactive waste from this and subsequent steps following the relevant regulations in your jurisdiction (*see* **Note 8**).
2. Transfer polypeptides from the gel onto a PVDF membrane (*see* **Note 11**). Most standard semidry or tank blot setups and protocols will be suitable (e.g., perform semidry transfer for 75 min at 100 V in SDS running buffer containing 20% methanol).
3. Stain PVDF membrane with Coomassie stain solution for 2–5 min with agitation on a platform rocker. Remove the Coomassie stain solution.
4. Wash the membrane with destain solution 2–3 times for 5 min each with agitation.
5. Air-dry membrane until completely dry (*see* **Note 12**).
6. Document membrane with a camera, imaging system, or scanner. The image of the membrane obtained at this step should be similar to the bottom panel of Fig. 1, clearly showing the characteristic four core histone bands around 15 kDa (three in the case of Xenopus histones due to virtually identical size of H2A and H2B).
7. In a fume cabinet, spray the membrane with EN^{3}HANCE spray (Perkin Elmer). Make sure to coat the membrane evenly with the solution. Let sit for 5 min, and then repeat two times.
8. Air-dry membrane until completely dry (*see* **Note 13**).

9. Expose to autoradiography film for 12–96 h in an autoradiography cassette in a −80 °C freezer (*see* **Note 14**). Bend one corner of the film to mark its orientation relative to the membrane.
10. Develop film in an X-ray developer.
11. Optional: To obtain quantitative information, cut out the Coomassie-stained bands of interest from the membrane with a scalpel or razor blade.
12. Place each individual band into a scintillation vial and add liquid scintillation cocktail.
13. Perform scintillation counting of all samples as well as vials containing only scintillation cocktail (for background count determination) in a scintillation counter suitable for ^{3}H.

3.3 Detection of Methylated Peptides by SDS-PAGE and ^{3}H Fluorography

This protocol outlines the steps required to detect methylation of histone peptides if radioactive ^{3}H-SAM has been used for the HMT reaction described in Subheading 3.1. It is applicable to all reactions performed using peptide substrates. To avoid loss of peptide in blotting, the SDS polyacrylamide gel used to resolve samples, rather than a membrane, is treated with autoradiography enhancer solution and exposed to an autoradiography film for detection of ^{3}H-SAM incorporation.

1. Resolve samples on a 10% SDS polyacrylamide gel. Run the gel until the dye front migrated about halfway into the resolving gel. This should result in a sharp band for the peptides migrating close to the dye front. Dispose of radioactive waste from this and subsequent steps following the relevant regulations in your jurisdiction (*see* **Note 8**).
2. Stain the gel in Coomassie staining solution for 30–60 min with agitation.
3. Destain gel with several washes of destain solution (15–30 min each) until the bands of interest are clearly visible and background is nearly clear.
4. Document the gel with a camera, imaging system or scanner.
5. Incubate the gel in ENLIGHTNING solution for 30 min. It is sufficient to use just enough to cover the gel.
6. Lift gel onto Whatman paper that has been presoaked in water.
7. Cover with clear wrap.
8. Dry on a gel dryer set to 50–55 °C until dry (1–1.5 h). Higher temperatures may increase the risk of cracks in the gel.
9. Remove clear wrap for optimal sensitivity.
10. Expose to autoradiography film for 12–96 h in an autoradiography cassette in a −80 °C freezer (*see* **Note 14**). Bend one

corner of the film to mark its orientation relative to the membrane.

11. Develop the film in an X-ray developer.

3.4 Detection of Methylated Peptides or Histones Using Antibodies

Using antibodies as a detection method is advantageous over radioactivity for a few reasons. First, albeit not as sensitive as radioactive detection, it is safer and easier to perform than radioactive HMT assays. Secondly, antibodies can be used to reveal the product specificity (monomethylation, dimethylation, or trimethylation) of a methyltransferase [5, 8]. A few caveats of using antibodies for HMT assays is that they are costly and there is a need to know which lysine of the histone is modified by a specific methyltransferase (otherwise, many antibodies would have to be used to "screen" for the modified lysine). Also, antibodies are less suitable when using native histones as substrates, as many modifications are already present and can mask the signal generated in the HMT assay.

1. Resolve samples by SDS-PAGE as described in Subheading 3.2 **step 1** (for histones) or Subheading 3.3 **step 1** (for peptides).
2. Transfer to nitrocellulose or PVDF membrane by wet or semi-dry transfer (*see* **Note 15**).
3. Quickly wash membrane by covering it in TBS-T solution. Let it sit for a few minutes. Remove the TBS-T.
4. Proceed to soak the membrane into the blocking solution. Gently shake the membrane in the blocking solution (add enough to cover the membrane) for 30–60 min at room temperature using a laboratory platform rocker. Wash briefly with TBS-T solution.
5. Dilute the primary antibody to a working concentration using the antibody solution. The minimum volume needed of antibody solution (containing the primary antibody at the appropriate concentration) is just enough to cover the membrane. Gently shake the membrane for 1 h at room temperature or at 4 °C overnight using a laboratory platform rocker. Remove the solution.
6. Perform three washes of the membrane in TBS-T solution (10 min each) using the platform rocker.
7. Dilute the secondary antibody to its working concentration in antibody solution and add to the membrane. Gently shake the membrane for 1 h at room temperature using a laboratory platform rocker. Remove the solution.
8. Perform two washes of the membrane in TBS-T solution (10 min each) using the platform rocker. Perform a third and final wash of the membrane in TBS solution (10 min). Remove the TBS solution at the end of the third wash.

9. Prepare the ECL substrate as recommended by the manufacturer. Add the ECL substrate directly to the membrane making sure the substrate is present on the whole membrane. Let it sit for 1–5 min.
10. Cover the membrane with clear wrap.
11. Expose to X-ray film (*see* **Note 16**).
12. Develop the film in an X-ray developer.

3.5 Detection of Methylated Peptides and Histones by Using Radioactivity and Filter Assays

This section describes the use of a rapid detection method for measuring the enzymatic activity based on incorporation of radioactive ^{3}H-SAM. Contrary to other detection methods described in this chapter, this method does not rely on separating the product(s) of the HMT reaction on a gel. However, it cannot resolve whether multiple histones or other proteins are being methylated in the same HMT reaction, as all ^{3}H-SAM incorporated into peptides or proteins will be detected.

1. The reactions are stopped by spotting the reactions onto Whatman P-81 filter papers (~1.5 cm^2 each). The filter papers are dried for ~15 min at room temperature before proceeding to the next step. Each filter paper should be labeled, as they will be combined in the next step.
2. Unincorporated ^{3}H-SAM is removed from the filters by washing them in 250 ml (using a 1 l beaker) of BD solution in a 1 l beaker. A maximum of 20 filter papers per 250 ml of BD solution should be used. Wash three times for 30 min each. Gently rock the beaker using a rocking shaker. Do not use a magnetic bar when washing, as the filters will get damaged.
3. Dry the filter papers at room temperature.
4. Each filter paper is quantified separately by liquid scintillation counting.

3.6 Troubleshooting

Figure 1 shows an example of a successful HMT assay using radioactive detection. If no methyltransferase activity is detectable with the protocol described in Subheading 3.1 followed by any of the appropriate detection methods, several options for optimization are available. Radioactivity-based detection methods are usually most sensitive and may be required to detect activity of certain methyltransferases. Although the amounts stated should be sufficient in most cases, the amount of ^{3}H-SAM can be increased to enhance the signal. Exposure times for fluorography on autoradiography film can be extended to several days or even weeks.

The methylation reaction itself can be enhanced by (1) increasing incubation times up to overnight, (2) altering the reaction temperature (both above and below 30 °C), (3) increasing reaction pH, and/or by (4) increasing concentration of SAM (both unlabeled and ^3H labeled). If the SAM stocks have been stored for prolonged times, consider replacing them with fresh ones. In addition, as the final molar concentration of ^3H-SAM is often in the submicromolar range especially for ^3H-SAM preparations with high labeling density, supplementing the reactions with additional 10–20 μM unlabeled SAM may increase sensitivity in cases where the increase in enzymatic activity due to elevation of SAM concentrations closer to its Km for the enzyme is greater than the competition effect observed by adding an excess of unlabeled SAM.

The amount of methyltransferase can be increased as well, however, care should be taken to prevent introduction of significant amounts of salt, which might decrease activity especially on nucleosomes [9]. Moreover, potential inhibitory factors from bacteria or cell lysates may be introduced with increasing amounts of methyltransferase preparations as well. It is also possible that further factors might be required for activity; for instance the catalytic subunits of PRC2 do not exhibit activity without the other complex members [1, 10]. Furthermore, while increasing the amount of substrate might enhance activity, assaying different types of substrates should also be considered since the activity of a methyltransferase can vary depending on the substrate [10, 11]. Also, enzymatic activity might only be detected with specific histone isoforms or in the presence of other modifications [4].

4 Notes

1. Both lysine and arginine methyltransferases have been shown to be optimally active in alkaline buffers, with pH optima often around 9–10 [12, 13]. We therefore recommend slightly alkaline buffer conditions with pH 8.5 usually providing a good starting point for optimization.
2. For optimum sensitivity of the assay, ^3H-SAM with a high specific activity should be used (~3 TBq/mmol, if available). Both radiolabeled and unlabeled SAM are relatively prone to decay, and stock solutions may lose significant activity within 6 months or less of storage. Stock solutions of SAM must be kept in a strongly acidic buffer (e.g., 5 mM sulfuric acid pH 2, 10% ethanol), as it is unstable at neutral or alkaline pH. Storage at −80 °C is not recommended, as this accelerates decay compared to storage at −20 °C.
3. Providing detailed instructions for the purification of specific enzymes is beyond the scope of this protocol. However, the protocol described here should allow for the detection of methyltransferase activity from a variety of sources, ranging from highly purified recombinant or native enzymes and enzyme complexes to crude or fractionated cellular extracts.

Both *E. coli* and insect cell (Sf9) expression hosts have been successfully used to prepare active methyltransferases using protein affinity tags such as His, GST, or FLAG. As a general recommendation, it is likely easier to use purified recombinant proteins expressed in *E. coli* (or insect cells if there are issues with expression/folding in *E. coli*) than native enzymes extracted from plant tissues. Insect cell systems also allow for the expression of individual subunits and purification of reconstituted protein complexes. Regardless of source, it is usually advisable to keep the volume of enzyme preparation per reaction to a minimum in order to minimize carry-over of potential inhibitors of activity. Activity of many methyltransferases, especially when using nucleosomes as substrates, is also inhibited by salt [9]. Enzyme preparations should therefore ideally be provided in a low-salt buffer, for instance by dialyzing the final purification product into BC100 buffer (*see* Subheading 2.1). Dialysis will also remove elution agents such as FLAG peptide, even though the latter commonly does not interfere with activity. Enzyme preparations should be aliquotted and stored at −80 °C to preserve activity.

4. Although not always representative of the native conformation of histones in chromatin, peptides derived from histone sequences such as the N-terminal tail of histone H3 are commonly used as substrates in HMT assays. Histone peptides containing several well-characterized modifications are readily available from several commercial sources or can be custom-made by peptide synthesis services. Peptides should be centered around the residue to be methylated and of sufficient length (>20 residues) to include neighboring residues that might be required for substrate recognition. Unless containing native N- or C-termini, cap the peptides with acetyl (N terminus) and amide (C terminus) groups to maintain electrostatic properties of peptides. As the mass of lyophilized material is a poor measure for peptide amounts due to varying amounts of residual salts, it is highly advisable to add a C-terminal tyrosine residue to allow for quantification by UV spectroscopy. Alternatively, a biotin moiety can be introduced in the form of a C-terminal biotinylated lysine residue to enable quantification by western blotting with anti-biotin antibodies. When comparing activities between differentially modified peptides, accurate quantification and matching of peptide amounts is crucial to obtain meaningful results. Peptides should be resuspended to a final concentration of 10 mM in water (minimal amounts of TFA or ammonium hydroxide can be added to help solubilize the peptides if needed). Aliquot and store at −20 °C. Use protein low-bind tubes to avoid loss of peptide due to adhesion to tube walls. Consider filtering through a 0.22 μm low protein

binding filter to prevent degradation due to potential bacterial contamination. If TFA or ammonium hydroxide is present in peptide stock solutions, ensure that the pH of the HMT reaction is unchanged by spotting 0.5–1 μl of the reaction onto each field of a suitable pH strip and determine pH based on color change.

5. Histone preparations from various sources can also be used as substrates. Recombinant, unmodified histones are available from commercial sources or can be expressed and purified from *E. coli* using well established protocols [14, 15]. Recombinant histones containing lysine/arginine-to-alanine point mutations are commonly used to elucidate the site specificity of methyltransferases with unknown substrate specificity. Recombinant histones with defined, site-specific modifications can further be generated by cysteine alkylation approaches or by native chemical ligation [16]. A growing selection of modified histones is also available commercially. They can also be isolated from native sources such as chicken erythrocytes [17] or HeLa cells (also available from vendors). Histones can either be provided as histone monomers or as complexes, such as H3–H4 tetramers or complete histone octamers. Histones are often provided in buffers containing 2 M NaCl or KCl, which is required to maintain solubility and integrity of the highly charged histones in absence of DNA. Histone preparations can therefore be a significant source of salt in HMT reactions, potentially inhibiting activity of the enzyme to be tested [9]. It is important to keep in mind that, when using histone octamers, the low concentration of salt in HMT reactions promotes dissociation into H2A–H2B dimers and H3–H4 tetramers [18], which will therefore represent the major histone species in the reaction.

6. Histones from recombinant or native sources can be assembled into nucleosomal substrates using well-established salt dialysis-based protocols [15]. DNA templates for assembly are commonly plasmids containing repeats of the 601 nucleosome position sequence [19] interspaced with linkers of varying lengths. These plasmids can be used directly for assembly. If exact positioning is crucial, the stretch of repeats can be excised and separated from the plasmid backbone, for which uneven spacing is expected due to the absence of positioning sequences. For mononucleosome particles, 601 DNA can either be amplified by PCR or excised from plasmids. The degree of saturation of DNA templates with histones should be monitored and oversaturation avoided. When simply seeking to detect activity, undersaturation is usually acceptable as methyltransferases often do not require complete occupancy of DNA with histone octamers to display activity, although

specific modes of regulation might be missed [20]. Reconstituted chromatin is usually stored in TE buffer, which is ideally suited for use in HMT assays due to its low salt content. Of the options presented here for substrates in these assays, reconstituted chromatin most accurately reflects the native substrate of most methyltransferases that act on chromatin, however, others may not be active on such reconstituted templates at all.

7. The optimal reaction temperature depends on the methyltransferase and needs to be determined experimentally. For example, *Neurospora crassa* Dim-5 is optimally active at 10 °C, remains 50% active at 30 °C, but is essentially inactive at 37 °C [13]. 30 °C should yield activity for a wide range of enzymes; however, the growth temperature of plants may provide a useful starting point for plant-derived histone methyltransferases or other plant enzymes.
8. When using radioactive detection methods, it is advisable to collect samples at the bottom of tubes by brief centrifugation in a tabletop centrifuge (10 s at full speed) before opening the tubes after all mixing and incubation steps in order to minimize the risk of radioactive contamination. Moreover, be cautious to properly handle and dispose of all tubes, pipette tips, blotting paper, and other consumables that have been in contact with ^{3}H-SAM. Buffers from SDS-PAGE and blotting transfer likewise need to be handled with appropriate care and disposed of properly. Equipment such as gel tanks and blotting setups should be checked for contamination. All procedures need to be carried out in accordance with local rules for the use of ^{3}H in laboratories.
9. If the goal of the assay is primarily to detect activity, incubation times as long as overnight can be employed. Usually, 30 min to 4 h should be sufficient to observe activity. However, when aiming to determine kinetic parameters of the enzyme, shorter time points within the linear range of the reaction are required to accurately determine initial velocities.
10. Fifteen percentage gels are recommended as they provide good resolution of all four core histones while still allowing adequate transfer of higher molecular weight methyltransferases in subsequent blotting steps. Both 18% and 12.5% gels may be used as well.
11. It is imperative to use PVDF membranes as nitrocellulose membranes are incompatible with the subsequent staining and scintillation cocktail spraying steps. Remember to presoak PVDF in methanol before use. Methanol for this purpose can be stored and reused.
12. Importantly, letting the PVDF membrane dry completely increases the signal-to-background ratio of the Coomassie-

stained bands and should be done before taking a picture. Air-drying takes ~30 min, but the process can be sped up with a hair dryer or laboratory heat gun, as long as these can be run at low temperature or cool mode to not heat up the membrane. Carefully hold the membrane with forceps and use a low fan speed to keep the membrane from flying away and becoming damaged.

13. A hair dryer can be used to speed up the drying process.
14. Exposing the film at −80 °C increases the sensitivity of detection, but it can be performed at ambient temperature as well.
15. For peptides, make sure to use a 0.2 μm or similar pore size membrane along with shorter transfer times to prevent transfer of peptides through the membrane. Staining the membrane by Ponceau can be helpful to ensure successful transfer of the peptide. This step may need to be optimized depending on the peptides used.
16. The strength of the signal obtained can vary based on a few parameters (e.g., efficiency of the HMT reaction, primary antibody concentration used). For these reasons, the time of exposure should be extended if no signal is observed on the film after a 1–2 min exposure.

Acknowledgments

We want to thank Jean-François Couture (University of Ottawa) for his advice on some of the assays described and Chantal LeBlanc for her helpful comments on this chapter. Work in the Voigt lab is supported by the Wellcome Trust and the Royal Society (joint Grant Ref. 104175/Z/14/Z, Sir Henry Dale Fellowship to PV) and has received funding from the European Research Council (ERC) under the European Union's Horizon 2020 research and innovation programme (ERC-STG grant agreement No. 639253). The Wellcome Trust Centre for Cell Biology is supported by core funding from the Wellcome Trust (Grant Ref. 092076).

References

1. Rea S, Eisenhaber F, O'Carroll D, Strahl BD, Sun ZW, Schmid M, Opravil S, Mechtler K, Ponting CP, Allis CD, Jenuwein T (2000) Regulation of chromatin structure by site-specific histone H3 methyltransferases. Nature 406 (6796):593–599. doi:10.1038/35020506
2. Alvarez-Venegas R, Pien S, Sadder M, Witmer X, Grossniklaus U, Avramova Z (2003) ATX-1, an Arabidopsis homolog of trithorax, activates flower homeotic genes. Curr Biol 13 (8):627–637
3. Jackson JP, Lindroth AM, Cao X, Jacobsen SE (2002) Control of CpNpG DNA methylation by the KRYPTONITE histone H3 methyltransferase. Nature 416(6880):556–560. doi:10.1038/nature731
4. Jacob Y, Bergamin E, Donoghue MT, Mongeon V, LeBlanc C, Voigt P, Underwood CJ,

Brunzelle JS, Michaels SD, Reinberg D, Couture JF, Martienssen RA (2014) Selective methylation of histone H3 variant H3.1 regulates heterochromatin replication. Science 343 (6176):1249–1253. doi:10.1126/science.1248357

5. Jacob Y, Feng S, LeBlanc CA, Bernatavichute YV, Stroud H, Cokus S, Johnson LM, Pellegrini M, Jacobsen SE, Michaels SD (2009) ATXR5 and ATXR6 are H3K27 monomethyltransferases required for chromatin structure and gene silencing. Nat Struct Mol Biol 16 (7):763–768. doi:10.1038/nsmb.1611
6. Chen Z, Grzybowski AT, Ruthenburg AJ (2014) Traceless semisynthesis of a set of histone 3 species bearing specific lysine methylation marks. Chembiochem 15 (14):2071–2075. doi:10.1002/cbic.201402313
7. Dawson PE, Muir TW, Clark-Lewis I, Kent SB (1994) Synthesis of proteins by native chemical ligation. Science 266(5186):776–779
8. Couture JF, Dirk LM, Brunzelle JS, Houtz RL, Trievel RC (2008) Structural origins for the product specificity of SET domain protein methyltransferases. Proc Natl Acad Sci U S A 105(52):20659–20664. doi:10.1073/pnas.0806712105
9. Stutzer A, Liokatis S, Kiesel A, Schwarzer D, Sprangers R, Soding J, Selenko P, Fischle W (2016) Modulations of DNA contacts by linker histones and post-translational modifications determine the mobility and modifiability of nucleosomal H3 tails. Mol Cell 61 (2):247–259. doi:10.1016/j.molcel.2015.12.015
10. Kuzmichev A, Nishioka K, Erdjument-Bromage H, Tempst P, Reinberg D (2002) Histone methyltransferase activity associated with a human multiprotein complex containing the Enhancer of Zeste protein. Genes Dev 16 (22):2893–2905. doi:10.1101/gad.1035902
11. Cao R, Wang L, Wang H, Xia L, Erdjument-Bromage H, Tempst P, Jones RS, Zhang Y (2002) Role of histone H3 lysine 27 methylation in Polycomb-group silencing. Science 298 (5595):1039–1043. doi:10.1126/science.1076997
12. Trievel RC, Beach BM, Dirk LM, Houtz RL, Hurley JH (2002) Structure and catalytic mechanism of a SET domain protein methyltransferase. Cell 111 (1):91–103
13. Zhang X, Tamaru H, Khan SI, Horton JR, Keefe LJ, Selker EU, Cheng X (2002) Structure of the Neurospora SET domain protein DIM-5, a histone H3 lysine methyltransferase. Cell 111 (1):117–127
14. Dyer PN, Edayathumangalam RS, White CL, Bao Y, Chakravarthy S, Muthurajan UM, Luger K (2004) Reconstitution of nucleosome core particles from recombinant histones and DNA. Methods Enzymol 375:23–44
15. Luger K, Rechsteiner TJ, Flaus AJ, Waye MM, Richmond TJ (1997) Characterization of nucleosome core particles containing histone proteins made in bacteria. J Mol Biol 272 (3):301–311. doi:10.1006/jmbi.1997.1235
16. Voigt P, Reinberg D (2011) Histone tails: ideal motifs for probing epigenetics through chemical biology approaches. Chembiochem 12 (2):236–252. doi:10.1002/cbic.201000493
17. Simon RH, Felsenfeld G (1979) A new procedure for purifying histone pairs H2A + H2B and H3 + H4 from chromatin using hydroxylapatite. Nucleic Acids Res 6(2):689–696
18. Eickbush TH, Moudrianakis EN (1978) The histone core complex: an octamer assembled by two sets of protein-protein interactions. Biochemistry 17(23):4955–4964
19. Thastrom A, Lowary PT, Widlund HR, Cao H, Kubista M, Widom J (1999) Sequence motifs and free energies of selected natural and non-natural nucleosome positioning DNA sequences. J Mol Biol 288(2):213–229. doi:10.1006/jmbi.1999.2686
20. Yuan W, Wu T, Fu H, Dai C, Wu H, Liu N, Li X, Xu M, Zhang Z, Niu T, Han Z, Chai J, Zhou XJ, Gao S, Zhu B (2012) Dense chromatin activates polycomb repressive complex 2 to regulate H3 lysine 27 methylation. Science 337 (6097):971–975. doi:10.1126/science.1225237

Chapter 21

Identification of Parent-of-Origin-Dependent QTLs Using Bulk-Segregant Sequencing (Bulk-Seq)

Nuno D. Pires and Ueli Grossniklaus

Abstract

Parent-of-origin effects play important roles in plant reproduction and are often mediated by epigenetic modifications at the histone or DNA level. However, the genetic basis underlying these modifications can be challenging to identify. Here, we describe an approach (Bulk-Seq) that can be used to map loci mediating parent-of-origin-dependent effects using whole-genome sequencing of pools of DNA.

Key words Bulk-segregant sequencing, Parent-of-origin effects, Seed development, Quantitative trait mapping, Segregation distortion, *Arabidopsis thaliana*

1 Introduction

Seed development is characterized by a complex set of interactions between paternal and maternal genomes [1–4]. Many of these interactions are mediated by parent-specific chromatin or DNA modifications, which are in turn shaped by genetic variation between genotypes. Here, we describe a bulk segregant sequencing (Bulk-Seq) protocol that can be used to genetically map parent-of-origin dependent effects using whole-genome sequencing. It was initially developed to identify paternal modifiers of the Arabidopsis *medea* (*mea*) mutant [5], which displays maternal seed abortion due to a misregulation of histone H3K27 trimethylation [6, 7].

The Bulk-Seq protocol allows for robust mapping of parent-of-origin quantitative trait loci (QTLs) without the need for extensive phenotyping and genotyping, requiring only a handful of sequencing libraries. The pooling strategy is particularly well suited to detect weak signals underlying complex polygenic traits (*see* **Note 1**). One advantage of Bulk-Seq over more classical bulk segregant mapping approaches (e.g., [8, 9]) is that it only requires one pool of plants to be selected (instead of generating two pools with extreme values). This makes the Bulk-Seq approach particularly well suited to the

Marian Bemer and Célia Baroux (eds.), *Plant Chromatin Dynamics: Methods and Protocols*, Methods in Molecular Biology, vol. 1675, DOI 10.1007/978-1-4939-7318-7_21, © Springer Science+Business Media LLC 2018

mapping of traits that affect seed fertilization, seed inviability or germination.

The Bulk-Seq strategy is described in Fig. 1. It involves creating a seed population (BC1) that inherits one set of chromosomes (W) from one parent and two segregating sets (W and B) from the other parent. These two genotypes (W and B) should have different effects in prefertilization or postfertilization fitness, so that they can be selected in BC1 individuals. Example scenarios could be alleles from genotype B leading to an increase in pollen fitness, to a maternal decrease of seed dormancy, to a paternal rescue of a maternal mutation, or to a paternal activation of a maternal reporter gene. DNA is then extracted from pools of selected BC1 seedlings (e.g., viable seeds, or fast germinating seeds), and whole-genome sequencing is used to identify genomic regions that exhibit biased transmission of W and B polymorphisms. Random segregation should result in an average B/W polymorphism ratio of 1:3 thoughout the genome, while chromosomal regions that contain a QTL for the trait of interest are expected to have a higher B/W ratio (up to a maximum of 1:1). A control BC1 population, containing individuals that were not selected (i.e., all the BC1 individuals), should be sequenced in parallel.

This protocol assumes a basic level of experience with command line scripting, including the installation of Unix programs and R packages (*see* **Note 2**). The code used to perform these analyses is available on GitHub at https://github.com/piresn/BulkSeq.

2 Materials

2.1 Plant Lines

This protocol requires two genotypes (W and B) that have different parent-of-origin specific effects on seed development. As an example (Fig. 1), the W and B genotypes are differentially transmitted to the progeny when crossed as males to a W^m female (*see* **Note 3**). Another example would be the maternal inheritance of B alleles resulting in the formation of fast-germinating seeds (while inheritance of W alleles produce highly dormant seeds). Any set of genotypes that differentially modifies the penetrance of a mutation or marker *m* expressed from the other parent can in principle be used (*see* **Note 4**).

2.2 DNA Extraction and Library Preparation

1. Liquid nitrogen.
2. DNA extraction kit (e.g., DNeasy Plant Mini Kit, Qiagen).
3. DNA quantification device (ideally, fluorometric-based method, e.g., Qubit Fluorometer, Life Technologies; but spectrophotometric alternatives are available, e.g., Nanodrop, Nanodrop Instruments).

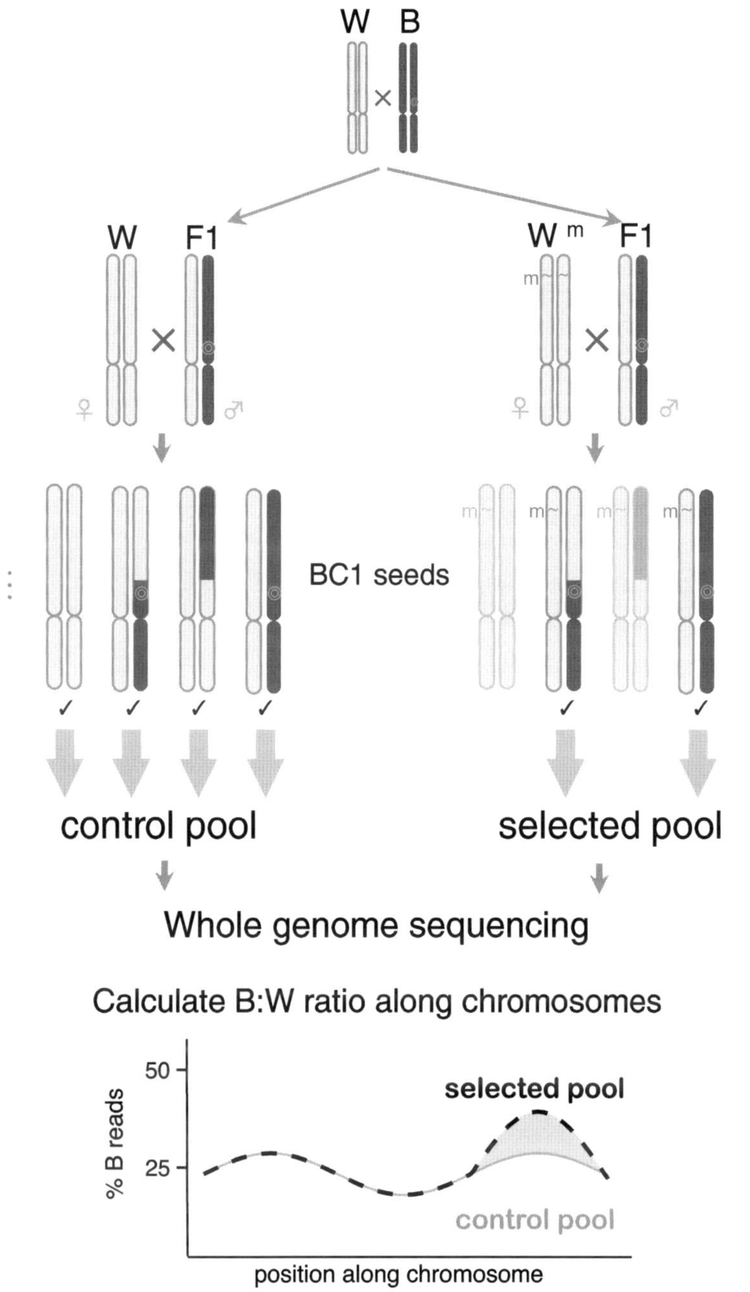

Fig. 1 Bulk-Seq strategy: in this scenario, *m* is a mutant in the W ("*white*") genetic background that causes seed abortion; the expression of *m* can be modulated by one or more alleles (*concentric circles*) present in the B ("*black*") genetic background. *m* could also be any trait (such as expression of a reporter gene) that is modified by B alleles from the other parent). The Bulk-Seq strategy can be used to genetically map the B modifier alleles. First, the W and B accessions are crossed to generate an F1 hybrid. This hybrid is then crossed to the mutant W^m: the progeny of this cross (BC1) will inherit a W^m chromosome from one parent, and different combinations of W and B alleles from the other parent (due to recombination during meiosis in F1 individuals). BC1 plants that inherit the modifier B alleles can be easily selected because they rescue the seeds produced by the W^m phenotype: DNA from pools of selected BC1 plants is extracted and sequenced; an increased ratio of B to W alleles along chromosomes can then be used to map the physical position of the B modifier alleles. The control pool population contains individuals that were not selected (in this case due to the absence of the genetic mark *m*). As an alternative scenario, one can also use other triggers (e.g., chemical or physical inductions) to differentiation the "selected" and "control" populations

4. DNA quantification kit if using fluorometric quantitation (e.g., Qubit dsDNA HS Assay kit, Life Technologies).
5. 3 M NaOAc, pH 5.2.
6. Isopropanol.
7. 70% ethanol.
8. TE buffer: 10 mM Tris–HCl (pH 8.0), 0.1 mM EDTA.
9. Library preparation kit (e.g., TruSeq DNA Sample Prep Kit v2, Illumina).

2.3 Read Mapping and SNP Calling

The following software programs and libraries are required:

1. FastQC 0.11.3.
2. cutadapt 1.8.3.
3. Samtools 1.2 (using htslib 1.2.1).
4. Bowtie 2 2.2.9.
5. R 3.3.1.
6. ggplot2 2.1.0.
7. zoo 1.7–13.
8. Python 3.4.0.
9. Bio 1.65.

3 Methods

3.1 Generation of Mapping Populations

1. Generate F1 hybrid lines by reciprocally crossing the accessions W and B (*see* **Note 5**).
2. Cross the F1 hybrids (as a male or female, depending on the trait analyzed) with a W^{m} parent and with a W parent (control). This will generate a "BC1 selected" population and a "BC1 control" population (*see* **Note 6**). Repeat these crosses to generate replicate BC1 populations.
3. Harvest and air-dry the BC1 seeds at room temperature for a minimum of 3 weeks.

3.2 Plant Selection and DNA Extraction

1. Select viable BC1 individuals, e.g., 2 weeks after germination (*see* **Note** 7) from the BC1 selected pool, and random individuals from the BC1 control pool.
2. Harvest one small leaf per plant and collect in a 1.5 ml tube. Pool 50 leaves (from 50 individuals) per tube and freeze in liquid nitrogen (*see* **Note 8**). The material can be stored at −80 °C for several weeks before DNA extraction.
3. Grind the leaf material to a fine powder, extract the DNA using the DNeasy Plant Mini Kit following the manufacturer's protocol, and elute with 150 μl water.

4. Quantify DNA using a fluorometric quantitation assay (e.g., Qubit dsDNA HS Assay kit) (*see* **Note 9**). One to two micrograms total DNA per extraction should be obtained at this stage.
5. If multiple extractions were performed in **step 3**, pool the DNA from the separate extractions in equal amounts to obtain a total of 3 μg DNA per pool.
6. Add 1/10 volumes of NaOAc 3 M and 1 volume of isopropanol (e.g., 50 μl sodium acetate and 500 μl of isopropanol for a 500 μl DNA sample). Mix well, and incubate for 15 min at room temperature. Centrifuge for 30 min at 10,000 × *g* at 12 °C and discard the supernatant.
7. Wash with 1 ml 70% ethanol, centrifuge for 1 min at 10,000 × *g*, discard the supernatant and let the tubes dry.
8. Resuspend the pellet in 25 μl TE buffer.
9. Quantify the DNA. The final concentration should now be 100–150 ng/μl.
10. Samples can be stored for several days at 4 °C, several weeks at −20 °C, or for several months at −80 °C.

3.3 Library Preparation and Sequencing

There are currently many sequencing options available, from cheap home-made solutions to specialized commercial kits. A simple option is to prepare libraries using the TruSeq DNA Sample Prep Kit v2 (Illumina) and sequence the DNA samples with single reads on an Illumina HiSeq 2500 platform (*see* **Note 10**). Read coverage should take into account the minor allele frequency expected: to be useful, SNPs should have at least one read retrieved from the minor allele in the control or selected pool samples. Therefore, if the expected minor allele frequency is 25%, one would like most SNPs to be covered by at least 4 reads; assuming that read coverage approximately follows a Poisson distribution, a median coverage of 11× would result in 99% of the SNPs being covered by four or more reads. In practice, however, lower coverages can also be used, since reads from neighbouring SNPs will be combined.

3.4 Read Mapping and SNP Calling

The code to perform these steps is available in the file mapping.sh at https://github.com/piresn/BulkSeq (*see* **Note 11**).

1. Check the quality of the reads using FastQC (http://www.bioinformatics.babraham.ac.uk/projects/fastqc).
2. Trim and discard low quality reads using cutadapt [10] (*see* **Note 12**).
3. Check read quality using FastQC and adjust the parameters of the last step if necessary (e.g., increase the -q quality trimming parameter).

4. Retrieve a reference genome sequence (e.g., the TAIR10 genome release from www.arabidopsis.org) (*see* **Note 13**) and index it using Bowtie2 [11].
5. Align reads to the indexed genome using Bowtie2 and the SAMtools package [12].
6. Index and sort the BAM alignment.
7. Call SNPs with mpileup and BCFtools (SAMtools) and save as a VCF file (*see* **Note 14**).

3.5 Create SNP Matrix File

The code to perform these steps is available in the file snpfile.R at https://github.com/piresn/BulkSeq (*see* **Note 15**).

1. Retrieve known SNPs from public datasets such as 1001genomes.org [13].
2. Combine the separate datasets, remove common SNPs between the two genotypes and create a SNP matrix file (snpm.txt) with the following format:

```
pos genot ref alt
chr1_956 Ler-1 C T
chr1_10904 Ler-1 A T
chr1_37388 Ler-1 G T
...
chrM_191621 Cvi_0 C T
chrM_195625 Cvi_0 A -
chrM_355145 Cvi_0 T -
chrM_355155 Cvi_0 A -
```

The first column contains the physical position of the SNP, the second identifies the genotype having an alternative allele (here Ler-1 or Cvi-0), the third and fourth columns contain the reference (here Col-0) and alternative base, respectively.

3.6 Calculate Allele Frequencies Along Chromosomes

An R script (cleanCounts.R) to perform the following steps is available at https://github.com/piresn/BulkSeq.

1. Combine the calculated allele frequencies (VCF file) with the SNP matrix (snpm.txt). Filter out SNPs where the reference sequence is ambiguous (*see* **Note 16**), and the ones where the predicted SNPs do not match the polymorphism identified in the sequenced samples.
2. SNPs with very high counts (often artifacts that arise during library preparation) can be removed by comparing the observed SNP distribution with a theoretical Poisson distribution. As an example, if the median coverage per SNP is 9 reads, then only 1% of the SNPs should have a coverage of more than 17 reads, and only 0.1% of the SNPs should have a coverage of more than 20 reads (assuming sequencing coverage is

approximately Poisson). Instead, a small subset of SNPs will often contain hundreds of reads. After comparing theoretical and actual distributions, we can set a threshold above which overrepresented SNPs are removed (e.g., all SNPs with more than 20 reads, the theoretical 0.1% percentile).

3. Calculate the median coverage frequency of each allele and confirm that it matches theoretical expectations (usually a 3:1 ratio).
4. The output from these steps is a table with allele frequencies in the following format:

```
CHR,POS,COV,TAIR10_BASE,LER_BASE,CVI_BASE,LER_COUNT,CVI_COUNT
chr1,502,16,T,NA,C,12,4
chr1,508,16,T,NA,C,12,4
chr1,657,7,C,NA,T,5,2
...
chr5,26974816,15,C,NA,G,0,15
chr5,26975144,12,T,NA,C,8,4
chr5,26975288,12,T,NA,C,0,12
```

The first two columns identify the physical position of the SNP, the COV field represents the total coverage (Ler-1 + Cvi-0 reads), TAIR10 is the reference allele (Col-0 accession), LER_BASE and CVI_BASE are the respective polymorphisms, LER_COUNT and CVI_COUNT represent the number of reads overlapping the respective allele).

3.7 Calculate Allele Frequencies Along Chromosomes

An R script (pool.R) to perform the following steps is available at: https://github.com/piresn/BulkSeq.

1. Combine two tables with Bulk-seq frequencies along common SNP positions (the control pool and selected pool datasets).
2. If an allele is not covered by any reads in either control or selected pool samples, this could be due to an incorrect SNP annotation. The safest solution is to remove these SNPs from the analysis. However, this is not recommended if the median coverage is low (*see* **Note 17**).
3. Allele counts from neighboring SNPs should be pooled together. This can be done along a sliding window with a width of a 5–100 SNP positions.
4. Calculate a statistic that reflects the divergence between the two datasets. A chi-square test, Fisher's exact test or a G statistic [14] will be appropriate in most cases. However, for highly polygenic traits where systematic differences between control and selected datasets are present along multiple positions, it may be more informative to simply calculate the enrichment of a set of alleles in one dataset relative to the

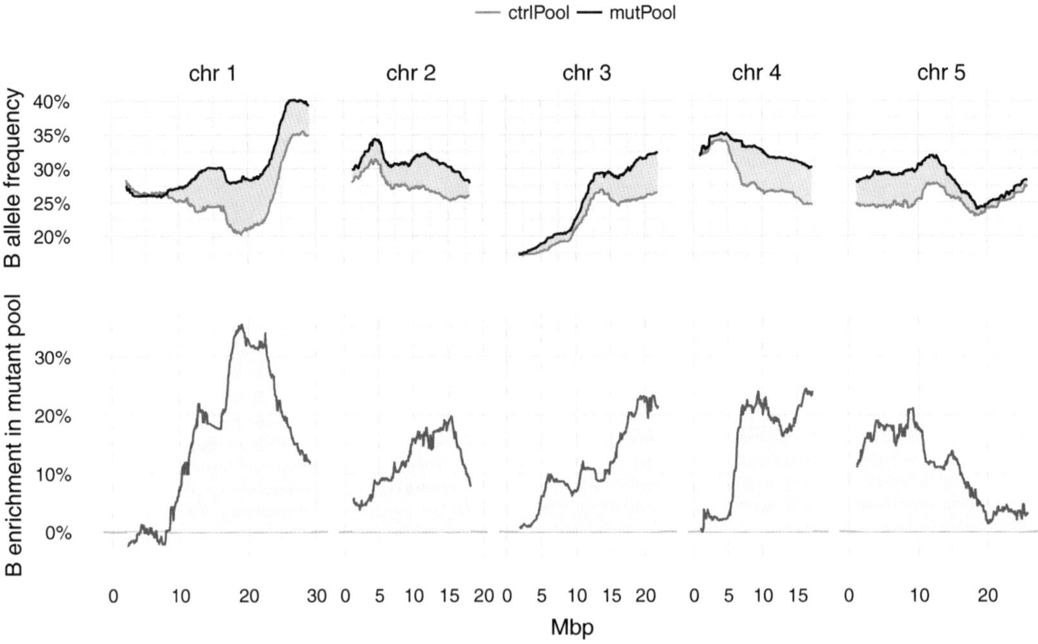

Fig. 2 Example of plots showing the relative frequency of B alleles (here Cvi-0) along chromosomes in the pools of mutant (selected) and control pools. The second plot shows the relative enrichment of B alleles in the mutant pool over the control pool

other (Fig. 2); this is the option demonstrated in the example code (*see* **Note 18**).

5. Individual allele ratios (and corresponding statistics) will probably be very noisy (i.e., have a high variance between neighboring SNPs). A median filter or weighted average can be used along sliding windows with a width of 5–100 SNPs to improve the signal-to-noise ratio (*see* **Note 19**).
6. Plot the results and identify the regions with an enrichment of B alleles (here Cvi-0) in the selected pool relative to the control pool.

4 Notes

1. When fine-mapping strong discrete traits, an alternative option to Bulk-Seq is to harvest dozens/hundreds of individual genotypes, and sequence individual sequences separately (multiplex) to identify the exact recombination points. Given the decreasing costs of whole-genome sequencing, variations or a combination of these two workflows may offer a convenient and inexpensive way to genetically characterize complex genetic and epigenetic traits.

2. We strongly recommend to read the documentation of each of the sequence mapping and analysis tools used; individual parameters should be adjusted according to the characteristics of each dataset (e.g., sequencing coverage and SNP density).
3. This is analogous to the transmission of Ler-1 and Cvi-0 as males when crossed with a Ler^{mea} female [5].
4. Hundreds of *A. thaliana* accessions have been extensively genotyped or sequenced. For exotic lines for which good SNP information is not available, one should sequence the parental W and B lines and manually perform SNP calling.
5. Maternal effects (including epigenetic effects and cytoplasmic inheritance) may significantly affect the development of subsequent generations. Therefore, it is advisable that half of the F1 hybrids individuals are obtained as WxB, and the other half as BxW.
6. The BC1 control population serves to control for differences in the transmission of paternal or maternal alleles that are inherent to these genotypes. These are often very strong and can be due to different factors such as pollen competition, reduced gametophytic viability or genetic incompatibilities. One can also use the Bulk-Seq specifically to identify such inherent differential transmission in the control crosses.
7. It is important that the germination and developmental stages of all BC1 plants is uniform.
8. While 50 is the recommend minimum number of individuals to sample, in practice it is recommended to sample more than this to minimize sampling effects. Ideally, this would be made as separate DNA extractions that can then be quantified and pooled in **steps 4** and **5**.
9. The Qubit fluorometer is a robust way to measure DNA concentrations. If it is not available, a spectrophotometer (e.g., NanoDrop) could be used instead, but care must be made to ensure that the samples are devoid of any contaminant RNA.
10. Genomic DNA will need to be fragmented mechanically or enzymatically at the start of library preparation. Refer to the manufacturer's protocols and recommendations (e.g., the TruSeq library preparation protocol recommends ultrasonication).
11. These examples use datasets available in the ArrayExpress database (www.ebi.ac.uk/arrayexpress) under accession number E-MTAB-5196 [5].
12. Alternatives to cutadapt include FASTX-Toolkit (http://hannonlab.cshl.edu/fastx_toolkit/) and trimmomatic [15].
13. To avoid mapping biases, one option is to use a reference genome sequence that N-masks the position of known

polymorphisms between the two genotypes. This can be done using the Python script GenomeSNPmask.py available at https://github.com/piresn/BulkSeq.

14. Instead of calling SNPs from the BAM alignment, another possibility is to use a program such as SNPsplit (http://www.bioinformatics.babraham.ac.uk/projects/SNPsplit/) to sort out the reads according to known SNP positions. The number of reads in each BAM file can then be calculated using BEDTools [16] and the ratio between control and selected pools quantified using EdgeR [17] or alternative suites.
15. This step must be customized to accommodate the format of the SNP datasets used. This example uses SNPs from Ler-1 and Cvi-0 accessions of *Arabidopsis* retrieved from the 1001 Genomes Consortium and SALK, respectively [13, 18].
16. Skip this step if the reference genome sequenced was N-masked.
17. If some values are left as zero, they may cause problems in downstream calculations; one should then introduce pseudocounts in all the fields (i.e., one read per allele).
18. If replicates are available (which is highly advisable), it is preferable to initially analyze each pair separately. If the results between replicates are indeed fond to be consistent, then different dataset can be merged based upon the sum of reads from common SNPs. Alternatively, a negative binomial model (as in EdgeR) can also be used to account for biological variation between replicates.
19. The width of the windows in this step and in the SNP pooling step should be adjusted in parallel: different combinations of window sizes should be tested to identify the ones that provide the best signal-to-noise ratio.

References

1. Raissig MT, Baroux C, Grossniklaus U (2011) Regulation and flexibility of genomic imprinting during seed development. Plant Cell 23:16–26. doi:10.1105/tpc.110.081018
2. Pires N (2014) Seed evolution: parental conflicts in a multi-generational household. Biomol Concepts 5:71–86
3. García-Aguilar M, Gillmor CS (2015) Zygotic genome activation and imprinting: Parent-of-origin gene regulation in plant embryogenesis. Curr Opin Plant Biol 27:29–35. doi:10.1016/j.pbi.2015.05.020
4. Piskurewicz U, Iwasaki M, Susaki D et al (2016) Dormancy-specific imprinting underlies maternal inheritance of seed dormancy in *Arabidopsis thaliana*. Elife 5:e19573
5. Pires ND, Bemer M, Müller LM et al (2016) Quantitative genetics identifies cryptic genetic variation involved in the paternal regulation of seed development. PLoS Genet 12:e1005806. doi:10.1371/journal.pgen.1005806
6. Grossniklaus U, Vielle-Calzada J-P, Hoeppner MA, Gagliano WB (1998) Maternal control of embryogenesis by *MEDEA*, a *Polycomb* group gene in *Arabidopsis*. Science 280:446–450. doi:10.1126/science.280.5362.446
7. Makarevich G, Leroy O, Akinci U et al (2006) Different polycomb group complexes regulate common target genes in Arabidopsis. EMBO

Rep 7:947–952. doi:10.1038/sj.embor.7400760

8. Takagi H, Abe A, Yoshida K et al (2013) QTL-seq: rapid mapping of quantitative trait loci in rice by whole genome resequencing of DNA from two bulked populations. Plant J 74:174–183. doi:10.1111/tpj.12105
9. Yang Z, Huang D, Tang W et al (2013) Mapping of quantitative trait loci underlying cold tolerance in rice seedlings via high-throughput sequencing of pooled extremes. PLoS One 8:e68433. doi:10.1371/journal.pone.0068433
10. Martin M (2011) Cutadapt removes adapter sequences from high-throughput sequencing reads. EMBnetjournal 17:1
11. Langmead B, Salzberg SL (2012) Fast gapped-read alignment with Bowtie 2. Nat Methods 9:357–359
12. Li H, Handsaker B, Wysoker A et al (2009) The sequence alignment/map format and SAMtools. Bioinformatics 25:2078–2079
13. Alonso-Blanco C, Andrade J, Becker C et al (2016) 1,135 genomes reveal the global pattern of polymorphism in *Arabidopsis thaliana*. Cell 166:481–491. doi:10.1016/j.cell.2016.05.063
14. Magwene PM, Willis JH, Kelly JK (2011) The statistics of bulk segregant analysis using next generation sequencing. PLoS Comput Biol 7:e1002255. doi:10.1371/journal.pcbi.1002255
15. Bolger AM, Lohse M, Usadel B (2014) Trimmomatic: a flexible trimmer for Illumina sequence data. Bioinformatics 30(15):2114–2120
16. Quinlan AR, Hall IM (2010) BEDTools: a flexible suite of utilities for comparing genomic features. Bioinformatics 26:841–842. doi:10.1093/bioinformatics/btq033
17. Robinson MD, McCarthy DJ, Smyth GK (2010) edgeR: a Bioconductor package for differential expression analysis of digital gene expression data. Bioinformatics 26:139–140. doi:10.1093/bioinformatics/btp616
18. Schneeberger K, Ossowski S, Ott F et al (2011) Reference-guided assembly of four diverse *Arabidopsis thaliana* genomes. Proc Natl Acad Sci U S A 108:10249–10254. doi:10.1073/pnas.1107739108

Chapter 22

QTL^{epi} Mapping in *Arabidopsis thaliana*

Kathrin Lauss and Joost J.B. Keurentjes

Abstract

While DNA sequence variation is known to be a major driver of phenotypic divergence, epigenetic variation has long been disregarded. One reason for that was the lack of suitable tools. The creation of epigenetically divergent but otherwise largely isogenic Arabidopsis populations has now alleviated some of these constraints. Epigenetic recombinant inbred line (epiRIL) populations allow for examining the effects of epigenetic variation on phenotypes. In addition, epiRILs enabled the development of epigenetic quantitative trait locus (QTL^{epi}) mapping, an approach to identify causal epigenetic factors. Here, we describe the successive steps of QTL^{epi} mapping in a broad sense, from the creation of epigenetically divergent populations to the identification of causal genes underlying particular phenotypes in Arabidopsis.

Key words epiRILs, QTL^{epi} mapping, Phenotyping, Epigenotyping

1 Introduction

Intraspecific natural variation (from here on termed natural variation) is defined as the wide phenotypic and adaptive diversity occurring in nature within a single plant species and caused by genetic differences. Natural variation has been effectively exploited to associate genotypic divergence with phenotypic trait variation and has been recognized as a valuable source for agricultural crop improvement [1].

One way to identify causal loci underlying quantitative trait variation is by using segregating populations for genetic mapping to detect quantitative trait loci (QTLs) [1, 2]. These approaches make use of sequence polymorphisms between lines that are used as genetic markers to label the parental origin of genomic fragments in progeny of crosses between those lines. Screening a large population of lines with a varying composition of "marked" parental genomic fragments basically allows monitoring co-occurrence, or cosegregation, of particular fragments with a specific phenotypic trait value. Widely used populations for QTL mapping are recombinant inbred lines (RILs). RILs are commonly generated by

Marian Bemer and Célia Baroux (eds.), *Plant Chromatin Dynamics: Methods and Protocols*, Methods in Molecular Biology, vol. 1675, DOI 10.1007/978-1-4939-7318-7_22, © Springer Science+Business Media LLC 2018

crossing genotypically (and phenotypically) divergent inbred lines and propagation by single seed descent until homozygosity to obtain a set of immortal lines harboring different genomic introgressions from their parents. However, classical RIL populations do not take into account epigenetic variation.

Variation in epigenetic marks includes DNA methylation (at cytosine residues) and histone modifications, which are known to regulate gene expression in a wide range of eukaryotic organisms, including plants [3–5]. It has been suggested to distinguish between transgenerational epigenetic memory of gene expression states, which is inherited across generations, and transient epigenetic changes induced by developmental or environmental stimuli [6, 7]. Examples for transgenerational epigenetic memory in plants are epialleles: allele variants that consist of an identical DNA sequence but which are differentially transcribed due to differences in their epigenetic modifications [8–10]. Epialleles being causative for particular phenotypes have been described in various plant species [11–15]. For example, the ***Flowering Wageningen*** (*FWA*) locus in Arabidopsis occurs in two epiallelic forms: a repressed state associated with extensive DNA methylation and a demethylated, highly transcribed state that causes a late flowering phenotype [15]. Hence, natural variation is, in addition to genetic factors, also regulated by epigenetic factors. That said, non-Mendelian segregation patterns of specific epigenetically regulated phenotypes can also be observed [16]. For instance, epialleles can undergo so called paramutation or transchromosomal methylation and demethylation events during hybridization [16, 17]. Thereby, one epiallele acquires the epigenetic profile (often DNA hypermethylation) of the other epiallele. Epialleles and paramutation phenomena thus exemplify the involvement of epigenetic variation in shaping plant phenotypes.

Currently, resources for studying and quantifying epigenetic variation and its association with phenotypic variation are rapidly emerging. Epigenetic recombinant inbred lines (epiRILs), which have similar DNA sequence but are polymorphic in terms of their epigenetic profile, are among the most informative of those tools. Analogous to RILs being used to identify causal genomic loci explaining trait variation, epiRILs can be used to detect causal loci in the epigenome (QTL^{epi} mapping).

Here, we explain the use of epiRILs for the identification of epigenetic patterns contributing to quantitative trait variation. First, the generation of epiRIL lines and the available epiRIL populations are discussed. We then explain how the phenotyping of these lines can be performed, and give details on the descriptive statistics of the phenotypic variation. The protocol continues with the generation of epigenetic linkage maps, and the actual QTL^{epi} mapping, for which the linkage map is coupled to the phenotypic data.

Finally, we present approaches that can be used to confirm the QTL^{epi}, and to fine-map the causal region to eventually identify the DMR that causes the phenotypic differences.

2 Materials

2.1 Epigenetic Recombinant Inbred Lines (epiRILs)

In order to study epigenetic variation and its influence on traits independent of genetic variation, epigenetic recombinant inbred lines (epiRILs) are a powerful experimental system. An epiRIL population is conceptually the same as a recombinant inbred line (RIL) population. The main difference is that RILs acquired a mosaic of different DNA sequence introgressions from their parents while epiRILs acquired a mosaic of epigenetic patterns from their otherwise isogenic parents. A RIL population is generated by creating a hybrid from two genetically (and phenotypically) distinct inbred lines of interest followed by single seed descent from the F_2 generation onward until at least the F_8 generation. This process results in an accumulation of several recombination events across the chromosomes and through the many generations of inbreeding those lines reach (near) full homozygosity. The creation of epiRILs is very similar, except that the inbred lines used as parents are isogenic apart from one parent carrying a mutation with a strong effect on epigenetic profiles. Several rounds of backcrossing and selfing until the F_8 result in a population of genetically nearly identical lines, which are segregating for epigenetic variation. Nonetheless, remaining sources of genetic variation in epiRILs could be transposable elements that lost their repressive epigenetic marks and became active again [18]. Therefore, in epiRILs it is useful to distinguish (phenotypic or molecular) effects induced by transposon insertions from those induced by epigenetic variation.

2.1.1 Confirmed Strategies to Create epiRIL Populations

To date, two epiRIL populations have been generated in *Arabidopsis thaliana* [19, 20]. The major aim of creating these populations was to generate lines displaying variation in heritable DNA methylation profiles, which can be used to study the effects of epigenetic variation. The two epiRIL populations were created in the genetic background of the reference accession Columbia (Col-0), using lines carrying mutations in *METHYLTRANSFERASE 1* (*MET1*) or *DECREASE IN DNA METHYLATION 1* (*DDM1*). Loss of MET1 leads to widespread loss of DNA methylation and affects the redistribution of repressive histone marks [21, 22], while loss of DDM1, a chromatin remodeler that is required for maintaining DNA-methylation, results in approximately 70% reduction in DNA methylation, particularly at transposable elements and repeats [23].

Steps:

1. Creation of F_1 from a wild-type and a DNA methylation mutant parent. To generate the *met1*-derived epiRIL population, homozygous *met1–3* mutant lines were crossed to Col-0 wild-type plants, creating F_1 plants with "epi"-heterozygous hypomethylated regions. Similarly, for the *ddm1*-derived epiRILs, homozygous *ddm1–2* mutant lines were crossed to Col-0 also resulting in F_1 plants with "epi" heterozygous hypomethylated regions.
2. Curing of the mutant allele. For both populations, individuals homozygous for wild type DDM1 or MET1 in the segregating F_2 were selected for inbreeding until the F_8 (Fig. 1). Curing of the mutant allele is essential to avoid the induction and accumulation of novel DNA methylation polymorphisms (alongside developmental defects) during the rounds of selfing [22–24]. Importantly, despite reestablishing the functional MET1 and DDM1 proteins, the DNA methylation polymorphisms that recombined in the F_1 generation remain largely unaffected [19, 20].
3. Successive rounds of selfing (for the *ddm1*-derived epiRILs one round of backcrossing and then selfing) up to the F_8. The F_2 plants were subsequently selfed, or backcrossed and selfed until generation eight (F_8) to create the *met1*-derived and *ddm1*-derived epiRILs, respectively. This one round of backcrossing of the F_1 to wild type Col-0 (Fig. 1) during the generation of the *ddm1*-derived epiRILs reduced the amount of methylome divergence but at the same time allowed the creation of a larger population of lines. The *met1*-derived epiRILs, lacking this round of backcrossing, displayed unstable phenotypes and many could not be propagated by selfing due to accumulation of detrimental phenotypic effects or reduced fertility [19]. Stable parental epialleles (wild-type or mutant-derived) are inherited in a Mendelian fashion in the course of selfing, just as during normal RIL generation. This means that the frequency of heterozygosity (here "epiheterozygosity") is declining by 50% in each subsequent epiRIL generation, resulting in <1% probability of epiheterozygosity at any specific locus in the F_8 (Fig. 1). A population of epiRILs thus consists of individual lines each containing a mosaic of homozygous wild type and demethylated regions (DMRs) derived from the respective initial crossing parents (Fig. 1).

Interestingly, methylation polymorphisms for a few loci were still detected even in the F_8 or F_9 generation of *met1*-derived epiRILs in addition to de novo establishment of nonparental epialleles. This probably reflects cases of transchromosomal (de)methylation [17] and/or progressive de novo methylation events [18].

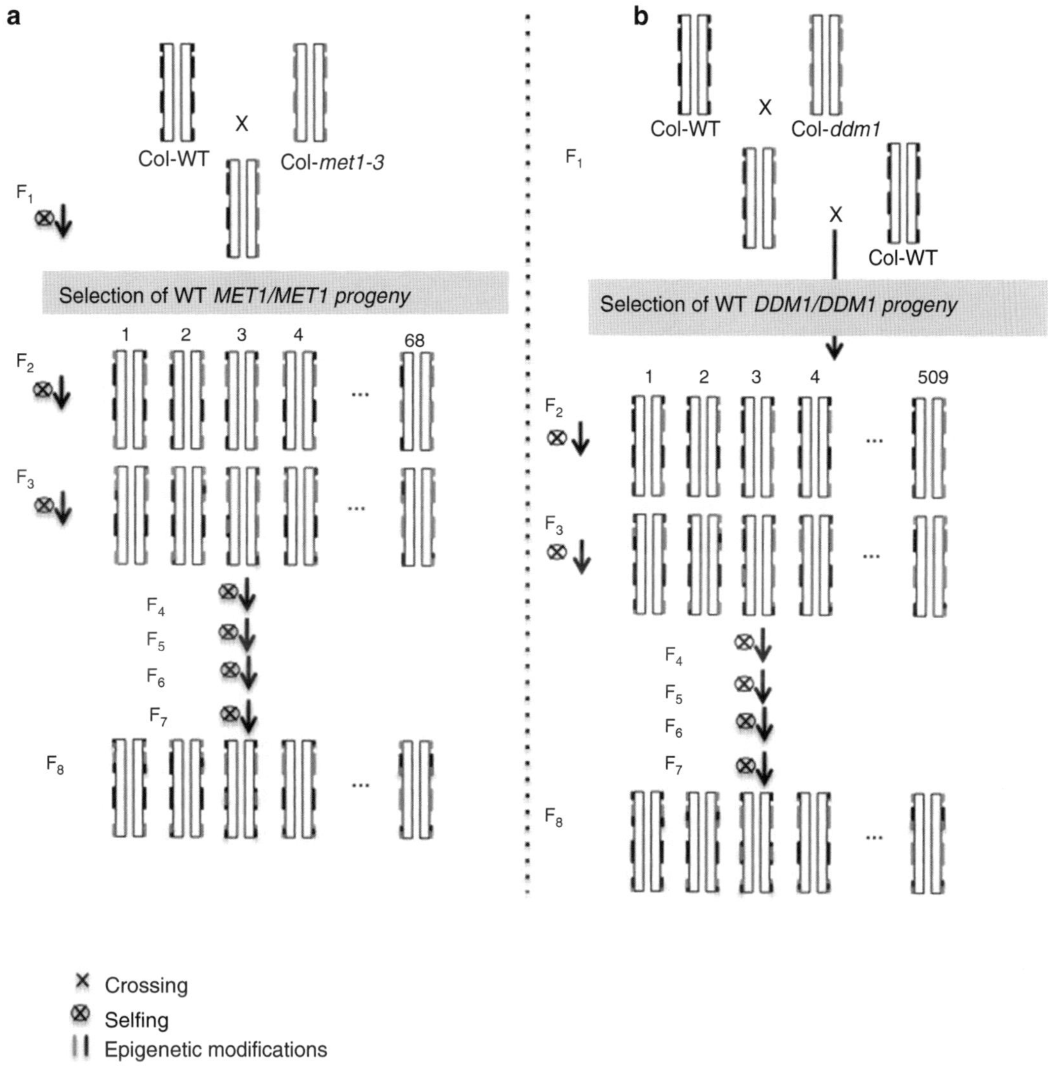

Fig. 1 Construction of epiRILs. (**a**) Construction of the *met1*-derived epiRILs: A Col-0 line was crossed to a *met1–3* mutant line (MET1 = METHYLTRANSFERASE 1) to create F1 hybrid plants. Selfed F1 plants gave rise to a segregating recombinant F2 population from which 96 individuals, homozygous for the wild type MET1 locus, were selected for an additional six generations of selfing by single seed descent. Initially 96 epiLines were created but only 68 could be advanced to the F8. By the F8 generation, lines reached (near) epihomozygosity. (**b**) Construction of the *ddm1*-derived epiRILs: A Col-0 line was crossed to a *ddm1–2* mutant line (DDM1 = DECREASE IN DNA METHYLATION 1). The resulting F1 was backcrossed as female parent to a Col-0 line. Subsequently, about 500 progeny plants homozygous for the wild type DDM1 allele were selected and propagated through six more rounds of selfing, generating a population of ~500 (near) epihomozygous different epiRILs by the F8. *Lines* are depicted schematically as one chromosome

Although it has yet to be determined to what extent such processes affect the currently available epiRILs, QTL^{epi} mapping procedures are challenged with this potential for non-Mendelian inheritance patterns (*see* **Note 1**).

The available epiRILs show phenotypic variation in various traits, confirming the concept of DNA methylation affecting quantitative traits [20, 25]. For specific purposes, it is possible to generate other sets of epiRILs in Arabidopsis following any of the explained strategies using different mutants that effect DNA methylation or any other epigenetic feature. However, the process is costly and time consuming and in order to be able to create a physical and/or genetic map of methylation markers, it is necessary to perform extensive epigenotyping.

2.1.2 Retrieving epiRIL Seed Stocks

The approximately 500 ***ddm1***-derived epiRILs are publicly available at the Arabidopsis Stock center of INRA Versailles (http://publiclines.versailles.inra.fr/). Furthermore, for subsets of those epiRILs whole genome methylation data [26] and DNA sequence data [27] is publicly available (123 and 52 lines, respectively). In case a larger population size is desired any additional lines to be included still need to be epigenotyped (*see* below).

The ***met1***-derived epiRILs are available from their founders upon request [19], although no methylome data are publicly available yet for this population.

2.2 Software for Statistics and QTL^{epi} Mapping

Different software packages are available to perform the statistical analysis on the phenotyping data, to create linkage maps and to execute the (^{epi})QTL mapping. For example:

1. For statistical analysis of phenotyping data:

   R is a freely available programming environment for data analysis. Once installed, any R package can be downloaded and used within this environment. An introduction to getting familiar with R can be found here: https://cran.r-project.org/doc/contrib/usingR.pdf.

   (a) R (free), https://www.r-project.org; https://cran.r-project.org.

   (b) R/moments (free), https://cran.r-project.org/web/packages/moments/index.html.

   (c) R/aod (free), https://cran.r-project.org/web/packages/aod/index.html.

   (d) R/car (free), https://cran.r-project.org/web/packages/car/index.html.

   (e) GenStat (commercial), https://www.vsni.co.uk/software/genstat/.

2. To create linkage maps:

   (a) R/qtl (free), http://www.rqtl.org.

   (b) MAPmaker (free), http://archive.broadinstitute.org/ftp/distribution/software/mapmaker3/.

   (c) JOINMAP (commercial), https://www.kyazma.nl/index.php/JoinMap/.
   (d) Mapchart (free), http://www.wur.nl/en/show/Mapchart-2.30.htm.
3. For (^{epi})QTL mapping:
   (a) MapQTL (commercial), https://www.kyazma.nl/index.php/MapQTL/.
   (b) R/qtl (free), http://www.rqtl.org.
   (c) QTL cartographer (free), http://statgen.ncsu.edu/qtlcart/.

3 Methods

To identify traits linked to heritable DNA methylation variation, a QTL^{epi} mapping analysis can be performed. QTL^{epi} mapping aims at identifying the epigenetic profiles at genomic loci that underlie a phenotypic trait of interest.

Before starting, it is important to carefully select the mapping population. Criteria that should be taken into account are, among others, a reasonable population size (which determines resolution and power of the mapping) and segregation of the trait of interest within the population. A good indication of the segregation of traits and the necessary population size can be obtained from phenotypic analyses of the population parents, which can be achieved prior to population analysis. Substantial trait variation might reflect large effect loci and suggests differential regulation in the parents, which will segregate in the epiRIL offspring.

3.1 Designing and Performing an Arabidopsis Phenotyping Experiment

1. Monitoring parental lines (optional). The difference in trait values between the parents of an epiRIL population can be a good indication of the involvement of epigenetic factors in the regulation of a trait (*see* also Subheading 3.2.1). In general, the majority of the epiRILs should display intermediate trait values relative to their parents, but extreme phenotypes (outside the parental values) can occur (*see* **Note 2**).
2. Ensure stable conditions in the growth facility. In order to obtain phenotyping results that are consistent within and across experiments, it is crucial to keep the growing environment as stable as possible [28] (*see* **Note 3**). In most facilities, conditions like humidity, temperature and day/night cycle are standardly monitored and kept stable. Additionally, if possible, it is recommended to measure carbon dioxide levels and light intensity within the chamber and ensure that all plants receive the same amount of water and nutrients (*see* **Notes 4** and **5**).

3. Propagation of epiRILs. To avoid the so-called "seed batch effects," lines to be phenotyped should be propagated simultaneously under the same growing conditions (*see* **Note 6**) for at least one generation.
4. Experimental replication. Homozygous lines allow for including replicates of genetically identical lines. This provides a better estimate of the line specific trait value and cancels out random biological within-line-variation. Ideally, published data from the same lines can be used to estimate the level of variation in the population and to determine the number of replicates needed. *See* for instance http://powerandsamplesize.com/ for power calculations to determine the sample size.
5. Population size vs. replicate number. Growing and phenotyping several replicates per line is advisable, as it will reveal within-line variation and thereby considerably enhance the reliability of the measurements and the subsequent QTL mapping (*see* **Notes** 7 and **8**).
6. Randomization. Randomize the replicates of all lines, including the parents of the population and possibly their F_1, throughout the growth/greenhouse chamber (*see* **Note 9**).
7. Stratification of seeds. To ensure uniform germination, stratify the seeds for 3–5 days in the dark before sowing (*see* **Note 10**).
8. Quantification of phenotypic traits. Sow seeds and phenotype traits of interest in a systematic manner. For Arabidopsis that can be for example physiological traits like biomass, photosynthesis or time to flowering, or molecular traits like gene transcription, metabolomics or proteomics. However, every other trait that can be accurately quantified and which segregates in the population is admissible for QTL^{epi} mapping.

3.2 Phenotypic Data Analysis

3.2.1 Descriptive Statistics of the Phenotypic Variation

Quantitative traits display a continuous distribution of trait values, measured in the different (epi)genotypes studied. Therefore, parameters of descriptive statistics, i.e., moments of trait value distributions, are routinely determined to summarize features of phenotypic datasets. To estimate whether a trait will segregate in the mapping population, the founding parents of the population can be analyzed first (*see* **Note 11**).

1. Determination of the population mean and variance. The population mean (μ) is determined from all individuals of a certain mapping population, as opposed to the line mean, which describes only the trait value of an individual epigenotype of a population (i.e., of the different replicas from one line). The standard deviation (SD) describes the extent of trait value variation in the population, i.e., between-line variation, which can be quite large for quantitative traits in plants due to transgressive segregation of multiple $QTLs^{epi}$. The SD of replicate

measurements of isogenic individuals provides an estimate of within-line variation independent of epigenetic variation, e.g., due to biological variation or measurement error. Both between- and within-line variation is used in the calculation of trait heritability (see below). The standard error of the mean (SEM), which is derived from the SD and the sample size, is often determined as an additional measure of variation and shows how well the determined mean represents the population.

2. Calculation of the coefficient of variation (CV). The variation of trait values in a population can be well described with the SD. However, for comparing variation between different epigenotypes, across experiments or between different traits, it is useful to calculate the coefficient of variation (CV). The CV is the ratio between the SD and the mean ($CV = SD/\mu$) and thus shows the trait variability in a population in relation to the population mean.

3.2.2 Determine the Distribution of the Phenotypic Data

Most QTL-analysis software tools apply parametric tests and, therefore, assume normal distributions of trait values in the population. However, in specific cases, the distribution of trait values can deviate substantially from normality. Although QTL analysis is quite robust against deviations from normal distribution, especially for large populations, transformation of data might considerably increase mapping power and reduce false positive QTL detection in the case of nonnormal distribution. It is therefore important to check the distribution of your data before starting the QTL analysis, and to correct for nonnormal distributed data if necessary. We present some approaches below.

1. Check the data distribution for normality:

   One way to check if data points are normally distributed is by visualizing them in a normal probability plot (or Q-Q plot). Hereby, the data points are compared with a theoretical normal distribution. In case of a normally distributed dataset, all (or most) data points will be based on the central line. This can be done in R with the functions qqnorm() and qqline() or the qqPlot() function from the R/car package. The latter immediately draws confidence limits around the reference line making interpretation easier. If the graphical output is inconclusive, a hypothesis test for normality like the Shapiro–Wilk test can be performed. The null hypothesis here is that the data is normally distributed, which means that a large *p* value (>0.05) indicates a normal distribution. This test can be done in R with the function shapiro.test().

2. Check the data distribution for normality and identify the type of possible deviations. The most common deviations from

normal distributed data are bimodal (two peaks), skewed (nonsymmetric), and platykurtic or leptokurtic (heavy- vs light-tailed) distributions. Skewness and kurtosis of distributions can be calculated, for example, with the R/moments package. Bimodal distributions are often caused by the segregation of a single large-effect QTL, while the level of kurtosis depends on the number of small-effect QTLs and the level of residual non-(epi)genetic biological variation. Skewness can be caused by biological restrictions on one side of the spectrum, e.g., flowering does usually not occur before a specific developmental stage is reached but can be considerably delayed. Also, the choice of the unit of measurement can cause skewed distributions, e.g., trait values expressed as percentages cannot exceed the lower and upper boundaries of 0% and 100%, respectively, and these extreme classes, therefore, often result in a spike in the data set.

3. Correct the data distribution. Most statistical packages, including the R-package, offer several modules for transformation of data. The approach will depend on the type of deviation:
   (a) A probit transformation can be performed in the case of skewed data due to limitations in the units of measurement. This can, for instance, be performed in R/aod using the glm function.
   (b) Other nonnormal distributions benefit from a LOG or SQRT transformation. Data sets with strong positive skewness and leptokurtic distributions are usually LOG-transformed, especially if trait values range several orders of magnitude.
   (c) SQRT transformations may improve the normality of negatively skewed and platykurtic distributions.

3.2.3 Heritability and Explained Variance

In QTL$^{(epi)}$ mapping, heritability (H^2) estimates are useful to properly interpret the proportion of heritable variation explained by all the QTLs$^{(epi)}$. The contribution of a single QTL(^{epi}) to the phenotypic trait variance (i.e., the statistical quantification of variation) is often expressed as explained variance, which is usually in the output of QTL analysis software, but can also be assessed by ANOVA.

The total variance (V_p) in a quantitative trait can be described as (epi)genetic variance (V_g) plus environmental variance (V_e), plus the variance derived from the interaction between (epi)genetic and environmental factors (V_{ge}). The latter is often derived from subtracting V_g and V_e from V_p.

$$V_p = V_g + V_e + V_{ge}.$$

While V_g can be assigned to specific genetic loci, V_e is random and constitutes biological variation, experimental and measurement error. Using ANOVA, V_g can be calculated from between-line variation, whereas V_e is calculated from within-line variation. To determine the latter adequately, multiple replicate measurements of isogenic population individuals are necessary, although this is not a strict necessity for QTL mapping. When no replicate measurements of epiRILs are available, the within-line variation of the parents of the population can be used as a proxy for V_e, assuming equal variance in the different epiRILs of the population, which is another prerequisite of parametric tests. Genetic variance can be further split into additive (variance from additive gene effects), dominant (variance from dominant gene action), and epistatic variance (variance from interaction between genes).

Two specific types of heritability can be estimated, broad-sense heritability (H^2) and narrow-sense heritability (h^2).

Broad-sense heritability (H^2)

In the broad sense, heritability is estimated using the total genetic variance (V_g) divided by the total phenotypic variance (V_p) of a population of genetically diverse lines.

$$H^2 = V_g/V_p.$$

In an experimental setup broad-sense heritability estimates indicate how much of the observed phenotypic variation can be explained by genetic or in this case epigenetic factors.

Narrow-sense heritability (h^2)

In the narrow sense, heritability is measured by the genetic variance due to additive effects (from all loci influencing the trait) divided by the total phenotypic variance.

$$h^2 = V_a/V_p.$$

Since dominance cannot be estimated in homozygous populations, such as epiRILs, broad-sense heritability is usually calculated.

3.3 QTL^{epi} Mapping

3.3.1 Creating an Epigenetic Linkage Map

The first step to link phenotypic variation in an epiRIL population to corresponding epigenetic loci is the creation of a linkage map. In RILs, genomic polymorphisms like single nucleotide polymorphisms (SNPs) are used as physical and genetic markers to determine which genomic fragment is derived from which parent. In epiRILs, differentially methylated regions (DMRs) that are stably inherited over generations serve the same purpose.

1. Identification of DMRs. To examine which DMRs are stably inherited over generations, it is necessary to perform genome-wide DNA methylation analysis on the founder parents, determine the DMRs present and compare that data with DMRs still detectable in the selected epiRILs. This can, for example, be done with bisulfite sequencing (*see* for protocols Chapters 2–4 [29–31]).

2. Transgenerational inheritance of epigenetic variation. Determine which DMRs have been stably inherited from the parents, e.g., using R-code. When the methylome data has been generated, it is possible to determine for each line, which region is derived from which parent. The methylome of each line has to be screened for DMRs that are not stably inherited (i.e., were not present in one of the parents). These have to be excluded from the data.
3. Calculating marker density. To perform QTL mapping accurately, it is important that the marker density is sufficiently high and that the spacing between markers is consistent in order to detect all cross overs. The principle assumption is that each locus on the genome is in linkage disequilibrium with one or more markers. Previously published epiRIL linkage maps reported a coverage of >80% of the Arabidopsis genome and a marker spacing of approximately 3.5 centiMorgan (cM) [27], where one cM refers to one recombination event per 100 meiosis events (*see* **Note 12**), which is sufficient to generate a saturating map.
4. Create a linkage map. DMRs that are stably inherited can be used as physical and genetic markers to create a map using freely available software packages like MAPmaker [32] or commercial packages such as JOINMAP [33]. Both packages offer elaborate user manuals, but briefly the following steps are considered:
   - Exclusion of lines with many missing data points.
   - Exclusion of markers with many missing data points.
   - Exclusion of redundant markers.
   - Generating a genetic map using a mapping function (e.g., Haldane or Kosambi).
   - Selecting a core set of markers for a representative map.

3.3.2 Mapping Epigenetic Quantitative Trait Loci (QTL^{epi}s)

Once an epigenetic linkage map has been created for all selected epiRILs, QTL^{epi} analysis can be performed similar to normal QTL analysis (Fig. 2). Commonly used software is MapQTL, R/qtl or QTL cartographer. The latter two are freely available and informative tutorials exist for all of them. For QTL^{epi} analyses, R/qtl is most commonly used, and in particular the scanone function, which assumes the presence of a single QTL. For this, the phenotyping data for the trait of interest is needed from the selected epiRILs. The segregation of trait values between lines is then compared to the segregation of epigenetic markers per locus by stepwise screening along their chromosomes in windows, and association scores are plotted on the epiRIL linkage maps created before.

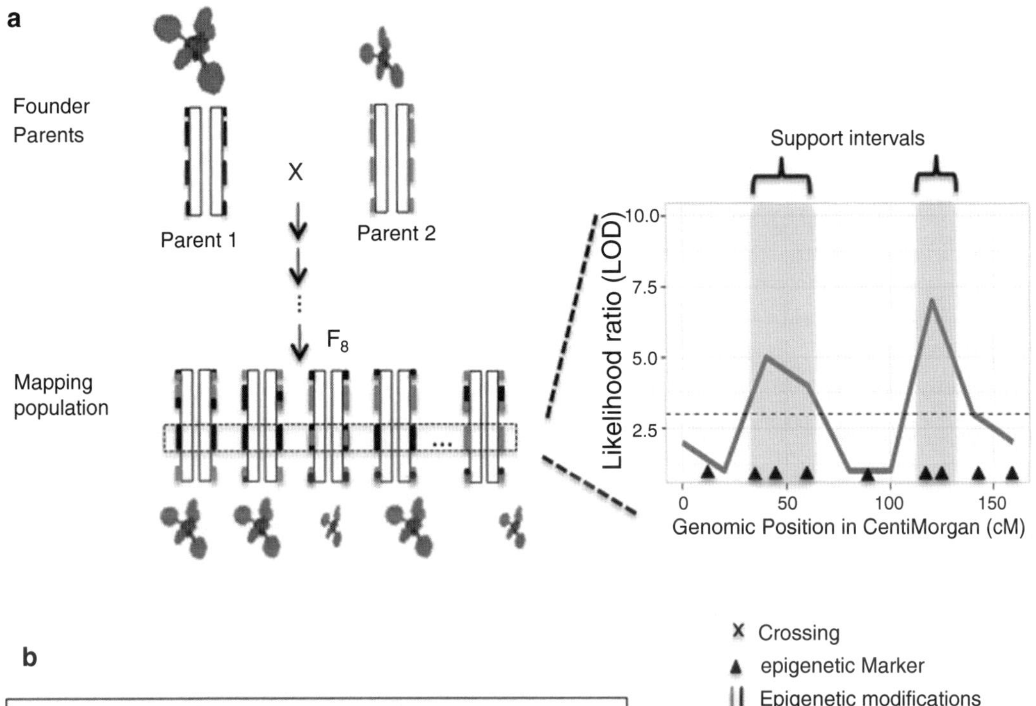

b

1. Genome-wide 5mC analysis of founder parents and mapping population → linkage map
2. Generate phenotypic data from mapping population and parents
3. epiQTL mapping → test each epigenomic position for co-occurrence with phenotype

Fig. 2 QTLepi mapping. (**a**) Schematic depiction of QTLepi mapping. Two parental lines were chosen that are isogenic, except one has a mutation affecting the epigenome. The parental lines differ phenotypically in the trait for which QTLepis should be mapped (leaf area). The F_8 generation is the mapping population. The parental lines and the mapping population are subjected to genome-wide analysis of stable epigenetic markers, which are in turn used to create a linkage map. Phenotypic data on the trait of interest is generated and QTLepis are mapped by determining the LOD (logarithm of the odds) at all genomic intervals. The *dotted line* represents the threshold for a significant QTL^{epi} and the shading indicates the support intervals around the peaks. (**b**) Stepwise approach of QTLepi mapping

1. Determination of LOD-scores per marker using interval mapping. Most QTL mapping approaches are based on interval mapping [34] which equally can be used for QTL^{epi} mapping. Interval mapping tests for the presence of a QTL every 2 cM or less between neighboring markers. At each locus, the method calculates the LOD (logarithm of the odds) score, which indicates the probability of a QTL at this position. If the LOD score exceeds a specific significance threshold, e.g., determined by permutation (see below), the presence of a QTL^{epi} in this

region is more likely to be caused by epigenetic effects than by random chance. Note that the LOD score is logarithmic, so a LOD score of three indicates that it is a thousand times more likely that linkage is due to an epigenetic effect than due to random chance.

Calculation LOD:

H1 : presence of a QTL H0 : absence of QTL.
LOD score = log10 (L(data |H1)/L(data |H0)).

2. Determination of significance thresholds by permutation testing. In order to account for the genome-wide QTL search, genome-wide significance of LOD scores is usually determined by permutation testing [35]. Basically, permutation testing compares the observed LOD scores with the maximum LOD score that could be obtained by chance with permuted data (*see* **Note 13**). Ideally, LOD significance thresholds should correspond to a false positive rate below 5%, which is often in the range of LOD 2–5. Permutation testing can be performed together with QTL mapping in R/qtl.
3. Localization of detected QTLs. To determine the most plausible genomic location of the detected QTL, confidence or support intervals are frequently constructed. For this, LOD scores of flanking loci around the maximum LOD score position are determined, and if they surpass a certain threshold, the genomic location of the markers is included in the interval. Usually, two-LOD support intervals, meaning loci that reach a LOD score of the maximum LOD minus two units, are included. Genes, or any other genetic factors, located within this support interval can be considered as candidates explaining the observed variation caused by the QTLs^{epi}.
4. Multiple QTL mapping. If there is more than one QTL segregating in the plant population, which can be easily deduced from the interval mapping, the mapping approach can be adjusted to account for the effect of each individual QTL. One option for an adjusted mapping approach is composite interval mapping (CIM) or multiple QTL Mapping (MQM) where multiple marker regression analysis is combined with interval mapping. Basically, this means that the marker that is most strongly linked to the detected QTL is assigned as a cofactor, which absorbs the variation introduced by that QTL, and thus reduces residual variation for tests at other marker positions and increases the power to detect additional QTLs.
5. Explained variance. The H^2 estimates are useful to properly interpret the proportion of heritable variation explained by all segregating QTLs^{epi}. However, a (multiple) regression model

considering the nearest linked methylation marker can be applied to get an estimate of the contribution of each of the detected QTL^{epi} to the total heritable trait variation. If heritability is high but the detected QTLs together explain just a part of the heritable variation, i.e., missing heritability, this means that there are more QTLs segregating that did not pass the significance threshold, possibly because of small-effect size and limited power. In addition, epistasis of two or more QTLs, whether detected or not, might explain considerable proportions of the total heritability.

3.4 Follow-Up Analysis: Confirmation, Fine-Mapping (Cloning), Validation and Functional Characterization

3.4.1 Confirmation of a QTL^{epi}

One strategy to confirm the presence of a QTL is mendelizing a QTL in an introgression line, i.e., a near isogenic line (NIL) harboring only the QTL interval of interest. A NIL is produced by several rounds of backcrossing a donor line (harboring the QTL) to a recurrent parental line, reducing the genomic introgression of the donor parent until only the QTL interval is present in an otherwise homogeneous recurrent background. After fixing the introgression by selfing, this line can be phenotyped and compared with the recurrent line. Principally, the same is possible for a QTLepi, with the distinction that the DNA methylation profile of the introgression has to be traced by epigenotyping. Since transchromosomal (de)methylation events may happen at the introgressed fragment, additional genotyping for a genetic marker (if available) is recommended, even though a DMR is causal for the QTL^{epi}.

Steps to generate an epiNIL (Fig. 3):

1. Cross an epiRIL containing the QTL introgression (and preferably as few as possible additional introgressions) with the recurrent wild-type parent used to generate the epiRIL population.
2. Propagate the F_1 by selfing to create a segregating F_2.
3. Epigenotype the progeny for all possible segregating introgressions (*see* **Note 14**). Two methods to determine methylation profiles at particular genomic positions are commonly used: endonuclease (McrBC) digestions followed by real-time PCR on marker positions and Targeted Bisulfite Sequencing.
   - Endonuclease digestions followed by real-time PCR.

     This technique is useful when a large quantity of (pooled) plants has to be assayed. This method is based on the endonuclease McrBC, which cleaves DNA containing methylated cytosines on one or both strands while it does not act on unmethylated DNA. Hence, a digest followed by quantitative (q)PCR on a marker region will allow quantifying the methylation at that position. To interpret the assay, controls need to be taken along that have the targeted position methylated and unmethylated (that can be for example the recurrent parents of the mapping population).

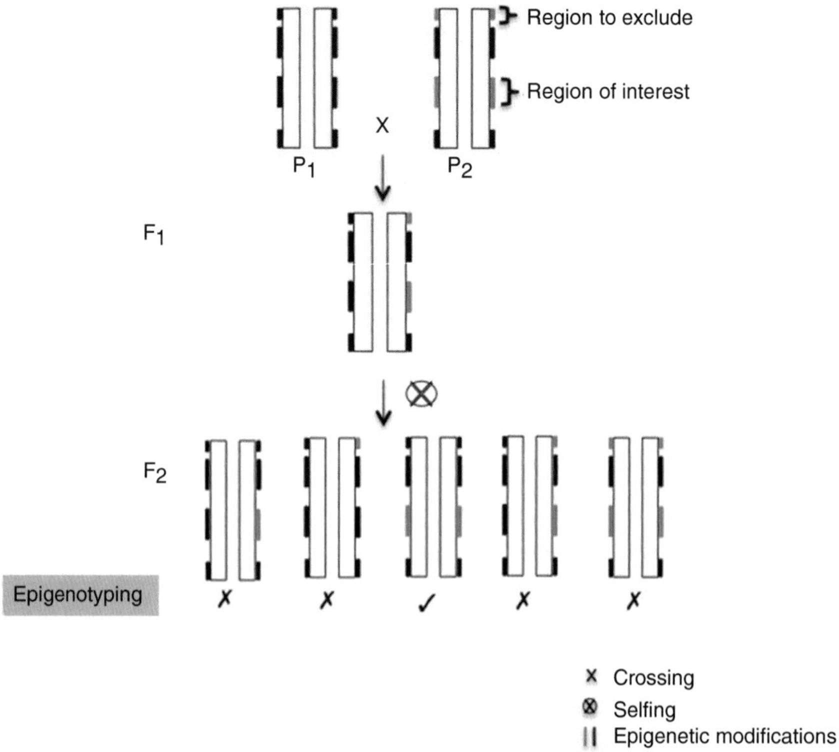

Fig. 3 Creation of an epigenetic near isogenic line (epiNIL) to confirm QTL^{epi} effects. The recurrent parent (P_1) is crossed to the epiRIL harboring the QTL region (P_2). The F_1 is selfed and the resulting F_2 individuals are epigenotyped for the desired introgression. Epigenotyping should also be used to exclude introgressions unrelated to the QTL regions

- The second method, targeted bisulfite sequencing, is recommended if a better quantification of the methylation profile and/or a higher resolution of a longer DNA stretch is desired. The method starts with bisulfite treatment of the genomic DNA, which results in conversion of unmethylated cytosines into uracil while methylated cytosines remain unaltered. Subsequently, the region of interest is amplified by PCR using unbiased degenerated primers (a primer mix that considers that each cytosine in the targeted region has the potential to be converted). The purified PCR product can either be sequenced directly or can be cloned into a vector with positive clones subjected to Sanger-sequencing.

Both methods establish the level of methylation of a specific genomic region and as such determine the descent of that region.

4. Select the epiNIL, which harbors only the QTL introgression (*see* **Note 15**), and compare its phenotype with the recurrent population parents.

3.4.2 Fine-Mapping and Functional Characterization

The next step, after identification and confirmation of a QTL^{epi} region, is to determine the candidate genes in this region, which introduce trait variation through differential methylation. In addition, transposon insertions need to be excluded as causes for the QTL effect.

1. Excluding transposons as causal loci. Genome-sequencing of a subset or the entire mapping population can reveal shared transposon insertions that could be causal. To date, genome-resequencing data is already publicly available for 73 of the *ddm1*-derived epiRILs [27]. Once transposon insertions in the QTL region have been identified in a subset of epiRILs, the remaining population can be screened by PCR for the same insertions.
2. Identification of candidate genes. Generally, QTL regions can be quite large (~ 1–2 Mbp) and, therefore, may still contain hundreds of genes. Hence, it may be useful to assign a priori putative target genes for further study.
   - Check the annotation of the genes for genes that may be involved in the trait of interest.
   - Check for the methylation status in the recurrent parents. Is there a DMR present in the locus?
   - Determine the gene expression levels of putative candidates. Differential methylation patterns in the mapping population imply gene expression differences, which can be analyzed by quantitative Reverse transcription-PCR (RT-PCR). In Arabidopsis, both methylation at promoter regions but also gene body methylation can affect gene transcription [36–38] (*see* **Note 16**).
3. Functional analysis of candidate genes and the causal DMR. To further confirm the causality of candidate genes, targeted gene knockdowns or knockouts can be analyzed. However, independent proof for the effect of DNA methylation at particular positions should be derived from cloning experiments, complementation or epigenetically modifying targets (region X methylated vs nonmethylated). Cloning of causal epigenomic loci is still challenging, as the DNA methylation profile might be altered during the process. However, methods to impose particular methylation profiles at loci of interest are available. For instance, transgenes can be used that induce the production of small RNAs, which in turn induce DNA methylation at their target site [39, 40]. Also, techniques for epigenomic editing are rapidly emerging. These methods rely often on specific DNA recognition domains fused to a catalytic domain of a chromatin-modifying enzyme, allowing targeting of the desired chromatin modification to any locus of

interest [41]. Examples for such techniques are CRISPR-Cas, Zinc finger proteins or transcription-activator-like effectors (TALEs) [41].

3.5 Examples of Research Applications of epiRIL Populations

epiRILs are a highly effective tool to assess the contribution of epigenetics to plant phenotypes. To date, they have been used to determine the contribution of epigenetic variation to quantitative traits [27], phenotypic plasticity in response to stress [42] or heterosis [43]. Excitingly, breeding strategies in canola crop plants already resulted in epiLines, which have improved energy use efficiency and drought resistance compared to their isogenic counterparts [44, 45]. This is an example of artificial selection of certain epigenetic states, which enhance physiological plant characteristics. The next logical step here is defining the relevant genomic positions by approaches like QTL^{epi} mapping. This indicates a huge potential of epigenetic variation in improving crop plants.

4 Notes

1. One way to ensure that non-Mendelian inheritance patterns are not affecting mapping approaches is by thoroughly monitoring of epigenetic profiles of mapping populations (and propagated lines) either locus-specific or genome-wide and using only stable positions as markers.
2. These extreme phenotypes are termed transgressive phenotypes [46]. Transgression is observed frequently, particularly in intraspecific crosses and when many opposite-effect QTLs segregate in the population. The distribution of transgressive trait values typically displays a leptokurtic shape. Possible explanations for transgressive segregation are the accumulation and action of complementary-effect genes, but also epistatic interactions or over-dominance may contribute. The transgression and parental values of a trait therefore provide meaningful information on the level of epigenetic regulation and the epigenetic architecture of that trait.
3. Phenotyping can be done in both growth and greenhouse chambers. However it should be realized that in the greenhouse environmental conditions are subject to more fluctuations and it is generally more difficult to keep a stable environment there.
4. Varying distances between plants and light source are common in growth chambers and can lead to plants being exposed to different light intensities. Basic (inexpensive) light sensors will allow to determine strongly affected areas and to design an experimental setup avoiding them.

5. If no automated watering system is available, watering manually with the exact same amounts is recommended.
6. Different handling of the parent lines can cause seed batch effects [47]. If the parental lines of seed batches have been propagated in the same way this step is not necessary.
7. When designing experiments with limited numbers, e.g., when growth space is limited, experiments are expensive or many different treatments are compared, it is usually better to increase the size of the mapping population first before including replicates of identical lines. This will provide a higher resolution while the locus effect is replicated in different lines. Unless there are large segregation distortions, the allele frequency of each locus is centered on 50% in the population.
8. Strong epistatic effects, inherently involved in quantitative trait regulation, increase the complexity in an experimental setup. The detection of epistasis requires increased numbers of replicates and population sizes.
9. Randomization will level out phenotypic effects caused by plant position in the growth chamber. *See* https://www.randomizer.org for a free resource to generate random numbers or assign individuals to experimental conditions.
10. Stratification will limit differences in germination.
11. The analysis of the founder parents can precede population measurements, but it is advised to analyze the parents in the same experimental settings as the derived population. Strong differences in trait values between the parents are indicative of epigenetic causal variation and might provide leads to the sign, strength and number of QTLs^{epi} segregating in the progeny. However, a lack of trait variation between the parents does not necessarily mean an absence of relevant epigenetic modifications. Multiple QTLs^{epi} with opposite roles might cancel out each other's effect in parental lines, but lead to transgression in population individuals (*see* **Note 2**). To obtain accurate estimates of parental trait values, it is recommended to include sufficient replicates in the analyses, especially if replicates of population individuals cannot be included in the experimental design. In the latter case, within line variation of the parents can then be used as a proxy to estimate heritability in the segregating population (see below). For this, it is advised that the parental lines are grown in parallel with the population.
12. Recombination during meiotic cell division results in the decay of linkage disequilibrium and as such the resolution of the population and the number of markers needed depends on the number of meiotic events. The resolution of a mapping population in turn depends on the crossing setup for its creation. For example an extra round of back-crossing, as in the

ddm1-derived epiRIL population stabilizes wild-type introgressions but also results in more crossovers than in the *met1*-derived epiRIL population. As a guideline, in classical Arabidopsis RILs, 5 cM correspond to approximately 1 megabasepair (Mbp) and recombination frequency is one to two cross overs per chromosome per meiosis. Taking fixation due to inbreeding into account, this means effectively two recombination events per chromosome in (epi)RILs. In small mapping populations, typically less than 200 individuals, this means that a spacing of 5 cM between markers, i.e., ~100–200 markers in Arabidopsis, should be sufficient to generate a saturating map.

13. The number of permutations lies often between 1000 and 10,000.
14. In order to account for recombination it is wise to use one marker at each end of the introgression.
15. If the desired epiNIL is not identified perform another round of backcrossing or screen more F_2 individuals.
16. For gene expression analysis it should be considered that fluctuations might occur across different tissues and also during the day (diurnal variation). Consequently, for comparative reasons it is important to profile the same tissue which was harvested at the same time of the day from all individuals.

Acknowledgment

K.L. was supported by the Centre for Improving Plant Yield (CIPY), which is part of the Netherlands Genomics Initiative and the Netherlands Organization for Scientific Research.

References

1. Alonso-Blanco C, Koornneef M (2000) Natural variation, an underexploited resource of genetic variation for plant genetics. Trends Plant Sci 5:22–29
2. Weigel D (2012) Natural variation in Arabidopsis: from molecular genetics to ecological genomics. Plant Physiol 158:2–22. doi:10.1104/pp. 111.189845
3. Law J, Jacobsen SE (2010) Establishing, maintaining and modifying DNA methylation patterns in plants and animals. Nat Rev Genet 11:204–220. doi:10.1038/nrg2719
4. Bologna NG, Voinnet O (2014) The diversity, biogenesis, and activities of endogenous silencing small RNAs in Arabidopsis. Annu Rev Plant Biol 65:473–503. doi:10.1146/annurev-arplant-050213-035728
5. Xu C, Tian J, Mo B (2013) siRNA-mediated DNA methylation and H3K9 dimethylation in plants. Protein Cell 4:656–663. doi:10.1007/s13238-013-3052-7
6. Springer NM (2013) Epigenetics and crop improvement. Trends Genet 29:241–247. doi:10.1016/j.tig.2012.10.009
7. Richards EJ (2011) Natural epigenetic variation in plant species: a view from the field. Curr Opin Plant Biol 14:204–209. doi:10.1016/j.pbi.2011.03.009
8. O'Malley RC, Ecker JR (2012) Epiallelic variation in Arabidopsis thaliana. Cold Spring Harb

Symp Quant Biol 77:135–145. doi:10.1101/sqb.2012.77.014571

9. Weigel D, Colot V (2012) Epialleles in plant evolution. Genome Biol 13:249. doi:10.1186/gb-2012-13-10-249
10. Finer S, Holland ML, Nanty L, Rakyan VK (2011) Review article the hunt for the epiallele. Environ Mol Mutagen 52(1):1–11. doi:10.1002/em.20590
11. Stam M, Belele C, Dorweiler JE, Chandler VL (2002) Differential chromatin structure within a tandem array 100 kb upstream of the maize b1 locus is associated with paramutation. Genes Dev 16:1906–1918. doi:10.1101/gad.1006702
12. Bender J, Fink GR (1995) Epigenetic control of an endogenous gene family is revealed by a novel blue fluorescent mutant of Arabidopsis. Cell 83:725–734
13. Henderson IR, Jacobsen SE (2008) Tandem repeats upstream of the Arabidopsis endogene SDC recruit non-CG DNA methylation and initiate siRNA spreading. Genes Dev 22:1597–1606. doi:10.1101/gad.1667808
14. Quadrana L, Almeida J, Asís R et al (2014) Natural occurring epialleles determine vitamin E accumulation in tomato fruits. Nat Commun 5:3027. doi:10.1038/ncomms5027
15. Soppe WJ, Jacobsen SE, Alonso-Blanco C et al (2000) The late flowering phenotype of fwa mutants is caused by gain-of-function epigenetic alleles of a homeodomain gene. Mol Cell 6:791–802
16. Hövel I, Pearson NA, Stam M (2015) Cis-acting determinants of paramutation. Semin Cell Dev Biol. doi:10.1016/j.semcdb.2015.08.012
17. Greaves IK, Groszmann M, Ying H et al (2012) Trans chromosomal methylation in Arabidopsis hybrids. Proc Natl Acad Sci U S A 109:3570–3575. doi:10.1073/pnas.1201043109
18. Marí-Ordóñez A, Marchais A, Etcheverry M et al (2013) Reconstructing de novo silencing of an active plant retrotransposon. Nat Genet 45:1029–1039. doi:10.1038/ng.2703
19. Reinders J, Wulff BBH, Mirouze M et al (2009) Compromised stability of DNA methylation and transposon immobilization in mosaic Arabidopsis epigenomes. Genes Dev 23:939–950. doi:10.1101/gad.524609
20. Johannes F, Porcher E, Teixeira FK et al (2009) Assessing the impact of transgenerational epigenetic variation on complex traits. PLoS Genet. doi:10.1371/journal.pgen.1000530
21. Finnegan EJ, Peacock WJ, Dennis ES (1996) Reduced DNA methylation in Arabidopsis thaliana results in abnormal plant development. Proc Natl Acad Sci U S A 93:8449–8454. doi:10.1073/pnas.93.16.8449
22. Mathieu O, Reinders J, Čaikovski M et al (2007) Transgenerational stability of the Arabidopsis epigenome is coordinated by CG methylation. Cell 130:851–862. doi:10.1016/j.cell.2007.07.007
23. Kakutani T, Jeddeloh J, Richards EJ (1995) Characterization of an Arabidopsis thaliana DNA hypomethylation mutant. Nucleic Acids Res 23:130–137
24. Kakutani T, Jeddeloh JA, Flowers SK et al (1996) Developmental abnormalities and epimutations associated with DNA hypomethylation mutations. Proc Natl Acad Sci U S A 93:12406–12411. doi:10.1073/pnas.93.22.12406
25. Roux F, Colomé-Tatché M, Edelist C et al (2011) Genome-wide epigenetic perturbation jump-starts patterns of heritable variation found in nature. Genetics 188:1015–1017. doi:10.1534/genetics.111.128744
26. Colomé-Tatché M, Cortijo S, Wardenaar R et al (2012) Features of the Arabidopsis recombination landscape resulting from the combined loss of sequence variation and DNA methylation. Proc Natl Acad Sci 109:16240–16245
27. Cortijo S, Wardenaar R, Colomé-Tatché M et al (2014) Mapping the epigenetic basis of complex traits. Science 343:1145–1148. doi:10.1126/science.1248127
28. Massonnet C, Vile D, Fabre J et al (2010) Probing the reproducibility of leaf growth and molecular phenotypes: a comparison of three Arabidopsis accessions cultivated in ten laboratories. Plant Physiol 152:2142–2157. doi:10.1104/pp.109.148338
29. Chen Y-R, Sheng Y, Zhong S (2017) Profiling DNA methylation using bisulfite sequencing (BS-Seq). In: Bemer M, Baroux C (eds) Plant chromatin dynamics: methods and protocols. Springer, New York, NY. doi:10.1007/978-1-4939-7318-7_2
30. Edelmann S, Scholten S (2017) Bisulphite sequencing using small DNA amounts. In: Bemer M, Baroux C (eds) Plant chromatin dynamics: methods and protocols. Springer, New York, NY. doi:10.1007/978-1-4939-7318-7_3
31. Kishore K, Pelizzola M (2017) Identification of differentially methylated regions in the Arabidopsis thaliana genome. In: Bemer M, Baroux C (eds) Plant chromatin dynamics: methods and protocols. Springer, New York, NY. doi:10.1007/978-1-4939-7318-7_4

32. Paterson AH, Lander ES, Hewitt JD et al (1988) Resolution of quantitative traits into Mendelian factors by using a complete linkage map of restriction fragment length polymorphisms. Nature 335:721–726. doi:10.1038/335721a0
33. van Ooijen JW (2011) Multipoint maximum likelihood mapping in a full-sib family of an outbreeding species. Genet Res 93:343–349. doi:10.1017/S0016672311000279
34. Lander ES, Botstein D (1989) Mapping Mendelian factors underlying quantitative traits using RFLP linkage maps. Genetics 121:185–199
35. Churchill GA, Doerge RW (1994) Empirical threshold values for quantitative trait mapping. Genetics 138:963–971
36. Zilberman D, Gehring M, Tran RK et al (2007) Genome-wide analysis of Arabidopsis thaliana DNA methylation uncovers an interdependence between methylation and transcription. Nat Genet 39:61–69. doi:10.1038/ng1929
37. Takuno S, Gaut BS (2012) Body-methylated genes in arabidopsis thaliana are functionally important and evolve slowly. Mol Biol Evol 29:219–227. doi:10.1093/molbev/msr188
38. Dubin MJ, Zhang P, Meng D et al (2015) DNA methylation in Arabidopsis has a genetic basis and shows evidence of local adaptation. elife 4:e05255. doi:10.7554/eLife.05255
39. Rajeevkumar S, Anunanthini P, Sathishkumar R (2015) Epigenetic silencing in transgenic plants. Front Plant Sci 6:693. doi:10.3389/fpls.2015.00693
40. McGinnis KM (2010) RNAi for functional genomics in plants. Brief Funct Genomics 9:111–117. doi:10.1093/bfgp/elp052
41. Kungulovski G, Jeltsch A (2015) Epigenome editing: state of the art, concepts, and perspectives. Trends Genet 32:101–113. doi:10.1016/j.tig.2015.12.001
42. Kooke R, Johannes F, Wardenaar R et al (2015) Epigenetic basis of morphological variation and phenotypic plasticity in Arabidopsis thaliana. Plant Cell 27:337–348. doi:10.1105/tpc.114.133025
43. Dapp M, Reinders J, Bédiée A et al (2015) Heterosis and inbreeding depression of epigenetic Arabidopsis hybrids. Nat Plants 1:15092. doi:10.1038/nplants.2015.92
44. Hauben M, Haesendonckx B, Standaert E et al (2009) Energy use efficiency is characterized by an epigenetic component that can be directed. Proc Natl Acad Sci U S A 106:20109–20114. doi:10.1073/pnas.0908755106
45. Verkest A, Byzova M, Martens C et al (2015) Selection for improved energy use efficiency and drought tolerance in canola results in distinct transcriptome and epigenome changes. Plant Physiol. doi:10.1104/pp.15.00155
46. Rieseberg LH, Archer M, Wayne RK (1999) Transgressive segregation, adaptation and speciation. Heredity (Edinb) 83(Pt 4):363–372. doi:10.1038/sj.hdy.6886170
47. Meyer RC, Toerjek O, Becher M, Altmann T (2004) Heterosis of biomass production in Arabidopsis. establishment during early development. Plant Physiol 134:1813–1823. doi:10.1104/pp.103.033001.hybrid

Part III

Spatial Chromatin Organization

Chapter 23

A Compendium of Methods to Analyze the Spatial Organization of Plant Chromatin

Aline V. Probst

Abstract

The long linear chromosomes of eukaryotic organisms are tightly packed into the nucleus of the cell. Beyond a first organization into nucleosomes and higher-order chromatin fibers, the positioning of nuclear DNA within the three-dimensional space of the nucleus plays a critical role in genome function and gene expression. Different techniques have been developed to assess nanoscale chromatin organization, nuclear position of genomic regions or specific chromatin features and binding proteins as well as higher-order chromatin organization. Here, I present an overview of imaging and molecular techniques applied to study nuclear architecture in plants, with special attention to the related protocols published in the "Plant Chromatin Dynamics" edition from Methods in Molecular Biology.

Key words Nuclear architecture, Fluorescence in situ hybridization, 3C techniques, Fluorescence immunolabeling, Microscopy

1 Introduction

Within the nucleus, the functional control center of the cell, DNA needs to be tightly packaged, while at the same time preserving controlled access to the cellular machinery. Given the small nuclear diameter in relation to the stretched size of a chromosome, it is evident that coordinated DNA packaging is of high importance for genome organization and function. To achieve this, nuclear DNA is organized together with histone proteins into chromatin, which can adopt distinct functional states characterized by differential enrichment in DNA methylation, histone marks and histone variants [1, 2]. Beyond this linear conception of chromatin, the rather complex and dynamic organization of chromatin within the three-dimensional (3D) space of the nucleus contributes to genome function.

Indeed, within the nucleus, chromosomes are not intermingled, but each chromosome occupies its own territory [3]. In mammalian cells, within the territory, chromatin forms topologically associating

Marian Bemer and Célia Baroux (eds.), *Plant Chromatin Dynamics: Methods and Protocols*, Methods in Molecular Biology, vol. 1675, DOI 10.1007/978-1-4939-7318-7_23, © Springer Science+Business Media LLC 2018

domains (TADs, [4]) and functionally relevant long-range interactions through loops that can bring specific regions, far away on the physical chromosome map, into close proximity in nuclear space. Such has been shown for promoters and enhancers [5, 6] or for multiple transcriptionally active genes that come together in transcription factories [7]. Transcriptionally silent or low expressed genes are found in proximity to the nuclear lamina in so called Lamina-associated domains (LADs) [8, 9]. Finally, nucleolar-associated domains (NADs), which are rather gene poor and silent chromatin domains, preferentially localized in close proximity to the nucleolus [10–12]. While coordinated gene expression is a major output influenced by chromosome architecture, other processes such as DNA replication [13] and repair [14] are equally affected by chromatin organization. In plants, much remains to be learned on the formation and dynamics of chromosome territories as well as the functional organization of the genome in LADs, TADs and local gene loops. Details of the available techniques to investigate nuclear architecture in plants will be discussed here. A summary of the different methods, including technical details, applications, and limitations, is presented in Table 1.

2 In Situ Labeling Techniques and Microscopy Imaging Approaches

One of the first methods used to assess the organization of DNA in the nucleus was colorimetric staining with an acidophilic dye and observation under light microscopy. This enabled, for example, the distinction of differential chromatin packaging in moss nuclei corresponding to heterochromatin and euchromatin [80] and to develop the theory of chromosome territories in mammalian cells [81]. Since then, chromatin analyses largely benefited from the advent of fluorescence microscopy imaging relying on fluorescent reporter protein tags (green fluorescent proteins and variants thereof), and the development of antibodies enabling specific immunolabeling of chromatin components or chromatin modifications [82–84] (Fig. 1, Table 1).

2.1 DNA and Chromatin Staining to Investigate Nuclear Architecture

Technical considerations. Global chromatin organization can be imaged after staining of DNA and chromatin with intercalating agents or other colorants; examples are PicoGreen or DRAQ5 for life cell imaging or Hoechst and DAPI for fixed cells. Imaging of DNA-stained nuclei allows to determine the size and shape of the nucleus, as well as distinguishing decondensed chromatin regions, which are lightly stained, from highly condensed heterochromatic regions that appear intensely colored (Fig. 1). The number, size, and distribution of these heterochromatic regions relative to the nuclear periphery or the nucleolus can be assessed in 2D on spread

Table 1
Comparison of the different 3D chromatin analysis methods used in plants

Method	Technical details	Applications	Limitations
Assessment of Global Nuclear Organization – Generally provides single-cell information and a measurement of cell-to-cell variation			
Whole chromatin staining	• DNA staining of whole nuclei and imaging by fluorescence microcopy (Chapters 29 and 33 [15, 16])	• Information on size and shape of nuclei and nucleolus, qualitative and quantitative heterochromatin parameters such as position, number, and distribution of heterochromatin domains [17, 18] • Cell-type specific analyses (in whole-mount tissue) or bulk analyses (isolated nuclei) [17]	• Requires intact nuclei, efficient DNA and chromatin labeling (challenging in whole mount) • Requires high-resolution 3D imaging to resolve submicrometer structures • Challenges in 3D image analysis: e.g., segmentation of nuclei and heterochromatin domains (Chapter 31 [19, 20])
Analysis of Nanoscale Chromatin Organization – Generally provides single-cell information and a measurement of cell-to-cell variation			
Whole chromatin, subnuclear domains, chromatin protein foci staining	• Cryofixation and freeze substitution, tissue sectioning, uranyl acetate chromatin staining, transmission electron microscopy (TEM) imaging (Chapter 34 [21])	• Unique nanoscale level of resolution (5–8 nm)Ultrastructural analyses of subnuclear domains, chromatin fibers • Qualitative and quantitative • Applicable to a broad range of materials • Cell-type specific (providing cell identification on tissue section)	• 2D information (3D reconstructions based on serial imaging of consecutive sections is challenging) • Uranyl acetate has a strong affinity, but is not specific to chromatin (stains also nucleoplasmic and nucleolar particles, of which abundance varies depending on cell type) • Requires specific, not widely available imaging instruments and expertise

(continued)

Table 1
(continued)

Method	Technical details	Applications	Limitations
Fluorescence in situ hybridization (FISH) *– Generally provides single-cell information and a measurement of cell-to-cell variation*			
Repeat-FISH	• Detection of repetitive sequences using directly or indirectly labeled DNA probes (Chapter 27 and 32 [22, 23]) or LNA-DNA mixmer probes (Chapter 28 [24])	• Describes organization of repeats (telomeres, centromeres, TE repeat families on metaphase chromosomes) (Chapter 30 [25]) and their localization relative to other nuclear structures [26] • Analysis of heterochromatin/repeat family organization (functional and evolutionary studies) • Somatic chromosome karyotyping [27] • FISH in whole mount tissue: provides single cell information for cells in tissue context • FISH in bulk nuclei (intact or spread/flattened) allows 2D/3D quantifications based on large sample size	• Direct labeling not always suitable • Loss of cell-type specific information in bulk preparations (3D or spread/flattened) • Robust and homogenous staining is technically challenging in whole-mount tissue and often requires optimization for a specific tissue (Chapters 27 and 28 [22, 24]), [28–34]; applicable on thick tissue sections [35–37] • Chromatin integrity can be compromised upon denaturation process (heat/formamide treatment)
BAC-FISH	• Detection of specific chromosomal regions or whole chromosome territories using directly or indirectly labeled probes generated from chromosome-specific BACs containing unique sequences [38]	• Spatial information of chromosome territory organization in single cells [30, 38–41] • Physical mapping of genomic contigs on pachytene spreads: facilitates genome assembly and annotation [42–44], evolutionary, comparative analyses of chromosome structure [45–48]	

Oligo-FISH	• Detection of specific chromosomal regions using sets of short directly labeled oligonucleotides	• Same as above	
Gene-FISH	• Padlock-FISH, relying on rolling circle amplification [49] • Directly or indirectly labeled gene probes (Chapters 27 and 32 [22, 23])	• Detection of a single gene and its position within nuclear space • Can be combined with automated 3D analysis of gene positioning (Chapters 25 and 27 [22, 50])	• Same as above • Technically difficult due to probe size with level of fluorophores close to detection threshold
GISH	• Detection of genomic sequences using whole DNA of another organism as probe	• Analysis of individual chromosomes, chromosomal segments, or the genomes of natural and artificial hybrids [51–54]	• Closely related genome might not be distinguished • Technically challenging for organisms with small genomes [52]
Visualization of lacO tagged transgenes with LacI-FP fusions *– Generally provides single-cell information and a measurement of cell-to-cell variation*			
LacI–LacO system	• Based on the bacterial operator–repressor system Lac repressor–lac operator (LacI–LacO) [55, 56] • Quantification of colocalization of tagged alleles in nuclear space [55] or proximity to nuclear compartments • Quantitative measurements of chromatin mobility *in planta* [57]	• Suitable for live imaging • Preservation of tissue integrity—No technical bias due to fixation • Can be used to functionally test the impact of spatial localization on gene expression [58]	• Requires the establishment of transgenic plants containing the LacI–LacO system • Risk of artificially inducing nonrandom organization of loci tagged with LacO arrays [59]
Immunolocalization/fluorescently tagged proteins *– Generally provides single-cell information and a measurement of cell-to-cell variation*			
Immunolocalization for whole chromatin/chromosomes, subnuclear domains/chromatin fiber analysis	• Chromatin immunostaining using specific antibodies (Chapters 24, 25, and 27) or use of photoactivatable fluorophores • Combinable with super-resolution microscopy imaging (Chapter 31 [19]) [60]	• Quantitative (relative levels) and qualitative (nuclear distribution) • Whole-mount tissue: comparative analysis between cell types • Isolated nuclei: allows for bulk, large-scale quantitative analyses; easy implementation	• Requires specific antibodies working robustly under standard immunostaining conditions or specific blinking antibodies/fluorophores for super-resolution • Loss of cell-type specific information in bulk preparations (3D or spread/flattened)

(continued)

Table 1
(continued)

Method	Technical details	Applications	Limitations
		• Potential to assess distribution at nanoscale level (20–120 nm depending on imaging platform) using super-resolution microscopy [61] • Multicolor imaging possible (multiple fluorophores/targets) • 3D image datasets	• Robust and homogenous staining is technically challenging in whole-mount tissue and often requires optimization for a specific tissue • For super-resolution: Restricted to thin preparations (isolated cells/nuclei), requires specific, not widely available imaging instruments and expertise, computationally intensive image reconstruction and analysis
Imaging of fluorescently tagged proteins	• Imaging of fluorescently tagged proteins using fluorescence microscopy (Chapter 31 [19])	• Does not require antibodies • Potential to assess protein dynamics in life imaging and in different tissues (Chapter 26 [62]) • Potential to assess distribution at nanoscale level using super-resolution microscopy (see above)	• Requires generation of transgenic plants • Protein localization may be altered due to protein fusion with fluorescent tag
Chromatin conformation capture techniques (3C, 4C, Hi-C) *See also technical review Chapter* 15 *– The readout is an average view of interactions among multiple cells and cell types*			
3C	• Detection of pairwise interaction between two loci in close vicinity in nuclear space: proximity allows for cross-linking and ligation-based cloning of interacting fragments (Chapters 14 and 15 [63, 64]) [65], analyzed by quantitative PCR starting from candidate viewpoint and interacting region: one versus one	• High resolution level detection of sequences involved in intrachromosomal folding/interchromosomal association [55, 66, 67]	• Requires prior knowledge of putatively interacting genomic regions • Rare interactions may be missed

4C	• 3C but interaction clones containing one viewpoint are screened by deep sequencing: one-versus-all [68]	• High resolution analysis of 3D genome organization: intramolecular model of interactions [69] • No requirement for a priori knowledge on interacting regions	• Probabilistic model of 3D organization: the readout is an average view of interactions among multiple cells and cell types • 3D model in relative space. No information on absolute 3D position within nuclear space
Hi-C	• 3C but interaction clones are blindly sequenced using NGS: all-versus-all [68]	• Provides global chromatin interactome [70–74] • Template for de novo genome assembly	• As before • Large datasets, challenging bioinformatics analysis • Difficulty to distinguish between specific and nonspecific interactions • Potential bias due to nonuniform distribution of restriction sites in the genome • Does not provide information where in the nucleus the interaction takes place
Genome profiling of chromatin protein-DNA associations – The readout is generally an average view of interactions among multiple cells and cell types, but tissue-specific promoters can be envisaged for DamID			
DamID (DNA adenine methyltransferase identification)	• DNA adenine methyltransferase fused to the protein of interest is stably expressed in plants; genomic regions in contact with the protein of interest will be methylated and can be identified by sequencing	• Whole genome approach without a priori • Mapping of binding sites of chromatin binding proteins [75, 76]	• Requires generation of transgenic lines with very low expression of Dam fusions and control lines [76] as artifacts can arise from uncontrolled methylation
ChIP-Seq (Chromatin Immunoprecipitation)	• Chromatin bound by a specific protein is isolated by immunoprecipitation and the immunoprecipitated DNA sequenced using NGS	• Genome-wide mapping of histone modifications, histone variants, chromatin binding proteins	• Requires specific and high affinity antibodies working robustly under immunoprecipitation conditions • Does not provide information where in the nucleus the interaction takes place

(continued)

Table 1
(continued)

Method	Technical details	Applications	Limitations
Identification of genomic regions associated with subnuclear structures – The readout is an average view of interactions among multiple cells and cell types			
Identification of nucleolus-associated chromatin domains (NADs)	• Isolation of nucleoli by fluorescence activated cell sorting (FACS) [77, 78], chromatin purification and sequencing by NGS [12] (Chapter 7 [79])	• FACS allows for high purity/ selectivity • De novo identification of NADs [12] • Robust, high-throughput: suitable for functional and comparative analyses	• Average view of multiple cells and cell types • Requires transgenic lines with nucleolus-localized fluorescent proteins for isolation by FACS • Limited amount of isolated material as FACS-based

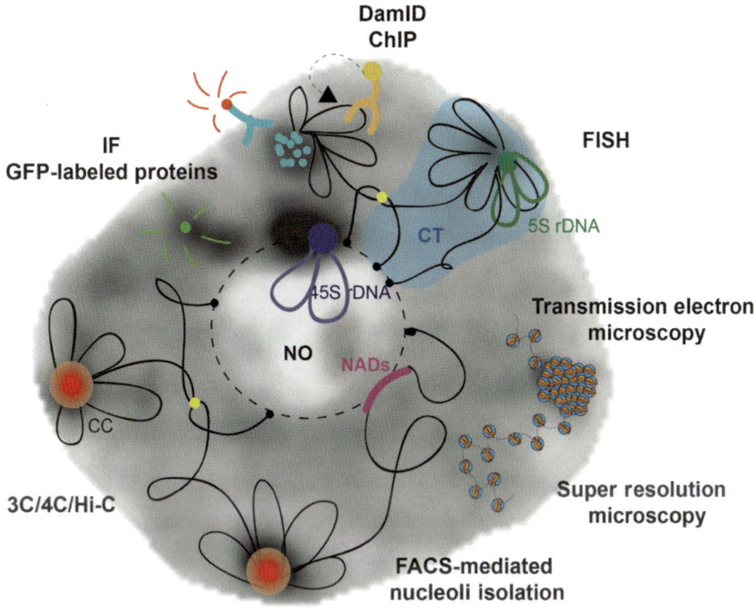

Fig. 1 Schematic representation of the different techniques described in this section overview to address nuclear organization exemplified for Arabidopsis. The model plant Arabidopsis comprises a small genome of about 135 Mb organized in five chromosomes, on which repetitive sequences and transposons are concentrated within the centromeric and pericentromeric regions (*red circles*) that include the 5S rDNA repeats (*green*). The 45S rRNA genes form long tandem repetitive arrays termed nucleolus organizer regions (NORs, *blue*). The mostly heterochromatic repetitive regions of the genome cluster in DAPI-bright chromocenters (CC) and have been studied extensively by fluorescence in situ hybridization (*FISH*). Immunofluorescence staining (*IF*) can be used to visualize histone modifications (here shown as *light blue dots* enriched at a chromocenter) with specific antibodies and *GFP-labeled chromatin binding proteins* can be imaged directly under a fluorescence microscope. The *DamID* technique that relies on chromatin-binding proteins coupled to a DNA adenine methyltransferase (*orange*) and Chromatin Immunoprecipitation (*ChIP*) have been applied to study the genome-wide enrichment of chromatin proteins. Genomic regions in close proximity of the nucleolus (NO), termed nucleolus-associated domains (NADs, *pink*) are identified by amplification or sequencing of genomic regions associated with *nucleoli* that were *isolated by FACS*. Finally, chromatin conformation capture (*3C*) techniques, in particular *Hi-C* have been used to reveal intrachromosomal interactions between centromere proximal regions of the chromosomes (centromere-proximal loops) and pericentromeres confirming the rosette like organization observed by FISH [85]. Furthermore, frequent interchromosomal interactions (such as the interactive heterochromatic islands or KNOT engaged regions, *yellow points*) implicating more distal chromosomal regions have been detected [71, 72]. Telomeres (*black points*) are clustered around the nucleolus. Drawings are superposed on a maximum Z projection of an Arabidopsis 3D nucleus from cotyledon epidermis (pavement cell) that is stained with a DNA intercalating-agent (DAPI). Euchromatin appears as *light grey*, chromocentres as *dark grey* foci and the nucleolus (NO) as a *white* region at the center of the nucleus. Chromatin density distribution in the nucleus can be analyzed by *transmission electron microscopy* or *super-resolution microscopy*

(flattened) nuclei or in 3D after segmentation of nuclei and heterochromatic domains (Table 1, Chapter 33 [15]).

Applications. Applications in plants include the characterization of nuclear size and shape and heterochromatin organization in wild-type *Arabidopsis thaliana* (Arabidopsis) plants and different mutants known to affect protein components of the nuclear periphery [17, 86].

2.2 Fluorescence In Situ Hybridization (FISH) Approaches to Detect Specific Genomic Regions Within the Nucleus

Technical considerations. To study the colocalization frequency of genomic loci or their localization relative to nuclear compartments (nucleolus, nuclear periphery, and nuclear bodies) these loci can be specifically labeled using *fluorescence in situ hybridization (FISH)*. FISH generally involves tissue fixation or isolation of nuclei, permeabilization followed by denaturation of DNA to permit the subsequent hybridization of the labeled probes to their complementary sequences (*see* also Chapter 30 [25]). Isolated, spread (flattened) nuclei are widely used for 2D FISH to study the organization of specific genomic loci or repetitive regions along chromosomes or relative to subnuclear structures [87, 88]. By contrast, FISH on intact nuclei can provide information on the 3D distances between different loci or between loci and nuclear compartments (*see* also Chapter 27 [22]). The 3D organization of chromatin in plant nuclei at interphase has been thoroughly characterized in Arabidopsis through several well-established and more recent techniques (Fig. 1, Table 1). Depending on the biological question, FISH probes can consist of specific genomic repeats (repeat-FISH), bacterial artificial chromosomes (BAC) clones covering 50–100 kb chromosomal regions (BAC-FISH), short oligonucleotides (oligo-FISH), PCR-based amplicons covering specific loci (gene-FISH), or whole genomic DNA (genomic in situ hybridization (GISH)). FISH probes can be directly labeled either by terminal conjugation of oligonucleotides with a fluorophore or by incorporation of nucleotides conjugated to fluorophores. Alternatively, probes can be indirectly labeled through incorporation of nucleotides conjugated to for example biotin or digoxigenin. While direct labeling offers the benefit of direct microscopic observation following hybridization washes, it suffers from poor incorporation of fluorescently labeled nucleotides [89]. By contrast, indirect labeling requires subsequent detection by antibodies coupled to fluorophores, which allows for signal amplification, but may increase background staining. Signal enhancement can also be achieved by amplifying the copy number of the DNA target to allow for more probes to bind. A recent study reported on FISH detection of single-copy genes in Arabidopsis with the use of padlock probes and in situ rolling circle amplification [49].

FISH in plants is mostly carried out on nuclear spreads from ethanol–acetic acid-fixed tissue [26] or isolated nuclei from paraformaldehyde (PFA)-fixed tissue [90, 91] that preserve the three-dimensional structure. Plant cells are surrounded by cell walls and exhibit natural fluorescence, which renders many of the sample preparation procedures generally employed in animal cells more difficult. Many approaches, in particular the study of nuclei within their tissue context, still face the challenge of limited probe and antibody accessibility and background noise. Nevertheless, different protocols have been developed including cryosections [92], acrylamide embedded [28, 29] or fixed whole-mount tissues, which are expected to reveal new information on the three-dimensional organization of chromosomal domains in a manner that conserves positional information of the cell within the tissue (*see* also Chapters 25, 27, and 32 of this issue [15, 22, 50]).

An inherent drawback of all FISH techniques is the requirement for denaturation of DNA mostly achieved by heat and formamide treatment, which can cause changes in nuclear morphology and chromatin organization. Low-temperature FISH procedures may offer an interesting alternative preserving chromosome integrity [93] but have, however, not been tested in plant cells yet.

Alternatively, the localization of specific loci in nuclear space can be studied using LacO arrays inserted at the locus of interest that are then bound by coexpressed LacI proteins coupled to fluorescent proteins [55, 56, 94], (Table 1). More recently, also in vivo labeling of endogenous genomic sequences has been realized in living Arabidopsis plants using transcription activator-like effectors (TALEs) coupled to fluorescent proteins [95]. Furthermore, the CASFISH procedure that relies on CRISPR-associated caspase 9 (Cas9) complexes may be used to detect repetitive sequences [96, 97].

Applications. By no means exhaustive, a few illustrative examples of FISH procedures applied in plants include: repeat-FISH, which has been used for the cytogenetic characterization of different species (e.g., by mapping rDNA loci [98]) or to study the organization of repetitive sequences in nuclear space on spread (flattened) nuclei [99] or in whole-mount sections [30]. BAC-FISH was used to position specific genomic regions relative to each other or to heterochromatic domains and to construct comprehensive cytogenetic maps using pachytene chromosomes revealing genome rearrangements [45, 88], to paint chromosomes to investigate chromosomal territory arrangements in interphase [30, 38, 39, 100] or to study chromatin condensation in different cell types [31]. Multiple oligonucleotide probes mapping to the same gene have been combined in Arabidopsis to visualize nascent and mature mRNA transcripts of *FLOWERING LOCUS C (FLC)* and to quantify mRNA expression in a cell-specific manner [101, 102]. While the visualization of nascent transcripts reveals the position of the gene locus, this requires that the gene is transcriptionally active.

To detect a genomic locus irrespective of its transcriptional state, gene sequences have been amplified using in situ rolling circle amplification to increase targets for the subsequent FISH probes. This technique has been applied to reveal the repositioning of light-inducible genes toward the nuclear periphery upon activation [49]. Finally, GISH, which uses total genomic DNA as probe [51], is applicable even for small plant genomes [52] and allowed characterization of wheat–barley introgression lines [53].

2.3 Immunolabeling to Analyze the Qualitative and Quantitative Distribution of Chromatin Modifications and Chromatin Proteins

Technical considerations. While FISH visualizes genomic regions, the nuclear distribution of specific proteins, post-translational modifications of histones or DNA methylation can be analyzed after detection by specific antibodies in *immunofluorescence staining*. Nuclei are fixed with PFA to preserve the protein components of chromatin and permeabilized to allow access of the specific primary antibodies and their subsequent revelation by fluorescently labeled secondary antibodies. Immunofluorescence staining techniques can be combined with FISH to study the enrichment of specific proteins such as histone modifications at a specific locus [103] (*see* Chapters 24, 25, and 27 [22, 50, 104]). Furthermore, chromatin components such as histones and histone variants can be expressed as fusion proteins with fluorescent proteins to image localization or to investigate their dynamics over time [105, 106] (*see* also Chapter 26 [62]). Interesting alternatives to antibodies specific for methylated DNA are the newly developed fluorescent dynamic sensors of DNA methylation (DYNAMETs), which rely on DNA methylation binding proteins coupled to fluorescent proteins [107].

Applications. Immunofluorescence on isolated nuclei has been applied to study the distribution of histone modifications during the cell cycle [90], to investigate the interdependence of DNA methylation and histone modifications [83, 108, 109], the stability of the epigenome over generations [110] or to characterize mutants deficient in the setting of DNA or histone methylation [111]. Fluorescently tagged proteins have helped for example to reveal the transmission of histone H3 variants through generations [112], the reprogramming of chromatin organization during the somatic to reproductive cell fate transition [113] and to follow the dynamics of centromere position over time [105].

3 Molecular Biology Techniques to Assess Nuclear Architecture

More recently, techniques have been developed that probe DNA–DNA or chromatin protein–DNA interactions. Coupled to next-generation sequencing (NGS) these techniques can provide a genome-wide, but cell population-averaged view of these interactions.

3.1 Chromatin-Conformation Capture Approaches to Study the Spatial Organization of Chromatin in the Nucleus

Technical considerations. Intrachromosomal and interchromosomal interactions between different loci can be studied using *chromatin-conformation capture* (3C) based assays that rely on formaldehyde crosslinking of genomic regions situated in spatial proximity [65, 68] (Fig. 1, Table 1). After cross-linking, DNA is digested with restriction enzymes and interacting regions are ligated and amplified by PCR or quantified by sequencing. The original 3C approach investigates the interaction between two loci, requiring prior knowledge on genomic loci that are potentially situated in close proximity in nuclear space (*see* Chapters 14 and 15 [63, 64]). Derivatives of this method coupled to sequencing (such as Hi-C, [114]) can now generate genome-wide interacting maps. The power of this technique resides in the simultaneous study of multiple interactions occurring simultaneously in the nucleus. A general drawback of 3C techniques is that they provide a static, cell population averaged view and do not consider cell-to-cell variability within a population. To overcome this issue, Hi-C has been carried out on single cells [115]. Technical biases might be introduced through the unequal distribution of restriction enzymes cutting sites in the genome. Therefore, DNA can be digested using DNase I [116] or micrococcal nuclease [117], thereby increasing the resolution up to the nucleosomal level. It needs to be kept in mind that the 3C-based techniques provide information on the spatial proximity of genomic regions, but do not deliver any information about the subnuclear compartment where such interactions occur. Such spatial information may be gained from new technical developments such as Chromatin Interaction Analysis by Paired-End Tag Sequencing (ChIA-PET-seq) that combines chromatin-conformation capture and chromatin immunoprecipitation (ChIP) [118] (reviewed in Chapter 14 [63]), but which has not yet been adapted to plants.

Applications. In plants, 3C experiments have been applied for example in maize to reveal epiallele-specific establishment of chromatin loops at the maize *b1* locus [66, 67] or the formation of gene loops at the *FLC* locus [119]. More recently, 4C has been used to analyze genome-wide interactions of individual loci [69] and Hi-C techniques at different levels of resolution have been implemented to study genomic interactions in the wild-type Arabidopsis genome and to understand how these local or long-range interactions are affected in mutants with altered nuclear size, heterochromatin organization, or specific epigenetic marks [70–74].

3.2 DamID and ChIP to Profile Chromatin Protein-DNA Associations

Technical considerations. To provide information on the proximity of a particular chromatin region to a specific nuclear compartment such as the nuclear lamina, the *DamID* technique has been developed (Fig. 1, Table 1). DamID relies on the detection of adenines methylated by a DNA adenine methyltransferase (Dam) fused to chromatin binding proteins [120] or nuclear lamina components

[75, 121]. Similar to Chromatin Immunoprecipitation (ChIP) coupled to sequencing, DamID allows for probing protein–DNA interactions in a genome-wide manner. While ChIP-Seq is widely used to map chromatin-binding proteins (*see* Chapter 5 [122]), DamID might be advantageous for specific biological questions as it allows visualizing more transient interactions (Table 1).

Applications. In plants, ChIP coupled to qPCR or sequencing has been extensively applied to map histone modifications [1, 2, 123], histone variants [124–126], or chromatin binding proteins [127, 128]. DamID was successfully used to map genome-wide binding sites of the chromatin binding protein LIKE HETERO-CHROMATIN PROTEIN-1 (LHP1) [76, 129].

Finally, the isolation of specific subnuclear structures such as nucleoli has been realized using ***Fluorescence activated cell sorting*** and applied to define DNA sequences by NGS that reside in close spatial proximity to nucleoli [12, 77] (*see* Chapter 7 [79]).

4 Conclusions and Perspectives

Each of the different approaches described here has its intrinsic bias, technical advantages or downsides (Table 1). Depending on the scientific question and the availability of tools, equipment and resources for downstream analyses, the appropriate approach should be carefully chosen.

The imaging and detection of FISH probes, antibodies or GFP-fusion proteins depends on the choice of appropriate imaging devices. For 2D nuclear spreads, standard epifluorescence microscopes are sufficient, while the detection of FISH signals, immunofluorescence signals or fluorescent proteins in 3D nuclear space requires the acquisition of multiple Z-stacks with confocal microscopes or epifluorescence microscopes equipped with structural illumination devices. Super-resolution microscopy techniques such as structured illumination microscopy (SIM) and photoactivated localization microscopy (PALM) have now been used in plants [61, 130]. In situ labeling together with microscopy imaging and image processing allow for single-cell level quantifications and offer the general advantage to provide information on chromatin arrangement at the single cell level within a tissue, or within a population of isolated cells (Table 1, *see* also Chapter 31 [19]). Hence these approaches are invaluable for their assessment of cell-to-cell variation in chromatin organization. On the other hand, microscopy is an inherently low throughput approach, due to manual preparation of individual samples, image recording and analysis. Automated systems, such as deep imaging systems, which allow high-content, high-resolution screening, are now being developed and can reduce analysis times in the future [131].

The power of the different techniques lies in their complementarity as interactions determined by 3C methods or a localization deduced from DamID experiments could be validated by 3D FISH experiments. Some technical challenges remain to be overcome to facilitate the study of plant nuclear architecture such as the further implication of live cell imaging approaches to investigate the dynamics of nuclear organization. In this respect, the development of methods to study the dynamics of proteins in the nucleus similar to the SNAP-Tag technology [132] will be important; photoconvertible fluorescent proteins [133] might be an interesting step in this direction. As the role of noncoding RNAs in genome organization [134] is becoming more and more recognized, it will also be interesting to further develop existing RNA-FISH protocols [102, 135]. As discussed above, a major limit of the 3C-based technologies is that they provide an average view of the cell population, which may mask specific interactions. While single cell Hi-C approaches have been carried out in mammalian cells, using a single cell suspension [115], similar approaches cannot be applied in plants. Enriching a particular cell population by INTACT technology [136] (*see* also Chapter 8 [137]) prior to 3C based techniques might be an alternative to improve the detection of functionally relevant details of genome organization in plants. Finally, the use of spatial statistics and stochastic geometry to model 3D nuclear organization in plants will contribute to a better understanding of plant nuclear organization (*see* for example Chapter 29) [16, 86].

Most of our information on genome organization is derived from work in Arabidopsis, which has a small genome, forms conspicuous chromocenters and adopts a rosette like organization. The genomes of many important crop species such as rye, barley, wheat and oat, however, are far richer in transposable elements and other repetitive sequences and were shown to rather adopt a Rabl chromosome arrangement, in which centromeres cluster on one and telomeres on the other side of the nucleus [138]. Much remains therefore to be discovered concerning the 3D arrangement of these large and highly repetitive genomes.

Acknowledgments

I thank S. Desset and C. Tatout for critical reading and C. Baroux for helpful editorial suggestions. This work was supported by ANR grant 'SINODYN' ANR-12 ISV6 0001, the Centre National de la Recherche Scientifique, the Institut National de la Santé et de la Recherche Médicale and the University Clermont Auvergne.

References

1. Roudier F, Ahmed I, Bérard C et al (2011) Integrative epigenomic mapping defines four main chromatin states in Arabidopsis. EMBO J 30:1928–1938
2. Sequeira-Mendes J, Aragüez I, Peiró R et al (2014) The functional topography of the Arabidopsis genome is organized in a reduced number of linear motifs of chromatin states. Plant Cell. doi:10.1105/tpc.114.124578
3. Cremer T, Cremer C (2001) Chromosome territories, nuclear architecture and gene regulation in mammalian cells. Nat Rev Genet 2:292–301
4. Dixon JR, Selvaraj S, Yue F et al (2012) Topological domains in mammalian genomes identified by analysis of chromatin interactions. Nature 485:376–380. doi:10.1038/nature11082
5. de Laat W, Duboule D (2013) Topology of mammalian developmental enhancers and their regulatory landscapes. Nature 502:499–506. doi:10.1038/nature12753
6. Lopes R, Korkmaz G, Agami R (2016) Applying CRISPR-Cas9 tools to identify and characterize transcriptional enhancers. Nat Rev Mol Cell Biol 17:597–604. doi:10.1038/nrm.2016.79
7. Osborne CS, Chakalova L, Brown KE et al (2004) Active genes dynamically colocalize to shared sites of ongoing transcription. Nat Genet 36:1065–1071. doi:10.1038/ng1423
8. Guelen L, Pagie L, Brasset E et al (2008) Domain organization of human chromosomes revealed by mapping of nuclear lamina interactions. Nature 453:948–951. doi:10.1038/nature06947
9. Pickersgill H, Kalverda B, de Wit E et al (2006) Characterization of the Drosophila melanogaster genome at the nuclear lamina. Nat Genet 38:1005–1014. doi:10.1038/ng1852
10. Németh A, Conesa A, Santoyo-Lopez J et al (2010) Initial genomics of the human nucleolus. PLoS Genet. doi:10.1371/journal.pgen.1000889
11. van Koningsbruggen S, Gierliński M, Schofield P et al (2010) High-resolution whole-genome sequencing reveals that specific chromatin domains from most human chromosomes associate with nucleoli. Mol Biol Cell 21:3735–3748. doi:10.1091/mbc.E10-06-0508
12. Pontvianne F, Carpentier M-C, Durut N et al (2016) Identification of nucleolus-associated chromatin domains reveals a role for the nucleolus in 3D organization of the A. thaliana genome. Cell Rep 16:1574–1587. doi:10.1016/j.celrep.2016.07.016
13. Pope BD, Ryba T, Dileep V et al (2014) Topologically associating domains are stable units of replication-timing regulation. Nature 515:402–405. doi:10.1038/nature13986
14. Seeber A, Gasser SM (2017) Chromatin organization and dynamics in double-strand break repair. Curr Opin Genet Dev 43:9–16. doi:10.1016/j.gde.2016.10.005
15. Desset S, Poulet A, Tatout C (2017) Quantitative 3D analysis of nuclear morphology and heterochromatin organization from whole mount plant tissue using nucleus. In: Bemer M, Baroux C (eds) Plant chromatin dynamics: methods and protocols. Springer, New York, NY. doi:10.1007/978-1-4939-7318-7_33
16. Arpon J, Gaudin V, Andrey P (2017) A method for testing random spatial model on nuclear object distributions. In: Bemer M, Baroux C (eds) Plant chromatin dynamics: methods and protocols. Springer, New York, NY. doi:10.1007/978-1-4939-7318-7_29
17. Poulet A, Duc C, Voisin M et al (2017) The LINC complex contributes to heterochromatin organisation and transcriptional gene silencing in plants. J Cell Sci 130:590–601
18. Poulet A, Arganda-Carreras I (2014) NucleusJ: an ImageJ plugin for quantifying 3D images of interphase nuclei. Bioinformatics 3:1144–1146
19. Baroux C, Schubert V (2017) Microscopy and image processing tools to analyse plant chromatin – practical considerations. In: Bemer M, Baroux C (eds) Plant chromatin dynamics: methods and protocols. Springer, New York, NY. doi:10.1007/978-1-4939-7318-7_31
20. Poulet A, Arganda-Carreras I, Legland D et al (2015) NucleusJ: an ImageJ plugin for quantifying 3D images of interphase nuclei. Bioinformatics 31:1144–1146. doi:10.1093/bioinformatics/btu774
21. Fabrice TN, Cherkezyan L, Ringli C, Baroux C (2017) Transmission electron microscopy imaging to analyse chromatin density distribution at the nanoscale level. In: Bemer M, Baroux C (eds) Plant chromatin dynamics:

methods and protocols. Springer, New York, NY. doi:10.1007/978-1-4939-7318-7_34

22. Bey TD, Koini M, Fransz PF (2017) Fluorescence in situ hybridization (FISH) and immunolabeling on 3D preserved nuclei. In: Bemer M, Baroux C (eds) Plant chromatin dynamics: methods and protocols. Springer, New York, NY. doi:10.1007/978-1-4939-7318-7_27
23. Ashenafi MS, Baroux C (2017) Automated 3D gene position analysis using a customized Imaris plugin. In: Bemer M, Baroux C (eds) Plant chromatin dynamics: methods and protocols. Springer, New York, NY. doi:10.1007/978-1-4939-7318-7_32
24. Simon L, Probst AV (2017) High-affinity LNA/DNA mixmer probes for detection of chromosome-specific polymorphisms of 5S rDNA repeats in *A. thaliana*. In: Bemer M, Baroux C (eds) Plant chromatin dynamics: methods and protocols. Springer, New York, NY. doi:10.1007/978-1-4939-7318-7_28
25. Bačovský V, Hobza R, Vyskot B (2017) Technical review: cytogenetic tools for studying mitotic chromosomes. In: Bemer M, Baroux C (eds) Plant chromatin dynamics: methods and protocols. Springer, New York, NY. doi:10.1007/978-1-4939-7318-7_30
26. Lysak M, Fransz P, Schubert I (2006) Cytogenetic analyses of Arabidopsis. Methods Mol Biol 323:173–186. doi:10.1385/1-59745-003-0:173
27. Kato A, Lamb JC, J a B (2004) Chromosome painting using repetitive DNA sequences as probes for somatic chromosome identification in maize. Proc Natl Acad Sci U S A 101:13554–13559. doi:10.1073/pnas.0403659101
28. She W, Grimanelli D, Baroux C (2014) An efficient method for quantitative, single-cell analysis of chromatin modification and nuclear architecture in whole-mount ovules in Arabidopsis. J Vis Exp:1–9. doi:10.3791/51530
29. Howe ES, Murphy SP, Bass HW (2013) Three-dimensional acrylamide fluorescence in situ hybridization for plant cells. In: Pawlowski WP, Grelon M, Armstrong S (eds) Plant meiosis methods in protocols. Humana, Totowa, NJ, pp 53–66
30. Berr A, Schubert I (2007) Interphase chromosome arrangement in Arabidopsis thaliana is similar in differentiated and meristematic tissues and shows a transient mirror symmetry after nuclear division. Genetics 176:853–863. doi:10.1534/genetics.107.073270
31. Costa S, Shaw P (2006) Chromatin organization and cell fate switch respond to positional information in Arabidopsis. Nature 439:493–496
32. Bauwens S, Van Oostveldt P, Engler G, Van Montagu M (1991) Distribution of the rDNA and three classes of highly repetitive DNA in the chromatin of interphase nuclei of Arabidopsis thaliana. Chromosoma 101:41–48. doi:10.1007/BF00360685
33. Raissig MT, Gagliardini V, Jaenisch J et al (2013) Efficient and rapid isolation of early-stage embryos from Arabidopsis thaliana seeds. J Vis Exp 7:e50371. doi:10.3791/50371
34. Gernand D, Rutten T, Varshney A, et al (2005) Uniparental chromosome elimination at mitosis and interphase in wheat and pearl millet crosses involves micronucleus formation, progressive heterochromatinization, and DNA fragmentation. 17:2431–2438. doi:10.1105/tpc.105.034249
35. Wegel E, Koumproglou R, Shaw P, Osbourn A (2009) Cell type-specific chromatin decondensation of a metabolic gene cluster in oats. Plant Cell 21:3926–3936. doi:10.1105/tpc.109.072124
36. Santos AP, Wegel E, Allen GC et al (2006) In situ methods to localize transgenes and transcripts in interphase nuclei: a tool for transgenic plant research. Plant Methods 2:18. doi:10.1186/1746-4811-2-18
37. Prieto P, Moore G, Shaw P (2007) Fluorescence in situ hybridization on vibratome sections of plant tissues. Nat Protoc 2:1831–1838. doi:10.1038/nprot.2007.265
38. Pecinka A, Schubert V, Meister A et al (2004) Chromosome territory arrangement and homologous pairing in nuclei of Arabidopsis thaliana are predominantly random except for NOR-bearing chromosomes. Chromosoma 113:258–269
39. Baroux C, Pecinka A, Fuchs J et al (2016) Non-random chromosome arrangement in triploid endosperm nuclei. Chromosoma 126:115–124. doi:10.1007/s00412-016-0578-5
40. Berr A, Pecinka A, Meister A et al (2006) Chromosome arrangement and nuclear architecture but not centromeric sequences are conserved between Arabidopsis thaliana and Arabidopsis lyrata. Plant J 48:771–783. doi:10.1111/j.1365-313X.2006.02912.x
41. Schubert V, Kim YM, Schubert I (2008) Arabidopsis sister chromatids often show complete alignment or separation along a 1.2-Mb euchromatic region but no cohesion "hot spots". Chromosoma 117:261–266. doi:10.1007/s00412-007-0141-5

42. Sun J, Zhang Z, Zong X et al (2013) A high-resolution cucumber cytogenetic map integrated with the genome assembly. BMC Genomics 14:461. doi:10.1186/1471-2164-14-461
43. Mandáková T, Lysak MA (2016) Chromosome preparation for cytogenetic analyses in Arabidopsis. Curr Protoc Plant Biol. John Wiley & Sons, Inc, Hoboken, NJ, pp 43–51
44. Doležel J, Vrána J, Cápal P et al (2014) Advances in plant chromosome genomics. Biotechnol Adv 32:122–136. doi:10.1016/j.biotechadv.2013.12.011
45. Fransz P, Linc G, Lee C-R et al (2016) Molecular, genetic and evolutionary analysis of a paracentric inversion in Arabidopsis thaliana. Plant J. doi:10.1111/tpj.13262
46. Mandáková T, M a L (2008) Chromosomal phylogeny and karyotype evolution in x=7 crucifer species (Brassicaceae). Plant Cell 20:2559–2570. doi:10.1105/tpc.108.062166
47. Kellogg EE a, Bennetzen JJL (2004) The evolution of nuclear genome structure in seed plants. Am J Bot 91:1709–1725. doi:10.3732/ajb.91.10.1709
48. Heslop-Harrison JS, Schwarzacher T (2011) Organisation of the plant genome in chromosomes. Plant J 66:18–33
49. Feng C, Qiu Y, Buskirk EK Van, et al (2014) Light-regulated gene repositioning in Arabidopsis. Nat Commun 5:1–9. doi:10.1038/ncomms4027
50. She W, Baroux C, Grossniklaus U (2017) Cell-type specific chromatin analysis in whole-mount plant tissues by immunostaining. In: Bemer M, Baroux C (eds) Plant chromatin dynamics: methods and protocols. Springer, New York, NY. doi:10.1007/978-1-4939-7318-7_25
51. Silva GS, Souza MM (2013) Genomic in situ hybridization in plants. Genet Mol Res 12:2953–2965. doi:10.4238/2013.August.12.11
52. Ali HBM, Lysak MA, Schubert I (2004) Genomic in situ hybridization in plants with small genomes is feasible and elucidates the chromosomal parentage in interspecific Arabidopsis hybrids. Genome 47:954–960. doi:10.1139/g04-041
53. Türkösi E, Cseh A, Darkó É, Molnár-láng M (2016) Addition of Manas barley chromosome arms to the hexaploid wheat genome. BMC Genet 17:871. doi:10.1186/s12863-016-0393-2
54. Younis A, Ramzan F, Hwang YJ, Lim KB (2015) FISH and GISH: molecular cytogenetic tools and their applications in ornamental plants. Plant Cell Rep 34:1477–1488. doi:10.1007/s00299-015-1828-3
55. Rosa S, De Lucia F, Mylne JS et al (2013) Physical clustering of FLC alleles during polycomb-mediated epigenetic silencing in vernalization. Genes Dev 27:1845–1850. doi:10.1101/gad.221713.113
56. Matzke AJ, Watanabe K, Winden J et al (2010) High frequency, cell type-specific visualization of fluorescent-tagged genomic sites in interphase and mitotic cells of living Arabidopsis plants. Plant Methods 6:1–11
57. Kato N, Lam E (2001) Detection of chromosomes tagged with green fluorescent protein in live Arabidopsis thaliana plants. Genome Biol 2:RESEARCH0045.
58. Smith S, Galinha C, Desset S et al (2015) Marker gene tethering by nucleoporins affects gene expression in plants. Nucleus 1034:00–00. doi:10.1080/19491034.2015.1126028
59. Pecinka A, Kato N, Meister A et al (2005) Tandem repetitive transgenes and fluorescent chromatin tags alter local interphase chromosome arrangement in Arabidopsis thaliana. J Cell Sci 118:3751–3758
60. Ricci MA, Manzo C, García-Parajo MF et al (2015) Chromatin fibers are formed by heterogeneous groups of nucleosomes in vivo. Cell 160:1145–1158. doi:10.1016/j.cell.2015.01.054
61. Schubert V, Weisshart K (2015) Abundance and distribution of RNA polymerase II in Arabidopsis interphase nuclei. J Exp Bot 66:1687–1698. doi:10.1093/jxb/erv091
62. Rosa S (2017) Measuring dynamics of histone proteins by photobleaching in Arabidopsis roots. In: Bemer M, Baroux C (eds) Plant chromatin dynamics: methods and protocols. Springer, New York, NY. doi:10.1007/978-1-4939-7318-7_26
63. Grob S, Cavalli G (2017) A Hitchhiker's guide to chromosome conformation capture. In: Bemer M, Baroux C (eds) Plant chromatin dynamics: methods and protocols. Springer, New York, NY. doi:10.1007/978-1-4939-7318-7_14
64. Weber B, Jamge S, Stam ME (2017) 3C in maize and Arabidopsis. In: Bemer M, Baroux C (eds) Plant chromatin dynamics: methods and protocols. Springer, New York, NY. doi:10.1007/978-1-4939-7318-7_15
65. Dekker J, Rippe K, Dekker M, Kleckner N (2002) Capturing chromosome conformation. Science 295:1306–1311
66. Louwers M, Splinter E, van Driel R et al (2009) Studying physical chromatin

interactions in plants using chromosome conformation capture (3C). Nat Protoc 4:1216–1229. doi:10.1038/nprot.2009.113

67. Louwers M, Bader R, Haring M et al (2009) Tissue- and expression level-specific chromatin looping at maize b1 epialleles. Plant Cell 21:832–842. doi:10.1105/tpc.108.064329
68. Sati S, Cavalli G (2016) Chromosome conformation capture technologies and their impact in understanding genome function. Chromosoma 126:33–44. doi:10.1007/s00412-016-0593-6
69. Grob S, Schmid MW, Luedtke NW et al (2013) Characterization of chromosomal architecture in Arabidopsis by chromosome conformation capture. Genome Biol 14: R129. doi:10.1186/gb-2013-14-11-r129
70. Moissiard G, Cokus SJ, Cary J et al (2012) MORC family ATPases required for heterochromatin condensation and gene silencing. Science 336(6087):1448–1451
71. Grob S, Schmid MW, Grossniklaus U (2014) Hi-C analysis in Arabidopsis identifies the KNOT, a structure with similarities to the flamenco locus of Drosophila. Mol Cell 55 (5):678–693. doi:10.1016/j.molcel.2014.07.009
72. Feng S, Cokus SJ, Schubert V et al (2014) Genome-wide Hi-C analyses in wild-type and mutants reveal high-resolution chromatin interactions in Arabidopsis. Mol Cell 55:694–707. doi:10.1016/j.molcel.2014.07.008
73. Wang C, Liu C, Roqueiro D et al (2015) Genome-wide analysis of local chromatin packing in *Arabidopsis thaliana*. Genome Res 25:246–256. doi:10.1101/gr.170332.113
74. Liu C, Wang C, Wang G et al (2016) Genome-wide analysis of chromatin packing in *Arabidopsis thaliana* at single-gene resolution. Genome Res 26(8):1057–1068. doi:10.1101/gr.204032.116
75. Cléard F, Karch F, Maeda RK (2014) DamID as an approach to studying long-distance chromatin interactions. In: Graba Y, Rezsohazy R (eds) Hox genes – methods and protocols. Springer, New York, NY, pp 279–289
76. Germann S, Juul-Jensen T, Letarnec B, Gaudin V (2006) DamID, a new tool for studying plant chromatin profiling in vivo, and its use to identify putative LHP1 target loci. Plant J 48:153–163
77. Pontvianne F, Blevins T, Chandrasekhara C et al (2013) Subnuclear partitioning of rRNA genes between the nucleolus and nucleoplasm reflects alternative epiallelic states. Genes Dev 27:1545–1550
78. Pontvianne F, Boyer-Clavel M, Sáez-Vásquez J (2016) Fluorescence-activated nucleolus sorting in Arabidopsis. In: Németh A (ed) The nucleolus – methods and protocols Springer New York, NYs, pp 203–211
79. Carpentier M-C, Picart-Picolo A, Pontvianne F (2017) A method to identify nucleolus-associated chromatin domains (NADs). In: Bemer M, Baroux C (eds) Plant chromatin dynamics: methods and protocols. Springer, New York, NY. doi:10.1007/978-1-4939-7318-7_7
80. Heitz E (1928) Das heterochromatin der Moose. Jb Wiss Bot 69:728–818
81. Cremer T, Cremer C (2006) Rise, fall and resurrection of chromosome territories: a historical perspective. Part II. Fall and resurrection of chromosome territories during the 1950s to 1980s. Part III. Chromosome territories and the functional nuclear architecture: experiments and. Eur J Histochem 50:223–272
82. Voss U, Larrieu A, Wells DM (2013) From jellyfish to biosensors: the use of fluorescent proteins in plants. Int J Dev Biol 57:525–533. doi:10.1387/ijdb.130208dw
83. Soppe WJ, Jasencakova Z, Houben A et al (2002) DNA methylation controls histone H3 lysine 9 methylation and heterochromatin assembly in Arabidopsis. EMBO J 21:6549–6559
84. Jasencakova Z, Soppe WJJ, Meister A et al (2003) Histone modifications in Arabidopsis-high methylation of H3 lysine 9 is dispensable for constitutive heterochromatin. Plant J 33:471–480
85. Fransz P, de Jong JH, Lysak M et al (2002) Interphase chromosomes in Arabidopsis are organized as well defined chromocenters from which euchromatin loops emanate. Proc Natl Acad Sci U S A 99:14584–14589
86. Andrey P, Kiêu K, Kress C et al (2010) Statistical analysis of 3D images detects regular spatial distributions of centromeres and chromocenters in animal and plant nuclei. PLoS Comput Biol 6:e1000853
87. Koornneef M, Fransz P, de Jong H (2003) Cytogenetic tools for Arabidopsis thaliana. Chromosom Res 11:183–194
88. Fransz PF, Armstrong S, De Jong JH et al (2000) Integrated cytogenetic map of chromosome arm 4S of A. thaliana: structural organization of heterochromatic knob and centromere region. Cell 100:367–376

89. Tasara T, Angerer B, Damond M et al (2003) Incorporation of reporter molecule-labeled nucleotides by DNA polymerases. II. High-density labeling of natural DNA. Nucleic Acids Res 31:2636–2646. doi:10.1093/nar/gkg371
90. Jasencakova Z, Meister a WJ et al (2000) Histone H4 acetylation of euchromatin and heterochromatin is cell cycle dependent and correlated with replication rather than with transcription. Plant Cell 12:2087–2100
91. Bowler C, Benvenuto G, Laflamme P et al (2004) Chromatin techniques for plant cells. Plant J 39:776–789
92. Tirichine L, Andrey P, Biot E et al (2009) 3D fluorescent in situ hybridization using Arabidopsis leaf cryosections and isolated nuclei. Plant Methods 5:11
93. Winkler R, Perner B, Rapp A et al (2003) Labelling quality and chromosome morphology after low temperature FISH analysed by scanning far-field and near-field optical microscopy. J Microsc 209:23–33. doi:10.1046/j.1365-2818.2003.01101.x
94. Rosin FM, Watanabe N, Cacas JL et al (2008) Genome-wide transposon tagging reveals location-dependent effects on transcription and chromatin organization in Arabidopsis. Plant J 55:514–525
95. Fujimoto S, Sugano SS, Kuwata K et al (2016) Visualization of specific repetitive genomic sequences with fluorescent TALEs in Arabidopsis thaliana. J Exp Bot 67:6101–6110. doi:10.1093/jxb/erw371
96. Deng W, Shi X, Tjian R et al (2015) CASFISH: CRISPR/Cas9-mediated in situ labeling of genomic loci in fixed cells. Proc Natl Acad Sci U S A 112:11870–11875. doi:10.1073/pnas.1515692112
97. Dreissig S, Schiml S, Schindele P et al (2017) Live cell CRISPR-imaging in plants reveals dynamic telomere movements. Plant J:0–3. doi:10.1111/tpj.13601
98. Garcia S, Kovařík A, Leitch AR, Garnatje T (2016) Cytogenetic features of rRNA genes across land plants: analysis of the Plant rDNA database. Plant J. doi:10.1111/tpj.13442
99. Probst A V, Fransz PF, Paszkowski J, Mittelsten Scheid O (2003) Two means of transcriptional reactivation within heterochromatin. Plant J 33:743–749
100. Schubert V, Rudnik R, Schubert I (2014) Chromatin associations in Arabidopsis interphase nuclei. Front Genet 5:1–11. doi:10.3389/fgene.2014.00389
101. Rosa S, Duncan S, Dean C (2016) Mutually exclusive sense–antisense transcription at FLC facilitates environmentally induced gene repression. Nat Commun 7:13031. doi:10.1038/ncomms13031
102. Duncan S, Olsson TSG, Hartley M et al (2016) A method for detecting single mRNA molecules in Arabidopsis thaliana. Plant Methods 12:13. doi:10.1186/s13007-016-0114-x
103. Chaumeil J, Augui S, Chow JC, Heard E (2008) Combined immunofluorescence, RNA fluorescent in situ hybridization, and DNA fluorescent in situ hybridization to study chromatin changes, transcriptional activity, nuclear organization, and X-chromosome inactivation. Methods Mol Biol (Clifton, NJ) 463:297–308
104. García-Aguilar M, Autran D (2017) Localisation of chromatin marks in Arabidopsis early embryos. In: Bemer M, Baroux C (eds) Plant chromatin dynamics: methods and protocols. Springer, New York, NY. doi:10.1007/978-1-4939-7318-7_24
105. Fang Y, Spector DL (2005) Centromere positioning and dynamics in living Arabidopsis plants. Mol Biol Cell 16:5710–5718
106. Rosa S, Ntoukakis V, Ohmido N et al (2014) Cell differentiation and development in Arabidopsis are associated with changes in histone dynamics at the single-cell level. Plant Cell 26:4821–4833. doi:10.1105/tpc.114.133793
107. Ingouff M, Selles B, Michaud C, et al (2017) Live-cell analysis of DNA methylation during sexual reproduction in Arabidopsis reveals context and sex-specific dynamics controlled by noncanonical RdDM. 1–12. doi:10.1101/gad.289397.116.GENES
108. Tariq M, Saze H, Probst A V et al (2003) Erasure of CpG methylation in Arabidopsis alters patterns of histone H3 methylation in heterochromatin. Proc Natl Acad Sci U S A 100:8823–8827
109. Mathieu O, Probst A V, Paszkowski J (2005) Distinct regulation of histone H3 methylation at lysines 27 and 9 by CpG methylation in Arabidopsis. EMBO J 24:2783–2791. doi:10.1038/sj.emboj.7600743
110. Mathieu O, Reinders J, Caikovski M et al (2007) Transgenerational stability of the Arabidopsis epigenome is coordinated by CG methylation. Cell 130:851–862
111. Jacob Y, Feng S, LeBlanc CA et al (2009) ATXR5 and ATXR6 are H3K27 monomethyltransferases required for chromatin structure and gene silencing. Nat Struct Mol Biol 16:763–768. doi:10.1038/nsmb.1611

112. Ingouff M, Rademacher S, Holec S et al (2010) Zygotic resetting of the HISTONE 3 variant repertoire participates in epigenetic reprogramming in Arabidopsis. Curr Biol 20:2137–2143
113. She W, Grimanelli D, Rutowicz K et al (2013) Chromatin reprogramming during the somatic-to-reproductive cell fate transition in plants. Development 140(19):4008. doi:10.1242/dev.095034
114. Lieberman-Aiden E, van Berkum NL, Williams L, et al (2009) Comprehensive mapping of long-range interactions reveals folding principles of the human genome. Science 326:289 LP-293.
115. Nagano T, Lubling Y, Stevens TJ et al (2013) Single-cell Hi-C reveals cell-to-cell variability in chromosome structure. Nature 502:59–64. doi:10.1038/nature12593
116. Ma W, Ay F, Lee C et al (2015) Fine-scale chromatin interaction maps reveal the cis-regulatory landscape of human lincRNA genes. Nat Methods 12:71–78. doi:10.1038/nmeth.3205
117. Hsieh THS, Weiner A, Lajoie B et al (2015) Mapping nucleosome resolution chromosome folding in yeast by micro-C. Cell 162:108–119. doi:10.1016/j.cell.2015.05.048
118. Choy J, Fullwood MJ (2017) Deciphering noncoding RNA and chromatin interactions: multiplex chromatin interaction analysis by paired-end tag sequencing (mChIA-PET). In: Ørom UA (ed) Enhancer RNAs - methods and protocols. Springer, New York, NY, pp 63–89. doi:10.1007/978-1-4939-4035-6_7
119. Crevillén P, Sonmez C, Wu Z, Dean C (2013) A gene loop containing the floral repressor FLC is disrupted in the early phase of vernalization. EMBO J 32:140–148. doi:10.1038/emboj.2012.324
120. van Steensel B, Delrow J, Henikoff S (2001) Chromatin profiling using targeted DNA adenine methyltransferase. Nat Genet 27:304–308. doi:10.1038/85871
121. van Steensel B, Henikoff S (2000) Identification of in vivo DNA targets of chromatin proteins using tethered dam methyltransferase. Nat Biotechnol 18:424–428. doi:10.1038/74487
122. Desvoyes B, Vergara Z, Sequeira-Mendes J et al (2017) A rapid and efficient ChIP protocol to profile chromatin binding proteins and epigenetic modifications in Arabidopsis. In: Bemer M, Baroux C (eds) Plant chromatin dynamics: methods and protocols. Springer, New York, NY. doi:10.1007/978-1-4939-7318-7_5
123. Gendrel AV, Lippman Z, Yordan C et al (2002) Dependence of heterochromatic histone H3 methylation patterns on the Arabidopsis gene DDM1. Science 297:1871–1873
124. Stroud H, Otero S, Desvoyes B et al (2012) Genome-wide analysis of histone H3.1 and H3.3 variants in Arabidopsis thaliana. Proc Natl Acad Sci 109:5370–5375
125. Wollmann H, Holec S, Alden K et al (2012) Dynamic deposition of histone variant H3.3 accompanies developmental remodeling of the Arabidopsis transcriptome. PLoS Genet 8:28–31. doi:10.1371/journal.pgen.1002658
126. Yelagandula R, Stroud H, Holec S et al (2014) The histone variant H2A.W defines heterochromatin and promotes chromatin condensation in Arabidopsis. Cell 158:98–109. doi:10.1016/j.cell.2014.06.006
127. Turck F, Roudier F, Farrona S et al (2007) Arabidopsis TFL2/LHP1 specifically associates with genes marked by trimethylation of histone H3 lysine 27. PLoS Genet 3:e86
128. Veluchamy A, Jegu T, Ariel F et al (2016) LHP1 regulates H3K27me3 spreading and shapes the three-dimensional conformation of the Arabidopsis genome. PLoS One:1–25. doi:10.1371/journal.pone.0158936
129. Zhang X, Germann S, Blus BJ et al (2007) The Arabidopsis LHP1 protein colocalizes with histone H3 Lys27 trimethylation. Nat Struct Mol Biol 14:869–871. doi:10.1038/nsmb1283
130. Wang C-JR (2013) Analyzing maize meiotic chromosomes with super-resolution structured illumination microscopy. In: Pawlowski WP, Grelon M, Armstrong S (eds) Plant meiosis – methods and protocols. Humana, Totowa, NJ, pp 67–78
131. Roukos V, Misteli T (2014) Deep imaging: the next frontier in microscopy. Histochem Cell Biol 142:125–131. doi:10.1007/s00418-014-1239-5
132. Clément C, Vassias I, Ray-Gallet D, Almouzni G (2016) Functional characterization of histone chaperones using SNAP-tag-based imaging to assess de novo histone deposition. In: Enzymology RMBT-M (ed) Enzyme epigenetics, part A, Academic, New York, NY, pp 97–117
133. Mathur J, Radhamony R, Sinclair AM et al (2010) mEosFP-based green-to-red photoconvertible subcellular probes for plants.

Plant Physiol 154:1573–1587. doi:10.1104/pp.110.165431

134. Rodriguez-Granados NY, Ramirez-Prado JS, Veluchamy A et al (2016) Put your 3D glasses on: plant chromatin is on show. J Exp Bot 67 (11):3205–3221. doi:10.1093/jxb/erw168
135. Pontes O, Li CF, Nunes PC et al (2006) The Arabidopsis chromatin-modifying nuclear siRNA pathway involves a nucleolar RNA processing center. Cell 126:79–92
136. Deal RB, Henikoff S (2011) The INTACT method for cell type-specific gene expression and chromatin profiling in Arabidopsis thaliana. Nat Protoc 6:56–68
137. Morao AK, Caillieux E, Colot V, Roudier F (2017) Cell type-specific profiling of chromatin modifications and associated proteins. In: Bemer M, Baroux C (eds) Plant chromatin dynamics: methods and protocols. Springer, New York, NY. doi:10.1007/978-1-4939-7318-7_8
138. Rabl C (1885) Über Zellteilung. Morphol Jahrb 10:214–330

Chapter 24

Localization of Chromatin Marks in Arabidopsis Early Embryos

Marcelina García-Aguilar and Daphné Autran

Abstract

During early embryo development, profound changes in chromatin structure and regulation take place. It is difficult to study these changes in plant embryos however, largely because of their relative inaccessibility, which impedes the application of current epigenomic and biochemistry protocols. To circumvent this issue and to analyze the epigenetic status of the embryo at both the cellular and subcellular level, we describe here a simple method to immunolocalize chromatin marks in whole mount early Arabidopsis embryos, either within maternal tissues or isolated from seeds. We show that this protocol can be combined with fluorescent protein markers, allowing for the simultaneous detection of several chromatin components and/or cell fate markers. This new protocol will facilitate deciphering the epigenetic circuits controlling early embryogenesis in plants.

Key words Embryos, Chromatin, Histone modifications, Transcriptional activity, Whole-mount immunolocalization, Fluorescent protein markers

1 Introduction

In sexual organisms, fertilization unites female and male gametes to form the zygote, which undergoes several rounds of cell division to begin embryogenesis. This developmental phase is a theatre of profound chromatin changes, which serve different purposes. First, the very specialized chromatin of parental gametes needs to change its structure to be combined in the new zygote nuclei; second, chromatin reprogramming is important to initiate zygotic transcriptional activation; and third, it contributes to resetting and/or properly reestablishing inherited parental epigenetic marks, while also allowing totipotency. Subsequently, chromatin control is crucial for definition of early cell fate lineages. These mechanisms have been well studied in animal systems, and numerous recent experiments have significantly improved our understanding of the dynamics of histone modifications, nucleosome

Marian Bemer and Célia Baroux (eds.), *Plant Chromatin Dynamics: Methods and Protocols*, Methods in Molecular Biology, vol. 1675, DOI 10.1007/978-1-4939-7318-7_24, © Springer Science+Business Media LLC 2018

positioning and composition, and DNA methylation during embryonic development [1].

In plants, a large body of work has begun to decipher the hormonal and genetic networks controlling the highly stereotyped divisions characteristic of early embryogenesis in the model eudicot *Arabidopsis thaliana* (Arabidopsis) [2]. By contrast, the epigenetic control of plant embryo development remains poorly understood, though epigenetics has recently been considered in the context of plant reproductive development and transgenerational inheritance [3, 4]. This lack of knowledge about epigenetic regulation of early embryogenesis derives from an essentially technical difficulty, that is, the small size and relative inaccessibility of early embryos embedded in maternal seed tissues, making epigenomic studies highly challenging—although some pioneer protocols exist, *see* for instance Chapter 8 of this book, for a protocol on how to isolate specific root cell types nuclei to study chromatin marks in ChIP-seq [5]. To circumvent this problem, genetic and in situ approaches have been used in several studies in both Arabidopsis and maize. Immunolocalization studies of histone marks, RNA Pol II status and fluorescent histone markers have shown that chromatin status is highly dynamic in plant gametes and zygotes [6].

Building on previously published protocols ([7–13]; *see* also Chapter 25 of this book [14]), we propose a simple method to immunolocalize histone tail modifications and Pol II status on whole-mount tissues of early Arabidopsis embryos. The two variations of the method (*see* Fig. 1 for a summary of the workflow), can be applied to either: (1) whole early seeds to study the zygote and 1 or 2-cell embryos (Protocol A, Subheadings 3.1.1–3.1.9) or (2) embryos isolated/popped out from seed tissues, which is possible from the 2- to 4-cell stage onward (Protocol B, Subheadings 3.2.1–3.2.8). Protocol A presented here allows monitoring chromatin in very early embryos as well as the surrounding endosperm and seed coat nuclei. This provides a whole picture of early seed development. By contrast, Protocol B focuses on isolated embryos, facilitating antibody penetration and imaging. This protocol is suitable to address the chromatin changes associated with patterning events during early embryogenesis. Here, we describe the immunolocalization of primary antibodies targeting RNA Pol II (Fig. 2) and histone H3 modifications (Fig. 3). This involves sample fixation, cell wall digestion, mounting in acrylamide, and membrane permeabilization, followed by the immunolabeling steps and counterstaining. Using either protocol A or B, the immunolocalization can be extended to other histone (H2, H4) modifications, and more broadly to antibodies targeting epigenetic effectors or even transcription factors and cellular components such as the cytoskeleton, as long as specific antibodies are available. Finally, we show that immunolocalization of histone marks can be well

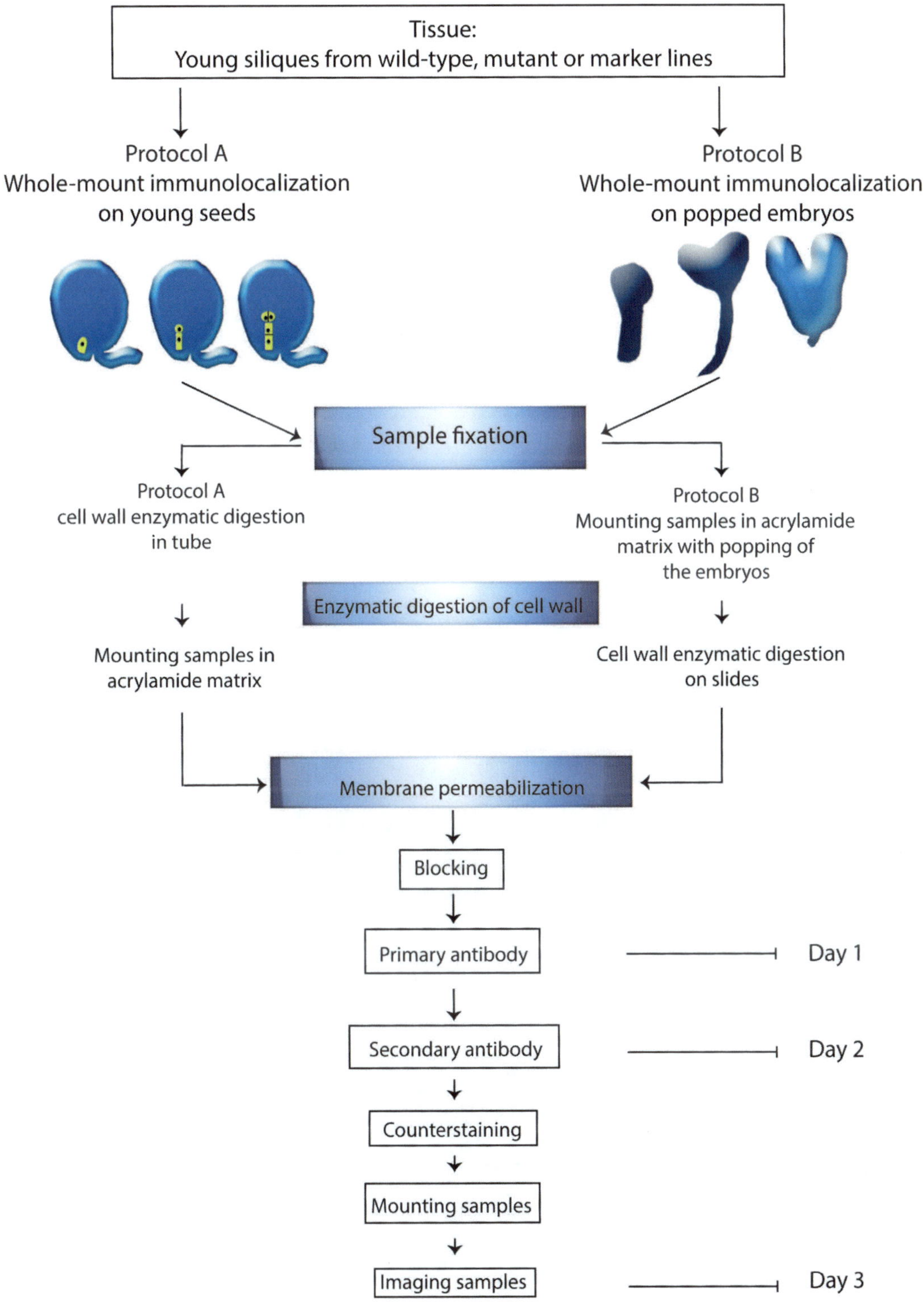

Fig. 1 Overview of the workflow for chromatin mark immunolocalization in whole mount early seeds (*Protocol A*) or whole mount isolated embryos (*Protocol B*). The specific and common steps of the two protocols are presented. We estimate 3 days laboratory work for both protocols

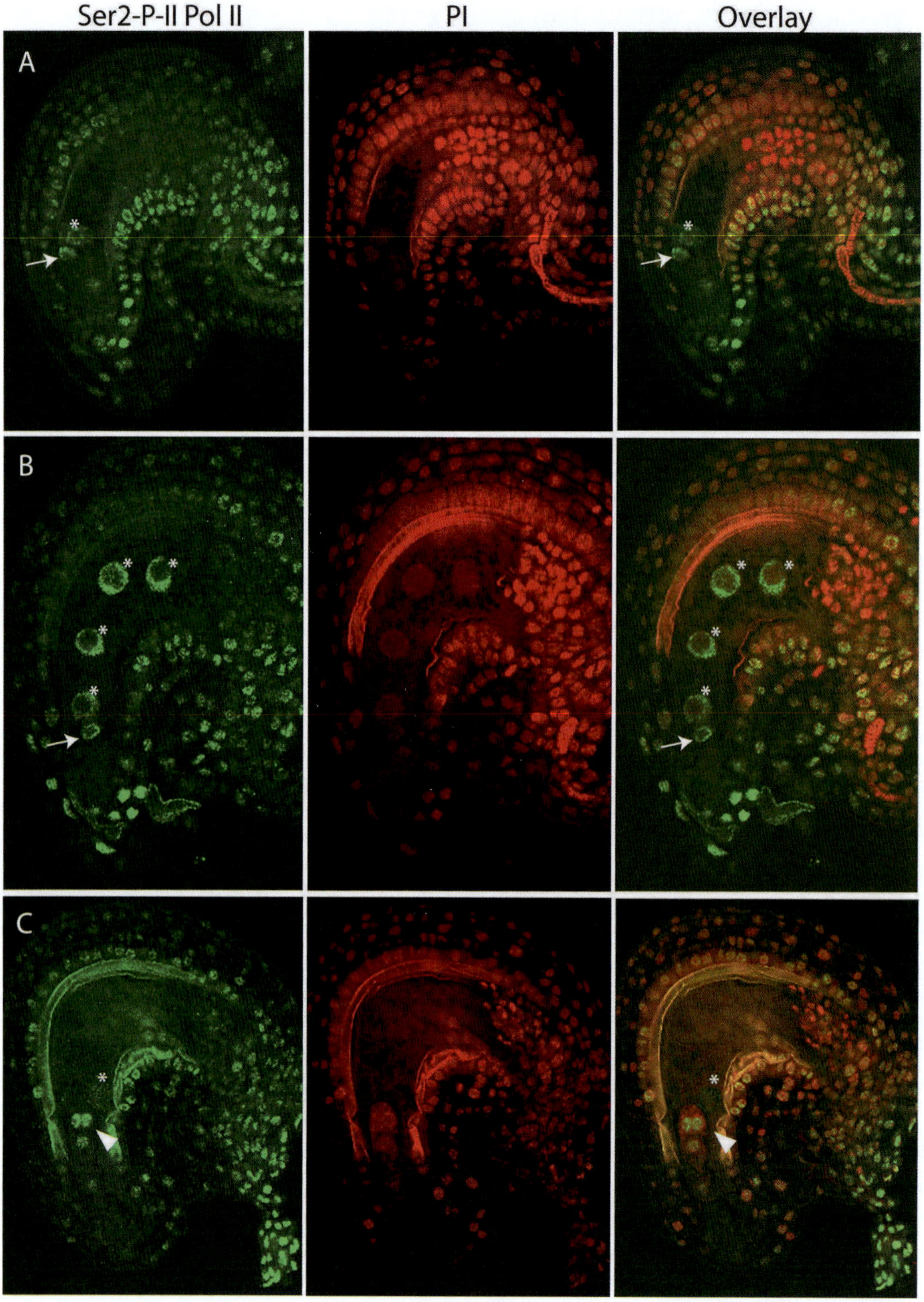

Fig. 2 Examples of immunolocalization of chromatin marks in whole mount early seeds (*Protocol A*). Localization of the transcriptionally active form of RNA polII in young seeds of Arabidopsis. Early seeds of Arabidopsis *med13/gct* mutant line (L*er*) were subjected to immunolocalization to detect the phosphorylated

combined with the analysis of fluorescent protein markers, increasing the possibilities to explore the epigenetic control of early embryo lineages in plants.

2 Materials

2.1 Equipment

1. 2, 1.5 and 0.5 ml eppendorf tubes.
2. Coplin jars (cleaned with sterile water and dry with ethanol).
3. Slide incubation chamber: plastic box with a wet paper towel (use sterile water, humidify the paper only, no excess of liquid), place two sticks (plexiglas rods or broken plastic pipettes or handmade polystyrene blocks) to elevate the microscope slides.
4. Adhesion microscope glass slides, or classic slides manually coated with poly-L-lysine: after cleaning with ethanol; using a clean small painting brush, slides are brushed with a small amount of poly-L-lysine solution, to deposit a homogenous and thin layer. After coating, place the slides in a vertical position and let poly-L-lysine dry at room temperature (*see* **Note 1**).
5. Razor blades.
6. Orbital shaker.
7. Dissection stereomicroscope.
8. Epifluorescence and/or confocal microscope for imaging.

2.2 Reagents and Solutions

1. Sterilized water [0.22 μm filtered, deionized (resistivity of 18.2 MΩ cm at 25 °C)].
2. 1× PBS.
3. Buffer 1: 1× PBS, 0.2% Triton X-100.
4. Buffer 2: 1× PBS, 2% Triton X-100.
5. Fixative: freshly prepared 4% Paraformaldehyde in 1× PBS (from powder or from 10% stock solution, prepare under the fume hood) mixed with 2% Triton X-100. Keep on ice or at 4 °C.

Fig. 2 (continued) version of the RNA pol II CTD repeat YSPTSPS (phospho S2) [H5]. (**a**) Seed containing a zygote and two endosperm nuclei. (**b**) Seed showing undivided zygote and four free endosperm nuclei. (**c**) Abnormal seed showing the 2-cell embryo and undivided central cell above the embryo proper. Symbols: *asterisk*, endosperm nuclei; *arrow*, zygote; *arrowhead*, 2-cell embryo. DNA was counterstained with propidium iodide (PI). Images are single optical sections acquired on an inverted Zeiss LSM 510 META confocal laser scanning microscope, using a 40× objective. Fluorescence was excited with Argon/2 and DPSS 561 lasers. Images were recorded using HFT488/561, BP575-615 (propidium iodide), BP 500-530 (Alexa 488) filters in a single channel configuration

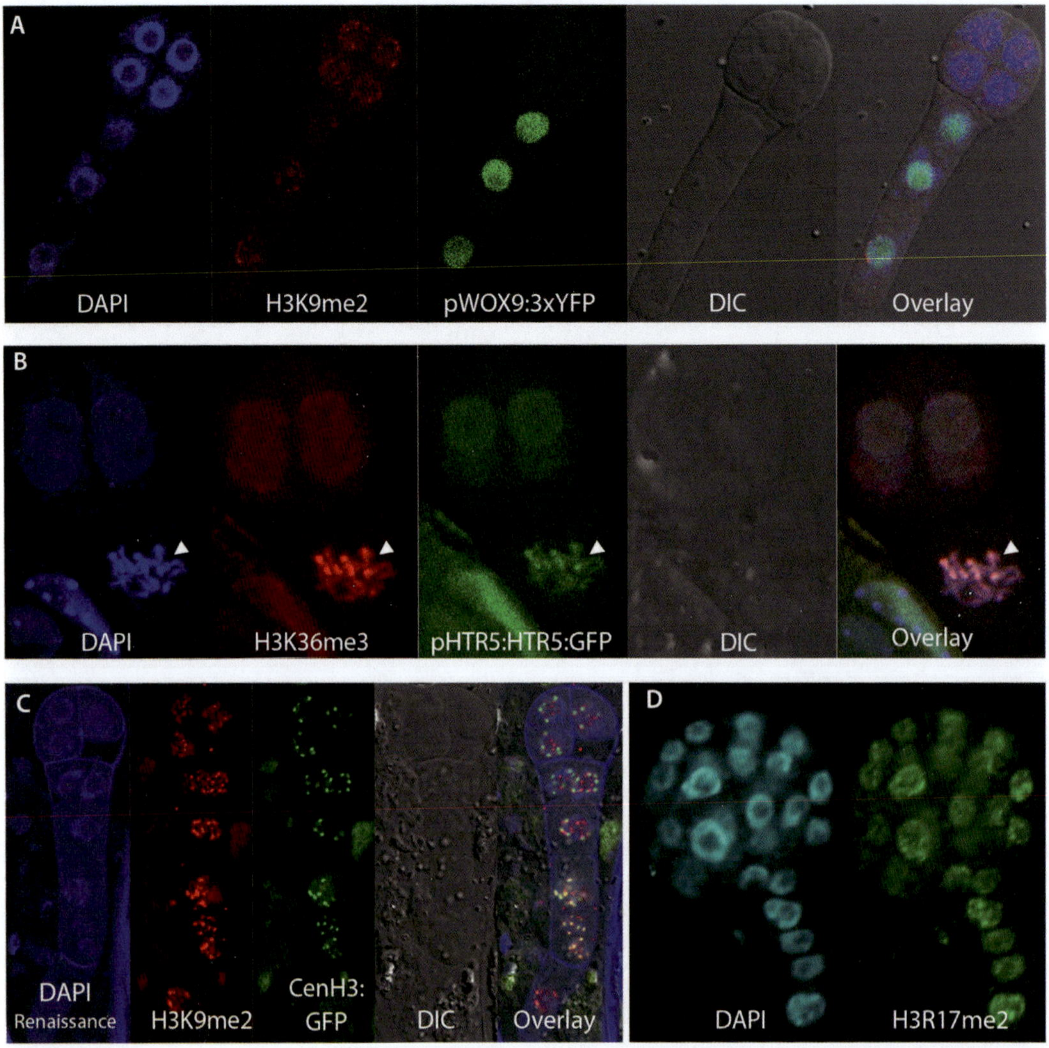

Fig. 3 Examples of immunolocalization of chromatin marks in whole mount isolated embryos (*Protocol B*). (**a**) Localization of the heterochromatin mark H3K9me2 (detected here with Alexa 568 conjugated secondary antibody—*red channel*) in an 8-cell stage embryo expressing the suspensor fate marker pWOX9:3XnYFP (*green channel*). DNA was counterstained with DAPI (*blue channel*)) and embryo cells were visualized using DIC (Differential Interference Contrast). (**b**) Localization of the euchromatin mark H3K36me3 (*red channel*) in embryo transitioning between the 2- and 4-cell stage, and expressing a GFP reporter of the histone H3.3 variant HTR5 (*green channel*). The signals colocalize along the chromosomes condensed during mitosis in the first suspensor nucleus (*arrow head*). (**c**) Localization of H3K9me2 in 8-cell stage embryo expressing a GFP reporter of the centromeric histone CenH3. Pericentromeric H3K9me2 foci are visible around the centromeres in the overlay, together with noncentromeric heterochromatic foci. DNA was counterstained with DAPI and cell walls were counterstained with SCRI Renaissance 2200 (*blue channel*). (**d**) Localization of H3R17me2 (detected here with Alexa 488 conjugated secondary antibody—*green channel*) in an early globular embryo. This histone modification is involved in transcriptional activation in animals. DNA was counterstained with DAPI (*cyan channel*). Images are single optical sections (**a**, **c**) or maximum intensity 3D projections (**b**, **d**), acquired in sequential mode on a Leica SP8 resonant confocal laser scanning microscope, equipped with 405, 488, and 561 nm lasers and HyD hybrid detectors, using a 63× NA 1.4 oil immersion objective (except in **d**: 40× NA 1.3 oil immersion objective)

6. Enzyme mix: 1% driselase, 0.5% cellulase, and 1% pectolyase w/v in water. To prepare 10 ml of enzymes mix, weigh 0.1 g driselase, 0.05 g cellulose, and 0.1 g pectolyase enzymes powder. Use a 15 ml Corning tube to mix the enzymes and add 8 ml of distilled-sterilized water. Stir gently to dissolve. Maintain the tube on ice to let the spume to reliquefy before adjusting to a final volume of 10 ml. Aliquot by 500 μl and store at −20 °C. Refreezing/thawing aliquots must be avoided, as it leads to enzymatic activity loss.
7. 15% acrylamide–bisacrylamide 19:1 (dilution from 40% acrylamide–bisacrylamide 19:1).
8. Catalysts for polyacrylamide gel polymerization: 25% Ammonium persulfate (dissolve powder in water. Can be stored 1 week at 4 °C, older solution can compromise polymerization) and tetramethylethylenediamine (TEMED) solution or 20% sodium bisulfite (dissolve powder in water, can be stored 1 week at 4 °C).
9. Propidium iodide (PI): stock solution 1000× at 1 μg/μl.
10. DAPI: stock solution 1000× at 10 μg/μl.
11. SCRI Renaissance 2200, aqueous stock solution.
12. Mounting media: antifading mounting reagent, 50% glycerol (diluted in ultrapure water).

2.3 Antibodies

The antibodies used to produce the data presented here are listed below. Alternatives are available but high purity grade and specificity (no cross-reactivity) should be verified (*see* **Note 2**).

Primary antibodies:

Anti-RNA pol II CTD repeat YSPTSPS (phospho S2), ab5095 Abcam. Rabbit polyclonal.

Anti-H3K9me2 ab1220 Abcam. Mouse monoclonal.

Anti-H3K36me3 ab9050 Abcam. Rabbit polyclonal.

Anti-H3R17me2(as) #39710 Active Motif. Rabbit polyclonal.

Secondary antibodies:

Alexa Fluor 488 goat anti-rabbit IgG (H + L), A11008, Molecular probes.

Alexa Fluor 488 goat anti-mouse IgG (H + L), A-11029, Invitrogen.

Alexa Fluor 568 donkey anti-rabbit IgG (H + L), A10042, Invitrogen.

Alexa Fluor 568 goat anti-mouse IgG (H + L), A-11004, Invitrogen.

3 Methods

We present here two alternative but similar protocols (Fig. 1): Protocol A refers to immunolocalization in embryos within the early seed, from zygote to 1–2 cell stages. Protocol B refers to immunolocalization in embryos at later stages, which are big enough to be pressed out of the seeds (see below), facilitating treatments, antibody penetration and imaging by separating the embryo from endosperm nuclei and integuments (*see* **Note 3**).

Either protocol can be followed, according to the developmental stage of the embryo to be studied, and/or depending on whether the embryo should be studied with intact endosperm and maternal seed coat or not. Protocols A and B have many steps in common, with specific variations that are detailed below.

3.1 Protocol A: Immunolocalization in Whole-Mount Early Seeds

DAY 1.

3.1.1 Sample Fixation

The aim is to stabilize the cellular and subcellular architecture of the tissue, while preventing the activity of endogenous degrading enzymes to preserve protein antigens. We use here paraformaldehyde, which is a strong cross-linking reagent. Contrary to formalin (i.e., formaldehyde solution), paraformaldehyde does not contain methanol, which can damage many epitopes. Neither glutaraldehyde (a good fixative to preserve cellular structures, but with autofluorescence), nor organic solvents such as acetone (which is not compatible with fluorescent proteins) are recommended.

Because antigens can be damaged by fixation, care should be given to not fix longer than indicated, as this could result in loss of signal. At the same time, imperfect fixation will lead also to signal loss, due to protein degradation and lack of structural integrity (*see* **Note 5**).

Selection of embryos at a certain stage is based on silique length and position. This synchronization may vary according to growth conditions and genotype; and can be determined beforehand by clearing embryos sampled from a gradient of siliques. In our growth conditions, early-fertilized seeds containing zygotes to two-celled embryos are isolated from siliques positioned close to the most recent open flower in the Arabidopsis inflorescence. These siliques typically still contain withered floral organs such as petals and stamens.

1. To isolate young seeds, fix carpel on a slide using doubled sided tape, to avoid any movement of the carpel and facilitate carpel dissection. Using an insulin needle (or fine #5 forceps) slice

both carpels longitudinally along the septum, then adhere the carpel valves to the tape to expose the young seeds.

2. Remove the seeds from the carpel (*see* **Note 4**).
3. Transfer them in a 1.5 ml Eppendorf tube containing freshly made fixative solution. Incubate 2 h with gentle shaking (50 rpm on orbital shaker) (*see* **Note 5**).
4. To remove the fixative solution, centrifuge the samples for 15–20 s at RCF 24,000 [× *g*].
5. Remove the fixative by pipetting, taking care not to touch the seed pellet.
6. Rinse the seeds in Buffer solution 1, centrifuge as in **step 4** and wash again in renewed Buffer solution 1 for 10 min with gentle shaking on ice (*see* **Note 6**).

3.1.2 Enzymatic Digestion of Cell Wall

The aim is to render the cell wall matrix permeable to antibodies and staining reagents. This is a critical step of the experiment. Under-digestion will lead to poor penetration of the reagents, while over-digestion may lead to chromatin degradation and loss of tissue integrity. The enzymes used for this step often come from almost crude extracts of plant pathogens, and each enzyme mix newly prepared should be calibrated to determine the adequate duration of digestion (*see* **item 6** of Subheading 2.2 and **Note 7**).

1. Centrifuge the samples again and remove the buffer by pipetting.
2. Thaw the enzyme mix aliquot, mix by flicking the tube to homogenize the suspension.
3. Add 30 μl of enzyme mix to the fixed tissue, and gently tap the bottom of the tube to separate the pellet of seeds.
4. Incubate the samples at 37 °C for 2 h (*see* **Note 7**).
5. To stop enzymatic digestion, centrifuge the tubes and pipet the enzyme mix.
6. Wash twice in Buffer 1, to completely eliminate enzyme mix residues (*see* **Note 6**).

3.1.3 Mounting the Samples in an Acrylamide Matrix on Slide

The goal is to maintain the samples at a fixed position on the microscope slide, into a porous system that allows reagent penetration. As such, this system is useful for the analysis of any small sample, such as early seeds as presented here, but also microsporocytes, pollen grains or ovules of Arabidopsis, for instance. The concentration of acrylamide and bisacrylamide regulates the pore size, which is important in the selection of molecules that go through the acrylamide matrix. Because acrylamide polymerization is fast, this step requires prompt manipulation (*see* **Notes 8–10**).

1. Prepare 250 μl aliquots of 15% acrylamide–bisacrylamide in 0.5 ml tubes, one tube per slide.

   All the following steps are performed under the stereomicroscope.
2. Centrifuge the samples 30 s at RCF 6000 [× *g*], and remove 1.3 ml of Buffer 1, maintaining the tubes on ice.
3. Pipet 50 μl of Buffer solution 1 containing the seeds from the bottom of the tube and transfer on an adhesive slide.
4. Drain the excess of Buffer solution 1 around the seeds: gently incline the slide, pipet the excess solution, but never let the seeds completely dry.
5. Add 2 μl of 25% APS and 1 μl TEMED to the 15% acrylamide tube, mix quickly (without producing air bubbles that can inhibit polymerization) and add 13 μl of this activated acrylamide mix over the seeds.
6. Immediately place a 22 × 22 mm coverslip on the acrylamide drop/seeds. This step must be done very quickly, as activated acrylamide takes only few seconds to polymerize (*see* **Notes 8–10**).
7. Let acrylamide pads polymerize at room temperature for at least 30 min. A slight retraction of the acrylamide pads along coverslip borders indicates that polymerization is done.
8. Remove the coverslip using a razor blade: insert the edge of a razor blade under a corner of the coverslip and toggle it to flip the coverslip.
9. Slides are placed in 50 ml 1× PBS or Buffer 1 in a Coplin jar (*see* **Note 6**).

3.1.4 Membrane Permeabilization Treatment

In this step, the lipids of the cell membranes are destabilized, to allow penetration of the antibodies to detect their target proteins. Detergents such as TritonX-100 (used here) and Tween 20 are excellent for dissolving cell membranes. However, since they are nonselective in nature, they can dissolve proteins along with lipids when used at high concentrations or for longer amount of time, affecting target protein detection. Washes after permeabilization are thus an important step.

1. Incubate the slides in a Coplin Jar in Buffer solution 2, on ice with gentle shaking on an orbital shaker, for 3 h.
2. Wash the slides 2× 10 min in 1× PBS to remove excess of Triton, which can create overly stringent conditions for immunobinding.

3.1.5 Blocking

The goal of blocking is to prevent nonspecific binding of the antibodies by coating all proteins in the sample. After blocking, the primary antibodies will surpass the blocked proteins to bind their cognate ligands, reducing the background signal. We use bovine serum albumin (BSA), for which antibodies have low affinity. For optimal results, freshly made BSA solution should be used.

The samples are blocked in 1% BSA in 1× PBS for 1 h at 37 °C in a Coplin jar with gentle shaking on an orbital shaker (*see* **Note 11**).

3.1.6 Immunolabeling with Primary Antibodies

This step is crucial as it consists in the detection per se of the target chromatin marks. Primary antibody will diffuse into the prepared tissues, and bind its specific antigen. Because antibodies are large molecules, this reaction takes time; and the thicker the tissue, the longer it takes. Here, for this whole-mount sample protocol, we use overnight incubation at 4 °C, to limit protein degradation. The specificity and quality of the primary antibody is essential for a successful experiment (*see* **Note 2**). If using a batch of antibody for the first time, a dilution series will help optimizing the signal to background ratio (*see* **Note 12**). A crucial step is the extensive washing of the unbound primary antibody to reduce unspecific binding of the secondary antibody (**steps 3** and **4**) (*see* **Note 13**).

1. Remove the 1% BSA solution.
2. Add 40 μl of adequate primary antibody dilution in 1× PBS, on the acrylamide pad (*see* **Note 12**). Place a clean coverslip over the acrylamide pads to avoid evaporation, and incubate in a wet chamber overnight a 4 °C.

   DAY 2.

3. Remove the coverslip and transfer the slides to Buffer 1 in a clean Coplin jar.
4. Incubate three times for 1 hour each in fresh Buffer 1, on ice, with gentle shaking on an orbital shaker (*see* **Note 13**).

3.1.7 Immunolabeling with Secondary Antibodies

The aim is to detect primary antibody using a fluorescently labeled secondary antibody. Several secondary antibodies will bind to each primary antibody, resulting in an amplified signal. Importantly, this amplification is linear, thus, quantification of secondary antibody signal correctly reflects the amount of target.

Secondary antibody raised against the species used to produce the primary antibody must be selected (*see* Subheading 2.3). In addition, preabsorbed secondary antibodies are chosen to prevent cross-reactivity between species.

We routinely use a 1:400 dilution for all secondary antibodies, however this can be adjusted for the specific target and sample, to optimize the signal and reduce background (*see* **Note 14**). Because secondary antibody signal detection depends on covalently bound fluorochromes, an important point is to carry out all incubations in darkness to avoid fluorochrome fading.

1. Drain the slides by maintaining them vertical on a piece of absorbent paper, to remove Buffer 1 (without drying the pads).
2. As for primary antibody incubation, pipet 40 μl per acrylamide pad of 1:400 dilution of the secondary antibody in Buffer 1 onto the pads and place a coverslips to avoid evaporation (*see* **Note 14**).
3. Incubate the samples in a humid chamber overnight at 4 °C, in the dark.

DAY 3, *For all the following steps, samples must be kept in the dark as much as possible.*

4. To remove uncoupled secondary antibody, take off the coverslips, and follow the same procedure as for primary antibody washes (*see* **Note 13**).

3.1.8 Counterstaining

The objective is to stain the DNA of each nucleus of the embryo or early seed samples. This facilitates identification of embryo/early seed structures and stages, provides a reference for specific chromatin mark quantification, and also indicates mitotic events. We counterstain routinely with either DAPI or PI, the choice depends on the fluorochrome combination and on the microscope to be used (*see* **Note 15**). For many nucleic acid dyes, fluorescence intensity increases significantly after binding; however, extensive washing is required to avoid fluorescence background.

To counterstain the samples with PI (1 μg/ml), or DAPI (10 μg/ml) diluted in Buffer 1 (*see* **Note 15**), incubate with gentle shaking on ice for 30 min in the dark. Then wash at least 30 min in Buffer 1, renewing the buffer every 5–10 min.

3.1.9 Mounting

The aim is to coat each acrylamide pad with a thin layer of a nondehydrating reagent with adequate optical properties (*see* below), to protect tissue structures, antibody binding and the fluorescence of the signals, while allowing imaging of the samples.

We obtained good results by mounting the samples either in 50% glycerol (refractive index: 1.39, best imaged with a water immersion lens) or in hardening antifading glycerol-based reagent (*see* **Note 16**) (refractive index: 1.47, to be imaged ideally with a glycerin immersion lens (RI 1.47) or alternatively an oil immersion lens (RI 1.51). Glycerol mounting is adequate when imaging can be performed a few days after mounting (to avoid long term storage, leading to signal fading), while hardening antifading reagents allow longer storage (*see* **Note 16**).

To mount in glycerol:

1. Remove a slide from the Coplin jar and place it under a stereomicroscope, add 20 μl of 50% glycerol.
2. Place a 22 × 22 mm coverslip slowly to avoid air bubbles and seal the slide with transparent nail polish.

To mount in hardening antifading reagent:

1. Drain the slides to remove buffer. Try to remove as much buffer as possible since buffer mixing with the mounting reagent will impair hardening and modify the refractive index, thereby decreasing image quality.
2. Place a drop of reagent on each acrylamide pad, and spread with a 22 × 22 mm coverslip, exerting a light pressure. Hardening at 4 °C overnight, in the dark, is required before sealing with nail polish. Optimal results are obtained after 3 days of complete hardening at 4 °C in the dark (*see* **Note 16**).

3.2 Protocol B: Immunolocalization in Isolated Embryos

DAY 1.

3.2.1 Sample Fixation

The aim is to stabilize the cellular and subcellular architecture of the tissue, while preventing the activity of endogenous degrading enzymes. We use here paraformaldehyde, which is a strong cross-linking reagent. Contrary to formalin (i.e., formaldehyde solution), paraformaldehyde does not contain methanol, which can damage many epitopes. Neither glutaraldehyde (a good fixative to preserve cellular structures, but with autofluorescence), nor organic solvents such as acetone (which is not compatible with fluorescent proteins) are recommended.

Because antigens can be damaged by fixation, care should be given to not fix longer than indicated, as this could result in loss of signal. At the same time, imperfect fixation will lead also to signal loss, due to protein degradation and lack of structural integrity (*see* **Note 5**).

Selection of embryos at a certain stage is based on silique length and position. This synchronization may vary according to growth conditions and genotype; and can be determined beforehand by clearing embryos sampled from a gradient of siliques. In our growth conditions and using Arabidopsis thaliana *Col_0* siliques ranging from 0.4 cm (2–4 cells embryos) to 1 cm (early globular embryos) in length are used.

1. Siliques are placed on a microscope slide and cut longitudinally along the dehiscence line, then transferred to fixative on ice.
2. Fix on ice for 2 h–3 h with gentle shaking (50 rpm on orbital shaker) (*see* **Note 5**)
3. To remove the fixative solution, the siliques are rinsed in Buffer 1 and washed twice in Buffer 1 on ice (*see* **Note 6**).

3.2.2 Mounting the Samples in an Acrylamide Matrix on Slide

The goal is to maintain the samples at a fixed position on the microscope slide, into a porous system allowing reagent penetration. As such, this system is useful for the analysis of any small sample, like early seeds as presented here, but also microsporocytes, pollen grains or ovules of Arabidopsis, for instance. The concentration of acrylamide and bisacrylamide regulates the pore size, which is important in the selection of molecules that go through the acrylamide matrix. Because acrylamide polymerization is fast, this step requires prompt manipulation (*see* **Notes 8–10**).

1. Prepare aliquots of 250 μl of 15% acrylamide–bisacrylamide 19:1 in 0.5 ml tubes, one tube per slide.

   All the following steps for mounting samples are done under the stereomicroscope.
2. On an adhesive slide, remove the seeds from the siliques (*see* **Notes 4** and **17**). Discard the siliques valves. Distribute the seeds or small groups of seeds over a 22 × 22 mm surface, to avoid overlapping and allow each seed to be surrounded by acrylamide and trapped in the matrix.
3. Before complete evaporation of the remaining buffer, activate the acrylamide mix by adding 2 μl of 25% APS and 1 μl TEMED to the 15% acrylamide tube, mix quickly and pipet between 13 μl (2–4 cell stage embryos) and 17 μl (16–32 cells stage embryos) on the seeds. This volume depends on seed size. Too little acrylamide will favor bubbles, while too much acrylamide will not allow popping out the embryos.
4. Before acrylamide polymerization, quickly place a 22 × 22 mm coverslip on the acrylamide drop, and immediately exert a light pressure on each seed/group of seeds using a tweezers or needle, monitoring under the stereomicroscope (*see* **Notes 8–10** and **18**). Globular stage embryos can be visualized when popping out, but younger stages are hardly visible under the stereomicroscope. Usually, endosperm leakage is visible, and a slight expansion of seeds indicates that popping occurred.
5. Let the acrylamide pads polymerize at room temperature for at least 30 min. A slight retraction of the acrylamide pads along the coverslip's borders indicates that polymerization has occurred.
6. Remove the coverslip using a razor blade: insert the edge of a razor blade under a corner of the coverslip and toggle it to flip the coverslip.
7. Place the slides in 50 ml of 1× PBS in a Coplin jar (*see* **Note 6**).

3.2.3 Enzymatic Digestion of Cell Walls and Membrane Permeabilization

The aim is to render the cell wall matrix permeable to antibodies and staining reagents. This is a critical step of the experiment. Underdigestion will lead to poor penetration of the reagents, while overdigestion may lead to chromatin degradation and loss of tissue integrity. The enzymes used for this step often come from almost crude extracts of plant pathogens, and each enzyme mix newly prepared should be calibrated to determine the adequate duration of digestion (*see* **item 6** of Subheading 2.2 and **Note** 7).

Membrane permeabilization can be achieved by destabilizing of the cell membrane lipids, to allow penetration of the antibodies to detect intracellular antigens. Detergents such as Triton X-100 (used here) and Tween 20 are excellent for dissolving cell membranes. However, since they are nonselective in nature, they can dissolve proteins along with lipids when used at high concentrations or for longer amount of time, affecting antigens detection. Washing after permeabilization is thus an important step.

1. Drain the slides by holding them vertical on a piece of absorbent paper, to remove buffer (without drying the pads).
2. Pipet 50 μl of enzyme mix directly on each pad, cover with a coverslip to avoid evaporation, and incubate for 1 h (or according to the required incubation time for each specific enzyme mix, *see* **Note** 7), at 37 °C in a wet chamber.
3. Rinse and wash the slides 2 × 5 min in Buffer 1 in a Coplin jar to completely remove enzyme residues (*see* **Note 6**).
4. Incubate the slides in a Coplin Jar in Buffer 2, on ice with gentle shaking on an orbital shaker, for 2 h.
5. Wash the slides 2 × 10 min in 1× PBS to remove excess of Triton that can create overly stringency conditions for immunobinding.

3.2.4 Blocking (Optional)

Blocking can be used in Protocol B when background is observed. The goal of blocking is to prevent non-specific binding of the antibodies by coating all proteins in the sample. After blocking, the primary antibodies will surpass the blocked proteins to bind their cognate ligands, reducing the background signal. We use bovine serum albumin (BSA), for which antibodies have low affinity. For optimal results, freshly made BSA solution should be used.

Samples are blocked in 1% BSA in 1× PBS for 30 min to 1 h at 37 °C in a Coplin jar with gentle shaking on an orbital shaker (*see* **Note 11**).

3.2.5 Immunolabeling with Primary Antibodies

This step is crucial as it consists in the detection per se of the target chromatin marks. Primary antibody will diffuse into the prepared tissues, and bind its specific antigen. Because antibodies are large molecules, this reaction takes time; and the thicker the tissue, the longer it takes. Here, for this whole-mount sample protocol, we use overnight incubation at 4 °C, to limit protein degradation. The specificity and quality of the primary antibody is essential for a successful experiment (*see* **Note 2**). If using a batch of antibody for the first time, a dilution series will help optimizing the signal to background ratio (*see* **Note 12**). A crucial step is the extensive washing of the unbound primary antibody to reduce unspecific binding of the secondary antibody (**steps 3** and **4**) (*see* **Note 13**).

1. Drain the slides by maintaining them vertically on a piece of absorbent paper, to remove the buffer (without drying the pads).
2. Add 50 μl of adequate primary antibody dilution in 1 × PBS (*see* **Note 12**). Place a clean coverslip over the acrylamide pads to avoid evaporation.
3. Incubate in a wet chamber overnight a 4 °C.

   DAY 2.

4. To wash uncoupled primary antibody, remove the coverslip and rinse the slides five times 1–2 h in Buffer 1, in a Coplin jar on ice with gentle shaking on an orbital shaker (*see* **Note 13**).

3.2.6 Immunolabeling with Secondary Antibodies

The aim is to detect primary antibody using a fluorescently labeled secondary antibody. Several secondary antibodies will bind to each primary antibody, resulting in an amplified signal. Importantly, this amplification is linear, thus quantification of secondary antibody signal correctly reflects the amount of target.

Secondary antibody raised against the species used to produce the primary antibody must be selected (*see* Subheading 2.3). In addition, preabsorbed secondary antibodies are chosen to prevent cross-reactivity between species.

We routinely use a 1:400 dilution for all secondary antibodies; however, this can be adjusted for the specific target and sample, to optimize the signal and reduce background (*see* **Note 14**). Because secondary antibody signal detection depends on covalently bound fluorochromes, an important point is to carry out all incubations in darkness to avoid fluorochrome fading.

1. Drain the slides by maintaining them vertical on a piece of absorbent paper, to remove Buffer 1 (without drying the pads).
2. As for primary antibody incubation, pipet 50 μl of 1:400 dilution of the secondary antibody in Buffer 1 onto the pads (*see* **Note 14**).
3. Incubate the samples in a humid chamber overnight at 4 °C, in the dark.

   DAY 3, *For all the following steps, samples must be kept in the dark as much as possible.*

4. To remove uncoupled secondary antibody, take off the coverslips, and follow the same procedure as for primary antibody washes (*see* **Note 13**).

3.2.7 Counterstaining

The objective is to stain the DNA of each nucleus of the embryo or early seed samples. This facilitates identification of embryo/early seed structures and stages, provides a reference for specific chromatin mark quantification, and also indicates mitotic events. We counterstain routinely with either DAPI or PI, the choice depends on the fluorochrome combination and on the microscope to be used (*see* **Note 15**). For many nucleic acid dyes, fluorescence intensity increases significantly after binding; however, extensive washing is required to avoid fluorescence background. For isolated embryos, simultaneous cell wall counterstaining is of significant help for embryo visualization (*see* below, ref. 13 and Fig. 3c).

To counterstain the samples with PI (1 μg/ml), or DAPI (10 μg/ml) diluted in Buffer 1 (*see* **Note 15**), incubate with gentle shaking on ice for 30 min in the dark. Then wash at least 30 min in Buffer 1, renewing the buffer every 5–10 min.

Embryo cell wall counterstaining can also be combined with DNA staining, using for instance SCRI Renaissance 2200 dye [15] (Fig. 3c). We used a 1:2000 dilution in 1× PBS, with extensive washes to avoid background signal.

3.2.8 Mounting

The aim is to coat each acrylamide pad with a thin layer of a nondehydrating reagent with adequate optical properties (see below), to protect tissue structures, antibody binding and the fluorescence of the signals, while allowing imaging of the samples.

We obtained good results by mounting the samples either in 50% glycerol (refractive index: 1.39, to be best imaged with a water immersion lens) or in hardening antifading glycerol-based reagent (*see* **Note 16**) (refractive index: 1.47, to be imaged ideally with a glycerin immersion lens (RI 1.47) or alternatively an oil immersion lens (RI 1.51). Glycerol mounting is adequate when imaging can be performed a few days after mounting (to avoid long term storage, leading to signal fading), while hardening antifading reagents allow longer storage (*see* **Note 16**).

To mount in glycerol:

1. Remove a slide from the Coplin jar and place it under a stereomicroscope, add 20 μl of 50% glycerol.
2. Place a 22 × 22 mm coverslip slowly to avoid air bubbles and seal the slide with transparent nail polish.

To mount in hardening antifading reagent:

3. Drain the slides to remove buffer. Try to remove as much buffer as possible since buffer mixing with the mounting reagent will impair hardening and modify the refractive index, thereby decreasing image quality.
4. Place a drop of reagent on each acrylamide pad, and place above a 22 × 22 mm coverslip, exerting a light pressure.

Hardening at 4 °C overnight, in the dark, is required before sealing with nail polish. Optimal results are obtained after 3 days of complete hardening at 4 °C in the dark (*see* **Note 16**).

3.3 Imaging

The goal is to observe, register and possibly quantify the fluorescence signals of the secondary antibody fluorochrome, of the counterstaining dyes, and the fluorescent proteins.

Dependent on the level of intracellular details required and the available lens, observation of Arabidopsis seeds containing zygotes and early embryos (Fig. 2) require 40× or 63× objectives for nucleus and cell division details. 20× and 10× objectives are suitable for observations of seeds containing torpedo stage and bent cotyledon stage embryos, respectively, but will not allow for detection of intranuclear patterns. An immersion lens with high Numerical Aperture (NA) is preferred for sensitivity, with immersion type matching as close as possible the refractive index of the mounting medium.

According to the combination of fluorophores, samples are imaged using the user's confocal microscope adequate lasers and filters. There are many different possible ways of imaging, but common general pitfalls, which we briefly recapitulate here, must be avoided [16]. Cross-excitation should be verified by imaging separately each channel, and if required, imaging must be done in sequential mode. To optimize x,y,z resolution, the x,y format should fit the Nyquist criterion (crucial if deconvolution treatment is planned), and z-step sectioning should be optimized to achieve a cubic voxel size (crucial for 3D analysis).

If samples are for quantitative analysis of fluorescence signal, it is important to (1) avoid signal saturation by verifying all sections under a LUT (Look Up Table) mode, ensuring that no section contains either empty or saturating pixels; (2) set up, for all samples to be compared, identical illumination intensity (laser), emission window, gain, scanning speed, pixel size (zoom); (3) prefer a linear mode of photon acquisition (i.e., photo-counting mode) and avoid averaging (or set up identical averaging for all samples). Given the variability from slide to slide in immunolocalization experiments, it is advisable to have an internal reference for normalization (i.e., DAPI or PI DNA staining), and/or perform within-slide comparisons.

For image processing and quantitative analysis, either commercial or open-source software are available, with detailed tutorials online [17, 18].

3.4 Combining Immunolocalization and Fluorescent Protein Markers

Immunolocalization of chromatin marks allows the observation of epigenetic dynamics during early embryo development, as well as for any other tissue for which epigenomic protocols are currently not routine in plants.

In Fig. 3, we show that this procedure is compatible with fluorescent proteins, opening up multiple possibilities. As a proof of concept, we took advantage of several fluorescent markers for chromatin components to monitor their status simultaneously with immunolocalization of histone marks. In addition, we show that early embryo cell fate markers can be combined with histone mark localization, to start addressing the important, yet unexplored question of epigenetic control of early embryo lineages in plants.

The coupling of immunolocalization and fluorescent protein markers offers many combinations to be explored (Fig. 3). We obtained good results, with long lasting fluorescent protein signals, using the following histone markers lines: centromeric Histone H3 pWOX2:CenH3:GFP [19]; Histone H3.3 pHTR5:HTR5:GFP [20]; Histone H2B pRPS5a:H2B:tdTomato and pRPS5a:H2B:GFP [21]; Histone H2A pH2AW:H2AW:mRFP [22]. The line pTFL2:TFL2:GFP [23], a H3K27me3 reader, also showed satisfying maintenance of the nuclear GFP signal. For embryo cell fate markers, good results were obtained so far for transcription factor promoters WOX2, WOX8, and WOX9, controlling expression of triple YFP nuclear fusions [24]. However, an independent WOX9 gene fusion with a single GFP (pSTIMPY:GFP [25]) and no nuclear targeting, did not retain any GFP signal after fixation. Therefore, expression level seems to be a limiting factor and nuclear factors and chromatin proteins are likely better protected from damage by fixative and permeabilization treatment. In critical cases, a lower percentage of paraformaldehyde and detergent can be tried to help maintaining a fluorescent protein signal. Altogether, the protocol presented here represents a simple method for high-resolution analysis in early embryogenesis, and establishes the basis for future simultaneous monitoring of embryo membranes or cell wall, using fluorescent protein markers, to facilitate embryo staging and 3D imaging. Additional possibilities include the use of a new generation of antibodies, such as nanobodies, to improve antibody penetration and reduce protocol length by direct coupling to fluorophores as well as combination with clearing procedures compatible with fluorescent proteins [26, 27].

4 Notes

1. Using "ProbeOn plus" adhesion slides from FisherBiotech, we obtained good adhesion of the acrylamide matrix throughout the process. Alternatively, slides manually coated with poly-L-lysine can be used.
2. When possible, primary antibody batches are chosen according to specificity tests published by The Antibody Validation Database: http://compbio.med.harvard.edu/antibodies/.

As a general rule, if used for the first time, the specificity of the primary antibody should be checked ideally by performing immunolocalizations on both wild-type and mutant lines affected in the deposition of the chromatin mark studied. Specificity should be established also by testing for single band detection on a Western blot. However, many antibodies need a three dimensional conformation for efficient binding, and might not work in Western blot.

3. It is also possible to isolate 1–2 cell embryos; however, most of the times, the basal suspensor cell is lost. Therefore, we recommend protocol A for younger stages.
4. For wild-type Arabidopsis siliques and Arabidopsis mutants that do not cause seed abortion, dissection of 10–15 siliques provides enough seeds for 3–4 slide preparations. Depending on seed stage, between 50 and 70 seeds per acrylamide pad works well. An excess number of seeds per slide can produce overlapping material and inefficient embedding of the antibody.
5. Vacuum can be applied to improve penetration of the fixative if poor fixation is observed (resulting in low signal and brownish tissues). Set up the vacuum and break it into 2–3 steps, with 5–10 min between each break.
6. If required for practical reasons, a longer incubation time of fixed samples in Buffer 1 or 1× PBS (up to one night at 4 °C) is possible for most histone modification antibodies we tested. However, if loss of signal is experienced, long incubation at this step should be avoided as it increases the probability of protein degradation.
7. Since commercial enzymes are not pure, each lot of enzyme mix will have a specific activity. Therefore, when a new enzyme mix is prepared, it is crucial to perform a calibration experiment first, testing different incubation times. We obtained good results with enzymes purchased from Sigma (plant cell culture tested).
8. If acrylamide polymerization seems too fast for the manipulation time needed, TEMED can be replaced by 1 μl 20% sodium bisulfite. The timing of acrylamide polymerization may also differ between new stock solutions and older stock solutions, which polymerize slower because they are more oxygenated.
9. Low Melting Point Agarose matrix has been used successfully in other protocols [7, 12]. We prefer acrylamide properties to maintain 3D structures and for its transparency, resulting in better imaging.
10. Coverslip placement should be fast, while taking care to avoid air bubble formation in the acrylamide matrix, as these hamper the final imaging of the samples. Manipulating the timing of

acrylamide polymerization (*see* **Note 8**) is useful to avoid this problem.

11. Alternatively, to reduce the amount of BSA to be used (for instance for availability reasons), add 50 μl of 1% BSA directly on the acrylamide matrix, cover it with a clean coverslip and maintain the samples in a wet chamber.
12. Dilution should be determined empirically for each antibody, each batch and according to the samples tested. Testing 1:200 and 1:500 dilutions is a good starting point for most chromatin mark antibodies we tested.
13. If background signal is observed, increase the total washing time (up to all day long), renewing the buffer every hour.
14. To avoid secondary antibody aggregates, briefly spin the stock tube at low speed (ca. 6000 × *g*) before pipetting.
15. If using DAPI in combination with green fluorophores, confocal imaging must be performed using sequential scanning, to avoid cross excitation. In addition, DAPI can generate chromatic aberrations relative to green or red emitting fluorophores, even using an apochromatic lens. Such aberrations are less severe between PI (red) and green fluorophores.
16. We obtained good results with the antifading, hardening, mounting mediums Prolong Gold or Diamond from Molecular Probes. Prolong is compatible with fluorescent protein markers. Refractive index of Prolong changes while hardening, and stabilizes around 1.47 after 48 h, allowing storage of the slides at −20 °C. We typically use 4 °C for 48 h storage, and −20 °C for 10 day storage. We observed decreased fluorescence after longer storage.
17. Mounting seeds of homogenous stage (and size) is crucial for correct popping out of the embryos.
18. To pop out the embryos, the correct tool and pressure is determined empirically by each user. Too much pressure will break the coverslip and/or destroy the samples. Too little pressure will not pop out the embryo.

Acknowledgments

We are grateful to Stewart Gillmor, Daniel Grimanelli, and Caroline Michaud for critical reading of the manuscript. We thank S. Gillmor, M. Ingouff, T. Higashiyama, F. Berger, N. DeStorme, C. Baroux, T. Laux, D. Weigel, and the NASC/ABRC Arabidopsis stock centers for mutants and marker lines. M.G.A. is supported by CONACyT Ciencia Básica Grant No. 237480. D.A. is supported by the European Union's Horizon 2020 MSCA-RISE-2014 project "ProCrop" under grant agreement No. 645674.

References

1. Perino M, Veenstra GJ (2016) Chromatin control of developmental dynamics and plasticity. Dev Cell 38:610–620. doi:10.1016/j.devcel.2016.08.004
2. Palovaara J, de Zeeuw T, Weijers D (2016) Tissue and organ initiation in the plant embryo: a first time for everything. Annu Rev Cell Dev Biol 6:47–75
3. Feng S, Jacobsen SE, Reik W (2010) Epigenetic reprogramming in plant and animal development. Science 330:622–627. doi:10.1126/science.1190614
4. Heard E, Martienssen RA (2014) Transgenerational epigenetic inheritance: myths and mechanisms. Cell 157:95–109. doi:10.1016/j.cell.2014.02.045
5. Morao AK, Caillieux E, Colot V, Roudier F (2017) Cell type-specific profiling of chromatin modifications and associated proteins. In: Bemer M, Baroux C (eds) Plant chromatin dynamics: methods and protocols. Springer, New York, NY. doi:10.1007/978-1-4939-7318-7_8
6. Baroux C, Autran D (2015) Chromatin dynamics during cellular differentiation in the female reproductive lineage of flowering plants. Plant J 1:160–176. doi:10.1111/tpj.12890
7. Grimanelli D, García M, Kaszas E et al (2003) Heterochronic expression of sexual reproductive programs during apomictic development in *Tripsacum*. Genetics 165:1521–1531
8. Baroux C, Pecinka A, Fuchs J et al (2007) The triploid endosperm genome of Arabidopsis adopts a peculiar, parental-dosage-dependent chromatin organization. Plant Cell 6:1782–1794
9. Garcia-Aguilar M, Michaud C, Leblanc O et al (2010) Inactivation of a DNA methylation pathway in maize reproductive organs results in apomixis-like phenotypes. Plant Cell 10:3249–3267. doi:10.1105/tpc.109.072181
10. Pillot M, Baroux C, Vazquez MA et al (2010) Embryo and endosperm inherit distinct chromatin and transcriptional states from the female gametes in Arabidopsis. Plant Cell 22:307–320. doi:10.1105/tpc.109.071647
11. Autran D, Baroux C, Raissig MT et al (2011) Maternal epigenetic pathways control parental contributions to Arabidopsis early embryogenesis. Cell 145:707–719. doi:10.1016/j.cell.2011.04.014
12. She W, Grimanelli D, Baroux C (2014) An efficient method for quantitative, single-cell analysis of chromatin modification and nuclear architecture in whole-mount ovules in Arabidopsis. J Vis Exp 88:e51530. doi:10.3791/51530
13. Belcram K, Palauqui JC, Pastuglia M (2016) Studying cell division plane positioning in early-stage embryos. Methods Mol Biol 1370:183–195. doi:10.1007/978-1-4939-3142-2_14
14. She W, Baroux C, Grossniklaus U (2017) Cell-type specific chromatin analysis in whole-mount plant tissues by immunostaining. In: Bemer M, Baroux C (eds) Plant chromatin dynamics: methods and protocols. Springer, New York, NY. doi:10.1007/978-1-4939-7318-7_25
15. Musielak TJ, Schenkel L, Kolb M et al (2015) A simple and versatile cell wall staining protocol to study plant reproduction. Plant Reprod 28:161–169. doi:10.1007/s00497-015-0267-1
16. Brown CM (2007) Fluorescence microscopy-avoiding the pitfalls. Journal of Cell Science 120:1703–1705. doi:10.1242/jcs.022079
17. Pauvonic I, Baroux C (2014). Quantification of chromatin modifications in whole-mount plant tissue. http://www.bitplane.com/learning/quantification-of-chromatin-modifications-in-whole-mount-plant-tissue-tutorial#sthash.O9pCatVT.dpuf. Edited by 2014
18. Bankhead P (2016), ImageJ/FIJI handbook, 3D processing chapter. https://petebankhead.gitbooks.io/imagej-intro/content/chapters/multidimensional_processing/multidimensional_processing.html
19. De Storme N, Nur Keçeli B, Zamariola L et al (2016) CENH3-GFP: a visual mark for gametophytic and somatic ploidy determination in Arabidopsis thaliana. BMC Plant Biol 16:1. doi:10.1186/s12870-015-0700-5
20. Ingouff M, Rademacher S, Holec S et al (2010) Zygotic resetting of the HISTONE 3 variant repertoire participates in epigenetic reprogramming in Arabidopsis. Curr Biol 20:2137–2143. doi:10.1016/j.cub.2010.11.012
21. Maruyama D, Hamamura Y, Takeuchi H et al (2013) Independent control by each female gamete prevents the attraction of multiple pollen tubes. Dev Cell 25:317–323. doi:10.1016/j.devcel.2013.03.013
22. Yelagandula R, Stroud H, Holec S et al (2014) The histone variant H2A.W defines heterochromatin and promotes chromatin condensation in Arabidopsis. Cell 158:98–109. doi:10.1016/j.cell.2014.06.006

23. Nakahigashi K, Jasencakova Z, Schubert I et al (2005) The Arabidopsis heterochromatin protein1 homolog (TERMINAL FLOWER2) silences genes within the euchromatic region but not genes positioned in heterochromatin. Plant Cell Physiol 46:1747–1756
24. Breuninger H, Rikirsch E, Hermann M et al (2008) Differential expression of WOX genes meadiates apical-basal axis formation in the Arabidopsis embryo. Dev Cell 14:867–876. doi:10.1016/j.devcel.2008.03.008
25. Wu X, Dabi T, Weigel D (2005) Requirement of homeobox gene STIMPY/WOX9 for Arabidopsis meristem growth and maintenance. Curr Biol 15(5):436–440. doi:10.1016/j.cub.2004.12.079
26. Kurihara D, Mizuta Y, Sato Y et al (2015) ClearSee: a rapid optical clearing reagent for whole-plant fluorescence imaging. Development 142:4168–4179. doi:10.1242/dev.127613
27. Musielak TJ, Slane D, Liebig C et al (2016) A versatile optical clearing protocol for deep tissue imaging of fluorescent proteins in Arabidopsis thaliana. PLoS One 11:e0161107. doi:10.1371/journal.pone.016110

Chapter 25

Cell-Type Specific Chromatin Analysis in Whole-Mount Plant Tissues by Immunostaining

Wenjing She, Célia Baroux, and Ueli Grossniklaus

Abstract

Chromatin organization in eukaryotes is highly dynamic, playing fundamental roles in regulating diverse nuclear processes including DNA replication, transcription, and repair. Thus, the analysis of chromatin organization is of great importance for the elucidation of chromatin-mediated biological processes. Immunostaining coupled with imaging is one of the most powerful tools for chromatin analysis at the cellular level. However, in plants, it is sometimes technically challenging to apply this method due to the inaccessibility of certain cell types and/or poor penetration of the reagents into plant tissues and cells. To circumvent these limitations, we developed a highly efficient protocol enabling the analysis of chromatin modifications and nuclear organization at the single-cell level with high resolution in whole-mount plant tissues. The main procedure consists of five steps: (1) tissue fixation; (2) dissection and embedding; (3) tissue processing; (4) antibody incubation; and (5) imaging. This protocol has been simplified for the processing of multiple samples without the need for laborious tissue sectioning. Additionally, it preserves cellular morphology and chromatin organization, allowing comparative analyses of chromatin organization between different cell types or developmental stages. This protocol was successfully used for various tissues of different plant species, including *Arabidopsis thaliana*, *Oryza sativa* (rice), and *Zea mays* (maize). Importantly, this method is very useful to analyze poorly accessible tissues, such as female meiocytes, gametophytes, and embryos.

Key words Chromatin organization, Whole-mount immunostaining, Chromatin analysis, Chromatin modification, Nuclear organization, *Arabidopsis thaliana*, Rice, Maize, Gametophyte

1 Introduction

In eukaryotes, genomic DNA and its associated histones are compacted into a highly organized structure known as chromatin. In plants, cytosine residues in DNA can be methylated, while histone tails can be modified by methylation, acetylation, ubiquitination, phosphorylation, glycosylation, sumoylation, and other modifications, which in turn modulate chromatin structure. The organization of the chromatin in the nucleus is not static, but changes in a highly dynamic fashion. Dynamic changes of chromatin organiza-

Marian Bemer and Célia Baroux (eds.), *Plant Chromatin Dynamics: Methods and Protocols*, Methods in Molecular Biology, vol. 1675, DOI 10.1007/978-1-4939-7318-7_25, © Springer Science+Business Media LLC 2018

tion occur in response to developmental and environmental cues, such as the establishment of reproductive lineages [1, 2], floral transition [3], progression of root development [4], seed maturation [5], senescence [6, 7], and light-signaling [8–10]. The dynamic organization of chromatin within the nucleus is considered to have a key influence on the accessibility to DNA of the factors that regulate replication, transcription, and repair [11–13]. Over the past decades, studies of chromatin organization using newly developed tools, such as live cell imaging, the sequencing of chromatin immunoprecipitated fragments (ChIP-seq), and chromosome conformation capture-based methods, such as Hi-C, have led to a better understanding of chromatin regulatory processes at different levels in plants [14–16]. Although high-throughput sequencing approaches have made significant breakthroughs in chromatin research through genome-wide analyses of chromatin interactions [15, 16], immunostaining coupled with imaging is a complementary and powerful tool to study the organization of chromatin. Particularly, when performed in intact tissues, it provides the unique possibility to visualize chromatin modifications on the cellular level, i.e., in a cell type-specific manner.

Whole-mount tissue immunostaining has classically been impaired by limitations with respect to poor permeability of plant tissues by the reagents and poor cellular preservation of the tissue after harsh chemical treatments. To overcome these problems, we and other labs developed robust protocols to efficiently investigate chromatin organization in plants by immunostaining [17–21]. In a complementary manner to our previously published video protocol [19], we provide here a detailed step-by-step description of tissue preparation, tissue clarification, and staining with practical notes facilitating handling and offering possible alternatives at specific steps. The protocol is based on embedding of the dissected plant tissue in acrylamide on a Superfrost Plus slide [22]. The embedded samples undergo a series of treatments, including fixation, cell wall digestion, permeabilization, incubation with the antibodies, and DNA staining (Fig. 1). This method was proven to be efficient and reliable for chromatin analysis in differentiating *Arabidopsis* megaspore mother cells [1], pollen mother cells [2], the developing female gametophyte [23], and roots [24]. It has also been successfully used for chromatin analysis in the developing ovules of other plant species, including rice and maize (Fig. 2). The procedure preserves cellular morphology and produces homogeneous fluorescent signals, which facilitates quantification of fluorescent intensity at the single-cell level.

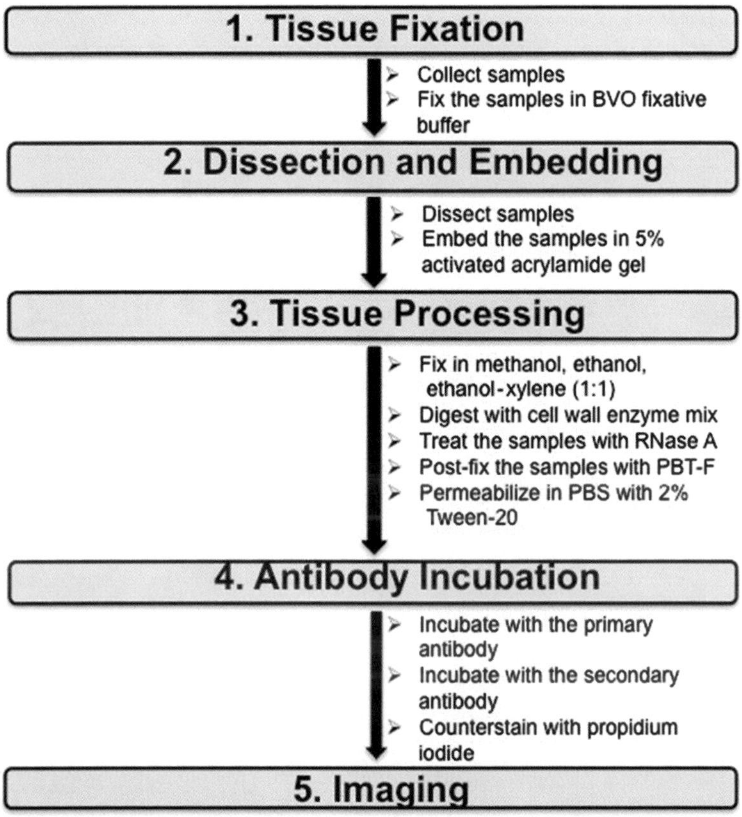

Fig. 1 Workflow of immunostaining in plants. This protocol has been successfully applied to roots, shoots, leaves, ovules, and other tissues from *Arabidopsis* as well as from other plant species. The procedure includes five major steps: (1) Tissue Fixation. Samples are collected and fixed in freshly made BVO buffer. (2) Dissection and Embedding. Tissues are dissected and embedded in 5% activated acrylamide. (3) Tissue Processing. The embedded samples are fixed in methanol, ethanol, ethanol–xylene, and then incubated with cell wall digestion enzyme mix at 37 °C for 1–2 h. After that, the samples are treated with RNase A at 37 °C for 1 h, and then fixed in freshly made PBT-F. Subsequently, permeabilization in PBS with 2% Tween 20 at 4 °C is performed. (4) Antibody Incubation. The samples are incubated first with the primary antibody, and then with the secondary antibody for 12-24 h at 4 °C. After the DNA is stained with propidium iodide, samples are mounted in anti-fade reagent. (5) Imaging. Images can be acquired by confocal laser scanning microscopy (CLSM) with appropriate laser and filter configurations

2 Materials

For all solutions, double deionized water (ddH_2O) is used unless specified otherwise.

2.1 Fixation

1. BVO Fixation Buffer: 2 mM EGTA, pH 7.5, 1% (v/v) formaldehyde (diluted from a 37 wt. % stock in H_2O), 10% DMSO, 1× PBS, 0.1% Tween 20 (prepare fresh 2 mL).

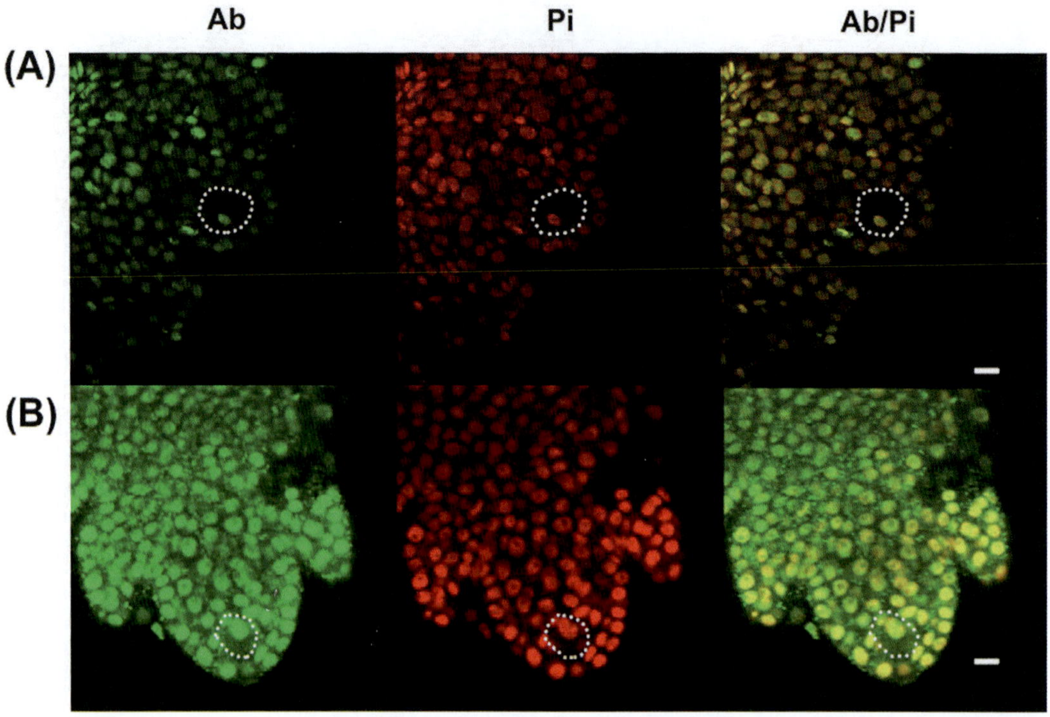

Fig. 2 Examples of whole-mount immunostaining in rice and maize ovules. (**a**) A wild-type rice ovule primordium was stained by the antibody against H3K4me3, and counterstained with propidium iodide (Pi). (**b**) Immunodetection of H3K4me2 in a wild-type mini-maize [25] ovule. The antibody (Ab) signal is *green*, Pi fluorescence is in *red*. The overlay of fluorescent signals from antibody and Pi staining is also shown (Ab/Pi). Megaspore mother cells are indicated by *white contours*. The images display single confocal sections. Scale bar: 40 μm

2. PBT: 1× PBS, 0.1% Tween 20 (prepare fresh 500 mL).
3. Shaker.

2.2 Dissection and Embedding

1. 30% acrylamide–bisacrylamide stock: 3 g acrylamide, 0.33 g bisacrylamide, 1× PBS (be careful, *see* **Note 1**, prepare 10 mL, keep at 4 °C for up to 6 months).
2. 5% acrylamide mix in PBS: 30% acrylamide–bisacrylamide, 1× PBS (prepare fresh 2 mL from 30% stock, make 200 μL aliquots, do not keep for more than 2 weeks at 4 °C).
3. 20% ammonium persulfate (APS): 0.2 g ammonium persulfate, 1 mL sterile water (prepare 10 mL, make 1 mL aliquots, keep at −20 °C for at least 6 months).
4. 20% sodium sulfite (NaPS): 0.2 g sodium sulfite, 1 mL sterile water (prepare 10 mL, make 1 mL aliquots, keep at −20 °C for at least up to 6 months).
5. Fine tungsten needles (*see* **Note 2**).
6. Superfrost Plus slides.

2.3 Tissue Processing

1. PBT.
2. PBT-F: 1× PBT, 2.5% (v/v) formaldehyde (prepare fresh 80 mL).
3. Cell wall-digestion enzyme mix: 0.5% (w/v) cellulase, 1% (w/v) driselase, and 0.5% (w/v) pectolyase (*see* **Note 3**, prepare 10 mL, make 1 mL aliquots and keep at −20 °C for at least up to 6 months).
4. Clean Coplin jars (three per experiment).
5. 80 mL methanol in a Coplin jar (keep covered under a fume hood for maximum 2 weeks).
6. 80 mL ethanol in a Coplin jar (keep covered under a fume hood for maximum 2 weeks).
7. 80 mL ethanol–xylene (1:1) in a Coplin jar (keep covered under a fume hood for maximum 2 weeks).
8. DNase-free RNase A (100 μg/mL, prepare 1 mL, make 100 μL aliquots and keep at −20 °C).
9. Moist chamber: plastic box with damp paper towel inside (use sterile ddH20) and plastic sticks used to elevate the slides above the wet background.
10. Flat-end forceps (1×).
11. Shaker.

2.4 Immunostaining

1. PBT.
2. Prolong Gold.
3. Propidium iodide (10 mg/mL, prepare 1 mL at 10 μg/mL).
4. Coplin jars.
5. Shaker.
6. Moist chamber.

3 Methods

The workflow is described in Fig. 1. Room temperature in our conditions corresponds to 19–23 °C. Perform all steps with formaldehyde in the flow hood.

3.1 Tissue Fixation

This step aims to preserve the cellular morphology and chromatin structure of the tissue.

1. Collect the plant tissue in a microfuge tube containing freshly made BVO fixative buffer on ice (*see* **Notes 4** and **5**).
2. Fix the tissue 30 min with gentle shaking at room temperature.

3. Spin the tubes containing the tissue in fixative in a bench top microcentrifuge 1 min at 400 × *g*.
4. Remove carefully the fixative buffer and add 1 mL of PBT, place the tubes on ice.

3.2 Dissection and Embedding

In this step, the tissue of analysis is released and embedded in an acrylamide gel to facilitate handling and homogeneous treatment in the subsequent steps.

1. Prepare five 1.5 mL tubes with each 200 μL of a freshly made 5% acrylamide mix.
2. Prepare five Superfrost Plus slides precleaned with 70% ethanol and labeled with a pencil.
3. Thaw one aliquot each of 20% APS and 20% NaPS, on ice.
4. Collect the dissected tissue by pipetting carefully with a cut-end tip or with forceps, depending on the size of the fragments, and place them on a clean slide; remove the excess liquid by pipetting.
5. Dissect the tissue on a Superfrost Plus slide under a stereomicroscope using fine tungsten needles (*see* **Notes 6** and **7**).
6. Quickly add and mix 12 μL NaPS and 12 μL APS with an aliquot of 200 μL acrylamide mix.
7. Add 30 μL of the activated acrylamide onto the dissected tissue.
8. Cover with a 20 mm × 20 mm coverslip, let polymerize at room temperature, 45–60 min (*see* **Note 8**).
9. Remove the coverslip using a razor blade. At this stage, the samples can be kept overnight at 4 °C in a Coplin jar containing 1× PBS.

3.3 Tissue Processing

This step aims to improve tissue clarity and permeability to the immunostaining reagents.

3.3.1 Tissue Clearing and Fixation (see Note 9)

Coplin jars are placed under the fume hood on paper towels. The slides are transferred from one Coplin jar to the other using flat-end forceps. Excess of liquid is allowed to drain before immersing the slide into the next solution. The setup in Coplin jars allows multiplex processing of eight slides in parallel. It is important to make sure the acryl pads do not dry between each step, while an excess of the solutions should not be carried over but drained on the tissue paper without touching the pad.

1. Incubate for 5 min in methanol.
2. Incubate for 5 min in ethanol.
3. Incubate for 30 min in ethanol–xylene (1:1).

4. Incubate for 5 min in ethanol.
5. Incubate for 5 min in methanol.
6. Incubate for 15 min in methanol and PBT (1:1), complemented with 2.5% Formaldehyde.
7. Rinse 2 × 10 min in PBT. At this stage, slides can be kept overnight at 4 °C. Cover the Coplin jar with a lid or Parafilm.

3.3.2 Cell Wall Digestion

1. Thaw an aliquot of the cell wall digestion mix on ice. Mix well by pipetting before use (*see* **Note 10**).
2. Take a slide from the Coplin jar, drain the excess of the liquid by placing it vertically on a paper towel.
3. Dry the backside of the slide with tissue paper; place it horizontally on a clean surface (clean bench or horizontal slide holder).
4. Add 100 μL of cell wall digestion mix over the acrylamide pad and cover with a 23 mm × 46 mm coverslip. Repeat for the other slides (*see* **Note 10**).
5. Incubate for 1–2 h at 37 °C in a moist chamber (*see* **Note 11**).
6. Carefully remove the coverslips and drain the excess of the enzyme mix on a tissue paper without touching the pad.
7. Wash the slides 2 × 5 min in PBT.

3.3.3 RNase A Treatment, Post-Fixation, and Permeabilization

1. Take a slide from the Coplin jar, drain the excess of the liquid as before.
2. Incubate each slide with 100 μL of RNase A at 100 μg/mL in PBS with 1% Tween 20 for 1 h at 37 °C in a moist chamber (*see* **Note 12**).
3. Wash the slides for 2 × 5 min in PBT.
4. Post-fix for 20 min in freshly made PBT-F.
5. Rinse the slides for 10 min in PBT.
6. Permeabilize for 1–2 h in PBS with 2% Tween 20 at 4 °C.
7. Rinse the slides for 2 × 5 min in PBT.

The slides can be kept at 4 °C at this stage for up to 2 weeks.

3.4 Immunostaining

Here, the antibodies are applied to the tissue to bind to the protein of interest. After incubation with the first and secondary antibody, extensive washing is required to wash off all unbound antibodies and thus reduce the background. The application of the antibodies takes place on the slide, which is then incubated in a moist chamber as described above. The washings are performed in Coplin jars. It is important to verify the conditions that allow specific, robust, and homogenous staining. In addition, specific considerations concerning antibody specificity, dilution, incubation times, and troubleshooting with regard to background signals, or troubleshooting with regard to the absence of signals is given in **Notes 13–17,** respectively.

1. Incubate each slide with 100 μL of primary antibody diluted in 1× PBS with 0.2% Tween 20 for 12–24 h at 4 °C (*see* **Notes 13–15**).
2. Wash the slides in PBT for 2–4 h at room temperature under gentle shaking.
3. Apply the secondary antibody 1:200 in 1× PBS + 0.2% Tween 20 for 12–24 h at 4 °C.
4. Wash the slides in PBT for 1 h at room temperature under gentle shaking.
5. Counterstain with 10 μg/mL propidium iodide in 1× PBS for 15 min, then rinse 15 min in 1× PBS under gentle shaking at room temperature (*see* **Note 16**).
6. Mount in anti-fading solution supplemented with 10 μg/mL propidium iodide. Let the mounting medium harden for 1 h before acquiring images by confocal laser scanning microscopy (CLSM) (*see* **Note 18**).

3.5 Imaging

1. Observe and acquire high-resolution images using CLSM. For long-time imaging or 3D series acquisition, the resonance-scanning mode is recommended, which allows better preservation of fluorescent signals (*see* **Notes 19** and **20**).

4 Notes

1. Acrylamide is a neurotoxin. Weigh out 0.33 g bisacrylamide and 3 g acrylamide under the hood. Dissolve the powder in 7 mL 1× PBS, and then adjust the volume to 10 mL with 1× PBS. Sterilize the solution by filtration (0.2 μm syringe type membrane filter), and store at 4 °C. The stock is stable for at least 6 months at 4 °C [24].
2. To sharpen tungsten needles, a Varta high energy 9 V battery was used. Remove the insulation from one wire to make an electrode, and immerse it in 10 n NaOH solution. Then, attach an alligator clip to the other end of the electric wire. Connect this end to the metal part of the tungsten needle holder. Hold the needle and submerge the end of the wire 1–2 cm into the solution for 5–15 s while moving it up and down slowly [26].
3. We experienced variability in staining homogeneity (ranging from no signal, signal in only part of the tissue, to 100% of the tissue showing staining), depending on the digestion time and the enzymatic activity (batch-specific, described by the

provider). It is recommended to produce a large amount of stock solution of the enzyme mix (e.g., 100 mL) and keep 1 mL aliquots at −20 °C. Each stock solution should first be tested on 1–2 slides before using it at large scale.

4. It is important to use young and healthy tissue. We found that tissue from old or stressed plants (drought, light, or pest stress) are not cleared and permeabilized efficiently, resulting in very poor staining.
5. The fixative should be freshly made. Samples can also be fixed in other fixatives such as 4% paraformaldehyde instead of BVO buffer. However, we didn't try other fixatives but 4% paraformaldehyde and BVO buffer. Fixation by 4% paraformaldehyde gives inconsistent signals or no staining. Furthermore, overfixation should be avoided, since it will produce weak fluorescence signals.
6. When dissection is not required (i.e., leaf, roots, floral organs), it is nevertheless recommended to embed small fragments. Care should be taken in order to not add too many fragments that may prevent good adhesion of the acrylamide pads and possibly lead to the formation of air bubbles.
7. Dehydration of the tissue during dissection leads to altered histological morphology and compromised staining signals. Thus, we recommend that the tissue should be covered by 1× PBS to avoid drying (less than 10 μL).
8. We strongly recommend 45 min for polymerization. If incomplete, this will lead to poor adhesion of the acrylamide gel and loss of the sample during the following steps.
9. The tissue can be fixed in 4% paraformaldehyde instead of BVO buffer (also *see* **Note 5**). This has the benefit of time saving by omitting the steps in Subheading 3.3.1, but the drawback of attenuating robustness of staining across samples, antibodies and the batch of cell wall digestion enzyme mix.
10. Depending on the purity of the enzyme extracts used to make the solution, some particles deposit at the bottom. It is important to mix well by pipetting before adding the solution onto the slide.
11. Cell wall digestion is a critical step for good penetration and homogenous distribution of the antibodies and staining reagents throughout the plant tissue. Limited digestion time and suboptimal enzyme concentrations result in weak and nonhomogenous staining signals. However, overdigestion may lead to damaged histological morphology. Thus, we recommend that different dilutions of the enzyme stock and digestion time should first be tested on 1–2 slides before using the stock at large scale.

12. This step is necessary when the preparation is counterstained with propidium iodide since this dye also binds to RNA molecules.
13. For immunostaining of histone modifications, select primary antibodies that have no cross-reactivities. It is advisable to verify the specificity of the antibodies in the open source database http://compbio.med.harvard.edu/antibodies/projects/1.
14. The optimal concentration of the primary antibody has to be tested using different dilutions (1:200, 1:500, and 1:1000) of the antibodies, especially for quantification analyses. We recommend to start with 1:200 dilution and increase the dilution to the maximum dilution (i.e., the lowest concentration) still enabling robust signals. In that way, the immunodetection is likely in a linear amplification phase favorable to analyses. Low dilutions (i.e., high concentration) correspond to a saturated detection level and hence are not suitable for quantification.
15. If immunostaining produces unspecific and high background noise, we recommend to block with 5% BSA in 1× PBS before applying the primary antibody. Increasing the duration of the washing steps to 1 h each is also a way to reduce background signals. Finally, the duration of the incubation with the antibody has to be determined on a case-by-case basis. In our hands, primary antibodies against modified histones were incubated for 12–24 h, while secondary antibodies were incubated for 24 h.
16. For counterstaining, the incubation duration with propidium iodide varies for different tissues and plant species. We find that 15 min incubation with propidium iodide works well for *Arabidopsis*, rice, and maize ovules. Besides propidium iodide, some other DNA dyes, such as 4′,6-diamidino-2-phenylindole (DAPI), can also be used. Compared to propidium iodide, DAPI penetrates more easily into the tissues. However, because DAPI has a broad fluorescent emission spectrum that overlaps with other fluorophores and a sequential scan will therefore be required. This sometimes complicates image acquisition.
17. If immunostaining produces a weak or absent signal, the incubation time and concentration of the antibodies can be increased. If DNA staining gives a weak or no signal, increasing the incubation time with propidium iodide should be tried.
18. We recommend to either acquire the images as soon as possible, or to keep the mounted samples at 4 °C in the dark to avoid fluorescence decay.
19. If several fluorochromes are used, it is advisable to verify the absence of cross talk between fluorochromes. If the cross talk between fluorochromes is present, set up a sequential scan.

Acquire transmission images separately and not simultaneously. For serial, three-dimensional image acquisition, we recommend to acquire images with the highest possible resolution in the x and y dimensions, and with 2-times oversampling in the z dimension (Nyquist's rule).

20. For quantitative measurements of chromatin modifications, we recommend to test the acquisition parameters, such as laser intensity, gain, pinhole, voxel size, and zoom factor at the beginning of the experiment to define a standard acquisition procedure that is to be strictly followed throughout the analysis of all slides.

Acknowledgments

This work was supported by the University of Zürich, the Swiss National Foundation and, in part, DuPont-Pioneer. We are thankful to Christof Eichenberger, Valeria Gagliardini, Arturo Bolanos, Matthias Philipp, Anja Frey, and Peter Kopf for general lab support, and to Karl Huwiler and Christian Frey for plant care.

References

1. She W, Grimanelli D, Rutowicz K et al (2013) Chromatin reprogramming during the somatic-to-reproductive cell fate transition in plants. Development 140:4008–4019
2. She W, Baroux C (2015) Chromatin dynamics in pollen mother cells underpin a common scenario at the somatic-to-reproductive fate transition of both the male and female lineages in *Arabidopsis*. Front Plant Sci 6:294
3. Tessadori F, Schulkes RK, van Driel R et al (2007) Light-regulated large-scale reorganization of chromatin during the floral transition in *Arabidopsis*. Plant J 50:848–857
4. Rosa S, Ntoukakis V, Ohmido N et al (2014) Cell differentiation and development in *Arabidopsis* are associated with changes in histone dynamics at the single-cell level. Plant Cell 26:4821–4833
5. van Zanten M, Koini MA, Geyer R et al (2011) Seed maturation in *Arabidopsis thaliana* is characterized by nuclear size reduction and increased chromatin condensation. Proc Natl Acad Sci U S A 108:20219–20224
6. Kolodziejek I, Koziol-Lipinska J, Waleza M et al (2007) Aspects of programmed cell death during early senescence of barley leaves: possible role of nitric oxide. Protoplasma 232:97–108
7. Ay N, Irmler K, Fischer A et al (2009) Epigenetic programming via histone methylation at *WRKY53* controls leaf senescence in *Arabidopsis thaliana*. Plant J 58:333–346
8. Tessadori F, van Zanten M, Pavlova P et al (2009) Phytochrome B and histone deacetylase 6 control light-induced chromatin compaction in *Arabidopsis thaliana*. PLoS Genet 5: e1000638
9. van Zanten M, Tessadori F, McLoughlin R et al (2010) Photoreceptors CRYPTOCHROME2 and phytochrome B control chromatin compaction in *Arabidopsis*. Plant Physiol 154:1686–1696
10. Bourbousse C, Mestiri I, Zabulon G et al (2015) Light signaling controls nuclear architecture reorganization during seedling establishment. Proc Natl Acad Sci U S A 112: E2836–E2844
11. Strahl BD, Allis CD (2000) The language of covalent histone modifications. Nature 403:41–45
12. Jenuwein T, Allis CD (2001) Translating the histone code. Science 293:1074–1080
13. Hübner MR, Eckersley-Maslin MA, Spector DL (2013) Chromatin organization and transcriptional regulation. Curr Opin Genet Dev 23:89–95

14. Rosa S, Shaw P (2013) Insights into chromatin structure and dynamics in plants. Biology 2:1378–1410
15. Park PJ (2009) ChIP-seq: advantages and challenges of a maturing technology. Nat Rev Genet 10:669–680
16. Liu C, Weigel D (2015) Chromatin in 3D: progress and prospects for plants. Genome Biol 16:170
17. Friml J, Benková E, Mayer U et al (2003) Automated whole mount localisation techniques for plant seedlings. Plant J 34:115–124
18. Sauer M, Paciorek T, Benková E et al (2006) Immunocytochemical techniques for whole-mount *in situ* protein localization in plants. Nat Protoc 1:98–103
19. She W, Grimanelli D, Baroux C (2014) An efficient method for quantitative, single-cell analysis of chromatin modification and nuclear architecture in whole-mount ovules in *Arabidopsis*. J Vis Exp 88:e51530
20. Escobar-Guzmán R, Rodríguez-Leal D, Vielle-Calzada JP et al (2015) Whole-mount immunolocalization to study female meiosis in *Arabidopsis*. Nat Protoc 10:1535–1542
21. Pasternak T, Tietz O, Rapp K et al (2015) Protocol: an improved and universal procedure for whole-mount immunolocalization in plants. Plant Methods 11:50
22. Bass HW, Marshall WF, Sedat JW et al (1997) Telomeres cluster *de novo* before the initiation of synapsis: a three-dimensional spatial analysis of telomere positions before and during meiotic prophase. J Cell Biol 137:5–18
23. Pillot M, Baroux C, Vazquez MA et al (2010) Embryo and endosperm inherit distinct chromatin and transcriptional states from the female gametes in *Arabidopsis*. Plant Cell 22:307–320
24. De Lucas PL, Turco GM et al (2016) Transcriptional regulation of *Arabidopsis Polycomb* repressive complex 2 coordinates cell type proliferation and differentiation. Plant Cell 28:2616–2631
25. McCaw ME, Wallace JG, Albert PS et al (2016) Fast-flowering mini-maize: seed to seed in 60 days. Genetics 204:35–42
26. Brady J (1965) A simple technique for making very fine, durable dissecting needles by sharpening tungsten wire electrolytically. Bull World Health Organ 32:143–144

Chapter 26

Measuring Dynamics of Histone Proteins by Photobleaching in Arabidopsis Roots

Stefanie Rosa

Abstract

Histone proteins play an important role in determining chromatin structure and gene expression. Studies in several systems have established that histones are in continuous turnover within the chromatin. It is therefore important to quantitatively measure the binding dynamics of these proteins in vivo. Photobleaching-based approaches such as fluorescence recovery after photobleaching (FRAP) are advantageous in that they are applied to living cells at a single cell level. In this chapter, I provide a detailed experimental protocol on how to perform histone FRAP experiments in *Arabidopsis thaliana* roots and how to analyze the most important parameters.

Key words Fluorescence recovery after photobleaching, Fluorescence microscopy, Green fluorescent protein, Half-time recovery, Mobile fraction, Histones, Arabidopsis

1 Introduction

Chromatin organization within the nucleus is extensively mediated by core and linker histones. The nucleosome, consisting of an octamer of core histones complexed with the DNA, has long been considered a stable building block of chromatin; however it has been shown that histone–DNA interactions are highly dynamic in nature [1–6].

Fluorescence recovery after photobleaching (FRAP) is a powerful experimental approach to study the dynamic behavior of proteins within living cells. In FRAP experiments, the desired protein is linked to a fluorescent protein suitable for expression and monitoring, such as green fluorescent protein (GFP), expressed in living cells and its recovery after bleaching is measured as an indicator of its mobility and dynamics. In addition to allowing a qualitative examination of histone distribution under traditional fluorescence microscopy, FRAP provides a quantitative analysis based on kinetic modeling methods. Time-lapse imaging of fluo-

Marian Bemer and Célia Baroux (eds.), *Plant Chromatin Dynamics: Methods and Protocols*, Methods in Molecular Biology, vol. 1675, DOI 10.1007/978-1-4939-7318-7_26, © Springer Science+Business Media LLC 2018

rescence recovery permits extraction of quantitative information regarding the protein's turnover rate, the size of mobile and immobile fractions, and the number of kinetically distinct populations of the protein.

FRAP is based on irreversibly damaging the emission of light from fluorescent molecules using high-power laser illumination. A typical FRAP experiment consists of three phases (Fig. 1a): (1) the initial fluorescence intensity is measured; (2) a short, high-intensity laser pulse is used to bleach fluorescent molecules that localize within a region of interest (ROI); (3) the fluorescence intensity is monitored within the ROI in a time-course series, providing information about the dynamics of the protein of interest. It is important to understand that photobleaching does not remove the proteins from the bleached region; what is measured is the exchange of bleached proteins within the bleached region for nonbleached molecules that diffuse from outside the bleached region, which in turn produces a recovery of fluorescence intensity. This recovery can be quantified by plotting the fluorescence intensity as a function of time within the defined ROI (Fig. 1b). Additional mathematical analysis and curve fitting allows extraction of

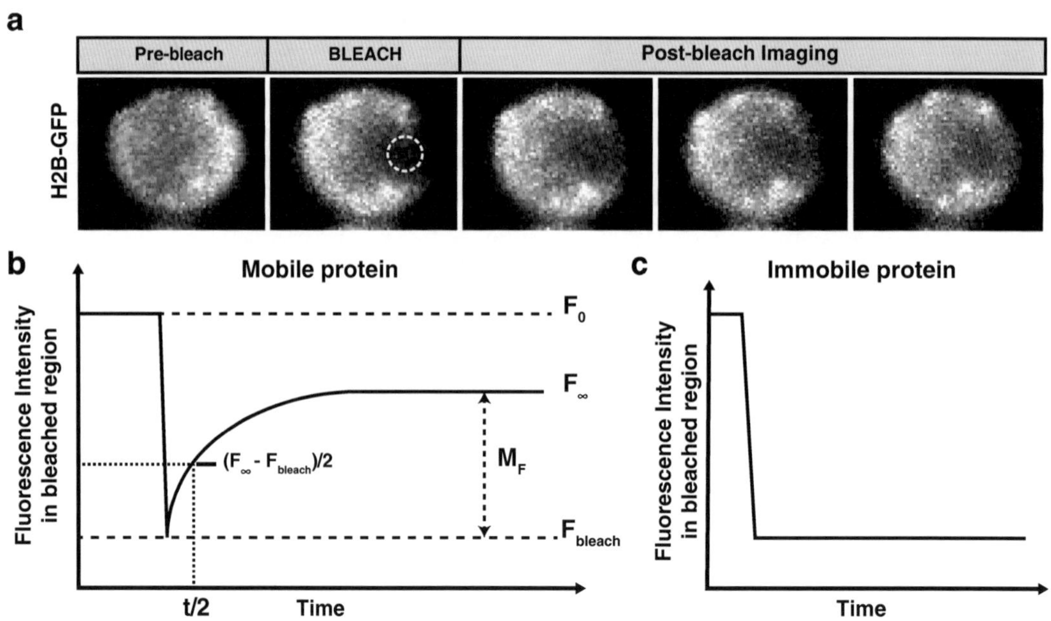

Fig. 1 The principle of fluorescence recovery after photobleaching. (**a**) Representative images of an Arabidopsis root nucleus expressing H2B-GFP indicating the different phases of a FRAP experiment. (**b**) Fluorescent intensity represented as a function of time. For the case of a mobile protein this enables the measurement of the half-time of recovery ($t_{1/2}$) and the mobile fraction (M_F). Fluorescence intensities before and immediately after bleaching are indicated (F_0 and F_{bleach}), as well as the fluorescence at the final end of the imaging time (F_{∞}). (**c**) Immobile proteins show no recovery

the main parameters such as mobile fraction and half-time of recovery.

The mobile fraction (M_F) is the proportion of bleached molecules that were replaced by nonbleached molecules within the ROI. M_F is calculated as the ratio of fluorescence intensity at the end of imaging time (F_∞) and the initial intensity (F_0) corrected by the bleaching depth (F_{bleach}) and background intensity. In a situation in which all the molecules are immobile (Fig. 1c), the bleached area remains bleached and $M_F = 0$. On the other hand, if all the molecules are mobile and free to diffuse, the fluorescence will reach the initial intensity, with $M_F = 1$ or 100%, when expressed in percentage. Finally, when both immobile and mobile fractions are present, the recovery curve reaches a plateau (Fig. 1b) and M_F will take values between 0 and 1, depending on the relative abundance of the two fractions. The half-time of recovery ($t_{1/2}$) is the time that it takes for the fluorescence intensity to reach half of its maximum value (at F_∞). The rate of recovery is often determined by the rate at which the bleached molecules dissociate from their binding sites, which in turn depends on the stability of the interaction with large or fixed molecules.

In this chapter I describe the basic requirements and methodology to perform FRAP as a method to assess mobility of histone proteins, particularly in Arabidopsis root tissues (*see* **Note 1**).

2 Materials

2.1 Plant Material and Reagents

1. Arabidopsis lines expressing a histone protein of interest tagged with GFP or another fluorescent reporter (*see* **Note 2**).
2. Phytagel (*see* **Note 3**).
3. Sucrose.
4. 5% bleach (hypochlorite) in dH_2O.
5. Plant growth chamber (*see* **Note 4**).
6. Laminar flow cabinet.
7. Chambered cover glass (e.g., Lab-Tek II cover glass chamber #1.5, Lab-Tek) (*see* **Note 5**).
8. Sterile razor blades.
9. Adhesive tape.
10. Growth medium: 4.4 g/L Murashige and Skoog medium (pH 5.8), supplemented with 1% w/v sucrose and 0.5% w/v Phytagel (*see* **Note 6**).

2.2 Microscope Setup

FRAP experiments can be carried out on most commercial confocal microscopes as well as on widefield microscopes. The basic requirements are: appropriate lasers and sets of filters for the fluorophores in use, high magnification lenses (×40 or ×63), strong lasers (for effective bleaching) and sensitive detectors (ideally HyD or GaAsP detectors, although PMTs also work). The microscope setup described here is currently used in our laboratory.

1. Inverted laser scanning confocal microscope (e.g., LSM 710, Zeiss).
2. Objective EC Plan-Neofluar 40×/1.3 Oil or another high numerical aperture oil immersion objective (*see* **Note 7**).
3. An argon 488 nm laser line is required for GFP (*see* **Notes 8** and **9**).

2.3 Software for Analysis

1. Software to extract the fluorescence intensity measurements from the time-lapse image series: for instance ImageJ software (NIH, Bethesda, MD, http://rsb.info.nih.gov/ij/) or alike.
2. Software to perform data analysis (mean intensity values, data plotting and curve fitting with best-fit equations): Microsoft Excel, GraphPad Software (La Jolla California USA, www.graphpad.com), MATLAB, or alike.

3 Methods

3.1 Plant Growth and Sample Preparation

The objective is to prepare healthy root tissue in a dedicated microscopic chamber (Fig. 2a) suitable for long term (up to overnight) time-lapse imaging (*see* **Note 5**) and with a setup allowing intact roots to be imaged.

1. Surface-sterilize seeds in 5% v/v sodium hypochlorite for 5 min and rinse three times in sterile distilled water.
2. Stratify the seeds by incubating for 2 days at 4 °C in darkness in 1.5 mL tubes.
3. Sterilize chambered cover glasses with 70% ethanol and leave to dry in a laminar flow cabinet.
4. Pour growth medium into the chambers.
5. When solidified, cut a small slice from the medium (~1 cm from the top) with a sterile razor blade in order to provide access to the coverslip.
6. Plate the seeds onto the top of the cut edge, close to the coverslip (Fig. 2a).
7. Grow the seedlings at 25 °C in continuous light (or other day–night conditions) in vertically oriented position for 3–5 days (*see* **Note 10**).
8. Seal the chamber with adhesive tape.

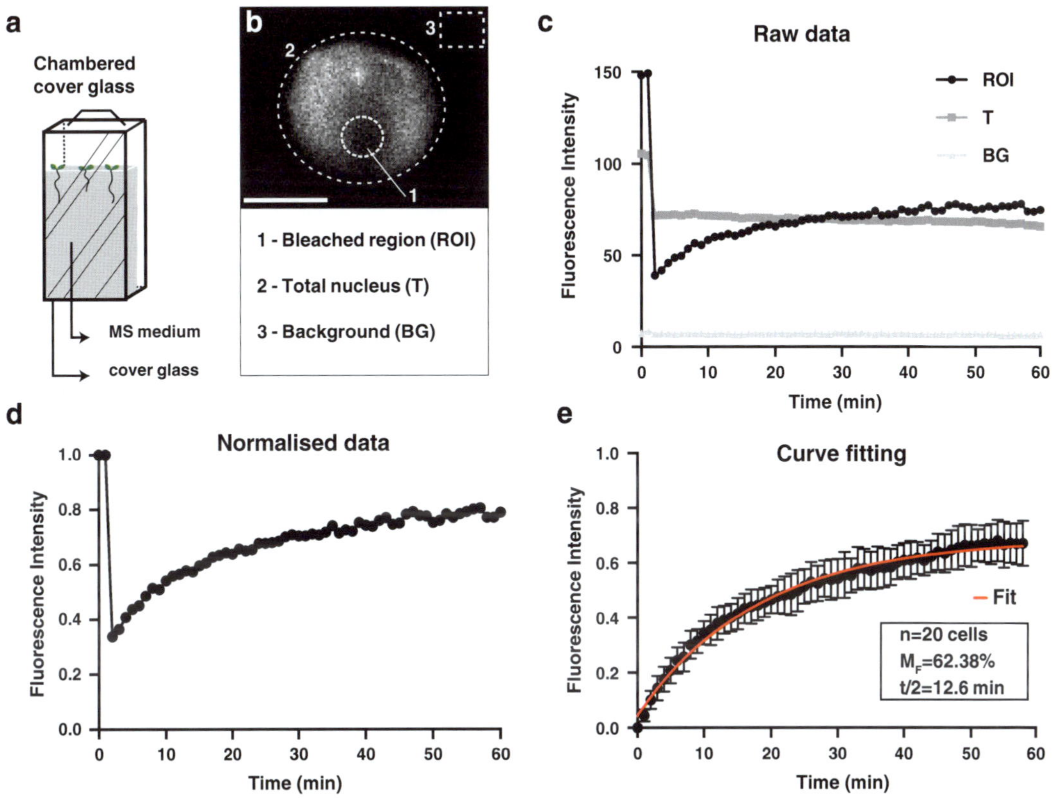

Fig. 2 FRAP data analysis workflow. (**a**) Schematic representation of the cover glass chamber used for plant growth and for imaging. (**b**) Representative image showing the selection of the three regions used for analysis (ROI, total nucleus, and background). Scale bar = 5 μm. (**c**) Fluorescence intensities extracted from the regions depicted in *b* during a FRAP experiment. (**d**) Recovery in the bleached region (ROI) shown in *c* after double normalization. (**e**) Fluorescence recovery curve averaged for 20 cells after normalization and correction for bleach depth. Error bars correspond to standard error of the mean (SEM). In *red*, the recovery curve corresponding to the best fit obtained from the experimental data

3.2 Image Acquisition

The objective is to produce time-lapse series following a bleaching event and covering the entire period of recovery (to determine the mobile vs immobile fraction) (*see* **Note 11**). The section below describes the settings required to image the full recovery of histone H2B. For proteins with unknown kinetics preliminary tests are necessary to determine the plateau stage and the speed of initial recovery, which will define the frequency of sampling (*see* **Note 12**).

1. Select an area of interest at a defined position relative to the root tip (meristematic/dividing, elongation, or differentiation zone) and a defined tissue layer (epidermis, cortex). Keep consistent throughout sampling (*see* **Note 13**).
2. Imaging settings: image acquisition should preserve the overall signal during the whole duration of imaging — use as low as

possible laser intensity, e.g., 2% and fast acquisition speed without averaging (*see* **Note 14**).

3. Image format: for a ~ 75 μm field of view, a format of 512 × 512 pixels with a 2–3× zoom factor is recommended providing a pixel size of ~0.1483 μm. Z-stacks of 15–20 slices depending on nucleus size with z-steps of 1 μm (*see* **Note 15**)
4. Pinhole setting: approx. 250 μm (*see* **Note 16**).
5. Bleaching settings: for bleaching, the laser intensity is increased to 100% (*see* **Note 17**) ROI size for bleaching: 1 μm in radius (*see* **Note 18**); 3–5 Iterations to achieve full bleach.
6. Experiment sequence: (1): Prebleach: One Z-stack before the bleach, (2) Bleach Pulse (*see* above), and (3) Post-Bleach: One Z-stack every 60 s for 60 min, starting immediately after the bleaching event (*see* **Notes 19** and **20**).

3.3 Image Processing

Image processing and mathematical modeling of fluorescence intensity curves are necessary steps to provide information on the dynamics of the protein of interest.

1. Collect a complete set of FRAP experiments (approx. 20 cells).
2. On ImageJ open each FRAP experiment file as a hyperstack (containing *z* and *t* dimensions) and apply a Z-stack maximum projection.
3. Nuclear movements, cell drift and tissue growth will likely occur during image acquisition. The nuclei must be aligned along the different time frames in order to proceed to quantification. This image registration process is done using StackReg plug-in of ImageJ (Plugin/Registration) (*see* **Note 21**).
4. Once images are aligned use the "ROI manager" tool (shortcut: "t") and add three regions of interests by drawing shapes (Fig. 2b, c) for the following elements:
   (a) bleached region (ROI);
   (b) total nucleus (T);
   (c) background (BG).

   Both ROI and T regions can be drawn by hand based on contrast of fluorescence. Because mean intensities over the ROI area are the values of interest, the precision of the outline for each area does not critically influence the result. These three ROIs are necessary for double normalization (Fig. 2d and Subheading 3.4).
5. Load the three regions onto the "ROI manager" window and select "Multi Measure" to obtain the results. The results table will be provided with the mean intensity values for each region

at the different time points. These measurements will be used later for normalizing the data.

6. Label the data in a fully comprehensive manner and store them in a text file (.csv) or Excel spreadsheet.

3.4 Data Analysis

Quantitative analysis of FRAP data requires normalization of recovery curves and several calculations should be performed before comparing FRAP measurements from different samples. For analysis of chromatin proteins we found that a double normalization [7] is the most appropriate method. This method takes into account the loss of signal due to bleaching during imaging and due to the bleach pulse at the bleaching region (corresponding to the fraction of the nucleus bleached). This method also normalizes the initial fluorescence intensity (F_0) to 1. For a double normalization the fluorescence intensity should be measured at three different regions mentioned above: the bleached region (ROI); the total nucleus (T) and a region outside the nucleus for background subtraction (BG). Normalization is then calculated as follows:

$$F_t = \frac{(T_{t=0} - BG)(ROI_t - BG)}{(T_t - BG)(ROI_{t=0} - BG)} \qquad (1)$$

Where,

- *Background subtraction*: is calculated at each time point t (BG).
- *Correction for the fluorescence signal lost due to bleaching and bleaching pulse*: calculated at each time point t.

$$\frac{(T_{t=0} - BG)}{(T_t - BG)}$$

- Normalization of $ROI_{t=0}$ to 1.

$$\frac{(ROI_t - BG)}{(ROI_{t=0} - BG)}$$

1. Export the raw intensity measurements of the three regions (ROI, T and BG) to Microsoft Excel and perform the double normalization according to Equation 1 (Eq. 1).
2. We additionally normalize for the variation on the bleaching depth as follows:

$$F_{t\ \mathrm{norm\ to\ bleachdepth}} = \frac{F_t - F_{\mathrm{bleach}}}{1 - F_{\mathrm{bleach}}}$$

Where F_t corresponds to the intensity of fluorescence at each time point t after double normalization, and F_{bleach} to the intensity of fluorescence immediately after bleaching.

3. Copy the normalized data to a program such as GraphPad Prism or alike and perform a curve fitting analysis (*see* **Note 22**). Use a nonlinear regression and fit to a single exponential function such as (*see* **Note 23**):

$$\mathrm{F(t)} = A \cdot \left(1 - e^{-kt}\right)$$

Where A is the mobile fraction and k the rate constant (Fig. 2e).

4. The mobile fraction (M_F) and half-life values ($t_{1/2}$) are determined by the software after curve fitting.
5. To statistically compare kinetics parameters and recovery curves between experiments (differing for instance in tissue treatment, genotype, cell type, etc.), fit the recovery of individual nuclei with an exponential function and determine the M_F and $t_{1/2}$ values. Then use those values to plot histograms and to assess the statistical significance between treatments.

4 Notes

1. In this protocol, I do not mention particular cells or tissues as the procedure can be applied to any cell type in the root. However, for cells located deep in the tissue, such as stem cells, multiphoton microscopy is necessary to bleach small and well-defined subcellular regions [8].
2. The fluorophore should ideally have the following characteristics: be bright to obtain a high signal-to-noise ratio; be photostable to prevent too much bleaching during imaging; be monomeric to avoid any unspecific associations between the tagged proteins.
3. Agar can also be used. However Phytagel produces more rigid media and is also more transparent, which is advantageous when growing plants vertically and for imaging plants directly in the growth media (in a cover glass chamber).
4. Light intensity, temperature, day–night regimes, and plant developmental stage should be, as much as possible, kept constant between experiments.
5. It is also possible to image root tips mounted on regular microscope slides and coverslips sealed with nail varnish or with double-sticky tape. Care should be brought to keep the tissue intact. Roots mounted in this way should not be imaged for longer than 1 h, as lack of oxygen will start affecting the sample conditions and therefore the FRAP recoveries.

6. Media containing Phytagel should be prepared fresh or stored as plates/chambers, as remelting will affect gelling properties. Agar media can be autoclaved and stored for remelting.
7. Using water as immersion medium can be inconvenient due to drying during the long time course of image acquisition. Oil or glycerol can be used, or alternatively, if available, any other immersion medium that more closely matches the refractive index of cells.
8. An alternative to photobleaching to measure dynamics of chromatin proteins is photoactivation [6, 8]. The advantage of using a photoactivatable tag is that much lower laser intensity is required for activation, thus substantially reducing photodamage to cells. Additionally, as the fluorescence of these proteins is seen only after photoactivation, newly synthesized non-photoactivated pools are not observed and do not complicate experimental results.
9. The laser line used should approximately match the excitation peak of the fluorescent tag used.
10. By placing the chambers vertically (or with a slight angle ~70°), the germinating roots will grow along the surface of the coverslip and can therefore be more easily imaged.
11. Alternatively, partial recovery can be acquired when comparing differences present only in the early phases of recovery.
12. There is no general rule and it is more a case of trial and error for each sample and system to obtain the optimal bleaching and imaging parameters.
13. Keep consistent when selecting the areas and cell types, as differences in recovery for the same protein between different stages of differentiation and cell types have been reported [6].
14. It is important to modulate the acquisition parameters in order to preserve the overall signal during the whole duration of imaging and to stay within the dynamic range of intensity values. Saturation of fluorescence intensity will result in inaccurate measurements of FRAP recovery.
15. The Z-stack is essentially used to correct for drifts along the Z-axis. However, for highly dynamic proteins, and short time courses, the Z-stack may be omitted and replaced by a larger pinhole.
16. The pinhole value should be adjusted according to each sample. For live imaging, the pinhole should be opened wide enough to acquire an adequate signal, while keeping the laser intensity low to minimize photobleaching during imaging and phototoxicity. Opening the pinhole too much will on the other hand affect the resolution by increasing the thickness of the

optical section. Thus the right balance should be found for each sample.

17. The laser intensity for the bleaching event should be set to achieve a sufficient decrease in fluorescence within the ROI. If one bleaching pulse is not sufficient to generate a clear bleached area, it is possible to apply several iterations (normally three to five) to achieve full bleach.
18. We chose the size of 1 μm in radius because it gives a relatively big and precise bleaching area. It is not essential to use this exact area. What is important is to keep the same bleaching area between experiments.
19. Cell and nuclei movement due to tissue growth may occur during image acquisition. This will sometimes lead to the cells of interest moving outside of the field of view. It is possible to avoid this by observing the samples during the imaging and manually correct for movements in x and y. Else, image registration for correcting drift is applied during processing.
20. To exclude the possibility that observed differences in fluorescence recovery reflect differences in rates of protein synthesis, the de novo expression rates of tagged histones in different samples can be measured by photobleaching entire cells and measuring the overall recovery of the fluorescence signal in the nucleus within the time corresponding to the FRAP experiment (e.g., ~1 h). Any recovery of fluorescence can be mostly attributed to newly synthesized tagged proteins.
21. This plugin aligns successive image frames by a cross-correlation procedure, which works easily and without operator intervention, as long as movements are not too large. More details, including screen shots, are available on the developers' website (http://bigwww.epfl.ch/thevenaz/stackreg/).
22. Discard data sets that exhibit highly variable background fluorescence intensities.
23. Multicomponent equations (such as double exponential functions) should be used when multiple populations of molecules, with differing diffusion rates, are present. This will be evident when a single exponential equation does not provide a good fit to the data.

Acknowledgments

This work was supported by the grant SFRH/BD/23202/2005 from the Portuguese Fundação para a Ciência e a Tecnologia (Portugal) and 3.3-GRO/1162118STP from the Humboldt Foundation (Germany). I thank Prof. Peter Shaw for helpful comments on the manuscript.

References

1. Kimura H, Cook PR (2001) Kinetics of core histones in living human cells: little exchange of H3 and H4 and some rapid exchange of H2B. J Cell Biol 153:1341–1353
2. Dion MF, Kaplan T, Kim M, Buratowski S, Friedman N, Rando OJ (2007) Dynamics of replication-independent histone turnover in budding yeast. Science 315:1405–1408
3. Jamai A, Imoberdorf RM, Strubin M (2007) Continuous histone H2B and transcription-dependent histone H3 exchange in yeast cells outside of replication. Mol Cell 25:345–355
4. Henikoff S (2008) Nucleosome destabilization in the epigenetic regulation of gene expression. Nat Rev Genet 9:15–26
5. Meshorer E, Yellajoshula D, George E, Scambler PJ, Brown DT, Misteli T (2006) Hyperdynamic plasticity of chromatin proteins in pluripotent embryonic stem cells. Dev Cell 10:105–116
6. Rosa S, Ntoukakis V, Ohmido N, Pendle A, Abranches R, Shaw P (2014) Cell differentiation and development in Arabidopsis are associated with changes in histone dynamics at the single-cell level. Plant Cell 26:4821–4833
7. Phair RD, Gorski SA, Misteli T (2004) Measurement of dynamic protein binding to chromatin in vivo, using photobleaching microscopy. Methods Enzymol 375:393–414
8. Rosa S, Shaw P (2015) Two-photon photoactivation to measure histone exchange dynamics in plant root cells. Bio-protocol 5:e1628

Chapter 27

Fluorescence In Situ Hybridization (FISH) and Immunolabeling on 3D Preserved Nuclei

Till David Bey, Maria Koini, and Paul Fransz

Abstract

The spatial distribution of genes in the nucleus emerges as an important factor in gene regulation and epigenetics. The position of loci relative to each other, to nuclear landmarks such as the nucleolus and chromocenters, as well as to chromatin proteins is therefore highly interesting. With *fluorescent in situ hybridization* (FISH) specific DNA sequences can be stained and antibodies allow the detection of specific proteins. Here, we present two protocols that preserve the 3D structure of nuclei. With whole-mount FISH, specific sequences can be stained in intact tissues and, secondly, a combined immunolabeling and FISH protocol on acrylamide-embedded nuclei makes it possible to stain DNA sequences and proteins simultaneously.

Key words Whole mount FISH, Immunolabeling, Chromosomes, Nucleus, 3D structure, Acrylamide embedding

1 Introduction

Chromosome organization involves the efficient folding of a long chromatin polymer into a defined nuclear area, the chromosome territory. Chromosome folding is a precise, systematic and most efficient process, since at any moment during the cell cycle the nuclear program can be changed in response to signals, thereby affecting the accessibility of genes for activation or repression. In the past few decades the chromosome territory has gained increasing interest by research groups linking spatial organization to gene activity. Although individual chromosomes have been examined in mitotic and meiotic cells since the nineteenth century, our knowledge of chromosome folding has long been limited due to less specific staining techniques. The development of sequence-specific labeling techniques such as in situ hybridization in the late 1960s made it possible to identify specific RNA and DNA targets [1, 2].

The microscopic analysis of the spatial position of a gene, gene regions or an entire chromosome in tissue context or in intact

Marian Bemer and Célia Baroux (eds.), *Plant Chromatin Dynamics: Methods and Protocols*, Methods in Molecular Biology, vol. 1675, DOI 10.1007/978-1-4939-7318-7_27, © Springer Science+Business Media LLC 2018

nuclei is one of the major challenges, particularly in plants. The plant cell wall is notorious to impede access of large detection molecules (DNA probes, antibodies) to the chromosome targets and isolated nuclei become flat after mounting on the microscopic slide surface. During the past two decades, several studies reported a successful 3D detection of DNA targets in different tissues: maize meiotic tissue [3], Arabidopsis roots [4–6], ovule primordia [7] and leaf fragments [8] using *fluorescence in situ hybridization* (FISH). One of the most important issues in whole mount FISH protocols is the preservation of the tissue structure and nuclear morphology. This can be achieved when applying FISH to cryo-sections of frozen tissue [8] or to cells/organs embedded in a stable matrix such as acrylamide [7] or simply by keeping the tissue fragments in a tube throughout the FISH procedure [4, 6]. The accessibility of the DNA target for the labeled probes, however, often remains a major problem, since cells of different tissues have a different cell wall composition or thickness. Hence, protocols have to be adapted for each tissue and each species.

Here, we present a FISH protocol for different tissue fragments and organs and a combined FISH immunolabeling protocol for isolated nuclei. The whole-mount FISH protocol (Subheadings 3.2 and 3.3) is suitable for most tissues including roots, seedlings, leaves, flower buds and anthers of Arabidopsis and other species. This protocol is based on Bauwens et al. [4], for which we adapted some steps to increase accessibility of the DNA. Different tissues, e.g., root, leaf fragments, and flower buds, can be simultaneously hybridized in the same tube with the same hybridization mix. Moreover, the protocol allows the reuse of the hybridization mix with precious, labeled probes. However, the use of whole-mount samples is limited when using antibodies, since uniform penetration of larger tissue fragments by the proteins is difficult. Additionally, tissues contain numerous structures that introduce optical aberrations, resulting in lower image quality and less microscopic resolution. Therefore, the second protocol we present here combines FISH and immunolabeling to isolated nuclei, embedded in acrylamide (Subheadings 3.4 and 3.5). The acrylamide embedding procedure is based on Bass et al. [3], who examined maize meiocytes. The combined FISH immunolabeling protocol allows to monitor proteins and DNA target sequences simultaneously.

2 Materials

2.1 Preparation of a Fluorescently Labeled Probe

1. DNA of at least 1 μg/μl (*see* **Note 1**).
2. 1 mM stock solution of labeled dUTP in ultrapure water (*see* **Note 2**).

3. 1 mM stock solution of dATP, dCTP, and dGTP (1 mM of each dATP, dCTP, and dGTP in ultrapure water).
4. 1 mM stock solution of dTTP in ultrapure water.
5. 5× dye mix for Alexa Fluor dyes: 200 μM dATP, dCTP, and dGTP; 50 μM dTTP; 150 μM labeled dUTP, prepared from stock solutions (*see* **Note 2**).
6. 5× cyanine dye mix: 100 μM dATP, dCTP, and dGTP; 100 μM labeled dUTP, prepared from stock solutions (*see* **Note 2**).
7. DNA polymerase I (10 U/μl, *see* **Note 2**).
8. 5× nick translation mix (e.g., from Roche).
9. Labeling reaction mix for Alexa Fluor dyes: 10 μg DNA, 1× Alexa Fluor dye mix, 1× nick translation mix, DNA polymerase I (additional 40 U/20 μl reaction) in ultrapure water (*see* **Note 2**).
10. Labeling reaction mix for cyanine dyes: 10 μg DNA, 1× cyanine dye mix, 1× nick translation mix in ultrapure water (*see* **Note 2**).
11. 1 M ethylenediaminetetraacetic acid (EDTA).
12. Heat block.
13. 3 M NaOAc.
14. Ethanol.
15. Microcentrifuge.
16. Sterile, ultrapure water.

2.2 Preparation of Whole-Mount Samples

1. Methanol.
2. Ethanol.
3. Xylene.
4. 1× PBS:10 mM Na_2HPO_4, 2 mM KH_2PO_4, 2.7 mM KCl, and 137 mM NaCl, pH 7.4.
5. Fixative solution I: 1% formaldehyde, 10% dimethylsulfoxide, 1× PBS, 60 mM EGTA.
6. 0.2% Macerozyme R-10 (e.g., from Duchefa) in 1× PBS.

2.3 Whole-Mount FISH in a Tube

1. Microscopy slides, preferably precleaned (e.g., Superfrost plain and frosted slides from Thermo Fisher).
2. Coverslips of thickness #1 (size 20 × 20 mm).
3. Nail polish.
4. 1× PBS: 10 mM Na_2HPO_4, 2 mM KH_2PO_4, 2.7 mM KCl, and 137 mM NaCl, pH 7.4.
5. 0.1 mg/ml DNase-free RNase A in 1× PBS.
6. Labeled probe (*see* Subheadings 2.1 and 3.1).

7. 1× PBT:10 mM Na_2HPO_4, 2 mM KH_2PO_4, 2.7 mM KCl, and 137 mM NaCl, pH 7.4, 0.1% Tween 20.
8. 2× SSC: 30 mM sodium citrate, 300 mM NaCl, pH 7.
9. Formamide (molecular biology grade, deionized for hybridization solution only).
10. HB50 buffer: 50% formamide, 2× SSC, 50 mM sodium phosphate buffer pH 7.
11. Hybridization solution I (0.5–1.5 μg of labeled DNA for repeats or 2–5 μg of labeled DNA for unique sequences): 50% formamide, 2× SSC (*see* **Note 3**).
12. 5 μg/ml DAPI (4′,6-diamidino-2-phenylindole, Hoffmann-La Roche) in antifade mounting medium (*see* **Note 4**).
13. Heat block.

2.4 Preparation of Acrylamide-Embedded Nuclei

1. Microscopy slides (standard slides are sufficient as a support for the gel-bearing coverslip, while precleaned slides are preferred for sample mounting, e.g., Superfrost plain and frosted slides (e.g., from Thermo Fisher)).
2. Coverslips of thickness #1 (sizes: 20 × 20 mm, 24 × 30 mm, and 24 × 50 mm).
3. Razor blades.
4. Nail polish (*see* **Note 5**).
5. 1× PBS: 10 mM Na_2HPO_4, 2 mM KH_2PO_4, 2.7 mM KCl, and 137 mM NaCl, pH 7.4.
6. Fixative solution II: 4% formaldehyde, 1× PBS.
7. Ice-cold nuclei isolation buffer (NIB): 500 mM sucrose, 100 mM KCl, 10 mM Tris–HCl pH 9.5, 10 mM EDTA, 4 mM spermidine, and 1 mM spermine.
8. NIB with 1% Triton X-100.
9. Filter with 100–200 μm pore size (*see* **Note 6**).
10. Ice-cold acrylamide solution (5% acrylamide–bisacrylamide (29:1), 0.1% tetramethylethylenediamine (TEMED)) in NIB.
11. 10% ammonium persulfate (APS) in ultrapure water.

2.5 Combined FISH and Immunolabeling on Acrylamide-Embedded Nuclei

1. Coplin staining jars (from plastic or glass, choose smallest size that can hold all slides).
2. 1× PBS: 10 mM Na_2HPO_4, 2 mM KH_2PO_4, 2.7 mM KCl, and 137 mM NaCl, pH 7.4.
3. 0.1 mg/ml DNase-free RNase A in 1× PBS.
4. Parafilm.
5. Moist, dark chamber.

6. Fixative solution II: 4% formaldehyde, 1× PBS.
7. 1% bovine serum albumine (BSA) in 1× PBS.
8. Primary antibody.
9. Secondary antibody.
10. Ethanol.
11. Hybridization solution II: 0.5 μg of labeled DNA for repeats or 1–2 μg of labeled DNA for unique sequences): 50% formamide, 2× SSC, and 50 mM sodium phosphate buffer pH 7 (*see* **Note 3**).
12. Hot plate.

3 Methods

3.1 Preparation of a Fluorescently Labeled Probe

Success in a FISH experiment depends on a good preparation as well as on a well-labeled probe. We label probes directly with a fluorescent dye by nick translation, which yields sufficiently strong hybridization signals and a low background. If probes are labeled with biotin or digoxigenin, the signal can be amplified in subsequent immunodetection steps. This, however, poses problems of antibody penetration in whole-mount FISH and often leads to increased background.

1. Prepare 20 μl labeling reaction mix for Alexa Fluor dyes or for cyanine dyes depending on the label (*see* **Note 2**) on ice. Keep the solution in the dark at all steps.
2. Incubate at 16 °C for 2 h.
3. Run 1 μl of the mix on a 1% agarose gel while keeping the rest of the solution on ice.
4. If most fragments are in a range of 50–500 bp, stop the reaction by adding 0.5 μl of 1 M EDTA and incubating at 65 °C for 10 min. If the fragment size is larger, incubate the labeling mixture longer at 16 °C (*see* **Note 7**).
5. The probe can be used directly without purification. If it is necessary to concentrate the probe, continue with **steps 6–12** (*see* **Note 3**).
6. Add 2 μl of 3 M NaOAc and 3 μl of 100% ethanol to the labeling reaction mix.
7. Precipitate at −80 °C for 15 min.
8. Centrifuge at full speed (10,000 × *g*) for 30 min and discard supernatant.
9. Wash with 100 μl of 70% ethanol.
10. Centrifuge at full speed for 5 min and discard supernatant.

11. Air-dry the pellet.
12. Resuspend the pellet in ultrapure water (the volume depends on the desired concentration).

3.2 Preparation of Whole-Mount Samples

In a whole-mount sample, the nuclei are preserved in their tissue context, which is important when cell-specific nuclear structures are studied in relation to surrounding cells. The tissue needs to be cleared and permeabilized to allow probes to penetrate and to be visualized. The following protocol is based on [4], with adjustments for FISH on small, unique targets (single locus FISH). All steps can be performed in 50 ml tubes using 10 ml of solution at each step unless indicated otherwise. Solutions can be exchanged by decanting holding the sample back with a forceps, except for xylene containing solutions that should be removed by pipetting using filter tips. Incubation steps are at room temperature (20 °C). Minimalize agitation of the samples to maintain the integrity of the tissue.

1. Transfer fresh tissue (seedlings, roots, leaf fragments or flower buds) to fixation solution (*see* **Note 8**) and incubate for 30 min at room temperature, with the last 5 min under vacuum (make sure the lid of the tube is open).
2. Incubate 2× 10 min in methanol.
3. Incubate 2× 5 min in absolute ethanol.
4. Incubate in a 1:1 mixture of xylene and ethanol for 30 min at room temperature (*see* **Note 9**).
5. Optional step for single locus FISH and for older tissue: Incubate in xylene for 30 min at 50 °C.
6. Wash 2× 5 min in ethanol.
7. Wash 2× 5 min in methanol.
8. Wash 2× 5 min in 1× PBT.
9. Wash in 1× PBS. The fixed tissue samples can be stored in 1× PBS at 4 °C overnight (*see* **Note 10**).

3.3 Whole-Mount FISH in a Tube

This protocol is suited for FISH on repetitive targets (e.g., rDNA and centromeric repeats) as well as small unique targets (single locus FISH). In the latter case follow the indicated additional steps. Use the sample from **step 9** of Subheading 3.2 in the 50 ml tube. Unless mentioned otherwise, use 10 ml of solution.

1. Incubate the sample in 2 ml of 0.1 mg/ml RNase for 1 h at 37 °C.
2. Optional step for single locus FISH: Add additional macerozyme R-10 to a final concentration of 0.2% to the RNase solution before incubation (*see* **Note 11**).

3. Wash 2× 5 min in 1× PBT.
4. Postfix the sample in 1% formaldehyde in 1× PBT for 30 min at room temperature.
5. Wash 2× 5 min in 1× PBT.
6. Incubate the material in a 1:1 mixture of HB50 and 2× SSC for 30 min at room temperature.
7. Incubate in HB50 for 30 min at room temperature.
8. Transfer the tissue fragment (e.g., two 10-day-old seedlings or a leaf fragment of similar size) into 30 μl hybridization solution I in a 1.5 ml tube (*see* **Note 3**). The sample must be fully submerged in the hybridization solution. Incubate for 1 h at room temperature.
9. Wrap the tube in aluminum foil to protect the fluorescently labeled probe from light.
10. Denature for 4 min at 85 °C. Place the tube on ice for 3 min.
11. To allow hybridization, place the tube at 37 °C overnight. For single locus FISH the hybridization can take 1–2 days.
12. Remove the hybridization solution (*see* **Note 12**).
13. Wash the sample in 1 ml HB50 for 5 min.
14. Wash the sample stringently in 1 ml HB50 for 1 h at 42 °C (*see* **Note 13**).
15. Wash in 1 ml of 1× PBS for 20 min.
16. Place the sample into a drop of 1× PBS (*see* **Note 14**). Remove any liquid and pipet 8 μl DAPI (5 μg/ml in antifade mounting medium). Cover with a coverslip and seal with nail polish.
17. Examine the nuclei with confocal laser scanning microscopy (CLSM).

3.4 Preparation of Acrylamide-Embedded Nuclei

In this protocol isolated nuclei are embedded in a gel to preserve their three-dimensional structure while allowing easy access for probes and antibodies. Following previously used protocols for meiocyte embedding [3], a 5% acrylamide gel is used. The gel is poured on a coverslip rather than a microscopy slide to allow imaging from the side of the gel that is perfectly attached to the coverslip. To minimally disturb the isolated nuclei, all pipetting steps of nuclei suspensions should be performed with pipet tips of which the end is cut off by 5 mm.

1. Fix 0.5 g of young leaves/seedlings in 10 ml of fixative solution II under vacuum for 20 min (*see* **Note 8**).
2. In the meantime prepare ten slides on which the gels will be poured. Glue a clean 24 × 30 mm^2 coverslip by the corners with nail polish onto each microscopy slide (*see* Fig. 1a and **Note 15**).

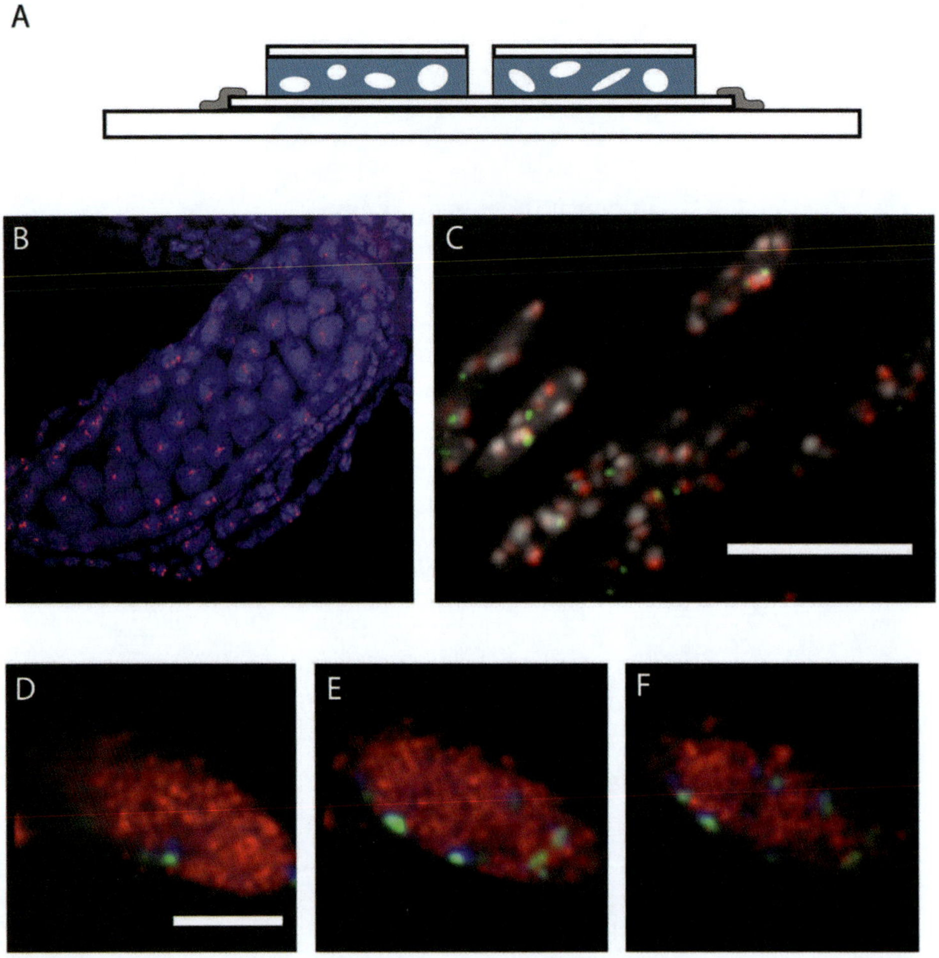

Fig. 1 Whole-mount FISH and combined FISH and immunolabeling on isolated nuclei. (**a**) Schematic overview of acrylamide preparation. A 24 × 32 mm coverslip for one sample or a 24 × 50 mm coverslip for two samples (depicted here) is glued onto a microscopy slide with nail polish. At the end of the preparation, the nuclei are suspended in the acrylamide gel that is sandwiched between this coverslip and a 20 × 20 mm coverslip (Schematic not drawn to scale). (**b**) Optical section of whole mount FISH to anthers of Arabidopsis with Cy3-labeled BAC pools (*red*) covering the major part of chromosome 4, counterstained with DAPI (*blue*). (**c**) Maximum intensity projection of whole-mount FISH to the *FT* locus with Cy3-labeled F19C20 BAC probe (*green*) and to the centromeric repeat with a Cy5-labeled pAL1 probe (*red*). The image shows phloem cells in the leaf. Scale bar 10 μm. (**d–f**) Series of optical sections of an acrylamide-embedded nucleus with centromeric repeats Cy3-labeled pAL1-Cy3 (*green*), active RNA polymerase II detected with Rabbit anti-RNAPIIS2P (*red*). Scale bar 5 μm. *See* **Note 24** for more information on the probes

3. Wash the plant material 3× 5 min in 1× PBS.
4. Place a petri dish on a metal block or plate on ice. Transfer the material to the petri dish and remove all 1× PBS (use a paper towel to drain excess solution). Keep the material on ice throughout the rest of the protocol unless otherwise noted.

5. Chop the material with a razor blade in 50 µl NIB with 1% Triton X-100 until a fine suspension is obtained (*see* **Note 16**).
6. Add 450 µl of ice-cold NIB to the suspension.
7. Filter through a nylon mesh (*see* **Note 6**) into a 1.5 ml tube.
8. Centrifuge with 500 × *g* for 3 min at 4 °C.
9. Discard the supernatant, add 500 µl of NIB, and gently resuspend the pellet.
10. Centrifuge with 500 × *g* for 3 min at 4 °C.
11. Discard the supernatant, add 100 µl of ice-cold acrylamide solution, and gently resuspend the pellet. It is important to keep the suspension on ice (*see* **Note 17**).
12. Add 1 µl 10% APS to the nuclei acrylamide solution and mix well by gently stirring the suspension with the pipet tip.
13. Pipet 12 µl of the suspension onto one of the 24 × 30 mm^2 coverslips prepared in **step 2** and cover the suspension with a 20 × 20 mm^2 coverslip. Avoid air bubbles that might destabilize the gel (*see* **Note 18**). Repeat this step for the remaining slides.
14. Let the gel polymerize for at least 60 min at room temperature.
15. The slides can be stored at 4 °C for up to a week.

3.5 Combined FISH and Immunolabeling on Acrylamide-Embedded Nuclei

We provide here a combined FISH and immunolabeling protocol on isolated nuclei embedded in acrylamide based on [9]. FISH or immunolabeling can also be performed independently by skipping steps as indicated. Use the preparation of acrylamide-embedded nuclei from **step 15** of Subheading 3.4. Perform all wash steps in Coplin jars at room temperature unless otherwise indicated.

1. Carefully lift up the 20 × 20 mm^2 coverslip on top of the gel with a razor blade (*see* **Note 19**).
2. Wash in 1× PBS for 5 min (*see* **Notes 20** and **21**).
3. Pipet 100 µl RNase (0.1 mg/ml 1× PBS) on the gel and cover with a 30 × 30 mm^2 piece of Parafilm. Incubate for 1 h at 37 °C in a moist chamber. This step can be skipped if an immunolabeling experiment without FISH is performed.
4. Wash in 1× PBS for 5 min (*see* **Note 21**).
5. If a FISH experiment is performed without immunolabeling skip **steps 6–14** and continue with **step 15**.
6. Postfix the nuclei by pipetting 100 µl of fixative solution II onto the gel. Incubate at room temperature for 10 min.
7. Wash in 1× PBS for 5 min (*see* **Note 21**).

8. Blocking step: cover the gel with 100 μl of 1% BSA in 1× PBS per gel and cover with a piece of Parafilm. Incubate at 37 °C for 30 min in a moist chamber.
9. Wash in 1× PBS for 5 min (*see* **Note 21**).
10. Prepare 50 μl of a 1:50–1:100 dilution of primary antibody in 1× PBS. Pipet the antibody solution onto the gel, cover it with Parafilm, and incubate in a moist chamber at 37 °C for 2 h or at 4 °C overnight (*see* **Note 22**).
11. Wash in 1× PBS three times for 20 min (*see* **Note 21**).
12. Prepare 50 μl of a 1:100–1:200 dilution of primary antibody in 1× PBS. Pipet the antibody solution onto the gel, cover it with Parafilm, and incubate in a moist chamber at 37 °C for 30 min (*see* **Note 23**).
13. Wash in 1× PBS three times for 10 min (*see* **Note 21**).
14. If an immunolabeling experiment is performed without FISH skip **steps 15–22** and continue with **step 23**.
15. Postfix the nuclei by pipetting 100 μl of fixative solution II onto the gel. Incubate at room temperature for 10 min.
16. Wash in 1× PBS for 5 min.
17. Dehydrate the sample through a series of 50%, 75%, and 100% ethanol. At each step pipet 100 μl onto the gel and incubate for 2 min. The ethanol solution with a lower concentration is pushed to the edges of the slide and can be removed with a paper towel.
18. Air-dry the slide. Remnants of liquid dilute the hybridization solution, which might lead to inconsistent results.
19. Add 10 μl hybridization solution II onto the gel and cover it with a 20 × 20 mm^2 coverslip (*see* **Note 3**). Keep the sample from this step on in the dark as much as possible, e.g., by covering the slide with aluminum foil.
20. Denature on a hot plate for 4 min at 90 °C.
21. Incubate in a prewarmed moist chamber at 37 °C for 3 h.
22. Wash in 1× PBS for 20 min (*see* **Note 13**).
23. Carefully remove the nail polish holding the 24 × 50 mm coverslip on the slide using a razor blade.
24. Pipet 20 μl of 5 μg/ml DAPI in anti-fade mounting medium on a clean slide. Place the coverslip with the gel onto the slide and seal with nail polish. Clean the coverslip before microscopic examination.
25. Examine the nuclei with CLSM.

4 Notes

1. Use pure and highly concentrated DNA for nick translation (above 1 μg/μl). Good results are achieved with midi-prep sized isolations. Backbones of plasmids and BACs do not need to be removed, as there is little homology to plant sequences.
2. We routinely use dUTP directly labeled with Alexa Fluor dyes (e.g., Alexa Fluor 488, Alexa Fluor 594, Thermo Fisher Scientific) and cyanine dyes (e.g., Cy3, Cy5, GE Healthcare) in FISH experiments. Alexa Fluor dyes are attached to dUTP with an alkynylamino linker, while Cy3/Cy5 are attached with an aminopropargyl linker. This influences their incorporation efficiency, requiring a higher DNA polymerase I concentration and addition of unlabeled dTTP for Alexa Fluor dyes [10].
3. With the protocols presented here optimal results are achieved when a directly labeled DNA probe is used. Repetitive targets can be well detected with a final probe concentration of 0.5 μg in 10 μl. Unique sequences in the order of 100 kb (BAC sized) require three times higher concentrations as the effective concentration of probe per sequence is lower. Detection of targets around and below 10 kb length may require probe concentrations around 5–10 times higher. Consider precipitating and resuspending the probe in a smaller volume. High probe concentrations do not lead to increased background. *See* also **Note 13**.
4. Vectashield (Vector Laboratories) is used but alternative mounting media are available. The refractive index of the mounting medium should match as much as possible to the one of the coverslip and of the immersion oil (usually 1.515).
5. Use clear, colorless "base coat" nail polish. "Top coat" varieties do not give sufficient adhesion for a coverslip on the slide to resist several wash steps.
6. A simple filter can be made from a 1 ml pipet tip and nylon mesh with a pore size of 100–200 μm. Cut off around 2 cm from the end of the tip. Melt the cut surface briefly in a burner flame and glue the nylon mesh to it. Remove any mesh extending over the edge of the tip.
7. The labeled probe should appear as a smear of fragments while unincorporated labeled dUTP appears as a band between 50 and 100 bp. Note that the size of labeled DNA differs from unlabeled DNA. An optimal range of probe length obtained by nick translation lies between 50 and 500 bp. Additionally, incorporation can be monitored by observing the loss of

unincorporated nucleotides and the increase in fluorescence of the labeled probe on a UV transilluminator.

8. Make sure that the tissue is properly submerged in the fixative. To facilitate submersion of trichome-rich leaves, add Tween 20 (0.1% final concentration). The tissue should be able to freely float in the solution. If more samples are used, increase all solution volumes accordingly.
9. This treatment clears the tissue and dissolves some cuticular waxes. Plants grown on soil possess a thicker cuticula than in vitro cultured plants on plate. Older leaves also have a pronounced cuticula compared to young leaves. Leaves older than 2 weeks benefit from an extended treatment. Shorten the incubation time in the mixture of xylene and ethanol (1:1) to 15 min and follow the additional xylene incubation step. The labeling efficiency for single locus FISH is also increased by this step.
10. Material can be stored longer (up to a week) but we have sometimes observed decreased efficiency of probe binding and sometimes degradation of the nuclear morphology. Especially for the detection of small targets it is advisable to use fresh preparations.
11. The concentration of macerozyme R-10 might need to be adjusted according to the material. Note that the activity of macerozyme R-10 may also vary between batches. It is advisable to make aliquots and store them at −20 °C. A test experiment (without probe) with a range of enzyme concentrations can be performed to evaluate the highest concentration for which the tissue still remains intact. For anther tissue of Arabidopsis, the entire immature flower bud needs to be digested with a mixture of 0.4% cytohelicase, 0.4% pectolyase and 0.4% cellulase (in 10 mM sodium citrate/citric acid, pH 4.5) prior to the RNase incubation. Anthers of tomato and Petunia have a thicker cell wall and need longer incubation or alternatively a higher (1%) concentration of the enzymes.
12. The hybridization solution can be stored at −20 °C and can be reused at least three times.
13. Unspecific binding of the probe results in an increase of background signal throughout the nucleus, often pronounced in the chromocenters. This can be caused by too short, aspecific probe molecules and can be overcome by more stringent washes at higher temperature. Instead, probe molecules that are too long result in unspecific staining along the nuclear periphery as they do not penetrate into the nuclei. In both cases preparing a new probe might be considered (*see* also **Note 7**).

14. In case of visualising meiocytes in anther tissue, the tissue should be covered by a 20 × 20 mm coverslip and gently tapped with the tip of a dissection needle. After submerging in liquid nitrogen the coverslip can be removed. After thawing, 8 μl of DAPI (5 μg/ml in Vectashield) is added. Cover with a coverslip and seal with nail polish.
15. In this setup the acrylamide gel remains directly attached to the coverslip. This prevents an uneven interface between gel and embedding medium, which introduces optical aberrations and attracts unspecific binding. The coverslip is fixed on a microscopy slide as a support to allow easy handling and washing in Coplin jars. If a 24 × 50 mm^2 coverslip is used instead of the 24 × 32 mm^2 coverslip, two gels can later be placed side-by-side. This allows handling of two different samples together in parallel.
16. Triton X-100 permeabilizes membranes and reduces the number of chloroplasts in the preparation. An additional Triton X-100 treatment after gel embedding is not necessary. Use a new razor blade and avoid scratching movements on the petri dish that can damage the nuclei.
17. At room temperature acrylamide polymerizes too quickly upon addition of APS, when preparing several slides. To prevent premature polymerization, use an ice-cold acrylamide solution and keep it on ice after resuspending the nuclei. Polymerization at 4 °C is delayed, but the resulting gel is opaque [11]. Therefore the acrylamide solution should be left at room temperature to fully polymerize once pipetted onto the slide.
18. The gel contains a low concentration of 5% acrylamide and is at most 30 μm thick. This allows probe and antibody molecules to diffuse quickly to the nuclei while giving enough stability to the gel. If larger or smaller gels and differently sized coverslips are preferred, adjust the volume accordingly.
19. Slide the razor blade carefully under one corner of the small coverslip on top of the gel. Then lever the top coverslip slowly up. The gel might stick to the top rather than to the bottom coverslip after longer storage. In this case rehydrate the gel in 1× PBS for 5 min first.
20. Washing before post-fixation is important to remove remaining monomeric acrylamide and polyamines. Both can participate in the formaldehyde crosslinking and block the access of antibodies to their targets.
21. Wipe the slide around the gel dry with a tissue after each wash. In the end, touch the edge of the gel carefully with a tissue to remove excess solution by capillary force. Otherwise the remaining liquid will dilute the solution in the next step.

22. Being more specific, monoclonal antibodies can result in lower background but might need overnight incubation at 4 °C. Extended incubation can also help in the detection of less abundant or inaccessible targets.
23. Negative controls should always be included if new antibody combinations are used to estimate the background staining. The specificity of the secondary antibody can be assessed by a staining without the respective primary antibody. Ideally, the specificity of the primary antibody is tested in a mutant or under a condition in which the target is not present.
24. Probes were directly labeled with Cy3-dUTP or Cy5-dUTP (GE Healthcare). The used antibodies are: rabbit antibody against RNAPIIS2P (Catalog no. 04-1571, Merck Millipore) and donkey antibody against rabbit labeled with Cy5 (Catalog no. 711-175-152, Jackson ImmunoResearch).

Acknowledgments

T.B. was financially supported by the European Commission Seventh Framework-People-2012-ITN (Project EpiTRAITS, GA-316965) and M.K. was financially supported by the Netherlands' Organization for Scientific Research (ALW2PJ/09053).

References

1. Pardue ML, Gall JG (1969) Molecular hybridization of radioactive DNA to the DNA of cytological preparations. Proc Natl Acad Sci U S A 64:600–604
2. Pardue ML, Gall JG (1970) Chromosomal localization of mouse satellite DNA. Science 168:1356–1358
3. Bass HW, Marshall WF, Sedat JW et al (1997) Telomeres cluster de novo before the initiation of synapsis: a three-dimensional spatial analysis of telomere positions before and during meiotic prophase. J Cell Biol 137:5–18
4. Bauwens S, Katsanis K, Montagu M et al (1994) Procedure for whole mount fluorescence in situ hybridization of interphase nuclei on Arabidopsis thaliana. Plant J 6:123–131
5. Costa S, Shaw P (2006) Chromatin organization and cell fate switch respond to positional information in Arabidopsis. Nature 439:493–496
6. Berr A, Schubert I (2007) Interphase chromosome arrangement in Arabidopsis thaliana is similar in differentiated and meristematic tissues and shows a transient mirror symmetry after nuclear division. Genetics 176:853–863
7. She W, Grimanelli D, Rutowicz K et al (2013) Chromatin reprogramming during the somatic-to-reproductive cell fate transition in plants. Development 140:4008–4019
8. Tirichine L, Andrey P, Biot E et al (2009) 3D fluorescent in situ hybridization using Arabidopsis leaf cryosections and isolated nuclei. Plant Methods 5:11
9. Pavlova P, Tessadori F, de Jong HJ et al (2010) Immunocytological analysis of chromatin in isolated nuclei. Methods Mol Biol 655:413–432
10. Kato A, Albert PS, Vega JM et al (2006) Sensitive fluorescence in situ hybridization signal detection in maize using directly labeled probes produced by high concentration DNA polymerase nick translation. Biotech Histochem 81:71–78
11. Gelfi C, Righetti PG (1981) Polymerization kinetics of polyacrylamide gels II. Effect of temperature. Electrophoresis 2:220–228

Chapter 28

High-Affinity LNA–DNA Mixmer Probes for Detection of Chromosome-Specific Polymorphisms of 5S rDNA Repeats in *Arabidopsis thaliana*

Lauriane Simon and Aline V. Probst

Abstract

Fluorescence in situ hybridization is a standard technique to visualize specific DNA sequences by hybridization with fluorescent probes and, most commonly, relies on DNA probes generated by nick translation. In this chapter, we describe the use of directly labeled LNA–DNA mixmer probes for the rapid detection of repetitive sequences on *Arabidopsis thaliana* nuclei spreads. We further demonstrate that due to the high thermal stability of the heteroduplexes and the resulting elevated binding affinity of LNA–DNA mixmer probes for their target DNA, these probes can be used to discriminate between repetitive sequences differing by only a few single nucleotide polymorphisms.

Key words *Arabidopsis thaliana*, Fluorescence in situ hybridization, LNA–DNA mixmer probes, Binding affinity, 5S rDNA, Nuclei spreads

1 Introduction

Fluorescence in situ hybridization (FISH), developed in the 1980s [1, 2], allows for the detection of specific DNA sequences on meiotic chromosome spreads or interphase nuclei [3] and therefore presents a powerful technique both for physical mapping of candidate sequences and for the study of chromosomal arrangements in a cell-specific manner.

In the model system *Arabidopsis thaliana*, FISH has widely been used to detect repetitive regions and to identify their chromosomal position [4, 5], but has also been applied to visualize whole chromosomes [6] or specific chromosomal domains [7]. FISH allowed both high-resolution physical mapping [7, 8], and investigating the impact of chromatin regulators on nuclear organization of repetitive sequences in interphase [9–11]. Most of the FISH techniques using plant tissues rely on DNA probes generated by nick-translation or random priming-based amplification, using

Marian Bemer and Célia Baroux (eds.), *Plant Chromatin Dynamics: Methods and Protocols*, Methods in Molecular Biology, vol. 1675, DOI 10.1007/978-1-4939-7318-7_28, © Springer Science+Business Media LLC 2018

Bacterial Artificial Chromosomes (BACs) as templates for euchromatic regions or cloned tandem repeats as templates for repetitive regions [12]. These DNA probes can be directly labeled by fluorophores or, alternatively, indirectly labeled via the incorporation of digoxigenin- or biotin-conjugated nucleotides, subsequently revealed by fluorescently labeled specific antibodies [12]. Immunodetection of digoxigenin- or biotin-conjugated nucleotides using several consecutive antibodies amplifies the signal, which is particularly useful for low or single-copy targets. However, the protocol is relatively time consuming and may complicate quantification and cause background staining.

An alternative to longer DNA probes are short oligonucleotide probes and, particularly, locked nucleic acids (LNA)–DNA mixmers. In these probes some nucleotides are replaced by LNA analogs, in which the ribose ring is “locked” by a methylene bridge between the 2′-O and the 4′-C [13, 14]. This methylene bridge constrains the LNA nucleosides in the ideal conformation for Watson-Crick binding, increasing thermal stability of heteroduplexes and subsequently binding affinity of LNA toward complementary DNA and RNA sequences [15], such that even single nucleotide mismatches might be discriminated [16]. LNA–DNA mixmers have been used to detect short DNA or RNA targets such as miRNAs [17, 18], as well as long noncoding RNAs and repetitive DNA sequences [19] in mammals. Compared to standard probes they might show further advantages due to the resistance to exonucleases and endonucleases [20, 21], which increases their stability. Furthermore, given that the LNA–DNA mixmers are short oligonucleotides that can be as short as 20 nucleotides, they might prove advantageous over classical probes in whole-mount preparations. Similar to DNA probes, the LNA–DNA mixmers can be coupled to digoxigenin or biotin for signal amplification or directly fluorescently labeled, the latter reducing background signals. A drawback is the elevated price of LNA–DNA mixmers compared to traditional methods; however, the amount of the probe provided is sufficient for about 500 hybridization reactions and costs corresponding to antibodies, fluorophores or nick-translation reagents are saved.

Here, we have used directly labeled LNA–DNA mixmer probes to detect the 5S rRNA genes in *A. thaliana*, which are highly repeated sequences (more than 1000 copies [22]). The majority of the 5S rRNA genes are clustered into three loci in the pericentromeric regions of chromosomes 3, 4, and 5 in the *A. thaliana* genome of the Columbia-0 (Col-0) accession [23]. Besides the elevated conservation of 5S rRNA gene sequences, the transcription termination sequence just downstream of the transcribed sequence shows a chromosome-specific DNA signature differing by only six nucleotides between the 5S rRNA genes of chromosome 4 and 5 [24]. We designed LNA–DNA mixmer probes specific to

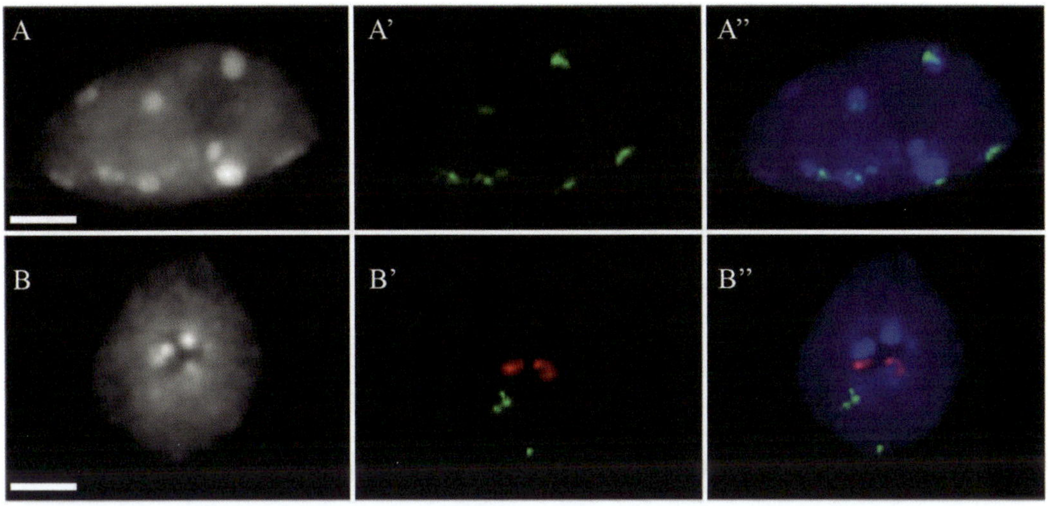

Fig. 1 Fluorescence in situ hybridization on cotyledon nuclei using directly labeled LNA–DNA mixmer probes. The probes were designed to detect all 5S rRNA genes (**A**, *green*) or specific for the 5S rRNA genes situated on chromosome 4 (**B**, *red*) and chromosome 5 (**B**, *green*). (**A**, **B**) Cotyledon nucleus stained with DAPI (*gray*); (**A'**, **B'**), FISH signals; (**A''**, **B''**) merged image, DNA in *blue*. Scale bar: 5 μM. (**A**) In a diploid nucleus, six 5S rDNA FISH signals are detected corresponding to the loci on chromosome 3, 4, and 5. The 5S rDNA repeats partially colocalize with the DAPI-bright chromocenters consisting mainly of centromeric and pericentromeric repetitive elements, but also form loops that emanate from the condensed chromatin domains. (**B**) The chromosome-specific probes recognize respectively the 5S rRNA genes on chromosome 4 (*red*) and on chromosome 5 (*green*)

the transcription termination sequence of 5S rRNA genes of chromosome 4 and 5, optimized hybridization and washing conditions and show that LNA–DNA mixmer probes can indeed be used to discriminate between the two 5S rDNA loci with distinct chromosomal location in nuclei spreads (Fig. 1).

This chapter describes a rapid FISH protocol including the preparation of nuclei spreads, hybridization and washing procedures adapted for short LNA–DNA mixmer probes targeting 5S rDNA.

2 Materials

2.1 Fixation of Arabidopsis Cotyledons and Tissue Digestion

1. Carnoy's fixative (*see* **Note 1**): 3:1 ethanol–glacial acetic acid, cold.
2. 2 mL microfuge tubes.
3. Citrate buffer: 10 mM sodium citrate/citric acid in distilled water, pH 4.5 (*see* **Note 2**).
4. Pectolytic enzyme mixture: 0.3% w/v pectolyase, 0.3% w/v cytohelicase, and 0.3% w/v cellulase (Sigma) in citrate buffer (*see* **Note 3**).

5. Fine forceps.
6. Six-well plate or small petri dish (35 × 10 mm).
7. Syringe needles.
8. Incubator at 37 °C.

2.2 Preparation of Nuclei Spreads

1. Cold 60% acetic acid.
2. Syringe needles.
3. Adhesive microscope slides (e.g., Star Frost, 76 × 26 mm, Knittel Glass).
4. Hot plate.
5. Cold Carnoy's fixative (*see* **Note 1**): 3:1 ethanol–glacial acetic acid, cold.
6. Coplin Jars.
7. 2× SSC: 0.3 M NaCl, 0.03 M sodium citrate, pH 7.
8. Pepsin: 100 mL of 100 μg/mL pepsin in 0.01 M HCl, prepared freshly.
9. 4% formaldehyde diluted from liquid 16% stock (*see* **Notes 4** and **5**) in 1× PBS (10 mM sodium phosphate, pH 7.0, 143 mM NaCl). At least 50 mL per Coplin jar is needed.
10. Deionized water.
11. Ethanol series: 70%, 90%, and 100%. At least 50 mL per Coplin Jar is required.
12. Antifade mounting medium (e.g., Vectashield, Vector Laboratories) containing 2 μg/mL DAPI (4′,6-diamidino-2-phenylindole).
13. 100 μg/mL DNase-free RNase A in 2× SSC.
14. Coverslips, 32 × 24 mm.
15. Incubator at 37 °C.
16. Moist chamber (*see* **Note 6**).

2.3 Hybridization and Washes

1. Directly labeled LNA–DNA mixmer probes (e.g., from Exiqon) (*see* **Note** 7), 100 μM in 1× PBS:
   - 5S rDNA probe detecting all 5S rRNA genes, "56FAM_CAAGCACGCTTAACT GCGGAGTTCTGAT".
   - 5S rDNA probe specific for chromosome 4, "TEX615_ACCAAAAAAAAAAAAAAAAAAAGAGGGATG" and
   - 5S rDNA probe specific for chromosome 5, "56FAM_AAAGGTTAAACATAAAAGAGGGATG".
2. Hybridization buffer (HB50): 50% deionized formamide, (*see* **Note 8**), 2× SCC, 50 mM Sodium phosphate buffer pH 7.
3. "HB50+": 20% dextran sulfate in HB50.

4. Hot plate (*see* **Note 9**).
5. Moist chamber.
6. Incubator at 55 °C.
7. Cover slips 32 × 24 mm.
8. 2× SSC: 0.3 M NaCl, 0.03 M sodium citrate, pH 7.
9. 0.75× SSC: Add 75 mL 2× SSC to 125 mL dH_2O.
10. 2× SSC-Tween: 0.1% Tween 20 in 2× SSC.
11. Antifade mounting medium (e.g., Vectashield, Vector Laboratories) containing 2 μg/mL DAPI (4′,6-diamidino-2-phenylindole).
12. Fluorescence microscope equipped with optical filters to detect DAPI, FITC/Alexa488/Fam, and Texas Red/Cy3 and a digital camera.

3 Methods

3.1 Fixation of Arabidopsis Cotyledons and Tissue Digestion

In this step the tissue is fixed and digested to release the nuclei. The procedure is described here for cotyledons from young seedlings, but other tissues can be used as well, although digestion times will need to be adjusted (*see* **Note 12**).

1. Sample Arabidopsis cotyledons from 10-day-old in vitro grown plants (*see* **Note 10**) and place up to 100 cotyledons into a 2 mL microtube containing Carnoy's fixative for fixation overnight at 4 °C. Per slide, 2–3 cotyledons are required.
2. The next morning (*see* **Note 11**), exchange the Carnoy's fixative 2–3 times by pipetting until the tissue becomes completely white. The more tissue was sampled, the more Carnoy's fixative exchanges will be needed.
3. Transfer about ten cotyledons into a six-well plate with forceps (enough for 3–4 slides).
4. Rinse the cotyledons twice with citrate buffer pH 4.5 at room temperature.
5. Remove the citrate buffer by pipetting and add the pectolytic enzyme mixture so that the mixture submerges the cotyledons (six-well plate can be slightly tilted to limit use of pectolytic enzyme mixture). Typically, 50–100 μL is enough to cover ten cotyledons.
6. Incubate in the incubator at 37 °C for about 1.5 h (*see* **Note 12**).
7. When the tissue is sufficiently digested (*see* **Note 13**), replace the enzyme mixture by 30 μL cold citrate buffer and proceed directly with the preparation of the nuclei spreads.

3.2 Preparation of Nuclei Spreads

In this step, the nuclei are released from the tissue, cleared from cytoplasm, and spread on microscope slides.

1. Use a needle to lacerate the cotyledons in the six-well plate until a fine suspension is obtained.
2. Pipette 10 μL of the suspension on a clean slide.
3. Place the slide on a hot plate set to 40 °C and add 10 μL of 60% acetic acid in drops around the suspension. Mix the suspension and acetic acid by gentle circular stirring with a needle (parallel to the glass surface to avoid any contact) for about 1 min (*see* **Note 14**). This step helps to dissolve the cytoplasm and facilitates nuclear adhesion to the slide. Avoid drying, and if needed, add more acetic acid while stirring.
4. Post-fix the preparation: Add in drops about 500 μL of cold Carnoy's fixative, kept on ice during the procedure, around the suspension and tilt the slide to mix both.
5. Wash the slide 10 min in 2× SSC in a Coplin jar at room temperature. All following steps are carried out in Coplin jars. Solutions are exchanged by emptying the Coplin jar, while making sure the slides do not glide out of the jar, and adding fresh solutions into the Coplin jar.
6. Digest the preparation with Pepsin for 3 min at 37 °C in an incubator (*see* **Note 15**).
7. Wash the slides with 2× SSC for 10 min at room temperature.
8. Post-fix the preparation in 4% formaldehyde (*see* **Note 4**) for 10 min.
9. Wash the slides with deionized water for 2 min.
10. Dehydrate the nuclei by incubating the slides in an ethanol series of 70%, 90%, and 100%, 2 min each.
11. Take the slides out of the Coplin jar and dry them vertically on tissue paper at room temperature for 10–20 min (the slides must be completely dry) (*see* **Note 16**).
12. Some slides can be sacrificed at this step to check the general quality of the preparation. Slides are mounted in antifade medium (e.g., Vectashield) containing DAPI and monitored under the fluorescence microscope (*see* **Note 17**).
13. Add 100 μL of RNase solution onto the slide and cover the nuclei preparations with a cover slip. Place the slide in a moist chamber and incubate for 1 h at 37 °C. This step removes RNA templates including 5S rRNA, which can cause background staining in the DNA-FISH experiment.
14. Place the slides in a Coplin jar and wash in 2× SSC for 5 min at room temperature.

15. Dehydrate the preparation by incubating the slides in a Coplin jar following an ethanol series of 70%, 90%, and 100%, 2 min each.
16. Take the slides out of the Coplin jar and dry them vertically at room temperature (the slides must be completely dry).

3.3 Hybridization and Washes

In this step, LNA–DNA mixmer probes are allowed to hybridize to denatured nuclear DNA. Excess of probes is removed through washes and the nuclear DNA is counterstained with DAPI before observation under the microscope.

1. Place the hot plate under the fume hood and heat it to 80 °C.
2. Dilute freshly one or two LNA-DNA mixmer probes to 0.1 μM in 15 μL HB50 and add 15 μL HB50+.
3. Pipette 30 μL of the hybridization mix per slide onto the nuclei spreads and cover with a coverslip. Avoid forming air bubbles.
4. Place the slides with the hybridization solution on the hot plate for 1 min. At this step chromosomes and probes are denatured; the time must be carefully monitored.
5. Hybridization: Transfer the slides in a moist chamber and incubate at 55 °C for 1 h (*see* **Note 18**).
6. At the same time, place the 2× SSC and 0.75× SSC wash solutions in closed Coplin jars in the 55 °C incubator.
7. Remove the coverslips by holding the slide vertically; the coverslip should glide down easily. Transfer to prewarmed 2× SSC and incubate for 5 min at 55 °C (*see* **Note 19**). The following steps are carried out in Coplin jars by exchanging the solutions.
8. Wash the slides 5 min in 0.75× SSC at 55 °C (*see* **Note 20**).
9. Wash the slides 5 min in 2× SSC at 55 °C.
10. Wash the slides 5 min in 2× SSC-Tween at room temperature.
11. Rinse the slide rapidly with deionized water.
12. Air-dry the slides.
13. Mount the slides in a drop of antifading medium containing DAPI (*see* **Note 21**).
14. Observe the slides under the fluorescence microscope (*see* **Note 22**).

4 Notes

1. Carnoy's fixative should be prepared freshly and kept on ice.
2. To prepare the citrate buffer 10× stock solution, add 0.1 M citric acid monohydrate into 0.1 M trisodium citrate dehydrate

until pH 4.5 is reached. The 10× stock can be stored at 4 °C and should be diluted into double distilled water prior to use.

3. We generally use the following products (C8274, C1794, P5936; all from Sigma) and prepare a stock solution containing 1% of each enzyme in citrate buffer, which is stored in small aliquots at −20 °C. Variations between batches may exist and digestion times may need to be adjusted for each new enzyme mixture batch.
4. Formaldehyde dilutions should be freshly prepared. Once opened, formaldehyde bottles should not be stored for longer periods of time to avoid concentration changes. We advice to use Formaldehyde supplied in small glass vials containing 10 mL of 16% formaldehyde (e.g., ThermoFisher) and to discard leftover dilutions.
5. Formaldehyde and formamide are toxic products in case of contact or inhalation. Dilutions and post-fixation steps as well as denaturation should be carried out under the fume hood using adequate protective equipment.
6. We typically place the slides horizontally into plastic slide storage boxes with a lid and a tissue imbibed with 2× SSC at the bottom. The slides should not be in contact with the imbibed tissue.
7. We have successfully used LNA–DNA mixmer probes between 20 and 25 nucleotides long with different fluorophores. The best positions of the LNA analogues in the probe sequence are determined by the provider [15, 16]. The probes are delivered in dry format and, upon arrival, should be diluted in 1× PBS to a stock solution of 100 μM. The stock solutions are stored in aliquots at −20 °C. While we present here results for the 5S rDNA probes, other repetitive sequences such as 180 bp centromeric repeats are ideal targets for LNA–DNA mixmer probes.
8. To avoid pH changes, deionized formamide solutions should be stored in small aliquots at −20 °C.
9. The temperature on the hot plate should be monitored and it is important that the surface is perfectly planar to allow uniform transfer of heat onto the slide.
10. Other tissues and other developmental stages such as leaves or flower buds can be used, but incubation times with the cell wall digestion mix should be adapted (*see* also **Note 11**). Flower buds contain both interphase nuclei, metaphase spreads as well as pollen mother cells at the pachytene stage, the latter having the advantage that chromosomes are synapsed and about 25 times longer than mitotic chromosomes [12].

11. The whole protocol can be carried out in 1 day, since the hybridization time is short (1 h) compared to standard DNA probes, which require an overnight hybridization step.
12. Depending on the tissue, the digestion time needs to be adjusted. For inflorescences the digestion time should be increased to 3 h.
13. Check if the tissue is sufficiently digested by gently squashing one cotyledon, which should release a white cloud corresponding to cells in suspension. If only fragments are released, incubate longer.
14. This is one of the most critical steps: acetic acid treatment is needed to clear and dissolve cytoplasmic components. Insufficient clearing creates clumps and layers of cellular fragments that will cause background during hybridization and DNA staining. The nuclei should appear smooth with a regular contour. Excess of treatment leads to nuclei with irregular contours and clumps in DNA staining. If the needle touches extensively the slide while stirring, nuclei might be scratched and damaged.
15. This step is optional, but helps to remove remaining cytoplasmic debris and improves access to the DNA probe. It is important to monitor precisely the incubation time and temperature. Possibly, the incubation time can be reduced when damaged nuclei are observed. Excess of treatment damages the chromatin and can lead to poor hybridization signals.
16. Slides can be stored at this step in a closed box at 4 °C.
17. The nuclei should appear isolated and not aggregated in clumps. No cytoplasm should remain around the nuclei. DAPI staining should reveal nuclei with smooth and regular contours, allowing to clearly distinguishing euchromatin (lightly stained) and heterochromatic regions (strongly stained). To avoid clumps of cells, increase digestion time and squash cotyledons more extensively.

    If excessive amounts of cytoplasm remain, the tissue digestion time can be adjusted and the density of cells reduced by using less starting material. Furthermore, during clearing, the amount of acetic acid can be increased and/or the stirring time prolonged (*see* **Note 14**). In addition, the time of pepsin treatment can be adjusted.
18. The temperature and the time of hybridization must be adjusted depending on the probe. When hybridization is performed with the 5S rDNA probe that detects all 5S rRNA genes, hybridization can be carried out at lower temperatures such as 37 °C. To achieve labeling specificity with the probes designed to hybridize exclusively to the 5S rDNA loci situated

on chromosome 4 or 5, hybridization should be performed at 55 °C.

19. It is important that the wash solutions have reached the required temperature.
20. If nonspecific binding of the probes is observed, the stringency of washing conditions can be increased: by lowering SSC concentrations, by adding Formamide to the wash buffers or by increasing the washing temperatures. Hybridization temperature can also be increased (*see* **Note 18**).
21. Coverslips can be sealed with nail polish and stored at 4 °C for several weeks.
22. For image acquisition, our typical microscope setup comprises a Leica SIM microscope, equipped with optical filters for DAPI, FITC, and TexasRed; a black and white digital CMOS camera (Hamamatsu, ORCA-Flash 4.0 V2) that produces 16-bit images; and Metamorph imaging software. To image the DNA stained by DAPI, the typical exposure time is 10 ms, for FISH signals in green and red channels, 100 ms is currently used. The appropriate exposure time should be adjusted for each channel. To create overlays of the black and white images acquired in the blue, green or red channel, the ImageJ "merge channels" option is used. Alternatively, images can be merged with Photoshop; however, care must be taken to monitor possible xy chromatic shifts.

References

1. Bauman J, Wiegant J, Borst P, van Duijn P (1980) A new method for fluorescence microscopical localization of specific DNA sequences by in situ hybridization of fluorochrome labelled RNA. Exp Cell Res 128:485–490
2. Langer-Safer PR, Levine M, Ward DC (1982) Immunological method for mapping genes on drosophila polytene chromosomes. Proc Natl Acad Sci U S A 79:4381–4385. doi:10.1073/pnas.79.14.4381
3. Levsky JM, Singer RH (2003) Fluorescence in situ hybridization: past, present and future. J Cell Sci 116:2833–2838. doi:10.1242/jcs.00633
4. Maluszynska J, Heslop-Harrison J (1993) Molecular cytogenetics of the genus Arabidopsis: in situ localization of rDNA sites, chromosome numbers and diversity in centromeric heterochromatin. Ann Bot 71:479–484. doi:10.1006/anbo.1993.1063
5. Maluszynska J, Heslop-Harrison JS (1991) Localization of tandemly repeated DNA sequences in *Arabidopsis thaliana*. Plant J 1:159–166. doi: 10.1111/j.1365-313X.1991.00159.x
6. Pecinka A, Schubert V, Meister A et al (2004) Chromosome territory arrangement and homologous pairing in nuclei of *Arabidopsis thaliana* are predominantly random except for NOR-bearing chromosomes. Chromosoma 113:258–269
7. Fransz PF, Armstrong S, de Jong JH et al (2000) Integrated cytogenetic map of chromosome arm 4S of *A. thaliana*: structural organization of heterochromatic knob and centromere region. Cell 100:367–376
8. Fransz P, de Jong JH, Lysak M et al (2002) Interphase chromosomes in Arabidopsis are organized as well defined chromocenters from which euchromatin loops emanate. Proc Natl Acad Sci U S A 99:14584–14589
9. Probst AV, Fransz PF, Paszkowski J, Mittelsten Scheid O (2003) Two means of transcriptional reactivation within heterochromatin. Plant J 33:743–749

10. Mathieu O, Jasencakova Z, Vaillant I et al (2003) Changes in 5S rDNA chromatin organization and transcription during heterochromatin establishment in Arabidopsis. Plant Cell 15:2929–2939
11. Mittelsten Scheid O, Probst AV, Afsar K, Paszkowski J (2002) Two regulatory levels of transcriptional gene silencing in Arabidopsis. Proc Natl Acad Sci U S A 99:13659–13662. doi:10.1073/pnas.202380499
12. Lysak M, Fransz P, Schubert I (2006) Cytogenetic analyses of Arabidopsis. Methods Mol Biol 323:173–186. doi:10.1385/1-59745-003-0:173
13. Koshkin AA, Singh SK, Nielsen P et al (1998) LNA (locked nucleic acids): synthesis of the adenine, cytosine, guanine, 5-methylcytosine, thymine and uracil bicyclonucleoside monomers, oligomerisation, and unprecedented nucleic acid recognition. Tetrahedron 54:3607–3630. doi:10.1016/S0040-4020(98)00094-5
14. Obika S, Nanbu D, Hari Y et al (1997) Synthesis of 2′-O,4′-C-methyleneuridine and -cytidine. Novel bicyclic nucleosides having a fixed C3, -endo sugar puckering. Tetrahedron Lett 38:8735–8738. doi:10.1016/S0040-4039(97)10322-7
15. Silahtaroglu AN, Tommerup N, Vissing H (2003) FISHing with locked nucleic acids (LNA): evaluation of different LNA/DNA mixmers. Mol Cell Probes 17:165–169. doi:10.1016/S0890-8508(03)00048-3
16. You Y, Moreira BG, Behlke MA, Owczarzy R (2006) Design of LNA probes that improve mismatch discrimination. Nucleic Acids Res 34:1–11. doi:10.1093/nar/gkl175
17. Song R, Ro S, Yan W (2010) In situ hybridization detection of microRNAs. Methods Mol Biol 629:287–294. doi:10.1007/978-1-60761-657-3_18
18. Søe MJ, Møller T, Dufva M, Holmstrøm K (2011) A sensitive alternative for microRNA in situ hybridizations using probes of 2′-O-methyl RNA + LNA. J Histochem Cytochem 59:661–672. doi:10.1369/0022155411409411
19. Probst AV, Okamoto I, Casanova M et al (2010) A strand-specific burst in transcription of pericentric satellites is required for chromocenter formation and early mouse development. Dev Cell 19:625–638. doi:10.1016/j.devcel.2010.09.002
20. Crinelli R (2002) Design and characterization of decoy oligonucleotides containing locked nucleic acids. Nucleic Acids Res 30:2435–2443. doi:10.1093/nar/30.11.2435
21. Frieden M, Hansen HF, Koch T (2003) Nuclease stability of LNA oligonucleotides and LNA-DNA chimeras. Nucleosides Nucleotides Nucleic Acids 22:1041–1043. doi:10.1081/NCN-120022731
22. Campell BR, Song Y, Posch TE et al (1992) Sequence and organization of 5S ribosomal RNA-encoding genes of Arabidopsis thaliana. Gene 112:225–228. doi:10.1016/0378-1119(92)90380-8
23. Fransz PF, Armstrong S, Alonso-Blanco C et al (1998) Cytogenetics for the model system *Arabidopsis thaliana*. Plant J 13:867–876
24. Cloix C, Tutois S, Yukawa Y et al (2002) Analysis of the 5S RNA pool in Arabidopsis thaliana: RNAs are heterogeneous and only two of the genomic 5S loci produce mature 5S RNA. Genome Res 12:132–144. doi:10.1101/gr.181301

Chapter 29

A Method for Testing Random Spatial Models on Nuclear Object Distributions

Javier Arpòn, Valérie Gaudin, and Philippe Andrey

Abstract

The cell nucleus is a structurally complex and dynamic organelle ensuring key biological functions. Complex relationships between nuclear structure and functions require a better understanding of the three-dimensional organization of the genome and of the subnuclear compartments. Quantitative image analysis coupled with spatial statistics and modeling is a relevant approach to address these questions. In this chapter, we describe a step-by-step procedure to process images and to test a spatial random model for the distribution of nuclear objects using chromocenters as an example. More elaborate models can be designed on the basis of the random model by introducing additional and more complex constraints to better fit observations and to question determinants of these spatial organizations.

Key words Image processing, Spatial descriptors, Spatial modeling, Random distribution, Nucleus, Chromocenters, Three-dimensional organization

1 Introduction

One of the major challenges in cell biology with major impacts on developmental biology is probably to better apprehend the genomic functions (transcription, replication, recombination, and DNA repair) in the spatial three-dimensional (3D) framework of the cell nucleus and their complex and dynamic relationships with the subnuclear compartments. A prerequisite is to depict the spatial organization of these compartments, such as chromocenters, nucleolus, or nuclear bodies [1], but also the spatial organization of the genome in the 3D nuclear space, to then further question the dynamics and determinants that drive these functional relationships.

On the one hand, high-throughput chromosome-conformation capture (Hi-C) techniques measure the average frequency of physical interactions between genomic regions over large sets of nuclei, allowing questioning the dynamics of the spatial genome organization. Resolution at the individual nucleus may

Marian Bemer and Célia Baroux (eds.), *Plant Chromatin Dynamics: Methods and Protocols*, Methods in Molecular Biology, vol. 1675, DOI 10.1007/978-1-4939-7318-7_29,

be expected in the future. On the other hand, but complementary to the Hi-C approaches, 3D imaging techniques unveil the absolute localization of target genomic sites or of nuclear compartments (hereafter collectively referred to as nuclear objects) in the 3D space of individual nuclei.

Image analysis can be used to extract quantitative information from acquired digital images. For example, Euclidean distances between nuclear objects or between objects and nuclear landmarks (such as the nuclear envelope) can be computed. Based on these quantitative measures, rules of spatial organization can be statistically investigated and evaluated. Such rules correspond either to mutual spatial interactions between nuclear objects or to spatial interactions between objects and nuclear landmarks such as the envelope. For example, one may want to test whether the location of a genomic region follows a specific localization in specific conditions or on the contrary can be considered as uniformly random within the nuclear space.

Up to now, tests for spatial rules of organization have mostly focused on radial distribution, for example to evaluate preferential peripheral positioning. Such tests are most generally achieved by comparing observed radial distributions to distributions expected under complete randomness. The radial distribution can be discretized, as in the so-called "shell" analysis based on counting the numbers of recorded locations within concentric shells of equal area or volume [2]. Conversely, the tests can be performed on the continuous radial distribution, as in the eroded-volume fraction approach [3].

Independently of the approach that is used (discrete or continuous), there are however two fundamental limitations on the use of classical radial analysis. First, analyzing the radial position alone encompasses an implicit projection of the 3D positions on the 1D radial direction. As a consequence, observing a random radial distribution is not sufficient to conclude to a completely random positioning of the objects of interest, as non-random spatial interactions and positioning rules can have been lost along the two other dimensions (Fig. 1a). Second, the theoretical null radial distribution expected under randomness is known only in the case of point positions. Hence, these approaches can only be applied to objects that are sufficiently small to be assimilated to points in acquired images (*i.e.*, object size below the optical resolution). Several nuclear compartments of interest, such as chromocenters or nuclear bodies, violate this assumption.

In this chapter, we describe a protocol from a methodological framework we are developing to overcome these limitations. Our methodology allows to evaluate a wide range of 3D spatial organization rules for nuclear objects of arbitrary and varying size [4]. The simplest spatial model to test is the complete randomness of object distribution, which allows to assess the existence of spatial heterogeneities or interactions. We describe here the step-by-step procedure that allows to implement this test. Because the method is conditioned by data extracted from the images, we detail the complete pipeline from image processing up to spatial analysis of object distribution within nuclei. Since appropriate image data is required for spatial analysis, we also provide recommendations about sample preparation and image acquisition.

The steps preceding spatial analysis are generally specific to the objects of interest. For example, specific staining protocols have to be used for different nuclear compartments or genomic regions. Here, we will consider the analysis of chromocenter distributions observed within DAPI-stained nuclei of ***Arabidopsis thaliana*** leaf cells as an illustrative example.

Conversely, the step-by-step spatial analysis method is a generic one and can be applied to question the rules that govern the spatial organization of different types of nuclear objects or domains (genomic regions labeled by FISH, fluorescently labeled subnuclear compartments, etc.). Even more, the methodology can be applied to many biological systems at different scales, from subcellular to histological organizations. It has been, for example, used to analyze the spatial organization of human P-bodies from 2D TEM images [5].

The completely random model is a simple reference model to initiate the analysis of spatial distributions. Additional constraints and parameters can be introduced according to research objects and questions to refine the spatial models into more sophisticated ones, which will better explain and get closer to the observed distributions of objects in the cell nucleus.

The step-by-step procedure for image processing and spatial analysis is sufficiently detailed to allow its implementation by any computer scientist or statistician in programming languages such as C/C++ or Java, or based on higher-level languages such as those from Matlab and R. We also provide pointers to existing implementations of some of the image processing steps in the popular ImageJ/Fiji software [6, 7]. Some implementation of the algorithms for the segmentation of nuclei and chromocenters from 3D DAPI images [4] are also available in the NucleusJ plugin [8] we developed for this software. The spatial analysis method for testing randomness will soon be made available as a plugin for the Free-D image data processing and analysis software [http://free-d.versailles.inra.fr; 9, 10].

2 Materials

2.1 Sample Preparation

To generate a population of nuclei, standardization of plant growth conditions is important as well as timing and tissue collection. Indeed, variations in nuclear parameters such as size or shape are expected with age, differentiation, tissue, cell type, or environmental cues [1]. The more homogeneous the material from biological replicate to biological replicate is, the more powerful the statistics will be. Until now most 3D studies in plants were performed on fixed material, either isolated or cell-sorted nuclei [4, 11, 12], nuclei in cryosections [13] or in nuclei from whole-mount samples [14]. A careful attention should be devoted to the fixation protocol to preserve chromatin structure without impacting on the relationships between nuclear objects and to preserve the nuclear shape and size without deformation such as flattening due to drying, coverslip application, etc. The analysis of nuclei in cryosections is a good compromise allowing protection against deformation and a reasonably good access for image acquisition. In cryosections, cell type identity can be taken into account. Live imaging using fluorescent markers of nuclear objects is another alternative [2, 15]. DAPI-staining needs to be well calibrated and reproducible to avoid saturation. Even so, variations in DAPI-staining intensities may be observed. Up to minor experimental fluctuations, it is recommended to standardize as much as possible the different steps from growth to image acquisition.

2.2 Image Acquisition

The acquisition conditions should be optimized according to microscope equipment and manufacturer's recommendations. To do so, an acquisition test followed by image analysis should be performed. Based on the results, acquisition parameters will be adjusted. Several rounds of tests may be required, but are mandatory to get accurate quantification and modeling. Depending on the microscope and the objective, parameters such as gain, offset, and voxel size in the XYZ directions need to be standardized to have as much as possible similar acquisition conditions from session to session (*see* **Note 1**). The constant gain should be set so as to avoid intensity saturation over the collection of acquired nucleus images.

Importantly, a good signal-to-noise ratio is essential to accurately extract the nuclear objects. Extraction of DAPI-labeled chromocenters from DAPI-labeled nucleoplasm is a difficult case, due to the fact that nuclear background and chromocenters are acquired in the same confocal channel. The specific segmentation method required for this case is described below.

2.3 Digital Images

The input material for image processing are 3D gray level images in which voxel values correspond to digitized DAPI fluorescence intensities (Fig. 1b). Images should be properly calibrated, *i.e.*, the physical dimensions of the voxels along the XYZ directions should be available (*see* **Note 2**).

Any standard image file format is appropriate for the procedures described below. However, it is recommended to stick to the proprietary file format of the acquisition system (Zeiss's LSM,

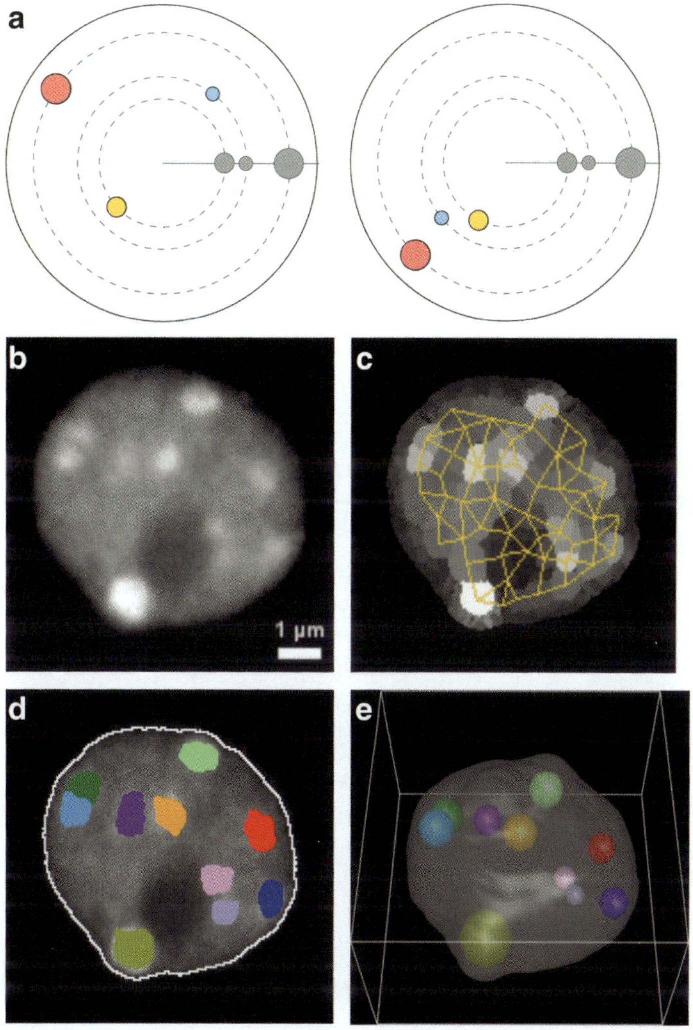

Fig. 1 Analysis and processing of 3D images of nuclei. (**a**) Radial analysis entails a projection of object positions from the 3D nuclear space onto the 1D radial dimension. The two displayed configurations, which are actually different in 3D, appear identical. Actual 3D organizations are lost. (**b**) Section through a 3D confocal image stack of an isolated DAPI-stained *A. thaliana* leaf cell nucleus. (**c**) Intermediate step in chromocenter segmentation. The nucleus is partitioned into many regions and operators are applied on the region adjacency graph to detect chromocenter regions. (**d**) Output of the segmentation pipeline (*White*: detected nucleus boundary; *Colors*: detected chromocenters). (**e**) Input data for the spatial analysis pipeline (*Gray*: nuclear envelope; *Colors*: chromocenter equivalent spheres)

Leica's LIF, etc.) whenever possible (*see* **Note 3**). If required, images converted to alternative file formats (such as plain TIF) can be used as well. Batch conversion from most of microscope proprietary formats to TIF can, for example, be performed under ImageJ/Fiji [6, 7] using the LOCI-BioFormats plugin [16].

2.4 Software

The Fiji image processing software [6, 7] is available at https://fiji.sc/. Installation instructions are found at https://fiji.sc/#download. An implementation of the Gaussian gradient is provided in the FeatureJ plugin available from https://imagescience.org/meijering/software/featurej/ with installation instructions. The watershed transform and the calculation of the region adjacency graph are available from the MorphoLibJ plugin [17]. Installation instructions are found at http://imagej.net/MorphoLibJ#Installation.

Statistical methods and tests are available in the R software [18] available from https://www.r-project.org/. The `ks.test()` method of the `dgof` package provides a permutation test for discrete goodness-of-fit tests.

3 Methods

3.1 Image Processing

We describe in this section how the information required for testing spatial randomness of object distribution can be extracted from acquired 3D images. This information encompasses the binary mask of the nucleus and the masks of the objects of interest. The procedure for segmenting chromocenters was specifically designed for the extraction of these compartments from images of DAPI-stained nuclei [4]. The procedure for segmenting nuclei from these images is largely generic and may be useful with other nuclear stains.

We consider here the simple situation where each image contains a single nucleus only. If required, the cropping of images containing several nuclei can easily be performed manually using, for example, the ImageJ/Fiji software (select the ***Rectangular selection tool***, then draw a rectangle around the nucleus, then run ***Image*** → ***Crop***). Cropping can also be automated. However, the description of the corresponding algorithms falls beyond the scope of this chapter.

3.1.1 Segmentation of the Nucleus

The first image processing step is the segmentation of the input image in order to obtain a binary mask of the nucleus. An automated intensity thresholding method such as Otsu's algorithm [19] will generally give close to, but not completely satisfactory results for this purpose. Thresholds computed with such automatic methods may indeed need to be corrected. For example, Otsu's method is sensitive to the relative proportion of image occupied by object versus background, meaning that for a given nucleus, the automatically computed threshold value depends on the margin around the nucleus. Consequently, the acquisition field-of-view and any subsequent cropping around the nuclei may impact the size, and to a lesser extent the shape, of the segmented nuclei.

We have thus designed a procedure for the automated segmentation of DAPI-stained nuclei that is robust to the relative size of the nuclei within their images [4]. For each nucleus, the procedure is the following:

1. Compute a threshold value using Otsu's method, and determine the corresponding nuclear binary mask (Fiji: *Image* → *Adjust* → *Threshold*, select *Otsu* and *Stack histogram* options, then click *Apply* and uncheck the *Calculate threshold for each image* option).
2. Using the binary mask, compute the average m and the standard-deviation s of the intensity values within the nucleus in the original intensity image (*Plugins* → *MorphoLibJ* → *Analyze* → *Intensity Measurements 2D/3D*).
3. Set the actual intensity threshold to $m - 2s$ and update the nuclear binary mask accordingly.
4. Apply mathematical morphology operators [20] such as hole filling (*Plugins* → *MorphoLibJ* → *Fill Holes*) and binary opening and closing (*Plugins* → *MorphoLibJ* → *Morphological filters 3D*) to correct potential artifacts in the obtained binary mask. Holes may indeed occur (for example, due to the presence of the nucleolus) as well as shape irregularities (such as bumps due to the intensity halo of chromocenters close to nuclear periphery).

3.1.2 Segmentation of Chromocenters

Important fluctuations of intensity values within the nucleus make simple thresholding inappropriate for the segmentation of chromocenters. However, correct segmentation of chromocenters can be achieved using the following procedure, which is based on the observation that chromocenters are characterized by a positive local contrast with their immediate surrounding [4].

1. Compute the 3D Gaussian gradient of the original image stack (*Plugins* → *FeatureJ* → *FeatureJ Edges*).
2. Run the watershed algorithm [20, 21] on the resulting 3D image, restricting the algorithm to run over the domain defined by the binary mask of the nucleus (*Plugins* → *MorphoLibJ* → *Segmentation* → *Classic Watershed*). This generates a complete partitioning of the nucleus into R regions, some of which correspond to actual chromocenters while others belong to the remaining nucleoplasm or to the nucleolus (Fig. 1c).
3. Compute the region adjacency graph (RAG) of these regions: each node of the graph corresponds to a region, and each edge connects two regions that are adjacent (Fig. 1c). The RAG can be obtained using *Plugins* → *MorphoLibJ* → *Analyze* → *Region Adjacency Graph*.

4. For each region i, compute its size V_i and determine the set $\mathcal{N}_i$ of its adjacent regions. Initialize m_i to the average intensity value of the region.
5. Update the $m_1, \ldots, m_R$ values by applying a morphological closing operation on the region adjacency graph. This operation consists first in a dilation step, where m_i is replaced by the largest m value in the neighborhood of region i, followed by an erosion step, where the updated m_i is replaced by the lowest value in the neighborhood. The effect of this operation is to assign to dark regions values from adjacent regions with higher values. The purpose is mainly to prevent obtaining spuriously large contrast values at the following step for non-chromocenter regions that touch the nucleolus.
6. Compute the contrast C_i of each region i according to:

$$C_i = m_i - \sum_{j=1}^{|\mathcal{N}_i|} V_j m_j \Big/ \sum_{j=1}^{|\mathcal{N}_i|} V_j$$

7. Compute a contrast threshold, mark all regions with contrast above the threshold as chromocenters, and discard the remaining regions. The contrast threshold can be determined manually or automatically, for example applying Otsu's algorithm to the set $\{C_1, \ldots, C_R\}$.

3.1.3 Quantitative Analysis

From the segmentation masks of the nucleus and of chromocenters (Fig. 1d), several morphological and photometric measurements can be extracted (*Plugins* → *MorphoLibJ* → *Particle Analysis 3D* and *Intensity Measurements 2D/3D*). Sizes can be quantified using volume, surface area, and axis lengths parameters. Shapes can be quantified using sphericity and length ratios between different axes, providing elongation and flatness measures. Fluorescence signals can be quantified using total or relative intensity, providing information about the relative heterochromatin fraction.

3.2 Testing Randomness of Object Distribution in a Single Nucleus

We begin with the statistical testing of complete spatial randomness for a pattern of objects observed in a given nucleus. In practice, analyzing a single nucleus is rarely of interest; however, it provides the basis for testing randomness over a population of nuclei, a procedure which is described in the next section. Complete spatial randomness of objects is a model in which objects are uniformly and independently distributed in the (nuclear) space, conditioning on the spatial constraints due to their sizes (no intersection with nuclear envelope or other objects).

3.2.1 Material

The input material is composed of (i) a triangular mesh representing the envelope of the nucleus and (ii) the number N, the individual positions $p_1, \ldots, p_N$ and the individual sizes of objects (Fig. 1e). The triangular mesh is computed from the binary mask of the nucleus using the Marching Cubes algorithm or one of its variants [22]. The individual positions are given by the centroids of the segmented objects. Sizes should be given as equivalent spherical radii, *i.e.*, the radii of spheres with the same volumes. The equivalent radius of an object with volume V is given by $\left(\frac{3V}{4\pi}\right)^{\frac{1}{3}}$. The coordinates of the vertices over the triangular mesh and the radii should be expressed in the same physical distance units (*see* **Note 4**).

3.2.2 Quantitative Spatial Descriptors of Object Distribution

To quantify the distribution of objects using spatial descriptors we use distance functions as in standard analysis of point patterns [23]. In the following sections, we assume that one of the following three common distance functions is used.

The *F*-function, also known as the void fraction or first contact distribution function, is related to the spacing between objects. It is formally defined as the cumulative distribution function (CDF) of the distance X between an arbitrary position within the nucleus and the closest object: $F(x) = P(X < x)$. In practice, it is estimated by first defining a set of L positions $\{q_1, q_2, \ldots, q_L\}$ within the nucleus and then by computing the distance to the nearest object of each of these positions (Fig. 2a). The estimation $\widehat{F}(x)$ of $F(x)$ is the observed proportion of distances inferior to x:

$$\widehat{F}(x) = \frac{1}{L}\sum_{k=1}^{L} \mathbf{1}_{d(q_k,\eta(q_k))<x}$$

where $\mathbf{1}_E$ is the indicator function of event E and $\eta(q_k)$ designates the closest object position from q_k. The positions $q_1, \ldots, q_L$ can be taken uniformly at random within the nucleus or according to a regular grid, for example placing a position at the center of each voxel (*see* **Note 5**).

The *G*-function is the cumulative distribution function of the distance Υ between any object and its nearest neighbor (Fig. 2b): $G(y) = P(\Upsilon < y)$. In practice, it is estimated by the empirical distribution of Υ:

$$\hat{G}(y) = \frac{1}{N}\sum_{i=1}^{N} \mathbf{1}_{d(p_i,\eta(p_i))<y}$$

where $\eta(p_i)$ designates the position of the closest neighbor of object at p_i.

The *H*-function is the cumulative distribution function of the distance Z between each object and any other one (Fig. 2c): $H(z) = P(Z < z)$. In practice, it is estimated by the empirical distribution of Z:

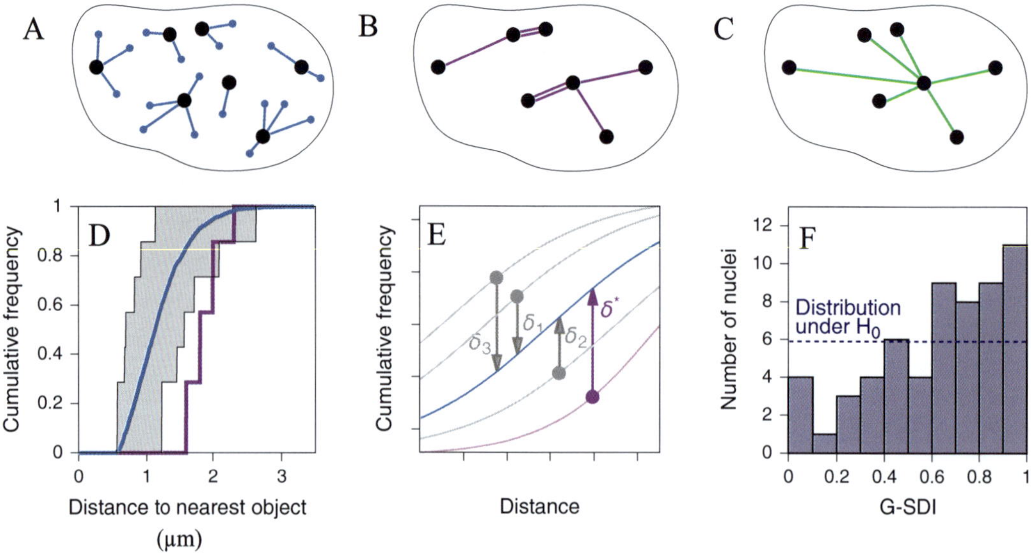

Fig. 2 Statistical spatial analysis. (**a**-**c**) Distance measurements used in spatial descriptors. Function *F* relies on distances between arbitrary positions and their nearest objects (**a**), function *G* on the distances between each object and its nearest neighbor (**b**), and function *H* on all object inter-distances (**c**). (**d**) Results obtained on one nucleus with function *G* (*Magenta*: observed CDF; *Blue*: average CDF under the random model; *Gray*: 95% envelope under the model). (**e**) Maximum signed distance between individual CDFs and model average (*Blue*: average CDF under the random model; *Magenta*: observed CDF; *Gray*: individual CDFs from Monte-Carlo simulations of the random model). (**f**) Distribution of the SDI computed using function *G* in a population of *A. thaliana* isolated leaf cell nuclei (*Dotted line*: uniform distribution expected under the random model)

$$\hat{H}(z) = \frac{2}{N(N-1)} \sum_{i=1}^{N} \sum_{j=i+1}^{N} \mathbf{1}_{d(p_i, p_j) < z}$$

3.2.3 Null Distribution Functions

For any of the *F*, *G*, or *H* descriptors, the null distribution function is the expected CDF under the completely random spatial model. The analytic expression of this distribution is generally unknown because of the arbitrary shape of the nucleus and because of the arbitrary and distinct object sizes. Monte-Carlo simulations of the completely random model are thus used to estimate this null distribution. It is essential that the actual nuclear shape, number and sizes of objects are taken into account in this process. The procedure is the following:

1. Generate *M* patterns of randomized object positions, conditioning on observed nuclear shape, number of objects and object sizes. Each pattern is obtained by randomly positioning a first object within the nuclear envelope, taking care that its distance to the nuclear border is at least as large as its equivalent radius. Then a second object is randomly positioned, avoiding intersection not only with nuclear border but also with the first

object. The process is repeated until all objects are placed (*see* **Note 6**).

2. For each random pattern, estimate the spatial descriptor (F, G, or H) as described in Subheading 3.2.2.
3. Average the resulting M cumulative distribution functions to obtain the estimated null distribution ($\widehat{F}_0, \hat{G}_0$ or $\hat{H}_0$) under the random model (Fig. 2d).

3.2.4 Test of Goodness-of-Fit

The goodness-of-fit of the completely random model to an observed nuclear pattern is tested by evaluating the statistical significance of the difference between the observed CDF and the null CDF. For a two-sided test, the null hypothesis is that the objects obey a completely random distribution and the alternative hypothesis is that they follow any other spatial distribution. The procedure is the following:

1. Compute the distance x^* where the observed and model average CDFs are maximally distant. Taking function F as an example, we have:

$$x^* = \underset{x \geq 0}{\operatorname{argmax}} |\widehat{F}(x) - \widehat{F}_0(x)|$$

2. Compute the signed maximal distance δ^* (Fig. 2e) between the two curves at position x^*:

$$\delta^* = \widehat{F}(x^*) - \widehat{F}_0(x^*)$$

3. Generate a second set of M independent completely random Monte-Carlo patterns and compute the signed maximum distances $\delta_1, \delta_2, \ldots, \delta_M$ between their corresponding CDFs and the null CDF as above for the observed pattern (Fig. 2e).
4. Sort in increasing order the $M + 1$ values $\delta^*, \delta_1, \delta_2, \ldots, \delta_M$ and determine the rank r of the empirical difference δ^* in this sorted set.
5. Compute the p-value of the two-sided test as [24]:

$$p\text{–value} = 2 \times \min\left\{\frac{r+1}{M+1}, 1 - \frac{r}{M+1}\right\}$$

3.3 Testing Randomness on a Population of Nuclei

Generally, a population of nuclei is analyzed to get robust results. Testing the goodness-of-fit of the model individually on each nucleus using the procedure described in the previous section encompasses a loss of statistical power compared to a single test over the whole data set. We thus recommend not to perform individual tests and to favor a global test to evaluate whether a spatial model fits with a population of observed nuclear patterns. One difficulty is that nuclei vary in size, shape, as well as in object number and individual object sizes. Normalization with regard to these fluctuations is thus required when testing over a population. We have designed a method in which each nucleus is used as its own reference, thus achieving implicitly the required normalization [4].

This global test relies on the observation that, under the null hypothesis, the Spatial Distribution Index (SDI) defined by

$$\mathrm{SDI} = 1 - \frac{r}{M+1}$$

is uniformly distributed between 0 and 1. The test is performed by computing the SDI for each nucleus and by comparing the distribution of SDI values in the population to the uniform distribution (Fig. 2f).

Classical one-sample distribution tests for continuous distributions, such as the Kolmogorov-Smirnov test, cannot be used for testing SDI uniformity because the SDI is by construction a quantity that takes discrete values by steps of size $\frac{1}{M+1}$. Therefore, the uniformity of the SDI distribution should be tested using procedures providing the equivalent of the Kolmogorov-Smirnov test for discrete distributions. The function `ks.test` in the package `dgof` [25] for the R software [18] provides such a procedure.

Depending on the homogeneity of the studied population and on the size of the effect (*e.g.*, departure from a spatial model) one wants to be able to detect, the required number of nuclei in the population will vary. However, based on our experience, we suggest sample sizes should be at least of 50–100 nuclei from different biological replicates.

3.4 Interpretation of the Test

In case the null hypothesis is rejected, the distribution of the SDI provides directions to interpret the actual distribution of objects. For any of the spatial descriptors, low SDI values indicate by construction that the distances measured on the observed data are smaller than those expected under the completely random model. For functions G and H, this would correspond to smaller inter-distances between objects, thus revealing a significant trend towards spatial proximity between objects. The same conclusion would be reached with function F in the case of high SDI values. Alternatively, high SDI values for functions G and H, and low SDI values for function F, would point to regular object distributions.

Applying the whole procedure to *A. thaliana* isolated leaf cell nuclei produced skewed SDI distributions towards high values for functions G and H, and towards low values for function F, thus revealing a more regular organization of chromocenters than expected under randomness (Fig. 2f) [4].

4 Notes

1. For image processing and spatial analysis, the physical voxel size (image calibration) is a more important property of acquired digital images than their dimensions (number of voxels). Voxel size should be set based on the optical resolution, which itself depends on the objective properties (Numerical Aperture) and the emission wavelength (recommended voxel sizes are 2–3 times smaller than resolution, and in any case not larger). The actual image size is actually not relevant. Ideally, it would be automatically adjusted based on the voxel size and the field-of-interest (hence, object size). Unfortunately, microscope software generally adjust the voxel size based on the field-of-view and predefined image dimensions.
2. The spatial calibration (physical size of a voxel in each of the three directions, generally expressed in nm or μm) is automatically stored in the image files by most acquisition software. However, as different software is generally used for the purpose of image visualization, processing, or analysis, it is recommended to check that spatial calibration is correctly read.
3. Conversions between image file formats should be avoided as much as possible. First, most of the metadata about imaging setup, that are automatically stored by acquisition software under proprietary formats, are generally lost during this conversion. Second, this increases the house-keeping and data-management burden—not mentioning the increased storage capacity that is required.
4. Theoretically, coordinates and sizes could also be expressed in voxels without affecting the results of spatial analysis. However, due to the anisotropic resolution of optical microscope between the axial and lateral resolutions, this is not recommended as it would imply a deformation (typically, flattening) of the nuclear envelope along the axial direction.
5. When random evaluation positions are used, their number/density should be set large enough to minimize the sampling effect on the estimated $\widehat{F}$. In practice, one can progressively increase the density of evaluation positions until $\widehat{F}$ becomes insensitive to sampling.
6. It may happen that there is no more room to place an additional object once some objects have already been placed. Therefore, several attempts are made to place at random an object until a valid position is found or until a predefined maximum number of attempts have been reached. In the latter case, all objects already positioned in the current pattern are removed and the process is restarted from the first object. It

may also be useful to randomize the ordering of the objects before the generation of each new random pattern.

Acknowledgements

JA was supported by a PhD fellowship provided by the European Commission Seventh Framework-People-2012-ITN project Epi-TRAITS (Epigenetic regulation of economically important plant traits, no-316965). The IJPB benefits from the support of the LabEx Saclay Plant Sciences-SPS (ANR-10-LABX-0040-SPS).

References

1. Del Prete S, Arpón J, Sakai K, Andrey P, Gaudin V (2014) Nuclear architecture and chromatin dynamics in interphase nuclei of *Arabidopsis thaliana*. Cytogenet Genome Res 143 (1–3):28–50
2. Fang Y, Spector DL (2005) Centromere positioning and dynamics in living *Arabidopsis* plants. Mol Biol Cell 16(12):5710–5718
3. Ballester M, Kress C, Hue-Beauvais C, Kiêu K, Lehmann G, Adenot P, Devinoy E (2008) The nuclear localization of WAP and CSN genes is modified by lactogenic hormones in HC11 cells. J Cell Biochem 105(1):262–270
4. Andrey P, Kiêu K, Kress C, Lehmann G, Tirichine L, Liu Z, Biot E, Adenot P-G, Hue-Beauvais C, Houba-Hérin N, Duranthon V, Devinoy E, Beaujean N, Gaudin V, Maurin Y, Debey P (2010) Statistical analysis of 3D images detects regular spatial distributions of centromeres and chromocenters in animal and plant nuclei. PLoS Comput Biol 6(7): e1000853
5. Ernoult-Lange M, Baconnais S, Harper M, Minshall N, Souquere S, Boudier T, Bénard M, Andrey P, Pierron G, Kress M, Standart N, le Cam E, Weil D (2012) Multiple binding of repressed mRNAs by the P-body protein Rck/p54. RNA 18(9):1702–1715
6. Schneider CA, Rasband WS, Eliceiri KW (2012) NIH Image to ImageJ: 25 years of image analysis. Nat Methods 9(7):671–675
7. Schindelin J, Arganda-Carreras I, Frise E, Kaynig V, Longair M, Pietzsch T, Preibisch S, Rueden C, Saalfeld S, Schmid B, Tinevez J-Y, White DJ, Hartenstein V, Eliceiri K, Tomancak P, Cardona A (2012) Fiji: an open-source platform for biological-image analysis. Nat Methods 9:676–682
8. Poulet A, Arganda-Carreras I, Legland D, Probst AV, Andrey P, Tatout C (2015) NucleusJ: an ImageJ plugin for quantifying 3D images of interphase nuclei. Bioinformatics 31(7):1144–1146
9. Andrey P, Maurin Y (2005) *Free-D*: an integrated environment for three-dimensional reconstruction from serial sections. J Neurosci Methods 145(1–2):233–244
10. Biot E, Crowell E, Burguet J, Höfte H, Vernhettes S, Andrey P (2016) Strategy and software for the statistical spatial analysis of 3D intracellular distributions. Plant J 87(2):230–242
11. Pecinka A, Schubert V, Meister A, Kreth G, Klatte M, Lysak MA, Fuchs J, Schubert I (2004) Chromosome territory arrangement and homologous pairing in nuclei of *Arabidopsis thaliana* are predominantly random except for NOR-bearing chromosomes. Chromosoma 113(5):258–269
12. Baroux C, Pecinka A, Fuchs J, Kreth G, Schubert I, Grossniklaus U (2017) Non-random chromosome arrangement in triploid endosperm nuclei. Chromosoma 126(1):115-124
13. Tirichine L, Andrey P, Biot E, Maurin Y, Gaudin V (2009) 3D fluorescent in situ hybridization using Arabidopsis leaf cryosections and isolated nuclei. Plant Methods 5:11
14. She W, Baroux C (2014) Chromatin dynamics during plant sexual reproduction. Front Plant Sci 5:354
15. Lindhout BI, Fransz P, Tessadori F, Meckel T, Hooykaas PJJ, van der Zaal BJ (2007) Live cell imaging of repetitive DNA sequences via GFP-tagged polydactyl zinc finger proteins. Nucleic Acids Res 35(16):e107
16. Linkert M, Rueden CT, Allan C, Burel J-M, Moore W, Patterson A, Loranger B, Moore J, Neves C, MacDonald D, Tarkowska A, Sticco C, Hill E, Rossner M, Eliceiri KW, Swedlow JR (2010) Metadata matters: access to image data in the real world. J Cell Biol 189(5):777–782

17. Legland D, Arganda-Carreras I, Andrey P (2016) MorphoLibJ: integrated library and plugins for mathematical morphology with ImageJ. Bioinformatics 32(22):3532–3534
18. R Development Core Team (2015) R: a language and environment for statistical computing. R Foundation for Statistical Computing, Vienna, Austria. ISBN 3-900051-07-0
19. Otsu N (1979) A threshold selection method from gray-level histograms. IEEE Trans Syst Man Cybern 9(1):62–66
20. Soille P (2003) Morphological image analysis: principles and applications, 2nd edn. Springer, Berlin, Germany
21. Vincent L, Soille P (1991) Watersheds in digital spaces: an efficient algorithm based on immersion simulation. IEEE Trans Pattern Anal Mach Intell 13(6):583–598
22. Lorensen WE, Cline HE (1987) Marching cubes: a high resolution 3D surface construction algorithm. ACM Comput Graph 21 (4):163–169
23. Diggle PJ (2014) Statistical analysis of spatial and spatio-temporal point patterns, 3rd edn. Chapman and Hall/CRC Press, Boca Raton
24. Baddeley A, Diggle PJ, Hardegen A, Lawrence T, Milne RK, Nair G (2014) On tests of spatial pattern based on simulation envelopes. Ecol Monogr 84(3):477–489
25. Arnold TB, Emerson JW (2011) Nonparametric goodness-of-fit tests for discrete null distributions. R J 3(2):34–39

Chapter 30

Technical Review: Cytogenetic Tools for Studying Mitotic Chromosomes

Václav Bačovský, Roman Hobza, and Boris Vyskot

Abstract

Significant advances in chromosome preparation and other techniques have greatly increased the potential of plant cytogenetics in recent years. Increase in longitudinal resolution using DNA extended fibers as well as new developments in imaging and signal amplification technologies have enhanced the ability of FISH to detect small gene targets. The combination of fluorescence in situ hybridization with immunocytochemistry allows the investigation of cell events, chromosomal rearrangements and chromatin features typical for plant nuclei. Chromosome manipulation techniques using microdissection and flow sorting have accelerated the analysis of complex plant genomes. Together, the different cytogenetic approaches are invaluable for the unravelling of detailed structures of plant chromosomes, which are of utmost importance for the study of genome properties, DNA replication and gene regulation. In this technical review, different cytogenetic approaches are discussed for the analysis of plant chromosomes, with a focus on mitotic chromosomes.

Key words Cell synchronization, Chromosome spreads, Laser microdissection, Cytogenetics, FISH, Immunostaining, Antibodies

1 Introduction

The increasing understanding of genome structure has been accompanied by the development of cytogenetic techniques that are now fundamental for a broad range of research areas and applications as well as for guiding genome sequencing efforts. Comparative analysis and genome studies significantly contributed to our understanding of genome evolution and structure. Cytogenetics as one of biological disciplines is approaching to study single cells and the composition of cellular compartments, and focuses on nuclei and single chromosomes using various techniques, particularly in situ hybridization, among others. In this respect, plant cytogeneticists were among the earliest genome researchers many years before the first plant genome was sequenced [1].

Marian Bemer and Célia Baroux (eds.), *Plant Chromatin Dynamics: Methods and Protocols*, Methods in Molecular Biology, vol. 1675, DOI 10.1007/978-1-4939-7318-7_30, © Springer Science+Business Media LLC 2018

After the introduction of fluorescence in situ hybridization (FISH), which allows the localization of specific DNA fragments in the chromosomes, different technical advances resulted in a broad range of FISH applications using various sets of fluorochromes, chromosome preparations, and probe sets allowing easier karyotype visualization. Also in the postgenomic era, in situ hybridization is still the principal method for rapid chromosome identification and characterization. FISH is used as a tool for chromosome identification, karyotype comparative analyses and for discrimination of single chromosomes for subsequent flow sorting and laser microdissection.

Cytogenetic mapping of genomic loci, chromosomal segments, and whole chromosomes offers an efficient strategy for sequence localization and characterization of complicated chromosomal regions (e.g., repetitive). Physical maps are a useful tool for determining the exact order of genes and the relative distance between two loci. Comparative analysis of genes and/or loci of interest among related species can be used for phylogenetic studies as well as for revealing genome reorganization. The combination of FISH with other methods such as immunostaining, provide a direct approach to anchoring DNA sequences to specific features of euchromatin and heterochromatin. The physical localization of DNA sequences with associated proteins enhances the resolution ability of FISH. Molecular cytology enables better understanding of basic processes during cell division and in this way provides a unique framework for structural and functional genomics. The combination of cytogenetics and genomics enables in depth analysis of rare events that are otherwise hard to detect. In this technical review, we discuss the use of various FISH and immunostaining techniques along with tools for chromosome manipulation such as microdissection and flow sorting (Fig. 1). Examples from the dioecious species ***Silene latifolia***, which possesses heteromorphic sex chromosomes, are presented here to illustrate the different techniques. After discussing the basic principles of cell synchronization and chromosome preparation, we describe in detail the application of FISH and individual FISH variations. Finally, we discuss the use of immunolabeling which enhances the resolution of FISH and provides unique information about chromosome organization and chromatin dynamics.

2 Chromosome Preparation

The preparation of high-quality chromosomes is essential for reproducible in situ hybridization. Parts of the cytoplasm, cellular or cell wall material often reduce the hybridization signal and generate a high level of background noise. All chromosomes should be spread separately in a single layer so that the number and morphology can

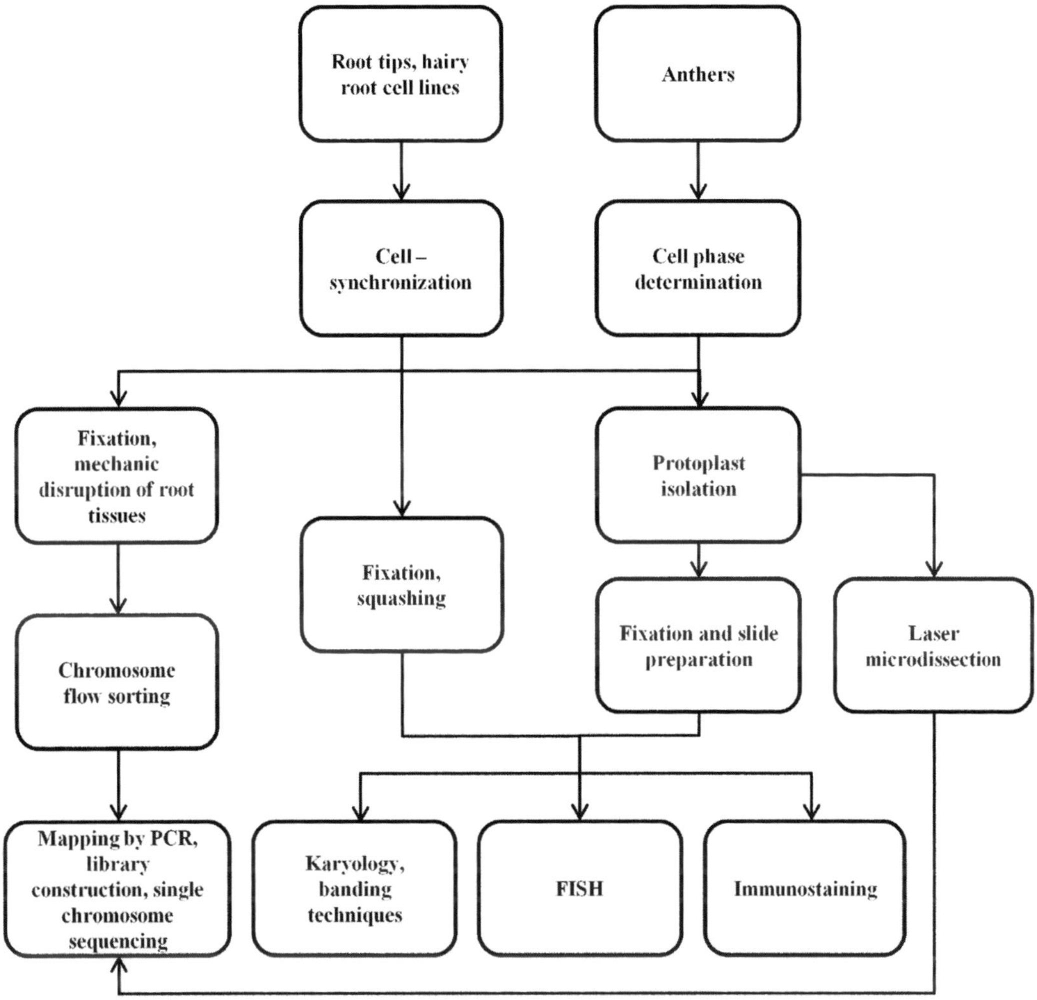

Fig. 1 Workflow diagram of common techniques for chromosome preparation and manipulation in plant cytogenetics. Each method is represented by a group of techniques, which are described in following chapters

be assessed. To achieve a high frequency of metaphases, it is necessary to accumulate dividing cells in a synchronous manner. In other words, the synchronization of plant material is crucial for metaphase chromosome preparation. In this chapter, a number of methods which provide valid results for subsequent in situ experiments are described.

2.1 Cell Cycle Stage Synchronization

In plant cytogenetics, chromosomes are normally obtained from root tips of living plants, small seedlings or hairy root cell lines. The number of well-spread metaphases is limited by the low frequency of synchronously dividing cells in a given root tip meristem. A variety of methods have been used to increase the rate of cell synchronization and the mitotic index in plants, using a wide range of compounds such as hydroxyurea (HU), mimosine, aphidicolin, and anti-tubulin substances like the alkaloid colchicine and, amiprophos-methyl (APM), nitrous oxide gas, benzamide designated RH-4032, and a novel phenylcyclohexene colchicine mimic RH-9472 [2, 3].

The synchronization itself is critically dependent on the state of dividing cells (the age of the meristem). Most of the methods are based on the arrest of meristematic cells in the G1/S phase using DNA synthesis inhibitors (e.g., HU) with minimal effect on cell viability. Treatment periods are generally longer than the length of the cell cycle in order to accumulate the majority of dividing cells. The time necessary for synchronization is species-dependent. Treatment prolongation leads to excessive cell death, nutrient starvation and causes chromosome breakages that can be accompanied by irreversible sister chromatid exchanges [4]. After removal of synthesis inhibitors, the arrested cells resume cycling in a synchronous manner.

The accumulation of cells in metaphase is achieved by anti-tubulin drugs that disrupt the mitotic spindle apparatus. The most common anti-tubulin drug is colchicine, although other chemically diverse alternatives are successfully used (e.g., dinitroanilines oryzalin and trifluralin, the phosphorothioamidate amiprophos-methyl, the benzamides pronamide and chlorpropham). In order to prepare longer (relaxed) pro-metaphase chromosomes with less condensed chromatin, a shorter incubation time and/or lower concentration of drugs is recommended. On the other hand, longer treatment often leads to chromosome decondensation and chromosome splitting into chromatids [5]. The length of the treatment correlates with the percentage of arrested cells in metaphase. Disruption of the spindle apparatus can also be substituted by cold treatment that leads to less condensed chromosomes but to a lower metaphase index. After chemical synchronization subsequent cold treatment is also recommended to increase the metaphase index and minimize the formation of "ball metaphases" leading to better separation of single chromosomes [6].

2.2 Chromosome Preparation Techniques

The release of intact plant chromosomes is a basic prerequisite for labeling using in situ hybridization. The preparation and staining of spread nuclei or intact metaphase chromosomes from plant cells is more challenging than from animal cells. This is largely due to their cell wall that constitutes an efficient barrier against the penetration of chemical reagents, dyes, DNA probes and antibodies. Plant tissues therefore need to be treated with a mix of cell wall digesting enzymes, typically comprising cellulose and pectin hydrolases among others. Great attention should be given to the calibration and to the effectivity of a new enzyme mix. It is critical to test each batch of enzymes first, because the activity of the crude or purified extracts varies between lots. There is no general protocol for chromosome preparation and a number of limiting factors should be considered, such as the amount of plant material and the cell phase or cell event to be studied. Plant materials can be highly variable and in particular the amounts of secondary metabolites in the cell wall differ, often as a result of differences in ploidy level. These metabolites hamper the treatment of plant tissues, leading to unsuccessful chromosome preparation. Some chromosome preparations therefore require additional permeabilization and prehybridization treatments, for example with detergents, proteases, or HCl treatment. Hence, optimization of the preparation procedure and the enzyme mix which lead to perfect digestion and chromosome spread is often required. Here, we describe common methods for chromosome preparation based on (1) squashing of intact tissues, or (2) dropping of digested tissues, or (3) drop-spreading of protoplast suspension.

1. In the squashing method, mitotic chromosomes with well-defined morphology are released and spread onto a microscopic slide into a single layer by gentle pressure on the digested tissue. This method is based on squeezing a small root tip cap (Fig. 2a, b) or fixed anthers (for meiotic chromosomes). Classically, this preparation was used in combination with acetocarmine staining to analyse the number and shape of metaphase chromosomes. Cells are spread in a drop of acetic

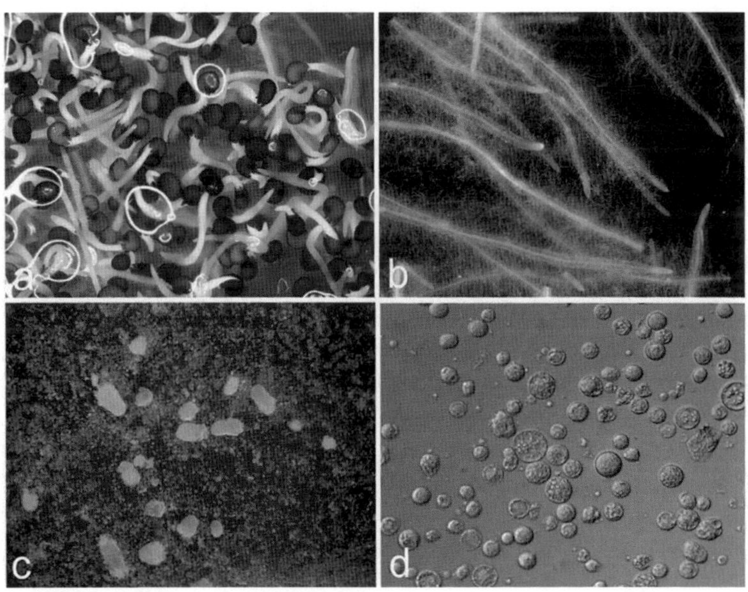

Fig. 2 Preparation of plant protoplasts for karyological analysis (*Silene latifolia*). (**a**) 2-day-old seedlings before protoplasting. (**b**) A hairy root culture induced by *Agrobacterium rhizogenes*. (**c**) Maceration of root tips in an enzyme mixture. (**d**) Released protoplasts before fixation

acid by gentle tapping on the cover slip or directly on the roots. A heat treatment enables clearing of the preparation from most of the cell wall and cytoplasm prior to freezing in liquid nitrogen or on dry ice. If any cytoplasm remains, subsequent washing steps in a series of acetic acid (low or medium concentration) can help, while preserving the chromosomes. If the chromosomes cannot be obtained by simple squashing due to the thickness or rigidity of the tissue, partial maceration of cell wall by digestion in an enzyme mix (cellulase, pectinase and cytohelicase) results in a satisfying preparation. Note that the mix may differ between plant materials, between different preparations, or after reuse. Using enzyme preparation, undesired nonmeristematic tissue is cut off and only meristem tips are transferred to the enzyme mix cocktail and digested, usually until they become soft (Fig. 2c). Following washes, root tips are fixed again in acetic acid and squashed. Chromosomes prepared by squashing are suitable for most experiments including single copy FISH. The advantage of this technique is that chromosomes can be prepared from cells at different cell cycle stages including meiotic chromosomes (pollen mother cells in anthers are highly synchronized; the appropriate bud/meiotic phase must be first determined). The squashing method does not require costly equipment, usually gives high quality metaphase spreads, and may be the only way of spreading large chromosomes. For additional information, several protocols are available [7, 8].

2. In situ experiments often use the dropping method also called air-drying technique or drop-spreading. This procedure was adapted for plants in 1994 [9], and later modified [10, 11]. The method relies on enzymatic digestion and disintegration of fixed plant meristematic tissue (Fig. 2a, b). After enzymatic digestion, the roots are washed in a low pH buffer and with absolute ethanol. The root tips are then transferred to freshly prepared fixative solution, disintegrated, and gently mixed. Dropping of the cell suspension can be done in high humidity conditions (e.g., humid box chamber). Chromosome spreads prepared by the dropping have well-defined morphology and structure, and need no further washing in acetic acid. Moreover, no further protease treatment is needed, as the chromosomes are completely released from the cells. Using this protocol, it is possible to obtain high quality chromosomes for reproducible FISH results on metaphase chromosomes. One caveat is that a large number of root tips are required in plant species with small meristems.

3. For the protoplast technique, protoplasts are derived from fresh tissue and therefore samples have to be processed immediately. The preparation starts after synchronization of young

seedlings (Fig. 2a) or hairy root cultures (Fig. 2b). Root tips are immersed in an enzyme solution to degrade the cell wall (cellulase, pectinase, and sorbitol (for equilibration of the osmotic value). The protoplast suspension (Fig. 2d) is filtered and resuspended in a hypotonic solution to swell the protoplasts and chromosomes. For further immunostaining, only a mild fixation in formalin is applied instead of acid fixatives [12]. For FISH, a protease or HCl treatment is additionally required to clear remaining cytoplasm, which produces background. For metaphase chromosome spread preparation, however, a considerable amount of starting plant material is needed to obtain protoplasts in a dense suspension. Further, protoplasts can be used also for preparing of interphase nuclei if they are cultivated in appropriate conditions [13]. Yet, upon appropriate scaling, the advantage of this method is a large number of well-spread metaphase plates as well as a large number of preparations for further experiments. On average, 40–50% of metaphases can be obtained if a standard protocol for synchronization as described above is followed.

3 Chromosome Manipulation: Flow Sorting and Laser Microdissection

Laser microdissection and flow sorting are powerful tools with a wide range of applications in cytogenetics and genomics. In particular, they simplify the analyses of large genomes by reducing sample complexity upon sequencing.

Although the chromosome sorting procedure is more than 40 years old, it has experienced a renewed interest due to high-throughput, whole-genome sequencing projects in recent years. Flow sorting is, in most aspects, an alternative to laser microdissection. Since only a minority of chromosomes can be distinguished based on size, chromosome-sorting has been long limited to downstream applications. Recently, two parallel methods have been developed to enhance discrimination of individual chromosomes. First, translocation deletion or alien addition chromosome lines are used as input material for sorting [14]. Second, specific DNA sequences using fluorescence in situ hybridization in solution (FISHIS) are used for proper identification of individual chromosomes during the sorting procedure [15]. Before flow sorting, chromosomes are released from cell populations highly enriched for metaphase cells, into a suitable isolation buffer. Chromosomes are categorized according to fluorescence intensity (relative DNA content) by DNA-specific fluorochromes or chromosome specific FISH probes (in the case of FISHIS). In contrast to microdissection, the analysis can be carried out in a short time with a large sample size and results in representative distributions of chromosomal DNA content termed flow karyotypes. The most widely used method is the preparation of chromosome samples from root tip meristems of young seedlings [5]. Contrary to cell cultures or artificial "hairy" root cultures, meristems are karyologically stable and meristem cells are easy to synchronize as described in Subheading 2.1.

The applications of flow sorted chromosomes are expanding along with advances in genomics. Historically, flow-sorted chromosomes were first used for physical mapping using DNA hybridization [16] and PCR. Later, flow sorted chromosomes were employed for small and large insert DNA (YAC, BAC) library construction [17]. Although separated chromosomes are routinely used for chromosome painting in animal models, chromosome painting using whole chromosome derived probes does not work in plants due to large numbers of various dispersed repeats in their genomes [18]. Combination of chromosome sorting with high-throughput next generation sequencing (NGS) technologies now facilitates the reduction of genome complexity in ongoing research in selected plant species [19]. This strategy also enables direct focus on genome and epigenome organization at the single chromosome level.

Laser microdissection, similar to flow sorting, allows for isolation of cells, subcellular compartments and chromosomes which can then be followed by DNA, RNA, and/or protein isolation. Early studies on isolation of a single chromosome from metaphase spreads used a simple mechanical micromanipulator [20]. Fully automated systems with laser-based manipulation were developed later. The main advantages of micromanipulation techniques are visual identification and control of a chromosome of interest (Fig. 3). Recent efforts to reduce the size of the microdissected

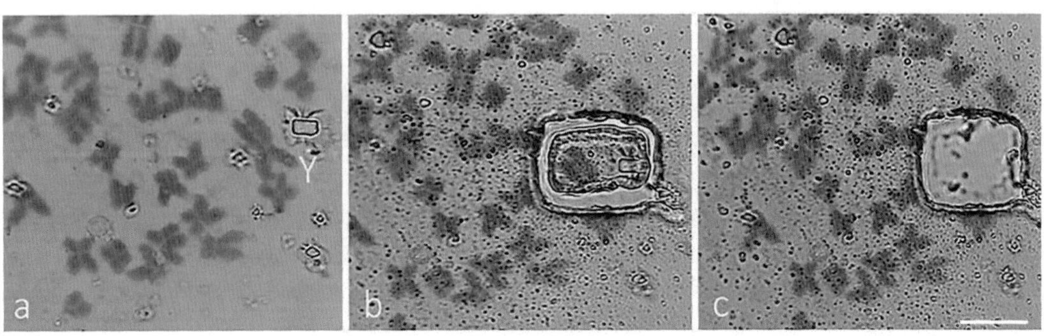

Fig. 3 Laser microdissection on *S. latifolia* metaphase chromosomes. Protoplast suspension was dropped on a polyethylene naphthalate membrane and stained with Giemsa. Chromosome spreads with sex chromosomes were investigated under an inverted microscope. (**a**) An example of selected Y chromosome. (**b**) The membrane was cut around the Y chromosome, and (**c**) the selected chromosome was catapulted by a single laser pulse into the cap of a PCR tube. The scale bar 10 μm

area led to the development of micromanipulators based on atomic force microscope nanolithography, enabling dissection of fragments as small as 0.4 μm [21]. In plants, laser microdissection at the chromosomal level was used mainly in the case of clearly distinguishable chromosomes (morphologically) such as sex chromosomes in spinach [22], in wild hop [23], rye B chromosomes [24], telo-chromosomes in wheat [25], and addition lines [26]. In addition to flow sorting, microdissection allows to separate subchromosomal regions such as individual chromosomal arms [12]. Since microdissection yields limited amounts of DNA, general use of this technique in molecular biology has depended on development of DNA amplification methods such as PCR [27]. There are three frequently used techniques for amplifying DNA from microdissected chromosomes: (a) adaptor mediated PCR (LA-PCR), (b) degenerated oligonuleotide-primed-PCR (DOP-PCR), and (c) whole-genome amplification using *Phi*29 polymerase (reviewed in [12]). Similar to flow sorting, key steps in chromosome microdissection are accumulation of a large number of metaphase chromosomes by synchronization of cell division and gentle fixation of chromosome material to avoid DNA damage. Laser microdissection also requires homogenous spread of chromosomes on specific synthetic inert membranes and permanent microscopic control of sample integrity and purity during the procedure (Fig. 3b, c). Applications of microdissected chromosomes are identical to those of flow sorted samples.

4 Fluorescence In Situ Hybridization

FISH (fluorescent in situ hybridization) is a very straightforward technique that involves hybridization of DNA molecules (probes) to their complementary sequences on chromosomal preparations. Probes of varying lengths are labeled directly by incorporation of fluorescent nucleotides or by incorporation of reporter molecules which are detected subsequently by antibodies or other affinity molecules attached to fluorochromes. Probes and targets are then visualized in situ by fluorescent or confocal microscopy. As a combined molecular and cytological approach, FISH enables to correlate DNA sequence with the structure and organization of plant nuclei and chromosomes, while retaining information at the single-cell level. Using various modifications it is possible to map the physical location of DNA sequences within the genome, to correlate linkage groups to specific chromosomes and to understand genome organization and the three-dimensional distribution of DNA sequences in interphase and meiosis. Compared to other techniques studying the structure of cells/nuclei, the advantage of FISH lies in longitudinal resolution (according to chromosome preparation), contrast, speed, and safety as well as multitarget localization [28]. Although this chapter is intended to describe various FISH techniques and methods, it is impossible to cover every modification. Therefore, the most used variants of FISH are listed below together with their recent applications.

4.1 FISH Detection of DNA Repeats

Plant genomes contain a large fraction of highly repetitive elements with an apparent random distribution across chromosomes. The sequences are generally divided into two classes: (1) tandem repeat units including ribosomal genes, telomere/subtelomere repeats or large blocks of satellites consisting of 160–180 bp and 320–360 bp long units, and (2) DNA transposable elements and retroelements with dispersed-like patterns [29]. Both classes are variable and their rapid evolution has led to changes in their abundance and repeat distribution. Owing to a high copy number and interspecies variability, the repetitive sequences may be very informative and serve as probes with robust signals for FISH, providing unique information about repetitive sequence organization that can hardly be obtained by other methods, e.g., repeat diversification [30] and accumulation of specific repeats in chromosomal domains [31]. FISH detection of DNA repeats can also be used to characterize the regulation of chromocenters [32].

4.1.1 Analysis of Transposable Elements

Transposable elements (TEs) can form up to 90% of the genome and their distribution is rather random [33]. However, some TEs are enriched in specific chromosomal locations such as Ty3-gypsy-like retroelements in the centromeres of *Beta* genus [34] or CRM clade retroelements which represent an active component of the centromeres in a wide range of angiosperm species [35]. A unique pattern of repeat distribution has also been found on the Y chromosome of *Silene latifolia* (Fig. 4a, b), where some elements like *Ogre* and *Copia* are accumulated, while others are absent [36]. Specific accumulation of *Athila* retroelements on the other hand, occurs predominantly on the X chromosome of *S. latifolia* [37]. These data show that the distribution and evolutionary dynamics of various repetitive elements differs [38].

The accumulation of TEs is influenced by chromatic features which may affect activation or reactivation of specific mobile elements. Heterochromatin is defined as a largely gene poor, epigenetically inactivated region, where some TEs appear to insert preferentially, e.g., the LTR retrotransposon *Athila* in *Arabidopsis.* TEs of different classes are often also found together in nuclear domains forming larger blocks of heterochromatin (reviewed in [39]). The cytogenetic analysis of chromosome arm 4S from *Arabidopsis* revealed a heterochromatic knob that may have been derived from an inversion event of pericentromeric sequences to an interstitial position. Interestingly, using probes to detect this knob, it was found that *Arabidopsis* ecotypes show polymorphisms for the presence of hk4S (one of the five regions located on the

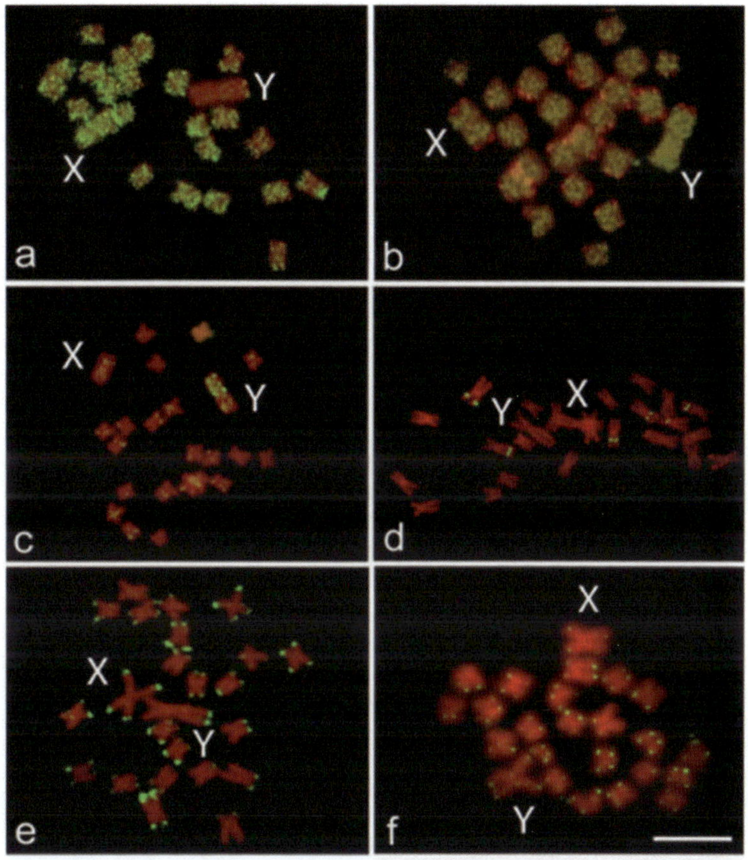

Fig. 4 Examples of fluorescence in situ hybridization on *S. latifolia* metaphase chromosomes (**a**–**f**). (**a**) Distribution of Ogre retroelements on the male metaphase (the missing signal on the Y). (**b**) Accumulation of Copia retroelements on the male metaphase. (**c**) Distribution of (CA)n microsatellite motif (the accumulation on the Y). (**d**) Hybridization with 5S-rRNA probe yields three signals per haploid genome. (**e**) Distribution of X.43.1 tandem subtelomeric repeat. (**f**) Telomeric repeat motif hybridizes on all chromosome ends. Bar indicates 5 μm (**a**–**f**)

short arm of chromosome 4) and that the knob has a different degree of chromatin condensation during cell division [40]. Another probe set for studying chromocenters in *Arabidopsis* revealed different levels of condensation of DNA repeats between leaf protoplasts, cultured cells and differentiated cells [13]. These studies indicate various structural dynamics in heterochromatin as well as different sequence composition in heterochromatic domains.

4.1.2 Tandem DNA Repeats

Tandem DNA repeats can be divided into rRNA repeat arrays, satellite, minisatellite, and microsatellite sequences [29]. All four classes are built from thousands of copied sequences in a tandem head-to-tail fashion. They are fast evolving and the monomer sequences are highly conserved [41]. As they are fast evolving, their distribution may be species specific (Fig. 4c). They are accumulated in centromeric, subtelomeric and telomeric regions or in regions with supressed recombination.

Due to the sequence organization and prominent signal bands, the repeat array sequences that are used in most FISH experiments are rRNA genes. rRNA include multigene families consisting of 18S–5.8S–25S rRNA, separated by two internal transcribed spacers (ITS) forming the 45S unit. The small subunits together with 5S are highly conserved even between distant species. Organization in long tandem arrays and high copy number at telomeric and subtelomeric regions correspond to heterochromatin and their position can provide useful information (Fig. 4d).

Other satellites form domains that are conserved between closely related species. Examples are the subtelomeric tandem repeated unit X-43.1 and the telomeric motif, which have similar distribution patterns in *S. latifolia* (Fig. 4e, f) as well as in other *Silene* species [36]. Similar results were found for centromeric satellites in the genus *Beta* [34]. Some of these satellites are similar within relatives, e.g., centromeric sequences from *Oryza sativa* can be used to study centromeres in other *Graminae* species using higher resolution chromosomal features. A number of centromeric sequences and anchored proteins have been described to date [42]. The use of repetitive sequences is important to fully understand the function and control of the centromere [43]. Furthermore, satellites play an indisputable role in the evolution of the centromere structure [44].

The accumulation of satellite sequences and their homology between related species in chromosomal domains make them an ideal tool to study specific events such as genomic turnover or chromosome aberration. The use of a mixture of different repetitive probes has been used for identification of individual chromosomes in a number of plant species, among which *Arabidopsis* [45], *Brassica* [46], maize [47], *Silene* species [48], *Pinus* [49], Norwegian spruce [50], and hemp [51].

4.2 BAC (Bacterial Artificial Chromosome) FISH

As the distribution of repetitive DNA sequences from the majority of plant genomes is dispersed, the identification of particular chromosomes is difficult. The use of probes derived from bacterial artificial chromosomes (BACs) is therefore usually preferred if the aim is to visualize a specific chromosome. BAC clones are typically 50–100 kb in size and are routinely used as FISH probes in species with small genomes and/or a limited number of repeat elements. However, when used for species with large genomes, the probes derived from individual BACs usually yield nonspecific cross-hybridization (Fig. 5a, b).

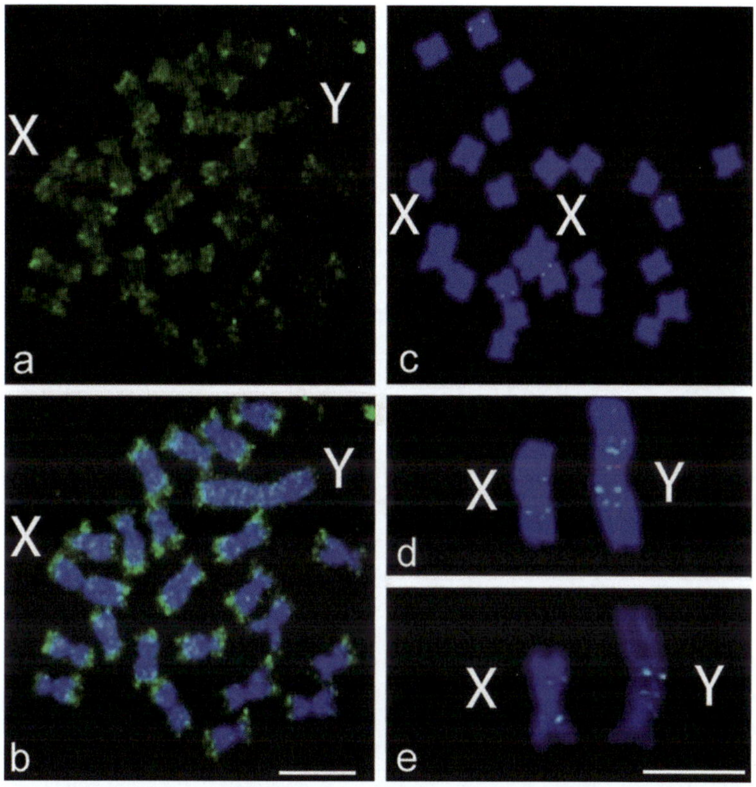

Fig. 5 FISH pattern of different BAC clones on *S. latifolia* chromosomes. (**a**) The pattern of a nonsupressed BAC probe (106D13) enriched for telomeric sequences labeled by fluorescence green (FITC) and (**b**) merged image of the nonspecific hybridization and stained with DAPI (4',6-diamidino-2-phenylindole). (**c**) Specific hybridization signal on chromosomes of *S. latifolia* using BAC clone 9B7 (*green*). (**d**) Comparative localization of BAC 9B7 (*green*) and BAC 7H5 (*red*). (**e**) Comparative localization of reference BAC 9B7 (*green*) and BAC 6B3 (*red*). The bars indicate 10 μm

The BAC FISH strategy is used for chromosome painting (CP) and chromosome comparative painting (CCP). CCP is based on cross-species hybridization of painting probes and therefore allows identification of homologous chromosomes and homologous regions shared among related species. CP and CPP are commonly used in crucifers (*Brassicaceae*), such as *A. thaliana* and relatives [52]. In crucifers combination of both approaches led to the identification and construction of an ancestral karyotype [53]. The usefulness of comparative cytogenetics was also demonstrated in *Orychophragmus* and *Zillineae* species by the detection of chromosome duplication and other chromosomal rearrangements that occurred after speciation [54]. CP and CPP may also be used to elucidate evolutionary mechanisms underlying the extant karyotypic variation [55, 56]. The use of CP in *Arabidopsis* revealed that chromosomes are arranged in a random fashion in differentiated cells [57]. However, CP and CPP rely on the availability of chromosome-specific BAC libraries which are arranged to cover whole chromosome(s) and small amounts of repetitive elements in the donor and target genomes as in *Arabidopsis*.

Also in other small-genome species, physical mapping of DNA sequences on chromosomes is possible using BAC derived probes. Pooled BACs are the most direct method for use, e.g., in rice [58] and stiff brome [59]. In plants possessing large genomes, a limited number of BAC clones yields a unique signal (Fig. 5c–e). Such BACs can serve as chromosome-specific cytogenetic DNA markers (CSCDM), which can be developed for individual chromosomes. A mixture of CSCDM markers can generate specific FISH signals that allow for identification of all chromosomes as demonstrated in sorghum [60]. This method is rather expensive if the number of analyzed BACs starts to be extensive. However, chromosome specificity can be increased by blocking dispersed repeats. The strategy using unlabeled total genomic DNA or DNA enriched for repetitive fractions (also called c_0t fractions) is known as competitive in situ suppression (CISS) hybridization. The CISS technique is based on isolation of c_0t-1 blocking DNA. After purification, the c_0t fraction is enriched for high or moderate copy number sequences and hence is more effective than total genomic DNA for prehybridizing and suppressing nonspecific signals [61]. CIS suppression has been successfully used, e.g., in maize, to map the short arm of chromosome 9 using a cosmid derived probe on pachytene chromosomes [62].

4.3 Genomic In Situ Hybridization

Genomic in situ hybridization (GISH) enables the analysis of parental genomes in a hybrid plant by application of the whole genomic DNA of individual parents [63]. Since 1989, GISH has become a powerful tool for comparison of interspecific and intergeneric hybrids and allopolyploid species as well as to study introgression, addition, and substitution lines. The principle of this technique is based on genome-specific dispersed repetitive sequences [28]. As the dispersed repeats evolve faster than genes, they enable differentiation of chromosomes from closely related species. Parental genomes can be distinguished using parental genomic DNA (gDNA) as a probe in cases where the dispersed repeats in each parental genome are highly diverged prior to interspecific hybridization or when the hybrid or allopolyploid is relatively recently formed. An important advantage of the GISH method is its ability to compare whole species genomes together. To discriminate between the two parental genomes, blocking c_0t-1 DNA can effectively prevent in situ probe hybridizing to common dispersed repetitive sequences and thereby enable distinction between parental chromosomes from more closely related species. Figure 6a demonstrates the use of c_0t-1 DNA in distinguishing parental genomes in a hybrid of *S. latifolia* and *S. viscosa*.

The GISH technique can also be used in phylogenetic studies. The combination of GISH and FISH with tandem repeat probes was applied to study the evolution of *Nicotiana* polyploid species [30]. Another development in the field of GISH phylogenetics is the measurement of differences in the intensity of hybridization signals using different gDNA probes of related species on one or several target species and hybrids [64].

4.4 Single- and Low-Copy FISH

For most plant species, the order and spatial localization of the genes on the chromosomes are based on genetic linkage maps. However, linkage map distances are not proportional to physical distances. In regions where recombination is suppressed (interstitial/proximal regions), the exact map locations of genetic markers and their relative positions cannot be unequivocally determined in most cases. This has been shown in barley [8] and tomato [65, 66]. For species with a large fraction of repeat elements, precise localization of target genes is very difficult due to problematic design of a gene/locus-specific probe which does not encompass repeat elements. Small DNA targets can solve this problem, but they are also often hard to detect using standard labeling systems (Fig. 6b). Four basic parameters are described in the text below that influence successful low copy gene localization: (1) type of chromosome preparation, (2) probe labeling system, (3) the length of the probe, and (4) the signal detection system.

1. For low copy FISH and small size probes, chromosomes should be intact with a well-defined morphology and clean from any debris which can interfere with probe penetration. Although the digestion of chromosomes with protease before FISH can be used to remove some of the contaminating

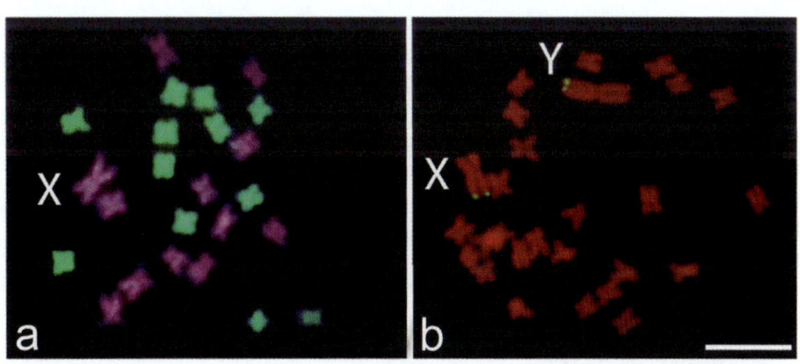

Fig. 6 Fluorescence in situ hybridization on *Silene* metaphase chromosomes. (**a**) Genomic in situ hybridization on an interspecific hybrid between the dioecious *S. latifolia* and hermaphrodite *S. viscosa*: the *red*-stained probe was the genomic DNA of *S. latifolia*, while the *green* probe was the genomic DNA of *S. viscosa*. (**b**) The localization of the gene DD44 gene on the sex chromosomes of the male genome of *S. latifolia*. The bar indicates 10 μm

nucleoplasm, the protease treatment also results in changes in chromosome structure. For this reason, in single copy in situ hybridization, the use of proteases should be avoided or used with caution. It was shown that some chromosomal DNA could be lost during in situ hybridization procedures resulting in low signal detection [67].

2. Essential in single copy FISH is also the probe preparation and the efficiency of incorporation of labeled nucleotides in the probe. The incorporation of different fluorochromes and brightness of probes were studied using various reaction parameters in direct nick translation [10]. This revealed that labeled-dNTPs can interfere with DNA polymerase I activity and that adding excess polymerase I can increase the efficiency of detecting probe signal. Another parameter that influences the brightness of the signal is the fluorescence quantum yield (QY). Fluorescence QY is the ratio of the number of fluorescence photons emitted to the number of photons adsorbed. The value is a measure of the relative extent to which these processes occur. The QY of most commonly used fluorochromes are described by their providers and a summary comparative list of available fluorochromes can be found online [68]. The highest QY is shown by Alexa Fluor 594 and Texas Red compared to FITC, Cy3, Cy5, or others. Texas Red was used successfully, e.g., in direct nick translation systems to localize targets as small as 2–3 kb in barley [8, 69].

3. A few approaches have been developed that allow the amplification of the signal from a small probe, such as rolling circle amplification [70] or click reaction [71] (*see* Subheading 4.7). One of these approaches is Tyr-FISH or tyramide amplification FISH. The Tyr-FISH uses horseradish peroxidase activity which catalyzes deposition of fluorophores directly adjacent to the immobilized enzyme. The Tyr-FISH involves (a) in situ hybridization, (b) signal amplification by streptavidin/antifluorochrome antigen-horseradish peroxidase, and (c) detection and imaging of the signal. The sensitivity of Tyr-FISH depends on the length of the probe or target sequence. This method is up to 100 times more sensitive than other existing techniques. The Tyr-FISH was successfully introduced into FISH technology in [17] and later modified in plants where targets of only 700 bp were localized on metaphase chromosomes of ***Allium cepa*** [72]. Although this technique is very promising, the Tyr-FISH has been used only in a few plant species such in rosa [73], wheat [74], oat [75], and barley [76]. The reason for neglecting the use of Tyr-FISH may lie in the difficulty to reach a satisfactory signal to noise ratio.

4. Significant improvements in single copy localization were reported using a cooled-charge-coupled device (CCD) camera.

Using CCD, the detection sensitivity can be increased 30-fold due to the integration of digital computer images and the intensity of fluorescent signals [77]. The detection limit is then not restricted by probe size, but by the signal that can be detected above the background noise (reviewed in [78]).

4.5 Fiber FISH

Fiber FISH is a FISH variant carried out on isolated and extended chromatin fibers. Fiber FISH thus enables the assessment of the effect of differences in probe length and the mapping of different probes relative to one another. With a resolution of ca 2.5–3.5 kb/μm, it is the only FISH variant suitable for high resolution physical mapping [79, 80]. The technique involves (a) isolation of interphase nuclei from leaves, (b) filtration of the cell lysate on nylon mesh membranes, (c) displacement of the isolation buffer by a storage buffer, (d) extension of the DNA fibers on microscope slides covered, e.g., by poly-L-lysine, and (e) in situ hybridization. The length of the extended DNA fibers is dependent on the stretching technique (fiber preparation) and nuclei isolation method. Longer fibers (in average 3 Mb long) can be obtained using mild stretching conditions [81] or using of nuclei in S phase. Although the fiber-FISH allows detection of small targets <1 kb, it may be difficult to distinguish a short fiber-FISH signal from background noise and therefore adjacent, longer reference probes should be used. Fiber FISH is therefore used in studies to detect genome rearrangements such as insertion or duplication in pericentromeric regions [82], organization of repetitive sequences in rice [83], in *Beta* species [34], and in *Arabidopsis* [80], DNA methylation in cotton [81], BAC positions in *Arabidopsis* [84], chloroplast DNA structure in higher plants [85], and transgenic DNA loci [86].

4.6 Oligo-Painting

Consistent identification of individual chromosomes relies mostly on the use of repetitive DNA (*see* Subheading 4.1) or large genomic DNA clones (*see* Subheading 4.2), and it is the basic prerequisite for efficient cytogenetic mapping. A new strategy for chromosome specific painting has been recently described [87]. This method allows either chromosome identification (both mitotic and meiotic chromosomes) or mapping of homologous chromosomes in related species [88]. Oligo-painting is based on the selection of 48-bp long oligonucleotides, which are nonoverlapping and unique to the target chromosome or to a chromosomal region. First, homopolymer sequences and repetitive motives are filtered out from the oligonucleotides in silico. After synthesis, the oligos are labeled via an RNA-intermediate and reverse transcription, using primers attached that contain a biotin or digoxigenin tag [89]. The RNA is then hydrolysed and the oligo-probes consisting of thousands of oligonucleotides are hybridized to chromosome spreads.

The superiority of this technique lies in its resolution and versatility. Compared to others (BACs and FISH detection of repetitive sequences), bulked oligo-probes can be designed to cover loci, a chromosome part, arm and even whole chromosome (s) using about 3–8 oligos per kb on metaphase chromosomes. Moreover, oligos from closely related species can be designed to have equal sequence similarity and thus produce similar FISH signal intensity. The probe design can therefore be customized for the goals of specific research projects. The limiting factor for developing bulked oligo-probes is the requirement of a reference genome from which sequences can be filtered out. If a reference genome is not available, probes can be designed on sequences of related species. Another disadvantage is the price of the oligo pool. According to [88], the price for the creation of a bulked oligo pool is about $ 1.500. On the other hand, synthesized oligo-libraries provide enough DNA template for more than one million FISH experiments.

4.7 FISH Resolution

Cytogenetic maps provide an efficient tool for sequence localization, validation of contig order and genes distances or characterization of regions that are subsequently analysed using next generation sequencing methods. The gene order and relative distance of a target gene and its proximal or distal distance to the centromere is then defined as longitudinal resolution. The ability to distinguish physical gene order (longitudinal resolution) differs for different chromosome preparation methods (e.g., protoplast technique/squashing or fiber preparation) as well as between chromosome types (meiotic or mitotic).

Conventional mitotic metaphase chromosome FISH (M-FISH) provides readily available material with a limited longitudinal resolution of about 2–10 Mb. Prometaphase chromosomes offer a longitudinal resolution of ca. 2 Mb. The highest resolution in somatic cells is provided by interphase nuclei (50–100 kb). Despite the lowest longitudinal resolution, M-FISH is still used in many studies for sequence mapping due the availability of metaphases, which are relatively easy to prepare.

The limitations of low resolution can be overcome by FISH on meiotic chromosomes. Pachytene chromosomes are longer than mitotic chromosome spreads by a factor 6–25 (reviewed in [78]). The advantage of meiotic preparations is their abundance in reproductive tissue/cells and the meiotic chromosomes are typically synchronized when isolated from pollen mother cells. The disadvantage is that they are not as abundantly available as somatic chromosomes.

An alternative for pachytene chromosomes are super-stretched chromosomes (70 kb longitudinal resolution) digested with proteinase K [90]. Stretched chromosomes offer fine FISH mapping in plant species where pachytene chromosomes are not suitable.

This technique requires chromosome-sorting analysis and is useful only for species in which all chromosomes can be distinguished. However, the highest longitudinal resolution is provided by fiber FISH.

Because the abundance of fluorochromes per site generally decreases when the resolution increases, the labeling system is also essential for a successful high-resolution experiment. The probes are usually labeled by a variety of methods which incorporate/add fluorochromes directly or incorporate/add a reporter molecule which is detected by addition of antibodies, such as biotin/streptavidin, digoxigenin/anti-digoxigenin, or specific antibodies such as for Tyr-FISH (*see* Subheading 4.4). A broad range of techniques is available to label the probes with a fluorochrome or reporter molecule, including nick translation, random priming, PCR, end labeling, and tailing. Each of these methods has its own limitations and label efficiencies [10]. Labeling efficiency is mostly affected by inhibition of already incorporated modified nucleotides in the probe sequence as demonstrated for Cy3-dUTP and Cy5-dUTP. Probes with an average length of 100 bp usually obtain 40–50 fluorochromes/kb, whereas probes with a length of 150–500 bp have a lower labeling efficiency with only 10–20 fluorochromes/kb [91]. Therefore, the use of the probe, the length of target sequence, and use of reporter molecules should be considered.

New approaches such as labeling through the copper-catalysed azide-alyne cycloaddition reaction (CuAAC) [26] or rolling-circle amplification of padlock probes [70] may considerably enhance the possibility to obtain sufficient signal with small probes at high resolution. CuAAC was already successfully applied to label long PCR products which were hybridized on plant chromosomes [71]. CuAAC probes can be used in immunohistochemistry and have the advantage that possible steric perturbation typical for short, heavily labeled probes is avoided. On the other hand, in situ detection of DNA molecules by target-primed rolling-circle amplification of padlock probes provides highly specific detection with minimal background and the entire reaction results in strong, discrete detection signals anchored to target sequences [70, 92].

Because, the list of modification which exists for each mentioned techniques is too broad, only some important factors affecting resolution were listed above. Other factors such as accessibility, DNA loss, and in situ renaturation of DNA target probe sequences are equally important as well as hybridization buffer and should be carefully checked during each experiment.

5 Immunostaining of Mitotic Chromosomes

Immunostaining techniques utilize polyclonal or monoclonal antibodies to recognize molecules of interest by a primary antibody.

The secondary antibody is usually conjugated with a fluorophore and detected by fluorescence microscopy. Hundreds of specific antibodies are commercially available enabling detection of corresponding proteins and other specific substances in both animal and plant samples. Here, we summarize some approaches useful in studies on the chromatin structure and function.

5.1 Detection of Modified Histones

Chemical modifications of nucleosomal histones influence chromatin properties, including their transcriptional competence. Histones undergo numerous posttranslational modifications such as acetylation, phosphorylation, methylation, poly-ADP ribosylation, and ubiquitinylation (Fig. 7). The targets for chemical modifications are especially additional amino groups of positively charged amino-acids (e.g., lysine and arginine) in N-terminal domains of H4, H3, H2A, and H2B. In some cases, a histone modification, like methylation, has a site-specific effect, e.g., methylation of lysines in positions 4 or 36 of histone H3 leads to gene activation, while methylation of lysines in positions 9 or 27 are typical markers for heterochromatin. Many transcriptional activators carry histone acetyltransferase activity, while corepressors possess histone deacetylase activity, confirming the importance of histone acetylation in transcription [93].

5.2 Detection of 5-Methylcytosine

DNA methylation is a very important epigenetic process controlling both structure and function. It is directed by DNA methyltransferases and often reflected by silencing of corresponding genes, transposable elements, and other repetitive sequences. The silencing is realized by binding of specific DNA methylation binding proteins. Probably the oldest cytogenetic technique used to study DNA methylation was in situ nick translation [94]. The principle is to cut DNA in chromosomes (compact chromosomes on a slide) by 5-mC (in)sensitive nucleases and to follow these nicks by polymerase reaction with labeled dNTPs. A

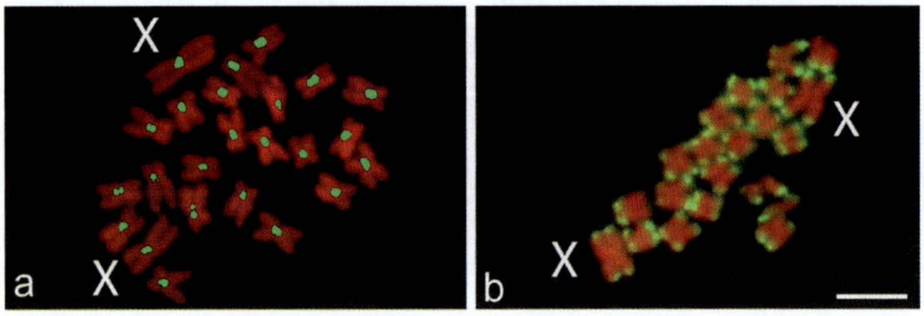

Fig. 7 Immunostaining with antisera against various forms of modified nucleosomal histones on *S. latifolia* metaphase chromosomes. (**a**) Antisera against histone H3S10ph stain only centromeres of metaphase chromosomes. (**b**) Antisera against histone H4K5ac illuminate gene-rich regions. The bar indicates 10 μm

better approach is indirect immunostaining with monoclonal antisera specific for 5-mC. These sera are used both in animal and plant research on both cytological and histological samples. In general, monoclonal antisera against 5-mC are bound to specific chromosome locations and are then visualized by labeled secondary antibodies. DNA methylation in the plant genome is common (up to 30% of cytosine residues are methylated) and, especially in the case of fluorescence, the methylation signal cannot be strictly quantified [95, 96].

5.3 Studies of DNA Replication Kinetics

DNA replication occurs in the S-phase of the cell cycle and it starts at thousands of replication origins. Replication origins of gene regions begin to replicate early in the S-phase (Fig. 8a), while heterochromatic regions replicate at the late S-phase (Fig. 8b). DNA replication kinetics is thus another epigenetic mechanism controlling gene function and silencing. Older cytogenetic experimental studies monitored DNA replication kinetics using short pulses of radioactively labeled thymidine which was detected by autoradiography. Radioactively labeled thymidine was later substituted by its analog, 5-bromo-2′-deoxyuridine (BrdU) detected by specific antibodies. The BrdU immunofluorescence signal is very clear, since there are no bromine epitopes in plant samples. Like anti-5-mC studies, the DNA epitope is studied, which enables strong fixation of samples (i.e., including acidic fixatives) allowing easier preparation of metaphase chromosomes. The problem, however, is that plant material (or dividing cells) are never perfectly synchronized. Careful microscopic analysis on abundant metaphases must be done to validly interpret the resulting data. A new thymidine derivate, 5-ethynyl-2′-deoxyuridine (EdU), has been introduced as a highly sensitive derivate permitting the detection

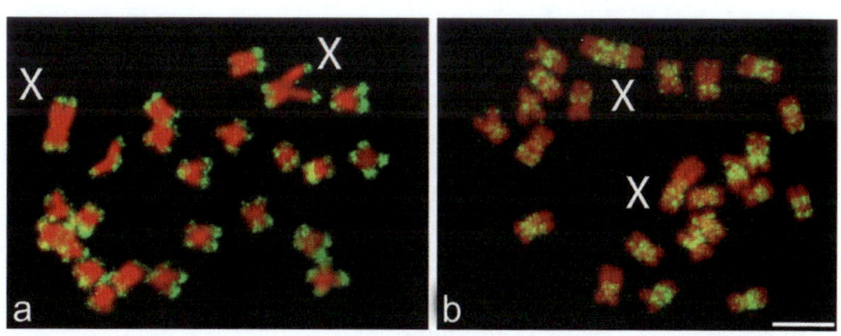

Fig. 8 Immunostaining with antisera against 5-bromo-2-deoxyuridine on *S. latifolia* metaphase chromosomes illuminates regions which incorporated BrdUrd during the S-phase. (**a**) BrdUrd applied at the beginning of the S-phase (early replicating regions are gene-rich), (**b**) BrdUrd applied at the end of the S-phase (late replicating regions largely contain heterochromatin). The bar indicates 10 μm

of DNA replication and DNA synthesis [97]. In contrast to BrdU and other derivates, samples are no longer fixed and DNA denaturation is not required, which yields in good structural preservation.

6 Conclusion

In this review, we discuss common principles of various methods for chromosome preparations in plants, FISH techniques and immunolabeling. Some of the most recent developments are described that may improve the standardly used methods for chromosome identification, gene localization, and visualization of specific chromatin features or cell events.

A variety of new sensitive methods are yearly described, which integrate plant cytogenetics with advanced techniques in high-resolution imaging, epigenetics, and genomics. Therefore, the right choice for each experiment depends mainly on the goals of the experiment (e.g., which resolution is required, what fragment size needs to be detected) and also the plant species (e.g., what material is studied, availability of reference genomes).

Acknowledgments

This research was supported by the Czech Science Foundation (grant 16-08698S). We greatly thank to Dr. Alexander Oulton for English revision of this chapter. We would like to thank to Dr. Jiri Siroky and Veronika Balounova for valuable comments.

References

1. Arabidopsis Genome Initiative (2000) Analysis of the genome sequence of the flowering plant Arabidopsis thaliana. Nature 408 6814):796-815.. doi:http://www.nature.com/nature/journal/v408/n6814/suppinfo/408796a0_S1.html
2. Cools T, Iantcheva A, Maes S, Van den Daele H, De Veylder L (2010) A replication stress-induced synchronization method for Arabidopsis thaliana root meristems. Plant J 64 (4):705–714. doi:10.1111/j.1365-313X.2010.04361.x
3. Halfmann RASD, Young DH (2007) Towards improved cell cycle synchronization and chromosome preparation. Method Cotton Sci 67:60–67
4. Andersson HC (1983) Hydroxyurea induces sister chromatid exchanges in G2: implications for the formation of chromosomal aberrations. Hereditas 98(1):61–64. doi:10.1111/j.1601-5223.1983.tb00578.x
5. Dolezel J, Cihalikova J, Lucretti S (1992) A high-yield procedure for isolation of metaphase chromosomes from root tips of Vicia faba L. Planta 188(1):93–98
6. Pan WH, Houben A, Schlegel R (1993) Highly effective cell synchronization in plant roots by hydroxyurea and amiprophos-methyl or colchicine. Genome 36(2):387–390. doi:10.1139/g93-053
7. Karafiátová M, Bartoš J, Doležel J (2016) Localization of low-copy DNA sequences on mitotic chromosomes by FISH. In: Kianian SF, Kianian PMA (eds) Plant cytogenetics: methods and protocols. Springer, New York, NY, pp 49–64. doi:10.1007/978-1-4939-3622-9_5
8. Karafiátová M, Bartoš J, Kopecký D, Ma L, Sato K, Houben A, Stein N, Doležel J (2013)

Mapping nonrecombining regions in barley using multicolor FISH. Chromosome Res 21 (8):739–751. doi:10.1007/s10577-013-9380-x
9. Martin R, Busch W, Herrmann RG, Wanner G (1994) Efficient preparation of plant chromosomes for high-resolution scanning electron microscopy. Chromosome Res 2(5):411–415. doi:10.1007/bf01552801
10. Kato A, Kato A, Albert PS, Vega JM, Kato A, Albert PS, Vega JM, Birchler JA (2006) Sensitive fluorescence in situ hybridization signal detection in maize using directly labeled probes produced by high concentration DNA polymerase nick translation. Biotech Histochem 81(2-3):71–78. doi:10.1080/10520290600643677
11. Aliyeva-Schnorr L, Ma L, Houben A (2015) A fast air-dry dropping chromosome preparation method suitable for fish in plants. J Vis Exp (106):e53470. doi:10.3791/53470
12. Hobza R, Vyskot B (2007) Laser microdissection-based analysis of plant sex chromosomes. In: Methods in cell biology, vol 82. Academic Press, New York, NY, pp 433–453. doi:10.1016/S0091-679X(06)82015-7
13. Tessadori F, Chupeau M-C, Chupeau Y, Knip M, Germann S, van Driel R, Fransz P, Gaudin V (2007) Large-scale dissociation and sequential reassembly of pericentric heterochromatin in dedifferentiated Arabidopsis cells. J Cell Sci 120(7):1200–1208. doi:10.1242/jcs.000026
14. Khlestkina EK (2014) Current applications of wheat and wheat–alien precise genetic stocks. Mol Breed 34(2):273–281. doi:10.1007/s11032-014-0049-8
15. Giorgi D, Farina A, Grosso V, Gennaro A, Ceoloni C, Lucretti S (2013) FISHIS: fluorescence in situ hybridization in suspension and chromosome flow sorting made easy. PLoS One 8(2):e57994. doi:10.1371/journal.pone.0057994
16. Lebo RV (1982) Chromosome sorting and DNA sequence localization. Cytometry 3 (3):145–154. doi:10.1002/cyto.990030302
17. Raap AK, van de Corput MPC, Vervenne RAM, van Gijlswijk RPM, Tanke HJ, Wiegant J (1995) Ultra-sensitive FISH using peroxidase-mediated deposition of biotin- or fluorochrome tyramides. Hum Mol Genet 4 (4):529–534. doi:10.1093/hmg/4.4.529
18. Schubert I, Fransz PF, Fuchs J, de Jong JH (2001) Chromosome painting in plants. Methods Cell Sci 23(1):57–69. doi:10.1023/a:1013137415093
19. Doležel J, Vrána J, Šafář J, Bartoš J, Kubaláková M, Šimková H (2012) Chromosomes in the flow to simplify genome analysis. Funct Integr Genomics 12(3):397–416. doi:10.1007/s10142-012-0293-0
20. Chambers R, Sands HC (1923) A dissection of the chromosomes in the pollen mother cells of tradescantia virginica L. J Gen Physiol 5 (6):815–819
21. Di Bucchianico S, Poma AM, Giardi MF, Di Leandro L, Valle F, Biscarini F, Botti D (2011) Atomic force microscope nanolithography on chromosomes to generate single-cell genetic probes. J Nanobiotechnol 9:27–27. doi:10.1186/1477-3155-9-27
22. Deng C-l, R-y Q, Cao Y, Gao J, S-f L, Gao W-j, Lu L-d (2013) Microdissection and painting of the Y chromosome in spinach (Spinacia oleracea). J Plant Res 126(4):549–556. doi:10.1007/s10265-013-0549-3
23. Yakovin N, Divashuk M, Yakovin N, Razumova O, Soloviev A, Karlov G (2014) Use of laser microdissection for the construction of Humulus japonicus Siebold et Zuccarini, 1846 (Cannabaceae) sex chromosome-specific DNA library and cytogenetics analysis. Compar Cytogenet 8(4):323. doi:10.3897/CompCytogen.v8i4.8473
24. Sandery MJ, Forster JW, Macadam SR, Blunden R, Jones RN, Brown SDM (1991) Isolation of a sequence common to A- and B-chromosomes of rye (Secale cereale) by microcloning. Plant Mol Biol Rep 9(1):21–30. doi:10.1007/bf02669286
25. Liu B, Segal G, Vega JM, Feldman M, Abbo S (1997) Isolation and characterization of chromosome-specific DNA sequences from a chromosome arm genomic library of common wheat. Plant J 11(5):959–965. doi:10.1046/j.1365-313X.1997.11050959.x
26. Stein N, Ponelies N, Musket T, McMullen M, Weber G (1998) Chromosome microdissection and region-specific libraries from pachytene chromosomes of maize (Zea mays L.) Plant J 13(2):281–289. doi:10.1046/j.1365-313X.1998.00033.x
27. Ludecke H-J, Senger G, Claussen U, Horsthemke B (1989) Cloning defined regions of the human genome by microdissection of banded chromosomes and enzymatic amplification. Nature 338(6213):348–350
28. Kato A, Vega JM, Han F, Lamb JC, Birchler JA (2005) Advances in plant chromosome identification and cytogenetic techniques. Curr Opin Plant Biol 8(2):148–154. doi:10.1016/j.pbi.2005.01.014

29. Schwarzacher T (2003) DNA, chromosomes, and in situ hybridization. Genome 46 (6):953–962. doi:10.1139/g03-119

30. Lim KY, Kovarik A, Matyasek R, Chase MW, Clarkson JJ, Grandbastien MA, Leitch AR (2007) Sequence of events leading to near-complete genome turnover in allopolyploid Nicotiana within five million years. New Phytol 175(4):756–763. doi:10.1111/j.1469-8137.2007.02121.x

31. Kubat Z, Hobza R, Vyskot B, Kejnovsky E (2008) Microsatellite accumulation on the Y chromosome in Silene latifolia. Genome 51 (5):350–356. doi:10.1139/G08-024

32. Baroux C, Pecinka A, Fuchs J, Schubert I, Grossniklaus U (2007) The triploid endosperm genome of arabidopsis adopts a peculiar, parental-dosage-dependent chromatin organization. Plant Cell 19(6):1782–1794. doi:10.1105/tpc.106.046235

33. Wicker T, Sabot F, Hua-Van A, Bennetzen JL, Capy P, Chalhoub B, Flavell A, Leroy P, Morgante M, Panaud O, Paux E, SanMiguel P, Schulman AH (2007) A unified classification system for eukaryotic transposable elements. Nat Rev Genet 8(12):973–982

34. Gindullis F, Desel C, Galasso I, Schmidt T (2001) The large-scale organization of the centromeric region in Beta species. Genome Res 11(2):253. doi:10.1101/gr.162301

35. Neumann P, Navrátilová A, Koblížková A, Kejnovský E, Hřibová E, Hobza R, Widmer A, Doležel J, Macas J (2011) Plant centromeric retrotransposons: a structural and cytogenetic perspective. Mob DNA 2(1):4. doi:10.1186/1759-8753-2-4

36. Cermak T, Kubat Z, Hobza R, Koblizkova A, Widmer A, Macas J, Vyskot B, Kejnovsky E (2008) Survey of repetitive sequences in Silene latifolia with respect to their distribution on sex chromosomes. Chromosome Res 16 (7):961–976. doi:10.1007/s10577-008-1254-2

37. Kralova T, Cegan R, Kubat Z, Vrana J, Vyskot B, Vogel I, Kejnovsky E, Hobza R (2014) Identification of a novel retrotransposon with sex chromosome-specific distribution in silene latifolia. Cytogenet Genome Res 143(1-3):87–95

38. Hobza R, Kubat Z, Cegan R, Jesionek W, Vyskot B, Kejnovsky E (2015) Impact of repetitive DNA on sex chromosome evolution in plants. Chromosome Res 23(3):561–570. doi:10.1007/s10577-015-9496-2

39. Bennetzen JL (2000) The many hues of plant heterochromatin. Genome Biol 1(1): reviews107.101–reviews107.104

40. Fransz PF, Armstrong S, de Jong JH, Parnell LD, van Drunen C, Dean C, Zabel P, Bisseling T, Jones GH (2000) Integrated cytogenetic map of chromosome arm 4S of A. thaliana: structural organization of heterochromatic knob and centromere region. Cell 100 (3):367–376. doi:10.1016/S0092-8674(00)80672-8

41. Mehrotra S, Goyal V (2014) Repetitive sequences in plant nuclear DNA: types, distribution, evolution and function. Genomics Proteomics Bioinformatics 12(4):164–171. doi:10.1016/j.gpb.2014.07.003

42. Houben A, Demidov D, Gernand D, Meister A, Leach CR, Schubert I (2003) Methylation of histone H3 in euchromatin of plant chromosomes depends on basic nuclear DNA content. Plant J 33(6):967–973. doi:10.1046/j.1365-313X.2003.01681.x

43. Richards EJ, Dawe RK (1998) Plant centromeres: structure and control. Curr Opin Plant Biol 1(2):130–135. doi:10.1016/S1369-5266(98)80014-9

44. Melters DP, Bradnam KR, Young HA, Telis N, May MR, Ruby JG, Sebra R, Peluso P, Eid J, Rank D, Garcia JF, DeRisi JL, Smith T, Tobias C, Ross-Ibarra J, Korf I, Chan SWL (2013) Comparative analysis of tandem repeats from hundreds of species reveals unique insights into centromere evolution. Genome Biol 14(1): R10–R10. doi:10.1186/gb-2013-14-1-r10

45. Fransz P, Armstrong S, Alonso-blanco C, Fischer TC, Torres-ruiz RA, Jones G (1998) Cytogenetics for the model system Arabidopsis thaliana. Plant J 13(6):867–876. doi:10.1046/j.1365-313X.1998.00086.x

46. Koo D-H, Plaha P, Lim YP, Hur Y, Bang J-W (2004) A high-resolution karyotype of Brassica rapa ssp. pekinensis revealed by pachytene analysis and multicolor fluorescence in situ hybridization. Theor Appl Genet 109 (7):1346–1352. doi:10.1007/s00122-004-1771-0

47. Kato A, Lamb JC, Birchler JA (2004) Chromosome painting using repetitive DNA sequences as probes for somatic chromosome identification in maize. Proc Natl Acad Sci U S A 101 (37):13554–13559. doi:10.1073/pnas.0403659101

48. Široký J, Lysák MA, Doležel J, Kejnovský E, Vyskot B (2001) Heterogeneity of rDNA distribution and genome size in Silene spp. Chromosome Res 9(5):387–393. doi:10.1023/a:1016783501674

49. Hizume M, Shibata F, Matsusaki Y, Garajova Z (2002) Chromosome identification and comparative karyotypic analyses of four Pinus

species. Theor Appl Genet 105(4):491–497. doi:10.1007/s00122-002-0975-4
50. Vischi M, Jurman I, Bianchi G, Morgante M (2003) Karyotype of Norway spruce by multicolor FISH. Theor Appl Genet 107 (4):591–597. doi:10.1007/s00122-003-1306-0
51. Divashuk MG, Alexandrov OS, Razumova OV, Kirov IV, Karlov GI (2014) Molecular cytogenetic characterization of the dioecious cannabis sativa with an XY chromosome sex determination system. PLoS One 9(1): e85118. doi:10.1371/journal.pone.0085118
52. Lysak MA, Mandáková T, Lacombe E (2010) Reciprocal and multi-species chromosome BAC painting in crucifers (Brassicaceae). Cytogenet Genome Res 129(1-3):184–189
53. Mandáková T, Lysak MA (2008) Chromosomal phylogeny and karyotype evolution in x=7 crucifer species (Brassicaceae). Plant Cell 20(10):2559–2570. doi:10.1105/tpc.108.062166
54. Lysak MA, Cheung K, Kitschke M, Bureš P (2007) Ancestral chromosomal blocks are triplicated in brassiceae species with varying chromosome number and genome size. Plant Physiol 145(2):402–410. doi:10.1104/pp.107.104380
55. Lysak MA, Fransz PF, Ali HBM, Schubert I (2001) Chromosome painting in Arabidopsis thaliana. Plant J 28(6):689–697. doi:10.1046/j.1365-313x.2001.01194.x
56. Mandáková T, Lysak MA (2016) Painting of arabidopsis chromosomes with chromosome-specific BAC clones. In: Current protocols in plant biology. John Wiley & Sons, Inc., New York, NY. doi:10.1002/cppb.20022
57. Pecinka A, Schubert V, Meister A, Kreth G, Klatte M, Lysak MA, Fuchs J, Schubert I (2004) Chromosome territory arrangement and homologous pairing in nuclei of Arabidopsis thaliana are predominantly random except for NOR-bearing chromosomes. Chromosoma 113(5):258–269. doi:10.1007/s00412-004-0316-2
58. Ohmido N, Fukui K, Kinoshita T (2010) Recent advances in rice genome and chromosome structure research by fluorescence in situ hybridization (FISH). Proc Jpn Acad Ser B Phys Biol Sci 86(2):103–116. doi:10.2183/pjab.86.103
59. Jenkins G, Hasterok R (2007) BAC 'landing' on chromosomes of Brachypodium distachyon for comparative genome alignment. Nat Protoc 2(1):88–98. http://www.nature.com/nprot/journal/v2/n1/suppinfo/nprot.2006.490_S1.html
60. Kim J-S, Childs KL, Islam-Faridi MN, Menz MA, Klein RR, Klein PE, Price HJ, Mullet JE, Stelly DM (2002) Integrated karyotyping of sorghum by in situ hybridization of landed BACs. Genome 45(2):402–412. doi:10.1139/g01-141
61. Zwick MS, Hanson RE, Islam-Faridi MN, Stelly DM, Wing RA, Price HJ, McKnight TD (1997) A rapid procedure for the isolation of C0t-1 DNA from plants. Genome 40 (1):138–142. doi:10.1139/g97-020
62. Sadder MT, Ponelies N, Born U, Weber G (2000) Physical localization of single-copy sequences on pachytene chromosomes in maize (Zea mays L.) by chromosome in situ suppression hybridization. Genome 43 (6):1081–1083. doi:10.1139/gen-43-6-1081
63. Schwarzacher T, Leitch AR, Bennett MD, Heslop-Harrison JS (1989) In situ localization of parental genomes in a wide hybrid. Ann Bot 64(3):315–324. doi:10.1093/oxfordjournals.aob.a087847
64. Markova M, Vyskot B (2009) New horizons of genomic in situ hybridization. Cytogenet Genome Res 126(4):368–375
65. Peterson DG, Lapitan NL, Stack SM (1999) Localization of single- and low-copy sequences on tomato synaptonemal complex spreads using fluorescence in situ hybridization (FISH). Genetics 152(1):427–439
66. Szinay D, Chang S-B, Khrustaleva L, Peters S, Schijlen E, Bai Y, Stiekema WJ, Van Ham RCHJ, De Jong H, Klein Lankhorst RM (2008) High-resolution chromosome mapping of BACs using multi-colour FISH and pooled-BAC FISH as a backbone for sequencing tomato chromosome 6. Plant J 56 (4):627–637. doi:10.1111/j.1365-313X.2008.03626.x
67. Raap AK, Marijnen JGJ, Vrolijk J, van der Ploeg M (1986) Denaturation, renaturation, and loss of DNA during in situ hybridization procedures. Cytometry 7(3):235–242. doi:10.1002/cyto.990070303
68. G McNamara (2007) Fluorophore Table-Earthlink. http://home.earthlink.net/~fluorescentdyes/McNamara2007FluorophoresTable.xls. Accessed February 2, 2017
69. Aliyeva-Schnorr L, Beier S, Karafiátová M, Schmutzer T, Scholz U, Doležel J, Stein N, Houben A (2015) Cytogenetic mapping with centromeric bacterial artificial chromosomes contigs shows that this recombination-poor region comprises more than half of barley chromosome 3H. Plant J 84(2):385–394. doi:10.1111/tpj.13006

70. Feng C-M, Qiu Y, Van Buskirk EK, Yang EJ, Chen M (2014) Light-regulated gene repositioning in Arabidopsis. Nat Commun 5:3027. doi:10.1038/ncomms4027. http://www.nature.com/articles/ncomms4027#supplementary-information
71. Hesse S, Manetto A, Cassinelli V, Fuchs J, Ma L, Raddaoui N, Houben A (2016) Fluorescent labelling of in situ hybridisation probes through the copper-catalysed azide-alkyne cycloaddition reaction. Chromosome Res 24 (3):299–307. doi:10.1007/s10577-016-9522-z
72. Khrustaleva LI, Kik C (2001) Localization of single-copy T-DNA insertion in transgenic shallots (Allium cepa) by using ultra-sensitive FISH with tyramide signal amplification. Plant J 25(6):699–707. doi:10.1046/j.1365-313x.2001.00995.x
73. Kirov I, Van Laere K, De Riek J, De Keyser E, Van Roy N, Khrustaleva L (2014) Anchoring linkage groups of the rosa genetic map to physical chromosomes with tyramide-FISH and EST-SNP markers. PLoS One 9(4):e95793. doi:10.1371/journal.pone.0095793
74. Pérez R, de Bustos A, Jouve N, Cuadrado Á (2009) Localization of Rad50, a single-copy gene, on group 5 chromosomes of wheat, using a FISH protocol employing tyramide for signal amplification (Tyr-FISH). Cytogenet Genome Res 125(4):321–328
75. Sanz MJ, Loarce Y, Ferrer E, Fominaya A (2012) Use of tyramide-fluorescence in situ hybridization and chromosome microdissection for ascertaining homology relationships and chromosome linkage group associations in oats. Cytogenet Genome Res 136 (2):145–156
76. Stephens JL, Brown SE, Lapitan NLV, Knudson DL (2004) Physical mapping of barley genes using an ultrasensitive fluorescence in situ hybridization technique. Genome 47 (1):179–189. doi:10.1139/g03-084
77. Wiegant J, Ried T, Nederlof PM, van der Ploeg M, Tanke HJ, Raap AK (1991) In situ hybridization with fluoresceinated DNA. Nucleic Acids Res 19(12):3237–3241
78. Figueroa DM, Bass HW (2010) A historical and modern perspective on plant cytogenetics. Brief Funct Genomics 9(2):95–102. doi:10.1093/bfgp/elp058
79. Fransz PF, Alonso-Blanco C, Liharska TB, Peeters AJM, Zabel P, de Jong JH (1996) High-resolution physical mapping in Arabidopsis thaliana and tomato by fluorescence in situ hybridization to extended DNA fibres. Plant J 9(3):421–430. doi:10.1046/j.1365-313X.1996.09030421.x
80. Jackson SA, Wang ML, Goodman HM, Jiang J (1998) Application of fiber-FISH in physical mapping of Arabidopsis thaliana. Genome 41 (4):566–572. doi:10.1139/g98-093
81. Wang K, Zhang W, Jiang Y, Zhang T (2013) Systematic application of DNA fiber-FISH technique in cotton. PLoS One 8(9):e75674. doi:10.1371/journal.pone.0075674
82. Stupar RM, Lilly JW, Town CD, Cheng Z, Kaul S, Buell CR, Jiang J (2001) Complex mtDNA constitutes an approximate 620-kb insertion on Arabidopsis thaliana chromosome 2: implication of potential sequencing errors caused by large-unit repeats. Proc Natl Acad Sci 98(9):5099–5103. doi:10.1073/pnas.091110398
83. Dong F, Miller JT, Jackson SA, Wang G-L, Ronald PC, Jiang J (1998) Rice (Oryza sativa) centromeric regions consist of complex DNA. Proc Natl Acad Sci 95(14):8135–8140
84. Jackson SA, Cheng Z, Li Wang M, Goodman HM, Jiang J (2000) Comparative fluorescence in situ hybridization mapping of a 431-kb Arabidopsis thaliana bacterial artificial chromosome contig reveals the role of chromosomal duplications in the expansion of the brassica rapa genome. Genetics 156(2):833–838
85. Lilly JW, Havey MJ, Jackson SA, Jiang J (2001) Cytogenomic analyses reveal the structural plasticity of the chloroplast genome in higher plants. Plant Cell 13(2):245–254
86. Nakano A, Suzuki G, Yamamoto M, Turnbull K, Rahman S, Mukai Y (2005) Rearrangements of large-insert T-DNAs in transgenic rice. Mol Genet Genomics 273(2):123–129. doi:10.1007/s00438-005-1116-y
87. Beliveau BJ, Apostolopoulos N, Wu C-t (2001) Visualizing genomes with oligopaint FISH probes. In: Current protocols in plant biology. John Wiley & Sons, Inc., New York, NY. doi:10.1002/0471142727.mb1423s105
88. Han Y, Zhang T, Thammapichai P, Weng Y, Jiang J (2015) Chromosome-specific painting in cucumis species using bulked oligonucleotides. Genetics 200(3):771–779. doi:10.1534/genetics.115.177642
89. Murgha Y, Beliveau B, Semrau K, Schwartz D, Wu C-t, Gulari E, Rouillard J-M (2014) Combined in vitro transcription and reverse transcription to amplify and label complex synthetic oligonucleotide probe libraries. Biotechniques 58(6):301–307. doi:10.2144/000114298
90. Valárik M, Bartoš J, Kovářová P, Kubaláková M, De Jong JH, Doležel J (2004) High-resolution FISH on super-stretched flow-sorted plant chromosomes. Plant J 37

(6):940–950. doi:10.1111/j.1365-313X.2003.02010.x

91. Yu H, Chao J, Patek D, Mujumdar R, Mujumdar S, Waggoner AS (1994) Cyanine dye dUTP analogs for enzymatic labeling of DNA probes. Nucleic Acids Res 22(15):3226–3232
92. Nilsson M, Malmgren H, Samiotaki M, Kwiatkowski M, Chowdhary B, Landegren U (1994) Padlock probes: circularizing oligonucleotides for localized DNA detection. Science 265 (5181):2085–2088. doi:10.1126/science.7522346
93. Spencer VA, Davie JR (1999) Role of covalent modifications of histones in regulating gene expression. Gene 240(1):1–12. doi:10.1016/S0378-1119(99)00405-9
94. Vyskot B, Araya A, Veuskens J, Negrutiu I, Mouras A (1993) DNA methylation of sex chromosomes in a dioecious plant, Melandrium album. Mol Gen Genet 239 (1):219–224. doi:10.1007/bf00281621
95. Siroky J, Ruffini Castiglione M, Vyskot B (1998) DNA methylation patterns of Melandrium album chromosomes. Chromosome Res 6(6):441–446
96. Castiglione MR, Cremonini R, Frediani M (2002) DNA methylation patterns on plant chromosomes. Caryologia 55(4):275–282. doi:10.1080/00087114.2002.10797876
97. Salic A, Mitchison TJ (2008) A chemical method for fast and sensitive detection of DNA synthesis in vivo. Proc Natl Acad Sci 105(7):2415–2420. doi:10.1073/pnas.0712168105

Chapter 31

Technical Review: Microscopy and Image Processing Tools to Analyze Plant Chromatin: Practical Considerations

Célia Baroux and Veit Schubert

Abstract

In situ nucleus and chromatin analyses rely on microscopy imaging that benefits from versatile, efficient fluorescent probes and proteins for static or live imaging. Yet the broad choice in imaging instruments offered to the user poses orientation problems. Which imaging instrument should be used for which purpose? What are the main caveats and what are the considerations to best exploit each instrument's ability to obtain informative and high-quality images? How to infer quantitative information on chromatin or nuclear organization from microscopy images? In this review, we present an overview of common, fluorescence-based microscopy systems and discuss recently developed super-resolution microscopy systems, which are able to bridge the resolution gap between common fluorescence microscopy and electron microscopy. We briefly present their basic principles and discuss their possible applications in the field, while providing experience-based recommendations to guide the user toward best-possible imaging. In addition to raw data acquisition methods, we discuss commercial and noncommercial processing tools required for optimal image presentation and signal evaluation in two and three dimensions.

Key words Plant chromatin, Widefield microscopy, Confocal microscopy, Light sheet microscopy, Super-resolution microscopy, Electron microscopy, Resolution, Image processing, Image segmentation, Deconvolution, Signal quantification

1 Introduction

Since its first description in the late nineteenth century as a stainable substance in the cell nucleus, chromatin has gained a considerable notoriety in biological sciences given its central role in gene expression and heredity [1]. Chromatin is a highly organized, macromolecular complex composed of DNA wrapped around specific, structural proteins—the histones. An octamer of histone variants forms the basic unit of chromatin, the nucleosome. Nucleosomes are arranged as a bead-on-a-string structure which itself can be folded along higher degrees of complexity. The abundance and the subtype of linker histone proteins regulate the local and global chromatin compaction states [2]. In addition to realize an efficient DNA packaging, chromatin provides a versatile molecular platform

Marian Bemer and Célia Baroux (eds.), *Plant Chromatin Dynamics: Methods and Protocols*, Methods in Molecular Biology, vol. 1675, DOI 10.1007/978-1-4939-7318-7_31, © Springer Science+Business Media LLC 2018

to regulate cell functions based on the genetic and epigenetic information. Epigenetic modifications include biochemical alterations of histone proteins and DNA. Those influence the properties of chromatin in terms of macromolecular fluidity, but also the accessibility to enzymatic, multiprotein complexes involved in transcription, repair, or replication [3]. The combinatorial principles of epigenetic instructions and the basic organizing principles of chromatin in the nucleus are remarkably shared between the plant and animal kingdoms, although critical differences remain [4].

Chromatin composition and organization can be described at three different levels: the epigenomic, microscopic, and probabilistic level.

Epigenomic studies provide molecular profiles of histone and DNA modifications along the chromosomes and in relation to the nature of genomic loci (repeats, transposable elements, protein-coding, or intergenic regions). Because of technical requirements, these profiles are mostly a readout of tens of thousands of cells from a whole organ, tissue or mix of replicate tissues. Although arising, single-cell epigenomics are not yet readily applicable to all kinds of organs or tissues, and are especially cumbersome in plants due to the difficulty to isolate single cells while preserving their cellular identity. Nevertheless, several approaches are being developed that dissect epigenome profiles among distinct cell types [5–7] (and *see* Chapter 8, [8]).

Microscopic studies of chromatin are historically rooted in the nineteenth century where basophilic dyes were used on cell spread preparations to visualize chromatin arrangements in different tissues or organisms [1]. Today, microscopic investigations of chromatin are based on fluorescence, molecular-cytogenetic labeling techniques providing the description of single nuclei and chromosomes, as well as the spatial, three-dimensional (3D) arrangement of chromatin and its composition. In addition, the technical revolution in cellular imaging of the past decade offers an unprecedented, nanoscopic level of information allowing resolving chromatin arrangement at the nucleosomal scale. The associated, powerful image processing algorithms allow quantifying and describing the higher-order degree of chromatin fiber organization.

A probabilistic level of chromatin description is provided by emerging techniques determining the frequency of *cis* or *trans* molecular interactions between chromosomal regions or together with specific nuclear domains (e.g., the nucleolus, the nuclear periphery). The so-called TADs, LADs, and NADs (topological-, lamina-, and nucleolus-associated domains, respectively) among others, emerge from a mathematical, likelihood-based interpretation of interaction frequencies within a population of hundreds of thousands of cells. These approaches are invaluable to decipher the mystery of genome regulation. In fact, a refined, spatial organization of chromosomes involving local (subchromosomal) and global (nuclear-level) arrangements creates functional regulatory domains that have a profound impact on gene expression and developmental programs (for review *see* [9]).

This chapter focuses on modern imaging techniques that allow microscopic and nanoscopic in situ investigations of nuclear architecture, and chromatin organization and composition. These techniques strongly rely on the use of fluorescent compounds engineered to label chromatin: chemicals (e.g., DNA dyes such as DAPI, propidium iodide or Sytox Green), fluorescent proteins (FPs) (GFP and derivatives introduced as translational fusions to chromatin proteins-of-interest), and fluorescently labeled antibodies against native proteins or DNA probes. Labeling methods are available that are noninvasive, and can be applied to intact tissues, isolated nuclei or cells, hence bypassing the need of sectioned material. Together with the advent of increasingly sensitive and versatile microscopy imaging platforms, modern cytogenetic labeling methods enable chromatin and nuclear organization to be resolved in a whole tissue or organ context but also at the single cell level, in 3D and with a resolution down to single molecules.

The variety and large number of labeling methodologies and imaging procedures can be particularly confusing to scientists new in the field. Furthermore, some of the latest fluorescence microscopy imaging technologies rely on advanced bio-imaging and physical concepts, high-end light source modulators and detectors, and complex processing algorithms. Collectively, this can be particularly overwhelming in decision making as to which method and approach to use. This chapter aims to provide guidelines toward choosing the appropriate microscopy imaging method for analyzing nuclear and chromatin organization and composition in situ. We review different imaging systems and discuss how they were used in plant chromatin studies. These systems include widefield epifluorescence microscopy (Subheading 2), confocal microscopy (Subheading 3), light sheet fluorescence microscopy (Subheading 4), super-resolution microscopy (Subheading 5), and electron microscopy imaging (Subheading 6). Whenever relevant, we provide recommendations on image acquisition parameters to achieve best-possible images which are necessary for high-resolution quantification of chromatin organization. The range of possible applications by the different systems presented here is illustrated in Fig. 1. In addition, Subheading 7 discusses common image processing approaches as well as recent developments and their relevance for plant chromatin analyses at the microscopic and nanoscale levels. Finally, Subheading 8 proposes perspectives in microscopy imaging that could contribute significantly to plant chromatin research.

While numerous works are cited, we apologize in advance to the authors whose work was not included due to space constraints. Here, only basic principles of the referred imaging techniques are provided (Box 1). For detailed, technical descriptions of the imaging instruments, the reader is referred to specialized review articles, books, or online education materials (*see* **Note 1**). While we focus

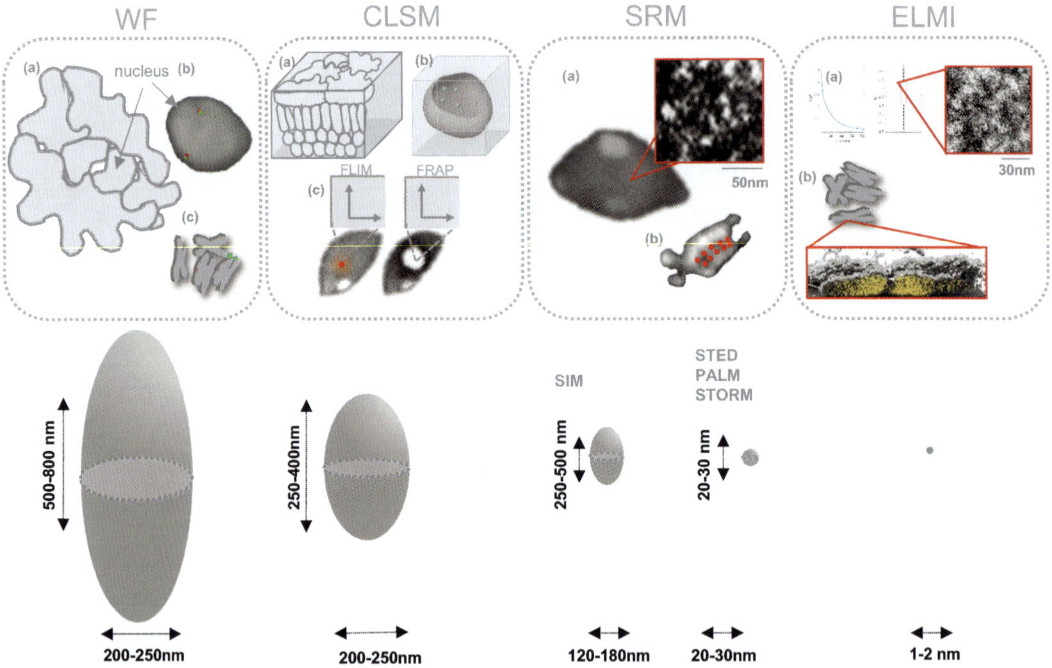

Fig. 1 Overview of microscopy systems applicable for plant chromatin imaging. Widefield (*WF*) imaging is suitable for rapid cell and tissue localization (*a*, leaf epidermal cells are represented as an example) of live chromatin markers, FISH staining of genomic regions in isolated nuclei (*b*) or chromosome spreads (*c*) (*red/green* signals). Confocal microscopy (*CLSM*) allows the 3D imaging of live chromatin reporters in whole-mount tissue (*a*, leaf section represented as an example), 3D imaging of FISH signals enabling 3D analyses (*b*, example image processing result: segmentation of the nucleus and nuclear signals), time-laps imaging of chromatin reporters to measure dynamic processes by FLIM or FRAP (*c*). Super-resolution microscopy (*SRM*) breaks the resolution limit of visible light imposed by its diffraction limit, thus allowing the nanoscale-level detection of chromatin organization using photolocalization of individual fluorophores (*a*, example of H3 immunolabeling in a leaf nucleus, GSD/STORM) or by using structured illumination imaging (SIM) (*b*, example of protein localization between chromatids). Electron Microscopy Imaging (*ELMI*) provides ultrastructural ranges of signal detection enabling nucleosomal scale analyses on nuclei from whole tissue sections (*a*, example of TEM preparation, uranyl acetate chromatin staining) or chromosome preparations (*b*, example of SEM preparation, *see* Fig. 5b). The lateral and axial resolutions provided by each instrument are indicated below each panel. The ranges are indicative and correspond to the best-possible resolutions but strongly depend on sample preparation and instrument setup (see main text for details). Images provided by C. Baroux: WF (*b*), CLSM (*b*, *c*), SRM (*a* and *b*, chromosome), and by F. N. Tohnyui (University of Zürich): ELMI (*a*, *b*). For ELMI (*b*, *inset*). *See* Fig. 5b for details

here on benefits and limitations for chromatin imaging in plant cells, several reviews give a general view of pitfalls and challenges associated with the different microscopy approaches ("beginners' guide," [10–12], or an in-depth overview of imaging equipment for plant cell imaging [13]). Moreover, many of the labeling techniques mentioned here are described in more detail in the dedicated method chapters of this book section. The reader is also referred to Tables 1 and 2 to gain an overview at a glance of the applications and limitations of the microscopy imaging techniques presented here and image processing/evaluation software, respectively.

Box 1 Microscopy Imaging Systems to Study Plant Chromatin:

A. Widefield (WF) Fluorescence

Widefield (WF) fluorescence microscopy relies on the illumination of the entire field-of-view of the sample preparation. Excitation of the fluorophores is enabled by a light source from which only a selective fraction of its spectrum is transmitted via a specific excitation filter inserted into a filter cube, or via LEDs or Lasers emitting specific wavelengths. The beam splitter and emission filter within the filter cube enable the selective transmission of the emitted fluorescence, so that nonspecific (e.g., auto) fluorescence is blocked. Thus, while filter cubes allow a specific range of light illumination/observation, they do not provide a highly specific nor tuneable profile; their selective bandwidth and attenuation capacity have to be carefully chosen based on the providers' characteristics and best-fit to the fluorophore to be observed (*see* **Note 1** and **11**). Furthermore, like for any imaging based on compound microscopes, the quality of the observation mainly depends on the objective. Important characteristics are the immersion medium (if not an air objective) and the corrections for spherical and chromatic aberrations that collectively allow optimizing the light collection. In addition, the numerical aperture (NA) of the objective defines the lateral and axial resolution (*see* **Note 9**). Finally, the characteristics of the digital camera should be carefully considered. The camera sensors vary in sensitivity, on an absolute scale but also across the visible range of wavelength (i.e., the sensors commonly used for transmission light microscopy may work very poorly in ranges of common fluorophores. In addition, their fabrics and design (CCD, EMCCD, ICCD, CMOS, *see* **Note 10**) have different properties relative to the size of the field of view, the pixel size, ability to capture low-level signals, levels of background signals (can be temperature dependent), signal integration times or dynamic ranges. In general, monochromatic cameras are preferred over color-cameras where the sensor area is split in RGB receptive pixel wells hence decreasing substantially the overall sensitivity [185]. In other words, not every digital camera is equally well suited for all fluorescence imaging experiments and first users are advised to seek for information and best suited camera beforehand. It is required that beginners in fluorescence microscopy should evaluate the optical features of the specimen for obtaining best possible informative images. It is worthwhile considering both fluorescence characteristics to choose the right light source and filter cube but also the imaging aims ahead of the experiment. Which level of details is required for an informative image (cellular/subcellular etc.)? What

(continued)

is the sample thickness and at which depth the signals have to be recorded? Are signals expected to be weak or strong?

B. Confocal Microscopy systems

Confocal Laser Scanning Microscopy (CLSM), Multiphoton CLSM (CLSM-MP), Confocal Spinning Disk Microscopy (CSDM)

Similar considerations concerning illumination source and objectives as described above apply to confocal microscopy imaging. Confocal imaging drastically improves the optical contrast based on the fundamental idea to place a hole the size of a pin ("pinhole") at the focal plane of the objective lens by excluding out-of-focus light. Proficient reviews and online tutorials describing this principle and implementations are available (*see* **Note 1**). Here we will focus on three main instruments useful to study plant chromatin: CLSM, CLSM-MP and CSDM. The main difference between CLSM and CSDM is the illumination and image recording strategy. In CLSM a laser of a specific wavelength is swept across the biological sample via controllable scanning heads to provide excitation in a point-by-point manner. Emitted fluorescence is then captured by photodetectors that will deliver a digitized image where one pixel corresponds to one scanned point. In such a system the image format is chosen by the user but the effective lateral resolution of the objective has to be taken into account (*see* **Note 9**). Detectors have varying sensitivity and background noise depending on the solution implemented by the provider, with the two main technologies being photomultiplier tubes (PMT) and avalanche photodiodes (APD). Modern detectors, though, combine the best of both (the so-called new generation and hybrid detectors) [186] and should be preferred for imaging if possible. Multiphoton CLSM uses a different illumination strategy. Instead of sending the energy required for excitation in a single photon with appropriate wavelength, two photons of about half the energy are sent on the focal point. The temporal resolution allows the energy to sum up to provide the optimal excitation energy. A benefit of this strategy is that higher wavelengths (of lower energy) are used that have superior properties in terms of tissue penetration depth, while causing less photodamage [187, 188]. Often, CLSM-MP is implemented with new generation detectors placed externally to the scanner box hence saving on distance with the effect to be much more efficient in collecting emitted fluorescence. This offers a more sensitive detection. Thus, CLSM-MP is well suitable for imaging of living cell nuclei deep within plant tissues. In CSDM the point illumination is multiplexed in such a way that several points of the image are illuminated at once through an array of microlenses arranged on a fast spinning disk. The image capture is often performed by high-sensitive digital sensors (CDD or sCMOS, *see* **Note**

(continued)

10), which offer superior output in terms of speed and sensitivity, and can play a critical role in live imaging or large sample acquisition when the fluorophore is fast photobleaching.

C. Light sheet fluorescence microscopy (LSFM)
LSFM, often also called single plane illumination microscopy (SPIM), relies on the principle that the sample is illuminated by a thin sheet of light instead of a single point (CLSM) or whole field of view (wide-field). The sample is imaged using regular objective lenses placed perpendicularly to the sheet by a fast, highly sensitive digital camera, thus enabling capturing fluorescence signals in a whole optical plane in a few milliseconds. Several detailed reviews are available [189]. LSFM enables the 3D analysis of specimens, in principle at a similar optical resolution as WF and confocal fluorescence microscopy. Lateral and spatial resolution strongly depend on the light sheet thickness which interplays with the useful field and depth of view [190]. Currently, the commercially available systems (using air or water immersion lenses with relatively low NA) do not offer, however, a submicrometric range of resolution suitable for fine-scale chromatin studies. Nevertheless, they offer the unprecedented possibility to generate images with isotropic signal distribution, which abolishes the problem of poor axial resolution in conventional, unidirectional imaging systems. Isotropy results from the combination of several 3D views of the same sample after rotation of the specimen in front of the objective or using multiple imaging objectives (multiview imaging) [189, 190]. Another asset is the high-speed optical sectioning capabilities with no out-of-focus excitation. As a consequence, light sheet methods induce less photobleaching and lower phototoxicity. Commercial LSFM systems are offered by the companies Carl Zeiss (Lightsheet Z.1), Leica Microsystems (Leica TCS SP8 DLS), Luxendo (InVi-SPIM, MuVi-SPIM) and LaVision Biotec (UltraMicroscope II).

D. Super-Resolution Microscopy (SRM)
Due to the diffraction limit of light as defined by Ernst Abbe and Rayleigh [191, 192] the spatial resolution of light microscopy including conventional fluorescence techniques is restricted, and reaches only ~200 nm laterally and ~600 nm in the axial dimension in biological specimens [173]. To overcome this restriction and to bridge the resolution gap between light and electron microscopy the so-called super-resolution (also referred to as optical nanoscopy) techniques have been developed and for which the main contributors Eric Betzig, William Moerner, and Stefan Hell received the Nobel Prize in Chemistry in 2014 [193]. They include Structured Illumination Microscopy (SIM), Photoactivated Localization Microscopy (PALM), Stochastic Optical Reconstruction Microscopy (STORM), Ground State Depletion followed by

(continued)

Individual Molecule return (GSDIM) and Stimulated Emission Depletion (STED). These "subdiffraction" methods can be divided into two different principles: (1) localization of individual fluorophores in the specimen with subdiffraction precision (PALM, STORM, GSDIM), and (2) structuring the illumination light to collect high spatial frequencies in the image that contain high resolution information (SIM, STED) [82]. SIM reaches a resolution of ~120 nm (Fig. 4a), PALM (Fig. 4b) and STORM ~20 nm, and STED up to ~20–50 nm.

Localization microscopy uses two distinct approaches depending on whether the diffraction limit is spatially or temporally broken. GSDIM (also commonly referred to GSD only) uses high-power lasers to bring the fluorophores into a long-lived "off state" (dark state), then single molecules stochastically return to a fluorescent state thereby temporally uncoupling the overall fluorescent population. In contrast, PALM and STORM rely on the activation of a limited subset of fluorophores at a given time point, so that the overall population is captured over sequential (time) imaging. STED rely on the application of multiple pulsed lasers with specificities regarding beam geometry and intensity that spatially deplete fluorophores not centered on the point spread function. STED uses a raster imaging mode similar to CLSM while GSD and PALM/STORM require recording 30,000–100,000 time frames (over 10–15 min) each capturing a subset of photons. Computational reconstruction uses specific algorithms taking into account photon intensity, frame correlation and number of events per frame among other factors to reconstruct the spatial distribution of the overall population of fluorophores.

In addition to noncommercial super-resolution microscope systems established in different research groups, since 2004 also commercial systems are produced by companies as Carl Zeiss, Nikon, Leica Microsystems and GE Healthcare Life. These systems have integrated image reconstruction software. Alternatively, open-source software such as Fiji provide different algorithms (for instance Thunderstorm), but those require a good technical understanding in order to fine-tune the different parameters.

All super-resolution techniques are based on imaging fluorescent molecules. Consequently, they are used to image labeled structures and molecules of interest. After fixation of a specimen, which should alter the native structures as little as possible, specific fluorescent affinity probes of preferably small size (FAB fragments of antibodies, nanobodies, snap-tags) may be used for labeling [91].

E. Electron Microscopy (ELMI)

Electron microscopes use a beam of accelerated electrons for illumination. The wavelength of an electron can be up to 100,000 times shorter than that of visible light photons. Therefore, electron

(continued)

microscopes achieve a higher resolution than light microscopes. They are applied to investigate the ultrastructure of a wide range of biological specimens including whole organisms, tissues, cell components and molecules.

Scanning electron microscopes (SEMs) produce images of a specimen by scanning it with a focused beam of electrons. This beam scans in a raster scan pattern and interacts with the atoms in the sample, producing electron signals containing information about the specimen's surface and composition. SEM achieves a resolution of ~1 nm.

Focused ion Beam Scanning Electron Microscopes (FIB/-SEMs) use focused ion beams for the sequential removal of material slices (FIB milling) followed by sequential SEM viewing steps in parallel. This allows the serial sectioning of biological specimens to investigate their interior virtually unachievable, and to reconstruct their ultrastructure in three dimensions.

During transmission electron microscopy (TEM) an electron beam is transmitted through an ultra-thin specimen. Instead of glass lenses of an optical light microscope TEMs use electrostatic and electromagnetic lenses to control the electron beam and to focus it. Due to the interaction of the electrons with the sample an image is formed. TEM can achieve a resolution of ~0.050 nm [194]. At smaller magnifications the TEM image contrast is induced by the absorption of electrons in the sample.

Correlative Light Electron Microscopy (CLEM) combines the efficiencies of light (including fluorescent) microscopy and electron microscopy. Light microscopy can provide widefield images of whole tissues and cells, but its resolution is limited. Electron microscopy provides a much higher resolution. CLEM combines the advantages of both techniques. This dual examination provides valuable complementary information. Cell structures and processes can be visualized in whole cell overview images with light microscopy. Then, for more ultrastructural details it can be zoomed in with electron microscopy.

2 Widefield Epifluorescence Microscopy

Widefield (WF) fluorescence microscopy relies on the illumination of the entire field-of-view of the sample preparation and the capture of the emitted light across an optical section which thickness is determined by the objective. As described in Box 1, successful imaging using WF requires awareness of the user on the best objective lens and filter cubes to be used for the biological question as well as of the camera sensitivity. Due to its simple implementation and widespread accessibility in most laboratories, WF epifluorescence microscopy is the preferred approach to analyze the

Table 1
Applications of different microscopy imaging platforms at a glance

Sample	Chromatin labeling	Type of chromatin analyses
Widefield and Confocal microscopy		
WF and confocal microscopy are applicable with the same sample type but deliver different datasets in terms of the format (3D with no sectioning for WF, 3D sectioning for confocal), resolution and depth of imaging (see text).		
Fixed, isolated nuclei Fixed, whole-mount tissue	Immunolabeling of modified histones, tagged chromatin protein, FISH, GISH labeling	*Qualitative analyses*: Cell-specific distribution, nuclear and subnuclear distribution (speckles, chromocenters, euchromatin, nuclear envelope, nucleolus, and nuclear bodies) *Quantitative analyses*: Heterochromatin content, intensity levels of histone modifications, histone variants, chromatin-binding proteins, etc.
Live tissue, cell culture	GFP reporter	Time-lapse imaging of chromatin motion, protein shuttling and loading, protein–protein interactions (BiFC, FRET and FLIM-FRET), protein turnover (FRAP)
Isolated chromosomes, chromatin fibers	FISH/GISH labeling, chromosome painting	Chromosomal distribution of genomic sequences/parental genomes, and karyotypes
Light sheet fluorescence microscopy		
To date no published application of LSFM imaging for chromatin studies—field under development		
Whole-mount organs, fixed and cleared	Immunolabeling and FISH	3D analyses of chromatin organization in whole organs: high-throughput imaging of multiple cells/cell-types
Whole-mount organs, alive	GFP reporter	3D/4D analyses with isotropic resolution in whole organs; possibly long-term live imaging of nuclei, chromatin, and chromatin proteins *in planta* Limited by low optical transparency of fresh tissues
Super-resolution microscopy		
Tissue sections	Immunolabeling and FISH	Identification of nuclei and nuclear components
Squash preparations of metaphase chromosomes	Immunolabeling and FISH	Chromatin/protein organization along condensed chromosomes, at

(continued)

Table 1 (continued)

Sample	Chromatin labeling	Type of chromatin analyses
(roots, shoots, flower buds, and anthers)		centromeres and (sub)telomeres; quantification and colocalization of proteins
Isolated nuclei and chromosomes	Immunolabeling and FISH	Chromatin and protein organization in interphase nuclei and condensed chromosomes, quantification and colocalization of proteins
Extended chromatin fibers	Immunolabeling and FISH	Arrangement of chromatin components along chromatin fibers
Electron microscopy		
Tissue sections	Uranyl acetate contrast staining; Nanogold-antibody labeling	*TEM*: euchromatin and heterochromatin distribution analyses with cell-type specific resolution in whole tissue sections; nanoscale level of chromatin organization through image signal distribution analyses *SEM*: surface imaging of chromatin structures in cross sections
Isolated nuclei and chromosomes	Platinum blue labeling of overall DNA, silver staining of overall protein, Nanogold-antibody labeling for specific proteins	*SEM* to analyze the ultrastructure and distribution of DNA-specific segments and proteins at the surface of nuclei and chromosomes; *FIB/SEM* to image the ultrastructure of chromatin and proteins in the interior of nuclei and chromosomes; SEs are used to image both DNA and protein together, BSEs to identify DNA only

For references to published work, please refer to the main text

distribution of reporter protein-tagged chromatin proteins within tissues. Typically, simple tissue preparations allow for answering the question whether the chromatin marker under study is ubiquitously expressed or whether it shows cell-specific distribution or accumulation patterns (Fig. 1). However, intact, often thick, tissue preparations are not appropriate to analyze the precise subnuclear distribution with WF, due to the strong light scattering and the capture of out-of-focus light, which blurs the subnuclear structures. Qualitative distribution analyses are preferentially performed on thin, dissected tissue preparations (e.g., epidermal peals, or

Table 2
Image processing software useful for plant nuclei image analyses (other than microscopy imaging built-in software)

Software	Application	URL	Reference
Resource overview			
Plant-image.analysis	Overview of image analysis software useful in plant biology	plant-image-analysis.org	[195]
FluoView	Digital image processing and analysis resources	fluoview.magnet.fsu.edu/resources/imageprocessing.html	
Open-source			
ImageJ/FiJi	*Basics*: image filtering, rendering, data visualization Advanced processing (customized plugins): 2D masks/segmentation, image registration, deconvolution, segmentation, super-resolution reconstruction, and motion tracking	https://fiji.sc/	[156, 196]
FluoRender	3D rendering, data visualization and analysis	www.sci.utah.edu/software/fluorender.html	[197]
Cell Profiler	cell detection and classification, quantification of numerous shapes, size, texture parameters—suited for nuclei analyses (2D images)	cellprofiler.org/	[198]
ChromDensityNano	Spatial chromatin density distribution analysis from TEM images	github.com/barouxlab/ChromDensityNano	[143, 144]
Commercial			
Imaris	3D, 4D rendering; nuclei and cell segmentation, export of quantitative measurements for segmented nuclei (shape, size, and signal intensities), 3D distance measurements, cell/nuclei tracking, and lineage analysis	www.bitplane.com	
Arivis	3D rendering, 4D visualization (large data), cell tracking, and quantitative measurements	www.arivis.com/en/imaging-science/	
Autoquant	Image deconvolution	www.mediacy.com/autoquantx3	
Huygens	3D visualization, image deconvolution/restoration, and basic analyses	svi.nl/HuygensSoftware	
Amira	3D rendering	www.fei.com/software/amira-for-life-sciences/	
Osiris	3D, 4D rendering, data visualization	www.osirix-viewer.com/	

gently squashed tissues), cell cultures, or isolated nuclei. Precise localization of distinct cellular compartments, such as chromocenters and the nucleolus (Fig. 2a), is best achieved with isolated nuclei or chromosome spreads which are optically more accessible, and not obstructed by the highly refractive cytoplasm and the cell wall, hence enabling a better resolution. The great benefit of using WF microscopy is the potential to collect a large amount of material, facilitating the scoring of nuclear patterns and quantification of fluorescence signals. Typically, hundreds of images can be easily collected from a few preparations to quantify the spatial distribution patterns of heterochromatin or specific genomic loci, chromosomal regions or whole chromosomes detected by fluorescence in situ hybridization (FISH) on isolated nuclei, metaphase plates or chromatin fibers (Fig. 3) (*see* also Chapter 23, [14]).

It should be noted that the quantification of signal intensities, as the prerequisite to measure the abundance of tagged protein or chromatin modifications, has to be done carefully, taken into account experimental biases (signal amplification in immunolabeling) and imaging parameters (illumination power, exposure time, camera gain). Some considerations are given below (*see* **Note 2**). Generally, unless the tissue analyzed is regularly flat and the cells are aligned in the same plane, cell-type specific quantifications within whole-mount tissue preparations may be better operated in 3D using confocal microscopy (see below).

Although WF microscopy is a robust, affordable approach for first-level investigations in whole tissue or for isolated material, the main limitation comes from the inherent optical design that captures out-of-focus light blurring subnuclear structures. To some extent, with samples of low complexity, post-acquisition image reconstruction using deconvolution algorithms can recreate an informative image where apparent blurry signals can be digitally cleaned into distinct, sharp structures, particularly in the lateral (x,y) dimension (*see* Subheading 7). However, WF is strongly limited in axial resolution, and for in-depth discrimination of distinct nuclei and chromosomes in multilayered tissues, confocal microscopy imaging should be preferred.

3 Confocal Laser Scanning Microscopy

Confocal imaging systems exclude out-of-focus light through the use of a pinhole in the optical patch (Box 1), resulting in a sharp image close to the maximum possible diffraction-limited resolution (*see* SRM in Box 1) in a given optical plane. In combination with a motorized-control adjusting the focal plane at different depth, it allows the generation of image stacks corresponding to optical sections along the sample depth which are the basis for 3D reconstructions. Box 1 describes several types of confocal imaging

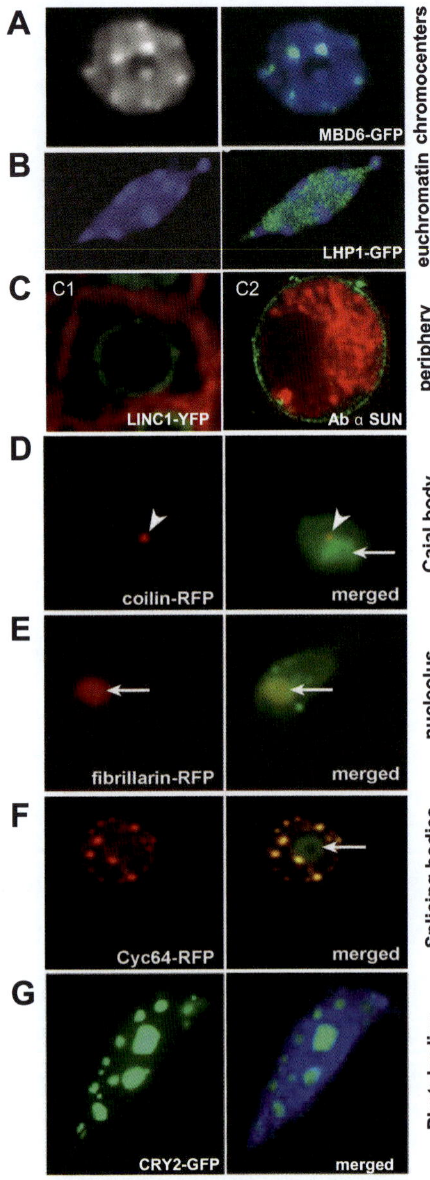

Fig. 2 Subnuclear localization of chromatin markers using WF and CLSM imaging on isolated nuclei or whole-mount plant tissues. Examples of high-resolution microscopy imaging of chromatin binding proteins or nuclear bodies in Arabidopsis (**a**, **b**, $\mathbf{c_1}$, **d**, **e**) and maize ($\mathbf{c_2}$). (**a**) Chromocenter localization of the methyl-cytosine binding protein AtMBD6-GFP detected by immunolocalization (anti-GFP) on isolated leaf nuclei. (**b**) Punctuate euchromatin distribution of a GFP tagged LIKE HETEROCHROMATIN PROTEIN1 fusion (LHP1-GFP) in a root hair nucleus (CLSM, live imaging). (**c**) Localization at the nuclear periphery, CLSM imaging: ($\mathbf{c_1}$) Arabidopsis LITTLE NUCLEI1 (LINC1)/CROWDED NUCLEUS1 (CRWN1) fused to YFP (*yellow*), live imaging with FM4-64 cell membrane counterstaining (*red*), intact floral tissue. ($\mathbf{c_2}$) Immunolabeled SUN proteins (*green*) in the nuclear envelope of isolated maize pollen meiocytes, with

systems varying in their illumination and detection method and referred to as point-scanning or spinning disk systems in a single or multiphoton implementation. All have in common the power of resolving submicrometer signals in the nucleus.

Achieving the best possible resolution of the system requires a careful setup of the imaging conditions that significantly influence the possibility to measure size, shape and spatial localization of signals. This includes evaluating tissue clearing media, choosing sample mounting and immersion media with similar refraction indices (to minimize the light refraction at the medium–coverslip interface), using high-quality cover glasses designed for confocal microscopy (controlled surface quality and thickness). Acquiring images with a format allowing a 2–3× oversampling (relative to the theoretical resolution of the objective and the moderate use of signal averaging) is also important. The summing up of signals in a photon counting mode should also be preferred over multiple averaging, particularly when fluorescence emission is weak, punctuate, or when images are to be used for quantifications (*see* **Note 2**). We describe here applications and technical considerations regarding high-resolution static imaging on stable sample preparations to analyze the spatial distribution of labeled chromatin or chromosomal components (Subheading 3.1), time-lapse imaging in fresh tissue to follow chromatin movements and dynamic changes of its composition in real-time (Subheading 3.2), and time-lapse imaging for quantifying chromatin mobility or molecular interactions (using fluorescence recovery after photobleaching (FRAP), fluorescence resonance energy transfer (FRET), and fluorescence lifetime imaging (FLIM) assays) (Subheading 3.3).

3.1 Static Imaging for High-Resolution, Quantitative Analyses of Chromatin Organization

Confocal imaging systems remain the platform of choice to record the 3D signal distribution in intact nuclei and/or in whole mount plant tissues. They allow high-resolution imaging down to the theoretical limit of optical resolution, particularly on fixed sample preparations where imaging time is not a limiting factor. Similar to WF microscopy, this allows to determine whether a tagged protein or genomic locus of interest localizes throughout the nucleoplasm or in a discrete compartment (e.g., nuclear bodies [15], nucleolus, heterochromatin foci or at the nuclear periphery [16]), and this in relation to the spatial nuclear morphology (Figs. 2 and 3). Another

Fig. 2 (continued) propidium iodide DNA counterstaining (*red*). (**d**, **e**) Colocalization of the GFP-tagged Arabidopsis DEAD box helicase, eIF4A-III (*green*, in merged pictures) with RFP-tags reporting on distinct nuclear compartments, live CLSM imaging in cell culture: Cajal body (**d**, COILIN-RFP), nucleolus (**e**, Fibrillarin-RFP), splicing speckles (**f**, Cyc64-RFP). (**g**) Immunolocalization in nuclear photobodies enriched with photoreceptors (CRY2-GFP, *green*; DAPI counterstaining, *blue*), imaging on isolated nuclei from etiolated seedlings exposed to blue light. Pictures reproduced with permission from [177] (**a**), [178] (**b**), [54] (**c_2**), [179] (**d–f**), [180] (**g**). (**c_1**), own picture; LINC1-GFP line: a gift from Eric Richards [181]

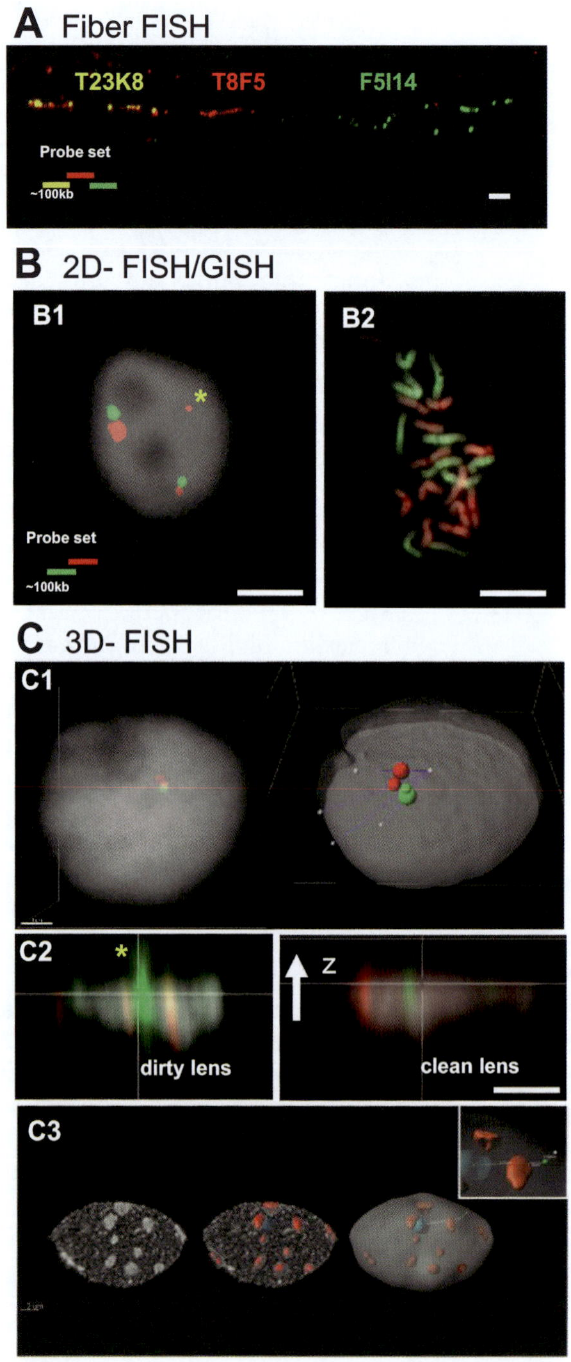

Fig. 3 High-resolution imaging of nuclear in situ hybridization signals. (**a**, **b**) 2D, widefield microscopy imaging of FISH signals on isolated chromatin fibers (**a**, BAC-DNA FISH), nuclei ($\mathbf{b_1}$, BAC-DNA FISH) and metaphase chromosomes ($\mathbf{b_2}$, GISH labeling of parental chromosomes sets in a *Primula* hybrid line, [182]; chromosome preparation by A. Guggisberg). (**c**) 3D, confocal microscopy imaging of FISH signals ($\mathbf{c_1}$, $\mathbf{c_2}$) in isolated leaf nuclei ($\mathbf{c_1}$, intact; $\mathbf{c_2}$, spread) and of DAPI

major asset is the possibility to detect several fluorophores sequentially (*see* **Note 3**) or simultaneously (typically up to four or more with high-end systems combining internal and external detectors).

Confocal imaging systems have also little constraints in the type of the biological sample, in particular compared to the very thin preparations required for WF or super-resolution microscopy (SRM) systems. Due to the highly light diffractive nature of plant tissues, however, the biological sample preparation plays a critical role. Image recording of nuclear structures may be performed on fresh tissues (*see* Subheading 3.1.1), fixed nuclei (*see* Subheading 3.1.2) or embedded isolated nuclei (*see* Subheading 3.1.3).

3.1.1 In Fresh, Living Tissue: Chromatin Signal Distribution Analyses and Protein Interaction Studies

Given the highly light diffractive nature of plant tissues [17], thicker samples will compromise signal intensity and resolution. Therefore, the imaging of nuclei in a cell layer far away from the coverslip is unlikely to provide submicrometric resolution. In these cases, the dissection of organs (e.g., leaf or cotyledon epidermis can be separated from underneath layers; root tip isolation; delicate opening of the carpel; anthers or seeds for accessing gametophytes or embryos) may be helpful. Multiphoton imaging is recommended for imaging in thick tissues as the higher wavelengths used have a superior tissue penetration (Box 1). The main benefit of imaging in fresh tissue is to observe tagged chromatin reporters *in planta* under native developmental and near-physiological conditions. The applications include localization studies with the possibility to quantify signal intensities (*see* Subheading 7), but also chromatin–protein and protein–protein interactions analyses.

These interactions are commonly elucidated using bimolecular fluorescence complementation (BiFC, *see* [18] for a protocol) or fluorescence resonance energy transfer (FRET). BiFC requires reliable controls to test against unspecific protein associations [19]. Yet it provides a powerful tool to verify or screen for novel protein associations in vivo and at tissue or cell-specific levels. Some examples include the demonstration that the flowering time control transcription factor CONSTANS physically interacts in the nucleus with the

Fig. 3 (continued) distribution showing conspicuous chromocenters (**c_3**, intact nucleus). (**c_2**) orthogonal section (*x,z*) across the nucleus shown in (**c_1**), shows an example of the dramatic effect of a dirty lens on axial signal resolution. High-resolution serial imaging allows for 3D image processing with signal *segmentation* (**c_1**, *right panel*; **c_3**, *middle* and *right panel*: *red*, chromocenters, *blue*, nucleolus, *grey*, nucleus) and *3D distance measurements* of signals relative to each other or to the nuclear periphery (thin lines **c_1** and **c_3**). (**c_3**) *Inset* shows a close-up with a green spot at the nuclear periphery for anchoring distance measurements along a vector intersecting with the center of the nucleus). Image segmentation and 3D distances are computed using Imaris. *See* also [45] for automated Imaris processing plugins. Pictures: **a**, **b**, **c_1**, **c_2**: by C. Baroux. **c_3**: preparation by M. Ashenafi. Scale bars: 2 μm

epigenetic readers MRG1/2 [20]). The modularity of the BiFC setup also enables the identification of associations between different chromatin modifying complexes that are not always unambiguously determined by in vitro approaches. For instance, this approach confirmed a suspected interaction between the transcription initiation complex and the histone modifier COMPASS in situ [21]. The simple imaging setup for BiFC also allows for screening assays aimed to identify novel chromatin protein associations. This approached unravelled novel interactions with the HISTONE DEACETYLASE COMPLEX1 (HDC1), including with linker histone variants (H1) and H3-binding proteins, but not H3 itself [22]. Due to the supposedly slow kinetics of fluorophore reconstruction, however, BiFC is often regarded as inappropriate to monitor short-lived or unstable complexes. In the later cases, FRET, or FLIM-FRET can be preferred approaches. Yet the difficulty to engineer a functional pair enabling a 1–10 nm proximity of the fluorophores, necessary for FRET activity, the relative complexity of the imaging setup and signal processing limits the application of these techniques [23]. However, successful examples demonstrate its power in elucidating complex chromatin protein arrangements in vivo. One key example is the finding of a direct binding of DET1, part of an ubiquitination complex that mediates protein degradation during light responses, with the nucleosomes in plant cells [24]. This led to the hypothesis that the ubiquitin degradation pathway plays an essential role in the chromatin remodeling that underlies photomorphogenesis. Upon weak or short molecular interactions, however, FRET measurements necessitate increased temporal resolution that is best provided by time-lapse imaging and quantification of fluorescence lifetime (FLIM-FRET, *see* Subheading 3.3).

3.1.2 In Fixed, Cleared Tissues: Quantitative, Single Cell-Level Imaging of Chromatin Labels

In the past years, several chemical treatments have been described [25–32] that improve the optical transparency by reducing light absorption and diffraction of fixed tissue, hence considerably enhancing the resolution power, which is particularly interesting for chromatin imaging. Further, they allow high-resolution imaging within the entire depth determined by the objective's working distance (*see* **Note 4**). A direct application is the probing of the chromatin status in different cell types within whole-mount tissues. Some clearing methods allow the preservation of the native fluorescence of fluorescent proteins, while others are better suited for immunostained samples.

Whole-mount tissue immunostaining protocols such as those provided for chromatin immunolabeling [33, 34] often contain chemical treatments improving the clarity. Hence, they do not require additional clearing steps as for fluorescent protein detection in fresh tissues. Typically, tissue clearing is improved by the removal of chlorophyll which is highly absorbent and diffractive by ethanol/

acetic acid or xylene/methanol and the depolymerization of the highly refringent cell wall by enzymatic digestions.

Whole-mount, in situ chromatin imaging offers the unprecedented benefit to address cell-specific composition and organization in tissue and organ context. Such approaches applied to ovule and seed preparations, which are particularly challenging for imaging, pioneered the finding that the plant chromatin undergoes specific dynamics in reproductive tissues. Histone variant composition, histone modifications and heterochromatin organization are dynamically remodeled in connection with epigenetic reprogramming during sporogenesis, gametogenesis, and embryogenesis, and endosperm development [35, 36]. For quantitative immunostaining assays, it is important to: (1) reduce the antibody concentration and incubation times to capture the linear phase of the immunodetection reaction, (2) normalize the chromatin labeling signals to DNA levels, and (3) set up the microscopy imaging with constant illumination, zoom, image format and detector mode between the samples (*see* also **Note 2**).

In addition, confocal imaging of whole-mount FISH offers to elucidate the cell-type and developmental stage-specific organization of target loci. Examples include the developmentally regulated locus decondensation in relation to cell-specific gene expression in roots [37] and parent-specific chromosomal behavior in wheat-pearl millet hybrid embryos [38]. In situ chromatin analyses are more challenging in crop plants often harboring thick tissues impeding on homogenous probe and reagent penetration. In these cases, tissue embedding and thick sectioning in slices capturing several cell layers offers a good solution allowing spatial, comparative analyses between cell types. Some illustrative examples include the study of the tissue-specific organization of genomic loci in wheat endosperm [39], the tissue-specific distribution of histone modifications in barley roots [40], ***Brachypodium*** seeds [41], and tomato tissues [42], and of tissue-specific chromatin modification underlying epigenetic differences between sexual and apomictic maize ovaries [43].

3.1.3 In Fixed Isolated and Spatially Preserved Nuclei

As described for WF imaging, nuclei isolation allows to produce a large amount of material suitable for large-scale quantification of chromatin patterns. However, 3D analyses require the preparation of spatially preserved nuclei retaining as much as possible their native form, for instance via embedding in a thin acrylamide matrix on slides (*see* Chapters 26, 31 and 32 [44–46]). Several methods have been described that allow the isolation of intact nuclei from plant tissues, either globally or using cell-sorting strategies. So far, flow-sorting based on DNA content and ploidy level with the help of FACS equipment is the most commonly used approach.

In addition, the recent development of nuclear tags to obtain tissue or cell-specific fractions [5–7, 47] (and Chapter 8 [8]) pave the way toward quantitative, high-resolution nuclear architecture studies at the single cell level.

3D imaging of isolated nuclei enables the highest possible resolution offered by CLSM with only little disturbance. In practice, it allows resolving signals laterally in the range of 200–250 nm depending on the numerical aperture (NA) of the objective and the wavelength applied (Fig. 1). 3D image stacks may be used in image processing for segmentation (*see* Subheading 7), which is the first step toward the analysis of the spatial distribution of signals within the nucleus [33, 48, 49] (and *see* Chapters 27, 32, and 33, [45, 50, 51]). This approach can be applied to identify chromocenters, which form conspicuous foci and can therefore be segmented digitally. Notably, 3D segmentation and spatial distribution analyses showed that *Arabidopsis* chromocenters are preferentially distributed at the nuclear periphery, and that they follow a specific dispersion pattern [49, 52, 53]. 3D imaging followed by 3D rendering enables also the reproduction of structures that are otherwise difficult to interpret in 2D. One example is the nuclear belt, formed by SUN proteins around the chromosomal bouquet during meiosis in maize [54].

3.2 Time-Lapse Imaging in Fresh Tissue to Follow Chromatin Dynamics in Real-Time

In combination with the development of expressed nucleus-specific protein markers, the fast time-lapse recording of dynamic events affecting chromatin under native conditions became feasible. For instance, chromatin mobility [55], centromere dynamics during the cell cycle [52, 56], real-time monitoring of DNA replication [57], histone variant incorporation [58] and variations in DNA methylation levels [59] were investigated using time-lapse imaging.

While most real-time imaging of live processes was performed by CLSM over 6–12 h, the affordable temporal resolution can be improved with spinning disk or multiphoton systems (*see* Box 1), but especially by light sheet microscopy (*see* Subheading 4). Long-term (>24 h) live chromatin imaging poses specific challenges where spatial and temporal resolution have to be balanced with the phototoxicity triggered upon cumulative illumination and the stability of the fluorophore. In addition, nuclear movements in living cells and tissue growth add another level of complexity impeding the nuclear tracking to follow chromatin processes (*see* also Subheading 7). Specific recommendations and issues for live cell imaging have already been discussed and provided in valuable guides [10, 12, 13].

The molecular stability, or residence time of chromatin complexes (histone variants or chromatin binding proteins) can be inferred from time-lapse imaging of FRAP of a localized region in the nucleus (*see* Chapter 24, [60]). The rationale behind this approach is that the recovery time provides an indication of the average residence time of the protein of interest (the shorter, the faster

3.3 Time-Lapse Imaging for Quantifying Chromatin Properties and Molecular Interactions (FRAP, FRET, and FLIM)

the fluorescence will be recovered in the bleached region). The difficulty in such an imaging approach is to balance the speed and frequency to capture the recovery events, while not impairing the cell viability (which will stop chromatin processes and fluorescence recovery) nor bleaching the rest of the nuclear fraction. Typically then, spatial resolution is reduced (smaller image format, *see* **Note 5** for an example) to keep an informative temporal resolution.

Visualizing chromatin–protein interactions *in planta* using FRET (*see* Subheading 3.1) can prove very difficult due to numerous local fluctuations of the nuclear environment influencing FRET emission and intensity. To circumvent this problem, one approach is to measure the decay of the emitted FRET fluorescence over time, following short excitation (instead of measuring its cumulative emission under constant illumination). This is the principle of time-domain, fluorescence life time imaging (FLIM). The FRET-FLIM approach [61] was successfully used in plants to demonstrate molecular associations among chromatin binding and/or chromatin regulator proteins. A few examples include interactions between specific members of the POLYCOMB REPRESSIVE 1 (PRC1) complex [62], pathogen effectors and the nucleosome or transcriptional regulators [63, 64], and between MADS-domain transcription factors [65]. The technique also allows to map internucleosome interactions in situ, and a further application is the characterization of local heterogeneities in the chromatin compaction along meiotic chromosomes as shown in worms [66, 67]. In addition, although not yet applied to plant cells, the FLIM-FRET approach provides a promising alternative to prove the DNA binding property of chromatin proteins with the DNA itself or with nascent RNA and DNA [68, 69]. It would be very interesting to apply these approaches to visualize spatial nuclear domain dynamics in plant cells, involving local chromatin arrangements, or short-lived protein complex interactions.

3.4 Limitations

It is relatively straightforward to perform 3D imaging with the currently available confocal imaging instruments. However, several imitations exist at the level of the imaging depth, the spatial and temporal resolution. The depth of imaging is impaired severely by the inherent opacity of plant tissues. When efficiently compensated by tissue clearing (*see* Subheading 3.1) confocal systems implemented with multiphoton excitation offer a significant improvement in the imaging depth. Due to the ability of higher wavelengths (in the 900–1300 nm range) to penetrate thick layers of living tissue with a higher efficiency multiphoton excitation provides a good solution to perform long-term, time-lapse imaging on live tissue in 3D.

In time-lapse imaging, a high imaging speed is essential to record fast events but also to prevent photobleaching (arising upon prolonged illumination with short, highly energetic wavelengths). In this respect, confocal systems equipped with a resonance scanning mode are advantageous over classical scanning systems (with a scanning speed up to 30 frames per second instead of 5–6, respectively). Yet point-scanning systems are notoriously slower than confocal spinning disk microscopy (CSDM) systems. The latter are better suited for fast, long-term nuclei imaging. Although CSDM offers a better temporal resolution, the NA is fixed by the spinning disk, hence limiting the spatial resolution.

For all systems the main limitations are posed by the theoretical optical resolution (constrained by the NA and the wavelength) and chromatic aberrations. Chromatic aberrations are particularly pronounced in the longitudinal axis (z) generating a highly anisotropic signal distribution (the typical spindle shape in z; *see* for example Fig. 3c), and hinder unambiguous 3D measurements without a robust image correction after the raw data acquisition. Isotropic images can be obtained from multiangle light sheet illumination microscopy (*see* Subheading 4), although currently available instruments have a diffraction-limited spatial resolution. An alternative to reconstruct the optimal spatial resolution from blurry images is to assign the signal coming from the focus/out-of-focus planes at acquisition (using compound detector arrays built as a multifaceted hexagonal fly eye) and recalculating the true focal signal. This principle has been implemented in the AiryScan detectors from the Carl Zeiss company.

4 Light Sheet Fluorescence Microscopy

Live imaging is essential to understand the dynamics of biological processes in cells, tissues and whole organisms. However, the use of conventional and confocal fluorescence microscopy for this purpose is restricted. Instead, in light sheet fluorescence microscopy, fluorophore bleaching and excitation-induced phototoxic effects are reduced via optical sectioning during the excitation process. Thus, the spatial long-term observation of biological processes under physiological conditions becomes feasible [70].

To perform long-term light sheet microscopy, special specimen preparation protocols allowing imaging under physiological conditions are required. For live imaging of developing *Arabidopsis* seedlings and tissues, especially adapted sample preparation methods and tools were developed. Von Wangenheim et al. provided guidelines and protocols to design live-imaging experiments for *Arabidopsis* shoots and roots [71]. Noncommercial light sheet systems have customized hardware, software, and specific sample loading approaches requiring local experts to run the platforms. In the recent years, however, the first commercial light-sheet microscopes (*see* Box 1) have become available and offer a relatively user-friendly imaging and image processing [71–73]. But progress is still awaited to improve the versatility in sample handling, especially for accommodating *in planta* imaging, and for the robustness of image processing (3D image registration). For imaging with the Lightsheet Z.1 (Carl Zeiss GmbH), Ovecka et al. published a protocol to observe cotyledons, leaves, roots and hypocotyls in developing *Arabidopsis* seedlings placed in a solid medium within glass capillaries or fluorinated ethylene propylene tubes [72]. A multisample *Arabidopsis* growth and imaging chamber (MAGIC) providing near-physiological imaging conditions and allowing high-throughput experiments has been constructed by de Luis Balaguer et al. [73] for the use with the Lightsheet Z.1. With this 3D printed device the authors quantified the number of cell divisions and demonstrated that the plants underwent cell divisions 416 times longer than with the glass capillary system supplied with the Lightsheet Z1.

To optimize the raw data acquisition and process the data, specific methods have been developed to improve image contrast, spatial resolution and signal strength, and to minimize scattering artifacts [74, 75]. In addition to commercial software solutions, open access platforms are available to improve the handling of light-sheet microscopy raw data [76].

The usefulness of light-sheet microscopy to image cell nuclei and divisions has been demonstrated in *Arabidopsis* plants expressing transgenic green (GFP) or yellow fluorescent protein (YFP) constructs. Sena et al. investigated the dynamics and spatial orientation of cell divisions in growing roots expressing the histone 2B (H2B) protein fused to YFP [77]. Novak et al. analyzed the dynamics of microtubule plus-end-tracking proteins fused to GFP in interphase nuclei and mitotic cell divisions of meristematic root cells [78]. Maizel et al. uncovered the growth rhythm of lateral roots by quantifying the contribution of cell elongation to the early morphogenesis of lateral root primordia, and additionally investigated the cell plate formation during cytokinesis [79]. The accumulation of the small GTPase RabA1d labeled by GFP was proven in cytokinesis during the cell plate formation in root meristems [80].

Although light-sheet microscopy has not yet been applied to investigate plant chromatin dynamics in detail and has so far only been applied in *Arabidopsis*, it is expected that this technique will become broader applied in the near future. Light sheet imaging can advance plant chromatin studies for two reasons: (1) the low phototoxicity and noninvasive design allowing long term imaging of nuclear processes and (2) the generation of isotropic 3D images, i.e., the resolution is equally good in all three dimensions. For (1), however, the high complexity of tissue organization and nuclear movements and changes over time requires the development of robust image processing algorithms enabling nuclei tracking and the extraction of nuclear shape and fluorescence levels (segmentation). For (2), the requirement is to image under multiple angles relative to the objective. Following acquisition, images are processed by fusing the sharpest part of each angle (image registration) into a single reconstructed image where the poor axial resolution disappears. This is of particular interest to resolve nuclear domains *in planta* and over time. This, also, requires intensive computational processing, for which open source plugins and commercial software are available [81]. Nevertheless, the optical resolution is limited by the NA of the objectives.

5 Super-Resolution Microscopy

Super-resolution microscopy (SRM, also referred to as optical nanoscopy) bridges the resolution gap between WF/Confocal light microscopy and electron microscopy. These "subdiffraction" imaging methods reach down to 20–50 nm and 50–100 nm in lateral and axial resolution, respectively (*see* Box 1) and can be divided into two different principles: (1) structuring the illumination light to collect high spatial frequencies in the image that contain high resolution information (SIM, STED) and (2) localization of individual fluorophores in the specimen with subdiffraction precision (PALM, STORM) [82]. Both structured illumination and localization microscopy may be used for fixed material but also for imaging living cells. Depending on the different super-resolution techniques and the imaging tasks to be performed (e.g., quantification or colocalization of molecules) the specimen preparations have to be adapted.

Excellent reviews describing and comparing the different super-resolution microscopy methods already exist [83–90]. These techniques were applied successfully in cell biology [91] with fixed and live specimens from both prokaryotes and eukaryotes and helped to discover new structures also within plant chromatin [92].

Coltharp and Xiao provided in a review the required technical background to choose the appropriate super-resolution method for designing and analyzing imaging experiments [89], and Dame and Tark-Dame summarized the results achieved by combining chromosome conformation capture and optical nanoscopy techniques to investigate chromatin [90]. Due to the fixation procedure required before applying immunostaining and FISH, structural artifacts within the specimen may occur. However, Markaki et al. demonstrated that by appropriately adapted 3D-FISH the key characteristics of the nuclei are preserved, and that the applied 3D-SIM yields new insights into the functional nuclear organization [93].Plant cell imaging is more challenging than imaging of prokaryotic and animal/human tissues, due to the varying refractive indices of plant cell organelles, which induce spherical aberrations and light scattering [87]. Nevertheless, the applications of SRM in plant chromatin research are also increasing [92].

Overall, SRM approaches remain relatively challenging in sample preparations especially for PALM/STORM and STED, and require technical expertise to optimize image acquisition and processing. For this reason, and together with the relatively high costs of the instruments, they are still not commonly used to study plant nuclei and chromatin. Nevertheless, the emergence of commercially available solutions offering the imaging instruments with a user-friendly platform and integrated image processing software, together with increasing scientific reports should encourage the plant chromatin community to exploit the unique benefits of these exciting new imaging techniques due to the supreme achievable resolution in light microscopy. In the following sections we describe and evaluate the applicability of the different super-resolution methods in plant chromatin research and imaging.

5.1 SIM

3D-SIM image reconstructions of nuclei from mammalian species indicated a sponge-like architecture of chromatin [83, 93, 94]. A similar organization has been observed by SIM for plant chromatin within interphase nuclei of *Arabidopsis* and rye [95–97]. Compared to WF followed by deconvolution, SIM imaging clearly delivers increased information about the substructural arrangement of chromatin within nuclei (Fig. 4a), but also the structures along condensed chromosomes could be analyzed in more detail [98]. Thus, looped chromatin fibers were identified by SIM at centromeres and chromosome arms, as shown for barley [99], rye [100] and pea [101].

Due to the absence of cytoplasm, isolated chromosomes and flow-sorted nuclei are well suitable to perform immunostaining and FISH followed by 3D-SIM [101, 102]. However, also the SIM imaging of squash preparations may achieve the doubling of resolution as demonstrated on monocentric and holocentric plant chromosomes [100, 103–105]. 3D-SIM image stacks are also the prerequisite for the colocalization and relative quantification of fluorescently labeled chromatin proteins and DNA interacting factors. This has been demonstrated for transcription elongation factors and RNA polymerase II within the euchromatin of *Arabidopsis* interphase nuclei [95, 106, 107] and by colocalizing centromere-specific DNA repeats, histones, and kinetochore proteins of the holocentric plant *Rhynchospora pubera* [104, 108].

The major advantage of SIM is that most preparation and labeling protocols and fluorophores used for widefield fluorescence microscopy are applicable without modification, thus allowing high-throughput experiments. In addition to proper tissue fixation, the use of high-quality glass slides and especially coverslips (e.g., Marienfeld high precision coverslips) are important to reach the possible resolution of up to ~120 nm by using a 488 nm excitation.

SIM raw data acquisition and image stack calculation is relatively fast. By using for example a Zeiss ELYRA PS.1 microscope system,

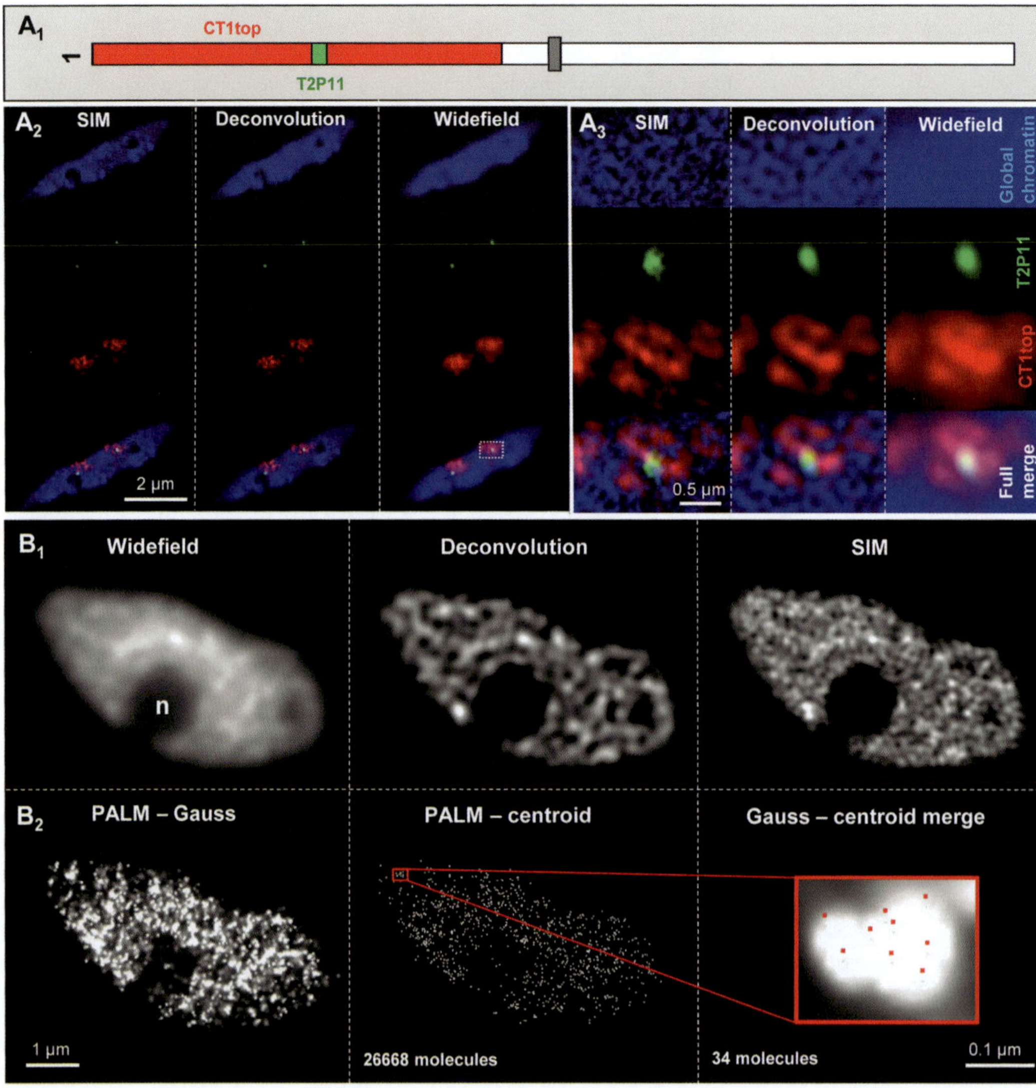

Fig. 4 Super-resolution microscopy. (**a**) Chromatin organization in a differentiated 4C *A. thaliana* leaf nucleus analyzed by SIM. (**a$_1$**) The top arm of chromosome 1 was stained in *red* with a ~14.2 Mb BAC contig and a ~85 kb segment (BAC T2P11) therein (*green*). (**a$_2$**) After FISH and SIM the organization of the euchromatic arm regions of both homologous chromosome territories (CT1top) becomes visible beyond the diffraction limit of light in comparison to the resolution achieved by deconvolution and widefield microscopy. (**a$_3$**) The enlarged region (*rectangle* in **a$_2$**) shows clearly the increased resolution compared to widefield and deconvolution imaging. (**b**) Distribution and quantification of active RNA polymerase II (phosphorylated at serine 2) molecules within the euchromatin of a differentiated 4C *A. thaliana* leaf nucleus analyzed by PALM (**b$_2$**) in comparison to SIM, widefield and deconvolution microscopy (**b$_1$**). After immunostaining with specific antibodies, SIM and PALM were performed consecutively. Compared to widefield and deconvolution microscopy SIM clearly delivers an increased resolution beyond the diffraction limit of light illustrating that RNAP II is distributed network-like within the euchromatin of the nucleus, but absent from the nucleolus (n) (**b$_1$**). (**b$_2$**) After 3D-PALM the resolution becomes further increased when visualized as Gauss distribution (*left*). Centroid visualization allows the exact localization and counting of single molecules (*middle*). The nucleus with a *z* extension of 1.72 μm contains in total 26,668 molecules which were counted in 43 slices of 40 nm. The merged Gauss-centroid view (*right*) represents the localization of nine single molecules in a single slice of a RNAPII cluster (*red rectangle*) containing in total 34 molecules in five slices with a total *z* range of 0.2 μm

the acquisition of raw image stacks of up to 30 slices at a distance of ~100 nm takes ~4–5 min in a sequential three-color experiment. These raw stacks are the basis to calculate 3D-SIM image stacks within ~20 s (*see* for example Fig. 4a) [102], and they may be used for 3D-movie rendering by open source or commercially available software (*see* Table 2 and [96, 101, 109, 110]).

It is also important to notice that correct instrument settings and optimal imaging conditions are required to avoid the generation of artificial signals resulting in apparent structures and shapes during the SIM calculation that do not exist in the biological sample. Especially overexposure should be avoided. However, open-source [111] and commercial software have been developed to identify and avoid such SIM data deficiencies.

5.2 PALM/STORM/GSDIM

Localization microscopy (PALM, STORM, GSDIM, *see* Box 1) works on the principle that fluorescence emission is temporally uncoupled between single molecules. One image integrated over a few nanoseconds captures only a fraction of the fluorophores following a stochastic, spatially resolvable distribution (*see* Box 1). The cumulative integration of all captured images reaches a circa fivefold higher resolution than SIM. Thus, single molecules may be identified, counted, and colocalized in single- and even two-color experiments, in 2D or 3D recording using common fluorophores.

Fluorophores. For instance, the fluorescent dyes Alexa488 and Cy5 proved to be efficient to label enzymes in *Arabidopsis* nuclei for PALM imaging [102, 112]. Yet for GSDIM imaging, other fluorophores particularly in the far-red spectrum (Alexa647) seem to provide interesting properties such as high quantum yield per switching event and low duty cycle (by being longer in the nonemitting state than the emitting state) [113] and prove suitable for immunolocalization of histone H3 in isolated Arabidopsis nuclei (Baroux, unpublished).

Sample preparation. Compared to preparations used for SIM the specimen preparations for PALM are a bit more challenging with delicate technical requirements at the level of sample preparation. The computationally intensive raw data acquisition and post-acquisition image reconstruction are also more time-consuming. To achieve reliable results, a high labeling density and efficient photoactivation are required [114]. Thus, the preparation of specimens directly onto clean coverslips has to be performed [102]. 2D-PALM may be performed under HILO (*see* **Note 6**), Epi and TIRF illumination [115]. 3D-PALM allows capturing *z* ranges of up to 2.0 μm. This range might be extended by using classical *z*-scans. TIRF, and for best performance also HILO illumination, requires the positioning of the specimen very close to the coverslip surface.

PALM using fluorescent proteins can be performed in buffers like PBS or Hepes to induce optimal fluorescently blinking molecules. Acquiring raw data sets (~30,000 frames) takes ~15 min, followed by a calculation procedure of also ~15 min by using commercial systems like the ELYRA PS.1 from Carl Zeiss. GSDIM/STORM imaging is, however, best performed in buffers with an enzymatic oxygen scavenging and small-molecule photostabilizing system that suppress the prevalence of the triplet dark state, thus enhancing stable molecule blinking [116].

Live imaging. Labeling of DNA-associated proteins and direct labeling of DNA allow a better understanding of chromatin structures. Flors and colleagues demonstrated that SYTO, YOHO and PicoGreen dyes are applicable to reach a high labeling density of photoswitchable fluorophores crucial to perform super-resolution imaging of DNA and chromatin using single-molecule localization methods in fixed and even living vertebrate cells [117, 118]. However, live PALM/STORM imaging has not been reported so far in plant cells; one main hindrance is the thickness of the plant tissues, their high light refraction under the illumination conditions and the poor sample viability under conditions required for this type of imaging.

Instruments. Noncommercial super-resolution microscope setups [119] and commercial systems such as the Elyra PS.1 [102] allow producing SIM and PALM/STORM data. Thereby, the observed SIM structures can be combined with precise PALM/STORM single molecule localization and counting. This clearly increases the information obtained from the specimens under investigation (Fig. 4b).

5.3 STED

STED imaging has a raster implementation similar to CLSM. The application of multiple lasers with distinct beam geometries (to produce fluorescence depletion in a ring structure centered on the optimal PSF focus point) increases, however, the scanning time by a factor of tenfold. This implies to work on stable, fixed biological specimen with robust (yet common) fluorophores [120]. As for the other SRM techniques, STED imaging imposes to work on isolated nuclei instead of complex plant tissues. It is thus well suited for the 3D imaging and analysis of immunolabeled proteins, chromatin modifications and FISH signals. To our knowledge there is to date no report on chromatin imaging in plant cells using STED systems. Yet its user-friendly implementation using commercially available solutions (becoming less expensive) offers a good alternative to structured and single molecule localization microscopy (see above), that are challenging in implementation and image processing.

6 Electron Microscopy

Different electron microscopy (ELMI) techniques such as transmission (TEM) scanning (SEM) and focused ion beam/scanning electron microscopy (FIB/SEM) have been applied to investigate the ultrastructure of plant chromatin. Compared to fluorescence light microscopy, ELMI imaging allows to visualize cell structures at a resolution of 1–2 nm (Fig. 1). However, it is challenging to label and localize DNA and proteins specifically, both for TEM and SEM. Additionally, correlative microscopy combining fluorescence and electron microscopy have been performed.

Most electron microscopy methods for analyzing plant chromatin were developed and applied by Gerhard Wanner (Germany), Kiichi Fukui (Japan), and their coworkers. Especially for SEM specimen preparation techniques and with it achieved ultrastructural observations were already summarized by Wanner and Schroeder-Reiter [121].

6.1 Scanning Electron Microscopy (SEM)

Already in the 1990s of the last century, special preparation techniques to analyze the spatial ultrastructure of plant chromosomes were developed. *Vicia faba* chromosomes were isolated by micromanipulation and from suspensions [122], or from standard squash preparations of mitotic and meiotic chromosomes from root tips (e.g., from barley, wheat, and rye [123–125]). The combination of enzyme maceration, treatment in acetic acid, and osmium impregnation using thiocarbohydrazide proved to be effective to examine the 3D ultrastructure of interphase nuclei and mitotic chromosomes [125].

For the visualization of DNA, chromosomes can be stained with Zirconium dichloride oxide, hydrochloric acid [126] or with platinum blue [127]. SEM imaging using secondary electrons (SE) shows both DNA and protein together, whereas backscattered electron (BSE) signals monitor the DNA content separately (Fig. 5a_1, a_2, a_4, a_5). Proteins can be visualized by staining with aqueous silver nitrate solution or with an aqueous solution of colloidal silver containing elementary silver [128] (Fig. 5a_3).

The development of Nanogold markers for immunodetection significantly improved the labeling efficiency and specificity for specific proteins to be detected by SEM [129–131] (Fig. 5a_6). In addition, FluoroNanogold markers are applicable to perform correlative light and electron microscopy after FISH [132, 133] (Fig. 5a_5) and immunostaining [134] (Fig. 5a_7). Beside analyzing the 3D structure of chromatin along the chromosome arms various investigations focused on the identification and organization of components forming monocentromeres and holocentromeres [103, 129–131] (Fig. 5a_1, a_2, a_3, a_6, a_7). Based on DNA and protein staining followed by SEM, the "dynamic matrix model" for plant metaphase chromosomes was established, proposing that the chromosomes are mainly composed of DNA packed in "chromomeres" (coiled solenoides) and a dynamic matrix formed by parallel protein fibers [128, 135] (Fig. 5a_8).

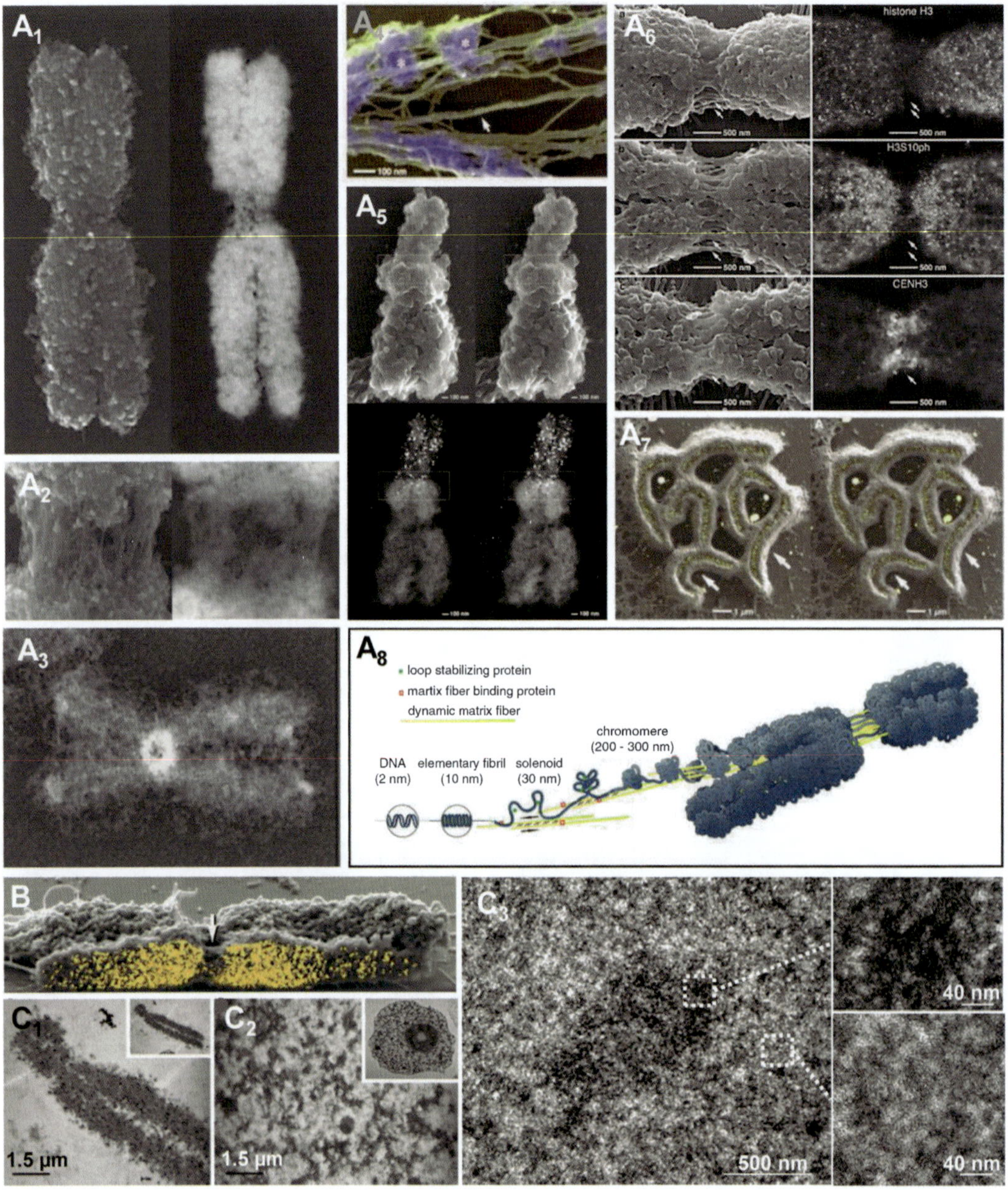

Fig. 5 Electron microscopy. (**a**) *SEM.* ($\mathbf{a_1}$) Scanning electron micrographs of a barley metaphase chromosome stained for DNA with platinum blue. The secondary electron (SE) image (*left*) shows both DNA and protein, whereas the backscattered electron (BSE) signal (*right*) monitors DNA content, which is high in areas corresponding with the chromomeres, lower in the centromeric region and between the sister chromatids [135]. ($\mathbf{a_2}$) Barley centromere structure at higher magnification stained for DNA with platinum blue. The comparison of the SE image (*left*) with the BSE image (*right*) indicates that at the centromere, the chromomeres border exposed parallel fibers (*left*) which contain less DNA than surrounding areas [121]. ($\mathbf{a_3}$) Barley metaphase chromosome stained with silver compounds showing a preferential protein accumulation at the centromeres and telomeres [121]. ($\mathbf{a_4}$) Detail of a decondensed barley metaphase chromosome stained with platinum blue for DNA (*purple*). The chromomeres show an intense DNA signal

In addition to interphase nuclei and mitotic chromosomes, the 3D ultrastructure of chromatin has also been investigated during meiosis. Although in rye the basic structural elements (30 nm fibers, chromomeres) were observed during meiosis, it has been proven that in general the chromosome surface is smoother than during mitosis [136, 137].

6.2 Focused Ion Beam/Scanning Electron Microscopy (FIB/SEM)

Although the 3D structure of spatial specimens can be visualized by SEM via stereo pairs (Fig. 5a_5, a_7), signals detected via BSE from markers localized in the chromosome interior are often blurry and faint due to electron scattering and absorption. Thus, the high resolution imaging of their spatial distribution is limited. Focused ion beam (FIB) milling (i.e., the sequential removal of material slices) in combination with SEM can be applied to investigate the chromatin packaging in the chromosome interior. This technique was helpful to clarify the inner chromatin structure of holocentric *Luzula* [138] chromosomes, and identified the 3D distribution of histone variants in the (peri)centromeres of monocentric barley chromosomes [130, 139] (Fig. 5b). In addition to compact

Fig. 5 (continued) (*asterisks*), whereas the chromosome matrix fibers (*yellow*) show either no, or very weak BSE signals (*arrow*) [135]. (**a_5**) Detection of 45S rDNA (NOR) at a peg-like terminal constriction on a metaphase chromosomes of the plant genus *Oziroë*. The 3D SEM micrographs (stereo pairs) show both chromosome topography (*upper* SE pair) and the spatial distribution of the hybridized enhanced gold signals labeling the NOR (*lower* BSE pair, DNA staining with platinum blue). The signals can be detected from different depths; diffuse signals originate from the interior or opposite side of the condensed pegs [132] (**a_6**) Topography (SE images, *left column*) and specific, but deviating label distribution (BSE images, *right column*) at the (peri) centromeres of somatic barley metaphase chromosomes immunolabeled by Nanogold Fab′-fragments for histone H3, phosphorylated histone H3 at ser10 (H3S10ph) and CENH3. The centromeres contain parallel fibrils [131]. (**a_7**) Holocentric *Lucula* metaphase chromosomes immunolabeled (FluoroNanogold enhanced with silver) for H2AThr120ph, which is predominantly localized within the longitudinal centromeric groove [183]. (**a_8**) Based on SEM imaging using DNA- and protein-specific labels a "dynamic matrix model" for plant metaphase chromatin composition and condensation has been proposed [183] (reproduced with permission of Karger AG (**a_1, a_4, a_8**); Elsevier Inc. (**a_2, a_3**) and Springer-Verlag (**a_5, a_6, a_7**). (**b**) *FIB/SEM*. Selected image out of a total of 128 images of FIB/SEM sections of a barley chromosome labeled by phosphorylated histone H3 at serine 10 (H3S10ph). The backscattered electron image (*yellow*) indicating H3S10ph is superimposed on the secondary electron image (*grey*). The direct viewing of the chromosome interior reveals that H3S10ph is accumulated around the centromere. The lateral fibrils at the centromere (*arrow*) are unlabeled [139]. (**c**) *TEM*. (**c_1**) *Vicia faba* metaphase chromosome and (**c_2**) nucleus stained with primary antibodies against phosphorylated histone H3, followed by secondary gold-labeled anti-rabbit IgG and subsequent gold enhancement. The metaphase chromosome shows evenly distributed immunosignals, whereas within the nucleus clusters appear. Inserted pictures show the corresponding chromosome and nucleus at lower magnification [142]. (**c_3**) Chromatin organization within an *Arabidopsis* root nucleus. The sample was prepared according to the high pressure freezing and freeze substitution method, followed by uranyl acetate contrast staining (*see* Chapter 34 of this issue). The enlarged regions (*rectangles*) show heterochromatin (*top*; chromocenter, *darker contrast*) surrounded by euchromatin (*bottom*; *lighter contrast*) (**b**, **c_1**, **c_2** reproduced with permission of Karger AG; **c_3**: preparation by N.T. Fabrice [144])

chromatin regions inside of the barley chromosomes, also chromatin free cavities were detected [130]. However, by using an ionic liquid technique enabling a sample preparation without dehydration, drying and coating, and thereby providing a sample that is closer to native conditions, such cavities could not be observed within barley metaphase chromosomes in FIB/SEM investigations [140, 141].

6.3 Transmission Electron Microscopy (TEM)

TEM has been used only in a few studies to investigate plant chromatin. Gold-labeled antibodies and subsequent gold enhancement proved to be effective to identify the distribution of phosphorylated H3 histones in *Vicia faba* metaphase chromosomes and nuclei after immunostaining [142] (Fig. 5c_1, c_2). Similarly, using Nano-gold antibodies, the ribosomal DNA of barley has been localized within nucleoli by in situ hybridization [125]. In addition, TEM imaging offers an unprecedented, ultrastructural level of details on chromatin organization. While chromatin patterns were formerly quantified based on the distribution of grey level intensities along the image, modern computational processing tools brought a renewed interest in TEM images as a source of information at the nanoscale level of chromatin organization [143]. Simple uranyl acetate contrast staining on plant tissue samples prepared by high pressure freezing and freeze substitution (Fig. 5c_3) allow quantifying the degree of chromatin compaction and periodicity using a user-friendly processing interface (*see* Chapter 34, [144]). This paves the way toward elucidating the structural role of chromatin regulators or the fine-scale chromatin state transitions upon environmental or developmental cues at the nucleosomal level. Thus, this approach proves highly complementary to fluorescence-based microscopy imaging approaches.

7 Image Processing

Resolving submicrometer signals in the nucleus requires a careful setup of the imaging conditions. Several issues are often underappreciated, but they critically influence the possibility to measure the spatial size and localization of signals. This includes matching mounting and immersion media (to minimize the light diffraction at the media–coverslip interface), adjusting the image format allowing a 2–3× oversampling (relative to the theoretical resolution of the objective) suitable for post-acquisition image deconvolution (*see* **Note 7**), and the moderate use of signal averaging (*see* **Note 8**).

The purpose of biological image processing is to obtain qualitative and quantitative data from signal distributions that are otherwise difficult or impossible to obtain by visual inspection, particularly on 3D or 3D time (4D) image series [145]. Several algorithms embedded in open source or commercially available software (Table 2) are available. They compute distinct tasks ranging from simple analytics (plots of signal distribution profiles), image visualization (noise filtering, volume rendering) and segmentation (2D or 3D masks generating digital objects based on signal distribution) to more complex processes leading to image restoration (deconvolution), motion drift correction (registration), or image reconstruction from complex spatial and temporal signal patterns (super-resolution microscopy). These tasks can be automated to process a large number of image data (batch processing), but this usually requires customized scripts and high-end computing environments.

We present below some examples of image processing that can be of practical use for rendering or analyzing chromatin signal distribution in plant cells. We do not cover super-resolution image reconstruction, which is specific for each imaging system and covered by specialized commercial software, reviews, and user forums.

7.1 Processing 2D Images

The processing of 2D images involves deblurring, intensity level quantification, spatial distribution analyses, motion drift correction and time-lapse tracking. A first level of analysis is a qualitative assessment of signal distribution, chromosomal structure or arrangement. Here, image deblurring via deconvolution can prove useful to resolve subnuclear structures (see below). A second level of analysis involves the scoring of signal pattern categories (e.g., distribution of a labeled protein in heterochromatin, or euchromatin, or both). It is a rapid and informative approach especially when comparing different cell types or genotypes. A third level of analysis considers signal intensities within regions of interest (ROI) and requires manual or automated isolation of ROIs used to compute intensity distributions. A fourth level is specific to time-lapse image processing requirements, with drift correction, registration, and object tracking. These different approaches are further detailed below.

7.1.1 Noise Reduction by Simple Filtering Approaches

Signal noise can often be identified as a salt-and-pepper dotty pattern or as residual, apparently random, and low-intensity signals drastically contrasting with major, conspicuous, and strong intensity objects. Noise can have several origins: experimental (e.g., unspecific antibody binding creating low-intensity dotty patterns) or technical (e.g., at the electronical level: spontaneous photoelectric conversion in camera sensors, amplifying noise with photomultiplier tubes, a problem largely corrected in modern devices). This should be carefully evaluated, however, against the possibility of a biological origin (i.e., where an apparent weak, random signal corresponds in fact to rare target localization) using experimental and imaging controls (samples without primary antibodies or tagged proteins etc.). Imaging noise can be removed by simple filters available in most processing software, which apply either linear or nonlinear signal processing approaches such as the common smoothing Gaussian filter and the edge-preserving median filter, respectively.

Noise reduction also proves very efficient in high-throughput imaging where spatial resolution is compromised for the benefit of speed of acquisition: recently, novel algorithms have been developed, allowing for superior and fast corrections for analyzing punctuate, chromosomal signals in yeast cell nuclei in large fields of view [146].

7.1.2 Deconvolution to Deblur Images

WF microscopy often suffers from out-of-focus light blurring, originating from subnuclear structures below 1 μm. Deconvolution algorithms aim at restoring an image closer to reality. It considers a mathematical function of signal distortion (point spread function, PSF) to ultimately compute an image free of optical aberration, comprising only the predicted source point of signal (*see* **Note 7**). 2D deconvolution is effective to improve the sharpness of nuclear speckles, chromatin signal patterns, chromosomal structures, and FISH signals (examples in Fig. 6a, b). It is important to consider carefully the parameters of image deconvolution (image format, wavelength, and refractive index of the immersion medium) as well as properly estimating the signal to noise ratio (SNR) and trying a different number of iterations to avoid producing shape artifacts. 3D deconvolution can efficiently restore for instance FISH signals down to punctuate structures consistent with the predicted size of targeted chromosomal domains (Fig. 6c). Yet the algorithms often do not entirely restore isotropic (spherical) signals (Fig. 6c, close-up, axial view). Another example is the revelation of submicrometric punctuate domains of varying intensity within chromocenters in DAPI-stained nuclei, normally shown as uniformly stained (blurred) domains, after 3D deconvolution (Fig. 6d).

7.1.3 Image Segmentation to Quantify Nuclear Signal Shapes and Intensities

The relative amount of a nuclear protein can be inferred from measuring the intensity of the signals recorded. This, however, requires a careful setup of the experimental, imaging and acquisition conditions (*see* **Note 2**) to avoid common pitfalls such as saturated signal levels (due to excess of antibody molecules or overexposure) and enable an internal normalization with a second, invariant marker (e.g., DNA staining). It should also be noted that signal intensity quantifications should not be done on images on which a noise reduction (smoothing) or deblurring preprocessing has been applied. These operations transform the image information, and are not suitable for comparative quantifications. Care should also be taken when quantifying 2D images generated from axial projections of 3D images. Typically, 3D image rendering software offers maximum intensity projections (MIP) that give an esthetically pleasing result, but should be prohibited for quantification. Indeed, as the name suggests, the projection considers only the highest intensity pixel in a given column (*z* axis). Instead, intensity sum projections should be preferred.

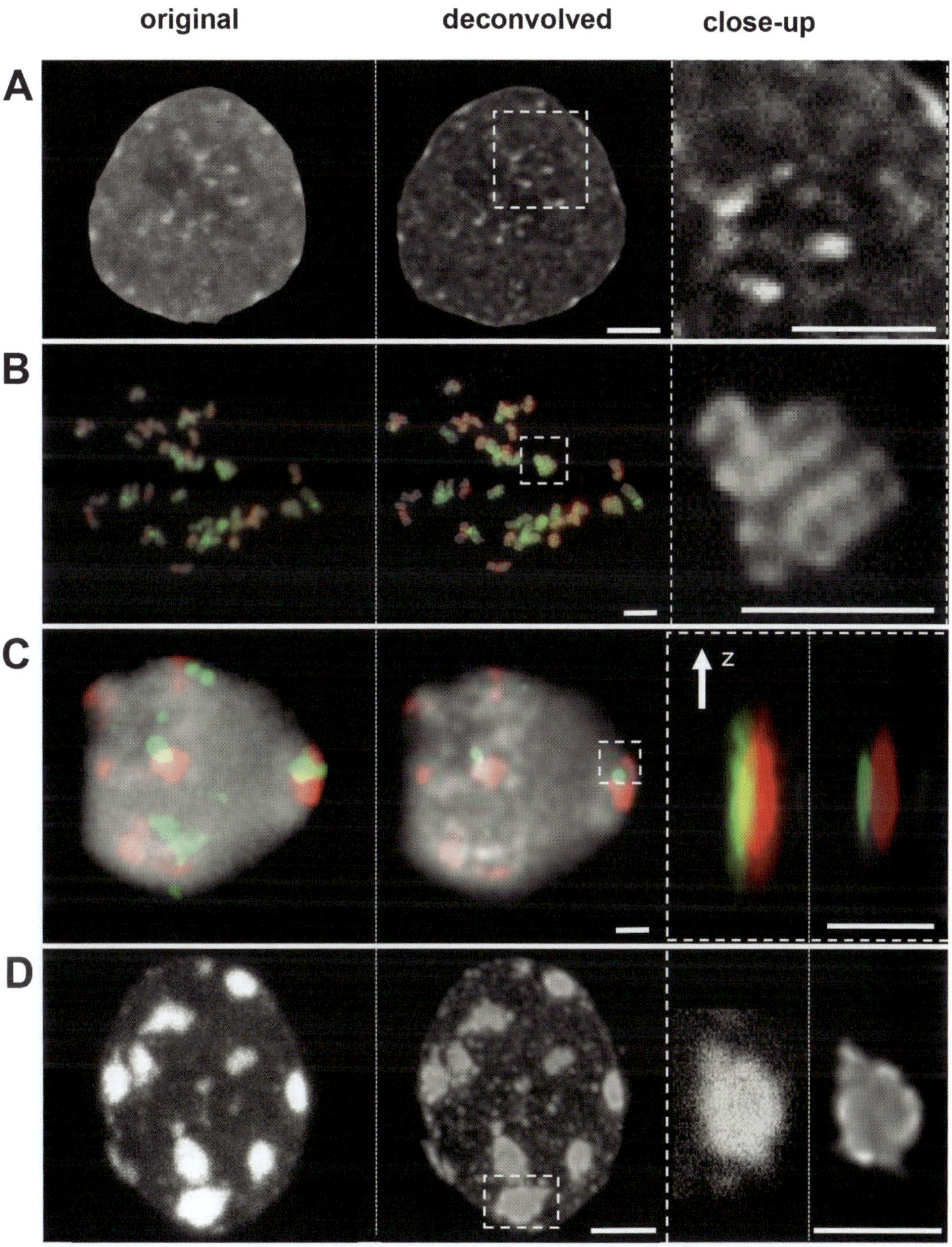

Fig. 6 Deconvolution removes out-of-focus blur to the benefit of image interpretation. Examples of deconvolution-based restoration of 2D, widefield (**a**, **b**) and 3D, confocal (**c**, **d**) microscopy images. (**a**) DAPI stained isolated, spread nucleus from Arabidopsis seed endosperm [184] (CB, own image). Restoration allows visualizing contrasted compaction domains in euchromatin (close-up). (**b**) Metaphase chromosome plate from *Primula egaliksensis* root, stained for its respective parental genomes by GISH [182] (chromosome preparation kindly provided by A. Guggisberg). (**c**) Isolated leaf nucleus, centromeric (*red*) and 45SrDNA

Isolated nuclei typically produce a large sample size suitable for various quantifications. When combined with immunostaining, for instance, the relative levels of the antibody signals normalized over DNA signals can be extracted from either hand-drawn ROIs (e.g., in Fiji using the ROI and Measure Tools, Table 2) or from automated masks running from customized plugins and considering intensity thresholds for defining nuclei boundaries. The latter for instance is commonly used to measure the relative heterochromatin fraction (RHF) on spread nuclei [147] and allows to confirm quantitative changes in heterochromatin organization during physiological or developmental transition or confirm the role of specific chromatin modifier mutants in maintaining heterochromatin (reviewed in [148–150]).

7.1.4 Spatial Signal Distribution Analyses

Spatial signal distribution analyses can be used to quantify chromatin density distributions and patterns of chromatin compaction. Ultrastructural microscopy approaches such as super-resolution (*see* Subheading 5) or electron microscopy (*see* Subheading 6) provide a nanoscale resolution of chromatin structures. These methods enable to investigate 2D and 3D patterns of chromatin organization. For instance, computing spatial autocorrelation functions on TEM images of nuclei evaluates chromatin signals with similar densities and their spatial distribution (Fig. 5c$_3$). This mathematical approach provides a proxy measure of periodicity versus dispersion of nanoscopic, local compact chromatin foci within euchromatin [143], which correlates with transcriptional competence [151]. This approach is well suited to analyze fine-scale chromatin organization in plant cells and provides novel opportunities for analyzing genetic mutants and chromatin dynamics during development. A user-friendly analysis interface is described in Chapter 34, [144]). Spatial signal localization analyses have also been used to resolve the distribution of histone variants to decrypt the nucleosomal level of chromatin compaction, using SRM imaging. Notably, Ricci et al. [152] demonstrated that the chromatin of pluripotent mammalian cells is composed of small and dispersed compact nucleosomal arrays compared to the large and more proximal arrays in differentiated cells, and that linker histones play a critical role in establishing these nanoscopic patterns. This work is inspiring both in the experimental (imaging) and modeling design for investigating the plant chromatin structure at finer scale, and in

Fig. 6 (continued) (*green*) FISH, DAPI counterstaining (*grey*). Deconvolution restores the fine-scale DNA (DAPI) distribution and reduces the chromatic aberration along the *z*-dimension (close up: link, nonrestored; right, restored) (**d**) DAPI-stained, isolated, intact (3D) leaf nucleus. Deconvolution reveals the fine-scale distribution of chromatin domains with varying compaction levels in euchromatin and in the chromocenters (close-up: link, nonrestored; right, restored). Image kindly provided by M. Ashenafi. Scale bars: 1 μm

relation to plant cell reprogramming during differentiation and responses to environmental cues.

7.1.5 Nuclei and Particle Tracking in Time-Lapse Movies

Time-lapse movies generated on WF systems, or 3D imaging systems after projection (see below), produce simple 2D sequences amenable to measure chromatin properties (e.g., by FRAP, FLIM-FRET, *see* Subheading 3), chromatin mobility or dynamic changes in nuclear organization during a biological process (*see* Subheadings 2 and 3). Yet specific challenges are present. Notably, the biological sample can drift within or away from the field of view over time, either due to tissue growth or to changes in the medium over prolonged illumination (i.e., due to temperature-induced convections). In addition, nuclei are extremely mobile in the cell, with movements ranging from gentle agitation to intracellular migration and shape changes [153]. Light, temperature, and mechanical stimuli influence nuclear movements [154]. From a practical point of view, when performing FRAP analyses for instance, these movements can be reduced by imaging in chambers with temperature control set at ~20 °C. For time-lapse analyses requiring intensity level measurements over time in a region of interest, it is important to correct drift and agitation using an ***image registration*** process [12, 155]. This can be achieved by using the open source image processing software ***Fiji*** (image registration plugins) [156].

Studying nuclear movements or the mobility of intranuclear foci (e.g., chromatin foci, nuclear speckles) requires algorithms able to track the fluorescence signal, based on object detection and connection between time frames (for example *see* [57, 157]. Again, in addition to commercial software (e.g., Imaris from Bitplane) several open-source algorithms are available to visualize graphically the direction, amplitude and speed of movements.

7.2 3D Processing and Quantifications

Similar to 2D images, 3D images can be preprocessed to reduce noise (filtering), to correct for optical aberrations (deconvolution) or motion drifts prior to downstream analyses. Algorithms are available that compute the same principle as described above (Table 2). Hence they will not be discussed here again. Specific challenges also arise for downstream analyses where the segmentation of nuclear signals in 3D-reconstructed image stacks is computationally challenging.

3D image analysis in nuclear organization studies involves different parameters that are biologically relevant, such as (1) signal number and intensity (2) signal domain shape, size and texture, and (3) signal domain position in the nuclear space relative to different nuclear compartments. The prerequisite to 3D quantification is to create a digital, 3D object capturing a region with fluorescence signals of interest. This process of isolating a group of pixels based on common features is called segmentation. It is one of the most difficult tasks in image processing. Several algorithms exist varying on the mathematical treatment of the nucleus [33, 48, 49] and the distribution of pixel intensities in the image. For the most basic approaches only thresholding (background subtraction) may be used. Other approaches are based on similarities between neighboring segments (growing approach), local contrasts in intensity maxima, or intensity distribution gradients (watershed methods), among others [158]. These approaches have a different sensitivity to noise or heterogeneous staining that they may produce different signal boundaries in lower intensity regions depending on the algorithm. Thus, it is worth considering more than one method to select the most effective one for the purpose of the study. Customized plugins are available in Fiji or open source codes in Imaris, among other options (*see* Table 2). Practical examples concerning chromatin and nuclear analyses are described below (*see* also Chapters 27, 32, 33 [45, 50, 51]).

For instance, the *Arabidopsis* chromocenters (conspicuous heterochromatin foci) are distinguishable by their contrasted signal intensities upon DNA staining. Algorithms were developed enabling their automated segmentation in 3D in *Arabidopsis* cells and allowing to study their number, size and shape in different cell types and mutants [48, 159]. In addition, their spatial distribution was mathematically modeled using spatial statistic approaches to describe their preferential peripheral localization as well as their spatial dispersion [49, 52, 53]. Another example is nuclei segmentation within a complex 3D image representing a whole-mount tissue stained by immunolabeling. It proved a powerful approach to quantify histone modifications and uncovered their specific dynamics with single cell level precision in *Arabidopsis* ovules [33] (and online tutorial [160]).

The emergence of user-friendly interfaces to implement these image processing plugins offer novel possibilities that will benefit the plant chromatin community. Although they have not been extensively used so far, they provide powerful tools to quantify the spatial distribution of nuclear speckles and bodies, chromatin protein complexes (e.g., Polycomb bodies and transcription factories) and genomic loci [15]. In combination with physiological, genetic and mathematical modeling approaches, they have the potential to unravel the functional organization of the plant nucleus.

Finally, the future of image processing applied to plant biology strongly relies on 4D image analysis. The time dimension brings another level of complexity, but of course an invaluable source of information regarding the evolution of the shape, size and number of nuclear compartments, of chromatin composition and the nature of chromatin dynamics. Despite the advent of 4D processing and cell tracking software [161, 162], including some specifically adapted to plant cell shapes [163, 164], motivated efforts are still needed to apply and develop them further to unravel nuclear dynamics in its entire complexity during *in planta* processes.

8 Perspectives

8.1 Moving the Plant Chromatin Field Toward More Super-Resolution Imaging and Image Processing-Based Analytics

The number and efficiency of optical instruments for biological imaging has been dramatically improved during the past decades. This is partly due to the technological progress in optical compounds, electronic sensors and composite material capturing light with increased sensitivity, as well as fine-motorized control of miniature components. But it is also largely based on novel theoretical developments enabling to reconstruct biological images from diverse forms of light manipulation (emission, absorption, diffraction together with improved optics) and on the development of better specimen preparations. This makes the choice for the correct imaging instrument even more difficult for the biologist. However, the imaging possibilities are often guided by the availability of instruments and expertise. Particularly, while CLSMs became routine instruments, ELMI and SRM have so far principally be operated by a few experts, as both implementation and image processing remain challenging, although some systems, such as STED and SIM, become affordable.

In terms of nuclear architecture and chromatin organization studies, imaging with high or super-resolution is unavoidable to elucidate the functional organization and dynamics of (sub)nuclear domains, chromosomal, nucleosomal and protein complex arrangements. High-end imaging of the nucleus in isolated cells or intact tissues remains only of little use without advanced image processing. Processing of image stacks is not only a requirement to visualize 3D patterns understandable for our brain, it also delivers the basis for quantitative analytics and mathematical modeling.

8.2 Novel Optical Label-Free Imaging Approaches

Meanwhile, optical imaging solutions beyond the classical systems relying on fluorescent labeling were developed. Numerous strategies allow the imaging of cells and nuclei without exogenous labels, solely relying on the intrinsic emission, absorption or light refraction properties of the material. For instance, digital holographic microscopy (DHM) enables to map refractive index differences in the nucleus [165]. Further, a sample-rotation-based tomography application of DHM proved to be able to provide spatial, isotropic images of refractive index maps in animal and plant cells and nuclei [166]. These imaging techniques merit to be further investigated to potentially reveal dynamic differences in biophysical properties of the chromatin along cellular differentiation, physiological or environmental responses.

Furthermore, recent developments in biophysics established mathematical interpretations of light diffraction patterns through living cells in spectroscopic microscopy enabling to infer nanoscale, structural patterns of the chromatin [167]. This approach, although perhaps still a bit theoretical, offers a new readout of chromatin images. Thus, the density and distribution of macromolecular complexes (chromatin) can be quantified and modelled. Altogether, these label-free imaging and image analysis techniques are promising to approach chromatin organization studies under a new angle, reconsidering its macromolecular and fractal organization in situ and at the nanoscale.

8.3 High-Throughput, Semiautomated Imaging

Another important field of chromatin investigation is connected to high-throughput, semiautomated imaging and image processing of plant nuclei, of serial tissue preparations or isolated cells. This will allow to perform genetic screens toward the identification of novel chromatin/nuclear structure regulators, but also to score for biological diversity in developmental and physiological responses at the nuclear level. The availability of semiautomated fluorescence microscopes and image analysis pipelines for cell/nuclei classification and analysis is the basis for such investigations. The power of such an approach is remarkably illustrated by the identification of regulators, via a high-throughput FISH-based RNAi screen, controlling gene positioning within the nucleus in mammalian cells [168]. This and other studies also confirm that there is a need to bridge cell/nuclear biology with system biology approaches to capture cell-to-cell variability and to describe better the rules underlying nuclear organization patterns.

8.4 Toward Customized Live-Imaging Reporters of Chromosomal Loci

More than a decade ago, the successful time-lapse imaging of a tagged genomic locus to follow its intra-nuclear dynamics was published [55]. However, due to the experimental, imaging and image processing challenges, the expectation to image chromosomal regions and specific loci "a la carte" to elaborate spatiotemporal maps of the nucleus underlying cellular processes and plant development has not yet been fulfilled. Nevertheless, significant progress toward reaching these aims can be observed. Improved microscopy imaging techniques are now evident, image processing is also constantly improving, and gene targeting technologies offer now a versatile solution to produce customized live-reporters for chromosomal loci. *Arabidopsis* lines expressing FP tags labeling various genomic repeats have been recently reported using either Zn-finger [169] or TALE effectors [170]. The new CRISPR-Cas9 technology seems to be most promising to label genomic loci as shown in human cells [171], which has meanwhile been adapted for plants [172].

9 Notes

1. Several microscopy imaging providers, filter cube and digital camera manufacturers provide interactive platforms and online articles with basic and advanced explanations on the underlying technologies and important considerations for optimized imaging, e.g., Science Lab (www.leica-microsystems.com/science-lab/), Olympus Microscopy Resource Center (http://www.olympusmicro.com/primer), Zeiss Campus (www.zeiss-campus.magnet.fsu.edu), and MicroscopyU (https://www.microscopyu.com).

   In addition, the *Handbook for Biological Confocal Microscopy* [173] provides an exhaustive, updated educational resource.
2. ***General considerations and practical tips for recording images suitable for signal quantification.*** A general advice is to consider signal normalization against the amount of DNA/chromatin. This can be useful for instance in fast cycling or endoreduplicated cell types where the level of the studied chromatin protein or modification may be correlated in a stoichiometric manner with the genome content. Dual imaging of the GFP- or antibody-tag against an internal reference should be implemented using for instance an invariant histone (H2B-RFP) or DNA counterstaining (e.g., propidium iodide or DAPI). In addition, signal levels should be adjusted to avoid saturation and the images acquired for comparative quantification should be taken under the same settings. In practice, the following issues should be carefully studied: (1) Always use the same image format, magnification and digital zoom, illumination settings between samples (confocal systems: laser power and laser transmission, WF, LSM, SRM systems: light source intensity, exposure time). (2) Since illumination quality can change over time, consider a dual marker and measure two channels for ratio measurements (normalization). (3) Use a linear gain, as little as possible gain (risks of signal distortion) or no gain at all (photon-counting mode when available), enquire on the providers' specification. (4) Adjust illumination/detection to obtain a distribution of signal intensity across a large dynamic range—avoid signal saturation, or, opposite, capturing a low intensity profile—in both cases contrast will be lost: use a LookUpTable (LUT) coloring mode and/or intensity histogram plot. For 8-bit images, the distribution should exploit the 0–256 range. For 16-bit images, intensities rarely map until the maximum possible level (65,536). (5) Mind photobleaching under long exposure time or excess of signal averaging, and prefer a moderate-to-high power illumination/short exposure/no signal averaging (photon counting mode) to low

power illumination/long integration. Signal sum-up without gain should be preferred over averaging amplified signals in CLSM systems. (6) Save images in uncompressed TIF format (never use JPEG for quantification).

3. Sequential scanning of multiple channels is preferred over simultaneous acquisition when the fluorophores show cross talk. The latter is determined by testing whether fluorophore 2 emission is detected under the excitation wavelength of fluorophore 1. Sequential scanning mode is available nowadays for most confocal instrument driving software. It should be designed in a way that in each sequence the software only has to activate/deactivate an excitation line/detector (i.e., the emission windows should be predefined for each channel in the same way for all sequences).
4. The working distance (WD) of the objective represents the distance between the front lens and the coverslip when the objective is in focus. Most immersion objectives for fluorescence microscopy have a WD ranging from 100 to 350 μm depending on the NA and magnification. 60× or 63× objectives often have a WD 1.5× to 2× larger than 100× objectives, with a comparable NA, and are thus preferable for imaging nuclei on whole-mount or thick preparations.
5. Example: we use here an objective with NA 1.3, offering a lateral resolution of ~150 nm and axial resolution of ~276 nm at 488 nm excitation. The nucleus is scanned over a square field of view of 10 × 10 μm. A 1024 × 1024 pixel image format produces an image with a pixel size of 10,000 nm/1024 = ~10 × 10 nm. This is 15-fold smaller than the optically resolved pixel (150 nm), i.e., 15-fold oversampling. According to the Nyquist–Shannon sampling theorem, a two to threefold oversampling is recommended to reconstruct two close signals within the resolvable distance [174]. Here, threefold oversampling means a pixel size of 50 × 50 nm. This is achieved by scanning our field of view with a format five times smaller than previously, i.e., ca. a 200 × 200 format. In practice, a standard 256 × 256 format is appropriate. Similarly, if our field of view is two times bigger (i.e., 20 × 20 μm), a 512 × 512 image scan format will give us pixels of 20 μm/512 = 39 nm side, which correspond to a ~3.8-fold oversampling, and a 400 × 400 pixel image format would provide a 50 × 50 nm (20 μm/400) pixel, i.e., threefold oversampling.
6. Highly Inclined and Laminated Optical sheet (HILO) microscopy is a method for fluorescent single-molecule imaging inside cells. The illumination by a highly inclined and thin beam increases the image intensity and decreases the background intensity, yielding a signal–background ratio about

eight-fold greater than that of epi-illumination. The high ratio yields clear single-molecule images in 3D, enabling to visualize and quantify molecular dynamics, interactions, and kinetics in cells [175].

7. Image deconvolution is a post-acquisition computational process releasing the image from out-of-focus signals. The out-of-focus signal is distributed around the point of emission in a circular area of decreasing intensities further away from the point (airy disk). This distribution is described by the point spread function (PSF) projecting a bell-shape curve centered on the focus point of highest intensity. Deconvolution takes into account either a measured or an estimated PSF (non-blind, adaptive blind methods, respectively), the NA of the objective, the excitation and emission wavelengths and the refractive index of the lens immersion medium into which light is transmitted. Deconvolution operates in a user-defined iterative manner to restore the image to the estimated real point-signal distribution. An excess of iterations can, however, lead to image distortion and thus requires careful inspection of the results by the user. For optimal deconvolution, the image should be acquired with slight oversampling, i.e., acquiring 2–3 times more pixels in the x, y, z dimensions, according to the Shannon–Nyquist sampling theorem [173, 174]. Thus for instance, if the ideal image format matching the optical resolution (provided by the objective's NA) is $500 \times 500 \times 600$ pixels in x, y, z dimensions with each pixel being $150 \times 150 \times 300$ μm in x, y, z size, a twofold oversampling will require increasing the image format to 1000×1000 pixels in x, y and doubling the amount of z-sections by halving the z-step. Several open source (e.g., *Fiji* [156]) and commercial deconvolution software (e.g., Huygens, svi.nl; and AutoQuant, Media Cybernetics, among others) are available.

8. Signal averaging can introduce a nonlinear relationship between signal intensities across the image, particularly in low gain or photon-counting detector mode (which provide a raw count of photons). In this case, particularly when the imaging instrument is equipped with a fast image capture or scanner (Resonance), a "summing up" mode where several images are arithmetically merged (signals are summed up for each pixel) will be preferable.

9. Lateral and axial resolutions (d) depend on the numerical aperture (not on the magnification) and the excitation wavelength. A good approximation following Abbe's formula is $d(xy) = \lambda/2\text{NA}$ and $d(z) = 2\lambda/\text{NA}^2$ so that typically for a 40× objective with NA 1.4, the lateral and axial resolution using a 488 nm excitation line will be 174 nm and 498 nm, respectively. However, a refined calculation of the theoretical limit of

resolution (i.e., where two points close to each other can be resolved) takes into account a limited diffraction (Rayleigh limit) and includes the NA of the condenser. In practice, the user is advised to verify the predicted resolution of the provider's objective for a given wavelength. This information can be calculated usually by a tool of the imaging software.

10. Two main types of camera sensors are available that correspond to digital sensors based on semiconductors: EMCCD sensors (Electron Multiplying Charge Coupled Device) and scientific CMOS sensors, (Complementary Metal–Oxide–Semiconductor). Roughly summarized, their main difference resides in the way the signal is amplified (per line in CCD and per pixel in sCMOS). The latter are nowadays considered as superior (in terms of speed, sensitivity and dynamic range) for biological imaging [176].
11. Several microscopy imaging providers, filter cube and digital camera manufacturers provide interactive platforms and online articles with basic and advanced explanations of the underlying technologies and important considerations for optimized imaging. For instances *see* **Note 1**.

Acknowledgments

CB is funded by the Swiss National Science Foundation (SNSF), the University of Zürich and SystemsX.ch. CB acknowledges the expert assistance and training provided by the Microscopy Facility of the University of Zürich particularly in TEM, LSM and SRM imaging (ZMB, Prof. Urs Ziegler, Jana Doehner, Dominik Haenni, Moritz Kirschmann, Andreas Kaech), Mariamawit Ashenafi for the 3D image file used in Fig. 3c$_3$. We thank Jörg Fuchs for flow sorting of nuclei, Martina Kühne and Andrea Kunze for slide preparation, Andreas Houben, Gerhard Wanner and Klaus Weisshart for critical reading of the manuscript, Marian Bemer for critical reading and suggestions.

References

1. Olins DE, Olins AL (2003) Chromatin history: a view from the bridge. Nat Rev Mol Cell Biol 4:809–814
2. Hergeth SP, Schneider R (2015) The H1 linker histones: multifunctional proteins beyond the nucleosomal core particle. EMBO Rep 16(11):1439–1453
3. Schneider R, Grosschedl R (2007) Dynamics and interplay of nuclear architecture, genome organization, and gene expression. Genes Dev 21(23):3027–3043
4. van Driel R, Fransz P (2004) Nuclear architecture and genome functioning in plants and animals: what can we learn from both? Exp Cell Res 296(1):86–90
5. Deal RB, Henikoff S (2011) The INTACT method for cell type-specific gene expression and chromatin profiling in Arabidopsis thaliana. Nat Protoc 6(1):56–68
6. Moreno-Romero J, Santos-Gonzalez J, Hennig L, Kohler C (2017) Applying the INTACT method to purify endosperm nuclei

and to generate parental-specific epigenome profiles. Nat Protoc 12(2):238–254

7. Kawakatsu T, Stuart T, Valdes M, Breakfield N, Schmitz RJ, Nery JR, Urich MA, Han X, Lister R, Benfey PN, Ecker JR (2016) Unique cell-type-specific patterns of DNA methylation in the root meristem. Nat Plants 2(5):16058
8. Morao AK, Caillieux E, Colot V, Roudier F (2017) Cell type-specific profiling of chromatin modifications and associated proteins. In: Plant chromatin dynamics, Methods in Molecular Biology. Springer, New York, NY
9. Gonzalez-Sandoval A, Gasser SM (2016) On TADs and LADs: spatial control over gene expression. Trends Genet 32(8):485–495
10. North AJ (2006) Seeing is believing? A beginners' guide to practical pitfalls in image acquisition. J Cell Biol 172(1):9–18
11. Lambert TJ, Waters JC (2017) Navigating challenges in the application of superresolution microscopy. J Cell Biol 216(1):53–63
12. Shaw SL (2006) Imaging the live plant cell. Plant J 45(4):573–598
13. Shaw SL, Ehrhardt DW (2013) Smaller, faster, brighter: advances in optical imaging of living plant cells. Annu Rev Plant Biol 64:351–375
14. Probst A (2017) A compendium of methods to analyse the spatial organization of plant chromatin. In: Bemer M, Baroux C (eds) Plant chromatin dynamics: methods and protocols, Methods in molecular biology. Springer, New York, NY. doi:10.1007/978-1-4939-7318-7_23
15. Mao YS, Zhang B, Spector DL (2011) Biogenesis and function of nuclear bodies. Trends Genet 27(8):295–306
16. Meier I (2009) Functional organization of the plant nucleus. In: Meier I (ed) Functional organization of the plant nucleus. Springer, Berlin, pp 1–8. doi:10.1007/978-3-540-71058-5_1
17. Cheng P-C (2010) Interaction of light with botanical specimens. In: Pawley JB (ed) Handbook for biological confocal microscopy, 3rd edn. Springer, New York, NY, pp 414–441
18. Ohad N, Yalovsky S (2010) Utilizing bimolecular fluorescence complementation (BiFC) to assay protein-protein interaction in plants. Methods Mol Biol 655:347–358
19. Horstman A, Tonaco IA, Boutilier K, Immink RG (2014) A cautionary note on the use of split-YFP/BiFC in plant protein-protein interaction studies. Int J Mol Sci 15 (6):9628–9643
20. Bu Z, Yu Y, Li Z, Liu Y, Jiang W, Huang Y, Dong AW (2014) Regulation of arabidopsis flowering by the histone mark readers MRG1/2 via interaction with CONSTANS to modulate FT expression. PLoS Genet 10 (9):e1004617
21. Song ZT, Sun L, Lu SJ, Tian Y, Ding Y, Liu JX (2015) Transcription factor interaction with COMPASS-like complex regulates histone H3K4 trimethylation for specific gene expression in plants. Proc Natl Acad Sci U S A 112(9):2900–2905
22. Perrella G, Carr C, Asensi-Fabado MA, Donald NA, Paldi K, Hannah MA, Amtmann A (2016) The histone deacetylase complex 1 protein of arabidopsis has the capacity to interact with multiple proteins including histone 3-binding proteins and histone 1 variants. Plant Physiol 171(1):62–70
23. Gadella TW Jr, van der Krogt GN, Bisseling T (1999) GFP-based FRET microscopy in living plant cells. Trends Plant Sci 4(7):287–291
24. Benvenuto G, Formiggini F, Laflamme P, Malakhov M, Bowler C (2002) The photomorphogenesis regulator DET1 binds the amino-terminal tail of histone H2B in a nucleosome context. Curr Biol 12(17):1529–1534
25. Hasegawa J, Sakamoto Y, Nakagami S, Aida M, Sawa S, Matsunaga S (2016) Three-dimensional imaging of plant organs using a simple and rapid transparency technique. Plant Cell Physiol 57(3):462–472
26. Kurihara D, Mizuta Y, Sato Y, Higashiyama T (2015) ClearSee: a rapid optical clearing reagent for whole-plant fluorescence imaging. Development 142(23):4168–4179
27. Warner CA, Biedrzycki ML, Jacobs SS, Wisser RJ, Caplan JL, Sherrier DJ (2014) An optical clearing technique for plant tissues allowing deep imaging and compatible with fluorescence microscopy. Plant Physiol 166 (4):1684–1687
28. Musielak TJ, Schenkel L, Kolb M, Henschen A, Bayer M (2015) A simple and versatile cell wall staining protocol to study plant reproduction. Plant Reprod 28 (3-4):161–169
29. Musielak TJ, Slane D, Liebig C, Bayer M (2016) A versatile optical clearing protocol for deep tissue imaging of fluorescent proteins in *Arabidopsis thaliana*. PLoS One 11(8): e0161107
30. Littlejohn GR, Gouveia JD, Edner C, Smirnoff N, Love J (2010) Perfluorodecalin enhances in vivo confocal microscopy resolution of *Arabidopsis thaliana* mesophyll. New Phytol 186(4):1018–1025

31. Littlejohn GR, Mansfield JC, Christmas JT, Witterick E, Fricker MD, Grant MR, Smirnoff N, Everson RM, Moger J, Love J (2014) An update: improvements in imaging perfluorocarbon-mounted plant leaves with implications for studies of plant pathology, physiology, development and cell biology. Front Plant Sci 5:140
32. Nagaki K, Yamaji N, Murata M (2017) ePro-ClearSee: a simple immunohistochemical method that does not require sectioning of plant samples. Sci Rep 7:42203
33. She W, Grimanelli D, Baroux C (2014) An efficient method for quantitative, single-cell analysis of chromatin modification and nuclear organization in whole-mount ovules in *Arabidopsis*. J Vis Exp (88):e51530
34. Escobar-Guzman R, Rodriguez-Leal D, Vielle-Calzada JP, Ronceret A (2015) Whole-mount immunolocalization to study female meiosis in *Arabidopsis*. Nat Protoc 10 (10):1535–1542
35. Pillot M, Baroux C, Vazquez MA, Autran D, Leblanc O, Vielle-Calzada JP, Grossniklaus U, Grimanelli D (2010) Embryo and endosperm inherit distinct chromatin and transcriptional states from the female gametes in *Arabidopsis*. Plant Cell 22(2):307–320
36. She W, Grimanelli D, Rutowicz K, Whitehead MW, Puzio M, Kotlinski M, Jerzmanowski A, Baroux C (2013) Chromatin reprogramming during the somatic-to-reproductive cell fate transition in plants. Development 140 (19):4008–4019
37. Costa S, Shaw P (2006) Chromatin organization and cell fate switch respond to positional information in *Arabidopsis*. Nature 439 (7075):493–496
38. Gernand D, Rutten T, Varshney A, Rubtsova M, Prodanovic S, Bruss C, Kumlehn J, Matzk F, Houben A (2005) Uniparental chromosome elimination at mitosis and interphase in wheat and pearl millet crosses involves micronucleus formation, progressive heterochromatinization, and DNA fragmentation. Plant Cell 17(9):2431–2438
39. Wegel E, Vallejos RH, Christou P, Stoger E, Shaw P (2005) Large-scale chromatin decondensation induced in a developmentally activated transgene locus. J Cell Sci 118 (Pt 5):1021–1031
40. Braszewska-Zalewska AJ, Wolny EA, Smialek L, Hasterok R (2013) Tissue-specific epigenetic modifications in root apical meristem cells of *Hordeum vulgare*. PLoS One 8 (7):e69204
41. Wolny E, Braszewska-Zalewska A, Kroczek D, Hasterok R (2015) In situ analysis of epigenetic modifications in the chromatin of Brachypodium distachyon embryos. Plant Signal Behav 10(5):e1011948
42. Bourdon M, Coriton O, Pirrello J, Cheniclet C, Brown SC, Poujol C, Chevalier C, Renaudin JP, Frangne N (2011) In planta quantification of endoreduplication using fluorescent in situ hybridization (FISH). Plant J 66(6):1089–1099
43. Garcia-Aguilar M, Michaud C, Leblanc O, Grimanelli D (2010) Inactivation of a DNA methylation pathway in maize reproductive organs results in apomixis-like phenotypes. Plant Cell 22(10):3249–3267
44. Bey TD, Koini M, Fransz P (2017) Fluorescence in situ hybridization (FISH) and immunolabeling on 3D preserved nuclei. In: Bemer M, Baroux C (eds) Plant chromatin dynamics: methods and protocols, Methods in molecular biology. Springer, New York, NY. doi:10.1007/978-1-4939-7318-7_27
45. Ashenafi M, Baroux C (2017) Automated 3D gene position analysis using a customized Imaris plugin: XTFISHInsideNucleus. In: Bemer M, Baroux C (eds) Plant chromatin dynamics: methods and protocols, Methods in molecular biology. Springer, New York, NY. doi:10.1007/978-1-4939-7318-7_32
46. She W, Baroux C, Grossniklaus U (2017) Cell-type specific chromatin analysis in whole-mount plant tissues by immunostaining. In: Bemer M, Baroux C (eds) Plant chromatin dynamics: methods and protocols, Methods in molecular biology. Springer, New York, NY. doi:10.1007/978-1-4939-7318-7_25
47. Marques-Bueno MM, Morao AK, Cayrel A, Platre MP, Barberon M, Caillieux E, Colot V, Jaillais Y, Roudier F, Vert G (2016) A versatile multisite gateway-compatible promoter and transgenic line collection for cell type-specific functional genomics in *Arabidopsis*. Plant J 85 (2):320–333
48. Poulet A, Arganda-Carreras I, Legland D, Probst AV, Andrey P, Tatout C (2015) NucleusJ: an ImageJ plugin for quantifying 3D images of interphase nuclei. Bioinformatics 31(7):1144–1146
49. Andrey P, Kieu K, Kress C, Lehmann G, Tirichine L, Liu Z, Biot E, Adenot PG, Hue-Beauvais C, Houba-Herin N, Duranthon V, Devinoy E, Beaujean N, Gaudin V, Maurin Y, Debey P (2010) Statistical analysis of 3D images detects regular spatial distributions of centromeres and chromocenters in animal and plant nuclei. PLoS Comput Biol 6(7): e1000853
50. Desset S, Poulet A, Tatout C (2017) Quantitative 3D analysis of nuclear morphology and heterochromatin organization from whole

mount plant tissue using NucleusJ. In: Bemer M, Baroux C (eds) Plant chromatin dynamics: methods and protocols, Methods in molecular biology. Springer, New York, NY. doi:10.1007/978-1-4939-7318-7_33

51. Arpon J, Gaudin V, Andrey P (2017) A method for testing random spatial model on nuclear object distributions. In: Bemer M, Baroux C (eds) Plant chromatin dynamics: methods and protocols, Methods in molecular biology. Springer, New York, NY. doi:10.1007/978-1-4939-7318-7_29
52. Fang Y, Spector DL (2005) Centromere positioning and dynamics in living *Arabidopsis* plants. Mol Biol Cell 16(12):5710–5718
53. de Nooijer S, Wellink J, Mulder B, Bisseling T (2009) Non-specific interactions are sufficient to explain the position of heterochromatic chromocenters and nucleoli in interphase nuclei. Nucleic Acids Res 37(11):3558–3568
54. Murphy SP, Gumber HK, Mao Y, Bass HW (2014) A dynamic meiotic SUN belt includes the zygotene-stage telomere bouquet and is disrupted in chromosome segregation mutants of maize (*Zea mays* L.) Front Plant Sci 5:314
55. Kato N, Lam E (2003) Chromatin of endoreduplicated pavement cells has greater range of movement than that of diploid guard cells in *Arabidopsis thaliana*. J Cell Sci 116 (Pt 11):2195–2201
56. Lindhout BI, Meckel T, van der Zaal BJ (2010) Zinc finger-mediated live cell imaging in Arabidopsis roots. Methods Mol Biol 649:383–398
57. Aki SS, Umeda M (2016) Cytrap marker systems for in vivo visualization of cell cycle progression in *Arabidopsis*. In: Caillaud M-C (ed) Plant cell division: methods and protocols. Springer, New York, NY, pp 51–57. doi:10.1007/978-1-4939-3142-2_4
58. Ingouff M, Hamamura Y, Gourgues M, Higashiyama T, Berger F (2007) Distinct dynamics of HISTONE3 variants between the two fertilization products in plants. Curr Biol 17(12):1032–1037
59. Ingouff M, Selles B, Michaud C, Vu TM, Berger F, Schorn AJ, Autran D, Van Durme M, Nowack MK, Martienssen RA, Grimanelli D (2017) Live-cell analysis of DNA methylation during sexual reproduction in Arabidopsis reveals context and sex-specific dynamics controlled by noncanonical RdDM. Genes Dev 31(1):72–83
60. Rosa S (2017) Measuring dynamics of histone proteins by photobleaching in *Arabidopsis* roots. In: Bemer M, Baroux C (eds) Plant chromatin dynamics: methods and protocols, Methods in molecular biology. Springer, New York, NY. doi:10.1007/978-1-4939-7318-7_26
61. Padilla-Parra S, Auduge N, Coppey-Moisan-M, Tramier M (2008) Quantitative FRET analysis by fast acquisition time domain FLIM at high spatial resolution in living cells. Biophys J 95(6):2976–2988
62. Molitor AM, Bu Z, Yu Y, Shen WH (2014) Arabidopsis AL PHD-PRC1 complexes promote seed germination through H3K4me3-to-H3K27me3 chromatin state switch in repression of seed developmental genes. PLoS Genet 10(1):e1004091
63. Le Roux C, Huet G, Jauneau A, Camborde L, Tremousaygue D, Kraut A, Zhou B, Levaillant M, Adachi H, Yoshioka H, Raffaele S, Berthome R, Coute Y, Parker JE, Deslandes L (2015) A receptor pair with an integrated decoy converts pathogen disabling of transcription factors to immunity. Cell 161 (5):1074–1088
64. Ramirez-Garces D, Camborde L, Pel MJ, Jauneau A, Martinez Y, Neant I, Leclerc C, Moreau M, Dumas B, Gaulin E (2016) CRN13 candidate effectors from plant and animal eukaryotic pathogens are DNA-binding proteins which trigger host DNA damage response. New Phytol 210 (2):602–617
65. Tonaco IA, Borst JW, de Vries SC, Angenent GC, Immink RG (2006) In vivo imaging of MADS-box transcription factor interactions. J Exp Bot 57(1):33–42
66. Lleres D, James J, Swift S, Norman DG, Lamond AI (2009) Quantitative analysis of chromatin compaction in living cells using FLIM-FRET. J Cell Biol 187(4):481–496
67. Lleres D, Bailly AP, Perrin A, Norman DG, Xirodimas DP, Feil R (2017) Quantitative FLIM-FRET microscopy to monitor nanoscale chromatin compaction in vivo reveals structural roles of condensin complexes. Cell Rep 18(7):1791–1803
68. Lorenz M (2009) Visualizing protein-RNA interactions inside cells by fluorescence resonance energy transfer. RNA 15(1):97–103
69. Cremazy FG, Manders EM, Bastiaens PI, Kramer G, Hager GL, van Munster EB, Verschure PJ, Gadella TJ Jr, van Driel R (2005) Imaging in situ protein-DNA interactions in the cell nucleus using FRET-FLIM. Exp Cell Res 309(2):390–396
70. Stelzer EH (2015) Light-sheet fluorescence microscopy for quantitative biology. Nat Methods 12(1):23–26
71. von Wangenheim D, Daum G, Lohmann JU, Stelzer EK, Maizel A (2014) Live imaging of

Arabidopsis development. Methods Mol Biol 1062:539–550

72. Ovecka M, Vaskebova L, Komis G, Luptovciak I, Smertenko A, Samaj J (2015) Preparation of plants for developmental and cellular imaging by light-sheet microscopy. Nat Protoc 10(8):1234–1247
73. de Luis Balaguer MA, Ramos-Pezzotti M, Rahhal MB, Melvin CE, Johannes E, Horn TJ, Sozzani R (2016) Multi-sample Arabidopsis growth and imaging chamber (MAGIC) for long term imaging in the ZEISS Lightsheet Z.1. Dev Biol 419 (1):19–25
74. Meinert T, Tietz O, Palme KJ, Rohrbach A (2016) Separation of ballistic and diffusive fluorescence photons in confocal Light-Sheet Microscopy of *Arabidopsis* roots. Sci Rep 6:30378
75. Royer LA, Lemon WC, Chhetri RK, Wan Y, Coleman M, Myers EW, Keller PJ (2016) Adaptive light-sheet microscopy for long-term, high-resolution imaging in living organisms. Nat Biotechnol 34 (12):1267–1278
76. Gualda E, Moreno N, Tomancak P, Martins GG (2014) Going "open" with mesoscopy: a new dimension on multi-view imaging. Protoplasma 251(2):363–372
77. Sena G, Frentz Z, Birnbaum KD, Leibler S (2011) Quantitation of cellular dynamics in growing *Arabidopsis* roots with light sheet microscopy. PLoS One 6(6):e21303
78. Novak D, Kucharova A, Ovecka M, Komis G, Samaj J (2015) Developmental nuclear localization and quantification of GFP-tagged EB1c in *Arabidopsis* root using light-sheet microscopy. Front Plant Sci 6:1187
79. Maizel A, von Wangenheim D, Federici F, Haseloff J, Stelzer EH (2011) High-resolution live imaging of plant growth in near physiological bright conditions using light sheet fluorescence microscopy. Plant J 68(2):377–385
80. Berson T, von Wangenheim D, Takac T, Samajova O, Rosero A, Ovecka M, Komis G, Stelzer EH, Samaj J (2014) Trans-Golgi network localized small GTPase RabA1d is involved in cell plate formation and oscillatory root hair growth. BMC Plant Biol 14:252
81. Preibisch S, Saalfeld S, Schindelin J, Tomancak P (2010) Software for bead-based registration of selective plane illumination microscopy data. Nat Methods 7(6):418–419
82. Rego EH, Shao L, Macklin JJ, Winoto L, Johansson GA, Kamps-Hughes N, Davidson MW, Gustafsson MG (2012) Nonlinear structured-illumination microscopy with a photoswitchable protein reveals cellular structures at 50-nm resolution. Proc Natl Acad Sci U S A 109(3):E135–E143
83. Schermelleh L, Heintzmann R, Leonhardt H (2010) A guide to super-resolution fluorescence microscopy. J Cell Biol 190 (2):165–175
84. Ball G, Parton RM, Hamilton RS, Davis I (2012) A cell biologist's guide to high resolution imaging. Methods Enzymol 504:29–55
85. Agrawal U, Reilly DT, Schroeder CM (2013) Zooming in on biological processes with fluorescence nanoscopy. Curr Opin Biotechnol 24 (4):646–653
86. Allen JR, Ross ST, Davidson MW (2014) Structured illumination microscopy for super-resolution. ChemPhysChem 15(4):566–576
87. Komis G, Samajova O, Ovecka M, Samaj J (2015) Super-resolution microscopy in plant cell imaging. Trends Plant Sci 20 (12):834–843
88. Nienhaus K, Nienhaus GU (2016) Where do we stand with super-resolution optical microscopy? J Mol Biol 428(2 Pt A):308–322
89. Coltharp C, Xiao J (2012) Superresolution microscopy for microbiology. Cell Microbiol 14(12):1808–1818
90. Dame RT, Tark-Dame M (2016) Bacterial chromatin: converging views at different scales. Curr Opin Cell Biol 40:60–65
91. Fornasiero EF, Opazo F (2015) Super-resolution imaging for cell biologists: concepts, applications, current challenges and developments. Bioessays 37(4):436–451
92. Schubert V (2017) Super-resolution microscopy - applications in plant cell research. Front Plant Sci 8:531
93. Markaki Y, Smeets D, Fiedler S, Schmid VJ, Schermelleh L, Cremer T, Cremer M (2012) The potential of 3D-FISH and super-resolution structured illumination microscopy for studies of 3D nuclear architecture. Bioessays 34(5):412–426
94. Markaki Y, Gunkel M, Schermelleh L, Beichmanis S, Neumann J, Heidemann M, Leonhardt H, Eick D, Cremer C, Cremer T (2010) Functional nuclear organization of transcription and DNA replication: a topographical marriage between chromatin domains and the interchromatin compartment. Cold Spring Harb Symp Quant Biol 75:475–492
95. Schubert V (2014) RNA polymerase II forms transcription networks in rye and *Arabidopsis* nuclei and its amount increases with

endopolyploidy. Cytogenet Genome Res 143 (1-3):69–77

96. Schubert V, Lermontova I, Schubert I (2013) The *Arabidopsis* CAP-D proteins are required for correct chromatin organisation, growth and fertility. Chromosoma 122(6):517–533
97. Ma W, Gabriel TS, Martis MM, Gursinsky T, Schubert V, Vrana J, Dolezel J, Grundlach H, Altschmied L, Scholz U, Himmelbach A, Behrens SE, Banaei-Moghaddam AM, Houben A (2016) Rye B chromosomes encode a functional Argonaute-like protein with in vitro slicer activities similar to its A chromosome paralog. New Phytol 213(2):916–928
98. Zakrzewski F, Schubert V, Viehoever P, Minoche AE, Dohm JC, Himmelbauer H, Weisshaar B, Schmidt T (2014) The CHH motif in sugar beet satellite DNA: a modulator for cytosine methylation. Plant J 78 (6):937–950
99. Ishii T, Karimi-Ashtiyani R, Banaei-Moghaddam AM, Schubert V, Fuchs J, Houben A (2015) The differential loading of two barley CENH3 variants into distinct centromeric substructures is cell type- and development-specific. Chromosome Res 23 (2):277–284
100. Demidov D, Schubert V, Kumke K, Weiss O, Karimi-Ashtiyani R, Buttlar J, Heckmann S, Wanner G, Dong Q, Han F, Houben A (2014) Anti-phosphorylated histone H2AThr120: a universal microscopic marker for centromeric chromatin of mono- and holocentric plant species. Cytogenet Genome Res 143(1-3):150–156
101. Neumann P, Schubert V, Fukova I, Manning JE, Houben A, Macas J (2016) Epigenetic histone marks of extended meta-polycentric centromeres of *Lathyrus* and *Pisum* chromosomes. Front Plant Sci 7:234
102. Weisshart K, Fuchs J, Schubert V (2016) Structured illumination microscopy (SIM) and photoactivated localization microscopy (PALM) to analyze the abundance and distribution of RNA polymerase II molecules in flow-sorted *Arabidopsis* nuclei. Bio Protocol 6(3): e1725. http://wwwbio-protocolorg/e1725
103. Heckmann S, Macas J, Kumke K, Fuchs J, Schubert V, Ma L, Novak P, Neumann P, Taudien S, Platzer M, Houben A (2013) The holocentric species *Luzula elegans* shows interplay between centromere and large-scale genome organization. Plant J 73 (4):555–565
104. Marques A, Ribeiro T, Neumann P, Macas J, Novak P, Schubert V, Pellino M, Fuchs J, Ma W, Kuhlmann M, Brandt R, Vanzela AL, Beseda T, Simkova H, Pedrosa-Harand A, Houben A (2015) Holocentromeres in *Rhynchospora* are associated with genome-wide centromere-specific repeat arrays interspersed among euchromatin. Proc Natl Acad Sci USA 112(44):13633–13638
105. Ribeiro SA, Vagnarelli P, Dong Y, Hori T, McEwen BF, Fukagawa T, Flors C, Earnshaw WC (2010) A super-resolution map of the vertebrate kinetochore. Proc Natl Acad Sci USA 107(23):10484–10489
106. Dürr J, Lolas IB, Sorensen BB, Schubert V, Houben A, Melzer M, Deutzmann R, Grasser M, Grasser KD (2014) The transcript elongation factor SPT4/SPT5 is involved in auxin-related gene expression in *Arabidopsis*. Nucleic Acids Res 42(7):4332–4347
107. Antosz W, Pfab A, Ehrnsberger H, Holzinger H, Köllen K, Mortensen S, Bruckmann A, Schubert T, Längst G, Griesenbeck J, Schubert V, Grasser M, Grasser K (2017) Composition of the *Arabidopsis* RNA polymerase II transcript elongation complex reveals interplay of elongation and mRNA processing factors. Plant Cell 29 (4):854–870
108. Marques A, Schubert V, Houben A, Pedrosa-Harand A (2016) Restructuring of holocentric centromeres during meiosis in the plant *Rhynchospora pubera*. Genetics 204 (2):555–568
109. Schubert V, Ruban A, Houben A (2016) Chromatin ring formation at plant centromeres. Front Plant Sci 7:28
110. Schubert V, Zelkowski M, Klemme S, Houben A (2016) Similar sister chromatid arrangement in mono- and holocentric plant chromosomes. Cytogenet Genome Res 149 (3):218–225
111. Ball G, Demmerle J, Kaufmann R, Davis I, Dobbie IM, Schermelleh L (2015) SIMcheck: a toolbox for successful super-resolution structured illumination microscopy. Sci Rep 5:15915
112. Schubert V, Weisshart K (2015) Abundance and distribution of RNA polymerase II in *Arabidopsis* interphase nuclei. J Exp Bot 66 (6):1687–1698
113. Dempsey GT, Vaughan JC, Chen KH, Bates M, Zhuang X (2011) Evaluation of fluorophores for optimal performance in localization-based super-resolution imaging. Nat Methods 8(12):1027–1036
114. Fernandez-Suarez M, Ting AY (2008) Fluorescent probes for super-resolution imaging in living cells. Nat Rev Mol Cell Biol 9 (12):929–943

115. Hedde PN, Nienhaus GU (2014) Super-resolution localization microscopy with photoactivatable fluorescent marker proteins. Protoplasma 251(2):349–362
116. Olivier N, Keller D, Gonczy P, Manley S (2013) Resolution doubling in 3D-STORM imaging through improved buffers. PLoS One 8(7):e69004
117. Flors C (2013) Super-resolution fluorescence imaging of directly labelled DNA: from microscopy standards to living cells. J Microsc 251(1):1–4
118. Flors C, Earnshaw WC (2011) Super-resolution fluorescence microscopy as a tool to study the nanoscale organization of chromosomes. Curr Opin Chem Biol 15 (6):838–844
119. Hamel V, Guichard P, Fournier M, Guiet R, Fluckiger I, Seitz A, Gonczy P (2014) Correlative multicolor 3D SIM and STORM microscopy. Biomed Opt Express 5 (10):3326–3336
120. Wurm CA, Kolmakov K, Göttfert F, Ta H, Bossi M, Schill H, Berning S, Jakobs S, Donnert G, Belov VN, Hell SW (2012) Novel red fluorophores with superior performance in STED microscopy. Optical Nanosc 1(1):7
121. Wanner G, Schroeder-Reiter E (2008) Scanning electron microscopy of chromosomes. Methods Cell Biol 88:451
122. Schubert I, Dolezel J, Houben A, Scherthan H, Wanner G (1993) Refined examination of plant metaphase chromosome structure at different levels made feasible by new isolation methods. Chromosoma 102 (2):96–101
123. Wanner G, Formanek H, Martin R, Herrmann RG (1991) High-resolution scanning electron-microscopy of plant chromosomes. Chromosoma 100(2):103–109
124. Martin R, Busch W, Herrmann RG, Wanner G (1994) Efficient preparation of plant chromosomes for high-resolution scanning electron microscopy. Chromosome Res 2 (5):411–415
125. Iwano M, Che FS, Takayama S, Fukui K, Isogai A (2003) Three-dimensional architecture of ribosomal DNA within barley nucleoli revealed with electron microscopy. Scanning 25(5):257–263
126. Jander G, Wendt H (1960) Lehrbuch der analytischen und präparativen anorganischen Chemie. Hirzel Verlag, Leipzig
127. Wanner G, Formanek H (1995) Imaging of DNA in human and plant chromosomes by high-resolution scanning electron microscopy. Chromosome Res 3(6):368–374
128. Wanner G, Formanek H (2000) A new chromosome model. J Struct Biol 132(2):147–161
129. Schroeder-Reiter E, Houben A, Wanner G (2003) Immunogold labeling of chromosomes for scanning electron microscopy: a closer look at phosphorylated histone H3 in mitotic metaphase chromosomes of Hordeum vulgare. Chromosome Res 11 (6):585–596
130. Schroeder-Reiter E, Perez-Willard F, Zeile U, Wanner G (2009) Focused ion beam (FIB) combined with high resolution scanning electron microscopy: a promising tool for 3D analysis of chromosome architecture. J Struct Biol 165(2):97–106
131. Houben A, Schroeder-Reiter E, Nagaki K, Nasuda S, Wanner G, Murata M, Endo TR (2007) CENH3 interacts with the centromeric retrotransposon cereba and GC-rich satellites and locates to centromeric substructures in barley. Chromosoma 116 (3):275–283
132. Schroeder-Reiter E, Houben A, Grau J, Wanner G (2006) Characterization of a peg-like terminal NOR structure with light microscopy and high-resolution scanning electron microscopy. Chromosoma 115(1):50–59
133. Schroeder-Reiter E, Sanei M, Houben A, Wanner G (2012) Current SEM techniques for de- and re-construction of centromeres to determine 3D CENH3 distribution in barley mitotic chromosomes. J Microsc 246 (1):96–106
134. Neumann P, Navratilova A, Schroeder-Reiter-E, Koblizkova A, Steinbauerova V, Chocholova E, Novak P, Wanner G, Macas J (2012) Stretching the rules: monocentric chromosomes with multiple centromere domains. PLoS Genet 8(6):e1002777
135. Wanner G, Schroeder-Reiter E, Formanek H (2005) 3D analysis of chromosome architecture: advantages and limitations with SEM. Cytogenet Genome Res 109(1-3):70–78
136. Zoller JF, Herrmann RG, Wanner G (2004) Chromosome condensation in mitosis and meiosis of rye (*Secale cereale* L.) Cytogenet Genome Res 105(1):134–144
137. Zoller JF, Hohmann U, Herrmann RG, Wanner G (2004) Ultrastructural analysis of chromatin in meiosis I + II of rye (*Secale cereale* L.) Cytogenet Genome Res 105(1):145–156
138. Heckmann S, Schroeder-Reiter E, Kumke K, Ma L, Nagaki K, Murata M, Wanner G, Houben A (2011) Holocentric chromosomes of Luzula elegans are characterized by a

longitudinal centromere groove, chromosome bending, and a terminal nucleolus organizer region. Cytogenet Genome Res 134 (3):220–228

139. Schroeder-Reiter E, Wanner G (2009) Chromosome centromeres: structural and analytical investigations with high resolution scanning electron microscopy in combination with focused ion beam milling. Cytogenet Genome Res 124(3-4):239–250
140. Dwiranti A, Lin L, Mochizuki E, Kuwabata S, Takaoka A, Uchiyama S, Fukui K (2012) Chromosome observation by scanning electron microscopy using ionic liquid. Microsc Res Tech 75(8):1113–1118
141. Hamano T, Dwiranti A, Kaneyoshi K, Fukuda S, Kometani R, Nakao M, Takata H, Uchiyama S, Ohmido N, Fukui K (2014) Chromosome interior observation by focused ion beam/scanning electron microscopy (FIB/SEM) using ionic liquid technique. Microsc Microanal 20(5):1340–1347
142. Houben A, Demidov D, Rutten T, Scheidtmann KH (2005) Novel phosphorylation of histone H3 at threonine 11 that temporally correlates with condensation of mitotic and meiotic chromosomes in plant cells. Cytogenet Genome Res 109(1-3):148–155
143. Cherkezyan L, Stypula-Cyrus Y, Subramanian H, White C, Dela Cruz M, Wali RK, Goldberg MJ, Bianchi LK, Roy HK, Backman V (2014) Nanoscale changes in chromatin organization represent the initial steps of tumorigenesis: a transmission electron microscopy study. BMC Cancer 14:189
144. Fabrice TN, Cherkezeyan L, Ringli C, Baroux C (2017) Transmission electron microscopy imaging to analyse chromatin density distribution at the nanoscale level. In: Bemer M, Baroux C (eds) Plant chromatin dynamics: methods and protocols, Methods in molecular biology. Springer, New York, NY. doi:10.1007/978-1-4939-7318-7_34
145. Meijering E, Carpenter AE, Peng H, Hamprecht FA, Olivo-Marin JC (2016) Imagining the future of bioimage analysis. Nat Biotechnol 34(12):1250–1255
146. Haider SA, Cameron A, Siva P, Lui D, Shafiee MJ, Boroomand A, Haider N, Wong A (2016) Fluorescence microscopy image noise reduction using a stochastically-connected random field model. Sci Rep 6:20640
147. Pavlova P, Tessadori F, de Jong HJ, Fransz P (2010) Immunocytological analysis of chromatin in isolated nuclei. Methods Mol Biol 655:413–432
148. Fransz P, ten Hoopen R, Tessadori F (2006) Composition and formation of heterochromatin in *Arabidopsis thaliana*. Chromosome Res 14(1):71–82
149. van Zanten M, Tessadori F, Peeters AJ, Fransz P (2012) Shedding light on large-scale chromatin reorganization in *Arabidopsis thaliana*. Mol Plant 5(3):583–590
150. Fransz PF, de Jong JH (2002) Chromatin dynamics in plants. Curr Opin Plant Biol 5 (6):560–567
151. Almassalha LM, Tiwari A, Ruhoff PT, Stypula-Cyrus Y, Cherkezyan L, Matsuda H, Dela Cruz MA, Chandler JE, White C, Maneval C, Subramanian H, Szleifer I, Roy HK, Backman V (2017) The global relationship between chromatin physical topology, fractal structure, and gene expression. Sci Rep 7:41061
152. Ricci MA, Manzo C, Garcia-Parajo MF, Lakadamyali M, Cosma MP (2015) Chromatin fibers are formed by heterogeneous groups of nucleosomes in vivo. Cell 160 (6):1145–1158
153. Chytilova E, Macas J, Sliwinska E, Rafelski SM, Lambert GM, Galbraith DW (2000) Nuclear dynamics in *Arabidopsis thaliana*. Mol Biol Cell 11(8):2733–2741
154. Higa T, Suetsugu N, Wada M (2014) Plant nuclear photorelocation movement. J Exp Bot 65(11):2873–2881
155. Qiu M, Yang G (2013) Drift correction for fluorescence live cell imaging through correlated motion identification. In: 10th Intern Symp Biomed Imaging, 7–11 April 2013. pp 452–455. doi:10.1109/ISBI.2013.6556509
156. Schindelin J, Arganda-Carreras I, Frise E, Kaynig V, Longair M, Pietzsch T, Preibisch S, Rueden C, Saalfeld S, Schmid B, Tinevez JY, White DJ, Hartenstein V, Eliceiri K, Tomancak P, Cardona A (2012) Fiji: an open-source platform for biological-image analysis. Nat Methods 9(7):676–682
157. Van Bruaene N, Joss G, Thas O, Van Oostveldt P (2003) Four-dimensional imaging and computer-assisted track analysis of nuclear migration in root hairs of *Arabidopsis thaliana*. J Microsc 211(Pt 2):167–178
158. Uchida S (2013) Image processing and recognition for biological images. Dev Growth Differ 55(4):523–549
159. Poulet A, Duc C, Voisin M, Desset S, Tutois S, Vanrobays E, Benoit M, Evans DE, Probst AV, Tatout C (2017) The LINC complex contributes to heterochromatin organisation and transcriptional gene silencing in plants. J Cell Sci 130(3):590–601

160. Paunovic I, She W, Baroux C (2013) http://www.bitplane.com/learning/quantification-of-chromatin-modifications-in-whole-mount-plant-tissue-tutorial
161. Faure E, Savy T, Rizzi B, Melani C, Stasova O, Fabreges D, Spir R, Hammons M, Cunderlik R, Recher G, Lombardot B, Duloquin L, Colin I, Kollar J, Desnoulez S, Affaticati P, Maury B, Boyreau A, Nief JY, Calvat P, Vernier P, Frain M, Lutfalla G, Kergosien Y, Suret P, Remesikova M, Doursat R, Sarti A, Mikula K, Peyrieras N, Bourgine P (2016) A workflow to process 3D+time microscopy images of developing organisms and reconstruct their cell lineage. Nat Commun 7:8674
162. Amat F, Lemon W, Mossing DP, McDole K, Wan Y, Branson K, Myers EW, Keller PJ (2014) Fast, accurate reconstruction of cell lineages from large-scale fluorescence microscopy data. Nat Methods 11(9):951–958
163. Fernandez R, Das P, Mirabet V, Moscardi E, Traas J, Verdeil JL, Malandain G, Godin C (2010) Imaging plant growth in 4D: robust tissue reconstruction and lineaging at cell resolution. Nat Methods 7(7):547–553
164. Bassel GW, Smith RS (2016) Quantifying morphogenesis in plants in 4D. Curr Opin Plant Biol 29:87–94
165. Chalut KJ, Ekpenyong AE, Clegg WL, Melhuish IC, Guck J (2012) Quantifying cellular differentiation by physical phenotype using digital holographic microscopy. Integrat Biol 4(3):280–284
166. Kus A, Dudek M, Kemper B, Kujawinska M, Vollmer A (2014) Tomographic phase microscopy of living three-dimensional cell cultures. J Biomed Opt 19(4):046009
167. Cherkezyan L, Zhang D, Subramanian H, Taflove A, Backman V (2016) Reconstruction of explicit structural properties at the nanoscale via spectroscopic microscopy. J Biomed Opt 21(2):025007–025007
168. Shachar S, Voss TC, Pegoraro G, Sciascia N, Misteli T (2015) Identification of gene positioning factors using high-throughput imaging mapping. Cell 162(4):911–923
169. Lindhout BI, Fransz P, Tessadori F, Meckel T, Hooykaas PJ, van der Zaal BJ (2007) Live cell imaging of repetitive DNA sequences via GFP-tagged polydactyl zinc finger proteins. Nucleic Acids Res 35(16):e107
170. Fujimoto S, Sugano SS, Kuwata K, Osakabe K, Matsunaga S (2016) Visualization of specific repetitive genomic sequences with fluorescent TALEs in *Arabidopsis thaliana*. J Exp Bot 67(21):6101–6110
171. Ma H, Naseri A, Reyes-Gutierrez P, Wolfe SA, Zhang S, Pederson T (2015) Multicolor CRISPR labeling of chromosomal loci in human cells. Proc Natl Acad Sci USA 112 (10):3002–3007
172. Dreissig S, Schiml S, Schindele P, Weiss O, Rutten T, Schubert V, Gladilin E, Mette M, Puchta H, Houben A (2017) Live cell CRISPR-imaging in plants reveals dynamic telomere movements. Plant J. doi:10.1111/tpj.13601
173. Pawley JB (2013) Handbook of biological confocal microscopy. Springer, New York, NY
174. Wilson T, Tan JB (1993) Three dimensional image reconstruction in conventional and confocal microscopy. Bioimaging 1(3):176–184
175. Tokunaga M, Imamoto N, Sakata-Sogawa K (2008) Highly inclined thin illumination enables clear single-molecule imaging in cells. Nat Methods 5(2):159–161
176. Beier HT, Ibey BL (2014) Experimental comparison of the high-speed imaging performance of an EM-CCD and sCMOS camera in a dynamic live-cell imaging test case. PLoS One 9(1):e84614
177. Zemach A, Li Y, Wayburn B, Ben-Meir H, Kiss V, Avivi Y, Kalchenko V, Jacobsen SE, Grafi G (2005) DDM1 binds *Arabidopsis* methyl-CpG binding domain proteins and affects their subnuclear localization. Plant Cell 17(5):1549–1558
178. Libault M, Tessadori F, Germann S, Snijder B, Fransz P, Gaudin V (2005) The *Arabidopsis* LHP1 protein is a component of euchromatin. Planta 222(5):910–925
179. Koroleva OA, Calder G, Pendle AF, Kim SH, Lewandowska D, Simpson CG, Jones IM, Brown JW, Shaw PJ (2009) Dynamic behavior of *Arabidopsis* eIF4A-III, putative core protein of exon junction complex: fast relocation to nucleolus and splicing speckles under hypoxia. Plant Cell 21(5):1592–1606
180. Yu X, Sayegh R, Maymon M, Warpeha K, Klejnot J, Yang H, Huang J, Lee J, Kaufman L, Lin C (2009) Formation of nuclear bodies of *Arabidopsis* CRY2 in response to blue light is associated with its blue light-dependent degradation. Plant Cell 21(1):118–130
181. Dittmer TA, Stacey NJ, Sugimoto-Shirasu K, Richards EJ (2007) LITTLE NUCLEI genes affecting nuclear morphology in *Arabidopsis thaliana*. Plant Cell 19(9):2793–2803
182. Guggisberg A, Baroux C, Grossniklaus U, Conti E (2008) Genomic origin and organization of the allopolyploid *Primula*

egaliksensis investigated by in situ hybridization. Ann Bot 101(7):919–927

183. Wanner G, Schroeder-Reiter E, Ma W, Houben A, Schubert V (2015) The ultrastructure of mono- and holocentric plant centromeres: an immunological investigation by structured illumination microscopy and scanning electron microscopy. Chromosoma 124 (4):503–517
184. Baroux C, Pecinka A, Fuchs J, Schubert I, Grossniklaus U (2007) The triploid endosperm genome of *Arabidopsis* adopts a peculiar, parental-dosage-dependent chromatin organization. Plant Cell 19(6):1782–1794
185. Käthner R, Zölffel M (2016) Light microscopy - technology and application. Süddeutscher Verlag onpact GmbH, Munich
186. Becker W, Su B, Holub O, Weisshart K (2011) FLIM and FCS detection in laser-scanning microscopes: increased efficiency by GaAsP hybrid detectors. Microsc Res Tech 74 (9):804–811
187. Feijo JA, Moreno N (2004) Imaging plant cells by two-photon excitation. Protoplasma 223(1):1–32
188. Benninger RK, Piston DW (2013) Two-photon excitation microscopy for the study of living cells and tissues. Curr Protoc Cell Biol. Chapter 4:Unit 4 11 11–24
189. Weber M, Huisken J (2011) Light sheet microscopy for real-time developmental biology. Curr Opin Genet Dev 21(5):566–572
190. Power RM, Huisken J (2017) A guide to light-sheet fluorescence microscopy for multiscale imaging. Nat Methods 14(4):360–373
191. Abbe E (1873) Beiträge zur Theorie des Mikroskops und der mikroskopischen Wahrnehmung. Arch Mikrosk Anat 9(1):413–468
192. Rayleigh L (1896) On the theory of optical images, with special reference to the microscope. Philos Mag 42:167–195
193. Van Noorden R (2014) Insider view of cells scoops Nobel. Nature 514(7522):286
194. Erni R, Rossell MD, Kisielowski C, Dahmen U (2009) Atomic-resolution imaging with a sub-50-pm electron probe. Phys Rev Lett 102 (9):096101
195. Lobet G, Draye X, Perilleux C (2013) An online database for plant image analysis software tools. Plant Methods 9(1):38
196. Schneider CA, Rasband WS, Eliceiri KW (2012) NIH Image to ImageJ: 25 years of image analysis. Nat Methods 9(7):671–675
197. Wan Y, Otsuna H, Chien C-B, Hansen C (2012) FluoRender: an application of 2D image space methods for 3D and 4D confocal microscopy data visualization in neurobiology research. IEEE Pacific Visualization Symposium [proceedings], pp 201–208
198. Carpenter AE, Jones TR, Lamprecht MR, Clarke C, Kang IH, Friman O, Guertin DA, Chang JH, Lindquist RA, Moffat J, Golland P, Sabatini DM (2006) CellProfiler: image analysis software for identifying and quantifying cell phenotypes. Genome Biol 7 (10):R100

Chapter 32

Automated 3D Gene Position Analysis Using a Customized Imaris Plugin: XTFISHInsideNucleus

Mariamawit S. Ashenafi and Célia Baroux

Abstract

Fluorescence in situ hybridization (FISH) is commonly used to visualize chromosomal regions or genomic loci within the nucleus, and can largely contribute to unraveling the link between structure and function in the nucleus. Three-dimensional (3D) analyses are required to best capture the nucleus' organizing principles, but the experimental setup and computational analyses are far from trivial. Here, we present a robust workflow for 3D FISH against repeats and single copy loci in embedded intact nuclei from Arabidopsis leaves. We then describe in detail the image acquisition, subsequent image deconvolution before 3D image processing, and the image reconstruction. We developed an automated batch image processing pipeline using a customized, open source plugin implemented in the Imaris environment.

Key words Fluorescence in situ hybridization, 3D FISH, Confocal imaging, Deconvolution, Automated image processing, 3D gene position, Imaris

1 Introduction

The nuclei of plant and animal cells at interphase share several organizing principles. This includes the presence of microscopically visible heterochromatic domains composed of genomic regions enriched in transcriptionally repressive chromatin modifications, and the occurrence of discrete chromosome territories in animal [1, 2] and plant [3–5] nuclei. Studies on yeast and animal model systems have shown that the nucleus has a functional organization. Notably, the position of certain genes relative to different nuclear compartments correlate with their expression status [6]. This suggested an influence of the 3D arrangement of chromatin on genome expression [7]. For instance, in mammalian cells, gene-rich chromatin tends to cluster away from the nuclear periphery and from the nucleolus. Active genes are also found on chromatin loops that protrude away from chromosome territories [6]. In yeast, the relocation of loci toward the nuclear periphery leads to gene silencing [8]. However, much less is known in plants. Yet a recent study

Marian Bemer and Célia Baroux (eds.), *Plant Chromatin Dynamics: Methods and Protocols*, Methods in Molecular Biology, vol. 1675, DOI 10.1007/978-1-4939-7318-7_32, © Springer Science+Business Media LLC 2018

on Arabidopsis reported the relocation of specific loci in response to light stimuli, which correlated with the induction of their expression [9].

In order to uncover whether plant nuclei also follow specific organizing rules with respect to gene position in the nucleus and expression, it is necessary to develop robust methods for probing and analyzing gene position in intact, three-dimensional (3D) nuclei. Here, we present a step-by-step protocol to generate and interpret data pertaining to the gene's position with regard to the nucleus periphery, the nucleolus and the chromocenters.

Fluorescence in situ hybridization (FISH) protocols have been developed in the past decades to allow the detection of repeat elements or chromosomal regions (using ~100 kb BAC probes) on isolated Arabidopsis nuclei. However, 3D FISH has proven difficult and only a few studies report successful applications in plants using whole-mount tissues [10]. In addition, single copy gene detection is also known as a challenging process likely due to the low density of fluorophores along the probes generated with classical labeling techniques. Oligo-painting [11] or padlock-type probes allow amplification of the target sites of immunolabeling [9], thereby increasing the signal. Here, however, a protocol is shown that uses a traditional nick translation-mediated incorporation of modified nucleotides and indirect immunolabeling that permits single copy gene detection. The probe template consists of a mix of PCR amplicons spanning a region of approximately 30 kb, thereby being large enough for high-confidence signal detection and allowing a specific location of targeted gene. The success of this method also relies on the embedding of isolated nuclei in a polyacrylamide matrix to preserve their 3D structure, while providing a sufficient optical clarity for high-resolution imaging. Image processing includes signal deconvolution, 3D reconstruction and threshold-based detection of signal maxima. For these steps we used the Huygens image deconvolution software (Huygens, Netherlands) and the 3D visualization and image processing software Imaris (Bitplane, CH), but alternative software packages are available. In addition, in order to analyze the position of FISH signals relative to the nuclear periphery, chromocenters and nucleolus, we developed a customized plugin implemented in Imaris. The methodology described in this chapter is summarized in the flowchart in Fig. 1.

2 Materials

General material for all DNA-FISH sections: 1.5 mL microcentrifuge tubes, microcentrifuge with temperature control, shaker, thermomixer, 15 mL tubes, and flat-end forceps.

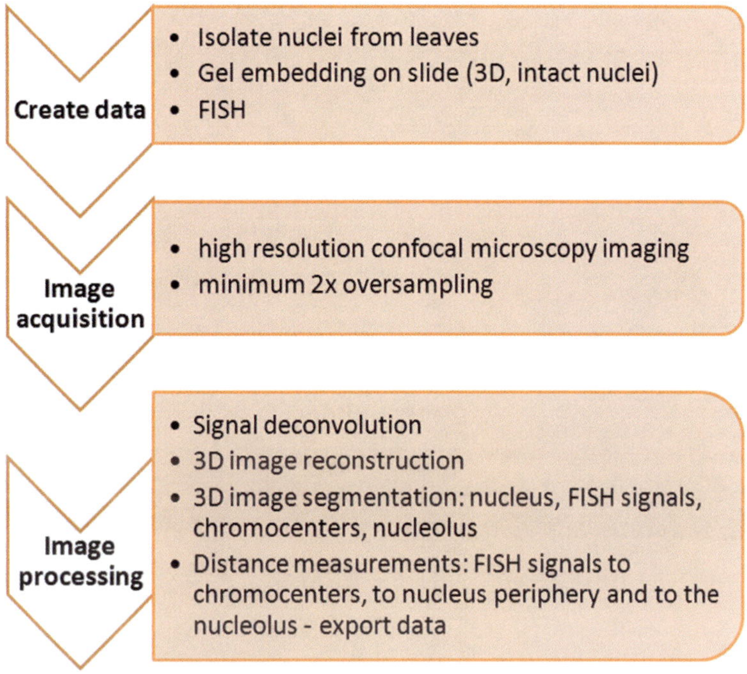

Fig. 1 3D gene position analysis flowchart

2.1 Nuclei Isolation

1. Arabidopsis plants (*see* **Note 1**).
2. Small petri dishes of about 3 cm diameter.
3. Large glass petri dish of about 80 mm × 15 mm.
4. 10× PBS:

   Solution A: Dissolve in 900 mL: 16.02 g Na_2HPO_4, H_2O and 73.84 g NaCl.

   Solution B: Dissolve in 200 mL: 2.76 g Na_2HPO_4, H_2O and 16.56 g NaCl.

   Adjust pH of Solution A to 7.0 with Solution B.
5. Buffer A: 4% Formaldehyde, 0.01% Triton in PBS.
6. Buffer B: 45 mM $MgCl_2$, 20 mM MOPS, 30 mM sodium citrate, 0.3% Triton X-100. Adjust to pH 7 using 0.1 N HCl (*see* **Note 2**).
7. 30 μm cell strainers.
8. Eppendorf tubes.
9. 2 mL Dounce tissue grinder set with large and small clearance pestles.
10. Single-edge carbon steel razor blade.

11. DAPI (4′,6-diamidino-2-phenylindole) dissolved in antifade mounting medium (Vectashield or similar), 1.5 μg/mL.
12. Fluorescence microscope with epifluorescence illumination.

2.2 Gel Embedding

1. Buffer C: 30% acrylamide, 3.3% bisacrylamide in PBS (*see* **Note 3**).
2. 20% sodium sulfite (NaS) in sterile ddH_2O.
3. 20% ammonium persulfite (APS) in sterile ddH_2O.
4. Superfrost microscope slides.

2.3 Probe and Slide Preparation

1. Phusion DNA polymerase with 5× Phusion buffer or equivalent.
2. 10 mM dNTPs.
3. Specific primers for the region of interest.
4. Thermocycler.
5. 70% EtOH.
6. 100% EtOH.
7. 3 M NaOAc.
8. 0.5 M EDTA pH 8.0.
9. 10 mL 0.5 M sodium phosphate buffer pH 7.0: 5.77 mL Na_2HPO_4 0.5 M and 4.23 mL NaH_2PO_4 0.5 M; check pH and adjust by adding some extra Na_2HPO_4 or NaH_2PO_4 if necessary.
10. Fluorimetric-based nucleic acid quantification device (*see* **Note 4**).
11. 20× SSC: 3 M NaCl, 0.3 M Na-citrate, adjust to pH 7 with 5 M HCl.
12. Digoxigenin (DIG) and/or Biotin (BIO) nucleic acid labeling kit.
13. HB50: 50% deionized formamide, 0.05 M sodium phosphate buffer pH 7, in 2× SSC (*see* **Note 5**).
14. DS: 20% dextran sulfate in HB50 (*see* **Note 5**).
15. FB: 1% formaldehyde in PBS (*see* **Note 6**).
16. 2.5 mg/mL Pepsin stock.
17. 0.1 M HCl stock (*see* **Note** 7).
18. Coplin jars.

2.4 Hybridization and Immunodetection

1. Heating blocks for microscope slides.
2. Digoxigenin (DIG) and/or Biotin (BIO) nucleic acid labeling kit.
3. HB50: 50% deionized formamide, 0.05 M sodium phosphate buffer pH 7, in 2×SSC (*see* **Note 6**).

4. Blocking buffer: 5% BSA, in 4× SSC (*see* **Note 8**).
5. 10× TN: 1 M Tris–HCl pH 7, 1.5 M NaCl.
6. TNB: 0.5% blocking reagent in 1× TN (*see* **Note 8**).
7. Primary antibodies: Mouse anti-DIG and Biotinylated Anti-Avidin (*see* **Note 9**).
8. Secondary antibodies: Alexa Fluor 488-coupled anti-mouse antibody and Avidin coupled with Texas-Red (*see* **Note 9**).
9. SF50: 50% deionized formamide in 2×SSC.
10. 4T: 0.05% Tween 20 in 4× SSC, pH 7.
11. TNT: 0.05% Tween 20 in 1× TN, pH 7.
12. FB: 1% formaldehyde in PBS (*see* **Note 6**).
13. Programmable hybridization incubator for microscope slides (*see* **Note 10**).
14. DAPI (4′,6-diamidino-2-phenylindole) dissolved in antifade mounting medium (Vectashield or similar), 1.5 μg/mL.
15. Transparent nail polish.

2.5 Image Acquisition and Processing

1. Confocal laser scanning microscope ideally equipped with a resonance scanner, new generation hybrid detectors and a 63× glycerol immersion, confocal-grade lens with chromatic and planar aberration corrections.
2. System Requirements: OS: Windows 7 or higher 64 bit or Mac 10.9 or higher (*see* **Note 11**).
3. Image deconvolution software (e.g., Huygens Professional, SVI, Netherlands).
4. Imaris (Bitplane AG, Switzerland).
5. Python 2.7 or higher (https://www.python.org/downloads/).
6. Imaris plugin: XTFISHInsideNucleus.py available on the open source Imaris repository (http://open.bitplane.com/tabid/235/Default.aspx?id=131).

3 Methods

3.1 Nuclei Isolation

Here, intact nuclei are isolated, while preserving their three-dimensional (3D) structure. The nuclei isolation protocol described in this section follows the procedure originally described in [12], with slight variations.

1. Prepare a fresh solution of Buffer A and keep it on ice.
2. Collect 5–6 healthy rosette leaves from Arabidopsis plants.
3. Place them in a small petri dish containing 10 mL Buffer A, make sure that all leaves are submerged.

4. Incubate with gentle shaking for 20 min at 4 °C
5. Rinse the leaves twice with PBS, remove the solution after the last wash.
6. Place the leaves in a large glass petri dish on ice (*see* **Note 12**).
7. Add immediately 200 μL Buffer B and chop finely with a razor blade, rapidly for about 30 s.
8. Repeat **step** 7 three times. Make sure you get a fine and homogenous suspension.
9. Transfer the suspension to a clean Dounce tissue grinder (*see* **Note 13**).
10. Grind with the large clearance pestle, by moving it gently up and down (five times maximum) (*see* **Note 13**).
11. Repeat **step 10** with the small clearance pestle (five times maximum) (*see* **Note 13**).
12. Filter the homogenate through two superimposed 30 μm cell strainers into a sterile 1.5 mL Eppendorf tube.
13. Centrifuge at 500 × *g* for 5 min at room temperature.
14. Discard the supernatant.
15. Add 100 μL of Buffer B, and gently resuspend the pellet by pipetting up and down with a cut-end tip.
16. Keep on ice.
17. Verify the quality of the preparation: add a 5 μL drop on a clean microscope slide. Add 5 μL of DAPI Vectashield. Add a coverslip and inspect with epifluorescence microscopy. The nuclei should show regular contours and chromocenters should be well visible.
18. Keep the rest of the suspension on ice. If the preparation quality is good, proceed immediately to gel embedding.

3.2 Gel Embedding

In this step, the nuclei are immobilized while their 3D structures are preserved. The protocol is scaled to capture ca 100 nuclei on a single slide, thus allowing to multiplex downstream FISH or immunostaining experiments. The nuclei extraction allows the preparation of 15 slides. In our hands, embedding in acrylamide pads on microscope slides gives better results than dried nuclei on slides. Acrylamide preparation and embedding is based on a protocol originally established in the Bass lab [13], and also shown in a video protocol [14]. Minor variations were introduced for FISH.

1. Place 15 clean microscope slides horizontally on a clean surface.
2. Prepare fresh solutions of 20% APS in ddH_2O and 20% NaS in ddH_2O.

3. In a 1.5 mL microcentrifuge tube add the following reagents on ice:

   166 μL PBS.

   34 μL Buffer C.

   70 μL nuclei suspension (use a 200 μL cut-end-tip).

   12 μL 20% APS.

   12 μL 20% NaS.

4. Mix gently by pipetting up and down with a 200 μL cut-end tip.
5. Add 15 μL of the sample solution to the center of each microscope slide with a cut-end tip.
6. Wait for 30 s before placing 18 × 18 mm cover slips.
7. Slides can be stored at 4 °C for 1 month, in dry, vertical racks.

3.3 Probe Preparation

We experienced robust signals with probes spanning 30 kb of a single-copy region. Here, the DNA probes are detected by indirect immunostaining of the modified nucleotides using commercially available antibodies. Alternative protocols use direct DNA labeling avoiding immunodetection. In our hands, however, the direct labeling did not yield reproducible results, while indirect labeling allowed for robust signals.

The FISH procedure essentially follows original protocols [15] with minor modifications. PCR amplification of the 30 kb DNA template was performed using the high-fidelity Phusion enzyme. Alternative probe labeling and detection may be used instead of this one (*see* Chapter 25 [16]).

The following proportions are set for probe preparation for one slide/one hybridization.

1. Set up the PCR reaction as described by the manufacterer (https://www.neb.com/protocols/1/01/01/pcr-protocol-m0530):

Component	20 μL reaction	Final concentration
Nuclease-free water	To 20 μL	
5× Phusion HF or GC buffer	4 μL	1×
10 mM dNTPs	0.4 μL	200 μM
10 μM forward primer	1 μL	0.5 μM
10 μM reverse primer	1 μL	0.5 μM
Template DNA	Variable	<250 ng
DMSO (optional)	(0.6 μL)	3%
Phusion DNA polymerase	0.2 μL	1.0 units/50 μL PCR

2. Run the PCR using the following program:

Initial denaturation	→	98 °C for 30 s
Number of cycles	→	25–35
Denaturation	→	98 °C for 5–10 s
Annealing	→	45–72 °C for 10–30 s
Extension	→	72 °C for 15–30 s per kb
Final extension	→	72 °C for 5–10 min
Hold	→	4–10 °C

3. Use 2 μL to verify that a specific amplicon is produced with 1% agarose gel electrophoresis.
4. Measure the DNA concentration of the PCR product using a fluorimetric-based nuclei acid quantification device, then calculate the volume needed for 900 ng (*see* **Note 4**).
5. If the PCR reaction does not yield ~1 μg, scale up to 50 μL reactions.
6. Transfer 900 ng into a fresh tube, add sterile standard ddH_2O to a final volume of 30 μL.
7. Proceed to the DNA precipitation: add 90 μL chilled 100% EtOH, 3 μL 3 M NaOAc, mix well, leave at −20 °C for at least 20 min and centrifuge at maximum speed in a benchtop microcentrifuge at 4 °C for 15 min. Remove the supernatant, briefly rinse the pellet with 70% EtOH and spin briefly at full speed, 4 °C for 1 min.
8. Remove the supernatant and resuspend in 16 μL sterile ddH_2O.
9. Add 4 μL DIG or BIO nucleic acid labeling mix, and gently pipette to mix.
10. Incubate for 90 min at 15 °C (*see* **Note 14**).
11. Stop the labeling reaction. Add 1 μL 0.5 M EDTA and incubate at 65 °C for 10 min. Spin briefly. At this point the probes can be stored at −20 °C until use.
12. Add 5 μL of each DNA probe (differently labeled probes either with DIG or BIO) and complete the volume with sterile ddH_2O up to 50 μL.
13. Precipitate with 150 μL 100% EtOH and 5 μL 3 M NaOAc following the same process as in **step** 7. Carefully dry the pellet on bench.
14. Add 10 μL HB50, and incubate at 42 °C for 10 min—do not pipette at this stage.

15. Add 10 μL DS, mix well by pipetting gently.
16. Denature probe for 15 min at 75 °C.
17. Immediately put on ice.

3.4 Slide Preparation

All steps are performed at room temperature (*see* **Note 15**) unless otherwise specified. For incubation at 38, 80, and 37 °C the heating blocks for microscope slides are used.

1. Label the slides with the FISH experiment number.
2. Immerse them in FB for 10 min for a mild fixation before the FISH treatment (*see* **Note 16**).
3. Rinse two times in PBS by exchanging the solution.
4. Drain the excess of liquid by holding the slide for a few seconds vertically.
5. Add 1 mL of 2.5 mg/mL pepsin in 100 mL 0.01 M HCl.
6. Add 100 μL of pepsin/0.01 M HCl on the gel pad and incubate for 1 min 25 s at 38 °C in a preheated slide heating block for microscope slides (*see* **Note** 17).
7. Transfer the slides to a Coplin jar filled with a PBS solution containing FB for 10 min.
8. Rinse two times in PBS.
9. Remove the last wash, remove as much liquid as possible, and keep the slides in the Coplin jar to air dry.

3.5 Hybridization

1. Cut small pieces of Parafilm that would cover the gel pads (around 18 × 18 mm) for all the slides.
2. Precondition the embedded nuclei with hybridization buffer: add 20 μL of HB50:DS (1:1), and cover the gel pads with Parafilm.
3. Incubate for 10 min at room temperature.
4. Remove the Parafilm, add 20 μL of probe and cover again with the same Parafilm.
5. Transfer the slides quickly to 80 °C in a preheated slide heating block for microscope slides and incubate for 2 min.
6. Hybridize for at least 16 h at 37 °C (*see* **Note 18**).

3.6 Immunodetection

All steps are performed at room temperature unless otherwise specified (*see* **Note 15**). For incubation at 37 and 42 °C, the heating blocks for microscopy slides are used.

1. Add 100 μL of SF50 and incubate for 5 min at 42 °C.
2. Remove the liquid by holding the slide vertically on a paper towel.

3. Add 100 μL of 2× SSC and incubate for 5 min at 42 °C, repeat **step 2**.
4. Add 100 μL of 4T and incubate for 5 min at 42 °C, repeat **step 2**.
5. Add 100 μL of blocking buffer for 30 min at 37 °C, repeat **step 2**.
6. Add 100 μL of 4T and incubate for 5 min, repeat **step 2**.
7. Add 100 μL of TNT, incubate for 5 min and repeat **step 2**. Repeat this step twice.
8. Add 100 μL of mouse anti-DIG in TNB (1:250) (for DIG nick translation) or Biotinylated Anti-Avidin in TNB (1:250) (for BIO nick translation), or 100 μL of TNB containing both antibodies (*see* **Note 19**).
9. Incubate for 90 min at 37 °C, repeat **step 2**.
10. Repeat **step 7**.
11. Add 100 μL of Alexa Fluor 488-coupled anti-mouse antibody (2.5:1000) in TNB (for DIG nick translation) or Avidin coupled with Texas-Red in TNB (1:1000) (for BIO nick translation), or 100 μL of TNB containing both antibodies (*see* **Note 19**).
12. Incubate for 90 min at 37 °C, repeat **step 2**.
13. Repeat **step 7**.
14. Counterstain with 10 μL of DAPI, add precision cover glass and seal with nail polish.
15. Leave for 10 min in the dark.
16. Slides can be stored vertically at 4 °C in the dark.

3.7 Image Acquisition

Objective: Obtain 3D images of individual nuclei with gene-FISH signals and counterstained with DAPI. We used a Leica TCS SP5 resonance CSLM, a 63 Glycerol immersion objective, NA 1.3, CS2 PL APO grade with corrective lenses for chromatic and planar aberrations. Other instruments with similar setup can be used. The important aspects are the following:

1. Image acquisition resolution affects the margin of error of all the distances and positions calculated by the plugin described under Subheading 3.10. The resolution of the images illustrated here is 20 nm × 20 nm × 80 nm, which is largely oversampled (*see* **Note 20**). This resolution is achieved using a pinhole opening of 1 airy unit, 20-fold zoom factor and 256 × 256 image format, and a z-step of 80 nm. An example of an acquired image is shown in Fig. 2.
2. Laser power and gain (if photomultiplier tubes, PMTs, are used) should be balanced to use the optimal dynamic range of

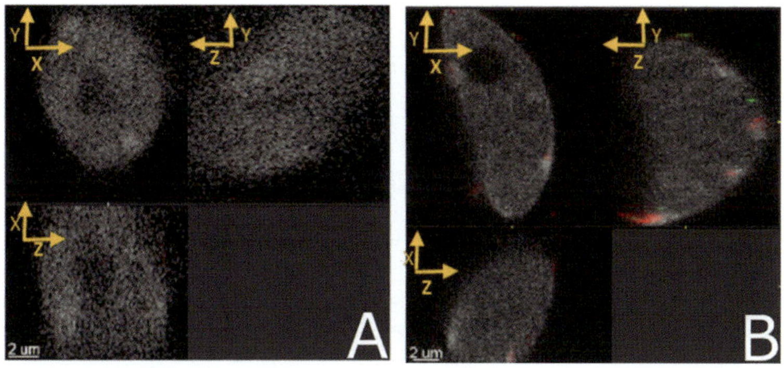

Fig. 2 Example of an acquired FISH image. (**a**) Image of nuclei after FISH without labeled probes. (**b**) Image of nuclei after FISH with labeled probes: CEN-repeat in the red channel and AT1G15690 in the green channel

pixel intensities, and avoid local saturation (using the Look Up Table (LUT) inspection tool) and bleaching.

3. Simultaneous three-channel acquisition should be controlled for the absence of cross-talk. Otherwise sequential scanning is recommended.
4. To speed up the acquisition of multiple nuclei from one preparation, automated z-series scans can be setup using a multiposition recorder grid following the provider's recommendation.
5. The objective lens should ideally allow for glycerol immersion (matching the refractive index (RI) of Vectashield). If available, use the lens correction for coverslip RI. Make sure to use a lens with confocal, fluorescence microscopy grade with best possible chromatic and spherical aberration corrections (particularly important for resolving the FISH signals in the *z* dimension).

3.8 Image Processing

The objective of the image processing part is to get numerical data out of digital image data. Processing consists of two steps: (1) image deconvolution and (2) image segmentation and distance computing. (1) Image deconvolution can in principle be achieved with any software, but we describe a workflow here using the Huygens platform. For alternative software, the users should follow the providers' instructions. (2) Image segmentation and the computing of FISH signal distances relative to the nuclear periphery, chromocenters, and nucleolus are achieved in the Imaris environment. While Imaris is a commercially available software often provided by academic microscopy facilities, the plugin used for this analysis is freely available in the Imaris Xtension open repository. We propose an automated batch processing in order to facilitate large data analysis, and to insure all images are processed in the

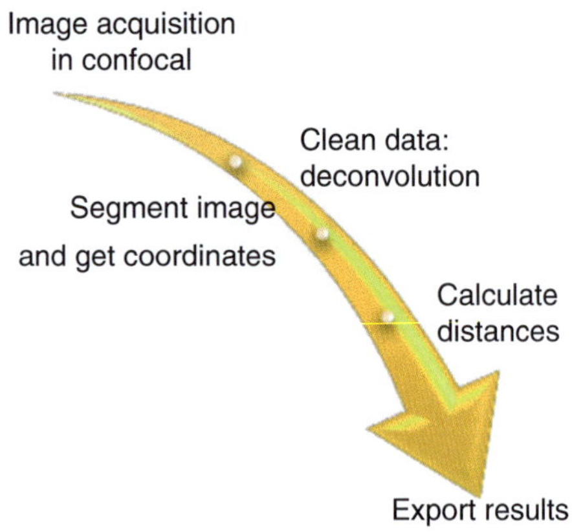

Fig. 3 Image processing flowchart

same manner but single image processing is possible and recommended the first time to become familiar with the process (described in the scheme in Fig. 3).

3.9 Deconvolution

1. Open one image in Huygens.
2. Set up the parameters for deconvolution using the Parameter editor (Edit menu or Right click on the image). The software should read the Meta data of the image and automatically detect the values of all the parameters, except for Lens immersion and Embedding medium. The user needs to set the values for these two parameters to glycerine 80% (this is the medium we used for lens immersion). Verify the values for all the other parameters. Validate the parameters ("Set all verified" and "Save") and save the template file corresponding to these parameters as *Temp1*. Validate ("Accept").
3. Set up the batch analysis mode (Batch tab), add a task. A step-by-step wizard opens, guiding you through the different steps. Validate each step via the arrow to (1) select the files to be processed using a browsing menu, (2) specify the deconvolution parameters to be used by selecting your formerly saved *Temp1* file, (3) adjust the parameters per channel by selecting the option "New Template." Table 1 shows the parameters recommended for each channel. Repeat this process for all channels.
4. Save the workflow and parameters you just set up as a template for future batch operations (Template tab: set the name of the template, for instance *Temp2* and save) and close the task ("Done").
5. Setup the output format and output location: in the Option tab, set Output format to "Imaris Classic." In the "Save

Table 1
Deconvolution parameters for huygens

Parameters	Values for all channels
Algorithm	Classic MLE
PSF mode	Theoretical
Max. iteration	40
Iteration mode	Optimized
Quality change threshold	0.05
Signal to noise ratio	15
Background mode	Lowest value
Background estimation radius	0.7
Relative background	0.0
Bleaching correction	If possible
Brick mode	Auto
PSFs per brick mode	Off
PSFs per brick, manual mode	1

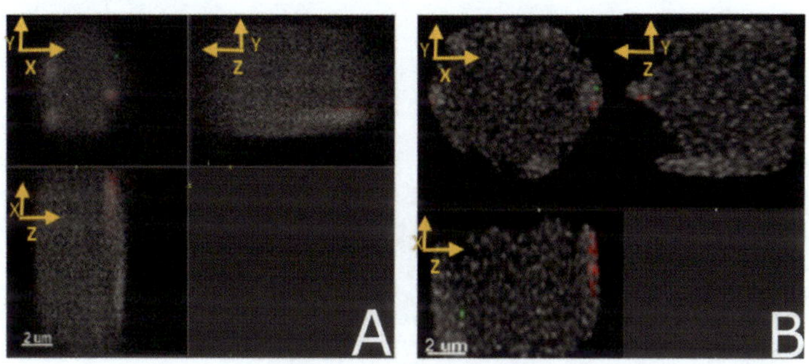

Fig. 4 Example of deconvolution output. (**a**) Raw image before deconvolution. (**b**) Image after deconvolution with Huygens. FISH signal: AT1G15690 in the red channel and AT5G25860 in the green channel

location" field, browse for the folder where all the resulting image files will be saved: *folder2*.

6. Start the deconvolution. An example of a deconvolution result is shown in Fig. 4.

3.10 Imaris Empowered Image Processing

The objective of our customized plugin is to (1) segment the image and (2) compute distances. Segmentation works on the principle of intensity thresholds and voxel size and uses processing modules implemented in Imaris. Segmentation produces four types of 3D objects: a nuclear surface, spots corresponding to FISH signals, surfaces corresponding to chromocenters and a nucleolus surface. For each FISH signal (spot), its distance to the nuclear surface (PF), to the nucleolus (NF) and to the closest chromocenter (CF) is calculated. All distances are normalized to the nuclear volume. The scheme of the analysis is described in Fig. 5.

The images produced by the plugin contain the features illustrated in Fig. 6. The figure shows that several mask channels are created and used to do the following segmentations: The DAPI

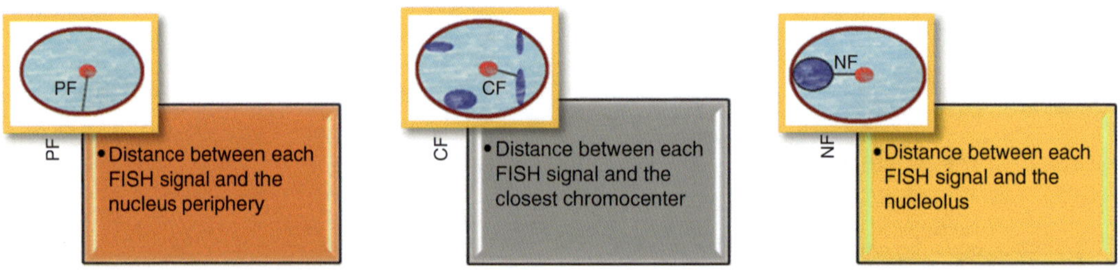

Fig. 5 Scheme of distances calculated with the plugin

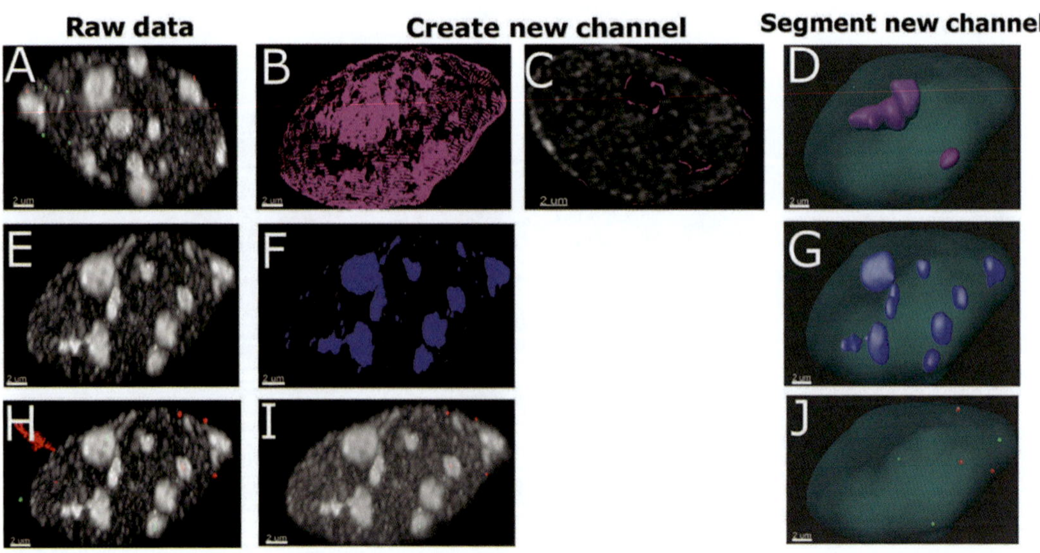

Fig. 6 Segmentation result. (**a**) Raw data: DAPI channel in gray. (**b**) Low intensity DAPI channel (LDI) in magenta, is created by the plugin to select voxels with intensity lower than average. 3D view. (**c**) DAPI and LDI channels, Z-slice view of the nucleoli. (**d**) Nucleus surface and the nucleoli are segmented using the DAPI and LDI channels. (**e**) Raw data: DAPI channel in gray. (**f**) Chromocenters channel, in purple, is created, by selecting voxels within the nucleus with intensity higher than average in the DAPI channel. (**g**) Using the latter, chromocenters are segmented. (**h**) Raw data: DAPI channel in gray, FISH signals: FISH_Ch0: AT1G15690 in the red channel and FISH_Ch1: AT5G25860 in the green channel. (**i**) High intensity FISH channels are created selecting voxels within the nucleus with intensity higher than average in each FISH channel; High intensity FISH_Ch0 in the red channel and High intensity FISH_Ch1 in the green channel. (**j**) High intensity FISH channels are segmented; AT1G15690 spots in red, AT5G25860 spots in green

channel is used to produce the nucleus surface (Fig. 6a and c), the Low intensity DAPI channel is used for the nucleolus surface (Fig. 6b and c), the Chromocenters channel is used for the Chromocenters surfaces (Fig. 6d and e), High intensity FISH channels are created for each FISH channel and are used to segment *FISH* spots (Fig. 6g and h). Options to fine tune segmentation are described in the Troubleshooting section.

The plugin produces four tables for each FISH channel:

FISHPeripheryDistanceTable_ChX.csv.

FISHNucleolusDistanceTable_ChX.csv.

FISHChromocenterDistanceTable_ChX.csv.

FISHPositionTable_ChX.csv.

Each table contains as many columns as there are images and as many rows as there are FISH spots. These tables are contained in the "XTFISHInsideNucleus_Result" folder. The latter is created by the plugin in the directory containing the images to be processed. For each image processed, an .ims file is created in this new folder.

Step by step

5. Download the two files: "XTFISHInsideNucleus.py" and "XTFISHInsideNucleus_Parameters.csv", from http://open.bitplane.com/tabid/235/Default.aspx?id=131. Both files are required to run the plugin. Save both files in the same folder: *folder1*.
6. Download and install python 2.7 or higher. Set installation directory to *pythonFolder* (in windows this is usually C:\Program Files).
7. Open Imaris, and under File/Preferences/Calculation/, set memory limit to 4 GB.
8. Go to the File tab and click on "Preferences" in the drop down menu.
9. In the "Preferences" window, click on "Custom Tools."
10. In the "Custom Tools" window, select "Browse" in the "Python application" field to locate the python.exe file: *pythonFolder.*
11. On the "XTension Folders" field, select "Add" and browse for *folder1*.
12. Select "Ok" on the bottom of the window.
13. In the top tool bar, select Image Processing/FISHInsideNucleus (on the bottom of the menu).
14. Set parameters:
    - Select "Batch of images" in Window1 (Fig. 7a) for batch processing (*see* **Note 21**).

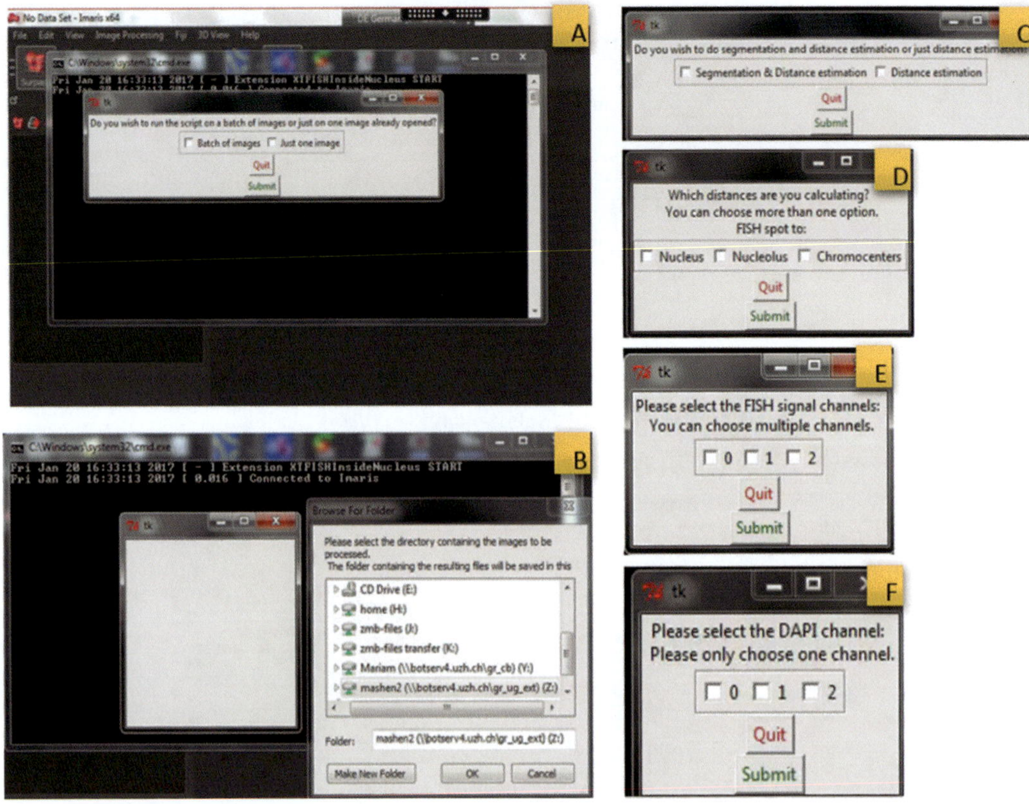

Fig. 7 Running the plugin. (**a**) Window1—Select if the plugin should process: "Batch of images" or "just one image." (**b**) Window2—If Batch of images is selected in Window1, browse for the folder (folder2) containing all the images to be processed. (**c**) Window3—if the plugin should perform segmentation, select "Segmentation & Distance estimation," if the segmentation has already been performed manually, select "Distance estimation." (**d**) Window4—To calculate the distance between FISH spots and nucleus periphery: select "nucleus," to the nucleolus: select "nucleolus," to the closest chromocenter: select "Chromocenters." It is possible to choose multiple options. (**e**) Window5—select FISH channels that should be segmented FISH spots. (This window pops up only if "Segmentation & Distance estimation" is selected in Window3). (**f**) Window6—select DAPI channel that should be segmented into nucleus, nucleolus and chromocenters. (This window pops up only if "Segmentation & Distance estimation" is selected in Window3)

- Browse for *folder2* in Window2 (Fig. 7b), the plugin will batch process the images contained in that folder.
- Select "Segmentation & Distance estimation" in Window3 (Fig. 7c), to run a two-step program: segmentation and distance calculation (*see* **Note 21**).
- Select options according to the distances desired (PF, NF or CF) in Window4 (Fig. 7d). For PF choose: "Nucleus," for NF: "Nucleolus," for CF: "Chromocenters." More than one option can be selected.

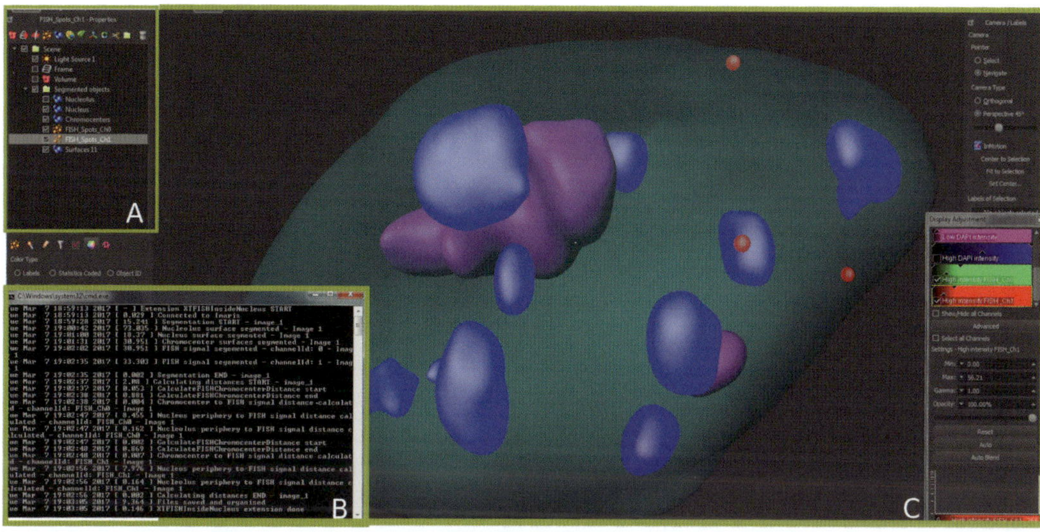

Fig. 8 Extension progress and result. (**a**) All instances (objects) created by the plugin are stored in the container "Segmented Objects." (**b**) While the plugin is running, a python terminal displays the progress of the plugin by logging in tasks that have been completed. (**c**) All objects created are displayed in the 3D viewer: magenta surface = nucleolus, red surfaces = chromocenters, red and green spots = FISH signals

- Select which channels are FISH channels in Window5 (Fig. 7e). More than one option can be selected.
- Select which channel is the DAPI channel in Window6 (Fig. 7f) (*see* **Note 21**).

While the plugin is running, the user can track the progress of the program by tracking the python terminal (Fig. 8b) that logs completed tasks, and by tracking the success of the segmentation process by viewing the objects in the 3D viewer (Fig. 8c).

Once the plugin has finished running, a window pops up, indicating that the program is completed, displaying the number of processed images.

All the produced files are saved in the folder XTFISHInside-Nucleus_Result that is created by the plugin in *folder2*.

3.11 Handling Segmentation Errors

It is recommended to go over all the .ims files produced to review the segmentation success, before analyzing the data produced. This can be done by a visual verification: check that the nucleus, chromocenters, and nucleolus surfaces and FISH spots correlate with the signals in the 3D image.

Troubleshoot option 1—for systematic segmentation error:

The segmentations are calibrated to our experiment. As a consequence, some parameters are set to a specific value. So in case the segmentations are erroneous for several images, the user can calibrate these values to fit their images, by using gradient values and doing several runs of the plugin. These values can be set in the "XTFISHInsideNucleus_Parmaeters.csv" file (*see* **Note 22**).

Troubleshoot option 2—for individual/random segmentation error:

An individual/random segmentation error corresponds to an inaccurate segmentation occurring on only one or a few images where the signal distribution and level does not allow an unambiguous thresholding by the plugin. One example: when the nucleolus is close to the nucleus surface, the plugin cannot distinguish between the voxels with low DAPI intensity inside the nucleolus and those outside the nucleus surface, as shown in Fig. 9a. An erroneous nucleolus segmentation is illustrated in Fig. 9b. To correct this, the segmentation must be done manually using the low DAPI intensity channel.

Manual segmentation steps:

1. Figure 9c Select the "create a surface" icon (1). Then select next (2).
2. Figure 9d Set the channel to Low DAPI Intensity channel (3), set smooth surface detail to 0.4 (4) and diameter of the largest sphere to 1.6 (5). Both values can be adjusted according to the quality of nucleolus segmentation.
3. Figure 9e Set the lowest intensity threshold (7) so that the entire nucleolus surface is segmented.
4. Figure 9f Select Volume and set a threshold of minimum value so that only the nucleolus surface is visible.

Same method shown here can be used for nucleus and chromocenter manual segmentation.

In order to perform the automated distance calculation on the corrected segmented image(s) follow the steps below:

1. If more than one image has been corrected, put all the modified images in one folder, *folder3*.
2. Run FISHInsideNucleus.
3. Set parameters:
   - Select "Batch of images" in Window1 (Fig. 6a) if there is more than one image otherwise select "Just one image."
   - Browse for *folder3* in Window2 (Fig. 6b).
   - Select "Distance estimation" in Window3 (Fig. 6c), to run a one-step program: distance calculation only.

The rest of the steps are the same as those described in the step-by-step section.

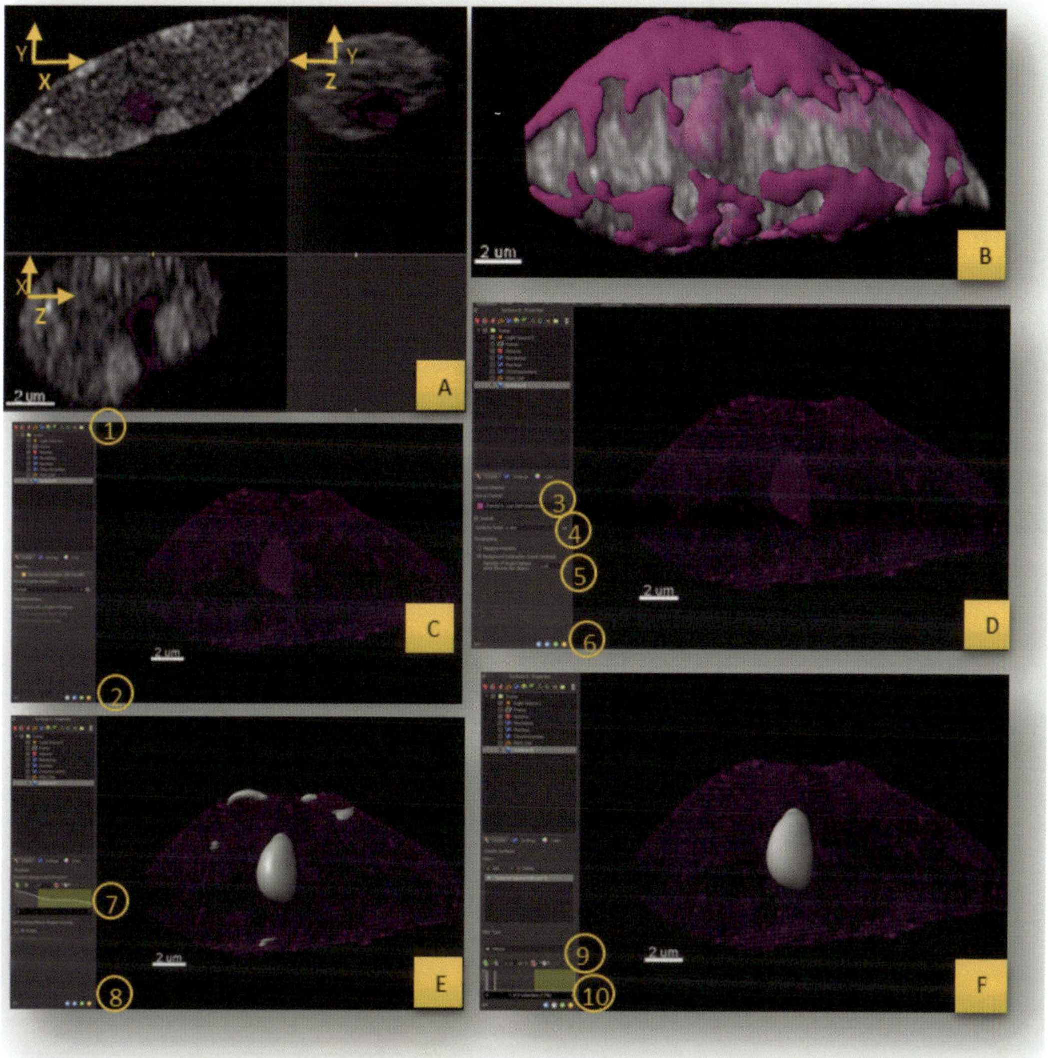

Fig. 9 Manual segmentation. (**a**) DAPI channel is in gray and LDI in magenta. (**b**) Erroneous nucleolus segmentation. (**c**) Create a new surface (*1*) and click on the next arrow (*2*). (**d**) Set the channel index (*3*), the smooth surface detail (*4*) and the diameter of the largest sphere, which fits into object (*5*) and click on the next arrow (*6*). (**e**) Set intensity threshold (*7*) and click on the next arrow (*8*). (**f**) Select volume and set a threshold of minimum value (*9*) and click on the next arrow (*10*)

4 Notes

1. After sterilization, the seeds are sown directly on soil. It is important to extract nuclei from healthy tissue (i.e., stressed, pigmented, or necrotic leaves should be avoided), from well-watered plants. To have a homogeneous population of nuclei, tissue should ideally be collected from the same leaf positions, at the same developmental stage, and around the same time of

the day, as these factors can influence chromatin organization. In this study, we used 35-days old Arabidopsis plants cultivated in a short-day growth chamber. The developmental stage of the plant should be adapted to the study conducted.

2. Prepare 10 mL aliquots of Buffer B and store at −20 °C up to 6 months.
3. After preparing 10 mL fresh Buffer C in a 50 mL bottle in the fume hood, sterilize by filtration and store in the bottle at 4 °C [13]. At this high concentration, this solution should be handled under the fume hood. When the concentration is lowered to 5%, the fume hood is no longer necessary.
4. Accurate quantification of double stranded DNA is critical. It is highly recommended to use a fluorimetric-based quantification discriminating single and double stranded DNA, and RNA. A Qubit system from Invitrogen or similar technology can be used. Alternatively, one can add a purification step after PCR, using a kit, and use the NanoDrop instead.
5. Prepare 10–20 mL of HB50/DS. Sterilize by filtration and prepare 50–100 μL aliquots, store at −20 °C up to 6 months. Use a fresh aliquot for each FISH.
6. FB can be stored at 4 °C for up to 6 months.
7. Prepare 10 mL 2.5 mg/mL pepsin stock, prepare 1 mL aliquots and store at −20 °C. Prepare 10 mL 0.1 M HCl stock and store at room temperature. Both preparations can be stored up to 6 months. Use a fresh aliquot of Pepsin for each FISH.
8. Prepare 10 mL Blocking buffer and TNB solution, sterilize by filtration, prepare 1 mL aliquots and store at −20 °C for up to 6 months. Use a fresh aliquot for each FISH.
9. Antibodies should be stored according to the manufacturer's instructions. We recommend preparing 5 μL aliquots, which should be stored as described by the manufacturer. In our experience, once an aliquot stored at −20 °C is used, it can be stored at 4 °C for 2 weeks only.
10. Instead of the heating block for microscope slides, one can use humid chambers with moist paper towels. However in our experience, the heating block produces a more controlled environment for our experiment and thereby allows replicable results.
11. Recommended computer system requirements: "RAM: 4 GB; 3 GHz CPU (Intel or AMD) dual core; Graphics: AMD Radeon R7 250 2 GB for Windows, nVidia GeForce GT 650M 1024 MB (slow triangle rendering) or Intel HD 4000 graphics or later (OS X only) for Mac; Monitor: 1280 × 1024

pixels or better; Mouse: 3 button wheel" http://www.bitplane.com/systemrequirements.aspx#sthash.yw2tN98V.dpuf.

12. In order to keep the nuclei intact, it is important not to leave the leaves or the suspension dry during **steps 6–8** of Subheading 3.1. Also, putting a layer of aluminum between the ice and the glass petri dish keeps ice from getting into the petri dish during **steps** 7 and **8**.
13. After each use of the Dounce tissue grinder, rinse with ddH_2O.
14. If the probe preparation is done for the first time for a particular DNA sequence, the nick translation reaction duration should be adjusted. To determine the proper conditions, load 2 μL of a test reaction on a 1% agarose gel after 10, 30, 60, and 90 min nick translation and more if necessary. It is important that most of the smear is found between 200 and 500 nt. Deviations from this will result in failed or suboptimal FISH signal.
15. The room temperature in our laboratory varies from 20 to 24 °C. Change in room temperature can introduce variability in FISH signal strength.
16. Please use the formaldehyde under the fume hood. The FB can be reused several times and can be stored at 4 °C up to 3 months.
17. Precise incubation duration is critical, underexposure of the nuclei to Pepsin solution would result in suboptimal FISH signal and an overexposure would result in damage of nuclei.
18. If FISH signal strength is too weak, one can try to increase the duration of the hybridization.
19. Antibody should be freshly prepared in TNB, be kept on ice in the dark and used only once. So during and after incubation with antibodies, the slides should be kept in the dark. We recommend using aluminum foil to cover the slides during **step 14** of Subheading 3.6.
20. The margin of errors of the distances calculated is indirectly proportional to the values of image acquisition resolution (the latter is the size of one voxel). It is therefore better to aim for a small voxel size. However the smaller the voxel size is, the longer one image acquisition will take. A good method for estimating the image acquisition resolution is to use the Nyquist calculator: https://svi.nl/nyquist/. The resolution can be adjusted using the zoom factor, the image format and z-step.
21. For these parameters, only one option should be selected. If the user selects more than one option, by default the first option will be the value of that parameter.

22. To have a better understanding of the meaning of these parameters, please look into XTFISHInsideNucleus_Manual.pdf that can be downloaded from http://open.bitplane.com/tabid/235/Default.aspx?id=131 /.

Acknowledgments

The authors thank Dr. Jana Döhner (Microscopy Imaging Facility, University of Zürich) for technical assistance with image acquisition and deconvolution; Dr. Peter Majer, Sacha Guyer, (Bitplane CH) for advice on the customized plugin; Drs. Wenjing She, Kostas Kritsas (University of Zürich), Imen Mestiri, and Fredy Barneche (IBENS Paris) for sharing experimental protocols as described here. This work was supported by a SystemsX iPhD grant (2014/235), the University of Zürich and the Baugarten Stiftung.

References

1. Zorn C, Cremer T, Cremer C, Zimmer J (1976) Laser UV microirradiation of interphase nuclei and post-treatment with caffeine. A new approach to establish the arrangement of interphase chromosomes. Hum Genet 35 (1):83–89
2. Cremer T, Cremer C, Schneider T, Baumann H, Hens L, Kirsch-Volders M (1982) Analysis of chromosome positions in the interphase nucleus of Chinese hamster cells by laser-UV-microirradiation experiments. Hum Genet 62 (3):201–209
3. Shaw PJ, Abranches R, Paula Santos A, Beven AF, Stoger E, Wegel E, Gonzalez-Melendi P (2002) The architecture of interphase chromosomes and nucleolar transcription sites in plants. J Struct Biol 140(1–3):31–38. doi:10.1016/S1047-8477(02)00537-3
4. Pecinka A, Schubert V, Meister A, Kreth G, Klatte M, Lysak MA, Fuchs J, Schubert I (2004) Chromosome territory arrangement and homologous pairing in nuclei of Arabidopsis thaliana are predominantly random except for NOR-bearing chromosomes. Chromosoma 113(5):258–269. doi:10.1007/s00412-004-0316-2
5. Berr A, Pecinka A, Meister A, Kreth G, Fuchs J, Blattner FR, Lysak MA, Schubert I (2006) Chromosome arrangement and nuclear architecture but not centromeric sequences are conserved between Arabidopsis thaliana and Arabidopsis lyrata. Plant J 48(5):771–783. doi:10.1111/j.1365-313X.2006.02912.x
6. Cremer T, Cremer C (2001) Chromosome territories, nuclear architecture and gene regulation in mammalian cells. Nat Rev Genet 2(4):292–301
7. Ruault M, Dubarry M, Taddei A (2008) Re-positioning genes to the nuclear envelope in mammalian cells: impact on transcription. Trends Genet 24(11):574–581. doi:10.1016/j.tig.2008.08.008
8. Andrulis ED, Neiman AM, Zappulla DC, Sternglanz R (1998) Perinuclear localization of chromatin facilitates transcriptional silencing. Nature 394(6693):592–595
9. Feng C-M, Qiu Y, Van Buskirk EK, Yang EJ, Chen M (2014) Light-regulated gene repositioning in Arabidopsis. Nat Commun 5:3027. doi:10.1038/ncomms4027
10. Bauwens S, Katsanis K, Van Montagu M, Van Oostveldt P, Engler G (1994) Procedure for whole mount fluorescence in situ hybridization of interphase nuclei on Arabidopsis thaliana. Plant J 6(1):123–131. doi:10.1046/j.1365-313X.1994.6010123.x
11. Beliveau BJ, Boettiger AN, Avendaño MS, Jungmann R, McCole RB, Joyce EF, Kim-Kiselak C, Bantignies F, Fonseka CY, Erceg J, Hannan MA, Hoang HG, Colognori D, Lee JT, Shih WM, Yin P, Zhuang X, Wu C-t (2015) Single-molecule super-resolution imaging of chromosomes and in situ haplotype visualization using Oligopaint FISH probes. Nat Commun 6:7147. doi:10.1038/ncomms8147
12. Bourbousse C, Mestiri I, Zabulon G, Bourge M, Formiggini F, Koini MA, Brown SC, Fransz P, Bowler C, Barneche F (2015) Light signaling controls nuclear architecture

reorganization during seedling establishment. Proc Natl Acad Sci U S A 112(21): E2836–E2844. doi:10.1073/pnas.1503512112

13. Howe ES, Murphy SP, Bass HW (2013) Three-dimensional acrylamide fluorescence in situ hybridization for plant cells. Methods Mol Biol (Clifton, NJ) 990:53–66. doi:10.1007/978-1-62703-333-6_6
14. She W, Grimanelli D, Baroux C (2014) An efficient method for quantitative, single-cell analysis of chromatin modification and nuclear architecture in whole-mount ovules in Arabidopsis. J Vis Exp 88:51530. doi:10.3791/51530
15. Lysak M, Fransz P, Schubert I (2006) Cytogenetic analyses of Arabidopsis. In: Salinas J, Sanchez-Serrano JJ (eds) Arabidopsis protocols. Humana Press, Totowa, NJ, pp 173–186. doi:10.1385/1-59745-003-0:173
16. Bey TD, Koini M, Fransz PF (2017) Fluorescence in situ hybridization (FISH) and immunolabeling on 3D preserved nuclei. In: Bemer M, Baroux C (eds) Plant chromatin dynamics: methods and protocols. Springer, New York, NY. doi:10.1007/978-1-4939-7318-7_27

Chapter 33

Quantitative 3D Analysis of Nuclear Morphology and Heterochromatin Organization from Whole-Mount Plant Tissue Using NucleusJ

Sophie Desset, Axel Poulet, and Christophe Tatout

Abstract

Image analysis is a classical way to study nuclear organization. While nuclear organization used to be investigated by colorimetric or fluorescent labeling of DNA or specific nuclear compartments, new methods in microscopy imaging now enable qualitative and quantitative analyses of chromatin pattern, and nuclear size and shape. Several procedures have been developed to prepare samples in order to collect 3D images for the analysis of spatial chromatin organization, but only few preserve the positional information of the cell within its tissue context. Here, we describe a whole mount tissue preparation procedure coupled to DNA staining using the PicoGreen® intercalating agent suitable for image analysis of the nucleus in living and fixed tissues. 3D Image analysis is then performed using NucleusJ, an open source ImageJ plugin, which allows for quantifying variations in nuclear morphology such as nuclear volume, sphericity, elongation, and flatness as well as in heterochromatin content and position in respect to the nuclear periphery.

Key words ImageJ, NucleusJ, Nucleus, 3D nuclear morphology, Heterochromatin, Chromocenters

1 Introduction

Light microscopy observations of stained tissue preparations have been traditionally applied to analyze nuclear organization including nuclear size and shape; and heterochromatin and nucleolus organization. These parameters were widely used in the diagnosis of human diseases [1]. New techniques in high-resolution microscopy imaging and computational processing of image series have enabled quantitative analyses of nuclear organization in three dimensions (3D) [2, 3]. 3D analyses have indeed highlighted the nonrandom organization of the human genome in the nuclear space, its organization in chromosome territories and the relationship between spatial organization of chromatin and gene transcription [4]. Nuclear organization analyses in plant tissues remains challenging due to technical difficulties at the level of sample staining and high-resolution imaging in thick tissues. Typically, the cell wall limits the

Marian Bemer and Célia Baroux (eds.), *Plant Chromatin Dynamics: Methods and Protocols*, Methods in Molecular Biology, vol. 1675, DOI 10.1007/978-1-4939-7318-7_33, © Springer Science+Business Media LLC 2018

penetration of DNA-intercalating agents, probes and antibodies. Furthermore, the cell contains specific components and organelles generating autofluorescence, which hampers imaging of the objects of interest [5]. Several methods allow staining of intact preparations of plant cells or tissues including acrylamide embedded tissue [6], cryosection [7], or whole-mount preparations [8]. In order to collect many nuclei from vegetative tissue, we have adapted the whole mount fixation method on vegetative tissues previously developed by Bauwens [8]. Nuclei are visualized using an intercalating agent incorporated into the DNA double helix to stain the nuclear DNA. We choose PicoGreen®, an ultrasensitive fluorescent nucleic acid stain, which in contrast to 4′,6-diamidino-2-phenylindole (DAPI) and propidium iodide (PI) has the property to penetrate into the nuclei of living plant tissues. Intercalating agents allow defining the nucleus as a distinct object within an image using DNA staining. In certain organisms, including Arabidopsis, such nuclear staining also reveals foci of higher intensity within the nucleus known as chromocenters corresponding to regions with dense and compact DNA domains within the nucleus (Fig. 1) [9].

We propose here a complete method, from the preparation of living or fixed samples up to the 3D analysis of nuclei images using the software NucleusJ [10] implemented as a plugin in the open source platform called ImageJ, one of the most popular image analysis platforms in Biology [11]. The method is divided in three major steps: (1) staining of nuclei within whole mount preparations of cotyledons with PicoGreen®, an intercalating agent preferred here over DAPI or PI for its efficient staining of both living and fixed tissues, (2) image acquisition of nuclei in 3D, and (3) 3D image processing. More specifically NucleusJ performs in two steps: First, the nucleus is defined through a segmentation step using a thresholding method initially defined by Otsu [12], which has been combined with a shape parameter called sphericity [10, 13] to optimize the selection of the threshold value. Second, segmentation of intranuclear objects such as chromocenters are performed using the watershed algorithm [13, 14], here adapted to objects in 3D.

2 Materials

Prepare all solutions using deionized water (dH_2O) and analytical grade reagents. Prepare and store all reagents at room temperature (unless indicated otherwise). Diligently follow all waste disposal regulations when disposing of hazardous waste. We do not add sodium azide to reagents.

2.1 Staining and Fixation

1. 9-day-old Arabidopsis seedlings (*see* **Note 1**).
2. Stereomicroscope (binocular).

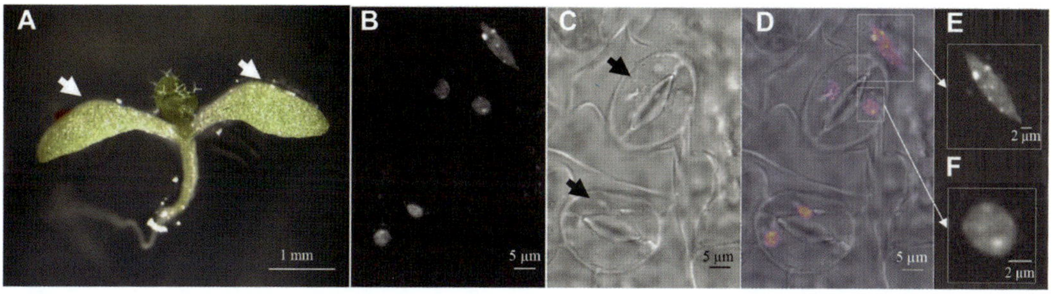

Fig. 1 Interphase nucleus from *Arabidopsis thaliana* cotyledon epidermis. (**a**) 9-day-old seedling used in the whole-mount protocol. The two cotyledons are indicated by *arrows*. (**b**) Z projection of PicoGreen® stained nuclei obtained from a fixed cotyledon. (**c**) Differential interference contrast (DIC) image from a fixed cotyledon. Two stomata made of two guard cells each are indicated by *arrows*. Pavement cells surround the stomata. (**d**) Overlay of PicoGreen® and DIC images. (**e**) Pavement cell nucleus and (**f**) guard cell nucleus are identified thanks to the DIC image and collected individually from original 3D images of the cotyledon epidermis (*white rectangles* in **d**). The nucleus is composed of a set of voxels displaying various gray levels and is delimited from the background (*black voxels*) by the limits of the staining. Chromocenters appear as regions of high fluorescence intensity within the nucleus (*light gray* to *white spots*)

3. Dissecting materials: forceps, scissors, and scalpel.
4. 1.5 mL microfuge tubes.
5. 1/10 PicoGreen® solution (Molecular Probes) (*see* **Note 2**): vortex the stock solution of PicoGreen® and dilute 10 μL of PicoGreen® into 90 μL of DMSO, vortex well. Store the aliquots at −20 °C.
6. Vacuum chamber.
7. dH_2O.
8. 35 mm ø petri dishes.

To stain nuclei in living tissues the following solutions are used:

9. 0.01% Triton X-100: 10 μL of 10% Triton X-100 in 10 mL dH_2O.
10. 1/400 PicoGreen®-Triton (*see* **Note 3**): dilute 12.5 μL of the 1/10 PicoGreen® solution into 487.5 μL of 0.01% Triton X-100. Mix well and protect from light (*see* **Note 4**).
11. Slides Superfrost, 24 × 60 mm coverslips of 0.17 mm thickness, double-sided adhesive tape.

To stain the nuclei in fixed tissues the following solutions are used:

12. 10× PBS: 1.3 M NaCl, 2.7 M KCl, 0.07 M Na_2HPO_4, and 0.03 M NaH_2PO_4 pH 7.2.
13. 0.25 M EGTA: Dissolve 9.5 g EGTA in 50 mL dH_2O, adjust to pH 7.5 with 1 N NaOH and adjust the volume to 100 mL with dH_2O.

14. FIX solution: 270 μL of 37% formaldehyde, 1 mL DMSO, 1 mL of 10× PBS, and 2.32 mL of 0.25 M EGTA, adjust to 10 mL with dH_2O.
15. 1/400 PicoGreen®-PBS (*see* **Note 3**): dilute 12.5 μL of the 1/10 PicoGreen® solution into 487.5 μL of 1× PBS. Mix well and protect from light (*see* **Note 4**).
16. Methanol and ethanol.
17. 1× PBS–glycerol (20:80).
18. Paper towel.
19. Slides Superfrost, 24 × 40 coverslips at 0.17 mm thickness and transparent nail polish.

2.2 Acquisition System

Several instruments can be used such as laser scanning confocal microscopy (CSLM), confocal spinning disk microscopy or Structured Illumination Microscopes (SIM). It is important, however, to use a high magnification objective lens (60–63×) with a high numerical aperture (NA) of 1.3–1.4. Refractive indexes $|(\eta)$ between the sample and the immersion medium ($\eta = 1.33$ to 1.52, respectively, for water and oil) should be as close as possible to avoid distortion especially in the *Z*-axis. However, the best compromise between high-NA oil- or water-immersion objectives and oil-, glycerol-, or water-immersion medium should be defined by the user in its operating system keeping in mind that distortion remain limited at low depth ($Z <$ 10–20 μm). The microscope should also be equipped with a high-end sensitive photodetector or digital camera (e.g., sCMO) with high-sensitivity (high quantum efficiency, low read noise) suited for low-level signal acquisition in the spectral range of the fluorescent probes.

2.3 Image Processing: Hardware and Software

- Computer environment with Intel's Core i5 or Core i7 processors and an allocated computer memory of 2–4 GB RAM (64-bit).
- Images deconvolution is a necessary step before processing by NucleusJ. From our experience, best results were achieved using the Huygens® software (Scientific Volume Imaging). For the images produced by our Leica SIM instrument (DM 6000 with Optigrid® module and MetaMorph software) deconvolution is already included.
- Fiji/ImageJ [11]. Website: https://imagej.nih.gov/ij/download.htmL
- NucleusJ plugin [10]. Website: http://imagejdocu.tudor.lu/doku.php?id=plugin:stacks:nuclear_analysis_plugin:start

3 Methods

3.1 Sample Preparation

1. Collect six cotyledons with a pair of scissors and forceps under the stereomicroscope.
2. Remove the roots and leaves with the forceps while holding the pair of cotyledons with another forceps (*see* **Note 5**) and place them in a microfuge tube containing dH_2O at room temperature. Go to Subheading 3.2 for living tissues or to Subheading 3.3 for fixed tissues procedures (*see* **Note 6**).

3.2 Staining of Nuclei in Living Tissues

1. Replace the dH_2O by 500 μL of 1/400 PicoGreen®-Triton. The cotyledons have to be completely immerged. Incubate at room temperature in the darkness for 15 min.
2. Place the open tubes on a rack into a vacuum chamber for 5 min and apply vacuum for 5 min at room temperature in the darkness; Incubate for 30 min (*see* **Note 7**).
3. Transfer the six pairs of cotyledons with a forceps into a 35 mm ø petri dish containing dH_2O.
4. Prepare a slide: stick double-sided adhesive tape at each extremity of a microscope slide leaving sufficient space to arrange the six cotyledons between them.
5. Collect one cotyledon from each pair using a scissor and put a drop of dH_2O on the cotyledon to prevent it from drying on the slide. Arrange the six cotyledons regularly spaced on the microscope slide between the two adhesive tapes.
6. Add a drop of dH_2O around the cotyledons.
7. Apply a 24 × 60 mm coverslip and gently press the extremities of the coverslip on the adhesive tape. Fill delicately the space between the slide and coverslip with dH_2O using a micropipette (*see* **Note 8**).
8. Discard the remaining cotyledons (*see* **Note 9**) and go immediately to Subheading 3.4.

3.3 Staining of Nuclei in Fixed Tissues

1. *Fixation*: replace the dH_2O by 1 mL of FIX solution (*see* **Note 10**). Verify that the cotyledons are completely immerged. Incubate for 25 min at room temperature and apply vacuum for 5 min at room temperature.
2. *Dehydration*: remove the FIX solution by pipetting and incubate three times 5 min in 1 mL methanol and then three times 5 min in 1 mL ethanol. The cotyledons should become white, repeat ethanol washes if not.
3. *Rehydration*: remove ethanol by pipetting and incubate three times 5 min in 1 mL 1× PBS.

4. *Staining*: remove 1× PBS by pipetting and incubate 30 min in 1 mL of 1/400 PicoGreen®-PBS. Then wash three times in 1 mL of 1× PBS.

3.3.1 Mounting

5. Transfer the cotyledons into a petri dish. For this, hold the tube upside down and gently tap the edge of the tube against the petri dish.
6. Prepare a slide with 100 μL PBS–glycerol.
7. Under the stereomicroscope: take each pair of cotyledons with a forceps and separate them with a scalpel.
8. Collect one cotyledon with the forceps, drain the excess of liquid by gently touching paper towel and quickly place it on the slide prepared with PBS: Glycerol.
9. Repeat with the other five cotyledons.
10. Transfer each immerged cotyledon from the slide onto a 24 × 60 mm coverslip with forceps and arrange the six cotyledons regularly. Cover them with 40 μL of PBS:glycerol. Take a new slide and put it on the coverslip: it adheres quickly by capillarity to the slide. This step is designed to minimize the volume of mounting medium between the sample and the coverslip.
11. Paint the contour of the coverslip with nail polish. This prevents drying and tissue drifting during imaging.
12. Discard the remaining cotyledons (*see* **Note 9**) and go to Subheading 3.4.

3.4 Image Acquisition

Here we used a SIM based on a Leica DM6000 B microscope equipped with an Optigrid® system to perform structured illumination microscopy, a HC PL APO 63×/1.4–0.6 oil-immersion lens and a Hamamatsu, digital CMOS camera, ORCA-Flash 4.0 V2 C11440-22CU, which allows capturing 16-bit data depth images of 4.0 megapixels (6.5 μm × 6.5 μm chip size). Using this optical system the lateral resolution (d*xy*) is 0.103 nm and the axial resolution (d*z*) is 0.200 nm (*see* **Note 11**).

This protocol does not describe the procedure to capture microscopy images as this is depending on the imaging platform; however, a few guidelines are given below.

1. Set the xy image format and z-step to allow for 2–3× oversampling, i.e., composed of voxel with *x*, *y*, *z* dimensions of ~100 × 100 × 200 nm (*xyz*) when using a 63×/1.4–0.6 oil objective here used at NA = 1.4 (*see* **Note 11**).
2. Set a field of view capturing one or several nuclei (in CSLM, the size of the field of view is depending on the zoom factor, which in turn influences the xy pixel size for a given image format (*see* **Note 11**).

3. Set the acquisition mode to capture an image with best possible dynamic range of pixel intensity distribution (*see* **Note 12**).
4. The cotyledon is not a flat object but instead displays an irregular surface, which complicates the automatic acquisition process. To image the cotyledon we recommend to start the acquisition outside the cotyledon epidermis and to acquire about 50–75 Z slices of 0.2 μm each.
5. Record a single plane image under transmission light using Differential Interference Contrast (DIC). This will later allow distinguishing between guard cells and pavement cells in the cotyledon epidermis (Fig. 1).
6. Collect six image stacks from at least three cotyledons coming from distinct seedlings capturing in total 150–200 nuclei.

3.5 ImageJ and NucleusJ Installation

This step of the protocol will be quickly performed for advanced users of ImageJ but may require some investment for inexperienced users. However, ImageJ is a versatile and user-friendly environment, which, once installed, allows many other applications than the one described in this protocol.

1. Download ImageJ at https://imagej.nih.gov/ij/download.htmL. Then proceed with the installation. Update ImageJ by selecting *Help/Update ImageJ* in the ImageJ bar menu (*see* **Note 13**).
2. Download NucleusJ and its dependencies (programs linked to NucleusJ) at http://imagejdocu.tudor.lu/doku.php?id=plugin:stacks:nuclear_analysis_plugin:start and save the files in the "plugins" folder of ImageJ on your computer. Restart ImageJ or simply apply the command *Help/Refresh Menus* before using NucleusJ for the first time.
3. Set up ImageJ memory at the minimum of 1700 MB by selecting *Edit/Options/Memory & Threads*. NucleusJ can be run under Windows, Mac or Linux systems with a 64-bit OS and a 64-bit version of Java (*see* **Note 14**).
4. Install the macro called "**Measure stack**" to be saved as a .txt file in the "macros" folder of ImageJ (*see* **Note 15**).

3.6 Image Preprocessing

From a complex image containing *n* nuclei, the first objective is to produce *n* stacks of images containing a single nucleus (Fig. 1). Individual nucleus image stacks should: (1) contain a complete (i.e., not trimmed) nucleus in all three dimensions, (2) show a high signal-to-noise ratio (i.e., low background signals, best monitored in each single plane image), (3) comprise at least one Z slice at each top/bottom side of the 3D image outside the nucleus and reporting on background signal, and (4) display a histogram of gray values distribution with best possible dynamic range (i.e., covering a maximal range of intensities from 0 to 63,536 for 16-bit images).

The protocol has been divided in (1) the creation of a Z projection and an overlay with the DIC image to orient the image and identify the nuclei depending on their cell type (2) the individualization (crop) of each nucleus from a large image containing 15–60 nuclei. At the end of this step, the user will have a collection of images containing a single nucleus per image suitable for NucleusJ analysis. Instead the user can select individual nuclei at the acquisition step (*see* Subheading 2.2).

3.6.1 Create a Z Projection and an Overlay with the DIC Image

1. Open Image J.
2. Open the image stack stained with PicoGreen®: *File/Open* (Fig. 2a) (*see* **Note 16**).
3. First, produce a Z projection from the image stack in order to easily differentiate the nuclei from the background by selecting *Image/stack/Z project* and choose *Max intensity.*
4. *Edit/Invert* to invert the Z projection.
5. *Image/Lockup Tables* and select *Fire* color (*see* **Note 17**).
6. *Image/Adjust/Brightness/Contrast* select *Auto* and adjust the best brightness-contrast to easily distinguish the nuclei as purple or brown full spots on a white or yellowish background.
7. Save the image Z projection: *File/Save As/Tiff.*
8. Open the DIC image stack: *File/Open.*
9. Second, produce an overlay between DIC and Z projection images by selecting *Image/Overlay/Add Image.* An example of such an overlay is given in Figs. 1d and 2b. Choose the name of the saved projection file in the open window, then select opacity to 50% and press OK.
10. *Edit/Save* the overlay image.
11. *Edit/Close* the Z projection file.
12. Finally, synchronize the original image stack and the overlay: *Analyze/Tools/Synchronize Windows* and then select *Synchronize All.* This step allows cropping the 3D stack based on a region of interest defined on the 2D overlay (see next step).

3.6.2 Select Single Nuclei to Create Stacks of Individualized Nuclei

1. The two files, overlay and stack, are open and synchronized (*see* Subheading 3.6.1) (Fig. 2a–c).
2. Create a *Work directory,* in which all the analysis will be stored, and a sub directory called *RawDataNucleus* (*see* **Note 18**).
3. Delimit a nucleus in the overlay using the *Rectangular* drawing tool in the ImageJ bar menu Fig. 2a, b).
4. Verify that the selected nucleus is totally included within the delimited volume in all slices and for the three axes by moving the depth scroll bar through the *z*-axis of the image stack. Be careful to discard incomplete nuclei in the *z*-axis or to readjust

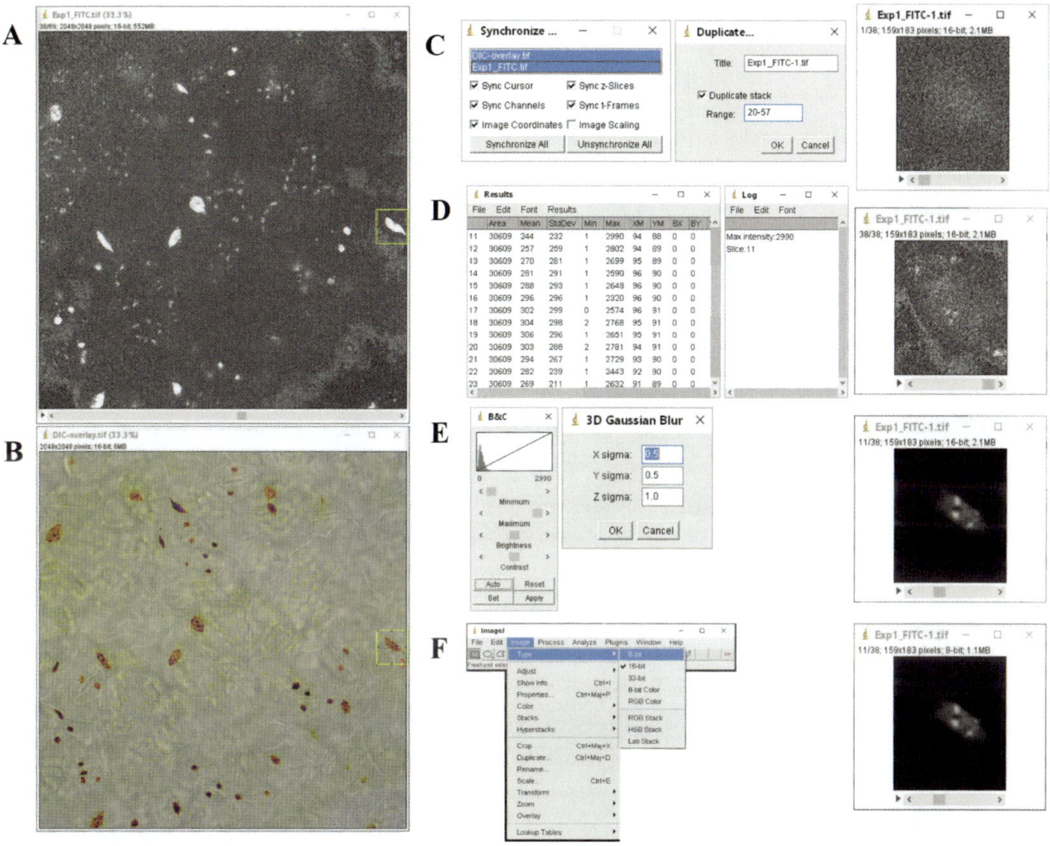

Fig. 2 Screenshots to select individual nuclei from large images with multiple nuclei. (**a** and **b**) Picogreen® (*top panel*) and DIC images are used to generate an overlay image (*bottom panel*). The rectangle tool (in *yellow* at the *right* of the images) is used to individualize (crop) nuclei. (**c**) The PicoGreen® and overlay images are synchronized and used to select individual nuclei from the PicoGreen® image. Individual nuclei are produced using the *Duplicate* function. (**d**) *MeasureStack* macro, which yields *Results* and *Log* pop-ups, allows identifying the Z slice containing the voxel of maximum intensity (slice 11 in this example). (**e**) Histogram is then optimized using the *Brightness/Contrast* tool and a *3D Gaussian Filter* is applied. (**f**) Image is then converted to 8-bit and saved as .tiff in the *RawDataNucleus* subdirectory. Images are given at the *right* of the figure at the various steps

the selected volume in the *x*, *y*-axes with the rectangle tool, as NucleusJ will not process incomplete nuclei.

5. Generate a subset of the original stack capturing a single nucleus (cropped image): identify the series of slices encompassing a specific nucleus, locate the top and bottom slices clearly beyond the nuclear boundaries, i.e., without signal above background, and duplicate the selected volume by using *Image/Duplicate* on the keyboard (Fig. 2c).
6. Determine the Z slice number containing the voxel of maximum intensity within the image stack in order to keep the maximum range of intensity in the subsequent steps. This can

be achieved using the "**Measure stack**" macro (*see* **Note 15**). To start the macro, select *Plugin/Macros/Run* and select the folder containing the macro. **Results** and **Log** windows will appear and indicate the Z slice of interest. Move to this slice with the scroll bar (Fig. 2d).

7. To correct the range of pixel intensity towards a distribution displaying the highest range of intensity values, the histogram is extended using *Image/Adjust/Brightness/Contrast*. For this, use the *Auto* button that will apply the maximum intensity in the histogram identified in **step 6** to all Z slices (Fig. 2e).
8. Apply a 3D Gaussian filter by selecting *Process/Filters/Gaussian Blur 3D*. From our experience with PicoGreen®, DAPI or Hoechst staining, best results are obtained when *X* is 0.5, *Y* is 0.5, and *Z* is 1 (Fig. 2e).
9. Convert the image to 8-bit by selecting *Image/Type/8-bit* (Fig. 2f) and save the cropped nucleus in the *RawDataNucleus*: *File/Save As/Tiff* (*see* **Note 19**).
10. Close the **Results** window. Looking at the overlay (Fig. 2b), identify the cell containing the nucleus to distinguish further between guard cells or pavement cells (*see* **Note 20**).
11. Repeat the **steps 3–10** for each nucleus of the image. The new folder *Work Directory/RawDataNucleus* will contain one image for each nucleus.

3.7 Image Processing and Analysis Using NucleusJ

NucleusJ is an ImageJ plugin that can be used for single or batch image analysis. Here only the batch mode is detailed. The analysis is performed in two steps: nuclear segmentation and then chromocenter segmentation. Segmented images are processed and stored in a specific subdirectory automatically created by the plugin in the main *Work directory*. Each step also produces quantitative parameters relative to nuclear morphology and heterochromatin organization, which are automatically saved as tabulated files in the main *Work directory*. The measurements delivered by NucleusJ related to Nuclear morphology are listed in Table 1 and include the volume of the nucleus (Vnuc in μm^3), the radius of a sphere of equivalent volume (ESR in μm), the total surface of the nucleus (SurfaceArea in μm^2), the elongation = length of longest axis/length of intermediate axis, the sphericity = $36\pi \times$ volume^2/surface area^3, the flatness = length of intermediate axis/length of shortest axis as well as two 2D parameters, which are the aspectRatio = length of longest axis/Length of small axis and the circularity = $4\pi \times$ surface area/Perimeter^2.

3.7.1 Nuclear Segmentation

1. Nuclear segmentation is first performed by selecting *Plugins/NucleusJ/Nucleus Segmentation (batch mode)* (*see* **Note 21**). A pop-up window called *Nucleus segmentation (batch)* appears (Fig. 3a).

Table 1
Nuclear morphology and heterochromatin organization parameters produced by NucleusJ

	Parameters	Abbreviations
Nuclear morphology	Volume of the nucleus	Vnuc
	Radius of a sphere of equivalent volume	ESR
	Total surface of the nucleus	SurfaceArea
	Length of longest axis/length of intermediate axis	Elongation
	$36\pi \times$ volume^2/surface Area^3	Sphericity
	Length of intermediate axis/length of shortest axis	Flatness
	Length of longest axis/length of small axis	AspectRatio
	$4\pi \times$ surface area/Perimeter^2	Circularity
Heterochromatin organization	Number of Chromocenters	NbCc
	Mean distance between Cc border and nuclear periphery/nucleus	DistanceBorderToBorderMean
	Mean distance between of Cc center and nuclear periphery/nucleus	DistanceBarycenterToBorderMean
	Mean volume of chromocenter/nucleus	VCcMean
	Total chromocenter volume/nucleus	VCcTotal
	Relative heterochromatic fraction (RHF) computed from the voxel volume	VolumeRHF
	Relative heterochromatic fraction (RHF) computed from the voxel intensity	IntensityRHF

2. Select the appropriate *RawDataNucleus* directory containing the images.
3. Select the Output Directory (*WorkDirectory*), in which the segmentation results will be saved.
4. Enter the voxel calibration corresponding to the optical system used for image acquisition. *x*, *y* and *z* values are set at 1 by default and expressed in *pixel* (*see* **Note 21**).
5. Define the minimum and maximum volume of the nucleus to be segmented. By default we are routinely using 15–2000 μm^3 but some mutants may have smaller or bigger nuclei.
6. Define the number of CPUs (Central Processing Units) used for image segmentation (*see* **Note 22**).
7. Press the Start button when complete. The image of the segmented nuclei will be processed and saved automatically in a new subdirectory created by the plugin and called *SegmentedDataNucleus* and located in the *WorkDirectory*.

3.7.2 Chromocenter Segmentation

Chromocenter segmentation is a two-step process, in which first NucleusJ automatically computes the intensity contrast of the regions detected by the 3D watershed and then the user manually

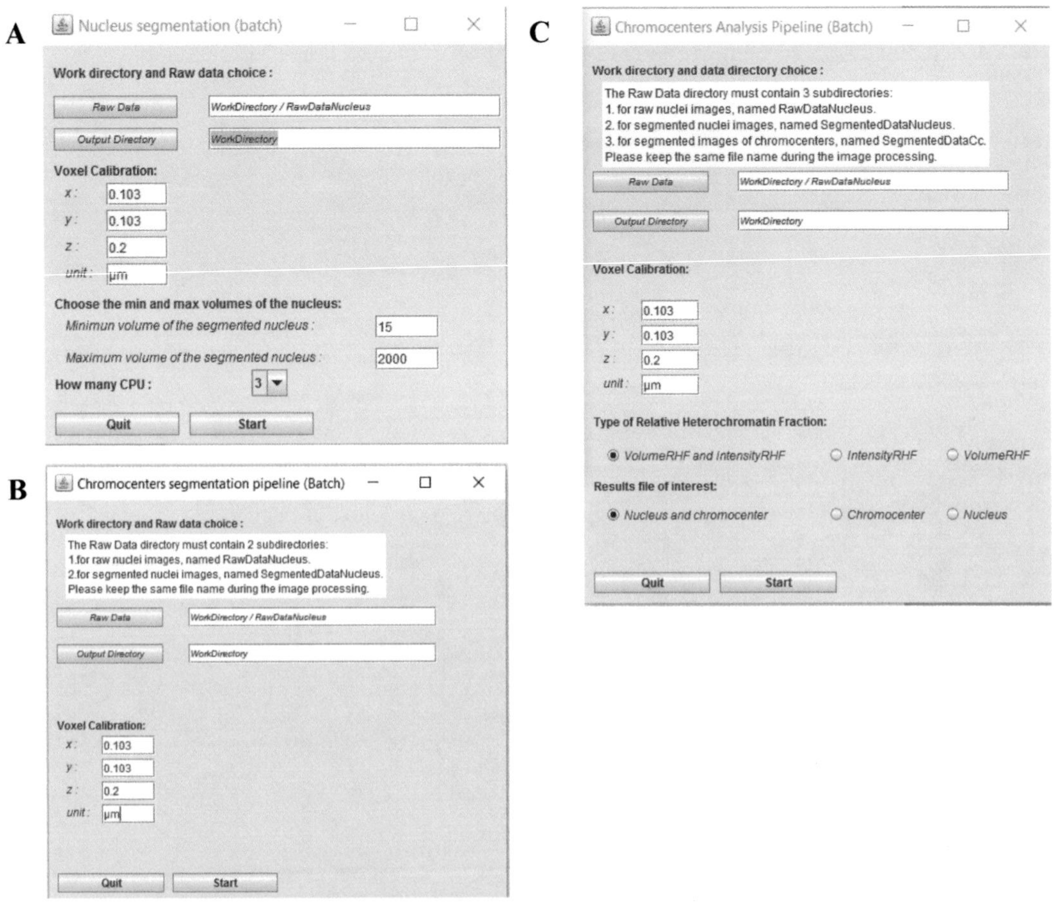

Fig. 3 Pop-up windows generated during NucleusJ processing and analysis

applies a threshold value to segment the chromocenters. Choosing the appropriate threshold is the most time consuming step of our pipeline and takes about 10–15 s per nucleus. The measurements delivered by NucleusJ related to chromocenters are listed in Table 1 and include the number of chromocenters (NbCc), the mean distance between Cc border and nuclear periphery/nucleus (DistanceBorderToBorderMean), the mean distance between of Cc centre and nuclear periphery/nucleus (DistanceBarycenterToBorderMean), the mean volume of chromocenter/nucleus (VCcMean), the total chromocenter volume/nucleus (VCcTotal), the relative heterochromatic fraction (RHF) computed from the voxel volume (VolumeRHF) and the relative heterochromatic fraction (RHF) computed from the voxel intensity (IntensityRHF).

1. *The intensity contrast* is performed by selecting *Plugins/NucleusJ/Chromocenters Segmentation (batch mode)* in the ImageJ bar menu. A pop-up window called *Chromocenters segmentation pipeline (Batch)* appears (Fig. 3b).

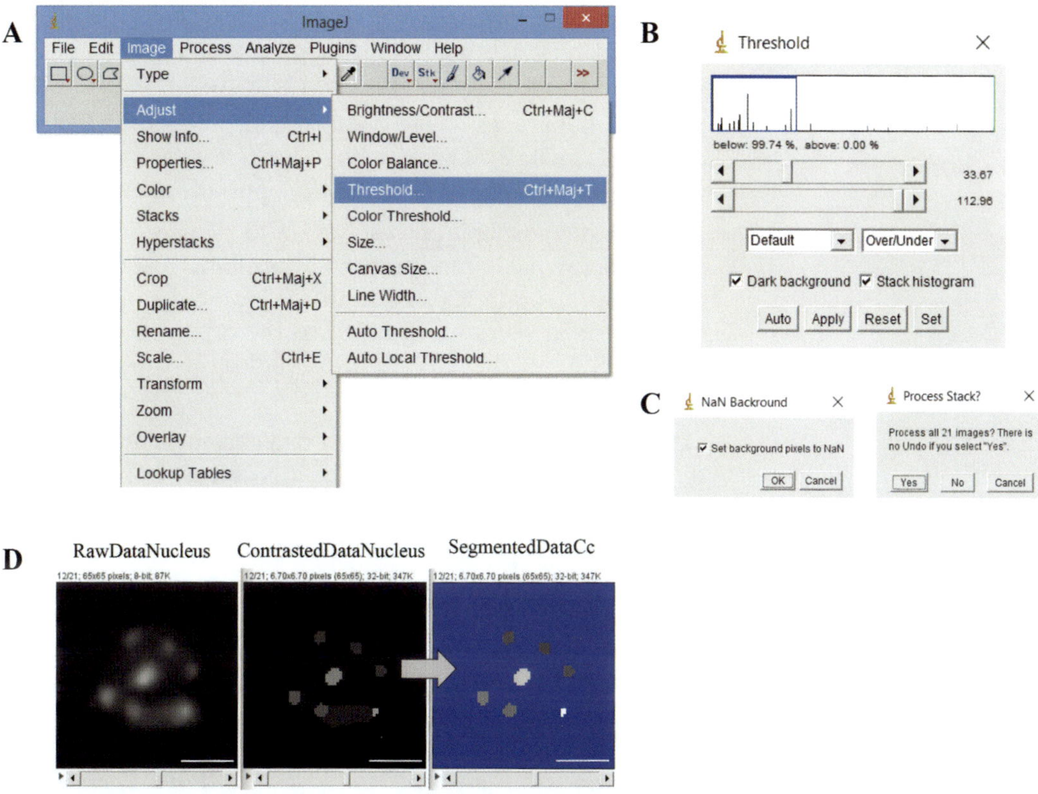

Fig. 4 Manual thresholding leading to segmented chromocenters. The segmented chromocenters are defined by the user in respect to the original image in order to retain only the most relevant signals corresponding to chromocenters. (**a**) The threshold tool from ImageJ is used during this process. (**b**) The threshold tool displays the histogram and allows for applying a threshold value (in *blue*) to all Z slices. Signal below this threshold value is considered as NaN. (**c**) The segmentation process is finalized through two successive pop-ups windows. (**d**) Example of segmentation of a raw nucleus is given at the *bottom* of the figure. Scale bar: 2 μm

2. Select the appropriate *RawDataNucleus* directory containing the images.
3. Select the Output Directory (*WorkDirectory*), in which the Image contrast results will be saved.
4. Enter the voxel calibration corresponding to the optical system (*see* **Note 21**).
5. Press the *Start* button when complete. The image contrast for each nucleus will be processed and saved automatically in a new subdirectory created by the plugin and called *ConstrastDataNucleus.*
6. To *segment the chromocenters* (Fig. 4), first create a new subdirectory called *SegmentedDataCc* to store the segmented chromocenters.
7. For each nucleus, open the raw and the corresponding image contrast.

8. Duplicate the image contrast using *Image/Duplicate* on the keyboard and synchronize the three open images with *Analyze/Tools/Synchronize Windows* and then select *Synchronize All*.
9. Select the duplicated image contrast and then *Image/Adjust/Threshold* (Fig. 4a). Check the box *Dark background* and *Stack histogram* and chose the *Over/Under* option (Fig. 4b).
10. To select the most appropriate threshold value, move the two cursors to the maximum value (right position) and then move the upper cursor to the left until the number of segmented chromocentres becomes identical to the one observed in the raw image. Apply the selected threshold to the Image contrast by clicking on *Apply*.
11. All the background voxels not part of a chromocenter are then converted to "nothing" (NaN: Not-a-Number) by selecting *set background pixels to NaN* of the pop-up window called *NaN background* (Fig. 4c). Apply this setting to all Z slices by selecting the *Yes* button in the *Process Stack?* pop-up window (Fig. 4c). The chromocenters are now segmented (Fig.4d).
12. Select *File/Save* and save the segmented chromocenters in the subdirectory *SegmentedDataCc* taking care to name the image as for the corresponding raw nucleus.

3.7.3 NucleusJ Analysis

1. Image analysis is performed by selecting *Plugins/NucleusJ/Chromocenters Analysis (batch mode)* in the ImageJ bar menu. A pop-up window called *Chromocenters Analysis Pipeline (Batch)* appears (Fig. 3c).
2. Select the appropriate *RawDataNucleus* directory containing the images.
3. Select the Output Directory, in which the Image contrast results will be saved.
4. Enter the voxel calibration corresponding to the optical system (*see* **Note 21**).
5. Press the Start button when complete. NucleusJ will generate and save two new files in the *Work directory*:
   - *NucAndCcParameters.tab*: contains quantitative parameters organized in a tabulated file. The list of the 15 parameters gained from the analysis is given in Table 1. From these data, the user can perform a statistical analysis by using the R package or other proprietary software as illustrated in [10].
   - *logError*: is a facultative file generated only when some nuclei are discarded from the segmentation analysis to alert the user.

4 Notes

1. The method can be applied to other tissues or developmental stages.
2. The dilutions indicated in the protocol are performed from the PicoGreen® reagent provided by the manufacturer as a liquid formulation where its concentration is not specified. The PicoGreen® has low solubility in dH_2O and we recommend using DMSO (dimethyl sulfoxide) to prepare the 1/10 dilution. Aliquots of 1/10 PicoGreen® in DMSO can be stored at −20 °C. Avoid exposure to light.
3. The optimal PicoGreen® dilution is 1/400 for the described protocol. Do not store PicoGreen® diluted in dH_2O. Avoid exposure to light.
4. Do not store; use only fresh solutions.
5. From our experience it is much easier to handle a pair of cotyledons than a single cotyledon.
6. Be careful not to process more than two samples of samples cotyledons at the same time, as image acquisition is time consuming.
7. For living material it is very important to respect the timing. From our experience, stained tissues become damaged 3 h after the start of PicoGreen® staining. Do not exceed 2.5 h between the start of staining incubation and the end of image acquisition.
8. Double adhesive tape is about 50 μm thick, which is also approximately the cotyledon thickness. It then provides an excellent mounting alternative and avoids further cotyledon movements. Before covering the samples with the coverslip, be careful to have sufficient dH_2O surrounding the cotyledons to avoid bubbles but not too much to avoid dH_2O leaking on the adhesive tape.
9. To avoid bias, we routinely observe one cotyledon per plant. If needed, the remaining cotyledons can be fixed and the staining step can be skipped.
10. To start at the same time many samples, dissect in dH_2O and replace dH_2O by FIX solution when all samples are ready. If the dissection time is more than 15 min, keep the tubes with dH_2O on ice until fixation.
11. In order to get the best results, images should be acquired at optimal resolution for the lateral resolution (dxy) $= 0.4\ \lambda em/NA$ and axial resolution (dz) $= 1.4\ \lambda em\ \eta/NA^2$ where the emission wavelength (λem) is 535 nm for PicoGreen®, the refractive index (η) is 1.518 for oil and the numerical aperture

(NA) is 1.4 when using a 63× oil-immersion objective. The sampling theory predicts that dividing these theoretical values of resolution by a factor of 2.3 improves the recovery of the best numeric data from the signal [5]. Note that such an over-sampling is suitable for a subsequent deconvolution analysis when a confocal microscope is used. Oversampling is already considered when acquisition is performed using a structured illumination microscope and in our condition corresponds to values given in **Note 21**.

12. When using SIM, check that the epifluorescence intensity of the image does not exceed the saturation threshold, i.e., that all the voxels of a given Z slice are below 65,535 (16 bits).
13. Commands in the ImageJ bar menu are indicated in italics and as grey underlined text.
14. A training Image dataset suitable for NucleusJ analysis can be downloaded at /media/WorkDirectory.zip, which contains a dataset of 77 nuclei from wild type and *crwn1 crwn2* mutant plants. *See* ref. 8 for more details.
15. This step generates a nucleus stack from 16 to 8 bit by keeping the maximum of the dynamic range. For each nucleus the slice containing the pixel with the maximum intensity has to be determined before applying the best dynamic to the other slices. This step is facilitated by the following macro:

```
macro "Measure Stack"{
saveSettings;
setOption("Stack position", true);
for (n=1; n&lt;=nSlices; n++) {setSlice(n);run("Measure");
}
restoreSettings;
// Find max intensity
max_intensity=0;
zpos=1;
 for (a=nResults()-1; a>=0; a-) {
 maxval = getResult("Max",a);
 if (maxval>max_intensity)
 {
 max_intensity = maxval;
 zpos = getResult("Slice",a);
 }
 }
 print("Max intensity:"+max_intensity+"\nSlice:"+ zpos);
}
```

Be careful that sometimes the pixel designed is an artifact outside the nucleus. In this case, discard the corresponding "nucleus" or search manually the maximum in the nucleus.

16. Images acquired with a Leica DM6000 B microscope are in a . tif format. For other formats, please see Bio-Formats documentation at the ImageJ website.
17. The display mode "Fire color" facilitates the detection of the boundaries of the nuclei in order to avoid the selection of uncomplete-imaged nuclei.
18. An example of directory and subdirectory organization created in the course of the NucleusJ analysis can be found in the NucleusJ documentation at http://imagejdocu.tudor.lu/doku.php?id=plugin:stacks:nuclear_analysis_plugin:start
19. The following macro can be run to perform this step:

```
run("Gaussian Blur 3D...", "x=0.5 y=0.5 z=1");
run("8-bit");
saveAs("Tiff")
```

20. Printed overlays can be used to label the nuclei with numbers and to assign the nucleus to a specific cell type (e.g., G for guard cell, and P for pavement cell).
21. If *Plugins/NucleusJ/Nucleus Segmentation & Analysis (batch mode)* is selected then only the nuclear segmentation is performed and will generate the segmented nuclei but also produce two tabulated files named **2DNucleiParameters.tab** and **3DNucleiParameters.tab** saved in the *Work directory.* The segmented nuclei are then saved into the *SegmentedDataNucleus* subdirectory. Optical calibration is dependent upon the microscope, objective, mounting medium and sample used in the experiment. In our experimental design, x is 0.103, y is 0.103, and z is 0.2, and the unit is μm.
22. In its first version, NucleusJ calculation demands a lot of memory. The best practice is to allocate as much CPUs as possible to NucleusJ and to close all other applications. For example we usually use three CPUs when four are available.

References

1. DeMay R (1996) The art and science of cytopathology. Chicago, American Society of Clinical Pathology
2. Croft JA, Bridger JM, Boyle S, Perry P, Teague P, Bickmore WA (1999) Differences in the localization and morphology of chromosomes in the human nucleus. J Cell Biol 145:1119–1131
3. Boyle S, Gilchrist S, Bridger JM, Mahy NL, Ellis JA, Bickmore WA (2001) The spatial organization of human chromosomes within the nuclei of normal and emerin-mutant cells. Hum Mol Genet 10:211–219

4. Fedorova E, Zink D (2008) Nuclear architecture and gene regulation. Biochim Biophys Acta 1783:2174–2184
5. Pawley JB, Masters R, Barry R (1996) Handbook of biological confocal microscopy, second edition. Opt Eng 35:2765–2766
6. Howe ES, Murphy SP, Bass HW (2013) Three-dimensional acrylamide fluorescence in situ hybridization for plant cells. Methods Mol Biol 990:53–66
7. Tirichine L, Andrey P, Biot E, Maurin Y, Gaudin V (2009) 3D fluorescent in situ hybridization using Arabidopsis leaf cryosections and isolated nuclei. Plant Methods 5:11
8. Bauwens S, Katsanis K, Van Montagu M, Van Oostveldt P, Engler G (1994) Procedure for whole mount fluorescence in situ hybridization of interphase nuclei on Arabidopsis thaliana. Plant J 6:123–131
9. Fransz PF, de Jong JH (2002) Chromatin dynamics in plants. Curr Opin Plant Biol 5:560–567
10. Poulet A, Arganda-Carreras I, Legland D, Probst AV, Andrey P, Tatout C (2015) NucleusJ: an ImageJ plugin for quantifying 3D images of interphase nuclei. Bioinformatics 31:1144–1146
11. Schneider CA, Rasband WS, Eliceiri KW (2012) NIH Image to ImageJ: 25 years of image analysis. Nat Methods 9:671–675
12. Otsu N (1979) A threshold selection method from greay-level histograms. IEE TransSys Man Cyber 9:62–66
13. Andrey P, Kiêu K, Kress C, Lehmann G, Tirichine L, Liu Z, Biot E, Adenot PG, Hue-Beauvais C, Houba-Hérin N, Duranthon V, Devinoy E, Beaujean N, Gaudin V, Maurin Y, Debey P (2010) Statistical analysis of 3D images detects regular spatial distributions of centromeres and chromocenters in animal and plant nuclei. PLoS Comput Biol 6:e1000853
14. Vincent L, Soille P (1991) Watersheds in digital spaces: an efficient algorithm based on immersion simulations. IEEE Trans Pattern Anal Mach Intell 13:583–598

Chapter 34

Transmission Electron Microscopy Imaging to Analyze Chromatin Density Distribution at the Nanoscale Level

Tohnyui Ndinyanka Fabrice, Lusik Cherkezyan, Christoph Ringli, and Célia Baroux

Abstract

Transmission electron microscopy (TEM) is used to study the fine ultrastructural organization of cells. Delicate specimen preparation is required for results to reflect the "native" ultrastructural organization of subcellular features such as the nucleus. Despite the advent of high-resolution, fluorescent imaging of chromatin components, TEM still provides a unique and complementary level of resolution capturing chromatin organization at the nanoscale level. Here, we describe the workflow, from tissue preparation, TEM image acquisition and image processing, for obtaining a quantitative description of chromatin density distribution in plant cells, informing on local fluctuations and periodicity. Comparative analyses then allow to elucidate the structural changes induced by developmental or environmental cues, or by mutations affecting specific chromatin modifiers at the nanoscale level. We argue that this approach remains affordable and merits a renewed interest by the plant chromatin community.

Key words Transmission electron microscopy, TEM, High-pressure freezing, Contrast staining, Nucleus, Euchromatin, Heterochromatin, Spatial chromatin density distribution, Nanoscale-level chromatin organization

1 Introduction

Before the advent of fluorescent molecular labels and the corresponding microscopy imaging techniques, nuclear organization was classically analyzed using transmission electron microscopy (TEM). TEM revealed common as well as distinct principles of chromatin domain organization, nucleolar morphology, and nuclear body distribution [1]. Although rarely employed nowadays for chromatin analyses, TEM enables a unique, nanoscale level of resolution. This level can neither be achieved by conventional fluorescence microscopy, constraint by the optical diffraction limit of ~200 nm, nor by super-resolution microscopy imaging (enabling down to 50–90 or 20 nm in the *xy* dimension depending on systems [2]. TEM-based analysis remains a valuable approach and

Marian Bemer and Célia Baroux (eds.), *Plant Chromatin Dynamics: Methods and Protocols*, Methods in Molecular Biology, vol. 1675, DOI 10.1007/978-1-4939-7318-7_34, © Springer Science+Business Media LLC 2018

has been successfully applied to reveal distinct nuclear features in plant cells upon developmental changes, environmental cues or genetic mutations. For instance, the distribution of heterochromatin foci is distinct between meristematic vs. differentiated cells in several plant species [3, 4]. Similarly, chromatin decondenses upon mycorrhizal fungal infection [5, 6] or upon exposure to pollutants such as cadmium [7]. However, because the staining procedure is not very specific (unless using immunogold labeling), TEM analyses do not replace—hence remain complementary to—conventional approaches consisting in imaging specific fluorescently labeled chromatin components (using DNA dyes, specific antibodies, or GFP-tagged chromatin proteins).

Specimen preparation at low temperatures as described below enables good ultrastructure preservation, allowing the quantification of chromatin organization at nanoscale. In former times, images were digitized and the intensity of grey values was captured in a serial manner across the image to draw intensity distributions and determine the proportion of condensed chromatin [8, 9]. Other feature-based quantifications were used to describe the heterogeneity, granularity, condensation, and margination of chromatin in plant nuclei [3, 10]. More recently, methods employing mathematical processing of the image signal distribution were developed to describe the periodicity of chromatin distribution and provide a measure of chromatin organization at the nanoscale level [11, 12]. The degree of organization described by the spatial distribution of local compact domains and the periodicity of such structures directly impacts on the access to the transcription machinery, and hence the transcriptional landscape [13]. Such analysis offers great promises to elucidate the impact of cellular differentiation, tissue treatment or genetic mutation on fine-scale chromatin organization at a level affecting transcription and, likely, epigenetic robustness, in plants.

Biological samples for TEM are fixed by physical means (e.g., high-pressure freezing: HPF) or chemically (e.g., fixation with glutaraldehyde and/or formaldehyde) to arrest cellular processes in as near-native state as possible. HPF is the most efficient fixation technique to date [14–16], and consists of rapidly freezing specimens ≤ 0.5 mm thick at the rate of 10,000 °C/s and around 2100 bar [17]. These samples then undergo freeze-substitution with an organic solvent (e.g., acetone or methanol, containing a chemical fixative, usually OsO_4 or glutaraldehyde) at low temperatures, usually −90 °C. Freeze substitution (FS) works on the basis that, at temperatures lower than its melting point, vitreous ice dissolves in the solvent and is replaced in the specimen by the solvent [18]. Freeze substituted samples are then raised to room temperature for resin embedding. Different types of resins are in use, but epoxy resins such as Epon generally offer better ultrastructural preservation for morphological studies. Thin sections about

70 nm thick are cut of embedded specimen and transferred onto electron microscope grids for contrast staining (usually with heavy metals such as uranium and lead) prior to TEM visualization. Here we present a step-by-step procedure for HPF/FS, Epon embedding, ultramicrotomy and contrast staining of Arabidopsis root specimens for TEM visualization and acquisition of images of nuclear chromosomal structures (euchromatin and heterochromatin). We confirm ultrastructural changes between cells in the meristematic and cell elongation zones of the Arabidopsis root as previously reported. The preparations allow to clearly distinguish the nucleolus, the euchromatin and heterochromatin compartments. In addition, the images allowed to quantify the distribution of chromatin densities (a proxy of chromatin compaction) at the nanoscale level based on electron density distribution. For this, we used an image signal processing based on spatial autocorrelation analyses; the analysis runs from a customized Matlab script previously reported [11]. We present a cost efficient and user-friendly graphical interface for running these analyses even for nonspecialists. The aim of such analyses is to quantify the degree of chromatin organization at the nanoscale level, which is poorly understood and yet invaluable in deciphering the functional organization of chromatin in cells.

2 Materials

Prepare all solutions using Milli-Q water unless stated otherwise. Store all solutions and reagents at the indicated temperature. Ensure that all waste disposal regulations are diligently followed when disposing of waste materials.

2.1 Sample Preparation

1. Murashige and Skoog medium including vitamins: ½ strength (2.2 g/l).
2. Phytagel or Bacto agar.
3. Square plastic petri dishes.
4. 4 °C dark chamber.
5. 22 °C growth chamber.
6. Arabidopsis seeds.

2.2 High-Pressure Freezing and Freeze Substitution

1. Scalpel.
2. Tweezers (fine-tipped straight and fine-tipped curved).
3. 1-Hexadecene: 92% grades.
4. Vacuum desiccator.
5. High-pressure freezing system and relevant accessories.
6. Fine and blunt forceps.

7. 6 mm aluminum specimen carriers with 150/150 μm and flat/300 μm recesses (*see* **Note 1**).
8. Liquid nitrogen.
9. Micropipettes (100 μl and 1 ml).
10. Acetone HPLC grade, water free (*see* **Note 2**).
11. Freeze substitution solution: 1% OsO_4 in water-free acetone. Stored in liquid nitrogen (be careful, *see* **Note 3**).
12. 2 ml Eppendorf tubes.
13. Freeze substitution system capable of progressive lowering of the temperature (e.g., Leica EM AFS2, Leica Microsystems).

2.3 Sample Embedding, Sectioning and Contrast Staining

1. Embedding Epon resin (*see* **Note 4**).

   Stock solution: Prepare 25 ml by adding 11.82 g Epon 812, 15.37 g durcupan ACM, 1.45 g dibutyl phthalate, and mix by inverting. Store the stock at RT.
   Embedding resin: Prepare 5 ml of Epon resin by adding 2.79 g stock solution, 2.38 g DDSA and 147 mg DMP 30 and mix by inverting gently.
2. Embedding molds (*see* **Note 5**).
3. Oven 60 °C.
4. Glass knife.
5. Diamond knife.
6. Syringe with 0.2 μm sterile filter attached.
7. Toluidine blue solution: 1 g Borax and 1 g Toluidine Blue O in 100 ml ddH_2O. Store at RT.
8. Formvar-coated electron microscope slot grid (*see* **Note 6**).
9. Electron microscope grid box.
10. Uranyl acetate: 2% in water. Store in dark at 4 °C (be careful, *see* **Note 7**).
11. Prepare lead citrate solution according to Reynolds [19]. Store at 4 °C (be careful, *see* **Note 8**).

2.4 Image Acquisition and Chromatin Density Distribution

1. Regular transmission electron microscope system operated at ≥80 kV acceleration voltage (e.g., CM 100, FEI, Eindhoven, The Netherlands) and equipped with a sensitive digital camera (e.g., Gatan Orius 1000 CCD camera, Gatan, Munich, Germany).
2. Image processing script at https://github.com/barouxlab/ChromDensityNano
3. Computational environment for image analysis: Fiji, Matlab, etc.

3 Methods

3.1 High-Pressure Freezing and Freeze Substitution

The aim of this step is to fix the tissue by high-pressure freezing and then to dehydrate at temperatures low enough to prevent the formation of ice crystals and ice crystal damage. Cryo-fixation proved superior to chemical fixation in preserving ultrastructural details particularly of chromatin. The tissue sample was collected from young seedlings grown in vitro, but the protocol below can be adapted to other tissues and organs types. Wild-type Arabidopsis seeds (ecotype Columbia-0) were sterilized, stratified for 3 days, and germinated vertically on ½ strength Murashige and Skoog medium (with 0.6% Phytagel w/v) in square plastic Petri dishes for 7 days as described [9].

1. Add 1-hexadecene to fill the 150 μm recess of the 6 mm specimen carrier.
2. Cut about 4 mm long Arabidopsis root sections using the scalpel and transfer into the carrier with tweezers (*see* **Note 9**).
3. Dip the flat side of the flat/300 μm carrier in 1-hexadecene and lay it on the specimen. Transfer the sandwich using tweezers into the bore of the dedicated specimen carrier cartridge and immediately freeze the set in the high-pressure freezer. HPF was performed in an EM HPM100 high-pressure freezer (Leica Microsystems). The frozen specimen cartridge is released in the liquid nitrogen Dewar (*see* **Note 10**).
4. Take out the cartridge, punch the specimen carrier sandwich out with liquid nitrogen-precooled tips of the fine-tip forceps. Store the specimen in liquid nitrogen until freeze substitution (*see* **Note 11**). The sandwiched carriers may fall apart in liquid nitrogen. Identify the one on which the specimen sticks and proceed with that one only.
5. Thaw the FS solution and transfer 1 ml portions into appropriately labeled 2 ml tubes. Precool the tubes to −90 °C in the freeze substitution system (e.g., Leica AFS 2 machine) and transfer the specimen plus carrier directly from liquid nitrogen into the tube. Run the freeze-substitution cycles as follows: 8 h at −90 °C, 6 h at −60 °C, 6 h at −30 °C, 1 h at 0 °C, with transition gradients of 30 °C per hour. The specimen mostly separates from the carrier after freeze substitution. Take away the carrier and rinse the specimen trice with water-free acetone, and proceed with embedding.

3.2 Sample Embedding, Sectioning, and Contrast Staining

Here the cryo-fixed and freeze substituted sample is embedded in Epon embedding resin and ultrathin sections are produced. Delicate handling is required for transferring the ultrathin sections onto the specimen electron microscope grid for downstream staining with an electron dense reagent (uranyl acetate). Ultrapure, sterile deionized water (e.g., Milli-Q water filtered through a 0.2 μm sterile filter) is used to fill the well of the diamond knife and for rinsing stained ultrathin sections.

1. Transfer the specimen into 50% v/v Epon resin/water-free acetone and incubate at RT for 2 h or overnight at 4 °C with gentle shaking.
2. Replace the solution with 100% embedding solution and incubate at 4 °C for 1 h with gentle shaking.
3. Add a few drops of embedding resin into the embedding mold to fill the bottom, and transfer the specimen with a toothpick (*see* **Note 12**) into the mold. Use fine-tip tweezers to carefully reorient the specimen (preferably under the binocular) according to the direction and/or area you intend to section. Then gently fill the mold with embedding medium and transfer the set for curing at 60 °C for 48 h. Curing can also be performed a bit longer, for example over the weekend.
4. After curing, trim away excess resin around the specimen with the glass knife to form a trapezoid shape. Collect 250 μm thick sections and stain with toluidine blue solution for quick observation with the light microscope to check if the zone of interest (sections containing the nuclei) is hit (*see* **Note 13**).
5. Cut 70 μm sections using the diamond knife and place serial sections onto formvar-coated electron microscope grids (*see* **Note 14**).
6. Place about 40 μl drops of 2% uranyl acetate (*see* **Note 15**) on Parafilm and float the grids with the specimen-side down on the drops for 15 min. Rinse the grids four times on drops of sterile-filtered Milli-Q water. Dry the grids by carefully sucking up the water with filter paper and let dry for about 10 min at room temperature or in a 37 °C incubator for 5 min (*see* **Note 16**).
7. Stain the section by placing the grids downward with the specimen side in contact with a drop of lead citrate for 10 min (*see* **Note 17**). Rinse four times on drops of sterile-filtered Milli-Q water. Dry the grids again by sucking up the water with filter paper, transfer them into the grid box, and let dry for about 2 h at room temperature or in a 37 °C incubator for 25 min (*see* **Note 18**).
8. The samples are ready for observation and assessment with the TEM. Alternatively, they can be stored in the EM grid box at room temperature for several years.

3.3 Image Acquisition

The aim is to record images of nuclei from stained sections at highest possible pixel resolution providing effective pixel size of 0.5–1.5 nm at 1× binning (Fig. 1) (*see* **Note 19**).

1. Before every imaging session, calibrate the instrument for the beam focus, beam intensity, and diffraction pattern according to the provider's recommendation.
2. Take an overview image of the tissue section containing nuclei, typically at a 7000-fold magnification.

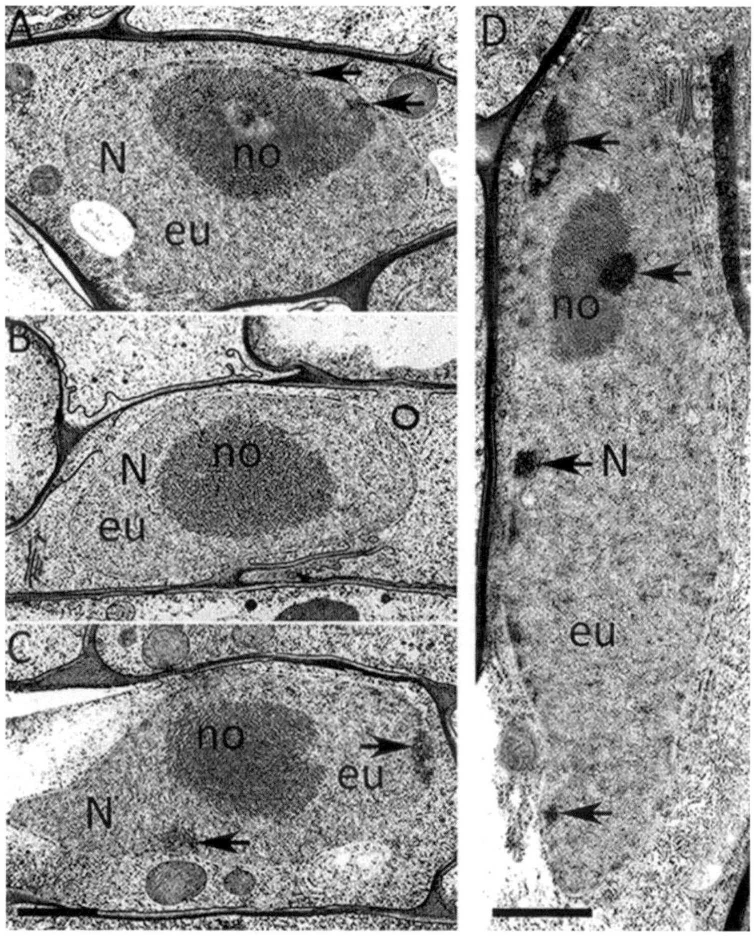

Fig. 1 Distinct chromatin distribution in the elongation zone of the Arabidopsis root. (**a–c**) Ultrastructure of nuclei of stele cells showing a large nucleolus (*no*), dispersed patches of electron dense heterochromatin like those indicated (*arrows*) and diffuse euchromatin (*eu*). (**d**) Typical nucleus of epidermal cells from the elongation zone showing the reduction of nucleolus size, patches of heterochromatin (*arrows*) associated with the nuclear membrane or with the nucleolus (*no*). The euchromatin is diffused and fills the majority of the remaining space. *N* nucleus. Scale bar = 1 μm

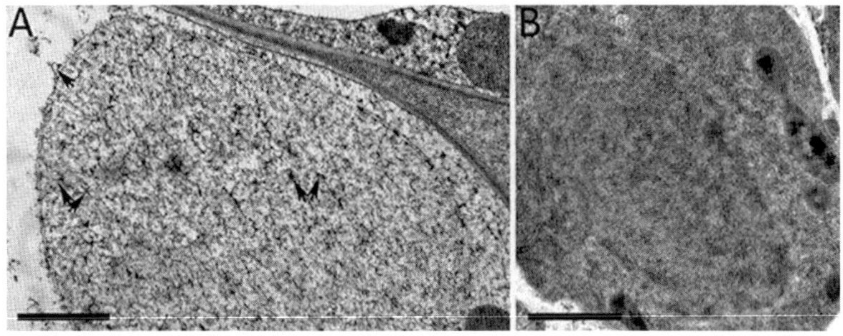

Fig. 2 Preparation Artifacts. (**a**) Ice crystal damage in the nucleus compromising the integrity of the chromatin. The *arrows* show clusters of cellular components that segregate during ice crystal damage altering the fine ultrastructural arrangement of cellular components. (**b**) Low contrast between cytoplasm and nucleus compromising the unambiguous identification of chromatin regions with distinct densities. Scale bar = 1 μm

3. Select a tissue layer of interest (e.g., epidermis, cortex, or inner cell types) and capture TEM fields encompassing a whole nucleus, typically at a magnification between 30,000 and 50,000-fold.
4. Inspect for the presence of possible water crystals and/or preparation artifacts in the chromatin (Fig. 2a). If the staining shows star-shaped or spiky structures, it is considered as artifact and the section should not be analyzed.
5. Verify that euchromatin and heterochromatin present a good contrast also in comparison to cytoplasmic areas. If the contrast does not allow unambiguous recognition of chromatin domains (Fig. 2b), it is possible to restain the sections.
6. For image acquisition at high magnification, make sure to match the zoom factor and corresponding pixel size to the pixel resolved on the camera chip, typically in the range of 0.5–1.5 nm (*see* **Note 19**).
7. Record 20–30 images per preparation. Include replicate sample preparation to exclude artifacts linked to one sample preparation (embedding, sectioning, staining....).
8. Save the images under the original software format (to keep the metadata of acquisition) as well as in uncompressed, 8-bits TIF files. Label the images with simple names tagged with a serial number to facilitate downstream image processing.

3.4 Chromatin Density Distribution Analysis

The aim of the image processing is to analyze the density distribution of the euchromatin compartment (Fig. 3). For this, first an image of density fluctuation is created, then a spatial autocorrelation function is run along consecutive image rotations. The *first step* consists of calculating and saving density autocorrelation functions

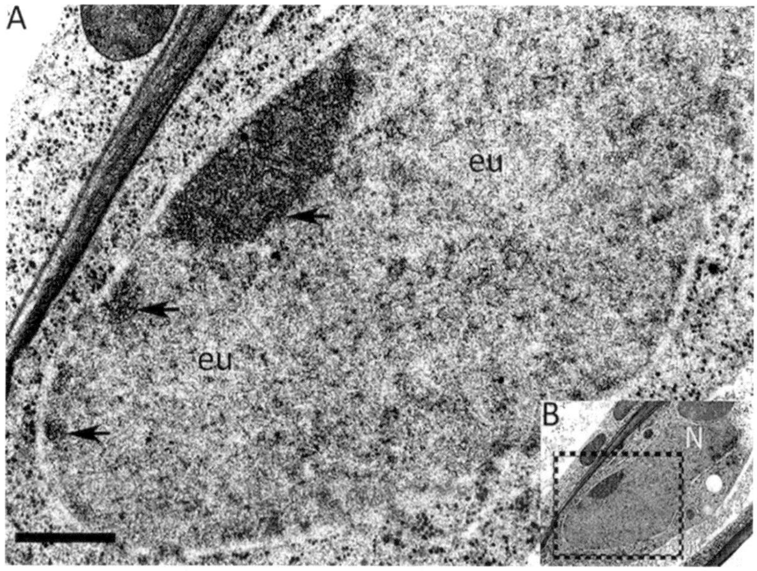

Fig. 3 Close-up electron micrograph of the nucleus. (**a**) A close-up view of a nuclear region (indicated in **b**) from the cell differentiation zone of the root. Heterochromatin (*arrows*) appears as electron dense masses near the nuclear membrane and euchromatin (*eu*) is more evenly stained with interspersed staining granules, probably corresponding to ribonucleoprotein particles. (**b**) Overview of the nucleus region featured in **a**. *N* nucleus. Scale bar = 500 μm

(ACFs) of images that need to be analyzed. The *second step* is generating a best-fit for the distribution of *D* (one of the three parameters describing the ACF based on the Whittle-Matern family of functions as described in [11]) across all images. This parameter evaluates the nature of the spatial heterogeneity of chromatin density. When the values of *D* are between 2 and 3, this implies that chromatin distribution is fractal with a fractal dimension *D* (the fractal nature of chromatin distribution has been repeatedly reported, [20–22]). In general, the greater the *D*, the higher the relative presence of larger structures in the image (described in detail in [11]). The interpretation for a fractal organization is that high values of *D* are manifested as larger and denser chromatin "clumps" and, at the same time, a larger accessible surface area of chromatin. The former factor increases repression of certain gene transcription, while the latter increases the overall transcription in the nucleus (*see* ref. 13). The analysis is meaningful when done on the euchromatin compartment only to avoid confounding effects of heterochromatin arrangements. The analysis is run using codes loaded in Matlab. This step-by-step manual allows users inexperienced with Matlab to run the analysis. Fine-tuning the graphics can be done by learning additional tools (Apps and command lines) in Matlab or by saving them as vector files for processing in a third-party program.

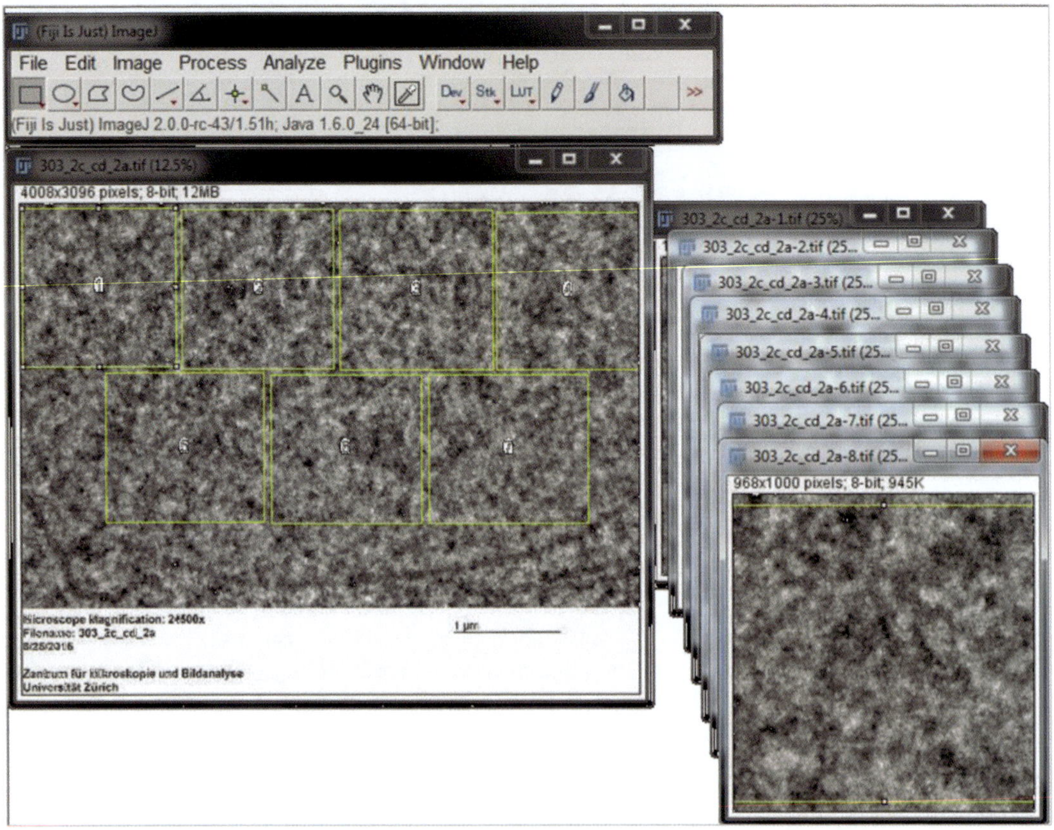

Fig. 4 Draw consecutive ROIs in euchromatin and duplicate

1. Open the images in Fiji. Make sure to open images taken at the same magnification.
2. Create a series of subset images corresponding to square or rectangular regions of interests capturing euchromatin regions only: for this draw a square ROI (*see* **Note 20**), duplicate (shortcut key "ctrl shift D").
3. Repeat this for several ROIs over one image (Fig. 4). Save all ROIs ("ctrl S" = save; "ctrl W" = close) into a separate folder under a logical numerical series (*see* **Note 21**).
4. Open Matlab.
5. Download the scripts from github: https://github.com/barouxlab/ChromDensityNano. The main GUI is *GUI_getACF_and_D.m* and to run properly it needs the functions *rotavg.m*, *Fit_loop.m*, *get_ACF.m*, and *get_D.m* saved in the same folder as the main GUI code. *When running the GUI for the first time Matlab may request to change the Matlab shadow folder, press "Change Folder".*
6. Open only *GUI_getACF_and_D.m* (*see* **Note 22**) and press Run. You will see the following window:

7. When running the script for the first time, it is advised to do so on a single picture, then 3–4 pictures before processing a large batch (computational time depends on the environment and file size).
8. First select the images (ROIs) that need to be analyzed: click on "Select directory with images".
9. In the editable window titled "Set image prefix" indicate the name of the group of ROI to analyze. For instance, all wild-type: load "wt_*.tif". The code loads the gray-scale ROI images. If only one image needs to be analyzed, enter its full name and do not include the asterisk sign. If all images of the folder are analyzed load "*".
10. In the editable window titled "Save data as" enter the name for the acf file (autocorrelation function) that will be generated. For instance, entering "wt" will create file "acf_wt" in the same directory as defined above.
11. In the editable window titled "R_{min} =" input the resolution of the images, in pixels. By default, this value is 1. However, if the images are blurry (*see* **Note 23**), R_{min} is the radius of an image of a point.
12. Checkbox "control for SNR": this operation aims at eliminating images with a bad signal to noise ratio. The threshold in pixels (the value of R_{min}) depends on the resolution at a given magnification and on the camera (*see* **Note 23**).
13. Run the code by clicking on the "Calculate ACF" button. It calculates an autocorrelation function (ACF) for each image and saves a file called acf_*NameDefinedByUser* with the density autocorrelation function that you will need for further processing.
14. When checkbox "Display each ACF" is enabled, the code will generate one figure per image showing from left to right: the original image, the 2D map of the autocorrelation function and the one-dimensional correlation function calculated from averaging the ACF for discrete radius r (*see* **Note 24**, Fig. 5).
15. Save one, several or all images for recording examples of individual ACF per ROI (Note that, however, the summary figure produced at the end of the process is the most informative, see below) The dimensions can be manually adjusted before saving, or processed in a third-party software after saving as a vectorized graphics (e.g., svg).
16. To de-activate the generation of individual figures (for n samples, n figures are generated), disable the checkbox "Display each ACF".
17. The saved file contains the user inputs for calculating the ACF (the value of R_{min}, flag for SNR control, the prefix of images

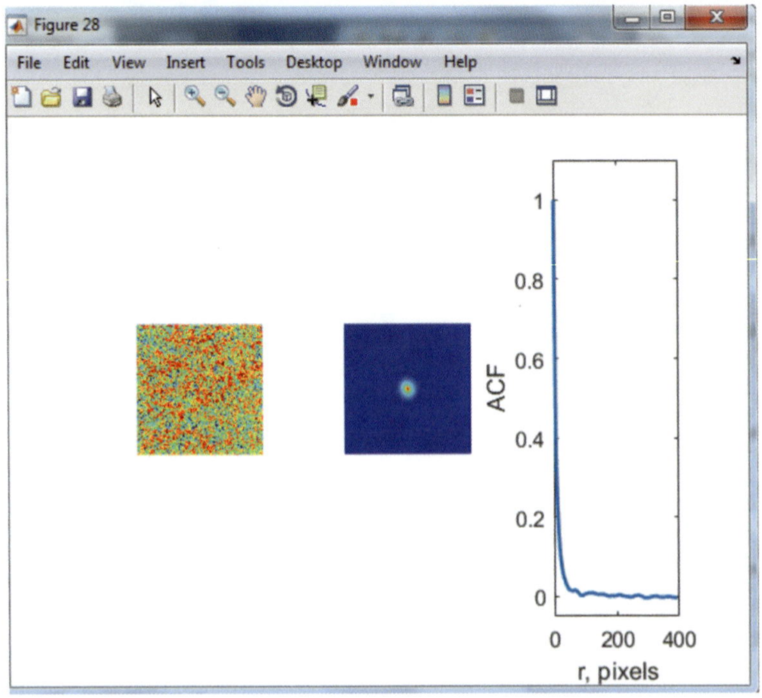

Fig. 5. Screenshot following the first step: evaluating the periodicity of signal dispersion using a spatial autocorrelation function. The original analyzed image on the *left*, its 2D autocorrelation in the *center*, and 1D ACF on the *right*

analyzed), as well as the outputs of the code: list of analyzed samples in the variable "sample", the calculated output matrix ACF (sample number in rows, and separation distance *r* in columns), and indices of images disqualified due to poor SNR in "outlier".

18. This is performed in the bottom panel titled "Calculate D".
19. Click on "Select ACF file", and select the file generated by the top panel of the code.
20. As above, set the value of R_{min} as well as the name of the file where the final data need to be saved.
21. Run the code by clicking on the "Calculate D" button. It generates (a) a plot showing the evolution of the autocorrelation as a function of "length scale" (*r*) which is a measure of discrete signal densities (a shallower decay means a higher density of larger length scales) and (b) a boxplot showing the median and sample distribution of *D* (higher *D* values indicate a higher density of larger length scales) (Fig. 6).
22. Enter the name for the dataset reporting *D* values. For instance, entering "wt" will create a file "D_wt". The final compiled data "D_ *NameDefinedByUser*" contains all data originally saved in the "acf_ *NameDefinedByUser*" (**step 13**)

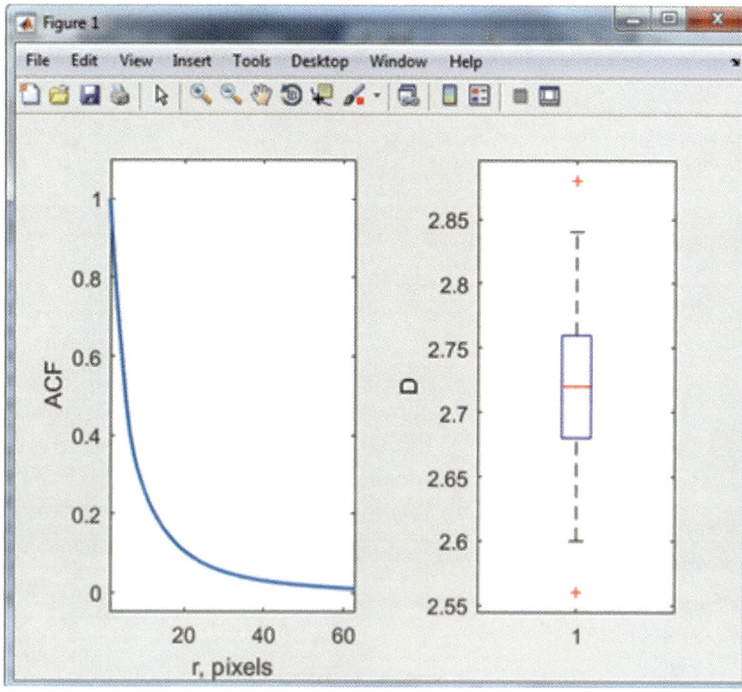

Fig. 6 Screenshot of the fitting process and quantification of chromatin mass density distribution. The average experimentally measured ACF for given length scales (*r*, *left*) and a boxplot of the calculated values of *D* (*right*) describing best the morphology of chromatin mass density distribution (see text for details)

along with the vector containing the fitted values of *D* and Ln (*see* **Note 24** and described in detail in [11].

23. Save the image. The dimensions can be manually adjusted before saving, or processed in a third-party software after saving as a vectorized graphics (e.g., svg).
24. For closing all the figure windows (for instance if then processing a different dataset), type "close all2" in the command line.
25. Initial statistical analysis of *D* values for different samples (e.g., genotypes, cell types, and tissue treatment) is performed by the last section of the GUI named "Compare ACF and D." Select two files previously created by the section "Calculate D" by clicking "Select D file 1" and "Select D file 2" respectively. Clicking "Compare D" will result in generating a figure with (a) average ACFs calculated for each group of images displayed together (the average ACF of the group identified by *D_file1* is in blue, that of *D_file2* is in red); (b) side-by-side boxplots of *D* values on each group with the *p*-value of a two-sided *t*-test assuming unequal variances comparing *D* values of the two groups displayed on top.

26. Further statistical analysis comparing *D* values for different samples (e.g., genotypes, cell types, tissue treatment) can be done in a third-party software (R, excel) by exporting the *D* values: double click on the "D_ *NameDefinedByUser*" saved in **step 22** (same folder as images analyzed), it is loaded in Matlab. Double click on the D variable, the values appear in a sheet and can be exported or simply copied.

4 Notes

1. Specimen carriers come in different arrangement modes which allow to achieve different cavity depth. Again, bearing the limitation posed by specimen thickness and considering the specimen thickness, cavity depth should be adjusted to achieve the lowest possible depth without squashing the specimen. High-pressure freezing is a fixation technique par excellence for preserving the fine ultrastructural details of a cell. However, there are limits to this technique including the thickness and water content of the specimen. High-pressure freezing would preserve ultrastructural details of specimens up to 500 μm thick [14] with increasing ultrastructural preservation achieved with decreasing sample thickness. Thus, specimens for high pressure freezing should have a thickness value below this limit, and where needed and if possible, specimens should be cut to smaller thicknesses. The Arabidopsis root has a diameter of 150–200 μm and, thus, is suitable for high pressure freezing. High water content in cells, a typical feature of most differentiating plant cells, can easily lead to ice crystal damage (the formation of crystal instead of vitreous ice, and the resulting segregation of cellular component leading to distortion of fine ultrastructural details). Ice crystal damage can be difficult to eliminate especially for thicker samples, and thus should be born in mind during ultrastructural analyses of high pressure frozen plant specimens.
2. High purity grade acetone is strongly recommended. The hydrophobic nature of Epon resin means that it would not mix with water and thus the specimens should be completely devoid of water to avoid artifactual damage during resin curing. Generally, we immerse acetone in 3 Å molecular sieve granules for at least 24 h before use.
3. Dissolve 500 mg OsO_4 crystals in 50 ml water-free acetone and store in liquid nitrogen. OsO_4 causes skin irritation, eye damage and can be fatal if inhaled. Avoid breathing fume and wear protective gloves, eye protection, and respiratory protection.
4. This Epon mixture is one that we have observed to penetrate deep into the tissue, and after curing, forms a hard enough and

yet least brittle block. This is highly beneficial during the subsequent trimming and sectioning of the specimen. Once the accelerator is added and the resin mixture is well mixed, proceed immediately to the final embedding. The resin begins to polymerize immediately after the accelerator is added and we strongly recommend that it should be used within the 60 min after, and not later. Curing for 48–60 h achieves the required rigidity; the block appears relatively weak if cured for much shorter at 60 °C.

5. Embedding molds come in different forms. It is recommended, for the choice of the mold, to consider the size and shape of the specimen, and particularly, the orientation of the specimen in the final block that would be suitable for subsequent trimming and sectioning.
6. Electron microscope grids generally have a shiny and a dark side. It is recommended to choose one of these sides for placing sections as this would guide in the subsequent contrast staining. Formvar-coated grids are commercially available. Alternatively, grids can be coated manually using 1% formvar solution as described [23] and stabilized with 8–10 nm thick carbon coating.
7. Dissolve 0.2 g uranyl acetate in 10 ml Milli-Q water. Uranyl acetate is toxic and potentially carcinogenic. Avoid inhaling or ingesting. Wear protective clothing and work in a fume hood assigned for radioactive substances. Dispose of the waste in appropriate container.
8. Prepare separately (a) 1.33 g lead nitrate in 15 ml Milli-Q water, (b) 1.76 g sodium citrate in 15 ml Milli-Q water, and (c) 0.4 g NaOH in 10 ml Milli-Q water. Add (a) and (b) and mix thoroughly until a homogenous milky solution is formed. Add 8 ml of (c). The solution immediately becomes clear. Add 12 ml of Milli-Q water, then gently mix and store aliquots of 10 ml at 4 °C. Lead citrate is toxic and causes serious eye damage. Avoid inhaling or ingesting. Wear protective clothing and work in a fume hood.
9. Hexadecene is a cryoprotectant and can significantly improve the quality of freezing. However, due to its highly hydrophobic nature, hexadecene may not readily mix with plant material. While we did not find it necessary for Arabidopsis root, degassing is recommended (for leave samples and others which do not submerge in 1-hexadecene) to submerge and get rid of air spaces trapped between the specimen and hexadecene. Air spaces are very poor conductors of heat and can impact freezing quality negatively.
10. Different version of HPF systems may work differently. It is recommended to follow the manufacturer's operational

manual. For a very detailed description for HPF with the EM HPM100 system, *see* ref. 16.

11. Frozen specimens should be stored and handled in liquid nitrogen until the start of freeze substitution to avoid ice crystal formation.
12. Arabidopsis roots become extremely fragile after freeze substitution. It is easier to transfer them with wooden toothpicks by gently curling around the specimen. The use of tweezers to pick up and transfer the specimen is not advised, as they easily fracture and damage it.
13. Trim away any excess resin around the specimen to make trapezoids shapes with smooth and perfectly parallel opposite sides. Where possible, limit the area of the section to the smallest possible. Block trimming can be done with an electronic block trimmer. However, this instrument may not be available. Manual trimming is the alternative where a very sharp glass knife is used. Rough edges and nonparallel opposite sides of the trapezoid shape due to blunt knives and/or brittle resin blocks can pose considerable difficulty during sectioning. It is important to know the average size of the cells in the specimen in order to estimate how deep to trim into the specimen block to hit nuclei. Cell density per mm varies considerably in Arabidopsis roots between the meristematic and differentiation zones. Nuclei can be seen under the light microscope after toluidine blue staining of semithin (about 250 μm thick).
14. With a small area, it is possible to collect many ultrathin sections onto the grid. And by analyzing serial sections, some level of depth information can be obtained.
15. Handling sections containing grids during staining is most efficient using the fine-tip curved tweezers. Uranyl acetate acts both as a fixative and a contrast stain. Uranyl acetate has the advantage of producing high electron density and image contrast enabling nucleic acid-containing regions to be stained in a highly preferential manner [24] as well as imparting a fine grain to the image. Uranyl acetate stains lipids, proteins and nucleic acids, and thus gives good contrasting of membranes, nucleic acids and nucleic acids containing protein complexes. Lead citrate binds to nucleic acids, lipids and proteins, but also interacts with and enhances the contrasting effect of uranyl acetate [25]. Some protocols recommend block staining of specimens with uranyl acetate, which would eventually shorten the post staining to only one step with lead citrate and also reduce the risk of damaging the ultrathin section by avoiding the additional washing and handling steps. When we stained specimens with 1% uranyl acetate for 1 h at room temperature after freeze substitution and omitted the uranyl acetate post

staining step, we observed a very weak contrasting of euchromatic and heterochromatic regions. Uranyl acetate and lead citrate solutions can contain precipitates which can be quite menacing as they can stick to ultrathin sections and obstruct visualization in the TEM. Thus, prior to use, spin down the solution for 3 min at 14,000 × *g* and only use the supernatant.

16. Drying uranyl acetate stained grids seems to stabilize the stain. Some authors recommend drying uranyl acetate stained grid for some hours before lead citrate staining. However, our shortened approach of drying by carefully sucking up with filter paper and placing the grid at 37 °C for about 5 min prior to lead staining gives very adequate contrasting.

17. The best contrasts were obtained when sections were stained with uranyl acetate for at least 15 min and lead citrate for 10 min. Staining for shorter time period (8 min of uranyl acetate and 5 min lead citrate) gave weak contrasts. The washing steps are just as critical as the centrifugation. Lead citrate staining is done in a CO_2-free environment to avoid it from reacting with CO_2 to form precipitates that can be menacing during visualization. Thorough washing on four drops of sterile filtered Milli-Q water can easily get rid of any precipitates. However, stained samples should not be left on each drop of water for more than 10 s, as the stain may redissolve in water.

18. It is forbidden for wet sections to be visualized in the TEM. The electron beams would burn any water droplet as well as the section around it. Stained, filter-paper-dried sections should be further air-dried for at least 2 h at RT or 25 min at 37 °C to get rid of any residual water before visualization in the TEM.

19. The practical resolution achieved by TEM imaging at ~100 kV is about 0.5 nm. However, several biological and technical factors may affect the resolution of the specimen ultrastructure provided in the image. First, the resolution of a biological structure depends on the method of sample preparation (fixation, dehydration/freeze substitution, embedding, and contrast staining), which may introduce some degree of distortion of the native arrangement of cellular components and alter the resolution. In our images, membrane bilayers with a thickness of approximately 10 nm were among the smallest cellular substructures visibly resolved, which provided a good indication of best-possible resolution from our preparation. Thus, we took into account Nyquist sampling theorem, which states that for the conversion of analogue to digital signals, the sampling rate must be greater than twice the highest frequency component of interest in the measured signal. We reasoned that images with pixel sizes between 0.5 and 1.5 nm would adequately sample regions of distinct electron densities and enable for the quantification of chromatin density distribution. We assume that images with pixel sizes below 0.5 nm would likely

contain too many pixels, with no additional relevant information and, conversely, pixel sizes above 1.5 nm may miss some fine details.

20. Dividing cells, by contrast are less suitable for such analyses due to a high density of euchromatic granules associated with transcriptional activity [26].
21. The shape and size of the ROI does not seem to affect significantly the measurements, at least in the range of 250 × 250–-1000 × 1000 pixels in our hands (<2% variation, data not shown). The user is however recommended to verify it on own images. Alternatively, a fixed ROI shape and size is applied to all images.
22. There are two possibilities: either save the ROI in a distinct folder per sample type (e.g., "wt" and "mutant") in which case the appropriate folder needs to be selected for each analysis; alternatively, ROIs can be saved in the same folder with a prefix "wt_1.tif", "wt_2.tif", "wt_3.tif", ... "mut_1.tif", "mut_2.tif", "mut_3.tif," and only the name needs to be changed in the second edible window: for selecting only wild-type images, type "wt_*.tif", or "mut_*.tif" for mutant images.
23. Place the folder with the scripts ideally in the Matlab folder locally. If placed however in a different folder, a prompt will appear when running the GUI the first time requesting to change the shadow folder path (validate "change folder"). The first time, run the script on a single image. If no image pops up, or if an error message appears in the command window tool, verify that the path of the selected input folder does not contain a typo, or that the name of the files (prefix, or all indicated with a * only) has no typo or space; eventually restart Matlab.
24. The pixel size depends on the magnification (each instrument has a parametric table of 1 mm equivalent for each magnification used) and the image resolution produced by the camera or scanner (in dpi). For instance, our imaging conditions and instruments (*see* Subheading 2) allowed a pixel size of 2 nm at 24,500× magnification.

Acknowledgment

This work was supported by The Swiss Initiative in Systems Biology SystemsX.ch (MechanX grant 145676 to CR and TNF), the Swiss National Science Foundation (grant 31003A_149974 to CB and the University of Zürich). We are grateful for the technical support of Dr. Andres Kaech and from the Center for Microscopy and Image Analysis of the University of Zürich.

References

1. Leitch AR (2000) Higher levels of organization in the interphase nucleus of cycling and differentiated cells. Microbiol Mol Biol Rev 64 (1):138–152
2. Thorn K (2016) A quick guide to light microscopy in cell biology. Mol Biol Cell 27 (2):219–222
3. Barlow PW (1985) Nuclear chromatin structure in relation to cell differentiation and cell activation in the cap and quiescent centre of Zea mays L. J Exp Bot 36(9):1492–1503
4. Olszewska MJ, Bilecka A, Kononowicz AK, Kolosziejczyk P (1988) Relationship between heterochromatin and repetitive DNA content and the dynamics of nuclear DNA endoreplication in root parenchyma cells. Biol Zent Bl 107 (3):311–326
5. Lingua G, D'Agostino G, Fusconi A, Berta G (2001) Nuclear changes in pathogen-infected tomato roots. Eur J Histochem 45(1):21
6. Fusconi A, Gnavi E, Trotta A, Berta G (1999) Apical meristems of tomato roots and their modifications induced by arbuscular mycorrhizal and soilborne pathogenic fungi. New Phytol 142(3):505–516
7. Tsai Y-C, Greco TM, Boonmee A, Miteva Y, Cristea IM (2012) Functional proteomics establishes the interaction of SIRT7 with chromatin remodeling complexes and expands its role in regulation of RNA polymerase I transcription. Mol Cell Proteomics 11(5):60–76
8. Nagl W, Cabirol H, Lahr C, Greulach H, Ohliger HM (1983) Nuclear ultrastructure: morphometry of nuclei from various tissues of Cucurbita, Melandrium, Phaseolus, Tradescantia, and Vicia. Protoplasma 115(1):59–64
9. Nagl W (1979) Nuclear ultrastructure: condensed chromatin in plants is species-specific (karyotypical), but not tissue-specific (functional). Protoplasma 100(1):53–71
10. Baluška F (1990) Nuclear size, DNA content, and chromatin condensation are different in individual tissues of the maize root apex. Protoplasma 158(1–2):45–52
11. Cherkezyan L, Stypula-Cyrus Y, Subramanian H, White C, Cruz MD, Wali RK, Goldberg MJ, Bianchi LK, Roy HK, Backman V (2014) Nanoscale changes in chromatin organization represent the initial steps of tumorigenesis: a transmission electron microscopy study. BMC Cancer 14(1):1
12. Belmont AS (1997) Nuclear ultrastructure: transmission electron microscopy and image analysis. Methods Cell Biol 53:99–124
13. Almassalha LM, Tiwari A, Ruhoff PT, Stypula-Cyrus Y, Cherkezyan L, Matsuda H, Cruz MAD, Chandler JE, White C, Maneval C (2017) The global relationship between chromatin physical topology, fractal structure, and gene expression. Sci Rep 7:41061
14. Wilson SM, Bacic A (2012) Preparation of plant cells for transmission electron microscopy to optimize immunogold labeling of carbohydrate and protein epitopes. Nat Protoc 7 (9):1716–1727
15. Matsko N, Mueller M (2005) Epoxy resin as fixative during freeze-substitution. J Struct Biol 152(2):92–103
16. Kaech A, Ziegler U (2014) High-pressure freezing: current state and future prospects. Methods Mol Biol 1117:151–171
17. Studer D, Humbel BM, Chiquet M (2008) Electron microscopy of high pressure frozen samples: bridging the gap between cellular ultrastructure and atomic resolution. Histochem Cell Biol 130(5):877–889
18. Feder N, Sidman RL (1958) Methods and principles of fixation by freeze-substitution. J Biophys Biochem Cytol 4(5):593–602
19. Reynolds ES (1963) The use of lead citrate at high pH as an electron-opaque stain in electron microscopy. J Cell Biol 17(1):208–212
20. Mirny LA (2011) The fractal globule as a model of chromatin architecture in the cell. Chromosom Res 19(1):37–51
21. Lebedev DV, Filatov MV, Kuklin AI, Islamov AK, Kentzinger E, Pantina R, Toperverg BP, Isaev-Ivanov VV (2005) Fractal nature of chromatin organization in interphase chicken erythrocyte nuclei: DNA structure exhibits biphasic fractal properties. FEBS Lett 579(6):1465–1468
22. Bancaud A, Lavelle C, Huet S, Ellenberg J (2012) A fractal model for nuclear organization: current evidence and biological implications. Nucleic Acids Res 40(18):8783–8792
23. Rowley JC, Moran DT (1975) A simple procedure for mounting wrinkle-free sections on formvar-coated slot grids. Ultramicroscopy 1 (2):151–155
24. Huxley HE, Zubay G (1961) Preferential staining of nucleic acid-containing structures for electron microscopy. J Biophys Biochem Cytol 11(2):273–296
25. Bernhard W (1969) A new staining procedure for electron microscopical cytology. J Ultrastruct Res 27(3–4):250–265
26. Hendzel MJ, Kruhlak MJ, Bazett-Jones DP (1998) Organization of highly acetylated chromatin around sites of heterogeneous nuclear RNA accumulation. Mol Biol Cell 9 (9):2491–2507.

Index

Marian Bemer and Célia Baroux (eds.), *Plant Chromatin Dynamics: Methods and Protocols*, Methods in Molecular Biology, vol. 1675, DOI 10.1007/978-1-4939-7318-7, © Springer Science+Business Media LLC 2018

D

E

F

G

Q

R

S

T

U

W